Hazardous Chemicals Handbook

Hazardous Chemicals Handbook

Second edition

Phillip Carson

PhD MSc AMCT CChem FRSC FIOSH

Head of Science Support Services, Unilever Research Laboratory,
Port Sunlight, UK

Clive Mumford

BSc PhD DSc CEng MIChemE

Consultant Chemical Engineer

Oxford Amsterdam Boston London New York Paris San Diego
San Francisco Singapore Sydney Tokyo

Butterworth-Heinemann
An imprint of Elsevier Science
Linacre House, Jordan Hill, Oxford OX2 8DP
225 Wildwood Avenue, Woburn, MA 01801-2041

First published 1994
Second edition 2002

British Library Cataloguing in Publication Data
A catalogue record for this book is available from the British Library

Library of Congress Cataloguing in Publication Data
A catalogue record for this book is available from the Library of Congress

ISBN 0 7506 4888 0

Typeset at Replika Press Pvt Ltd, Delhi 110 040, India
Printed and bound in Great Britain

Contents

Preface to the second edition

The aim of this book remains as for the first edition, namely to provide an initial point of ready reference for the identification of hazards and precautions for dangerous chemicals. It is targeted not only at those in the chemical and process industries, but also anyone likely to work with chemicals within industry and in the service sector, e.g. hospitals, universities, research laboratories, engineering, agriculture, etc. It embraces the entire life-cycle of chemicals during transport, storage, processing, marketing, use and eventual disposal and should appeal to chemists, occupational and environmental health practitioners and students, engineers, waste handlers, safety officers and representatives, and health care professionals. Clearly, more detailed texts or professional advice may need to be consulted for specific applications.

Since the first edition in 1994 there have been no significant changes in the fundamentals of chemistry, physics and toxicology upon which the safe handling of chemicals are based. There has, however, been some increase in knowledge relating to the chronic toxicological and potential environmental effects of specific chemicals, and in legislation and government guidelines. These are reflected in the second edition. In general, within the UK the predominant legislation relating to substances hazardous to health, the Control of Substances Hazardous to Health Regulations 1999 and its accompanying Approved Code of Practice, incorporate significant changes since the 1988 (and 1994) versions. There has been an increase in the controls applicable to the marketing and transportation of different classes of chemicals. Those applicable to major hazards have changed under the Control of Major Accident Hazard Regulations 1999. Other legislation has been introduced: e.g. the Confined Spaces Regulations 1997, the Reporting of Injuries, Diseases and Dangerous Occurrences Regulation 1995, the Health and Safety (Safety Signs and Signals) Regulations 1996, and the Pressure Systems Safety Regulations 2000 which is of importance to the scope of this text. Increased concern as to the possible environmental impacts of chemical discharges and disposal has been accompanied by more comprehensive legislation for control. General safety legislation was expanded by the introduction of various separate regulations in 1993, including that dealing with management of health and safety at work; workplace health, safety and welfare; workplace equipment; and personal protective equipment. These improvements are, in general, now reflected in industry.

The opportunity has been taken to improve each chapter and to update the information. The main changes include an expansion of the terminology in Chapter 2 and provision of an introduction to basic chemical principles for non-chemists in a new Chapter 3. Chapter 4 on Physicochemistry contains additional examples. Chapter 5 on Toxic chemicals has been enlarged and the table of hygiene standards updated. Chapters 6, 7 and 8 on Flammable chemicals, Reactive chemicals and Cryogens, respectively, have been updated and expanded. The scope of Chapter 9 on Compressed gases has been widened to include additional examples together with the basic techniques of preparing gases *in situ*. Chapter 10 summarizes techniques for monitoring air quality and employee exposure. It has also been expanded to provide guidance on monitoring of water and land pollution.

The chapter on Radioactive chemicals (Chapter 11) has been updated. Considerations of safety in design (Chapter 12) are presented separately from systems of work requirements, i.e. Operating procedures (Chapter 13). The considerations for Marketing and transportation of hazardous chemicals are now addressed in two separate chapters (Chapters 14 and 15). Chemicals and the Environment are now also covered in two chapters (Chapters 16 and 17) to reflect the requirement that the impact of chemicals on the environment should be properly assessed, monitored and controlled. Although a substantial contribution to atmospheric pollution is made by emissions from road vehicles and other means of transport, and this is now strictly legislated for, this topic is outside the scope of this text. Chapter 18 provides useful conversion factors to help with the myriad of units used internationally.

Whilst the hazards identified, and the principles and practice for the control of risks are universal, i.e. they are independent of location, in order to assist quick-reference an appendix of relevant contemporaneous UK legislation has been added as a guide together with a much-expanded Bibliography in Chapter 19. Finally, for convenience of use, the Index has been enlarged.

It is hoped that the improvements will help to achieve the objectives for which the text was originally conceived, i.e. to summarize in relatively basic terms the hazards associated with chemicals and how the ensuing risks can be controlled, and to provide sufficient detailed information to supplement that obtainable from suppliers, government publications, trade associations, and computerized data banks where recourse to specialized textbooks may be premature, difficult or unnecessary.

P.A.C.
C.J.M.

Preface to the first edition

The aim of this handbook is to provide a source of rapid ready reference to help in the often complex task of handling, using and disposing of chemicals safely and with minimum risk to people's health or damage to facilities or to the environment.

The range of chemicals and chemical mixtures in common use in industry is wide: it is obviously impossible to list them all in a concise handbook, or to refer to all their proprietary names. The approach here has been to avoid 'random listing' and to arrange by type of hazard, dealing with the most widely used substances and those properties and characteristics of behaviour that are directly relevant to common use and to compliance with safety legislation. Numerous sources not restricted to those in the Bibliography were searched for information and although not listed, to achieve conciseness, these are acknowledged. The multiplicity of data sources also means that minor variations occur due to differences in the procedures and methods for their determination; however they provide general guidance. Whilst the data quoted in this text has been carefully collated, its accuracy cannot be warranted. For this reason, and to avoid overlooking consideration of other chemical-specific hazards or location-dependent legislation, it is advisable to refer to a Chemical Safety Data Sheet before using any chemical. These are readily available from suppliers (e.g. in the UK under S.6 of the Health & Safety at Work etc. Act 1974). For exhaustive treatment of physical, toxicological, flammable/explosive and reactive properties, and the background to – and limitations of – their determination or prediction, the reader is referred to standard textbooks (see Bibliography) such as:

The Safe Handling of Chemicals in Industry (Carson and Mumford)
Dangerous Properties of Industrial Materials (Sax and Lewis)
Handbook of Reactive Chemical Hazards (Bretherick)
Handbook of Toxic and Hazardous Materials (Sittig)
Patty's Industrial Hygiene and Toxicology (Clayton and Clayton)

The identification, assessment, control and monitoring of chemical-related hazards and environmental pollution control are, of course, required under a wide range of statutory legislation, dependent upon the country concerned. For example, in the UK the Health and Safety at Work etc. Act 1974, the Control of Substances Hazardous to Health Regulations 1988, the Highly Flammable Liquids and Liquefied Petroleum Gases Regulations 1972, the Control of Pollution Act 1974 and the Environmental Protection Act 1990 are supplemented by a wide variety of other measures. Legislative controls tend to change frequently and it is important to ensure that a check is made on current requirements and constraints in any specific situation involving chemicals.

It is hoped that this book will prove valuable to safety advisers, environmental health officers, emergency services personnel, safety representatives and those engaged in the transport or disposal of wastes – in fact, to anyone involved with chemicals 'in the field', i.e. away from ready access to chemical safety data sheets, detailed texts, library facilities or computerized databanks. It also provides a useful summary for those who may need to make only passing reference to the hazardous properties and potential effects of chemicals, such as general engineering students and occupational health nurses.

P.A.C.
C.J.M.

1

Introduction

Industrial hazards cover a wide spectrum including fire and explosion, mechanical hazards (e.g. from moving machinery), electrical hazards, occupational exposures to ionizing and non-ionizing radiation, biological hazards (e.g. acute or chronic infections, parasitism, and toxic or allergic reactions to plant and animal matter), physical hazards (e.g. tripping, falling, impact from vehicles or falling objects) and ergonomical hazards (e.g. lifting or carrying heavy or awkward loads or from repetitive operations). Work-related stress can also lead to mental and physical ill-health. Different hazards may be associated with the manufacture, storage, transport, use, and disposal of chemicals. Environmental hazards, through persistent or accidental losses of chemicals, may also be related to these operations.

Working with pathogenic micro-organisms bears passing similarity to chemicals. Hence, in the UK micro-organisms are classified as *hazardous substances* under the Control of Substances Hazardous to Health Regulations and there is an accompanying Code of Practice. However, biological hazards arising from the working environment or from more specialized activities, e.g. working with pathogenic organisms in laboratories, are beyond the scope of this book. This text deals solely with occupational, industrial and environmental hazards associated only with chemicals. It includes fires and explosions since they inevitably involve chemical compounds.

Chemicals are ubiquitous as air, carbohydrates, enzymes, lipids, minerals, proteins, vitamins, water, and wood. Naturally occurring chemicals are supplemented by man-made substances. There are about 70 000 chemicals in use with another 500–1000 added each year. Their properties have been harnessed to enhance the quality of life, e.g. cosmetics, detergents, energy fuels, explosives, fertilizers, foods and drinks, glass, metals, paints, paper, pesticides, pharmaceuticals, plastics, rubber, solvents, textiles; thus chemicals are found in virtually all workplaces. Besides the benefits, chemicals also pose dangers to man and the environment. For example:

- Of the many industrial fires in the UK in 1997 each of some 411 cost more than £50 000 with total losses amounting to £186m. These spanned a wide range of industrial and related premises as shown in Table 1.1. The most common sources of ignition (see Chapter 6) that year are shown in Table 1.2.
- In the UK alone occupational health risks due to chemicals are illustrated by:
  - 152 incidents in 1998 involving supply and use of flammable gas with around 70% causing carbon monoxide poisoning and 30 fires/explosions;
  - 554 new cases of pneumoconiosis (excluding asbestosis) and 3423 assessed cases of bronchitis or emphysema (in coal miners) during the same period;
  - annually there are 4500 cases of work-related skin disease (80% contact dermatitis), ca 1500 cases of occupational asthma (mainly from solder flux, isocyanates, wood dust, spray painting, metal treatment, plastics), ca 200 cases of allergic rhinitis;
  - between 2% and 8% of all cancer deaths are of occupational origin;

Table 1.1 Breakdown of U.K. fires causing more than £50 000 damage in 1997

Occupancy	*No. of fires*	*Loss £000*	*% of total cost of all fires*
Agriculture, forestry and fishing	27	3282	2
Paper, printing and publishing	14	6480	3
Food, drink and tobacco	13	7235	4
Rubber and plastic	7	5371	3
Textiles, footwear and clothing	7	1894	1
Timber and wood products excluding furniture	6	1042	1
Chemicals and allied products	5	4543	2
Construction	5	515	–
Metal manufacture	4	871	–
Engineering	4	545	–
Other manufacturing industries	25	20 249	11
Retail distribution	27	15 021	8
Transport and communications	18	13 390	7
Wholesale distribution	8	26 250	14
Education	39	23 407	13
Recreational/cultural	30	6946	4
Clubs and public houses	19	4668	3
Cafes/restaurants	14	2431	1
Insurance, banking and business services	10	1730	1
Hotels/boarding houses	6	2516	1
Hospitals	4	925	–
Public admin./defence/law enforcement	3	430	–
Hostels/holiday camps	1	99	–
Homes for disabled	1	100	–
Domestic dwellings	63	9970	5
Other	26 434	518	14

Table 1.2 Accidental fires (UK) in 1997: sources of ignition

Ignition source	*No. of fires*	*% of all fires*	*Loss £000*	*% of total cost of all fires*
Electrical appliances	110	26.8	55 491	29.8
Smokers' materials	17	4.1	2138	1.1
Gas appliances (excluding blowlamps and welding)	10	2.4	2595	1.4
Blowlamps: all fuels	7	1.7	2176	1.2
Welding and cutting appliances	7	1.7	1340	0.7
Oil and petroleum appliances (excluding blowlamps and welding)	6	1.5	714	0.4
Unspecified appliances	5	1.2	578	0.3
Rubbish burning	4	1	417	0.2
Chimney, stovepipe and flue	2	0.5	303	0.2
Natural occurrence	2	0.5	215	0.3
Ashes/soot	1	0.2	50	0.1
Other	22	5.4	3789	2
Total	193	47	69 806	37.7

  - ca 1760 cases per year of acute poisonings and injuries from chemicals, the most common being from acids, caustic, and gases, with process operatives and tradesmen being at greatest risk;
  - an estimated 9000 cases of sick building syndrome per year.
- The UK Environment Agency deals with over 6000 oil pollution incidents each year. One estimate suggests that the chemical industry contributes to 50% of all air pollution with proportions approximating to sulphur dioxide (36%), carbon dioxide (28%), nitrogen oxides (18%), carbon monoxide (14%) and black smoke (10%). Motor spirit refining is responsible for ca 26% of emissions of volatile organic compounds to the atmosphere. In 1996 there were over 20 000 reports of water pollution incidents with 155 successful prosecutions.
- The EC produces in excess of 2 billion tonnes of waste each year. 414 million tonnes arise in the UK and a further 68 million tonnes of hazardous waste are imported. All wastes must be disposed of safely.

Society must strike a balance between the benefits and risks of chemicals. In the workplace it is a management responsibility to ensure practices control the dangers, and it is for employees to collaborate in implementing the agreed procedures. Management must also prevent uncontrolled environmental releases and ensure all wastes are disposed of safely and with proper regard for their environmental impact. The aims of this book are to raise awareness and to help users identify, assess and control the hazards of chemicals to permit optimum exploitation whilst minimizing the dangers.

The hazards of ‘chemicals’ stem from their inherent flammable, explosive, toxic, carcinogenic, corrosive, radioactive or chemical-reactive properties. The effect of exposure on personnel may be acute, e.g. in a flash-fire or due to inhalation of a high concentration of an irritant vapour. Alternatively, prolonged or intermittent exposure may result in an occupational disease or systemic poisoning. Generally acute effects are readily attributable; chronic effects, especially if they follow a long latency period or involve some type of allergic reaction to a chemical, may be less easy to assign to particular occupational exposures. The possible permutations of effects can be very wide and exposure may be to a combination of hazards. For example, personnel exposed to a fire may be subject to flames, radiant heat, spilled liquid chemicals and vapours from them, leaking gases, and the pyrolytic and combustion products generated from chemical mixtures together with oxygen deficient atmospheres. However, whether a hazardous condition develops in any particular situation also depends upon the physical properties of the chemical (or mixture of chemicals), the scale involved, the circumstances of handling or use, e.g. the processes involved and degree of containment, and upon the control measures prevailing, e.g. provision of control and safety devices, local exhaust ventilation, general ventilation, personal protection, atmospheric monitoring and systems of work generally.

Hazard recognition and assessment always start from a knowledge of the individual properties of a chemical. What this may include is exemplified by Table 1.3. Additional properties, including those in Table 1.4, are relevant to environmental hazards, e.g. relating to behaviour on spillage or emission, and determination of permissible levels for disposal to air, land or water systems. Other properties may be relevant, e.g. odour which can serve as an, albeit often unreliable, means of detection. (Refer to Table 5.12.)

An elementary introduction to chemistry is given in Chapter 3; this serves only to provide background and for more advanced consideration reference will be necessary to specific text books, e.g. as listed in the Bibliography. A brief discussion of the relevance of physicochemical principles to hazard identification is given in Chapter 4. Relevant toxic and flammable properties, and summaries of appropriate precautions to cater for them during handling, use and disposal, are provided in Chapters 5 and 6, respectively. Reactive hazards are discussed in Chapter 7. The special problems with cryogenic materials and chemicals under pressure, typified by compressed

Table 1.3 Comprehensive information possibly required for a hazardous chemical

Name of chemical; other names
Uses
General description of hazards
General description of precautions
  Fire-fighting methods
  Regulations
  Sources of advice on precautions

Characteristics: evaluate as appropriate under all process conditions
Formula (chemical structure)
Purity (identity of any contaminants), physical state, appearance, other relevant information
Concentration, odour, detectable concentration, taste

Physical characteristics

Molecular weight	Particle size; size distribution
Vapour density	Foaming/emulsification characteristics
Specific gravity	Critical temperature/pressure
Melting point	Expansion coefficient
Boiling point	Surface tension
Solubility/miscibility with water	Joule–Thompson effect
Viscosity	Caking properties

Corrosivity
Contamination factors (incompatibility), oxidizing or reducing agent, dangerous reactions

Flammability information

Flash point	Vapour pressure
Fire point	Dielectric constant
Flammable limits (LEL, UEL)	Electrical resistivity
Ignition temperature	Electrical group
Spontaneous heating	Explosion properties of dust in a fire
Toxic thermal degradation products	

Reactivity (instability) information

Acceleration rate calorimetry	Drop weight test
Differential thermal analysis (DTA)	Thermal decomposition test
Impact test	Influence test
Thermal stability	Self-acceleration temperature
Lead block test	Card gap test (under confinement)
Explosion propagation with detonation	JANAF
	Critical diameter
	Pyrophoricity

Toxicity information
  Toxic hazard rating
  Hygiene standard (e.g. OEL, TLV)
  Maximum allowable concentration (MAC)
  Lethal concentration (LC_{50})
  Lethal dose (LD_{50})
Biological properties
Exposure effects
  Inhalation (general)
  Respiratory irritation
  Ingestion
  Skin/eye irritation
  Skin and respiratory sensitization
  Mutagenicity
  Teratogenicity
  Carcinogenicity
Radiation information
  Radiation survey
  Alpha/beta/gamma/neutron exposure and contamination

gases, are dealt with in Chapters 8 and 9. The unique problems associated with radioactive chemicals are described in Chapter 11.

The foregoing relates mainly to normal laboratory or commercial quantities of chemicals. Additional considerations arise with those quantities of flammable, explosive, reactive, bulk toxic, or hypertoxic chemicals which constitute *major hazards*, i.e. which may pose a hazard to neighbouring factories, residents, services etc. or a more substantial potential risk to the environment. Within the UK the Control of Major Accident Hazards Regulations 1999 requires that the operator of any establishment where a dangerous substance listed in column 1 of Parts 2 or 3 of Schedule 1 (reproduced here as Tables 1.5 and 1.6) is present in a quantity equal to or greater than that listed in column 2 of those Parts shall notify the competent authority. Detailed procedures and precautions are then applicable to such sites depending partly upon whether they are 'lower tier' or 'upper tier', i.e. sites at which the quantity present is equal to or exceeds that listed in column 3. The special considerations with such installations are detailed in specialist texts noted in the Bibliography. In the UK the Planning (Hazardous Substances) Regulations 1992 also require the holder to obtain a 'hazardous substances consent' for any site on which it is intended to hold a bulk quantity of any of 71 substances above a 'controlled quantity' (Table 1.7).

Table 1.4 Typical data on hazards to the environment

Aquatic toxicity (e.g. to fish, algae, daphnia)
Terrestrial toxicity (to plants, earthworms, bees, birds)
Biotic degradation
Abiotic degradation
Photodegradation
Biochemical oxygen demand
Chemical oxygen demand
Hydrolysis as a function of pH
Bioaccumulation
Oil/water partition coefficient

To proceed to assess, and recommend control strategies for, any operation involving a mixture of chemicals – e.g. a chemical process, welding fume, mixed effluents – can be a complex exercise. It can rarely be solved by rigidly following a checklist, although checklists, examples of which are given in the various chapters, can provide useful guidelines. And although associated hazards are not covered here, the control of chemical hazards in the workplace cannot be achieved in isolation from a consideration of electrical, mechanical, ergonomic, biological and non-ionizing radiation hazards. Hence these must be included in any hazard analysis and control system.

To ensure that an operation is under control may necessitate environmental monitoring; this is summarized in Chapter 10. Principles of safe design are given in Chapter 12. General safety considerations, administration and systems of work requirements, including elementary first aid, are summarized in Chapter 13. For example, the recommended strategy is to include provision for appropriate first aid procedures within the system of work before specific chemicals are brought into use; to so order work practices that the risk of exposure is minimized; and in the event of an accident involving any but the most trivial injuries – with no foreseeable likelihood of complications or deterioration – to seek immediate medical assistance.

Additional considerations, e.g. relating to labelling, information supply and emergency procedures, arise when marketing and transporting chemicals. While – as with Chapter 13 and with control measures generally – what is required will vary with specific legislation and basic requirements are summarized in Chapters 14 and 15.

Table 1.5 Schedule 1 Part 2 of the COMAH Regulations Named Substances (Explanatory notes omitted)

Column 1 *Dangerous substances*	*Column 2* *Quantity in tonnes*	*Column 3* *Quantity in tonnes*
Ammonium nitrate (as described in Note 1 of this Part)	350	2500
Ammonium nitrate (as described in Note 2 of this Part)	1250	5000
Arsenic pentoxide, arsenic (V) acid and/or salts	1	2
Arsenic trioxide, arsenious (III) acid and/or salts	0.1	0.1
Bromine	20	100
Chlorine	10	25
Nickel compounds in inhalable powder form (nickel monoxide, nickel dioxide, nickel sulphide, trinickel disulphide, dinickel trioxide)	1	1
Ethylenimine	10	20
Fluorine	10	20
Formaldehyde (concentration ≥ 90%)	5	50
Hydrogen	5	50
Hydrogen chloride (liquefied gas)	25	250
Lead alkyls	5	50
Liquefied extremely flammable gases (including LPG) and natural gas (whether liquefied or not)	50	200
Acetylene	5	50
Ethylene oxide	5	50
Propylene oxide	5	50
Methanol	500	5000
4,4-Methylenebis (2-chloroaniline) and/or salts, in powder form	0.01	0.01
Methylisocyanate	0.15	0.15
Oxygen	200	2000
Toluene diisocyanate	10	100
Carbonyl dichloride (phosgene)	0.3	0.75
Arsenic trihydride (arsine)	0.2	1
Phosphorus trihydride (phosphine)	0.2	1
Sulphur dichloride	1	1
Sulphur dioxide	15	75
Polychlorodibenzofurans and polychlorodibenzodioxins (including TCDD), calculated in TCDD equivalent	0.001	0.001
The following CARCINOGENS:		
4-Aminobiphenyl and/or its salts, Benzidene and/or its salts, Bis(chloromethyl) ether, Chloromethyl ether, Dimethylcarbamoyl chloride, Dimethylnitrosoamine, Hexamethylphosphoric triamide, 2-Naphthylamine and/or its salts, 1.3-Propane sultone, 4-Nitrodiphenyl	0.001	0.001
Automotive petrol and other petroleum spirits	5000	50 000

Table 1.6 Schedule 1 Part 3 of the COMAH Regulations. Categories of substances and preparations not specifically named in Part 2. (Explanatory notes omitted)

Column 1 Column 2 *Categories of dangerous substances*	*Column 3* *Quantity in tonnes*	*Quantity in tonnes*
1. VERY TOXIC	5	20
2. TOXIC	50	200
3. OXIDIZING	50	200
4. EXPLOSIVE (Where the substance or preparation falls within the definition given in Notes 2a)	50	200
5. EXPLOSIVE (Where the substance or preparation falls within the definition given in Notes 2b)	10	50
6. FLAMMABLE (Where the substance or preparation falls within the definition given in Notes 3a)	5000	50 000
7a. HIGHLY FLAMMABLE (Where the substance or preparation falls within the definition given in Notes 3bi)	50	200
7b. HIGHLY FLAMMABLE (Where the substance or preparation falls within the definition given in Notes 3bii)	5000	50 000
8. EXTREMELY FLAMMABLE (Where the substance or preparation falls within the definition given in Notes 3c)	10	50
9. DANGEROUS FOR THE ENVIRONMENT in combination with risk phrases		
(i) R50: Very toxic to aquatic organisms	200	500
(ii) R51: Toxic to aquatic organisms and R43: may cause long-term adverse effects in the aquatic environment	500	2000
10. ANY CLASSIFICATION not covered by those given above in combination with risk phrases		
(i) R14: Reacts violently with water (including R15)	100	500
(ii) R29: In contact with water, liberates toxic gas	50	200

Table 1.7 Planning (Hazardous Substances) Regulations 1992
Hazardous substances and controlled quantities

Hazardous substance	*Controlled quantity*
Part A Toxic substances	
1. Acetone cyanohydrin (2-cyanopropan-2-ol)	200 t
2. Acrolein (2-propenal)	200 t
3. Acrylonitrile	20 t
4. Allyl alcohol (2-propen-1-ol)	200 t
5. Allylamine	200 t
6. Ammonia (anhydrous or as solution containing more than 50% by weight of ammonia)	100 t
7. Arsenic trioxide, arsenious (III) acid and salts	1 t
8. Arsine (arsenic hydride)	1 t
9. Bromine	40 t
10. Carbon disulphide	20 t
11. Chlorine	10 t
12. Ethylene dibromide (1,2-dibromoethane)	50 t
13. Ethyleneimine	50 t
14. Formaldehyde (>90%)	50 t
15. Hydrogen chloride (liquefied gas)	250 t
16. Hydrogen cyanide	20 t
17. Hydrogen fluoride	10 t
18. Hydrogen selenide	1 t
19. Hydrogen sulphide	50 t
20. Methyl bromide (bromomethane)	200 t
21. Methyl isocyanate	150 kg
22. Nickel tetracarbonyl	1 t
23. Nitrogen oxides	50 t
24. Oxygen difluoride	1 t
25. Pentaborane	1 t
26. Phosgene	750 kg
27. Phosphine (hydrogen phosphide)	1 t
28. Propyleneimine	50 t
29. Selenium hexafluoride	1 t
30. Stibine (antimony hydride)	1 t
31. Sulphur dioxide	20 t
32. Sulphur trioxide (including the sulphur trioxide content in oleum)	15 t
33. Tellurium hexafluoride	1 t
34. 2,3,7,8-Tetrachlorodibenzo-p-dioxin (TCDD)	1 kg
35. Tetraethyl lead	50 t
36. Tetramethyl lead	50 t
Part B Highly reactive substances and explosive substances	
37. Acetylene (ethyne) when a gas subject to a pressure ≤620 millibars above that of the atmosphere, and not otherwise deemed to be an explosive by virtue of Order in Council No 30,[(a)] as amended by the Compressed Acetylene Order 1947,[(b)] or when contained in a homogeneous porous substance in cylinders in accordance with Order of Secretary of State No 9,[(c)] made under the Explosives Act 1875.[(d)]	50 t
38. Ammonium nitrate and mixtures containing ammonium nitrate where the nitrogen content derived from the ammonium nitrate >28% of the mixture by weight other than: (i) mixtures to which the Explosives Act 1875 applies; (ii) ammonium nitrate based products manufactured chemically for use as fertilizer which comply with Council Directive 80/876/EEC;[(e)] or (iii) compound fertilizers.	500 t
39. Aqueous solutions containing >90 parts by weight of ammonium nitrate per 100 parts by weight of solution.	500 t
40. Ammonium nitrate based products manufactured chemically for use as fertilizers which comply with Council Directive 80/876/EEC and compound fertilizers where the nitrogen content derived from the ammonium nitrate >28% of the mixture by weight.	1000 t
41. 2,2-Bis(tert-butylperoxy)butane (>70%)	5 t
42. 1,1-Bis(tert-butylperoxy)cyclohexane (>80%)	5 t

Table 1.7 Cont'd

Hazardous substance	*Controlled quantity*
43. tert-Butyl peroxyacetate (>70%)	5 t
44. tert-Butyl peroxyisobutyrate (>80%)	5 t
45. tert-Butyl peroxyisopropylcarbonate (>80%)	5 t
46. tert-Butyl peroxymaleate (>80%)	5 t
47. tert-Butyl peroxypivalate (>77%)	5 t
48. Cellulose nitrate other than:	
(i) cellulose nitrate to which the Explosives Act 1875 applies; or	50 t
(ii) solutions of cellulose nitrate where the nitrogen content of the cellulose nitrate ≤12.3% by weight and the solution contains ≤55 parts of cellulose nitrate per 100 parts by weight of solution.	
49. Dibenzyl peroxydicarbonate (>90%)	5 t
50. Diethyl peroxydicarbonate (>30%)	5 t
51. 2,2-Dihydroperoxypropane (>30%)	5 t
52. Di-isobutyryl peroxide (>50%)	5 t
53. Di-n-propyl peroxydicarbonate (>80%)	5 t
54. Di-sec-butyl peroxydicarbonate (>80%)	5 t
55. Ethylene oxide	5 t
56. Ethyl nitrate	50 t
57. 3,3,6,6,9,9-Hexamethyl-1,2,4,5-tetroxacyclononane (>75%)	5 t
58. Hydrogen	2 t
59. Liquid oxygen	500 t
60. Methyl ethyl ketone peroxide (>60%)	5 t
61. Methyl isobutyl ketone peroxide (>60%)	5 t
62. Peracetic acid (>60%)	5 t
63. Propylene oxide	5 t
64. Sodium chlorate	25 t
65. Sulphur dichloride	1 t
Part C Flammable substances (unless specifically named in Parts A and B)	
66. Liquefied petroleum gas, such as commercial propane and commercial butane, and any mixtures thereof, when held at a pressure >1.4 bar absolute.	25 t
67. Liquefied petroleum gas, such as commercial propane and commercial butane, and any mixture thereof, when held under refrigeration at a pressure ≤1.4 bar absolute.	50 t
68. Gas or any mixture of gases which is flammable in air, when held as a gas.	15 t
69. A substance or any mixture of substances, which is flammable in air, when held above its boiling point (measured at 1 bar absolute) as a liquid or as a mixture of liquid and gas at a pressure >1.4 bar absolute.	25 t
70. A liquefied gas or any mixture of liquefied gases, which is flammable in air and has a boiling point <0°C (measured at 1 bar absolute), when held under refrigeration or cooling at a pressure ≤1.4 bar absolute.	50 t
71. A liquid or any mixture of liquids not included in entries 68 to 70 above, which has a flash point <21°C.	10 000 t

[a] S.R. & O. 1937/54.
[b] S.R. & O. 1947/805.
[c] S.R. & O. 1919/869.
[d] 1875 c.17.
[e] OJ No L250, 23.9.80, p. 7.

All chemical operations produce waste either as solid wastes (including pastes, sludge and drummed liquids), liquid effluents, or gaseous emissions (including gases, particulate solids, mists and fogs). Relevant data are summarized in Chapters 16 and 17.

Since data have been collated from a variety of sources, and tend to be presented in mixed units, and because rapid conversion of units is an advantage in many on-site situations, conversion tables are included in Chapter 18. Finally, since safety with chemicals cannot be addressed exhaustively in a handbook, selected sources of reliable current information on chemical hazards and their control are listed in Chapter 19.

2

Terminology

ACID A chemical compound whose aqueous solution turns blue litmus paper red, reacts with and dissolves certain metals to form salts, and reacts with bases to produce salts and water. They are capable of transferring a hydrogen ion (proton) in solution.

ACUTE Describes a severe and often dangerous condition in which relatively rapid changes occur.

ACUTE TOXICITY Adverse health effects occurring within a short time period of exposure to a single dose of a chemical or as a result of multiple exposures over a short time period, e.g. 24 hours.

AEROSOL A colloidal suspension of liquid or solid particles dispersed in gas.

AFFF, AQUEOUS FILM-FORMING FOAM Fire-fighting foam which flows on burning liquid as a film, providing rapid knock-down.

ALCOHOL-RESISTANT FOAM Foam for use against fires involving liquids miscible with water, e.g. alcohol, acetone.

ANION A negatively charged atom or group of atoms, or a radical which moves to the positive pole (anode) during electrolysis.

ANOXIA Deficient supply of oxygen to tissues.

ANTIBODY A modified protein circulating in the serum of an animal, synthesized in response to a foreign molecule (antigen) that has entered the body.

ANTIGEN A foreign substance (usually a protein) that stimulates formation of antibody.

ASPHYXIA The result of a diminished supply of oxygen to the blood and tissues and interference with the respiratory function. Simple anoxia may be caused by 'inert gases', e.g. nitrogen, and some flammable gases, e.g. methane. Toxic anoxia may be caused by certain substances, e.g. carbon monoxide and hydrogen cyanide, which interfere with the body's ability to transfer or utilize oxygen in the tissues. Rapid unconsciousness and death can occur in either case.

ASTHMA Periodic attacks of wheezing, chest tightness and breathlessness resulting from constriction of the airways.

ATOM The smallest unit of an element incapable of further subdivision in the course of a chemical reaction.

ATOMIC NUMBER The number of protons in an atomic nucleus.

ATOPY Hypersensitivity where tendency to allergy is inherited.

AUTO-IGNITION TEMPERATURE The minimum temperature required to initiate or cause self-sustained combustion of material in the absence of any external source of energy. (Values may change significantly with geometry, gas/vapour concentration, and presence of catalyst.) Any ignition source must be at a temperature of, or greater than, the ignition temperature of the specific substance.

BASE A chemical compound whose aqueous solution turns red litmus paper blue and is capable of accepting or receiving a proton from another substance. They react with acids to form salts and water.

BATNEEC Term used in the Environmental Protection Act and other legislation. Certain polluting processes are required to use the Best Available Techniques Not Entailing Excessive Cost (BATNEEC) to reduce the environmental impact of a prescribed process as far as possible. Environment Agency inspectors determine what constitutes BATNEEC for each application and are to change the definition as improved technologies or techniques become available.

BIOCHEMICAL OXYGEN DEMAND (BOD) Official term given to measure how polluting organic industrial effluent is when it is discharged into water. This effluent is feed for bacteria which consume oxygen, making it more difficult for plant and fish life to survive. The lower the BOD level, the less polluting the effluent.

BLACK LIST The Black List was introduced by the EC in Directive 76/464/EEC on dangerous substances released into water as list I. It contains substances selected mainly on the basis of their toxicity, persistence and accumulation in living organisms and in sediment.

BEST PRACTICABLE ENVIRONMENTAL OPTION (BPEO) Organizations may be encouraged to undertake systematic decision processes with a view to seeking the BPEO that provides the most benefit (or least damage) to the environment, at an acceptable cost.

BLEVE, BOILING LIQUID EXPANDING VAPOUR EXPLOSION Instantaneous release and ignition of flammable vapour upon rupture of a vessel containing flammable liquid above its atmospheric boiling point.

BLOWING AGENT Chemical liable to decomposition at low temperature to produce a large volume of gas.

CARCINOGEN An agent (whether chemical, physical or biological) capable of increasing the incidence of malignant neoplasms. Defined in Regulation 2 of the Control of Substances Hazardous to Health Regulations 1999 as:

(a) any substance or preparation which if classified in accordance with the classification provided for by Regulation 5 of the Chemicals (Hazard Information and Packaging for Supply) Regulations 1994 would be in the category of danger, carcinogenic (category 1) or carcinogenic (category 2) whether or not the substance or preparation would be required to be classified under those Regulations; or
(b) any substance or preparation:
    (i) listed in Schedule 1, or
    (ii) arising from a process specified in Schedule 1 which is a substance hazardous to health.

CATION A positively charged atom or group of atoms, or a radical which moves to the negative pole (cathode) during electrolysis.

CHEMICAL BOND Strong forces of attraction holding atoms together in molecules or crystalline salts.

CHLOROFLUOROCARBONS (CFCS) Organic substances containing chlorine and fluorine which were initially thought to be harmless and found extensive use, e.g., as propellants because they are largely non-flammable. Some CFCs have since been found to be one of the main sources of atmospheric ozone depletion and a greenhouse gas. Until recently they were used extensively as aerosol propellants, solvents, refrigerants and in foam making. Many countries have now agreed to eliminate CFCs as soon as possible.

CHLORINATED HYDROCARBONS Hydrocarbons containing chlorine atoms, e.g. trichloroethylene. Some of these chemicals accumulate in the food chain and do not readily degrade. Some plastics which contain certain chlorinated hydrocarbons release dioxins into the air, when burnt at low temperatures.

CHRONIC Occurring for a prolonged period.

CHRONIC TOXICITY Adverse health effects resulting from repeated daily exposures to a chemical for a significant period.

CLASS A FIRE A fire involving solids, normally organic, in which combustion generally occurs with the formation of glowing embers.

CLASS A POISON (USA) A toxic gas/liquid of such a nature that a very small amount of the gas, or vapour of the liquid, in air is dangerous to life.

CLASS B FIRE A fire involving liquids or liquefiable solids.

CLASS B POISON (USA) Any substance known to be so toxic that it poses a severe health hazard during transportation.

CLASS C FIRE A fire involving gases or liquefied gases in the form of a liquid spillage, or a liquid or gas leak.

CLASS D FIRE A fire involving metals.

CNS DEPRESSANT Substances, e.g. anaesthetics and narcotics, which depress the activity of the central nervous system. Symptoms following exposure include headache, dizziness, loss of consciousness, respiratory or cardiac depression, death.

CONFINED SPACE A space which is substantially, although not always entirely, enclosed and where there is a reasonably foreseeable risk of serious injury from hazardous substances or conditions within the space or nearby. The risks may include flammable substances; oxygen deficiency or enrichment; toxic gases, fume or vapour; ingress or presence of liquids; free-flowing solids; presence of excessive heat. For the purpose of the Confined Spaces Regulations 1997 a 'confined space' means any place, including any chamber, tank, vat, silo, pit, trench, pipe, sewer, flue, well or other similar space in which, by virtue of its enclosed nature, there arises a reasonably foreseeable specified risk.

CONTACT DERMATITIS Inflammation of the skin due to exposure to a substance that attacks its surface.

CONTROLLED WASTE All household, industrial or commercial waste of any quantity or description.

CORROSIVE A substance that chemically attacks a material with which it has contact (body cells, materials of construction).

COSHH (CONTROL OF SUBSTANCES HAZARDOUS TO HEALTH) The Control of Substances Hazardous to Health Regulations 1999 establish the responsibilities of employers with regard to all substances which pose a health hazard in the workplace.

CRYOGEN A substance used to obtain temperatures far below freezing point of water, e.g. $< -78°C$.

CVCE (CONFINED VAPOUR CLOUD EXPLOSION) Explosion of a gas or vapour which is initially 'confined' within a vessel, building, piping, etc.

DANGEROUS SUBSTANCES (UK) Defined substances which may be hazardous to the fire services in an emergency. (Dangerous Substances (Notification and Marking of Sites) Regulations 1990.)

Defined substances over which control is exercised for conveyance in all road tankers or in tank containers >3 m^3 capacity. (The Carriage of Dangerous Goods by Road Regulations 1996.)

Defined substances covered by a comprehensive system to inform consumers of potential dangers and to reduce the hazard when carried by road. The Chemical (Hazard Information and Packaging for Supply Regulations 1994).

Defined substances, including all toxic gases, all flammable gases, asbestos and most hazardous wastes, for which carriage in packages or in bulk is controlled. (The Carriage of Dangerous Goods by Road Regulations 1996).

DETONATION Explosion in which the flamefront advances at more than supersonic velocity.

DISCHARGE CONSENTS Permission to discharge trade effluent directly into controlled waters is given by the National Rivers Authority in the form of a discharge consent which will specify amounts and conditions. Discharges to public sewers are controlled by discharge consents by one of the ten Water Service Companies.

DUST Solid particles generated by mechanical action, present as airborne contaminant (e.g. <75 µm in size).

DUTY OF CARE The concept of the duty of care for waste is set out in Section 34 of the Environmental Protection Act (1990) which states that it is the duty of any person who imports, produces, carries, keeps, treats or disposes of controlled waste to keep that waste properly under control.

ECOTOXICOLOGY The study of toxic effects of chemical and physical agents on living organisms as well as human beings, especially on populations and communities within defined ecosystems.

ENDOTHERMIC REACTION A chemical reaction resulting in absorption of heat.

ENVIRONMENT AGENCY The Environment Agency provides a comprehensive approach to the protection and management of the environment by combining the regulation of land, air and water. Its creation is a major and positive step, merging the expertise of the National Rivers Authority, Her Majesty's Inspectorate of Pollution, the Waste Regulation Authorities and several smaller units from the Department of the Environment.

EPIDEMIOLOGY The study in populations of health factors affecting the occurrence and resolution of disease and other health-related conditions.

ERYTHEMA Reddening of skin, inflammation.

EXOTHERMIC REACTION A chemical reaction in which heat is released and, unless temperature is controlled, may lead to runaway conditions.

FIBROSIS Scarring, usually of lung tissue.

FIRE POINT The minimum temperature at which a mixture of gas/vapour and air continues to burn in an open container when ignited. The value is generally above the flash point.

FLAMMABLE RANGE The concentrations of flammable gas or vapour between the LEL and UEL at a given temperature.

FLASH POINT The lowest temperature required to raise the vapour pressure of a liquid such that vapour concentration in air near the surface of the liquid is within the flammable range, and as such the air/vapour mixture will ignite in the presence of a suitable ignition source, usually a flame. (Open cup values are approximately 5.5° to 8.3°C higher than the closed cup values.)

FOG (MISTS) Liquid aerosols formed either by condensation of a liquid on particulate nodes in air or by uptake of liquid by hygroscopic particles.

FUME Airborne solid particles (usually <0.1μm) that have condensed from the vapour state.

HAZARD The inherent property of a substance capable of causing harm (e.g. toxicity, radioactivity, flammability, explosivity, reactivity, instability). In a broader context anything that can cause harm, e.g. electricity, oxygen-deficiency, machinery, extreme temperature.

HAZARDOUS WASTE An unofficial class of industrial wastes which have to be disposed of with particular care. In the UK the closest definition is for 'special wastes'. Certain toxic organic wastes, such as PCBs, have to be burned in high-temperature incinerators.

HEAVY METALS A group of metals which are sometimes toxic and can be dangerous in high concentrations. The main heavy metals covered by legislation are cadmium, lead, and mercury. Industrial activities such as smelting, rubbish burning, waste disposal and adding lead to petrol increase the amount of toxic heavy metals in the environment.

HUMIDIFIER FEVER A flu-like illness caused by inhalation of fine droplets of water from humidifiers that have become contaminated.

HYDROCARBONS Organic compounds that contain only hydrogen and carbon. The major sources of hydrocarbons in the atmosphere are vehicle emissions (unburned fuel) and gas leaks. Contributes to acid rain.

HYGIENE STANDARD See OES, MEL, TLV.

INERTING Depression of the flammable limits of a flammable gas/vapour–air mixture by the addition of an inert gas, e.g. nitrogen, carbon dioxide, or similar mixtures, to render it non-flammable.

ION An isolated electron or positron, or an atom or molecule, which by loss or gain of one or more electrons has acquired a net electric charge.

IONIZING RADIATION The transfer of energy in the form of particles or electromagnetic waves of a wavelength of 100 nanometers or less or a frequency of 3×10^{15} hertz or more capable of producing ions directly or indirectly.

ISOTOPE One of two or more atoms having the same atomic number but different mass number.

IPC (INTEGRATED POLLUTION CONTROL) Under this new integrated approach to pollution control land, water and air are to be treated collectively rather than as separate environmental media. Industries are given consents to pollute with the effect on all three media being taken into consideration. IPC was introduced by the Environmental Protection Act (1990) and an EC system of Integrated Pollution, Prevention and Control is being introduced.

JET FIRE Fuel burning as a flame when flammable gas or vapour issues from a pipe, or other orifice, and burns on the orifice.

KINETICS The branch of physical chemistry concerned with measuring and studying the rates and mechanisms of chemical reactions.

LANDFILL Disposal of waste in the ground. This method is commonly used for both domestic waste and more hazardous chemical waste. Landfill sites used for difficult and potentially-dangerous wastes are now engineered, managed and monitored to prevent poisons leaking out.

LC_{50} The calculated concentration of a substance that causes death in 50% of a test population under prescribed conditions in a prescribed period of time (normally expressed as ppm or mg/m^3 for gases, mg/l for liquids).

LD_{50} The calculated dose of chemical (mg per kg body weight) causing death in 50% of test population. (The species of animal, route of administration, any vehicle used to dissolve or suspend the material, and the time period of exposure should be reported.)

LEACHATE Liquid that leaks from waste disposal sites. (In a broader sense liquid, e.g. solution, removed from a solid by a solvent, such as water.)

LEGIONNAIRES' DISEASE Infection caused by inhaling a fine spray of airborne water carrying *Legionella pneumophila* bacteria.

LEL (LOWER EXPLOSIVE, OR FLAMMABLE, LIMIT) The minimum concentration of a gas, vapour, mist or dust in air at a given pressure and temperature that will propagate a flame when exposed to an efficient ignition source. Generally expressed as % by volume for gases and vapours, and as mg/m^3 for mists or dusts.

LPG (LIQUEFIED PETROLEUM GAS) Petroleum gas stored or processed as a liquid in equilibrium with vapour by refrigeration or pressurization. The two LPGs in general use are commercial propane and commercial butane supplied to product specifications, e.g. BS 4250. (These, or mixtures thereof, comprise LPG for the purpose of the Highly Flammable Liquids and Liquefied Petroleum Gas Regulations 1972.)

MAJOR ACCIDENT An occurrence (including in particular, a major emission, fire or explosion) resulting from uncontrolled developments in the course of operation of any establishment and leading to serious danger to human health or the environment, and involving one or more dangerous substances. Requirements with respect to the control of major accident hazards involving dangerous substances apply to defined establishments under the Control of Major Accident Hazards Regulations 1999.

MASS NUMBER The sum of the number of protons and neutrons in the nucleus of an atom.

MEL, MAXIMUM EXPOSURE LIMIT (UK) The maximum concentration of an airborne substance (averaged over a reference period) to which employees may be exposed by inhalation under any circumstances. (Listed in 'Occupational exposure limits', EH40/–HSE.)

METAL FUME FEVER Non-specific, self-limiting illness resembling an attack of influenza caused mainly by exposure to fumes of zinc, copper, or magnesium and less frequently due to exposure to other metal fumes. Exposures occur from molten metals, e.g. in smelting, galvanizing, welding.

METALWORKING FLUID Fluid applied to a tool and workpiece to cool, lubricate, carry away particles of waste and provide corrosion protection. Generally comprising neat mineral oils, or water-based materials, or a mixture of the two. Fluids may also contain emulsifiers, stabilizers, biocides, corrosion inhibitors, fragrances and extreme pressure additives.

MINERAL OIL Oil derived from petroleum. Includes a wide range of hydrocarbons from light oils, kerosene and gas oils, to the heavier fuel and lubricating oils.

MOLECULES Groups of atoms held together by strong chemical forces and forming the smallest unit of a compound. The atoms may be identical, e.g. H_2 or different, e.g. H_2O.

MULTIPLE CHEMICAL SENSITIVITY An acquired disorder characterized by recurrent symptoms, referable to multiple organ systems, occurring in response to many chemically-unrelated compounds at doses far below those established in the general population to cause harmful effects. No single widely accepted test of physiologic function can be shown to correlate with symptoms.

MUTAGEN A chemical or physical agent that can cause a change (mutation) in the genetic material of a living cell.

NARCOSIS Drowsiness or sleepiness.

NATURAL GAS Flammable gas consisting essentially of methane with very minor proportions of other gases. Flammable limits approximately 5–15%. Odourized for commercial distribution within the UK.

NRA (NATIONAL RIVERS AUTHORITY) The National Rivers Authority were the body responsible for the management of water resources and the control of water pollution in England and Wales. They are now part of the Environment Agency.

ODOUR THRESHOLD The minimum concentration of a substance at which the majority of test subjects can detect and identify the substance's characteristic odour.

OES, OCCUPATIONAL EXPOSURE STANDARD (UK) The concentration of an airborne substance (averaged over a reference period) at which, according to current knowledge, there is no evidence that it is likely to be injurious to employees if they are exposed by inhalation, day after day. (Specified by HSC in Guidance Note EH40.)

OXIDIZING AGENT Compound that gives up oxygen easily or removes hydrogen from another compound. It may comprise a gas, e.g. oxygen, chlorine, fluorine, or a chemical which releases oxygen, e.g. a nitrate or perchlorate. A compound that attracts electrons.

OXYGEN DEFICIENCY Depletion of oxygen content in an atmosphere to below the normal 21%. Exposure to <18% must not be permitted. Concentrations 6% to 10% oxygen can lead to loss of consciousness.

OXYGEN ENRICHMENT Increase in oxygen content of air to above the normal 21%. Enrichment within a room to >25% can promote or accelerate combustion.

OZONE A reactive form of oxygen the molecule of which contains 3 atoms of oxygen. In the ozone layer it protects the earth by filtering out ultra-violet rays. At ground level, as a constituent of photochemical smog, it is an irritant and can cause breathing difficulties.

OZONE LAYER A thin layer of ozone that lies about 25 kilometres above the earth in the stratosphere. Forms a protective screen against harmful radiation by filtering out ultra-violet rays from the sun.

PARAOCCUPATIONAL EXPOSURE Exposure of workers to an airborne contaminant from a nearby process or operation not forming part of their jobs. Also termed 'neighbourhood exposure'.

PCBS (POLYCHLORINATED BIPHENYLS) Toxic synthetic chemicals with excellent heat resistance and low electrical conductivity properties. Now little used but considerable quantities remain in old electrical equipment. Produces dioxins and polychlorinated dibenzo-furans when burned below 1200°C. PCBs are toxic and bio-accumulative.

PERCUTANEOUS ABSORPTION Absorption via the skin, e.g. due to local contamination or a splash of chemical.

PERMIT-TO-WORK A document needed when the safeguards provided in normal production are unavailable and the manner in which a job is done is critical to safety. Identifies conditions required for safe operation.

PNEUMOCONIOSIS A group of lung diseases of a chronic fibrotic character due to the inhalation and retention in the lungs of a variety of industrial dusts. The main diseases are asbestosis, silicosis, coalworkers' pneumoconiosis and mixed-dust pneumoconiosis; less common pneumoconioses are associated with talc, clay or aluminium.

POOL FIRE A fire involving a flammable liquid spillage onto ground or onto water, or within a storage tank or trench. The pool size depends upon the scale and local topography. Fire engulfment and radiant heat pose the main risks.

PRACTICABLE Capable of being done in the light of current knowledge and invention.

PRESCRIBED DISEASE A disease prescribed for the purpose of payment of disablement benefit under the Social Security Act 1975 and the Social Security (Industrial Injuries) (Prescribed Diseases) Regulations 1985 and subsequent amendments. (Conditions due to physical agents, biological agents and miscellaneous conditions are classified in addition to conditions due to chemical agents.)

PRESCRIBED PROCESS Industrial process which requires an official authorization because of the likelihood of causing pollution under the provisions of the Environmental Protection Act (1990).

PRESCRIBED SUBSTANCE A substance controlled by Section I of the Environmental Protection Act (1990) because of its potential to pollute. Different substances are prescribed for release to different environmental media.

PRESSURE SYSTEM Defined in the Pressure System Safety Regulations 2000 as a system containing one or more pressure vessels of rigid construction, any associated pipework and protective devices; the pipework with its protective devices to which a transportable gas container is, or is intended to be, connected; or a pipeline and its protective devices which contains or is liable to contain a relevant fluid, but does not include a transportable gas container. Here 'relevant fluid' is steam; any fluid or mixture of fluids which is at a pressure of >0.5 bar above atmospheric pressure, and which fluid or a mixture of fluids is a gas, or a liquid which would have a vapour pressure of >0.5 bar above atmospheric pressure when in equilibrium with its vapour at either the actual temperature of the liquid or 17.5°C; or a gas dissolved under pressure in a solvent contained in a porous substance at ambient temperature and which could be released from the solvent with the application of heat.

PULMONARY OEDEMA Production of watery fluid in the lungs.

PYROPHORIC SUBSTANCE A material that undergoes such vigorous oxidation or hydrolysis (often with evolution of highly-flammable gases) when exposed to atmospheric oxygen or to water, that it rapidly ignites without an external source of ignition. This is a special case of spontaneous combustion.

REASONABLY PRACTICABLE The implication that the quantum of risk is balanced against the sacrifice or cost in terms of money, time and trouble necessary to avert that risk. If the risk outweighs the sacrifice or cost, additional precautions are necessary.

RECYCLING The use of materials, usually after further processing, which otherwise would be thrown away. Becoming common practice in industry, especially with expensive commodities such as chemical solvents although many products require a commercial subsidy in order to make recycling viable.

RED LIST The Red List was drawn up by the UK Government in 1989 in response to international conferences of the states bordering the North Sea. The aim was to reduce inputs of Red List substances by 50% by 1995, from 1985 levels. All those substances listed on the Red List are included on the EC Black List.

Authorizations to discharge Red List substances are dealt with by the Environment Agency under Integrated Pollution Control (IPC).

REDUCING AGENT A material that adds hydrogen to an element or compound; a material that adds an electron to an element or compound.

REPORTABLE DISEASE (UK) A disease which must be reported to the authorities when linked to specified types of work. (The Reporting of Injuries Diseases and Dangerous Occurrences Regulations 1995.)

RESPIRABLE DUST That fraction of total inhalable dust which penetrates to the gas exchange region of the lung (usually considered to be in the range 0.5 μm–7 μm).

RESPIRATORY SENSITIZER (ASTHMAGEN) A substance which can cause an individual's respiratory system to develop a condition which makes it 'over-react' if the substance is inhaled again. Such an individual is 'sensitized'; over-reaction is then likely to occur at concentrations of the substance which have no effect on unsensitized persons and lead to characteristic symptoms, e.g. rhinitis (a runny nose), conjunctivitis or in severe cases asthma or alveolitis.

RISK The likelihood that a substance will cause harm in given circumstances.

SAFE SYSTEM OF WORK A formal procedure resulting from systematic examination of a task to identify all the hazards. Defines safe methods to ensure that hazards are eliminated or risks controlled.

SEALED SOURCE A source containing any radioactive substance whose structure is such as to prevent, under normal conditions of use, any dispersion of radioactive substances into the environment, but it does not include any radioactive substance inside a nuclear reactor or any nuclear fuel element.

SENSITIZATION DERMATITIS Inflammation of the skin due to an allergic reaction to a sensitizer.

SENSITIZER A substance that causes little or no reaction in a person upon initial exposure but which will provoke an allergic response on subsequent exposures.

SICK BUILDING SYNDROME A group of symptoms more common in workers in certain buildings and which are temporarily related to working in them. Symptoms include lethargy, tiredness, headache; also sore/dry eyes, dry throat, dry skin, symptoms suggestive of asthma, blocked or runny nose. Cause is multifunctional but does include agents encountered in the workplace.

SMOKE Particulate matter (usually <0.5 μm in diameter) in air resulting usually from combustion, including liquids, gases, vapours and solids.

SOLVENTS Liquids that dissolve other substances. Chemical solvents are used widely in industry: e.g. by pharmaceutical makers to extract active substances; by electronics manufacturers to wash circuit boards; by paint makers to aid drying. Solvents can cause air and water pollution and some can be responsible for ozone depletion.

SPECIAL WASTE Controlled waste which is subject to special regulations regarding its control and disposal because of its difficult or dangerous characteristics. The UK definition of special waste is similar, but not identical, to the EC's hazardous waste.

SPONTANEOUS COMBUSTION Combustion that results when materials undergo atmospheric oxidation at such a rate that the heat generation exceeds heat dissipation and the heat gradually builds up to a sufficient degree to cause the mass of material to inflame.

STEAM EXPLOSION Overpressure associated with the rapid expansion in volume on instantaneous conversion of water to steam.

SUBSTANCE HAZARDOUS TO HEALTH As defined in Regulation 2 of the Control of Substances Hazardous to Health Regulations 1999,

(a) a substance which is listed in Part 1 of the approved supply list as dangerous for supply within the meaning of the Chemicals (Hazard Information and Packaging for Supply) Regulations 1994 and for which an indication of danger specified for the substance in Part V of that list is very toxic, toxic, harmful, corrosive or irritant;
(b) a substance for which the Health and Safety Commission has approved a maximum exposure limit or an occupational exposure standard;
(c) a biological agent;
(d) dust of any kind except of a substance within para. (a) or (b) above, when present at a concentration in air equal to or greater than:
 (i) 10 mg/m^3, as a time-weighted average over an 8-hr period, of total inhalable dust, or
 (ii) 4 mg/m^3, as a time-weighted average over an 8-hr period of respirable dust;
(e) a substance, not covered by (a) or (b) above, which creates a hazard to the health of any person which is comparable with the hazards created by substances mentioned in those sub-paragraphs.

SYNERGISTIC When the combined effect, e.g. of exposure to a mixture of toxic chemicals, is greater than the sum of the individual effects.

SYSTEMIC POISONS Substances which cause injury at sites other than, or as well as, at the site of contact.

TERATOGEN A chemical or physical agent that can cause defects in a developing embryo or foetus when the pregnant female is exposed to the harmful agent.

THERMODYNAMICS The study of laws that govern the conversion of one form of energy to another.

TLV-C, THRESHOLD LIMIT VALUE – CEILING (USA) A limit for the atmospheric concentration of a chemical which may not be exceeded at any time, even instantaneously in workroom air.

TLV-STEL, THRESHOLD LIMIT VALUE – SHORT TERM EXPOSURE LIMIT (USA) A maximum limit on the concentration of a chemical in workroom air which may be reached, but not exceeded, on up to four occasions during a day for a maximum of 15 minutes each time with each maximum exposure separated by at least one hour.

TLV-TWA, THRESHOLD LIMIT VALUE – TIME WEIGHTED AVERAGE (USA) A limit for the atmospheric concentration of a chemical, averaged over an 8-hr day, to which it is believed that most people can be exposed without harm.

TOTAL INHALABLE DUST Airborne material capable of entering the nose and mouth during breathing and hence available for deposition in the respiratory tract.

TOXIC WASTE Poisonous waste, usually certain organic chemicals such as chlorinated solvents.

TRADE EFFLUENT Any waste water released from an industrial process or trade premises with the exception of domestic sewage.

UEL, UPPER EXPLOSIVE (OR FLAMMABLE) LIMIT The maximum concentration of gas, vapour, mist or dust in air at a given pressure and temperature in which a flame can be propagated.

UVCE (UNCONFINED VAPOUR CLOUD EXPLOSION) Explosion which may occur when a large mass of flammable vapour, normally >5 tonnes, after dispersion in air to produce a mixture within the flammable range is ignited in the open. Intense blast damage results, often causing 'domino effects', e.g. secondary fires.

VALENCY The number of potential chemical bonds that an element may form with other elements.

VOCS (VOLATILE ORGANIC COMPOUNDS) Gaseous pollutants whose sources include vehicle emissions and solvents.

WDAS (WASTE DISPOSAL AUTHORITIES) Body responsible for planning and making arrangements for waste disposal in their area with the waste disposal companies and for providing household waste dumps under the Environmental Protection Act 1990.

3

General principles of chemistry

Introduction

This chapter provides a brief insight into selected fundamental principles of matter as a background to the appreciation of the hazards of chemicals.

Chemistry is the science of chemicals which studies the laws governing their formation, combination and behaviour under various conditions. Some of the key physical laws as they influence chemical safety are discussed in Chapter 4.

Atoms and molecules

Chemicals are composed of atoms, discrete particles of matter incapable of further subdivision in the course of a chemical reaction. They are the smallest units of an element. Atoms of the same element are identical and equal in weight. All specimens of gold have the same melting point, the same density, and the same resistance to attack by mineral acids. Similarly, all samples of iron of the same history will have the same magnetism. Atoms of different elements have different properties and differ in weight.

Atoms are comprised of negatively charged electrons orbiting a nucleus containing positively-charged protons and electrically-neutral neutrons as described in Chapter 11. The orbits of electrons are arranged in energy shells. The first shell nearest to the nucleus can accommodate two electrons, the second shell up to eight electrons, the third 18 electrons, and the fourth 32 electrons. This scheme is the 'electronic configuration' and largely dictates the properties of chemicals. Examples are given in Table 3.1.

Chemicals are classed as either elements or compounds. The former are substances which cannot be split into simpler chemicals, e.g. copper. There are 90 naturally-occurring elements and 17 artificially produced. In nature the atoms of some elements can exist on their own, e.g. gold, whilst in others they link with other atoms of the same element to form molecules, e.g. two hydrogen atoms combine to form a molecule of hydrogen. Atoms of different elements can combine in simple numerical proportions 1:1, 1:2, 1:3, etc. to produce compounds, e.g. copper and oxygen combine to produce copper oxide; hydrogen and oxygen combine to produce water. Compounds are therefore chemical substances which may be broken down to produce more than one element. Molecules are the smallest unit of a compound.

Substances such as brass, wood, sea water, and detergent formulations are *mixtures* of chemicals. Two samples of brass may differ in composition, colour and density. Different pieces of wood of the same species may differ in hardness and colour. One sample of sea water may contain more salt and different proportions of trace compounds than another. Detergent formulations differ

Table 3.1 Electronic configuration of selected elements

Element	*Symbol*	*Atomic (proton) number*	*No. electrons in 1st shell*	*No. electrons in 2nd shell*	*No. electrons in 3rd shell*	*No. electrons in 4th shell*
Hydrogen	H	1	1			
Helium	He	2	2			
Lithium	Li	3	2	1		
Beryllium	Be	4	2	2		
Boron	B	5	2	3		
Carbon	C	6	2	4		
Nitrogen	N	7	2	5		
Oxygen	O	8	2	6		
Fluorine	F	9	2	7		
Neon	Ne	10	2	8		
Sodium	Na	11	2	8	1	
Magnesium	Mg	12	2	8	2	
Aluminium	Al	13	2	8	3	
Silicon	Si	14	2	8	4	
Phosphorus	P	15	2	8	5	
Sulphur	S	16	2	8	6	
Chlorine	Cl	17	2	8	7	
Argon	Ar	18	2	8	8	
Potassium	K	19	2	8	8	1
Calcium	Ca	20	2	8	8	2
Scandium	Sc	21	2	8	9	2
Titanium	Ti	22	2	8	10	2
Vanadium	Va	23	2	8	11	2
Chromium	Cr	24	2	8	13	1
Manganese	Mn	25	2	8	13	2

between brands. It is possible to isolate the different chemical components from mixtures by physical means.

Periodic table

The number of protons plus neutrons in an atom is termed the mass number. The number of protons (which also equals the number of electrons) is the atomic number. When elements are arranged in order of their atomic numbers and then arranged in rows, with a new row starting after each noble gas, the scheme is termed the periodic table. A simplified version is shown in Table 3.2.

The following generalizations can be made. Period 2 elements at the top of Groups I to VII tend to be anomalous. Atomic and ionic radii decrease across a period but increase down a group. Elements in a period have similar electronic configurations and those in groups have the same outer electronic arrangements. Elements at the top of a group tend to differ more from the succeeding elements in the group than they do from one another. Metals are on the left of the table and non-metals on the right. Elements such as silicon and germanium are borderline. Elements become less metallic on crossing a period and more metallic on descending a group. The Group I elements are alkali metals with reactivity increasing from top to bottom of the table. So the exothermic reaction of potassium (K) with water is more vigorous than that of lithium (Li).

Electronegativities increase across a period to a maximum with Group VII, the halogens, for which reactivity decreases from top to bottom of the table. Elements in Group 0 are unreactive

Table 3.2 Periodic table of the elements

Group / Period	I	II											III	IV	V	VI	VII	0
1	H 1																	He 2
2	Li 3	Be 4											B 5	C 6	N 7	O 8	F 9	Ne 10
3	Na 11	Mg 12	Transition elements										Al 13	Si 14	P 15	S 16	Cl 17	Ar 18
4	K 19	Ca 20	Sc 21	Ti 22	V 23	Cr 24	Mn 25	Fe 26	Co 27	Ni 28	Cu 29	Zn 30	Ga 31	Ge 32	As 33	Se 34	Br 35	Kr 36
5	Rb 37	Sr 38	Y 39	Zr 40	Nb 41	Mo 42	Tc 43	Ru 44	Rh 45	Pd 46	Ag 47	Cd 48	In 49	Sn 50	Sb 51	Te 52	I 53	Xe 54
6	Cs 55	Ba 56	* 57–71	Hf 72	Ta 73	W 74	Re 75	Os 76	Ir 77	Pt 78	Au 79	Hg 80	Tl 81	Pb 82	Bi 83	Po 84	At 85	Rn 86
7	Fr 87	Ra 88	** 89–102															
***Lanthanide series**			La 57	Ce 58	Pr 59	Nd 60	Pm 61	Sm 62	Eu 63	Gd 64	Tb 65	Dy 66	Ho 67	Er 68	Tm 69	Yb 70	Lu 71	
****Actinide series**			Ac 89	Th 90	Pa 91	U 92	Np 93	Pu 94	Am 95	Cm 96	Bk 97	Cf 98	Es 99	Fm 100	Md 101	No 102		

and termed 'inert', or 'noble', gases. Elements 21–30, 39–48 and 72–80 are termed 'transition elements' whilst those between 57 and 71 are termed 'lanthanides' or 'rare earth' elements, and elements 89–102 the 'actinides'.

Valency

Atoms combine to form molecules or compounds by linking together using chemical bonds. The number of bonds that elements are able to produce is termed their valency. Chemical changes are most conveniently represented by use of symbols and formulae in chemical equations. Atoms of each element are represented by a symbol. The chemical formula of a compound is merely a composite of its constituent atoms together with numerical indications of the ratio of the combining elements. If a number of atoms of one kind are present in a molecule the number is indicated by a subscript. So carbon dioxide is CO_2, ammonia NH_3, nitric acid HNO_3 and ammonium nitrate is NH_4NO_3.

Chemical equations are used to describe reactions between compounds. The formulae of the reactants are written on the left-hand side of the equation and the formulae of the products on the right. If a number of molecules of one kind take part in the reaction the number is written as a coefficient in front of the formulae. The two sides of the equation must balance.

To illustrate, both hydrogen and chlorine have a valency of one. Elemental hydrogen consists of two hydrogen atoms linked to form a molecule of hydrogen written as H_2. Elemental chlorine comprises molecules of two atoms, Cl_2. One molecule of hydrogen can react with one molecule of chlorine to produce two molecules of hydrogen chloride:

$$H_2 + Cl_2 = 2HCl$$

Oxygen has a valency of two, nitrogen three and carbon four. Thus, elemental oxygen consists of molecules comprising two oxygen atoms. Because of their valencies, oxygen and hydrogen will co-react in a ratio of 1:2 respectively, i.e. one molecule of oxygen reacts with two molecules of hydrogen to form two molecules of water:

$$2H_2 + O_2 = 2H_2O$$

Whereas some atoms have only one valency, others have several, e.g. sulphur has valencies of two, four and six and can form compounds as diverse as hydrogen sulphide, H_2S (valency two), sulphur dioxide, SO_2 (valency four) and sulphur hexafluoride, SF_6 (valency six). Clearly some compounds comprise more than two different elements. Thus hydrogen, sulphur and oxygen can combine to produce sulphuric acid, H_2SO_4. From the structure it can be seen that hydrogen maintains its valency of one, oxygen two and sulphur is in a six valency state.

H—O   O
   S
H—O   O

Chemical bonds

Chemical bonds are strong forces of attraction which hold atoms together in a molecule. There are two main types of chemical bonds, viz. covalent and ionic bonds. In both cases there is a shift in the distribution of electrons such that the atoms in the molecule adopt the electronic configuration of inert gases.

Ionic bonds are formed by the transfer of electrons between atoms. For example, calcium has two outer electrons, whilst chlorine has seven. By transfer of its two outer electrons, one to each chlorine atom, the calcium atom becomes doubly positively charged (a cation), Ca^{++}, and has the stable configuration of inert argon. The chlorine atoms each having gained an electron become negatively charged (an anion), $2Cl^-$, also with the stable configuration of argon. The negatively-charged chlorine atoms then become electrostatically attracted to the positively charged calcium ion to form calcium chloride, $CaCl_2$.

Covalent bonds form when non-metallic atoms combine by sharing, rather than transferring, electrons. This is achieved by overlapping of their electronic shells. The overlapping region attracts both atomic nuclei and bonds the atoms. For example, hydrogen atoms have one electron. In the hydrogen molecule each atom contributes one electron to the bond thereby allowing each hydrogen atom control of two electrons giving it the electron configuration of the inert gas helium. In a water molecule, the oxygen atom, with six outer electrons, gains control of an extra two electrons supplied by two atoms of hydrogen and gives it the configuration of the inert gas neon.

Bonds may also be broken symmetrically such that each atom retains one electron of the pair that formed the covalent bond. This odd electron is not paired like all the other electrons of the atom, i.e. it does not have a partner of opposite spin. Atoms possessing odd unpaired electrons are termed ‘free radicals’ and are indicated by a dot alongside the atomic or molecular structure. The chlorination of methane (see later) to produce methyl chloride (CH_3Cl) is a typical free-radical reaction:

$$Cl_2 \overset{\text{UV light}}{=} 2Cl\bullet \text{ (Chlorine free radicals)}$$

$$Cl\bullet + CH_4 = HCl + CH_3\bullet \text{ (Methyl free radicals)}$$

$$CH_3\bullet + Cl_2 = CH_3Cl + Cl\bullet$$

When many molecules combine the macromolecule is termed a polymer. Polymerization can be initiated by ionic or free-radical mechanisms to produce molecules of very high molecular weight. Examples are the formation of PVC (polyvinyl chloride) from vinyl chloride (the monomer), polyethylene from ethylene, or SBR synthetic rubber from styrene and butadiene.

$$\underset{\text{Vinyl chloride}}{\underset{|}{CH}(Cl)=CH_2} + \underset{|}{CH}(Cl)=CH_2 + \underset{|}{CH}(Cl)=CH_2 = \underset{\text{Polyvinyl chloride}}{(-CH(Cl)-CH_2-)_n}$$

Hydrogen bound covalently to nitrogen, oxygen or fluorine may bond with another atom of one of these elements by a unique linkage known as the 'hydrogen bond'. Since the hydrogen nucleus cannot control more than two electrons this is not a covalent bond but a powerful dipole–dipole interaction, traditionally indicated by dotted lines. It occurs in water, hydrofluoric acid, undissociated ammonium hydroxide and many hydrates, e.g.:

$$H-O-H\cdots O(H)-H\cdots O(H)-H \qquad F-H\cdots F-H$$

Hydrogen bonding accounts for the abnormally high boiling points of, e.g., water, hydrogen fluoride, ammonia, and many organic compounds (see later) such as alcohols.

Oxidation/Reduction

Processes which involve oxygen as reactant are termed oxidation reactions. Some are beneficial, e.g. the burning of fuels in the body to produce energy, the manufacture of chemical intermediates, e.g. cyclohexanone in the manufacture of nylon. Some are detrimental such as the oxidation of food to make it rancid: anti-oxidants are added to retard this. Compounds that give up oxygen easily or remove hydrogen from another compound are classed as oxidizing agents. They may comprise a gas, e.g. oxygen, chlorine, fluorine, or a chemical which releases oxygen, e.g. a nitrate or perchlorate. Thus copper becomes oxidized when it reacts with oxygen to form copper oxide:

$$2Cu + O_2 = 2CuO$$

Reactions which involve the use of hydrogen as a reactant are termed reductions, e.g. the addition of a molecule of hydrogen across the unsaturated C=C in olefins to produce saturated alkanes. The material which adds hydrogen, or removes oxygen, is termed the reducing agent.

Clearly oxidation and reduction always occur together. The oxidation of copper by oxygen is accompanied by reduction of the oxygen to copper oxide. Similarly, whilst metal oxides such as CuO, HgO, SnO and PbO are reduced to the metal by hydrogen the latter becomes oxidized to water during the process:

$$CuO + H_2 = Cu + H_2O$$

The reactions are therefore termed redox reactions. Redox reactions can also be explained at the electronic level. Thus, the conversion of atomic copper to the oxide produces Cu^{++} ions as copper atoms have lost electrons. The reduction of lead oxide by hydrogen entails removal of electrons from the oxide to produce neutral metallic lead. Oxidizing agents are therefore chemicals which attract electrons, and reducing agents chemicals that add electrons to an element or compound. Using these definitions the presence of either oxygen or hydrogen is not required, e.g. since chlorine is a stronger oxidizing agent than iodine it displaces iodine from iodides:

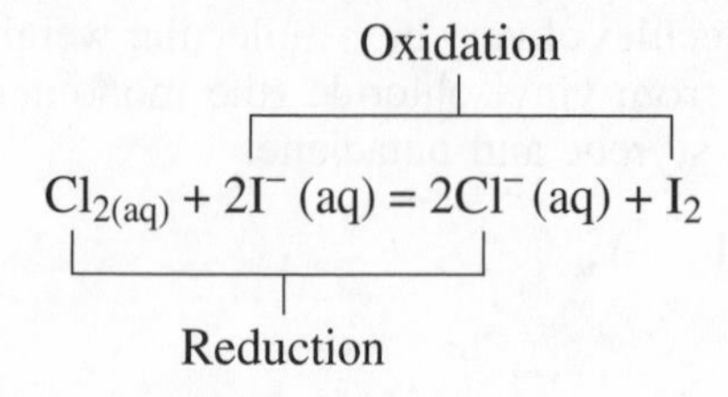

Physical state

Chemicals exist as gases, liquids or solids. Solids have definite shapes and volume and are held together by strong intermolecular and interatomic forces. For many substances, these forces are strong enough to maintain the atoms in definite ordered arrays, called crystals. Solids with little or no crystal structure are termed amorphous.

Gases have weaker attractive forces between individual molecules and therefore diffuse rapidly and assume the shape of their container. Molecules can be separated by vast distances unless the gas is subjected to high pressure. Their volumes are easily affected by temperature and pressure. The behaviour of any gas is dependent on only a few general laws based upon the properties of volume, pressure and temperature as discussed in Chapter 4.

The molecules of liquids are separated by relatively small distances so the attractive forces between molecules tend to hold firm within a definite volume at fixed temperature. Molecular forces also result in the phenomenon of interfacial tension. The repulsive forces between molecules exert a sufficiently powerful influence that volume changes caused by pressure changes can be neglected i.e. liquids are incompressible.

A useful property of liquids is their ability to dissolve gases, other liquids and solids. The solutions produced may be end-products, e.g. carbonated drinks, paints, disinfectants or the process itself may serve a useful function, e.g. pickling of metals, removal of pollutant gas from air by absorption (Chapter 17), leaching of a constituent from bulk solid. Clearly a solution's properties can differ significantly from the individual constituents. Solvents are covalent compounds in which molecules are much closer together than in a gas and the intermolecular forces are therefore relatively strong. When the molecules of a covalent solute are physically and chemically similar to those of a liquid solvent the intermolecular forces of each are the same and the solute and solvent will usually mix readily with each other. The quantity of solute in solvent is often expressed as a concentration, e.g. in grams/litre.

Important common physical properties related to these states of matter are summarized in Table 3.3.

Acids

Acids and bases (see later) are interrelated. Traditionally, acids are compounds which contain hydrogen and which dissociate in water to form hydrogen ions or protons, H^+, commonly written as:

Table 3.3 Important common physical properties

Gas
Density
Critical temperature, critical pressure (for liquefaction)
Solubility in water, selected solvents
Odour threshold
Colour
Diffusion coefficient
Liquid
Vapour pressure–temperature relationship
Density; specific gravity
Viscosity
Miscibility with water, selected solvents
Odour
Colour
Coefficient of thermal expansion
Interfacial tension
Solid
Melting point
Density
Odour
Solubility in water, selected solvents
Coefficient of thermal expansion
Hardness/flexibility
Particle size distribution/physical form, e.g. fine powder, flakes, granules, pellets, prills, lumps
Porosity

$$HX = H^+ + X^-$$

Examples include hydrochloric acid, nitric acid, and sulphuric acid. These are strong acids which are almost completely dissociated in water. Weak acids, such as hydrogen sulphide, are poorly dissociated producing low concentrations of hydrogen ions. Acids tend to be corrosive with a sharp, sour taste and turn litmus paper red; they give distinctive colour changes with other indicators. Acids dissolve metals such as copper and liberate hydrogen gas. They also react with carbonates to liberate carbon dioxide:

$$2HCl + Cu = CuCl_2 + H_2$$

$$2HCl + CaCO_3 = CaCl_2 + CO_2 + H_2O$$

Both acids and alkalis are electrolytes. The latter when fused or dissolved in water conduct an electric current (see page 55). Acids are considered to embrace substances capable of accepting an electron pair. Mineral acids have wide usage as indicated by Table 3.4.

Bases

Alkalis tend to be basic compounds which dissociate in water to produce hydroxyl ions, OH^- thus:

$$XOH = X^+ + OH^-$$

Specifically, an alkali is a hydroxide of one of the alkali or alkaline earth metals. Examples include the hydroxides of potassium, sodium, and calcium (where X is K, Na, and Ca, respectively).

Table 3.4 Industrial uses of mineral acids

Acid	*Main uses*
Hydrochloric acid (HCl)	Chemical manufacture, chlorine, food and rubber production, metal cleaning, petroleum well activation
Nitric acid (HNO_3)	Ammonium nitrate production for fertilizers and explosives, miscellaneous chemical production
Phosphoric acid (H_3PO_4)	Detergent builders and water treatment, foods, metal industries
Sulphuric acid (H_2SO_4)	Alkylation reactions, caprolactam, copper leaching, detergents, explosives, fertilizers, inorganic pigments, textiles
Hydrofluoric acid (HF)	Etching glass

The first two are very soluble in water but the last is less so. Weaker bases include ammonium hydroxide where X is NH_4. In fact every acid can generate a base by loss of a proton and the definition now includes any compound capable of donating electron pairs, e.g. amines. Bases turn litmus paper blue and show characteristic effects on other indicators. They are soluble in water, tarnish in air, and in concentrated form are corrosive to the touch. Common examples are given in Table 3.5.

Table 3.5 Common bases

Base	*Properties*
Sodium hydroxide (NaOH) (caustic soda)	White deliquescent solid. Sticks, flakes, pellets. Dissolution in water is highly exothermic. Strongly basic. Severe hazard to skin tissue
Potassium hydroxide (KOH) (caustic potash)	White deliquescent solid. Sticks, flakes, pellets. Dissolution in water is highly exothermic. Strongly basic. Severe hazard to skin tissue
Calcium hydroxide ($Ca(OH)_2$) (slaked lime)	White powder soluble in water yielding lime water. Alkaline
Ammonium hydroxide (NH_4OH) (aqueous ammonia solution)	Weakly alkaline. Emits ammonia gas. Severe eye irritant

Halogens

The halogens comprise fluorine, chlorine, bromine and iodine. They are remarkable for their similarity in chemical behaviour and for the gradation in physical properties. They are highly reactive and form covalent links with hydrogen, carbon, silicon, nitrogen, phosphorus, sulphur and oxygen, and with other halogen atoms. The halogen atoms also produce ionic bonds with metals. The elements tend to be toxic.

Fluorine is a greenish-yellow gas which condenses to a yellow liquid and pale yellow solid. It is the most chemically active of all the elements.

Chlorine (see page 280) is a heavy greenish-yellow gas with a choking smell. It combines directly with nearly all elements, exceptions being carbon, nitrogen and oxygen. It will react spontaneously with phosphorus, arsenic, antimony and mercury. Most metals will react with it when wet or on heating. It has a powerful affinity for hydrogen and will attack hydrocarbons.

Bromine is a dark red volatile liquid with a pungent odour. The vapour attacks the eyes and mucous membranes. It combines spontaneously and with deflagration with phosphorus, arsenic and potassium and with many other elements when warmed. It bleaches litmus and turns starch paper orange/yellow.

Iodine is a dark grey solid which is easily vaporized to a deep blue/violet vapour. It is sparing soluble in water but dissolves in aqueous potassium iodide to give a brown solution. It combines directly with many elements.

Metals

A metal is an electropositive element. There are over 70 metals in the earth's crust. Examples include copper, gold, iron, platinum, silver and tungsten. Chemically, in solution, a metal atom releases an electron to become a positive ion. In bulk metals are solids and tend to have high melting and boiling points (an exception is mercury). They are lustrous, relatively dense, malleable, ductile, cohesive and highly conductive to both electricity and heat.

Metals are crystalline in structure and the individual crystals contain positive metal ions. The outer valency electrons appear to be so loosely held that they are largely interspersed amongst the positive ions forming an electron cloud which holds the positive ions together. The mobility of this electron cloud accounts for the electrical conductivity. The crystal structure also explains the hardness and mechanical strength of metals whereas the elasticity is explained by the ability of the atoms and ions to slide easily over each other. Metals can be blended with other metals to produce alloys with specific properties and applications. Examples include:

- Brass (alloy of copper and zinc) used for ship's propellers, screws, wind instruments.
- Bronze (alloy of copper and tin) used for coins, medals, statues, church bells.
- Duralumin (alloy of aluminium, magnesium, copper and manganese) used for structural purposes, e.g. in aircraft construction.
- Nichrome (alloy of nickel, iron and chromium) used for heating elements.
- Solder (alloy of tin and lead) used for joining metals, e.g. in electrical circuits.
- Stellite (alloy of cobalt, tungsten, chromium, and molybdenum) used for surgical instruments.

(Variations in physical properties occur with changes in relative proportions.)

Group I metals are good conductors of heat and electricity and are so soft that they can be cut with a knife. As a result of their low specific gravities, Li, Na, and K float on water. They react vigorously with electronegative elements such as O, S and Cl. Indeed the ease with which the outer electron is detached from the atom explains their highly-reactive nature. This is exemplified by sodium which can only be handled if air is excluded, e.g. by nitrogen blanketing, or under vacuum, or submersed in oil.

Group IIA metals include Be, Mg, Ca, Sr, Ba and Ra which are grey, moderately-hard, high melting-point substances. Like the alkali metals they attack water to liberate hydrogen but with less vigour. The salts of the alkaline earths are generally less stable towards heat and water than those of alkali metals, and less water soluble.

Group IIB includes Zn, Cd and Hg. Zinc has some resemblance to magnesium but the other metals in the group have little in common. At room temperature mercury is unaffected by air, water or non-oxidizing agents whereas zinc is more reactive, albeit tempered by a protective hydroxide film, a property utilized in galvanizing.

Compounds tend to be covalent. Metals form complex ions and their oxides are only weakly basic. Mercury forms no hydride.

Aluminium is an extremely light, white metal and whilst hard is malleable and ductile. On exposure to air the metal forms a protective oxide film which reduces its reactivity. Its compounds tend to be covalent in nature: the sulphate is hydrolysed in solution and the trichloride is volatile.

Both tin and lead from Group IV can form valency two and four compounds. Two of the four outer electrons can behave as inert when the atoms are bivalent. Bivalent tin (stannous) derivatives are covalent whereas the nitrate and sulphate of bivalent lead (plumbous) are ionic. Some tetavalent compounds such as the hydrides and chloride are unstable, e.g.:

$$PbCl_4 + 2H_2O = PbO_2 + 4HCl$$

Whereas stannic oxide is neither oxidizing nor reducing, plumbic oxide is a powerful oxidizer.

Tin finds widespread use because of its resistance to corrosion, or as foil or to provide protective coats/plates for other metals. Properties of lead which make industrial application attractive surround its soft, plastic nature permitting it to be rolled into sheets or extruded through dies. In the finely-divided state lead powder is pyrophoric; in bulk form the rapidly-formed protective oxide layer inhibits further reaction. It dissolves slowly in mineral acids. Industrial uses include roofing material, piping, and vessel linings, e.g. for acid storage.

The transition metals Cr, Mn, Fe, Co and Ni possess bi- and trivalent states. Chromium is a hard, malleable, white metal capable of high polish and does not tarnish in air. It is used for plating steel. Together with nickel it is also used in grades of stainless steel. Manganese is a grey metal which decomposes water and dissolves in dilute acids. Its chief use is in steel to remove trace quantities of oxygen and sulphur and to produce tough steel. Iron is a white, soft, malleable, ductile magnetic metal when pure and is used mainly in steel production. It is attacked by oxygen or steam to produce an oxide, Fe_3O_4. When exposed to ordinary atmospheric conditions it becomes covered with rust, i.e. hydrated ferric oxide, $2Fe_2O_3.3H_2O$. Cobalt does not oxidize in air at room temperature but oxidizes slowly if heated to yield cobaltous oxide, CoO. It dissolves slowly in acids becoming passive in concentrated nitric acid. Nickel is silver grey, hard, malleable, capable of high polish and resistant to attack by oxygen at room temperature but yields the oxide on heating. It dissolves in dilute nitric acid but is rendered passive by the concentrated acid. It forms the volatile, toxic tetra carbonyl with carbon monoxide.

The metals copper, silver and gold from Group IX are sometimes termed coinage metals. They possess characteristic metallic lustre, take high polish and resist attack by air. They are extremely malleable and ductile and excellent conductors of heat and electricity. All are attacked by chlorine; copper alone is attacked by oxygen. None of the metals displace hydrogen from acids. Copper has a characteristic red colour. It is used for cooking utensils and wires in telegraphs, telephones, power lines, and electrical machinery. Silver is a lustrous, white metal capable of high polish. It is tough, malleable, ductile and an efficient conductor of heat and electricity. Whilst resistant to attack by oxygen, on exposure to air it is slowly covered with a black film of silver sulphide. Uses include electroplating, mirrors, silverware, and crucibles. Gold is a yellow, malleable, ductile metal which does not tarnish in air and is inert to any mineral acid. It reacts with halogens and aqua-regia (a mixture of hydrochloric and nitric acids in the ratio of 4:1).

Oxygen and sulphur

Oxygen is the first member of Group IV with six electrons in the outer shell. It is a colourless, tasteless and odourless gas which condenses to a blue liquid and freezes to a blue solid under cryogenic conditions (discussed in Chapter 8). The oxygen atom exerts covalent, ionic, or co-ionic characteristics. The principle types of compounds are those in which the oxygen atom

- exerts two ionic bonds by accepting two electrons from the same or different atoms, e.g. Ca^{++} O^{-};
- exerts two covalent bonds by sharing electron pairs, e.g. H_2O;
- exerts co-ionic character by combining with another atom which already has the inert configuration but of which at least one pair of electrons is unshared, e.g.

$$\begin{array}{l} Cl \diagdown \\ Cl — P^{+} — O^{-} \\ Cl \diagup \end{array}$$

Thus oxygen can feature in a wide variety of compounds including ozone, oxides, water, hydrogen peroxide, carbonates, nitrates/nitrites, etc. It comprises about 21% of normal air (by volume).

Sulphur molecules are S_8 and it can exist in several forms. Its compounds are more acidic than those of oxygen and it may assume covalency up to six. It forms a series of oxides and oxyacids of diverse chemistry. Combustion yields mainly SO_2, a cause of atmospheric pollution from sulphur-bearing fossil fuels.

Nitrogen, phosphorus, arsenic and antimony

None of these elements from Group V form cations of the type N^{+++++} due to loss of all five valency electrons. All the elements are strongly electronegative and readily form covalent bonds with other elements.

Nitrogen is a colourless, tasteless, odourless gas which is slightly soluble in water (see also page 296). It is non-toxic and inert and comprises about 79% of normal air (by volume). It neither burns nor supports combustion and at room temperature does not react with any substance. On heating, however, it combines with oxygen to produce nitric oxide NO, with hydrogen to produce ammonia NH_3, and with silicon to form silane SiH_4, with calcium carbide to form calcium cyanamide $CaCN_2$ and with metals such as lithium, calcium, barium, magnesium and aluminium to form the corresponding nitrides.

Phosphorus exists as white and red phosphorus. The former allotrope may be preserved in the dark at low temperatures but otherwise reverts to the more stable red form. The white form is a waxy, translucent, crystalline, highly-toxic solid subliming at room temperature and inflaming in air at 35°C, so it is handled under water. The red form is a reddish violet crystalline solid which vaporizes if heated at atmospheric pressure and condenses to give white phosphorus. The red form ignites in air at 260°C. Both are insoluble in water, and white phosphorus can be stored beneath it. Phosphorus forms a host of compounds such as phosphine, tri- and penta-halides, tri-, tetra- and penta-oxides, oxyacids including hypophosphorous, orthophosphorous and orthophosphoric acids.

Arsenic exists as grey, yellow and black forms of differing physical properties and susceptibilities towards atmospheric oxygen. The general chemistry is similar to that of phosphorus but whereas phosphorus is non-metallic, the common form of arsenic is metallic. Traces of arsenides may be present in metallic residues and drosses; these may yield highly toxic arsine, AsH_3, with water.

Antimony is a bluish white metal with good lustre but poor heat conducting ability. It is stable in air and resistant to dilute acids but attacked by halogens, sulphur, phosphorus and arsenic.

pH

The strength of acids and bases is measured on a pH (potential of hydrogen) scale:

$$pH = -\log_{10} [H^+]$$

The hydrogen ion concentration of a normal solution of a strong acid is about 1 gram-ion per litre and that of a typical strong base is 10^{-14} gram-ion per litre. Because of the vast range of possible concentrations it is convenient to use a logarithmic scale to express the hydrogen ion concentration of a solution. The symbol pH is used to denote the degree of acidity of a solution. Pure water which dissociates slightly to produce 10^{-7} gram-ions of H^+ per litre is taken as the standard of neutrality. Thus water has a pH of 7. Solutions of pH less than 7 are acidic and those greater than 7 are alkali. The pH of a solution can be determined electrically using a hydrogen or glass electrode and reference electrode (e.g. calomel electrode) or by chemical indicators. The pH scale is shown in Figure 5.1.

Salts

Acids and alkalis react with each other to produce salts and water, e.g.:

$$HCl + NaOH = NaCl + H_2O$$

Thus salts are compounds formed by replacement of hydrogen in an acid by a metal. Clearly non-metals can also be involved, e.g.:

$$NH_4OH + HCl = NH_4Cl + H_2O$$

Salts are non-volatile and in the fused state or in solution conduct an electric current. Many salts are hydrated in the solid state with water of crystallization.

These reactions are exothermic and must be carefully controlled if the reactants are concentrated, since the rates can be very rapid.

Organic chemistry

Carbon is in the same group in the periodic table as silicon, germanium, tin and lead. The electronic structures are characterized by the presence of four electrons in the external quantum shell. The elements, however, do not form ions of the type X^{++++} and compounds are covalent in the quadrivalent state. Lead and tin may be bivalent when lead forms ionic valencies. Carbon differs from the other elements in this group by forming an enormous range of compounds, the chemistry of which is a special discipline, organic chemistry. There are over a million known organic compounds, including sugar, starch, alcohol, resins and mineral oil. The versatility of carbon arises from:

- the stability of the compounds produced whether from electropositive elements such as hydrogen, or from electronegative elements such as oxygen or fluorine;
- the ability of carbon to covalently link with other carbon atoms with one, two or three bonds, e.g. $H_3C—CH_3$ (ethane), $H_2C{=}CH_2$ (ethylene), $HC{\equiv}CH$ (ethyne, or acetylene). These links may be in the form of chain or ring skeletons. Compounds comprising mainly carbon and hydrogen are termed hydrocarbons.

Tables 3.6 and 3.7 illustrate some of the key organic groupings. For convenience organic compounds can be classified as either aliphatic or aromatic.

Table 3.6 Examples of aliphatic organic structures

Structure	Name	Structure	Name
Carbon–carbon groups		**Carbon–nitrogen groups**	
>C–C<	Paraffin	>C–NH_2	Primary amine
>C=C<	Olefin	>C–N(H)–C<	Secondary amine
–C≡C–	Acetylene	>C–N(C<)–C<	Tertiary amine
		>C–NO_2	Nitro
Carbon–oxygen groups		>C–C≡N	Nitrile (or cyanide)
>C–OH	Alcohol	–C(=O)–NH_2	Amide
>C–O–C<	Ether	**Carbon–sulphur groups**	
>C–O–O–C<	Peroxide	>C–SH	Mercaptan
>C=O	Ketone	>C–SO_3H	Sulphonic acid
–C(=O)–Cl	Acyl chloride	(>C)₂S=O	Sulphoxide
–C(=O)–H	Aldehyde	(>C)₂S(=O)₂	Sulphone
–C(=O)–OH	Fatty acid		

Aliphatic compounds

Aliphatic compounds are straight chain or acyclic compounds and are characterized by addition and free-radical chemistry.

Table 3.7 Selected aromatic compounds

Structure	Name	Structure	Name
C_6H_6	Benzene	C_6H_5OH	Hydroxy benzene (Phenol)
$C_6H_5NH_2$	Amino benzene (Aniline)	$C_6H_5CH_3$	Methyl benzene (Toluene)
$C_6H_5CO_2H$	Benzoic acid	$CH_3C_6H_2(NO_2)_3$	Trinitro toluene (TNT)
$C_6H_5SO_3H$	Benzene sulphonic acid	$C_{10}H_8$	Naphthalene
C_6H_5Cl	Chlorobenzene	$C_{10}H_7NH_2$	Beta naphthylamine

Carbon/carbon compounds

Paraffins

Compounds containing only carbon and hydrogen are termed paraffins or alkanes. The general formula for these compounds is C_nH_{2n+2} where n is an integer. When only single bonds are present between carbon atoms they are classified as 'saturated'. Examples include, ethane, propane, and butane; the last two are common fuel gases:

CH_4	methane	(natural gas)
$CH_3—CH_3$	ethane	
$CH_3—CH_2—CH_3$	propane	commonly used as liquefied petroleum gas
$CH_3—CH_2—CH_2—CH_3$	butane	

$$\begin{array}{c} CH_2 \\ / \quad \backslash \\ CH_2 — CH_2 \end{array}$$ cyclopropane

The alkanes are almost insoluble in water, sodium hydroxide and sulphuric acid but soluble in non-polar solvents. The liquid density increases as the size of the paraffin molecule increases but tends to level off at 0.8, i.e. all alkanes are less dense than water; therefore they will float and spread as thin films on water. The boiling points and melting points increase as the number of carbon atoms rises. The physical properties of cyclic aliphatic hydrocarbons resemble those of the straight-chain counterparts, although the boiling points and densities of the cyclic compounds are somewhat higher. The strong carbon–carbon and carbon–hydrogen bonds render paraffins relatively unreactive and the few reactions they undergo require forcing conditions and tend to produce mixtures. On heating between 400 and 600°C they can undergo thermal degradation or 'cracking' to produce simpler alkanes, olefins and hydrogen; this can increase the flammable hazards.

Olefins

When carbon atoms are linked by a double bond the compounds are called olefins. Since these molecules contain less than the maximum quantity of hydrogen they are termed unsaturated. Examples include ethylene, propylene, and butylene. Note that the latter can exist in several forms:

$CH_2{=}CH_2$	ethylene
$CH_3CH{=}CH_2$	propylene
$CH_3{-}CH_2{-}CH{=}CH_2$	1-butylene
$CH_3{-}CH{=}CH{-}CH_3$	2-butylene
$(CH_3)_2C{=}CH_2$	iso-butylene

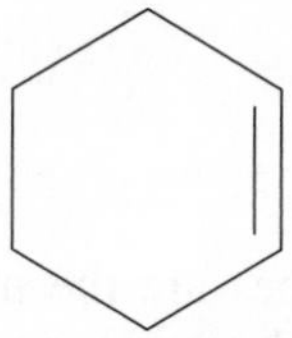

cyclohexene

Their physical properties are essentially those of the alkanes. It is the unsaturated linkages that dominate the chemistry and the main reaction is one of addition (e.g. hydrogen, halogen, and hydrogen halides) across the double bond to produce saturated compounds. This reactivity is utilized in the manufacture of long-chain polymers, e.g. polyethylene and polypropylene.

Acetylenes

Compounds with even less hydrogen to carbon than olefins are acetylenes or alkynes as exemplified by:

$HC{\equiv}CH$	acetylene
$CH_3C{\equiv}CH$	propyne
$CH_3CH_2C{\equiv}CH$	1-butyne
$CH_3C{\equiv}CCH_3$	2-butyne

Physical properties are similar to alkanes and the chemistry is dictated by the carbon triple bond. This bond is less reactive than the olefin double bond towards electrophilic reagents, but more reactive towards chemicals that are themselves electron rich. Some metals, e.g. copper, react to form metal acetylides. If allowed to dry out the heavy metal acetylides are prone to explode (Chapter 7).

Carbon/Halogen compounds

One or several hydrogen atoms in hydrocarbons can be substituted by halogen to produce alkyl halides. This significantly alters the toxicity, e.g. substitution of a chlorine atom in a hydrocarbon leads to an increase in the potential narcotic and anaesthetic effects. Because of the increased molecular weight, alkyl halides have considerably higher boiling points than the corresponding hydrocarbon. For a given alkyl group the boiling point increases with increasing atomic weight of halogen, with fluorides having the lowest boiling point and iodides the highest. Increasing the halogen content also reduces the ease with which some compounds undergo chemical or biological oxidation and hence they can accumulate in the environment. Some halogenated organic substances react with ozone in the upper atmosphere and deplete the planet of this gas which provides a protective shield against harmful ultra-violet light.

Some alkyl halides are toxic, e.g. trichloromethane or chloroform ($CHCl_3$) and tetrachloromethane or carbon tetrachloride (CCl_4). Progressive chlorination of hydrocarbons gives liquids and/or solids of increasing non-flammability, density, viscosity, solvent power and decreasing specific heat, dielectric constant and water solubility. So perchloroethylene, $CCl_2{=}CCl_2$, is a common dry-cleaning solvent and trichloroethylene, $CHCl{=}CCl_2$, is widely used in vapour degreasing of metal components. Despite being polarized molecules they are insoluble in water.

As with other groups, halogens can substitute hydrogen in organic compounds containing additional functional moieties such as carboxylic acids to form acid chlorides, e.g. acetyl chloride CH_3COCl. These are reactive acidic compounds liberating hydrochloric acid on contact with water.

Carbon/Nitrogen compounds

Of the organic compounds of sufficient basicity to turn litmus paper blue amines are the most significant. These compounds have trivalent nitrogen bonded directly to carbon by single bonds with the general formula RNH_2, R_2NH or R_3N where R is an alkyl or aryl group. The first are classed as primary amines, the next secondary amines and the last tertiary amines. The chemistry is influenced by the number of hydrogen atoms attached to the nitrogen.

Amines, like ammonia NH_3, are polar compounds and, except for tertiary amines, form intermolecular hydrogen bonds leading to higher boiling points than non-polar compounds of the same molecular weight, but lower boiling points than alcohols or acids. The smaller molecules, containing up to about six carbon atoms, dissolve in water. Aliphatic amines are similar in basicity to ammonia and form water-soluble salts with acids:

$$RNH_2 + HCl = RNH_3^+Cl^-$$

Nitriles, or alkyl cyanides, are compounds in which carbon is bound to nitrogen by triple bonds. They tend to be stable, neutral substances with pleasant smells and are less toxic than hydrogen cyanide. The smallest compounds are water soluble liquids and all are soluble in organic solvents.

Tertiary amines can be oxidized to form amine oxides in which the amino nitrogen atom is linked to a single oxygen atom. The resulting compounds are basic dissolving in water thus:

$$R_3N \rightarrow O + H_2O = [R_3N{-}OH]^+ \, OH^-$$

When nitrogen linked to two oxygen atoms is bound to carbon the compounds are termed nitroparaffins. When pure these compounds are colourless liquids with pleasant smells. They are sparingly soluble in water and most can be distilled at atmospheric pressure. The lower members are used as solvents for oils, fats, cellulose esters, resins, and dyes. Nitroparaffins are also used as raw materials for the synthesis of other chemicals such as pesticides, drugs, explosives, fuels (e.g. nitromethane in drag racing fuel). Some nitroparaffins are explosive as described in Chapter 7.

Carbon/Oxygen compounds

Compounds containing oxygen linked to a carbon and a hydrogen atom are termed alcohols. Simple examples are methyl and ethyl alcohol, CH_3OH and C_2H_5OH, respectively. Reactions of alcohols are characterized by the replacement of either the OH hydrogen atom or the entire OH group:

$$\underset{\text{methyl alcohol}}{2CH_3OH} + 2Na = \underset{\text{sodium methoxide}}{2CH_3ONa} + H_2$$

$$CH_3OH + HCl = \underset{\text{methyl chloride}}{CH_3Cl} + H_2O$$

Compounds in which oxygen bridges two carbon atoms are termed ethers, e.g. diethyl ether, $CH_3CH_2—O—CH_2CH_3$. This is used as a solvent and an anaesthetic. Generally, ethers are unreactive compounds but on standing they can react with atmospheric oxygen to produce explosive peroxides, e.g. diethyl peroxide, $CH_3CH_2—O—O—CH_2CH_3$.

Oxygen can link solely to carbon atoms by double bonds to form carbonyl compounds containing the C=O group. If the same carbonyl group is linked to another carbon the compounds are classed as ketones, if connected to a hydrogen atom they are aldehydes, and if connected to OH groups they are carboxylic acids. The C=O carbon can also be bonded to other atoms such as halogens, nitrogen and sulphur. The carbonyl group tends to dominate the chemistry of aldehydes and ketones since it can be oxidized to carboxylic acids, reduced to alcohols or undergo addition reactions. Carboxylic acids are acidic in nature, typically reacting with bases to form salts or with alcohols to produce often sweet-smelling esters:

$$\underset{\text{acetic acid}}{CH_3CO_2H} + \underset{\text{sodium hydroxide}}{NaOH} = \underset{\text{sodium acetate}}{CH_3CO_2Na} + \underset{\text{water}}{H_2O}$$

$$CH_3CO_2H + CH_3OH = \underset{\text{methyl acetate}}{CH_3CO_2CH_3} + H_2O$$

Carbon/Sulphur compounds

Since sulphur is in the same group as oxygen in the periodic table it replaces oxygen in organic structures to produce ‘thio’ analogues such as:

R—S—H	thio alcohols or alkyl thiols
R—S—R	thioethers or alkyl sulphides
R—C(=O)—SH	thioacids
R—C(=S)—SH	dithioacids

$$\begin{array}{c} S \\ \| \\ R\text{—}C\text{—}R \end{array} \quad \text{thioketones}$$

$$\begin{array}{c} R\text{—}C\text{—}H \\ \| \\ S \end{array} \quad \text{thioaldehydes}$$

With the exception of methanethiol, which is a gas, thiols are colourless, evil-smelling liquids. Their boiling points are lower than those of the corresponding alcohols, reflecting their reduced association and degree of hydrogen bonding between hydrogen and sulphur. For the same reason thiols are less water soluble than their oxygen counterparts. The chemistry of thiols resembles that of alcohols but they are more acidic, reflecting the stronger acidity of hydrogen sulphide over water, of which alcohols can be regarded as alkyl derivatives. They are used as feedstocks in rubber and plastics industries and as intermediates in agricultural chemicals, pharmaceuticals, flavours and fragrances. Because they react with rubber-containing materials selection of hose and gasket material is crucial.

Alkyl sulphides are the sulphur analogues of ethers from which they differ considerably in chemistry. They are unpleasant-smelling oils, insoluble in water but soluble in organic solvents. They tend to be comparatively inert. Mustard gas, $ClCH_2CH_2$—S—CH_2CH_2Cl, an oily liquid boiling at 216°C with a mustard-like smell, is highly poisonous and a vesicant, and for this reason found use in chemical warfare.

Alkyl sulphoxides occur widely in small concentrations in plant and animal tissues. No gaseous sulphoxides are known and they tend to be colourless, odourless, relatively unstable solids soluble in water, ethyl alcohol and ether. They are freely basic, and with acids form salts of the type $(R_2SOH)^+ X^-$. Because sulphoxides are highly polar their boiling points are high. Their main use is as solvents for polymerization, spinning, extractions, base-catalysed chemical reactions and for pesticides.

Thioketones and aldehydes readily polymerize to the trimer and isolation of the monomer is difficult.

Thioacids have a most disagreeable odour and slowly decompose in air. Their boiling points are lower than those of the corresponding oxygen counterparts and they are less soluble in water, but soluble in most organic solvents. An important dithioacid is dithiocarbonic acid (HO—CS_2H). Whilst the free acid is unknown, many derivatives have been prepared such as potassium xanthate giving a yellow precipitate of copper xanthate with copper salts:

$$\underset{\text{potassium hydroxide}}{KOH} + \underset{\text{carbon disulphide}}{CS_2} + \underset{\text{ethyl alcohol (ethanol)}}{C_2H_5OH} = \underset{\text{potassium xanthate}}{C_2H_5O\text{—}CS_2K} + H_2O$$

Unlike oxygen, sulphur can exist in higher valency states and as a result can be incorporated into organic structures in additional ways. Examples include:

$$\begin{array}{c} O \\ \| \\ R\text{—}S\text{—}R \\ \| \\ O \end{array} \quad \text{Alkyl sulphones}$$

$$RSO_3H \quad \text{Sulphonic acids}$$

Sulphones are colourless, very stable, water-soluble solids that are generally resistant to reduction. The most important sulphones are sulpholane (1) and sulpholene (2):

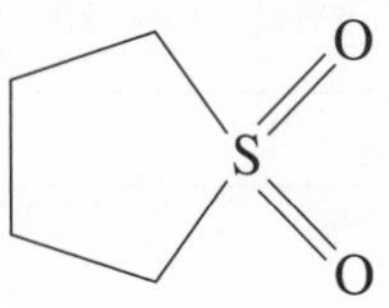

(1) 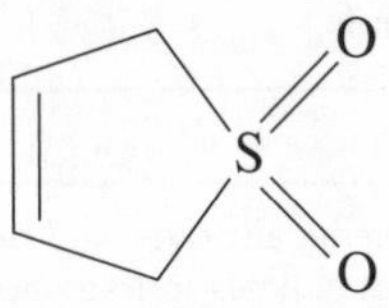 (2)

Compound 1 is used as a solvent in the food, paint, resin/plastics, soap and woodpulp/paper industries, and as a plasticizer. Compound 2 is used as an intermediate for the manufacture of hydraulic fluid additives and cosmetics. Sulphonal (2:2-bis(ethylsulphonyl)-propane), another important sulphone, is a colourless solid, stable to acids and alkalis, with hypnotic properties.

Sulphonic acids are water soluble, viscous liquids. Their acidity is akin to that of sulphuric acid; they form salts with bases but fail to undergo esterification with alcohols. Their properties vary according to the nature of R: some are prone to thermal decomposition. They are used as surfactants and in the dye industry; some have biological uses. 2-Amino-ethanesulphonic acid is the only naturally occurring sulphonic acid.

Aromatic compounds

Aromatic compounds are benzene and its derivatives and compounds that resemble benzene in their behaviour in a chemistry dominated by ionic substitution. Benzene has the formula C_6H_6 commonly written as the ring:

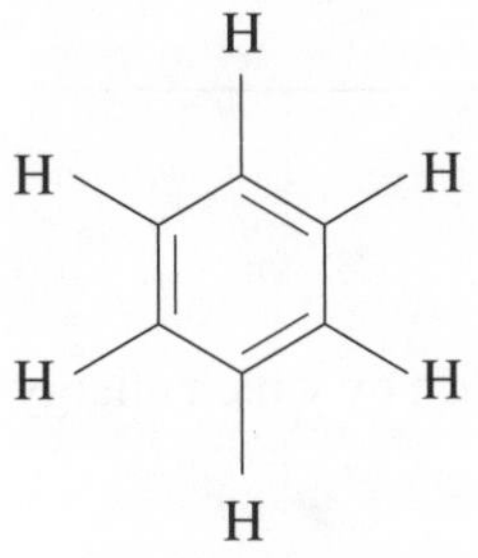

In reality all carbon atoms share equally the pool of electrons which constitute the double bonds and benzene resists addition across the double bonds which would otherwise destroy its unique structure and stability. Single or multiple hydrogen atoms can be substituted to form a host of derivatives containing similar functional groups to those above, e.g. saturated and unsaturated aliphatic chains, amino, carboxylic acidic, halogeno, nitro, and sulphonic acid groups as shown in Table 3.6.

Aromatic compounds find wide industrial use as exemplified by Table 3.8.

Benzene and alkylbenzenes possess low polarity with similar physical properties to hydrocarbons. They are insoluble in water but soluble in non-polar solvents such as ether. They are less dense than water (Table 6.1) and boiling points rise with increasing molecular weight (ca 20–30°C increment for each carbon atom). Since melting point depends not only on molecular weight but also on molecular shape, the relationship to structure is more complicated. Benzene itself is a colourless liquid boiling at 80°C and freezing at 5.4°C. It is highly flammable with a flash point of –11°C but with a narrow flammable range of 1.4–8%. It is acutely toxic producing narcotic effects comparable to toluene but it also poses chronic effects on bone marrow which may lead to anaemia or even leukaemia. Similar effects are not proven for pure toluene, but in the past commercial toluol was contaminated with benzene.

Table 3.8 Industrial uses of selected aromatic compounds

Compound	*Use*
Aniline (amino benzene) $C_6H_5NH_2$	Agrochemicals, dyes and pigments, pharmaceuticals, photographic chemicals, polymers, rubbers
Benzene C_6H_6	Production of ethyl benzene, cumene, cyclohexane, maleic anhydride, nitrobenzene, chlorobenzene, detergents
Benzene (and alkylbenzene) sulphonic acids $C_6H_5SO_3H$	Detergents, phenols, dyes
Benzoic acid $C_6H_5CO_2H$	Alkyl resins, chemical intermediate, oil drilling additive, medicines
Nitro benzene $C_6H_5NO_2$	Production of aniline, para-aminophenol, and dyes
Phenol (hydroxy benzene) C_6H_5OH	Phenolic resins as adhesives and for the car industry, caprolactam
Toluene (methyl benzene) $C_6H_5CH_3$	Automotive fuels additive, organic solvent, production of benzene, styrene, and terephthalic acid
Styrene (vinyl benzene) $C_6H_5CH{=}CH_2$	Polymers $(C_8H_8)_n$ for audio and video cassettes, carpet backing, domestic appliances, packaging, food containers, furniture, toys, vehicle parts
Xylenes (di methyl benzene) $C_6H_4(CH_3)_2$	Plasticizers, polymer fibres and resins, solvents

Combustion chemistry

In biological systems the oxidation of fuels by oxygen is a fundamental reaction by which energy is created, along with by-products such as water and carbon dioxide:

$$3O_2 + 2(\text{—}CH_2\text{—}) = 2CO_2 + 2H_2O + \text{ENERGY}$$

Anything that interferes with this mechanism in humans can result in reduced well-being or even death. Silicosis or asbestosis may impair oxygen transport from lungs to blood. Inhaled carbon monoxide may combine with haemoglobin and prevent it from carrying oxygen. Waste products may not be removed efficiently when kidney function is damaged by toxic chemicals such as mercury and phenolic substances. Fuels such as glucose may be prevented from entering cells due to inactivation of the necessary catalytic enzymes by, e.g., toxic metals or fluoracetate.

The oxidation of fuels is common outside living systems, combustion being the extreme example. Combustion is a chemical reaction between a fuel and usually oxygen, with the liberation of energy often as heat. A flame is produced when sufficient energy is liberated, usually in the visible range of the spectrum though some are in the infra-red or ultra-violet region and invisible. Chemical combustion processes are initiated by heat, light, sparks, etc. As the temperature of the combustible substance rises it reaches the ignition temperature specific to the material and to the pressure, and the combustion process begins. It spreads from the ignition source to the adjacent layer of gaseous mixture of fuel and oxidant (usually oxygen in air). In turn the burning layer ignites the next layer until equilibrium is reached between the total heat energies of reactants and products when the combustion process ceases. When the rate of heat lost from the mass is less than the heat liberated by the combustion process an explosion occurs. When combustion increases progressively such that the flame front advances supersonically, compression from the shockwave causes an increase in temperature and self-ignition of the fuel, i.e. detonation. The requirements for chemicals to burn are discussed in Chapter 6.

Most organic materials will burn; the smaller molecules may be highly flammable. In the simplest form carbon (e.g. charcoal) in the presence of a surplus of oxygen will produce carbon dioxide:

$$C + O_2 = CO_2$$

Usually, however, fuels are hydrocarbons and the products of combustion can be complex and depend upon the nature of the fuel, the amount of oxygen present, and the temperature. A great deal of energy is required to break carbon–carbon and carbon–hydrogen bonds such as the high temperatures of flames. Once the energy barrier is surmounted the subsequent chain of events proceeds readily with the evolution of energy, often sufficient to keep the combustion reaction in progress. Simple hydrocarbons in excess oxygen will produce carbon dioxide and water:

$$2C_2H_6 + 7O_2 = 4CO_2 + 6H_2O$$

If nitrogen or sulphur is present in the fuel then the mixture of combustion products may include oxides of these elements. In the absence of excess oxygen incomplete oxidation occurs to produce partially oxidized carbon compounds such as aldehydes, ketones, phenols, and carbon monoxide. Carbon monoxide is extremely toxic and some of the other compounds are respiratory irritants.

Since air comprises about 21% oxygen and 79% nitrogen, with traces of other gases, e.g. CO_2, complete combustion of methane (i.e. natural gas) in air can be represented as:

$$CH_4 + 2O_2 + 8N_2 = CO_2 + 2H_2O + 8N_2$$

This demonstrates how the oxygen is depleted resulting, as summarized in Chapter 6, in an irrespirable atmosphere rich in nitrogen. High temperature combustion may also result in the generation of oxides of nitrogen, NO_x, which are respiratory irritants.

Under certain conditions some inorganic materials will burn. Magnesium metal as powder or ribbon when heated to its melting point in oxygen burns to produce magnesium oxide, and in air to produce a mixture of magnesium oxide and magnesium nitride. Aluminium also burns in air at high temperatures to produce a mixture of the oxide and nitride. Dust explosion characteristics of various inorganic materials are included in Table 6.1.

Some materials such as oil-impregnated cotton and iron pyrites are prone to spontaneous combustion, whilst selected materials such as metal alkyls and metals in a finely divided state burn on immediate contact with water or air. These are termed ‘pyrophoric’. Examples and precautions for their control are described in Chapter 6.

Dangers arising from fires therefore include:

- Burns from heat radiation, or fire engulfment.
- Asphyxiation due to consumption of oxygen until the concentration is <18%.
- Poisoning from toxic combustion products. In chemical fires, particularly those involving mixtures, an extremely complex mixture of gases and particulates, e.g. smoke may be produced. The composition depends upon the initial compounds involved, the temperatures attained and the oxygen supply, and is hence often unpredictable. Some gaseous compounds may derive from thermal breakdown, i.e. pyrolysis, of the chemicals rather than oxidation as illustrated in Tables 3.9 and 3.10.
- Injury from collapse of weakened structures.
- Explosions (see Chapter 6).

Chemical reactivity

Enthalpy

Energy cannot be created or destroyed but is converted from one form to another. Thermodynamics is the study of energy transfer during reactions and on the work done by chemical systems. It defines the energy required to start a reaction or the energy given out during the process. This change in energy is denoted ΔU. During chemical reactions energy may be absorbed or liberated. If this is in the form of heat, and since most reactions are performed at constant pressure, this is termed enthalpy and denoted by H. The enthalpy change (or heat of reaction) is:

$$\Delta H = H2 - H1$$

where $H1$ is the enthalpy of reactants and $H2$ the enthalpy of the products (or heat of reaction).

When $H2$ is less than $H1$ the reaction is exothermic and ΔH is negative, i.e. temperature increases. When $H2$ is greater than $H1$ the reaction is endothermic and the temperature falls. The heat of reaction is usually expressed in the equation as ΔH, e.g.

$$2H_2\text{ (gas)} + O_2\text{ (gas)} = 2H_2O\text{ (liquid)}\ \Delta H\text{ (298K)} = -571.6\text{ kJmole}^{-1}$$

Data exist for the enthalpy of chemical reactions, formation of substances from their constituent elements, combustion, fusion, neutralization, solution, vaporization, etc.

Electrochemistry

When strips of reactive metals such as zinc are placed in water a potential difference, the electromotive force (emf), is set up; the metal becomes negatively charged due to the transfer of zinc ions to the solution and the build-up of electrons on the metal. The metal strips or rods are termed the

Table 3.9 Classes of pyrolysis products produced during fires

Chemical type involved in a fire	*Pyrolysis product*
Halogenated plastic	Polyaromatic hydrocarbons Aliphatics Substituted benzenes Halogenated aliphatics Dioxins and furans
Non-halogenated plastics	Polycyclic aromatic compounds Aliphatics Substituted benzenes Heavy metals
Halogenated chemicals	Polycyclic aromatic hydrocarbons Aliphatics Substituted benzenes Halogenated aliphatics Dioxins and furans
Non-halogenated chemicals	Polycyclic aromatic hydrocarbons Aliphatics Substituted benzenes
Metal-based pesticides	Range of organics Heavy metals

electrode and the potential difference is termed the electrode potential. The latter depends upon the identity of the metal, the temperature and the concentration of metal ions in solution. Thus copper is less reactive than zinc so that if the two metals are immersed in water and connected by a wire electrons will flow through the external circuit from Zn (cathode) to Cu (anode). The process is termed electrolysis and the arrangement by which chemical energy is converted into

Table 3.10 Compounds liberated from a range of materials during fires

Fire material		*CO*	*HCN*	*HCl*	*P_2O_5*	*Isocyanate*	*Irritants, e.g. acrolein*	*HF and HBr*	*PAHs*	*NO_x*
Chlorine	Emissions									
	On-site	NA	NA	NA	NA	NA	NA	NA	NA	NA
	Off-site									
Oil refineries (storage tanks)	Emissions	+	–	–	–	–	++	–	+++	–
	On-site	–	–	–	–	–	++	–		–
	Off-site	–	–	–	–	–	+			
Paints and solvents	Emissions	+++	–	++	+	++	++	–	++	–
	On-site		–	+	–	++	+	–		–
	Off-site	–	–	–	–	+	–	–	–	–
Petrol	Emissions	++	–	–	–	–	++	–	+	–
	On-site	–	–	–	–	–	+	–		–
	Off-site	–	–	–	–	–	–	–	–	–
Phosphorus	Emissions	+++	–	+	+++	++	++	–	++	–
	On-site		–	+	++	++	+	–		–
	Off-site	–	–	–	+	+	–	–	–	–
Plastics	Emissions	+++	+++	+++	+	++	++	+	++	++
	On-site			++	–	++	++			+
	Off-site	–	–	+	–	+	+	–		–
Resins and adhesives	Emissions	+++	++	+	–	++	++	+	++	++
	On-site			+	–	++	++			+
	Off-site	–	–	–	–	+	+	–	–	–
Rubber	Emissions	+++	+	+	–	–	+++/++	–	++	+
	On-site			+	–	–	++/+	–		
	Off-site	–	–	–	–	–	+	–	–	–
Upholstery (polyurethane)	Emissions	+++	+++	+++	–	++	++	+	++	++
	On-site			++	–	++	++			+
	Off-site	–	–	+	–	+	+	–	–	–
Vegetation (forests)	Emissions	+	–	–	–	–	+	–	+	+
	On-site	–	–	–	–	–	+	–		
	Off-site	–	–	–	–	–	–	–	–	–
Waste tips	Emissions	–	+	+	–	+	++	+	+	+
	On-site	–		+	–	+	+			
	Off-site	–	–	–	–	–	–	–	–	–

Key

NA	Not applicable since chlorine is the main risk
Emissions	Total emissions during a fire
On-site	Exposure of workers and emergency service personnel
Off-site	Exposure of the general public and the wider environment
–	Zero or little emission or exposure
+	Likely to be some emission or exposure
++	Likely to be low level emission or exposure
+++	Likely to be greatest emission or exposure

electrical energy is termed a 'chemical' or 'galvanic cell'. To standardize conditions emfs are determined at 25°C of cells containing a molar metallic electrode (i.e. a rod of metal immersed in a solution containing 1 gram-ion of the metallic ion per litre) opposed to a molar hydrogen electrode (i.e. a plate of platinum covered with a thin film of hydrogen to simulate a rod of hydrogen).

The ease with which an atom gains or loses electrons is termed the electronegativity of the element. Tabulation of the elements in order of ease by which they lose electrons is called the electrochemical series and is shown in Table 6.10. Chapter 4 explains the importance of this to the formation and control of corrosion, and Chapter 6 discusses the relevance to predicting reactivity of metals towards water and their potential to become pyrophoric.

Other industrial applications of electrolysis include extraction/purification of metals from ores, electroplating, and the manufacture of certain chemicals such as sodium hydroxide. In the latter, sodium chloride solution when electrolysed is converted to sodium hydroxide to produce chlorine at the anode and hydrogen at the cathode. Both of these gaseous by-products are collected for industrial use; chlorine is used in the production of bleach and PVC; hydrogen is used as a fuel, to saturate fats, and to make ammonia.

Rates of chemical reaction

Whereas thermodynamics describes the energy requirements of a reaction, the speed at which it progresses is termed kinetics. It is important to be able to control the rate of chemical reactions for commercial and safety reasons. If a reaction takes too long to progress the rate at which a product is manufactured would not be viable. Alternatively, if reactions progress too fast and 'runaway' out of control there could be dangers such as explosions. The rate at which reactions take place can be affected by the concentration of reactants, pressure, temperature, wavelength and intensity of light, size of particles of solid reactants, or the presence of catalysts (i.e. substances which alter the speed of reactions without being consumed during the reaction) or impurities. Catalysts tend to be specific to a particular reaction or family of reactions. Thus nickel is used to facilitate hydrogenation reactions (e.g. add hydrogen to C=C double bonds) whereas platinum is used to catalyse certain oxidation reactions. Sometimes care is needed with the purity of reactants since impurities can act as unwanted catalysts; alternatively, catalysts can be inactivated by 'poisoning'.

The effect of temperature on different types of reaction is shown in Figure 7.5.

For reactions which progress slowly at room temperature it may be necessary to heat the mixture or add a catalyst for the reaction to occur at an economically-viable rate. For very fast reactions the mixture may need to be cooled or solvent added to dilute the reactants and hence reduce the speed of reaction to manageable proportions. In general the speed of reaction

- doubles for every 10°C rise in temperature;
- is proportional to the concentration of reactants in solution;
- increases with decreased particle size for reactions involving a solid;
- increases with pressure for gas phase reactions.

Chapter 4 discusses reaction kinetics further.

4

Physicochemistry

Hazards from processes using chemicals are assessed on the basis of:

• Inherent chemical properties	Toxic, flammable/explosive, reactive, unstable
• Chemical form	Liquid, solid (briquette, flake, powder), gas, vapour, air-borne particulate (including mist, fume, froth, aerosol, dust)
• Quantity	In storage, held up in process stages, in the working atmosphere, as wastes, etc.
• Processing conditions	Use of high or low temperature, high pressure, vacuum or possible hazardous reactions (polymerization, oxidation, halogenation, hydrogenation, alkylation, nitration, etc.)

Hazards can often be foreseen from basic physicochemical principles, as summarized below.

Vapour pressure

The vapour pressure of a chemical provides an indication of its volatility at any specific temperature. As an approximation, the vapour pressure p' of a pure chemical is given by

$$\log_c p' = (A/T) + B$$

where A and B are empirically determined constants and T is the absolute temperature.

Hence the vapour pressure of a chemical will increase markedly with temperature.

For a component 'a' in a mixture of vapours, its partial pressure p_a is the pressure that would be exerted by that component at the same temperature if present alone in the same volumetric concentration. So with a mixture of two components, 'a' and 'b', the total pressure is

$$P = p_a + p_b$$

If an inert gas is also present, its pressure is additive:

$$P = p_a + p_b + p_{inert}$$

In an 'ideal mixture' the partial pressure p_a is proportional to the mole fraction y_a of the component in the gas phase:

$$p_a = y_a P$$

and this partial pressure is also related to the concentration in the liquid phase expressed as mole fraction x_a by:

$$p_a = p'_a x_a$$

where p'_a is the vapour pressure of component 'a' at the prevailing temperature. So, if all the components are miscible in the liquid phase the total pressure P of a mixture is

$$P = p'_a x_a + p'_b x_b + p'_c x_c$$

As a result:

- The flash point of any flammable liquid will be lowered if it is contaminated with a more volatile, flammable liquid.
- Application of heat to a flammable liquid (e.g. due to radiation or flame impingement in a fire, or because of 'hot work') can generate a flammable vapour–air mixture.
- Increase in temperature of a toxic liquid can create an excessive concentration of toxic vapour in air. This may occur as the result of an exothermic reaction.
- The pressure in the vapour space of an incompletely full, sealed vessel containing liquid cannot be reduced by partially draining off liquid.
- The pressure in an incompletely full container of liquid will increase with temperature and can, in the extreme, result in rupture due to over-pressurization unless adequate relief is provided. (This may occur following an uncontrolled exothermic reaction.) Alternatively, partial ejection of the contents can occur on opening.
- The composition of the vapour in equilibrium with a miscible liquid mixture at any temperature, e.g. on heating during distillation, will be enriched by the more volatile components. The composition of the liquid phase produced on partial condensation will be enriched by the less volatile components. Such 'fractionation' can have implications for safety in that the flammability and relative toxicity of the mixtures can change significantly.

Gas–liquid solubility

For a dilute solution, the partial pressure exerted by a dissolved liquid (a solute) 'a' in a liquid solvent is given by

$$p_a = Hx_a$$

where H is Henry's law constant for the system and x_a is the mole fraction of solute. A different value of H is applicable to each gas–liquid system.
As a result:

- The solubility of a gas generally decreases with any increase in temperature. So, if a solution in a closed receptacle is heated above the filling temperature during transport or storage, loss of gas can result on opening or liquid discharge.
- With a 'sparingly-soluble' gas a much-higher partial pressure of that gas is in equilibrium with a solution of a given concentration than is the case with a highly soluble gas.
- Exposure of a solution to any atmosphere will lead to the take-up, or release, of gas until equilibrium is eventually attained.
- Rapid absorption of a gas in a liquid in an inadequately-vented vessel can result in implosion, i.e. collapse inwards due to a partial vacuum.

Liquid-to-vapour phase change

Evaporation of liquid to form vapour is accompanied by a considerable increase in volume. For example, at atmospheric pressure one volume of water will generate 1600 volumes of steam. Similarly 4.54 litres of gasoline will yield 0.93 m^3 of neat vapour on complete vaporization. The reverse process, condensation, is accompanied by a considerable – and often rapid – decrease in volume. As a result:

- Contact of water with molten metals or salts or hot oil (above 100°C at atmospheric pressure) can result in a 'steam explosion', or a 'boil-over', with ejection of process materials. Similar effects occur with other volatile liquids.
- Evaporation of a relatively-small volume of liquid in an enclosed space can produce a flammable or toxic vapour hazard. Leakage, or spillage, of a chemical maintained as a liquid above its atmospheric boiling point by pressure (e.g. liquefied petroleum gases) or as a liquid by refrigeration (e.g. ammonia) can result in a sizeable vapour cloud.
- Sudden cooling of a vapour-filled vessel which is sealed, or inadequately vented, may cause an implosion due to condensation to liquid.
- Cooling of vapour in a vented vessel may cause sucking-back of process materials or ingress of air.
- Vaporization in enclosed containers can produce significant pressure build-up and explosion.

In addition to volume changes the effect of temperature is also important. Thus the specific latent heat of vaporization of a chemical is the quantity of heat, expressed as kJ/kg, required to change unit mass of liquid to vapour with no associated change in temperature. This heat is absorbed on vaporization so that residual liquid or the surroundings cool. Alternatively an equivalent amount of heat must be removed to bring about condensation. Thus the temperature above a liquefied gas is reduced as the liquid evaporates and the bulk liquid cools. There may be consequences for heat transfer media and the strength of construction materials at low temperatures.

Solid-to-liquid phase change

The phase change of a chemical from solid to liquid generally results in an expansion in volume. (Ice to water is one exception.) As a result:

- Ejection of liquid can occur from open pipelines when solid blockages are released by external heating, e.g. by steam. (This hazard is increased if pressure is applied upstream of the constricton.)
- Melting of solid in a sealed system may exert a significant internal pressure.

Density differences of gases and vapours

As an approximation, at constant pressure,

$$\text{density of a gas/vapour} \propto \frac{\text{relative molecular mass}}{\text{absolute temperature}}$$

Since few chemicals (e.g. hydrogen, methane, ammonia) have a molecular weight less than that

Table 4.1 Densities of some toxic gases and vapours relative to air at 20°C

	Density gas/ density of air	*Relative molecular mass*
Bromine vapour	5.54	160
Phosgene	3.43	99
Chlorine	2.46	71
Sulphur dioxide	2.22	64
Acrylonitrile vapour	1.84	53
Hydrogen cyanide vapour	0.94	27
Hydrogen fluoride vapour	0.69	20
Ammonia	0.59	17

of air, under ambient conditions most gases or vapours are heavier than air. For example, for common toxic gases refer to Table 4.1; for flammable vapours refer to Table 6.1. At constant pressure the density of a gas or vapour is, as shown, inversely proportional to the absolute temperature. As a result:

- On release, vapours heavier than air tend to spread (i.e. to 'slump') at low level and will accumulate in pits, sumps, depressions in ground, etc. This may promote a fire/explosion hazard, or a toxic hazard, or cause an oxygen-deficient atmosphere to form, depending on the chemical.
- Heavy vapour can remain in 'empty' vessels after draining out liquid and venting via the top with similar associated hazards.
- On release, vapours which are less dense than air at ambient temperature may tend to spread at low level when cold (e.g. vapour from liquid ammonia or liquefied natural gas spillages).
- Gases less dense than air may rise upwards through equipment, or buildings, and if unvented will tend to accumulate at high level. This is an important consideration with piped natural gas which tends to diffuse upwards from fractured pipes, open valves or faulty appliances. Hydrogen from leakages in use, e.g. from cylinders (refer to Table 9.12), or from electrolytic processes, e.g. battery-charging, rapidly diffuses upwards.
- Hot gases rise by thermal lift. Hence in the open air they will disperse. Within buildings this is a serious cause of fire escalation and toxic/asphyxiation hazards if smoke and hot gases are able to spread without restriction (or venting) to upper levels.
- A balanced flue can serve to effectively vent a combustion process in a gas-fired appliance, but must be sound in construction and unrestricted to avoid leaks.
- Once a gas or vapour has been mixed with air, it is the mean density of the mixture which is important (similar considerations arise when mixing other gases). The mean density of a gas mixture is given by:

$$p_{\text{mixture}} = \frac{p_g V_g + p_a V_a}{V_g + V_a}$$

where V_g, V_a are the volumes of gas and air, and p_g, p_a the densities of gas and air respectively. Clearly if V_a is large relative to V_g, or if p_g does not differ significantly from p_a, the value of p_{mixture} approximates to p_a. As a result:

- The density of air saturated with a chemical vapour may not differ significantly from that of air itself. Refer to Table 4.2. This is an important consideration when designing ventilation systems, i.e. both high- and low-level extract vents may be desirable.

Table 4.2 Relative densities of air saturated with selected chemicals at 25°C

	Relative density of saturated air (air at 25°C)
Benzene	1.21
Bromochloromethane	1.07
Carbon tetrachloride	1.65
Diisobutyl ketone	1.01
Nitroethane	1.04
Parathion	1.0

Density differences of liquids

The specific gravities (s.g.) of liquid chemicals vary widely, e.g. for the majority of hydrocarbon fuels s.g. <1.0 but for some natural oils and fats, chlorinated hydrocarbons, s.g. >1.0. Density is generally reduced by any increase in temperature. As a result:

- On heating up, thermal expansion of a liquid in sealed piping, equipment or a container may exert sufficient hydraulic pressure to cause rupture or failure. (Hence specific filling ratios are followed with containers, e.g. road tankers.)
- A lighter liquid can spread over and, if immiscible, remain on top of a denser liquid. Thus liquid fuels and many organic liquids will spread on water; this may result in a hazard in sumps, pits or sewerage systems and often precludes the use of water as a jet in fire-fighting.
- Stratification of immiscible liquids may occur in unagitated process or storage vessels.

Immiscible liquid–liquid systems

In a combination of two immiscible liquids, each exerts its own vapour pressure independently. The total pressure is then the sum of the vapour pressures,

$$P = p'_a + p'_b$$

Also if a solute C is present in solvent A when it is mixed with solvent B, some transfer will occur of C to B. Eventually equilibrium will be attained between the concentrations of C in each phase. For many dilute solutions this is expressed by

$$y = mx$$

where x is the mass (or mole) fraction of C in A, y is the mass (or mole) fraction of C in B and m is the partition coefficient. In concentrated solutions the equilibria are better represented by a distribution curve. As a result of these equilibria:

- The boiling point of a mixture of immiscible liquids can be significantly lower than that of either chemical, so violent boiling may occur unexpectedly on mixing them whilst hot.
- Partition of solute into a second immiscible liquid (e.g. water) may result in its release if the latter is subsequently exposed to air, e.g. in a sump or effluent drain.
- Trace contamination of an immiscible liquid can occur following accidental contact with another liquid even if the mutual solubilities are considered insignificant.

Vapour flashing

If a liquid near its boiling point at one pressure is 'let down' to a reduced pressure, vapour flashing will occur. This will cease when the liquid temperature is reduced, due to removal of the latent heat of vaporization, to a temperature below the saturation temperature at the new pressure.

As a result:

- Flashing of vapour containing entrained mist may occur on venting equipment or vessels containing volatile liquids. This may create a toxic or flammable hazard depending on the chemical; with steam the risk is of scalding. Rupture of equipment can produce a similar effect.
- Escapes or spillages of liquefied petroleum gas, or chlorine or ammonia, rapidly generate a vapour cloud.
- Loss of containment, e.g. due to a crack or open valve, from beneath the liquid level in a liquefied gas vessel is potentially more serious than if it occurs from the gas space because the mass flowrate is greater.
- Absorption of heat (auto-refrigeration) and consequent temperature reduction on flashing may have a serious effect on associated heat transfer media, upon the strength of materials of construction, and result in frosting at the point of leakage. Exposure of personnel carries a risk of frostbite.

Effects of particle or droplet size

Airborne particulate matter may comprise liquid (aerosols, mists or fogs) or solids (dust, fumes). Refer to Figure 5.2. Some causes of dust and aerosol formation are listed in Table 4.3. In either case dispersion, by spraying or fragmentation, will result in a considerable increase in the surface area of the chemical. This increases the reactivity, e.g. to render some chemicals pyrophoric, explosive or prone to spontaneous combustion; it also increases the ease of entry into the body. The behaviour of an airborne particle depends upon its size (e.g. equivalent diameter), shape and density. The effect of particle diameter on terminal settling velocity is shown in Table 4.4. As a result:

- All combustible solids can create a dust explosion hazard if dispersed in air as a fine dust within certain concentration limits. Refer to Table 6.2. The hazard increases with decreasing particle size.
- Particles in the respirable size range, i.e. about 0.5–7 μm, will, once dispersed, remain airborne for extended periods. Indeed since they are sensitive to slight air currents they may be permanently suspended.
- A dust cloud comprising a distribution of particle sizes soon fractionates, e.g. visible matter settles to the ground in a few minutes. Hence the size distribution of airborne particles may differ significantly with time and from that of the source material. (This is particularly relevant to occupational hygiene measurements involving toxic dust emissions.)

Surface area effects in mass transfer or heterogeneous reactions

The rate of mass transfer across a phase boundary or interface can be expressed by

$$N = K.\, A\, (\Delta C)_m$$

Table 4.3 Selected sources of dusts and aerosols

Particulate form	*Common sources*
Dust	• Crushing • Grinding • Milling • Sieving • Drying, e.g. spray, drum, fluidized bed drying • Powder mixing • Emptying bags of powder, or fines from bag filters • Powder transfer, e.g. conveyors • Brushing • Buffing
Aerosol	• Droplet generation by air pressure differential across the film, e.g. bursting of liquid film in bubbles, froth across bottle mouths and pipette tips • Formation of froth by mixing of gas and liquid in shake cultures, fermenters or test tubes • Forceful ejection of the contents of pipettes or syringe, especially if already containing gas bubbles • Vibration or twanging of hypodermic needles or platinum loops • Impact of a droplet on a liquid, e.g. from liquid falling under gravity • Separation of two moist surfaces, e.g. withdrawal of the plunger in a syringe barrel or stopper from a tube • Centrifugal forces, e.g. escape of liquid from a centrifuge • Sizzling, e.g. when a platinum loop or inoculation tool is flamed or plunged hot into liquid • Deliberate atomization in some humidification operations, gas scrubbing, spray drying, spray painting • Leakage of liquids transferred under pressure, e.g. from flange joints, pump glands • Misting during condensation (e.g. in distillations) gas absorption, vaporizations • Use of propellants, e.g. hand-held aerosol cans • Electrolytic processes • Rapid liquid discharge from an orifice, e.g. shower heads, taps

Table 4.4 Terminal velocities of particles of different sizes

Diameter (μm)	*Rate of fall* (m/s)
5000	9
1000	4
500	3
100	3×10^{-1}
50	75×10^{-3}
10	3×10^{-3}
5	75×10^{-5}
1	36×10^{-6}
0.5	10×10^{-6}
0.1	36×10^{-8}

(Particle density 1000 kg/m^3; air at 20°C, 100 kPa.)

where N is mass transferred/unit time
K is a mass transfer coefficient
A is the interfacial area
$(\Delta C)_m$ is the mean concentration gradient, representing the deviation from equilibrium. Hence the rate is directly related to coefficient K, which will generally increase with any increase in turbulence such as increased relative velocity between the phases or agitation; to the exposed surface area A; and to the concentration difference, whether it is a pressure or humidity differential or a solubility relationship. As a result:

- The rate of evolution of a toxic or flammable vapour from a liquid (e.g. in an open vessel, from a spillage or as a spray) is directly related to the exposed area. Therefore, the rate of vapour formation from solvent-impregnated rag, from solvent-based films spread over a large area, from foams or from mists can be many times greater than that from bulk liquid.
- All gas absorption processes are surface area dependent. Hence water fog may be an effective means of dealing with emissions of soluble gases, e.g. ammonia or hydrogen fluoride.
- The rates of gas–solid reactions are surface area dependent, so finely-divided metals, coal etc. may be prone to oxidation leading to spontaneous combustion. A combustible dust will burn much more rapidly than the bulk sold, and if dispersed in air cause a dust explosion (refer to Table 6.2).
- The rates of gas–liquid reactions are surface area dependent. Hence in the spontaneous combustion of oil impregnating fibrous thermal insulation on hot equipment, oxidation is facilitated by the large exposed surface area and, since the dissipation of heat is restricted, the temperature can rise until the oil ignites spontaneously.
- The important factors, on exposure to chemicals that are toxic by absorption via the skin, are the contact area and the duration of exposure (refer to Table 13.8).

Enthalpy changes on mixing of liquids

Mixing of two or more chemicals which have dissimilar molecular structures may be exothermic (liberating heat) or endothermic (absorbing heat). As a result:

- Unless controlled, the enthalpy release when some liquids are mixed may result in their ejection from equipment or, in the extreme, an explosion.

Critical temperatures of gases

Every gas has a critical temperature above which it cannot be liquefied by the application of pressure alone. The critical pressure is that required to liquefy a gas at its critical temperature. Data for common gases are given in Table 4.5. As a consequence:

- Liquefied gases may be stored fully refrigerated, with the liquid at its bubble point at near atmospheric pressure; fully pressurized, i.e. at ambient temperature; or semi-refrigerated with the temperature below ambient but the vapour pressure above atmospheric pressure. Of the gases listed in Table 4.6, all those with critical temperatures below ambient must be maintained under refrigeration to keep them in the liquid phase.

Table 4.5 Critical temperature and pressure data for common gases

	Critical Temperature (°C)	*Critical pressure* (bar)
Water (steam)	374	–
Sulphur dioxide	157	219
Chlorine	144	78
Ammonia	132	77.7
Nitrous oxide	39	–
Carbon dioxide	31.1	73.1
Oxygen	–119	50
Nitrogen	–147	33.7
Hydrogen	–240	12.9

- If the temperature remains constant, the pressure within any cylinder containing liquefied gas will remain constant as gas is drawn off (i.e. more liquid simply evaporates) so the quantity of gas remaining cannot be deduced from the pressure.

Table 4.6 Gases commonly stored in liquefied form

Gas	*Boiling point at 1 bar abs.* (°C)	*Liquid density at boiling point* (kg/m^3)	*Volume ratio gas (1 bar abs., 20°C) to liquid (at boiling point)*	*Vapour pressure at 38°C* (bar abs.)	*Critical temperature* (°C)
Can be stored without refrigeration					
Ethylene oxide	11	883	425	2.7	195
n-Butane	0	602	242	3.6	152
Butadiene (1, 3)	–4	650	280	4.1	152
Butylene (α)	–6	626	262	4.3	146
Isobutane	–12	595	241	5.0	135
Ammonia	–33	682	962	14.6	133
Propane	–42	582	315	13.0	97
Propylene	–48	614	347	15.7	92
Requires refrigeration					
Ethane	–89	546	436	–	32
Ethylene	–104	568	487	–	9
Methane (LNG)	–162	424	637	–	–82
Oxygen	–183	1140	860	–	–119
Nitrogen	–196	808	696	–	–147

Chemical reaction kinetics

The rate of chemical reaction is generally a function of reactant concentration and temperature. For many homogeneous reactions therefore, if they are exothermic,

$$\text{rate of generation} \propto e^{RT_r}$$

where R is the gas constant and T_r is the absolute temperature. If the heat is removed by forced convection to a coolant in a jacket or coil,

$$\text{rate of removal} \propto T_r - T_a$$

where T_a is the coolant temperature. Thus:

- Since the generation rate is exponential whereas the removal rate is linear, for any exothermic reaction in a specific reactor configuration a critical condition may exist, i.e. a value of T_r beyond which 'runaway' occurs.
- A reaction which is immeasurably slow at ambient temperature may become rapid if the temperature is raised.

With some reactions which have a significant rate at ambient temperature, e.g. catalysed reactions, oxidations or self-polymerization of certain polymers, severe hazards may be associated with an elevation in temperature.

Exothermic reactions require control strategies which may involve temperature control, dilution of reagents, controlled addition of one reagent, containment/venting and provision for emergencies. Refer to p. 248.

Many liquid phase or heterogeneous solid–liquid or gas–liquid reactions result in gaseous products or byproducts. These products may be toxic (refer to Table 5.1) or flammable (refer to Table 6.1), or result in overpressurization of any sealed container or vessel. Unless pressure relief is provided, relatively small volumes of reactants – the presence of which may not be expected – may generate sufficient gas pressure to rupture a container. The causes of pressure build-up may be:

- Reactions with water, e.g. reaction of phosphorus oxychloride with water to produce gaseous hydrogen chloride:

  $POCl_3 + H_2O \rightarrow H_3PO_4 + HCl$ (refer to Table 7.1)

- Electrolytic corrosion (see later).
- Reaction due to contaminants, e.g. in reusable containers or in transfer pipelines.
- Reaction with construction materials, e.g. nitric acid can produce nitric oxide gas on contact with copper in pipes or copper windings in motors of canned pumps:

  $3Cu + 2HNO_3 \rightarrow 3\ CuO + 2NO + H_2O$

- Slow decomposition of a chemical of limited stability. For example, the slow decomposition of 98–100% formic acid to gaseous carbon monoxide in a full 2.5 litre bottle would produce 7 bar pressure during one year at 25°C if unvented:

  $HCOOH \rightarrow CO + H_2O$

- Self-initiated reactions, e.g. pyruvic acid on storage can become oxidized by air (or airborne yeasts) to form sufficient gaseous carbon dioxide to overpressurize the container:

  $CH_3COCO_2H + [O] \rightarrow CH_3CO_2H + CO_2$

  Pyruvic acid should therefore be stored refrigerated with light and air excluded.

The precautions required can be any combination of those in Table 4.7.

Corrosion

Pure metals and their alloys interact gradually with the elements of a corrosive medium to form stable compounds and the resulting metal surface is considered to be 'corroded'. The corrosion reaction comprises an anode and an electrode between which electrons flow. Table 6.10 shows the

Table 4.7 Precautions applicable to reactions producing gaseous products or byproducts

Temperature control
Adequately sized pressure relief
Elimination of contaminants, including metallic residues, from process streams and equipment
Selection of materials of construction compatible with the chemical(s) in use, properly cleaned and passivated
Elimination of ingress of reactive chemicals, e.g. water, air
Date labelling and inventory control in storage
Cleaning and inspection of reusable containers, tankers etc, before refilling

anodic–cathodic series, or electrochemical series, for selected metals and for hydrogen (since the discharge of hydrogen ions takes place in most corrosion reactions). Metals above hydrogen in the series displace hydrogen more easily than do those below it. As a general rule, when dissimilar metals are used in contact with each other and are exposed to an electrically-conducting solution, combinations of metals should be chosen that are as close as possible to each other in the series. Coupling two metals widely separated in the series will generally produce accelerated attack on the more active metal. Often, however, protective passive oxide films and other effects will tend to reduce galvanic corrosion. Insulating the metals from each other can prevent corrosion. The dual action of stress and a corrodent may result in stress corrosion cracking or corrosion fatigue.

Corrosion may be uniform or be intensely localized, characterized by pitting. The mechanisms can be direct oxidation, e.g. when a metal is heated in an oxidizing environment, or electrochemical. Galvanic corrosion may evolve sufficient hydrogen to cause a hazard, due to:

- Formation of a flammable atmosphere with air in equipment or piping.
- Build-up of internal pressure within a weakening container.
- Production of atomic hydrogen as a species; this may penetrate metal to produce blistering or embrittlement.

The consumption of oxygen due to atmospheric corrosion of sealed metal tanks may cause a hazard, due to oxygen-deficiency affecting persons on entry.

Stresses may develop resulting from the increased volume of corrosion products, e.g. rust formation involves a seven-fold increase in volume.

Many salts are corrosive to common materials of construction, as demonstrated in Tables 4.8 and 4.9. Corrosion may be promoted, or accelerated, by traces of contaminants.

Whereas corrosion of metals is due to chemical or substantial electrochemical attack, the deterioration of plastics and other non-metals which are susceptible to swelling, cracking, crazing, softening etc. is essentially physicochemical rather than electrochemical.

Corrosion prevention

Corrosion prevention is achieved by correct choice of material of construction, by physical means (e.g. paints or metallic, porcelain, plastic or enamel linings or coatings) or by chemical means (e.g. alloying or coating). Some metals, e.g. aluminium, are rendered passive by the formation of an inert protective film. Alternatively a metal to be protected may be linked electrically to a more easily corroded metal, e.g. magnesium, to serve as a sacrificial anode.

Some corrosion-resistant materials for concentrated aqueous solutions and acids are given in Tables 4.10 and 4 .11. The resistance of some common polymers to organic solvents is summarized in Table 4.12. The attack process is accelerated by an increase in temperature. The chemical resistance of a range of common plastics is summarized in Table 4.13.

Table 4.8 Comparison of corrosion rates by solutions of salts

Salts (type and examples)	*Corrosion rates for listed construction materials*				
	Carbon steel	304 SS	316 SS	Alloy 400 (65 Ni-32Cu)	Nickel 200
Non-oxidizing non-halides					
Alkaline (pH >10)					
e.g. sodium carbonate	L	L	L	L	L
Neutral					
e.g. sodium sulphate	M	L-M; SCC	L-M	L	L
sodium nitrate	L-M; SCC	L; pits	L, pits	L	L
Acid					
e.g. nickel sulphate	S	L-M	L-M	M	M; pits
Non-oxidizing halides					
Neutral					
e.g. sodium chloride	M; pits	M; SCC; pits	M; SCC; pits	M	M
Acid					
e.g. zinc chloride	S	S; SCC; pits	S; SCC; pits	M	M
ammonium chloride	S	S; SCC; pits	M; SCC; pits	M	M-S
Oxidizing non-halides					
Neutral					
e.g. sodium chromate	L(1)	L	L	L	L
sodium nitrite	L(1)	L	L	L	L
potassium permanganate	M	M	M	M	M
Acid					
e.g. ferric sulphate	S	L	L	S	–
silver nitrate	S; SCC	M	M	S	S
Oxidizing halides					
Alkaline					
e.g. sodium hypochlorite	S	S; pits	S; pits	M-S; pits	M-S; pits
Acid					
e.g. ferric chloride	S; SCC	S; pits	S; pits	S	S
cupric chloride	S	S; pits	S; pits	S	S
mercuric chloride	S	S; SCC; pits	S; SCC; pits	S; SCC	S

L Low: <5 mpy, for all concentrations and temperatures <boiling
M Moderate: <20 mpy, perhaps limited to lower concentrations and/or temperatures
S Severe: >50 mpy
SCC Induces stress corrosion cracking
(mpy = mils per year: 1 mil = 0.001 in = 25.4 μm)
(1) Chemical acts as corrosion inhibitor if present in sufficient amounts, but may cause pitting if in lower amounts.

Force and pressure

Pressure is defined as force per unit area. (It may be expressed in a variety of units; refer to Chapter 18.) So, pressure × area = force.

As a result:

- A relatively small pressure can result in a very large force if it is applied over a large area. Inadequately vented atmospheric storage tanks may therefore rupture if, for some reason, e.g. a high inflow, they are subjected to a relatively low internal pressure. Large side-on structures, windows, etc. are particularly prone to damage from an explosion even at a significant distance from the epicentre.

Table 4.9 Selected inorganic salts highly corrosive to carbon steel (Corrosion rate >50 mpy)

Aluminium sulphate	Magnesium fluorosilicate
Ammonium bifluoride	Mercuric chloride
Ammonium bisulphite	Nickel chloride
Ammonium bromide	Nickel sulphate
Ammonium persulphate	Potassium bisulphate
Antimony trichloride	Potassium bisulphite
Beryllium chloride	Potassium sulphite
Cadmium chloride	Silver nitrate
Calcium hypochlorite	Sodium aluminium sulphate
Copper nitrate	Sodium bisulphate
Copper sulphate	Sodium hypochlorite
Cupric chloride	Sodium perchlorate
Cuprous chloride	Sodium thiocyanate
Ferric chloride	Stannic ammonium chloride
Ferric nitrate	Stannic chloride
Ferrous ammonium sulphate	Stannous chloride
Ferrous chloride	Uranyl nitrate
Ferrous sulphate	Zinc chloride
Lead nitrate	Zinc fluorosilicate
Lithium chloride	

- High-pressure equipment, the walls of gas cylinders, etc., are subjected to very high forces. Hence metallurgical integrity is vital.
- Implosion of glass, plastic or inadequately designed metal equipment can occur under partial vacuum conditions.

Stored pressure energy

Any gas stored or transferred under pressure represents an energy source. If piping or equipment failure occurs, energy is given out as the gas rapidly expands down to virtually atmospheric pressure. Elastic strain energy in the walls is also given out, but, by comparison, this is relatively small.

For a perfect gas the energy content is:

$$W = \frac{P_o V \ln P_o}{P_a}$$

where W is the energy content (joules), V is the container volume (m^3), and P_o and P_a are the internal and atmospheric pressure (Mpa), respectively. As a result:

- Pressure rupture of even a small piece of equipment may result in a serious explosion generating missiles travelling at high velocity. This is an important consideration when using laboratory glassware or industrial glass equipment and pipework.
- A pressure not registered on a gauge may be sufficient to result in an accident.
- Pneumatic testing of equipment should only be used when hydraulic testing is impracticable; appropriate safeguards, e.g. pre-inspection, reduction of the internal gas space, pressure relief, and use of a blast pit or enclosure, are necessary.
- High inventories of stored pressure energy constitute a major accident hazard.

Table 4.10 Corrosion-resistant materials for concentrated aqueous solutions

Material	*Acids*								*Alkalis*		*Salts*			*Organic solvents*		*Not recommended for*
	Oxidizing			*Reducing*			*Organic*				*Aqueous solutions*					
	Oleum	70–100% H_2SO_4	HNO_3 (conc.)	0–70% H_2SO_4	0–37% HCl	0.80% HF	Acetic acid	Formic acid	NaOH (caustic)	NH_4OH (ammonia)	NaCl	NaOCl	$FeCl_3$	Aliphatics	Aromatics	
Alloy C	L	L	L	R	L	L	R	R	R	R	R	R	L	R	R	Acid services >65°C, especially hydrochloric acid or acid solutions with high chloride contents
Tantalum	V	R	R	L	R	N	R	R	N	N	R	R	R	R	R	Hot oleum (>50°C), strong alkalis, fluoride solutions, sulphur trioxide
Glass/silicates	R	R	R	R	R	N	R	R	L	L	R	R	R	R	R	Strong alkalis, especially >54°C, distilled water >82°C, hydrofluoric acid, acid fluorides, hot concentrated phosphoric acid, lithium compounds >177°C, severe shock or impact applications
Carbon, impreg. with furan	N	L	L	–	–	L	–	–	R	R	R	R	R	–	–	Strong oxidizers, very strong solvents
Carbon, impreg. with phenolic	N		L	R	R	L	R	R	N	R	R	R	R	R	R	Strong alkalis, very strong oxidizers
FEP/TFE	R	R	R	R	R	R	R	R	R	R	R	R	R	R	R[(1)]	Molten alkali metals, elemental fluorine, strong fluorinating agents
Furan resin	N	N	V	R	L	N	L	L	R	R	R	V	R	R	L[(1)]	Strongly oxidizing solutions, liquid bromine, pyridine
Phenolic resin	N	L	V	R	R	L	L	R	N	N	R	N	L	R	L[(1)]	Strong alkalis or alkali salts, very strong oxidizers
Epoxy resin[(2)]	N	N	V	L	R	–	L	N	R	L	R	N	R	L	L[(1)]	Strong oxidizing conditions, very strong organic solvents

R Recommended for full range concentrations up to boiling or to temperature limit of (non-metallic) product form.
L Generally good service but limited in concentration and/or temperature.
V Very limited in concentration and/or temperature for service.
N Not recommended.
(1) See Table 4.12
(2) Epoxy hardener will strongly affect chemical resistance.

Table 4.11 Construction materials for use with strong acids

Acid	*Construction material*	*Important safety consideration*
Acetic	316 L stainless steel	Excess acetic anhydride in glacial acetic acid can accelerate corrosion; chloride impurities (ppm levels) can cause pitting and stress-corrosion cracking
	Copper/copper alloys	Not for highly oxidizing conditions
	Aluminium alloys	Sensitive to contaminants; requires very clean welding; attacked very rapidly in concentrations near 100% or with excess acetic anhydride
Formic	304 L stainless steel	For ambient temperature only
	Copper/copper alloys	Not for oxidizing conditions, including air
Hydrochloric	Rubber-lined steel (natural rubber)	Low tolerance for organic solvent impurities; temperature limited according to hardness of rubber; steel fabrication must be properly done
	Alloy C	Not for hot concentrated HCl
	Alloy B	Not for oxidizing conditions (test if reducing conditions are in doubt)
	Tantalum	Not for fluoride impurities
	Impervious graphite	Fragile
	Reinforced plastic	HCl may attack or permeate laminate; temperature limited (<65.5°C); requires excellent engineering design and fabrication quality
Hydrofluoric	Alloy 400	Not for oxidizing conditions (test if in doubt)
	Copper	Not for oxidizing conditions (test if in doubt); <65.5°C only.
	Cupro-nickel	Not for oxidizing conditions (test if in doubt); concentration and temperature slightly limited
	Carbon steel	Not below 60% concentration, depending on impurities
	Impervious graphite	Fragile; limit to below 60% concentration
	Polyvinylidene fluoride	–
Nitric	304 L stainless steel	Must use low-carbon (or stabilized grade) if welded; not for fuming acid concentrations above 65.5°C
	High-silicon iron	Casting only; limited shock resistance; only for concentrations above 45% if temperatures over 71°C
	Aluminium (e.g. 3003, 5052)	Mostly for over 95% concentration, not for below 85% concentration; requires very clean welding.
	Titanium	May ignite in red-fuming nitric acid if water is below 1.5% and nitrogen dioxide is above 2.5%
Oleum	Carbon steel	Not for 100–101% H_2SO_4 concentration; limited in temperature
	Glass-lined steel	Limited shock resistance
Phosphoric	316 L stainless steel	Must use low-carbon or stabilized grade if welded, up to 85% concentration and 93°C
Sulphuric	Carbon steel	Not for below 70% concentration; ambient temperatures only; flow velocities below 0.6 to 1.2 m/s
	Alloy 20 variations	Limited temperature at 65–75% concentration
	High-silicon iron	Castings only; limited shock resistance
	Chemical lead	Soft and suffers from erosion; creeps at room temperature; limit to below 90% concentration
	Glass	Limited shock resistance
	Alloy C	Better for reducing acid strengths (<60% concentration)
	Rubber-lined steel	For dilute not concentrated (oxidizing) strengths; temperature limited according to rubber hardness and acid concentration; steel fabrication must be properly done
	Brick linings with silicate mortar	Absorption of the corrosive by the masonry (use membrane substrate); poor properties in tension or shear (use in compression); many brick linings 'grow' in service but if used in archlike contours, growth merely increases compression

Table 4.12 Solvent resistance of polymers

Solvent	*Example*	*Thermosetting resins*					*Thermoplastics*						*Elastomers*		
		Epoxy (120–148°C)	Furan (120°C)	Phenolic (93°C)	Polyester FRP (104°C) (Bisphenol-A fumarate)	Vinyl ester FRP (93°C) (Low-temperature variety)	FEP/TFE (204/260°C)	Nylon 6/6 (93–120°C)	Polyethylene (65°C)	Polypropylene (107°C)	Polyvinyl chloride (60°C)	Polyvinylidene fluoride (135°C)	Butyl rubber (93°C)	Natural rubber (65°C)	Neoprene (93°C)
Alcohols	Methanol	G	E	G	G	G	E	P	E	E	E	E	E	E	G
Aldehydes	Formaldehyde	G	E	G	G	G	E	E	G	E	G	G	G	G	E
Aliphatics	Heptane	G	E	E	G	E	E	G	P	G	G	E	P	P	G
Aliphatic amines	Diethylamine	G	E	P	P	P	E	P	G	G	–	G	–	–	–
Aromatics (and derivatives)	Benzene	E	G	G	P	P	E	G	P	G	P	G	P	P	P
	Aniline	P	G	G	P	P	E	P	G	E	P	G	G	P	P
	Phenol	G	G	G	F	P	E	P	G	G	P	G	P	P	P
	Xylene	G	E	E	G	G	E	G	P	F	P	G	P	P	P
Chlorinated aliphatics	Trichloroethylene	G	G	E	G	P	E	G	P	P	P	E	P	P	P
	Ethylene chloride	G	G	G	P	P	E	–	P	F	P	E	P	P	P
Ketones	Acetone	G	G	P	G	G	E	G	P	G	P	P	G	P	P
Miscellaneous:															
Pyridine	Aromatics	P	P	P	P	P	E	F	F	E	P	F	P	P	P
Tetrahydrofuran	Aromatics	–	P	–	P	–	E	F	P	F	P	F	P	P	P
Furfural	Aromatics	G	E	P	G	P	E	–	P	P	P	F	–	–	–

E Recommended to maximum temperature of product form.
G Recommendation limited to somewhat lower temperature, or restricted in product from.
F Very limited recommendation; for ambient temperature only.
P Not recommended. Severe attack.
(Temperatures are approximate maxima.)

Expansion and contraction of solids

Hazardous situations can develop with changes in volume with temperature. The effect of temperature on gases and liquids is mentioned on pages 45 and 65, respectively.

The thermal expansion and contraction of solids can also have safety implications. For a given material the amount of its linear expansion, or contraction, in one direction is directly related to temperature and its original size (i.e. length, diameter, circumference). Thus:

$$\text{change in length} = \alpha L \Delta t$$

where α = thermal coefficient of linear expansion,
Δt = change in temperature, and
L = original linear measurement.

Table 4.14 lists the coefficients of expansion for selected materials of construction.

Thus, unacceptable stresses can arise in rigid construction materials in apparatus, equipment, piping, etc. if subjected to large temperature fluctuations. For example, conventional glass is prone to failure due to thermal shock.

Table 4.13 Chemical resistance of common plastics

Chemical	*Resins*						
	CPE	*LPE*	*PP*	*PMP*	*FEP/ ETFE*	*PC*	*PVC*
Acetaldehyde	GN	GF	GN	GN	EE	FN	GN
Acetamide, sat.	EE	EE	EE	EE	EE	NN	NN
Acetic acid, 5%	EE	EE	EE	EE	EE	EG	EE
Acetic acid, 50%	EE	EE	EE	EE	EE	EG	EG
Acetone	EE	EE	EE	EE	EE	NN	FN
Adipic acid	EG	EE	EE	EE	EE	EE	EG
Alanine	EE	EE	EE	EE	EE	NN	NN
Allyl alcohol	EE	EE	EE	EG	EE	EG	GF
Aluminium hydroxide	EG	EE	EG	EG	EE	FN	EG
Aluminium salts	EE	EE	EE	EE	EE	EG	EE
Amino acids	EE	EE	EE	EE	EE	EE	EE
Ammonia	EE	EE	EE	EE	EE	NN	EG
Ammonium acetate, sat.	EE	EE	EE	EE	EE	EE	EE
Ammonium glycolate	EG	EE	EG	EG	EE	GF	EE
Ammonium hydroxide, 5%	EE	EE	EE	EE	EE	FN	EE
Ammonium hydroxide	EG	EE	EG	EG	EE	NN	EG
Ammonium oxalate	EG	EE	EG	EG	EE	EE	EE
Ammonium salts	EE	EE	EE	EE	EE	EG	EG
n-Amyl acetate	GF	EG	GF	GF	EE	NN	FN
Amyl chloride	NN	FN	NN	NN	EE	NN	NN
Aniline	EG	EG	GF	GF	EE	FN	NN
Antimony salts	EE	EE	EE	EE	EE	EE	EE
Arsenic salts	EE	EE	EE	EE	EE	EE	EE
Barium salts	EE	EE	EE	EE	EE	EE	EG
Benzaldehyde	EG	EE	EG	EG	EE	FN	NN
Benzene	FN	GG	GF	GF	EE	NN	NN
Benzoic acid, sat.	EE	EE	EG	EG	EE	EG	EG
Benzyl acetate	EG	EE	EG	EG	EE	FN	FN
Benzyl alcohol	NN	FN	NN	NN	EE	GF	GF
Bismuth salts	EE	EE	EE	EE	EE	EE	EE
Boric acid	EE	EE	EE	EE	EE	EE	EE
Boron salts	EE	EE	EE	EE	EE	EE	EE
Brine	EE	EE	EE	EE	EE	EE	EE
Bromine	NN	FN	NN	NN	EE	FN	GN
Bromobenzene	NN	FN	NN	NN	EE	NN	NN
Bromoform	NN	NN	NN	NN	EE	NN	NN
Butadiene	NN	FN	NN	NN	EE	NN	FN
n-Butyl acetate	GF	EG	GF	GF	EE	NN	NN
n-Butyl alcohol	EE	EE	EE	EG	EE	GF	GF
sec-Butyl alcohol	EG	EE	EG	EG	EE	GF	GG
tert-Butyl alcohol	EG	EE	EG	EG	EE	GF	EG
Butyric acid	NN	FN	NN	NN	EE	FN	GN
Cadmium salts	EE	EE	EE	EE	EE	EE	EE
Calcium hydroxide, conc.	EE	EE	EE	EE	EE	NN	EE
Calcium hypochlorite, sat.	EE	EE	EE	EG	EE	FN	GF
Carbazole	EE	EE	EE	EE	EE	NN	NN
Carbon bisulphide	NN	NN	EG	FN	EE	NN	NN
Castor oil	EE	EE	EE	EE	EE	EE	EE
Cedarwood oil	NN	FN	NN	NN	EE	GF	FN
Cellosolve acetate	EG	EE	EG	EG	EE	FN	FN
Caesium salts	EE	EE	EE	EE	EE	EE	EE
Chlorine, 10% in air	GN	EF	GN	GN	EE	EG	EE
Chlorine, 10% (moist)	GN	GF	GN	GN	EE	GF	EG
Chloroacetic acid	EE	EE	EG	EG	EE	FN	FN

Table 4.13 Cont'd

Chemical	*Resins*						
	CPE	*LPE*	*PP*	*PMP*	*FEP/ ETFE*	*PC*	*PVC*
p-Chloroacetophenone	EE	EE	EE	EE	EE	NN	NN
Chloroform	FN	GF	GF	FN	EE	NN	NN
Chromic acid, 10%	EE	EE	EE	EE	EE	EG	EG
Chromic acid, 50%	EE	EE	EG	EG	EE	EG	EF
Cinnamon oil	NN	FN	NN	NN	EE	GF	NN
Citric acid, 10%	EE	EE	EE	EE	EE	EG	GG
Citric acid, crystals	EE	EE	EE	EE	EE	EE	EG
Coconut oil	EE	EE	EE	EG	EE	EE	GF
Cresol	NN	FN	EG	NN	EE	NN	NN
Cyclohexane	GF	EG	GF	NN	EE	EG	GF
Decalin	GF	EG	GF	FN	EE	NN	EG
o-Dichlorobenzene	FN	FF	FN	FN	EE	NN	GN
p-Dichlorobenzene	FN	GF	EF	GF	EE	NN	NN
Diethyl benzene	NN	FN	NN	NN	EE	FN	NN
Diethyl ether	NN	FN	NN	NN	EE	NN	FN
Diethyl ketone	GF	GG	GG	GF	EE	NN	NN
Diethyl malonate	EE	EE	EE	EG	EE	FN	GN
Diethylene glycol	EE	EE	EE	EE	EE	GF	FN
Diethylene glycol ethyl ether	EE	EE	EE	EE	EE	FN	FN
Dimethyl formamide	EE	EE	EE	EE	EE	NN	FN
Dimethylsulphoxide	EE	EE	EE	EE	EE	NN	NN
1,4-Dioxane	GF	GG	GF	GF	EE	GF	FN
Dipropylene glycol	EE	EE	EE	EE	EE	GF	GF
Ether	NN	FN	NN	NN	EE	NN	FN
Ethyl acetate	EE	EE	EE	EG	EE	NN	FN
Ethyl alcohol	EG	EE	EG	EG	EE	EG	EG
Ethyl alcohol, 40%	EG	EE	EG	EG	EE	EG	EE
Ethyl benzene	FN	GF	FN	FN	EE	NN	NN
Ethyl benzoate	FF	GG	GF	GF	EE	NN	NN
Ethyl butyrate	GN	GF	GN	FN	EE	NN	NN
Ethyl chloride, liquid	FN	FF	FN	FN	EE	NN	NN
Ethyl cyanoacetate	EE	EE	EE	EE	EE	FN	FN
Ethyl lactate	EE	EE	EE	EE	EE	FN	FN
Ethylene chloride	GN	GF	FN	NN	EE	NN	NN
Ethylene glycol	EE	EE	EE	EE	EE	GF	EE
Ethylene glycol methyl ether	EE	EE	EE	EE	EE	FN	FN
Ethylene oxide	FF	GF	FF	FN	EE	FN	FN
Fluorides	EE	EE	EE	EE	EE	EE	EE
Fluorine	FN	GN	FN	FN	EG	GF	EG
Formaldehyde, 10%	EE	EE	EE	EG	EE	EG	GF
Formaldehyde, 40%	EG	EE	EG	EG	EE	EG	GF
Formic acid, 3%	EG	EE	EG	EG	EE	EG	GF
Formic acid, 50%	EG	EE	EG	EG	EE	EG	GF
Formic acid, 98–100%	EG	EE	EG	EF	EE	EF	FN
Fuel oil	FN	GF	EG	GF	EE	EG	EE
Gasoline	FN	GG	GF	GF	EE	FF	GN
Glacial acetic acid	EG	EE	EG	EG	EE	GF	EG
Glycerine	EE	EE	EE	EE	EE	EE	EE
n-Heptane	FN	GF	FF	FF	EE	EG	FN
Hexane	NN	GF	EF	FN	EE	FN	GN
Hydrochloric acid, 1–5%	EE	EE	EE	EG	EE	EE	EE
Hydrochloric acid, 20%	EE	EE	EE	EG	EE	EG	EG
Hydrochloric acid, 35%	EE	EE	EG	EG	EE	GF	GF
Hydrofluoric acid, 4%	EG	EE	EG	EG	EE	GF	GF

Table 4.13 Cont'd

Chemical	*Resins*						
	CPE	*LPE*	*PP*	*PMP*	*FEP/ ETFE*	*PC*	*PVC*
Hydrofluoric acid, 48%	EE	EE	EE	EE	EE	NN	GF
Hydrogen	EE	EE	EE	EE	EE	EE	EE
Hydrogen peroxide, 3%	EE	EE	EE	EE	EE	EE	EE
Hydrogen peroxide, 30%	EG	EE	EG	EG	EE	EE	EE
Hydrogen peroxide, 90%	EG	EE	EG	EG	EE	EE	EG
Isobutyl alcohol	EE	EE	EE	EG	EE	EG	EG
Isopropyl acetate	GF	EG	GF	GF	EE	NN	NN
Isopropyl alcohol	EE	EE	EE	EE	EE	EE	EG
Isopropyl benzene	FN	GF	FN	NN	EE	NN	NN
Kerosene	FN	GG	GF	GF	EE	GF	EE
Lactic acid, 3%	EG	EE	EG	EG	EE	EG	GF
Lactic acid, 85%	EE	EE	EG	EG	EE	EG	GF
Lead salts	EE	EE	EE	EE	EE	EE	EE
Lithium salts	EE	EE	EE	EE	EE	GF	EE
Magnesium salts	EE	EE	EE	EE	EE	EG	EE
Mercuric salts	EE	EE	EE	EE	EE	EE	EE
Mercurous salts	EE	EE	EE	EE	EE	EE	EE
Methoxyethyl oleate	EG	EE	EG	EG	EE	FN	NN
Methyl alcohol	EE	EE	EE	EE	EE	FN	EF
Methyl ethyl ketone	EG	EE	EG	EF	EE	NN	NN
Methyl isobutyl ketone	GF	EG	GF	FF	EE	NN	NN
Methyl propyl ketone	GF	EG	GF	FF	EE	NN	NN
Methylene chloride	FN	GF	FN	FN	EE	NN	NN
Mineral oil	GN	EE	EE	EG	EE	EG	EG
Nickel salts	EE	EE	EE	EE	EE	EE	EE
Nitric acid, 1–10%	EE	EE	EE	EE	EE	EG	EG
Nitric acid, 50%	EG	GN	GN	GN	EE	GF	GF
Nitric acid, 70%	EN	GN	GN	GN	EE	FN	FN
Nitrobenzene	NN	FN	NN	NN	EE	NN	NN
n-Octane	EE	EE	EE	EE	EE	GF	FN
Orange oil	FN	GF	GF	FF	EE	FF	FN
Ozone	EG	EE	EG	EE	EE	EG	EG
Perchloric acid	GN	GN	GN	GN	GF	NN	GN
Perchloroethylene	NN	NN	NN	NN	EE	NN	NN
Phenol, crystals	GN	GF	GN	FG	EE	EN	FN
Phosphoric acid, 1–5%	EE	EE	EE	EE	EE	EE	EE
Phosphoric acid, 85%	EE	EE	EG	EG	EE	EG	EG
Phosphorus salts	EE	EE	EE	EE	EE	EE	EE
Pine oil	GN	EG	EG	GF	EE	GF	FN
Potassium hydroxide, 1%	EE	EE	EE	EE	EE	FN	EE
Potassium hydroxide, conc.	EE	EE	EE	EE	EE	NN	EG
Propane gas	NN	FN	NN	NN	EE	FN	EG
Propylene glycol	EE	EE	EE	EE	EE	GF	FN
Propylene oxide	EG	EE	EG	EG	EE	GF	FN
Resorcinol, sat.	EE	EE	EE	EE	EE	GF	FN
Resorcinol, 5%	EE	EE	EE	EE	EE	GF	GN
Salicylaldehyde	EG	EE	EG	EG	EE	GF	FN
Salicylic acid, powder	EE	EE	EE	EG	EE	EG	GF
Salicylic acid, sat.	EE	EE	EE	EE	EE	EG	GF
Salt solutions	EE	EE	EE	EE	EE	EE	EE
Silver acetate	EE	EE	EE	EE	EE	EG	GG
Silver salts	EG	EE	EG	EE	EE	EE	EG
Sodium acetate, sat.	EE	EE	EE	EE	EE	EG	GF
Sodium benzoate, 1%	EE	EE	EE	EE	EE	EE	EE

Table 4.13 Cont'd

Chemical	*Resins*						
	CPE	*LPE*	*PP*	*PMP*	*FEP/ ETFE*	*PC*	*PVC*
Sodium hydroxide, 1%	EE	EE	EE	EE	EE	FN	EE
Sodium hydroxide, 50% to sat.	EE	EE	EE	EE	EE	NN	EG
Sodium hypochlorite, 15%	EE	EE	EE	EE	EE	GF	EE
Stearic acid, crystals	EE	EE	EE	EE	EE	EG	EG
Sulphuric acid, 1–6%	EE	EE	EE	EE	EE	EE	EG
Sulphuric acid, 20%	EE	EE	EG	EG	EE	EG	EG
Sulphuric acid, 60%	EG	EE	EG	EG	EE	GF	EG
Sulphuric acid, 98%	EG	EE	EE	EE	EE	NN	NN
Sulphur dioxide, liq., 46 psi	NN	FN	NN	NN	EE	GN	FN
Sulphur dioxide, wet or dry	EE	EE	EE	EE	EE	EG	EG
Sulphur salts	FN	GF	FN	FN	EE	FN	NN
Tartaric acid	EE	EE	EE	EE	EE	EG	EG
Tetrachloromethane	FN	GF	GF	NN	EE	NN	GF
Tetrahydrofuran	FN	GF	GF	FF	EE	NN	NN
Thionyl chloride	NN	NN	NN	NN	EE	NN	NN
Titanium salts	EE	EE	EE	EE	EE	EE	EE
Toluene	FN	GG	GF	FF	EE	FN	FN
Tributyl citrate	GF	EG	GF	GF	EE	NN	FN
Trichloroethane	NN	FN	NN	NN	EE	NN	NN
Trichloroethylene	NN	FN	NN	NN	EE	NN	NN
Triethylene glycol	EE	EE	EE	EE	EE	EG	GF
Tripropylene glycol	EE	EE	EE	EE	EE	EG	GF
Turkey red oil	EE	EE	EE	EE	EE	EG	EG
Turpentine	FN	GG	GF	FF	EE	FN	GF
Undecyl alcohol	EF	EG	EG	EG	EE	GF	EF
Urea	EE	EE	EE	EG	EE	NN	GN
Vinylidene chloride	NN	FN	NN	NN	EE	NN	NN
Xylene	GN	GF	FN	FN	EE	NN	NN
Zinc salts	EE	EE	EE	EE	EE	EE	EE
Zinc stearate	EE	EE	EE	EE	EE	EE	EG

E 30 days of constant exposure cause no damage. Plastic may even tolerate exposure for years.
G Little or no damage after 30 days of constant exposure to the reagent.
F Some signs of attack after 7 days of constant exposure to the reagent.
N Not recommended; noticeable signs of attack occur within minutes to hours after exposure. (However, actual failure might take years.)
First letter: at room temperature.
Second letter: at 52°C.

Resins
CPE Conventional (low-density) polyethylene.
LPE Linear (high-density) polyethylene.
PP Polypropylene.
PMP Polymethylpentene.
FEP Teflon FEP (fluorinated ethylene propylene). Teflon is a Du Pont registered trademark.
ETFE Tefzel ethylene-tetrafluoroethylene copolymer. (For chemical resistance, see FEP ratings.) Tefzel is a Du Pont registered trademark.
PC Polycarbonate.
PVC Rigid polyvinyl chloride.

Table 4.14 Thermal coefficients of linear expansion

Material	*Coefficient of expansion per °C*
Aluminium	0.000025
Brass	0.000018
Bronze	0.000018
Copper	0.000017
Glass (Pyrex)	0.000003
Iron (cast)	0.000011
Lead	0.000029
Platinum	0.000009
Silver	0.000021
Steel (mild)	0.000012
Tin	0.000025
Zinc	0.000029

Table 4.15 Pressure increase of common liquids due to thermal expansion on 16.6°C temperature rise

Liquid	*P (psia)*
Acetic acid	3200
Acetone	3260
Aniline	5190
Benzene	3860
n-Butyl alcohol	2590
Carbon tetrachloride	3310
Methyl alcohol	3900
Petroleum (s.g. = 0.8467)	2340
Toluene	3340
Water	1100

Liquids can also exert pressure due to thermal expansion. Table 4.15 provides an indication of pressure increases due to temperature increases for selected common liquids in full containers or pipes. Serious accidents can arise unless the design of rigid plant items such as pipework takes into account the changes in volume of liquids with temperature fluctuation by the following or combinations thereof:

- Expansion bellows.
- Expansion joints.
- Expansion loops or offset legs.
- Positioning of equipment so as to exploit the inherent flexibility.
- Supports, guides, anchors, piles.
- Decrease in pipe wall thickness.

Consideration should be given to the effects of thermal expansion of liquids and pressure-relief valves installed unless:

- pipelines cannot become blocked or heated;
- liquid temperature is controlled;
- line is normally in service and can be vented and drained;
- one end of valve is fitted with check valve.

If pressure relief is important methods include:

- installation of pressure-relief valve;
- bypass;
- check valves;
- ensuring liquid is drained before blockage can occur.

5
Toxic chemicals

Introduction

Chemicals may be encountered as reactants, solvents, catalysts, inhibitors, as starting materials, finished products, by-products, contaminants, or off-specification products. They may vary from pure, single substances to complex proprietary formulations.

Exposures to chemicals may involve solids, liquids, or airborne matter as mists, aerosols, dusts, fumes (i.e. μm-sized particulates), vapours or gases in any combination. Many situations, e.g. exposure to welding fumes or to combustion products from fossil fuels, include mixtures both of chemicals and of physical forms. Quantification of exposure is then difficult.

An exposure to a specific chemical in relatively low concentrations over a period may result in chronic effects. At higher concentrations, the effects may be acute. Some chemicals produce local damage at their point of contact with, or entry into, the body; others produce systemic effects, i.e. they are transported within the body to various organs before exerting an adverse effect.

For a classification of airborne contaminants, refer to Table 5.1

Hazard recognition

The toxicity of a substance is its capacity to cause injury once inside the body. The main modes of entry into the body by chemicals in industry are inhalation, ingestion and absorption through the skin. Gases, vapours, mists, dusts, fumes and aerosols can be inhaled and they can also affect the skin, eyes and mucous membranes. Ingestion is rare although possible as a result of poor personal hygiene, subconscious hand-to-mouth contact, or accidents. The skin can be affected directly by contact with the chemicals, even when intact, but its permeability to certain substances also offers a route into the body. Chemicals accorded a 'skin' notation in the list of Occupational Exposure Limits (see Table 5.12) are listed in Table 5.2. Exposure may also arise via skin lesions.

Types of toxic chemicals

Irritant chemicals

Primary irritants cause inflammation. Inflammation is one of the body's defence mechanisms. It is the reaction of a tissue to harm which is insufficient to kill the tissue and is typified by

Table 5.1 Classification of airborne contaminants

Classification	*Sub-groups*	*Examples*
Irritants Have a corrosive or a vesicant (blistering) effect on moist or mucous surfaces. Concentration may be more important than duration of exposure. Animals and man react similarly.		
		Vapour, gases, mists
(a) Primary	Upper respiratory	Acrolein; sulphur dioxide, hydrogen chloride, chromic acid; formaldehyde.
	Upper and lower respiratory	Fluorine; chlorine; bromine; ozone; cyanogen chloride.
	Lower respiratory	Phosgene; nitrogen dioxide; arsenic trichloride.
	Skin	Inorganic acids (chromic, nitric); organic acids (acetic, butyric); inorganic alkalis (sodium hydroxide, sodium carbonate); organic bases (amines); organic solvents.
		Dusts Detergents; salts (nickel sulphate, zinc chloride); acids, alkalis, chromates.
(b) Secondary or allergens	Skin sensitizers	Epoxy-resins; picryl chloride; or chlor-2-4-dinitrobenzene; *p*-phenyl diamine.
	Respiratory sensitizers	Isocyanates; proteolytic enzymes; *p*-phenylene diamine; complex salts of platinum; cyanuric chloride.
Asphyxiants Exert an effect by interference with oxidation of tissue. Animals and man react similarly.	Simple anoxia caused by oxygen deficiency in inhaled air.	Carbon dioxide; methane; hydrogen; nitrogen; helium.
	Toxic anoxia caused by damage to the body's oxygen transport or utilization by adverse reaction of biologically active substances.	Carbon monoxide; cyanogen, hydrogen cyanide; nitrites; arsine; aniline, dimethyl aniline, toluidine; nitrobenzene; hydrogen sulphide (causes respiratory paralysis by impairment of oxygen utilization in the central nervous system).
Anaesthetics and narcotics Exert principal effects as simple anaesthesia, by a depressant action on the central nervous system.		Acetylene Olefins Ether Paraffins Aliphatic ketones Aliphatic alcohols Esters ↓ Decreasing anaesthetic action compared with other effects.
Systemic poisons Substances which cause injury at other than the site of contact.	Visceral organs in general	Many halogenated hydrocarbons and metals.
	Haematopoletic (i.e. blood-forming system)	Benzene; phenols.
	Nervous system	Carbon disulphide; methanol; phenol; *n*-hexane; methyl *n*-butyl ketone; organophosphorus compounds; tetra-alkyl lead compounds.

Table 5.1 Cont'd

Classification	*Sub-groups*	*Examples*
		Toxic inorganic substances e.g. Lead, manganese, cadmium, antimony, beryllium, mercury; arsenic; phosphorus; selenium and sulphur compounds, fluorides.
Respiratory fibrogens		*Fibrogenic dusts* e.g. Free crystalline silica, (quartz, tridymite, cristobalite), asbestos (chrysotile, amosite, crocidolite etc.), talc.
Carcinogens		
Cancer-producing agents	Skin	Coal tar pitch dust; crude anthracene dust; mineral oil mist; arsenic.
	Respiratory	Asbestos; polycyclic aromatic hydrocarbons; nickel ore; arsenic; bis-(chloromethyl) ether; mustard gas.
	Bladder/urinary tract	β-naphthylamine; benzidine; 4-aminodiphenylamine.
	Liver	Vinyl chloride monomer.
	Nasal	Mustard gas; nickel ore.
	Bone marrow	Benzene.
Inerts		*Gases* Simple asphyxiants Argon; methane; hydrogen; nitrogen; helium. *Particulates* e.g. cement, calcium carbonate.

constriction of the small vessels in the affected area, dilation of the blood vessels, increased permeability of the vessel walls, and migration of the white blood and other defensive cells to the invading harmful chemical. The aim is to concentrate water and protein in the affected area to 'dilute' the effect and wash away the chemical. Production of new cells is speeded up and contaminated surface cells are shed.

The respiratory system is the main target organ for vapour, gas or mist. Readily-soluble chemicals, e.g. chlorine or phosgene, attack the upper respiratory tract; less soluble gases, e.g. oxides of nitrogen, penetrate more deeply into the conducting airways and, in some cases, may cause pulmonary oedema, often after a time delay.

For example, sulphur dioxide is highly water soluble and tends to be absorbed in the airways above the larynx. Responses at various concentrations are summarized in Table 5.3. However, in the presence of particulate catalysts and sunlight, conversion to sulphur trioxide occurs and the irritant response is much more severe.

Other parts of the body are also vulnerable: the skin and eyes from direct contact/rubbing or from exposure to airborne material including splashes; the mouth and pharynx by ingestion of solid or liquid chemicals.

One effect of direct contact of liquid or solid, and less often vapour, with the skin is a contact irritant dermatitis. Some dusts can also act as primary irritants. Even chemically-inert dusts, e.g. from glass fibres, can induce a dermatitis due to abrasion; this is made worse if a reactive chemical, e.g. a synthetic resin binder, is also involved. Examples of primary irritants include acids; alkalis; defatting compounds, e.g. organic solvents, surfactants; dehydrating agents; oxidizing agents and reducing agents.

Table 5.2 Materials with an 'Sk' notation in list of Occupational Exposure Limits

Acrylamide
Acrylonitrile
Aldrin
Allyl alcohol
Aniline
Azinphos-methyl
Aziridine
Butan-1-ol
2-Butoxyethanol
n-Butylamine
γ-BHC (Lindane)
Bromoform
Bromomethane
Butan-2-one
2-sec-Butylphenol
Carbon disulphide
Carbon tetrachloride
Chlorinated biphenyls
2-Chlorobuta-1,3-diene
1-Chloro-2,3-epoxy propane
2-Chloroethanol
Chloroform
1-Chloro-4-nitrobenzene
Chlorpyrifos
Cresols, all isomers
Cumene
Cyanides
Cyclohexylamine
Diazinon
1,2-Dibromoethane
2,2′-Dichloro-4,4′-methylene dianiline (MDOCA)
1,3-Dichloropropene
Dichlorvos
Dieldrin
2-Diethylaminoethanol
Diethyl sulphate
Di-isopropylamine
N,N-Dimethylacetamide
N,N-Dimethylaniline
Dimethyl formamide
Dimethyl sulphate
Dinitrobenzene
2,4-Dinitrotoluene
1,4-Dioxane
Dioxathion
Endosulfan
Endrin
2-Ethoxyethanol
2-Ethoxyethyl acetate
Ethylene dinitrate
4-Ethylmorpholine
2-Furaldehyde (furfural)
Furfuryl alcohol
Glycerol trinitrate
Heptan-3-one
Heptan-2-one
Hexahydro-1,3,5-trinitro-1,3,5-triazine
Hexan-2-one
Hydrazine
Hydrogen cyanide
2-Hydroxypropylacrylate
2,2-Iminodi(ethylamine)
Iodomethane
Malathion
Mercury alkyls
Methacrylonitrile
Methanol
2-Methoxyethanol
2-Methoxyethyl acetate
(2-Methoxymethylethoxy) propanol
Methoxypropanol
2-Methyl-4,6-dinitrophenol
5-Methylhexan-2-one
4-Methylpentan-2-ol
4-Methylpentan-2-one
1-Methyl-2-pyrrolidone
N-Methyl-*N*,-2,4,6-tetranitroaniline
N-Methylaniline
Mevinphos
Monochloroacetic acid
Morpholine
Nicotine
4-Nitroaniline
Nitrobenzene
Nitrotoluene
Octachloronaphthalene
Parathion
Parathion-methyl
Pentachlorophenol
Phenol
Phorate
Piperidine
Polychlorinated biphenyls (PCB)
Propan-1-ol
Propylene dinitrate
Prop-2-yn-1-ol
Sodium fluoroacetate
Sulfotep
Tetrabromoethane
Tetraethylpyrophosphate
Tetrahydrofuran
Tetramethyl succinonitrile
Thallium, soluble compounds
Tin compounds, organic
Toluene
o-Toluidine
Tricarbonyl (eta-cyclopentadienyl) manganese
Tricarbonyl (methylcyclopentadienyl) manganese
Trichlorobenzene
1,1,2-Trichloroethane
Trichloroethylene
2,4,6-Trinitrotoluene
Xylene
Xylidine

Table 5.3 Typical effects of sulphur dioxide concentrations in air

Concentration (ppm)	*Response*
0.5–0.8	Minimum odour threshold
3	Sulphur-like odour detectable
6–12	Immediate irritation to nose and throat
20	Reversible damage to respiratory system
>20	Eye irritation
	Tendency to pulmonary oedema and eventually respiratory paralysis
10 000	Irritation to moist skin within a few minutes

In extreme cases irritant chemicals can have a corrosive action. Corrosive substances can also attack living tissue (e.g. to cause skin ulceration and, in severe cases, chemical burns with degradation of biochemicals and charring), kill cells and possibly predispose to secondary bacterial invasion. Thus whilst acute irritation is a local and reversible response, corrosion is irreversible cell destruction at the site of the contact. The outcome is influenced by the nature of the compound, the concentration, duration of exposure, the pH (see Figure 5.1) and also, to some extent, by individual susceptibility etc. Thus dilute mineral acids may be irritant whereas at higher concentrations they may cause corrosion.

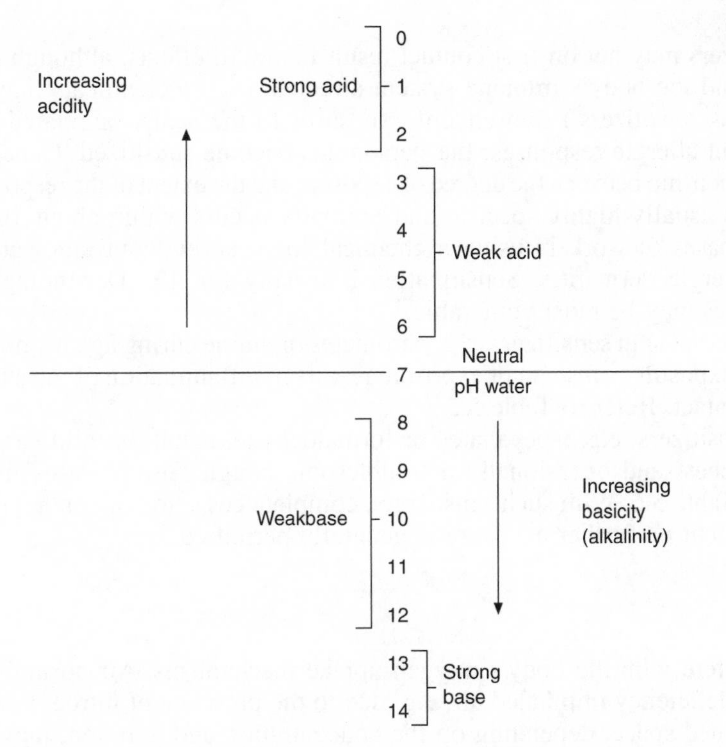

Figure 5.1 *The pH scale*

A summary of the more common corrosive chemicals is given in Table 5.4. Note that this includes many primary irritants, such as:

- Chemicals which give strong acid reactions, often on interaction with water, e.g. mineral acids. Some organic acids can also be corrosive. Phenolics can result in local anaesthesia so that the pain will be absent for a time, i.e. contact may go unheeded.
- Halogen compounds.
- Acid anhydrides/halides which react with water to form their parent acids.
- Common bases, which render aqueous solutions alkaline.
- Certain oxidizing/reducing compounds and salts which, in the form of solid (bulk or dust) or as solution, can produce irritation by thermal burns.

Strong acids and alkalis produce effects within moments: e.g. sulphuric and nitric acids quickly become hydrated by the water content of the skin/mucous membranes and combine with skin protein to form albuminates, sometimes with charring. Some substances, e.g. certain organotins or hydrofluoric acid, produce a more delayed reaction. Thus on the skin hydrofluoric acid produces an effect which varies, depending on concentration and duration of exposure, from mild erythema to severe burns and intense pain, sometimes delayed by several hours after the initial exposure. A tough white lump forms over the area of skin damage under which progressive destruction of cell tissue continues. Burns under the finger nails are notable in this respect because of the difficulties of treatment. Similarly, inhalation of the vapour can cause corrosion of the respiratory system and pulmonary oedema. If hydrofluoric acid is swallowed, burns to the mouth and pharynx can occur with vomiting and ultimate collapse.

Sensitizers

Generally sensitizers may not on first contact result in any ill effects, although cellular changes can be induced and the body's immune system affected. (Some chemicals may act as primary irritants as well as sensitizers.) Subsequent exposures to the same, or related, chemicals may bring about violent allergic responses: the person has become sensitized. Generally there is no mathematical relationship between the degree of exposure and the extent of the response. Sensitization to a compound is usually highly specific and normally occurs within about 10 days, although there have been cases of workers using a chemical for years without untoward effects before developing an allergic dermatitis. Sensitization is usually for life. Depending upon the toxic mechanism, atopics may be most vulnerable.

Thus with industrial skin sensitizers, e.g. chromates or amine curing agents, no effect is usually observed on first exposure; subsequent exposure results in inflammation of the skin, not restricted to the areas of contact. Refer to Table 5.5.

Respiratory sensitizers, e.g. isocyanates or formaldehyde, result, in mild cases, in a sense of tightness of the chest and occasionally a troublesome cough. Severe cases involve bronchial asthma. Refer to Table 5.6. With such sensitizers, complete cessation of contact is often followed by rapid recovery but no further exposure is generally permitted.

Asphyxiants

Asphyxiants interfere with the body's oxygen uptake mechanisms. Air normally contains 21% oxygen. Oxygen deficiency in inhaled air, e.g. due to the presence of nitrogen, argon, or carbon dioxide in a confined space, depending on the concentration and duration, may affect the body and ultimately cause death from simple anoxia (Table 5.7).

Table 5.4 Common corrosive chemicals

Acids and anhydrides	Acetic acid Acetic anhydride Acid mixtures Battery fluids Chloroacetic acid Chlorosulphonic acid Chromic acid Dichloroacetic acid Fluoroboric acid Fluorosilicic acid Hydrobromic, hydrochloric, hydrofluoric and hydroiodic acids Methacrylic acid Nitric acid	Nitrohydrochloric acid Perchloric acid Phenosulphonic acid Phosphorus pentoxide Propionic acid Selenic acid Spent acids Sulphamic acid Sulphuric acid and oleum (fuming sulphuric acid) Sulphurous acid Thioglycolic acid Trichloroacetic acid
Akalis	Ammonium hydroxide Potassium hydroxide (caustic potash) Quaternary ammonium hydroxides Sodium hydroxide (caustic soda)	
Halogens and halogen salts	Aluminium bromide and chloride Ammonium bifluoride and other bifluorides Antimony trichloride, pentachloride and pentafluoride Beryllium chloride Boron trichloride Bromine Chlorine Calcium fluoride Chromic fluoride Chromous fluoride Fluorine Iodine Iron chlorides (ferric chloride, ferrous chloride) Lithium chloride Phosphorus oxybromide and oxychloride (phosphoryl bromide and chloride)	Phosphorus sulphochloride (thiophosphoryl chloride) Phosphorus trichloride and pentachloride Potassium fluoride and bifluoride Potassium hypochlorite Pyrosulphuryl chloride Sodium chlorite Sodium fluoride Sodium hypochlorite Stannic chloride Sulphur chloride Sulphuryl chloride Thionyl chloride Titanium tetrachloride Vanadium dichloride Zinc chloride
Interhalogen compounds	Bromine trifluoride and pentafluoride Chlorine trifluoride Iodine monochloride	
Organic halides, organic acid halides, esters and salts	Acetyl bromide Acrylonitrile monomer Allyl chloride Allyl chloroformate Allyl iodide Ammonium thiocyanate Anisoyl chloride Benzhydryl bromide (diphenyl methyl bromide) Benzoyl chloride Benzyl bromide Benzyl chloride Benzyl chloroformate (benzyl chlorocarbonate) Butyl acid phosphate Chloracetyl chloride	*p*-Chlorobenzyl chloride Chloropropionyl chloride Dibromoethane (ethylene bromide) 1,2-Dichloroethane (ethylene chloride) Diisooctyl acid phosphate Ethyl chloroformate Ethyl chlorocarbonate Ethylene oxide Fumaryl chloride Iso-propylchloroformate Methyl chloroformate Methyl chlorocarbonate Propionyl chloride Sodium fluorosilicate

Table 5.4 Cont'd

Chlorosilanes	Allyl trichlorosilane	Hexadecyl trichlorosilane
	Amyl trichlorosilane	Hexyl trichlorosilane
	Butyl trichlorophenyl-trichlorosilane	Methyl trichlorosilane
	Cyclohexyl trichlorosilane	Nonyl trichlorosilane
	Dichlorophenyl trichlorosilane	Octadecyl trichlorosilane
	Diethyl trichlorosilane	Octyl trichlorosilane
	Diphenyl dichlorosilane	Phenyl trichlorosilane
	Dodecyl trichlorosilane	Trimethyl trichlorosilane
		Vinyl trichlorosilane
Miscellaneous corrosive substances	*Proprietary mixtures, e.g. cleaning, disinfecting, bleaching, degreasing solids or solutions, based on these chemicals are corrosive to a degree dependent upon dilution*	
	Ammonium sulphide	Hydrazine
	Benzene sulphonyl chloride	Hydrogen peroxide
	Benzyl dimethylamine	Organic peroxides
	Beryllium nitrate	Phenols
	Catechol	Silver nitrate
	Chlorinated benzenes and toluenes	Soda lime
	Chlorobenzaldehyde	Sodium aluminate
	Chlorocresols	Sodium amide
	Cresols	Sodium bisulphate
	Cyclohexylamine	Sodium bisulphite
	Dibenzylamine	Sodium chromate and dichromate
	Dichlorophenol	Sodium hydride
	Diethyl sulphate	Sodium pyrosulphate
	Diketene	Triethyltetramine
	Dimethyl sulphate	Tritolyl borate
	Hexamethylenediamine	

Table 5.5 Common industrial skin sensitizers

Coal-tar and its direct derivatives
Acridine
Anthracene
Carbazole
Cresol(1)
Fluorene
Naphthalene
Phenanthrene
Phenol(1)
Pyridine
Tar

Dyes
Amido-azo-benzene
Amido-azo-toluene
Aniline black
Auramine
Bismarck brown
Brilliant indigo, 4 G.
Chrysoidine
Crystal and methyl violet
Erio black
Hydron blue
Indanthrene violet, R.R.
Ionamine, A.S.
Malachite green
Metanil yellow
Nigrosine
Orange Y
Paramido phenol
Paraphenylendiamine
Pyrogene violet brown
Rosaniline
Safranine
Sulphanthrene pink

Dye intermediates
Acridine and compounds
Aniline and compounds
Benzanthrone and compounds
Benzidine and compounds
Chloro compounds

Table 5.5 Cont'd

Naphthalene and compounds
Naphthylamines
Nitro compounds

Explosives
Ammonium nitrate
Dinitrophenol
Dinitrotoluol
Fulminate of mercury
Hexanitrodiphenylamine
Lead styphnate
Picric acid and picrates
Potassium nitrate
Sensol
Sodium nitrate
Trinitromethylnitramine (Tetryl)
Trinitrotoluene

Insecticides
Arsenic compounds(1)
Creosote
Fluorides(1)
Lime(1)
Mercury compounds(1)
Nicotine
Organic phosphates
Petroleum distillates(1)
Phenol compounds(1)
Pyrethrum

Natural resins
Burgundy pitch
Copal
Dammar
Japanese lacquer
Pine rosin
Wood rosin

Oils
Cashew nut oil(1)
Coconut oil
Coning oils (cellosolves, eugenols)
Cutting oils (the inhibitor or antiseptic they contain)
Essential oils of plants and flowers
Linseed oil
Mustard oil(1)
Sulphonated tung oil

Photographic developers
Bichromates
Hydroquinone
Metol
Para-amido-phenol
Paraformaldehyde
Paraphenylendiamine
Pyrogallol

Plasticizers
Butyl cellosolve stearate
Diamyl naphthalene
Dibutyl tin laurate
Dioctylphthalate
Methyl cellosolve oleate
Methyl phthalylethylglycola
Phenylsalicylate
Propylene stearate
Stearic acid
Triblycol di-(2,ethyl butyrate)

Rubber accelerators and anti-oxidants
Guanidines
Hexamethylene tetramine
Mercapto benzo thiazole
Ortho-toluidine
Para-toluidine
Tetramethyl thiuram monosulphide and disulphide
Triethyl tri-methyl triamine

Synthetic resins
Acrylic
Alkyd
Chlorobenzols
Chlorodiphenyls
Chloro-naphthalenes
Chlorophenols
Cumaron
Epoxies
Melamine formaldehyde
Phenol formaldehyde
Polyesters
Sulphonamide formaldehyde
Urea formaldehyde
Urethane
Vinyl

Others
Enzymes derived from *B. subtilis*

(1) Compounds which also act as primary irritants.

Table 5.6 Some substances recognized as causing occupational asthma

Substance	*Examples of use*
Isocyanates	Plastic foam, synthetic inks, paints and adhesives
Platinum salts	Platinum refining workshops and some laboratories
Acid anhydride and amine hardening agents, including epoxy resin curing agents, e.g. ethylene diamine, triethylene tetramine	Adhesives, plastics, moulding resins and surfaces coatings
Fumes from the use of resin (colophony) as a soldering flux	The electronics industry
Proteolytic enzymes	Biological washing powders and the baking, brewing, fish, silk and leather industries
Animals, including insects and other arthropods or their larval forms	Research and educational laboratories, pest control and fruit cultivation
Dusts from barley, oats, rye, wheat or maize, or meal or flour made from such grain	The baking or flour milling industry or on farms
Antibiotics, e.g. cephalosporins, hydralizine, ampicillins, piperazine, spiramycin	Manufacture, dispensing
Cimetidine	Manufacture of cimetidine tablets
Wood dusts; some hardwoods (e.g. iroko, mahogany); some softwoods (e.g. western red cedar)	Furniture manufacture
Ispaghula powder	Manufacture of bulk laxatives
Castor bean dust	Processing
Ipecacuanha	Manufacture of ipecacuanha tablets
Azodicarbonamide	Blowing agent in the manufacture of expanded foam plastics for wallcoverings, floor coverings, insulation and packaging materials
Glutaraldehyde	Hospitals, laboratories, cooling tower systems and leather tanning
Persulphate salts and henna	Manufacture of hair care products and their application
Crustaceans	Fish and food processing industries
Reactive dyes	Dyeing, printing and textile industries
Soya bean	Soya bean processing and food industries
Tea dust	Tea processing and food industries
Green coffee bean dust	Coffee processing and food industries
Fumes from stainless steel welding	Stainless steel fabrication operations
Natural rubber latex	Latex gloves, adhesives, surgical apparatus and appliances
Water-mix metalworking fluids	Coolants in metalworking
Certain cyanoacrylates	Adhesives
Methyl methacrylate	Adhesives
Diazonium salts	Polymer manufacture
Paraphenylenediamine	Hair dyes and treatments
Formaldehyde	Preserving, resin and foam manufacture
Cobalt	Hard metal manufacture and tools
Nickel	Electroplating
Bromelein, papain	Meat tenderizing
Amylase	Flour improver
Triglycidyl isocyanurate	Polyester-based powder coatings
Azodicarbonamide	Plastics manufacture; flour improver
Butadiene diepoxide	Polymer manufacture
2,3-expoxy-1-propanol (glycidol)	Oil stabilizer

Table 5.7 Typical effects of depleted oxygen levels in air

Oxygen concentration (%)	*Effect*
16–21	No noticeable effect
12–16	Increased respiration, slight diminution of coordination
10–12	Loss of ability to think clearly
6–10	Loss of consciousness, death

Levels below 19.5% oxygen can have detrimental effects if the body is already under stress, e.g. at high altitudes. Exposures below 18% should not be permitted under any circumstance. Other chemicals, e.g. carbon monoxide, result in toxic anoxia due to damage of the body's oxygen transport or utilization mechanism.

Anaesthetics and narcotics

Anaesthetics and narcotics, e.g. hydrocarbons and certain derivatives such as the various chlorinated solvents or ether, exert a depressant action on the central nervous system.

Systemic poisons

Systemic poisons attack organs other than the initial site of contact. The critical organs are the kidneys, liver, blood and bone marrow.

Respiratory fibrogens

The hazard of particulate matter is influenced by the toxicity and size and morphology of the particle. Figure 5.2 gives typical particle size ranges for particles from various sources. The critical size of dust (and aerosol) particles is 0.5 to 7 μm, since these can become deposited in the respiratory bronchioles and alveoli. If dust particles of specific chemicals, e.g. silica or the various grades of asbestos, are not cleared from the lungs then, over a period, scar tissue (collagen) may build up; this reduces the elasticity of the lungs and impairs breathing. The characteristic disease is classified as pneumoconiosis. Common examples are silicosis, asbestosis, coal pneumoconiosis and talc pneumoconiosis.

An appreciation of the composition and morphology of the dust is important in the assessment of hazard. Thus, among silica-containing compounds, crystalline silicates and amorphous silicas (silicon dioxide) are generally not considered fibrogenic, whereas free crystalline silica and certain fibrous silicates such as asbestos and talcs can cause disabling lung diseases. Table 5.8 indicates the approximate free silica content of various materials; Table 5.9 lists a range of silica-containing materials according to type.

Carcinogens

Cancer is a disorder of the body's control of the growth of cells. The disease may be genetic or influenced by life style or exposure to certain chemicals, termed carcinogens. For a list of examples of human chemical carcinogens, and the relevant target organs, refer to Table 5.10.

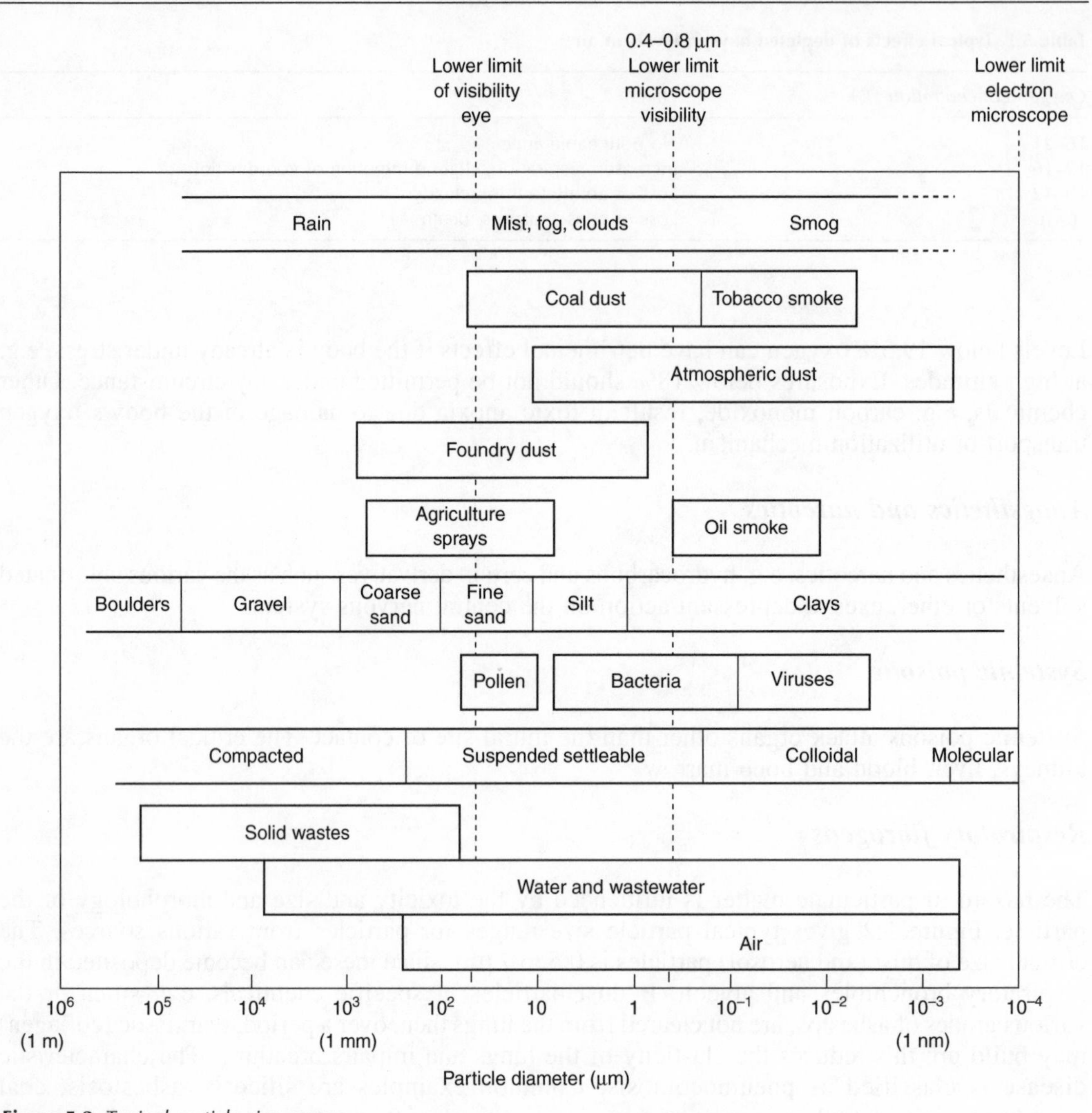

Figure 5.2 *Typical particle size ranges*

Table 5.8 Crystalline SiO_2 in various materials

Material	*Normal range crystalline SiO_2(%)*
Foundry moulding sand	50–90
Potteryware body	15–25
Brick and tile compositions	10–35
Buffing wheel dressings	0–60
Road rock	0–80
Limestone (agricultural)	0–3
Feldspar	12–25
Clay	0–40
Mica	0–10
Talc	0–5
Slate and shale	5–15

Table 5.9 Types of silica-containing dusts

Crystalline free silica (SiO_2, including microcrystalline forms)	
Chalcedony	A heat-resistant, chemically inert form of microcrystalline quartz. A decorative material. Rare in industry.
Chert	A microcrystalline form of silica. An impure form of flint used in abrasives.
Cristobalite	A crystalline form of free silica, extremely hard and inert chemically; very resistant to heat. Quartz in refractory bricks and amorphous silica in diatomaceous earth are altered to cristobalite when exposed to high temperatures (calcined). Cristobalite is extensively used in precision casting by the hot wax process, dental laboratory work, and certain speciality ceramics.
Flint	A microcrystalline form of native quartz, more opaque and granular than chalcedony. Used as an abrasive and in ceramics.
Jasper	A microcrystalline impure form of silica similar to chert. Used for decorative purposes. Rare in industry.
Quartz	Vitreous, hard, chemically-resistant free silica, the most common form in nature. The main constituent in sandstone, igneous rocks, and common sands.
Tridymite	Vitreous, colourless form of free silica. Formed when quartz is heated to 870°C (1598°F).
Tripoli (rottenstone)	A porous siliceous rock resulting from the decomposition of chert or siliceous limestone. Used as a base in soap and scouring powders, in metal polishing, as a filtering agent, and in wood and paint fillers. A cryptocrystalline form of free silica.
Amorphous free silica (Noncrystalline)	
Diatomaceous earth	A soft, gritty amorphous silica composed of minute siliceous skeletons of small aquatic plants. Used in filtration and decoloration of liquids, insulation, filler in dynamite, wax, textiles, plastics, paint, and rubber. Calcined and flux-calcined diatomaceous earth contains appreciable amounts of cristobalite, and dust levels should be the same as for cristobalite.
Silica gel	A regenerative absorbent consisting of the amorphous silica manufactured by the action of HCl on sodium silicate. Hard, glossy, quartz-like in appearance. Used in dehydrating and in drying and as a catalyst carrier.
Silicates (compounds made up of silicon, oxygen and one or more metals with or without hydrogen. Asbestos dust is the most hazardous (page 148). Others, e.g. talc, mica, vermiculite, have caused a silicatosis on prolonged exposure.)	
Asbestos	A hydrated magnesium silicate in fibrous form. The fibres are believed to be the more hazardous component of asbestos dust.
Clays	A great variety of aluminium–silicate bearing rocks, plastic when wet, hard when dry. Used in pottery, stoneware, tile, bricks, cements, fillers and abrasives. Kaolin is one type of clay. Some clay deposits may include appreciable amounts of quartz. Commercial grades of clays may contain up to 20% quartz.
Feldspar	Most abundant group of materials, composed of silicates of aluminium with sodium, potassium, calcium, and rarely barium. Most economically important mineral. Used for ceramics, glass, abrasive wheels, cements, insulation and fertilizer.
Fuller's earth	A hydrated silica–alumina compound, associated with ferric oxide. Used as a filter medium and as a catalyst and catalyst carrier and in cosmetics and insecticides.
Kaolin	A type of clay composed of mixed silicates and used for refractories, ceramics, tile and stoneware.
Mica	A large group of silicates of varying composition, but similar in physical properties. All have excellent cleavage and can be split into very thin sheets. Used in electrical insulation.
Portland cement	Fine powder containing compounds of lime, alumina, silica and iron oxide. Used as a construction material.
Silicon carbide (carborundum)	Bluish-black, very hard crystals. Used as an abrasive and refractory material.
Talc	A hydrous magnesium silicate used in ceramics, cosmetics, paint and pharmaceuticals, and as a filler in soap, putty and plaster.
Vermiculite	An expanded mica (hydrated magnesium-aluminium-iron silicate). Used in lightweight aggregates, insulation, fertilizer and soil conditioners, as a filler in rubber and paints, and as a catalyst carrier.

Table 5.10 Chemicals associated with cancer in humans (not all those listed are still in industrial use)

Chemicals and industrial processes which are carcinogenic for humans

Substance or process	*Site affected and type of neoplasm*
4-Aminobiphenyl	Bladder – carcinoma
Arsenic and certain compounds	Skin, lung, liver – carcinoma
Asbestos	Respiratory tract – carcinoma Pleura and peritoneum – mesothelioma Gastrointestinal tract – carcinoma
Auramine manufacture	Bladder – carcinoma
Benzene	Blood – leukaemia
Benzidene	Bladder – carcinoma
Bis (chloromethyl) ether and technical grade chloromethyl ether	Lung – carcinoma
Chlornaphazine	Bladder – carcinoma
Chromium and certain compounds	Lung – carcinoma
Diethylstilbestrol	Female genital tract – carcinoma (transplacental)
Haematite mining (underground)	Lung – carcinoma
Isopropanol manufacture (strong acid process)	Respiratory tract – carcinoma
Melphalan	Blood – leukaemia
Mustard gas	Respiratory tract – carcinoma
2-Naphthylamine	Bladder – carcinoma
Nickel refining	Respiratory tract – carcinoma
Soots, tars, and mineral oils	Skin, lung, bladder – carcinoma
Vinyl chloride	Liver – angiosarcoma Brain Lung – carcinoma Lymphatic system – lymphoma

Chemicals which are probably carcinogenic in humans

Substance	*Site affected (human)*
Acrylonitrile	Colon, lung
Aflatoxins	Liver
Amitrole	Various sites
Auramine	Bladder
Beryllium and certain compounds	Bone, lung
Cadmium and certain compounds	Kidney, prostate, lung
Carbon tetrachloride	Liver
Chlorambucil	Blood
Cyclophosphamide	Bladder, blood
Dimethylcarbamoyl chloride	
Dimethyl sulphate	Lung
Ethylene oxide	Gastrointestinal tract, blood
Iron dextran	Connective tissue
Nickel and certain compounds	Respiratory tract
Oxymetholone	Liver
Phenacetin	Kidney, bladder
Polychlorinated biphenyls	Skin, various sites
Thiotepa	Blood

Hazard assessment

Indicators of toxicity hazards include LD_{50}, LC_{50}, plus a wide range of *in vitro* and *in vivo* techniques for assessment of skin and eye irritation, skin sensitization, mutagenicity, acute and chronic dermal and inhalation toxicity, reproductive toxicology, carcinogenicity etc.

The LD_{50} is the statistically derived single dosage of a substance that can be expected to cause death in 50% of the sample population. It is therefore an indicator of acute toxicity, usually determined by ingestion using rats or mice, although other animals may be used. LD_{50} is also determined by other routes, e.g. by skin absorption in rabbits. The values are affected by species, sex, age, etc.

The LC_{50} is the lethal concentration of chemical (e.g. in air or water) that will cause the death of 50% of the sample population. This is most appropriate as an indicator of the acute toxicity of chemicals in air breathed (or in water, for aquatic organisms). Table 5.11 illustrates the use of LD_{50} values to rank the toxicity of substances.

Table 5.11 Toxicity rating system

Toxicity rating	*Commonly used term*	*LD_{50} Single oral dose for rats* (g/kg)	*4hr Vapour exposure causing 2 to 4 deaths in 6-rat group* (ppm)	*LD_{50} Skin for rabbits* (g/kg)	*Probable lethal dose for humans*
1	Extremely toxic	≤0.001	<10	≤0.005	Taste (1 grain)
2	Highly toxic	0.001–0.05	10–100	0.005–0.043	1 teaspoon (4 ml)
3	Moderately toxic	0.05–0.5	100–1000	0.044–0.340	1 oz (30 g)
4	Slightly toxic	0.5–5.0	1000–10 000	0.35–2.81	1 pint (250 g)
5	Practically non-toxic	5.0–15.0	10 000–100 000	2.82–22.6	1 quart (500 g)
6	Relatively harmless	>15.0	>100 000	>22.6	>1 quart

Hygiene standards

Hygiene standards are employed as indicators of risk to man from inhalation of toxic or nuisance chemicals at work.

Some indication of risk of employee exposure to airborne chemicals can be gauged from an analysis of the level of exposure for comparison with known human dose/response data such as those for carbon monoxide and hydrogen sulphide listed in Tables 5.31 and 5.32. More commonly the reference is to published hygiene standards based on human epidemiology, animal data and extrapolations from information on related chemicals, with built-in safety factors. Table 5.12 lists hygiene standards published annually by the American Conference of Governmental Industrial Hygienists (ACGIH), known as threshold limit values (TLV), and the UK equivalents published by the Health and Safety Executive (HSE), known as Occupational Exposure Limits (OELs). The table is a useful first point of reference but the original publications should be consulted for most up-to-date values, an indication of proposed changes, and more detailed guidance on their interpretation. It is also important to consult the latest documentation explaining the toxicological background to the values.

Table 5.12 Hygiene standards (*see* key and notes on pages 108–110)

Substance		*Threshold limit values (USA)* TWA (ppm)[a]	(mg/m³)[b]	STEL (ppm)[a]	(mg/m³)[b]		*Occupational exposure limits (UK)* TWA[c] (ppm)	(mg/m³)	STEL[c] (ppm)	(mg/m³)	*Air odour threshold* (ppm; v/v)
Acetaldehyde		–	–	C25	–		20	37	50	92	2000
Acetic acid		10	25	15	37		10	25	15	37	21
Acetic anhydride		5C	–	–	–		0.5	0.1	2	0.4	39
Acetone		500	–	750	–		750	1810	1500	3620	57
Acetone cyanohydrin		–	–	SKC4.7	–		–	–	–	–	–
Acetonitrile		40	67	60	101		40	68	60	102	0.23
Acetophenone		10	–	–	–		–	–	–	–	–
Acetylene		–	(d)	–	–		–	(d)	–	–	230
Acetylene dichloride, *see* 1,2-Dichloroethylene											
Acetylene tetrabromide		1	14	–	–						
Acetylsalicylic acid (aspirin)		–	5	–	–		–	5	–	–	
Acrolein		–	–	C0.1	–		0.1	0.25	0.3	0.8	0.61
Acrylamide	SK	–	0.03[e]	–	–	SK	–	0.3 MEL	–	–	
Acrylic acid	SK	2	5.9	–	–		10	30	20	60	110
Acrylonitrile	SK	2[e]	4.3[e]	–	–	SK	2	4.4 MEL	–	–	0.12
Adipic acid		–	5	–	–		–	–	–	–	–
Adiponitrile		2	–	–	–		–	–	–	–	–
Aldrin	SK	–	0.25	–	–	SK	–	0.25	–	0.75	
Allyl alcohol	SK	0.5	–	–	–	SK	2	4.8	4	9.7	1.8
Allyl chloride		1	3	2	6		–	–	–	–	0.84
Allyl glycidyl ether (AGE)		1	–	–	–	SK	5	24	10	47	
Allyl propyl disulphide		2	12	3	18		–	–	–	–	
α-Alumina, *see* Aluminium oxide											
Aluminium											
Total inhalable dust		–	–	–	–		–	10	–	–	
Respirable dust		–	–	–	–		–	4	–	–	
Metal dust		–	10	–	–		–	–	–	–	
Pyro powders, as Al		–	5	–	–		–	–	–	–	
Welding fumes, as Al		–	5	–	–		–	–	–	–	
Alkyls (NOC-d), as Al		–	2	–	–		–	–	–	–	
Aluminium oxides											
Total inhalable dust		–	10	–	–		–	10	–	–	
Respirable dust		–	–	–	–		–	4	–	–	
Aluminium salts (soluble)		–	2	–	–		–	2	–	–	
4-Aminodiphenyl	SK	–	(g)	–	–		–	–	–	–	
2-Aminoethanol, *see* Ethanolamine											
2-Aminopyridine		0.5	2	–	–		0.5	2	2	8	
Amitrole		–	0.2	–	–		–	–	–	–	
Ammonia		25	17	35	24		25	18	35	25	4.8

Ammonium chloride fume		–	10	–	20		–	10	–	20	
Ammonium perfluorooctanoate	SK	–	0.01	–	–		–	–	–	–	
Ammonium sulphamate		–	10	–	–		–	10	–	20	
Amosite, *see* Asbestos											
n-Amyl acetate		50	–	100	–		50	270	100	541	1800
sec-Amyl acetate		50	–	100	–		50	270	100	541	61000
Aniline and homologues	SK	2	7.6	–	–	SK	1	4	–	–	1.9
Anisidine (o-, p-isomers)	SK	0.1	0.5	–	–	SK	0.1	0.5	–	–	
Antimony and antimony compounds											
except stibine (as Sb)		–	0.5	–	–		–	0.5	–	–	
Antimony hydride (Stibine)		0.1	–	–	–		0.1	0.52	0.3	1.6	
Antimony trioxide production		(g)	–	–	–		–	–	–		
p-Aramid respirable fibres (1)							–	–	–		
ANTU		–	0.3	–	–		–	–	–	–	
Argon		–	(d)	–	–		–	(d)	–	–	
Arsenic elemental and inorganic		–	0.01[g]	–	–		–	–	–	–	
compounds As								(except arsine)			
Arsenic trioxide production		–	(e)	–	–		–	–	–	–	
Arsine		0.05	0.16	–	–		0.05	0.16	–	–	0.10
Asbestos, all forms[g]			0.1 Fibre/cc[h]			(j)					
Asphalt (petroleum) fumes		–	0.5	–	–		–	5	–	10	
Atrazine		–	5	–	–		–	10	–	–	
Azinphos-methyl	SK	–	0.2	–	–	SK	–	0.2	–	0.6	
Azodicarbonamide		–	–	–	–SEN	–	1.0	–	3 MEL		
Barium, soluble compounds, as Ba		–	0.5	–	–		–	0.5	–	–	
Barium sulphate											
Respirable dust		–	–	–	–	–	4	–	–		
Total inhalable dust		–	10	–	–	–	10	–	–		
Benomyl		–	10	–	–		–	10	–	15	
Benzene		0.5[g]	–	2.5	–		5	15 MEL	–	–	0.85
Benz(a)anthrene		(e)									
Benzenethiol		–	–	–	–		0.5	2.3	–	–	
Benzene-1,2,4-tricarboxylic acid 1,2-anhydride		–	–	–	C0.04	SEN	–	0.04	–	0.12	
Benzidine	SK	–	(g)	–	–		–	–	–	–	
p-Benzoquinone, *see* Quinone											
Benzo(b)fluoanthene		(e)									
Benzoyl peroxide		–	5	–	–		–	5	–	–	
Benzo(a)pyrene		–	(e)	–	–		–	–	–	–	
Benzotrichloride	SK	(e)	–	C0.1	–		–	–			
Benzoyl chloride		–	–	C0.5	–		–	–	–		

(1) *p*-Aramid respirable fibres are subject to an OES of 0.5 fibres/ml, 8 hr TWA

Table 5.12 Cont'd

Substance		*Threshold limit values (USA)* TWA (ppm)[a]	TWA (mg/m^3)[b]	STEL (ppm)[a]	STEL (mg/m^3)[b]		*Occupational exposure limits (UK)* TWA[c] (ppm)	TWA[c] (mg/m^3)	STEL[c] (ppm)	STEL[c] (mg/m^3)	*Air odour threshold* (ppm; v/v)
Benzyl acetate		10	–		–		–		–	–	
Benzyl butyl phthalate		–	–	–	–		–	5	–	–	
Benzyl chloride		1	5.2	–	–		0.5	2.6	1.5	7.9	23
Beryllium and compounds, as Be		–	0.002[g]	–	0.01		–	0.002	–		
Biphenyl		0.2	1.3	–	–		0.2	1.3	0.6	3.8	240
2,2-Bis(p-chlorophenyl)-1,1,1-trichloroethane		–	–	–	–		–	1	–	3	
see Diglycidyl ether											
Bis(2-dimethylamino-ethyl) ether		0.05	–	0.15	–		–	–	–	–	
Bis(2-ethyl hexyl) phthalate		–	–	–	–		–	5	–	10	
Bismuth telluride		–	10	–	–		–	10	–	20	
Se-doped		–	5	–	–		–	5	–	10	
Borates, tetra, sodium salts											
Anhydrous		–	1	–	–		–	1	–	–	
Decahydrate		–	5	–	–		–	5	–	–	
Pentahydrate		–	1	–	–		–	1	–	–	
Bornan-2-one		2	–	4	–		2	13	3	19	
Boron oxide		–	10	–	–		–	10	–	20	
Boron tribromide		–	–	C1	–		–	–	1	10	
Boron trifluoride		–	–	C1	–		–	–	1	2.8	
Bromacil		–	10	–	–		1	11	2	22	
Bromine		0.1	0.66	2	1.3		0.1	0.7	0.3	2	
Bromine pentafluoride		0.1	0.72	–	–		0.1	0.7	0.3	2.2	2.0
Bromochloromethane, *see* Chlorobromomethane											
Bromoethane, *see* Ethyl bromide											
Bromoform	SK	0.5	5.2	–	–	SK	0.5	5	–	–	0.39
Bromomethane, *see* Methyl bromide											
Bromotrifluoromethane		1000	6190	–	–		1000	6190	1200	7430	
1,3-Butadiene		2[e]	–	–	–		10	22 MEL	–	–	640
Butane		800	1900	–	–		600	1450	750	1810	0.29
Butanethiol, *see* Butyl mercaptan											
2-Butanone, *see* Methyl ethyl ketone (MEK)											
trans-But-2-enal		–	–	–	–		2	6	6	18	
2-Butoxyethanol (EGBE)	SK	20	–	–	–	SK	25	123 MEL	–	–	250
n-Butyl acetate		150	713	200	950		150	724	200	966	390
sec-Butyl acetate		200	950	–	–		200	966	250	1210	

tert-Butyl acetate		200	950	–	–		200	966	250	1210	
Butyl acrylate		2	–	–	–		10	53	–	–	290
n-Butyl alcohol	SK	–	–	C50	–	SK	–	–	50	154	60
sec-Butyl alcohol		100	303	–	–		100	308	150	462	38
tert-Butyl alcohol		100	303	–	–		100	308	150	462	2.1
Butylamine	SK	–	–	C5	–	SK	–	–	5	15	2.7
n-Butyl chloroformate		–	–	–	–		1	5.6	–	–	
tert-Butyl chromate, as CrO_3	SK	–	–	–	C0.1		–	–	–	–	
Butylated hydroxy toluene		–	10	–	–		–	10	–	–	
n-Butyl glycidyl ether (BGE)		25	133	–	–		25	135	–	–	
n-Butyl lactate		5	30	–	–		5	30	–	–	0.71
Butyl mercaptan		0.5	1.8	–	–		–	–	–	–	510
o-sec-Butylphenol	SK	5	31	–	–	SK	5	30	–	–	
p-tert-Butyltoluene		1	6	20	121		–	–	–	–	2.0
Cadmium dusts and salts, as Cd											
elemental		–	0.01[e]	–	–			–			
compounds, as Cd		–	0.002[e]	–	–			(except cadmium sulphide pigments)			
Cadmium oxide fume, as Cd		–	–	–	–		–	0.025 MEL		0.05 MEL	
Cadmium sulphide pigments, respirable dust, as Cd		–	–	–	–		–	0.03 MEL	–	–	
Caesium hydroxide		–	2	–	–		–	3	–	–	
Calcium carbonate		–	10[f]	–	–		–	10 (total inhalable dust) 4 (respirable dust)	–	–	
Calcium chromate		–	0.001[e]	–	–	–	–	–	–	–	
Calcium cyanamide		–	0.5	–	–	–		0.5	–	1	
Calcium hydroxide		–	5	–	–		–	5	–	–	
Calcium oxide		–	2	–	–		–	2			
Calcium silicate		–	10[f]	–	–		–	10 (total inhalable dust) 4 (respirable dust)	–	–	
Calcium sulphate		–	10[f]	–	–		–	10 (total (inhalable dust) 4 (respirable dust)	–	–	
Camphor, synthetic, *see* Bornan-2-one											
Caprolactam											
Dust		–	1	–	3		–	1	–	3	
Vapour		5	23	10	47		5	23	10	47	

Table 5.12 Cont'd

Substance		*Threshold limit values (USA)* TWA (ppm)[a]	TWA (mg/m³)[b]	STEL (ppm)[a]	STEL (mg/m³)[b]		*Occupational exposure limits (UK)* TWA[c] (ppm)	TWA[c] (mg/m³)	STEL[c] (ppm)	STEL[c] (mg/m³)	*Air odour threshold* (ppm; v/v)
Captafol	SK	–	0.1	–	–	SK	–	0.1	–	–	
Captan		–	5	–	–		–	5	–	15	
Carbaryl		–	5	–	–		–	–	–	–	
Carbofuran		–	0.1	–	–		–	0.1	–	–	
Carbon black		–	3.5	–	–		–	3.5	–	7	
Carbon dioxide		5000	9150	30 000	54 800		5000	9150	15 000	27 400	0.067
Carbon disulphide	SK	10	32	–	–	SK	10	32 MEL	–	–	92
Carbon monoxide		25	28	–	–		30	35	200	232	0.00050
Carbon tetrabromide		0.1	1.4	0.3	4.1		0.1	1.4	0.3	4	
Carbon tetrachloride	SK	5[e]	31[e]	10	62	SK	2	13	–	–	0.052
Carbonyl chloride, *see* Phosgene											
Carbonyl fluoride		2	5.4	5	13		–	–	–	–	
Catechol	SK	5	23	–	–		5	23	–	–	
Cellulose (pure)		–	10	–	–		–	10 (total inhalable dust)	–	20	
								4 (respirable dust)	–	–	
Chlordane	SK	–	0.5	–	–	SK	–	–	–	–	
Chlorinated camphene	SK	–	0.5	–	1		–	–	–	–	
Chlorinated diphenyl oxide		–	0.5	–	–		–	–	–	–	
Chlorine		0.5	1.5	1	2.9		0.5	1.5	1	2.9	3.2
Chlorine dioxide		0.1	0.28	0.3	0.83		0.1	0.3	0.3	0.9	0.011
Chlorine trifluoride		–	–	C0.1	0.38C		–	–	0.1	0.38	
Chloroacetaldehyde		–	–	C1	3.2C		–	–	1	3.3	
Chloroacetone	SK	–	–	C1	C3.8		–	–	–	–	
α-Chloroacetophenone		0.05	0.32	–	–		0.05	0.32	–	–	1.4
Chloroacetyl chloride		0.05	0.23	0.15	0.7		–	–	–	–	
Chlorobenzene		10	–	–	–		50	234	–	–	110
o-Chlorobenzylidene malononitrile	SK	–	–	C0.05	C0.39		–	–	–	–	
Chlorobromomethane		200	1060	–	–		200	1060	250	1340	0.50
2-Chloro-1,3-butadiene, *see* β-Chloroprene											
Chlorodifluoromethane		1000	3590	–	–		1000	3590	–	–	
Chlorodiphenyl (42% chlorine)	SK	–	1	–	–						
Chlorodiphenyl (54% chlorine)	SK	–	0.5	–	–	SK	–	0.1	–	–	
1-Chloro-2,3-epoxy propane, *see* Epichlorohydrin											

2-Chloroethanol, *see* Ethylene chlorohydrin											
Chloroethylene, *see* Vinyl chloride											
Chloroform		10[e]	49[e]	–	–		2	9.9	–	–	0.12
bis(Chloromethyl)ether		0.001[g]	0.005	–	–		0.001	0.005	–	–	
Chloromethyl methyl ether		[e]	[e]	–	–		–	–	–	–	
1-Chloro-4-nitrobenzene		–	–	–	–	SK	–	1	–	2	
1-Chloro-1-nitropropane		2	10	–	–		–	–	–	–	
Chloropentafluoroethane		1000	6420	–	–		1000	6420	–	–	
Chloropicrin		0.1	0.67	–	–		0.1	0.7	0.3	2	0.13
β-Chloroprene	SK	10	36	–	–		–	–	–	–	0.68
3-Chloropropene		–	–	–	–		1	3	2	6	
2-Chloropropionic acid		0.1	–	–	–		–	–	–	–	
o-Chlorostyrene		50	283	75	425		–	–	–	–	
Chlorosulphonic acid		–	–	–	–		–	1	–	–	
o-Chlorotoluene		50	259	–	–		50	264	–	–	150
2-Chloro-6-(trichloromethyl)pyridine, *see* Nitrapyrin											
Chlorpyrifos		–	0.2	–	–	SK	–	0.2	–	0.6	
Chromite ore processing (Chromate), as Cr		–	0.05[g]	–	–		–	–	–	–	
Chromium metal		–	0.5	–	–		–	0.5	–	–	
Chromium (III) compounds, as Cr		–	0.5	–	–		–	0.5	–	–	
Chromium (VI) compounds, as Cr											
Water soluble		–	0.05[g]	–	– }		–	0.05	–	–	
Water insoluble		–	0.01[g]	–	– }						
Chromyl chloride		0.025	0.16	–	–	–	–	–	–		
Chrysene		[e]	[e]	–	–		–	–	–	–	
Chrysotile, *see* Asbestos											
Clopidol		–	10	–	–		–	–	–	–	
Coal dust		–	2 (respirable fraction)	–	–		–	2 (respirable dust)	–	–	
Anthracite		–	0.4	–	–		–	–	–	–	
Bitumous		–	0.9	–	–		–	–	–	–	
Coal tar pitch volatiles, as benzene solubles		–	0.2[g]	–	–		–	0.14	–	–	
Cobalt as Co, elemental and inorganic compounds		–	0.02	–	–		–	0.1 (cobalt and compounds)	–	–	
Cobalt carbonyl as Co		–	0.1	–	–		–	–	–	–	
Cobalt hydrocarbonyl as Co		–	0.1	–	–		–	–	–	–	
Copper											
Fume		–	0.2	–	–		–	0.2	–	–	
Dusts and mists as Cu		–	1	–	–		–	1	–	2	
Cotton dust, raw		–	0.2[l]	–	–		–	2.5	–	–	

Table 5.12 Cont'd

Substance		Threshold limit values (USA) TWA (ppm)[a]	TWA (mg/m^3)[b]	STEL (ppm)[a]	STEL (mg/m^3)[b]		Occupational exposure limits (UK) TWA[c] (ppm)	TWA[c] (mg/m^3)	STEL[c] (ppm)	STEL[c] (mg/m^3)	Air odour threshold (ppm; v/v)
Cresol, all isomers	SK	5	22	–	–	SK	5	22	–	–	17 000 (m-cresol)
Cristobalite, *see* Silica, crystalline											
Crocidolite, *see* Asbestos											
Crotonaldehyde		–	–	C0.3	–		–	–	–	–	17 (trans)
Crufomate		–	5	–	–		–	–	–	–	
Cryofluorane (INN)		1000	7110	–	–		1000	7110	1250	8890	
Cumene	SK	50	246	–	–	SK	25	125	75	375	570
Cyanamide		–	2	–	–		–	2	–	–	
Cyanides as CN, *see* Hydrogen cyanide					–						
Cyanogen		10	22	–	–		10	22	–	–	
Cyanogen chloride		–	–	C0.3	C0.75		–	–	0.3	0.77	
Cyclohexane		300	1030	–	–		100	350	300	1050	25
Cyclohexanol	SK	50	206	–	–		50	208	–	–	0.15
Cyclohexanone	SK	25	100	–	–		25	102	100	408	0.88
Cyclohexene		300	1010	–	–		300	1020	–	–	0.18
Cyclohexylamine		10	41	–	–	SK	10	41	–	–	2.6
Cyclonite	SK	–	0.5	–	–	SK	–	1.5	–	3	
Cyclopentadiene		75	203	–	–		–	–	–	–	1.9
Cyclopentane		600	1720	–	–		–	–	–	–	
Cyhexatin		–	5	–	–		–	5	–	10	
2,4-D		–	10	–	–		–	10	–	20	
DDT (Dichloro-diphenyltrichloroethane)		–	1	–	–		–	1	–	3	
Decaborane	SK	0.05	0.25	0.15	0.75		–	–	–	–	
Demeton	SK	0.01	0.11	–	–		–	–	–	–	0.060
Diacetone alcohol		50	238	–	–		50	240	75	362	0.28
Dialkyl 79 phthalate		–	–	–	–		–	5	–	–	
Diallyl phthalate		–	–	–	–		–	5	–	–	
2,2′-Diaminodiethylamine		–	–	–	–	SK	1	4	–	–	
1,2-Diaminoethane, *see* Ethylenediamine											
Diammonium peroxydisulphate as S_2O_8		–	–	–	–		–	1	–	–	
Diatomaceous earth, *see* Silica, amorphous											
Diazinon	SK	–	0.1	–	–	SK	–	0.1	–	0.3	
Diazomethane		0.2[e]	0.34	–	–		carcinogen	–	–	–	
Diborane		0.1	0.11	–	–		0.1	0.1	–	–	2.5
Dibromodifluoromethane		–	–	–	–		100	872	150	1310	

1,2-Dibromoethane, *see* Ethylene dibromide											
2-*N*-Dibutylamino-ethanol	SK	0.5	3.5	–	–		–	–	–	–	
Dibutyl phenyl phosphate	SK	0.3	3.5	–	–		–	–	–	–	
Dibutyl phosphate		1	8.6	2	17		1	8.7	2	17	
Dibutyl phthalate		–	5	–	–		–	5	–	10	
6,6′-Di-tert-butyl-4-4′-thiodi-*m*-cresol		–	–	–	–		–	10	–	20	
Dichloroacetylene		–	–	C0.1	C0.39		–	–	0.1	0.4	
o-Dichlorobenzene	SK	25	152	50	153		–	–	50	306	0.30
p-Dichlorobenzene		10	31	–	–		25	153	50	306	0.18
3-3′-Dichlorobenzidine		–	(e)	–	–	–	–	–	–	–	
1,4-Dichloro-2-butene		0.005[e]	–	–	–		–	–	–	–	
Dichlorodifluoromethane	SK	1000	4950	–	–		1000	5030	1250	6280	
1,3-Dichloro-5,5-dimethyl hydonton		–	0.2	–	0.4		–	0.2	–	0.4	
1,1-Dichloroethane		–	–	–	–		200	823	400	1650	
1,2-Dichloroethane, *see* Ethylene dichloride											
1,1-Dichloroethylene *see* Vinylidene chloride											
1,2-dichloroethylene		200	793	–	–		200	806	250	1010	17 (trans)
Dichloroethyl ether	SK	5	29	10	58		–	–	–	–	0.049
Dichlorofluoromethane		10	42	–	–		10	43	–	–	
Dichloromethane		50	175	–	–		100	350 MEL	300	1060	
1,1-Dichloro-1-nitroethane		2	12	–	–		–	–	–	–	
2,2′-Dichloro-4-4′-methylene dianiline (Mb OCA)		–	–	–	–	SK	–	0.005 MEL	–	–	
Dichlorophenoxyacetic acid, *see* 2,4-D											
1,2-Dichloropropane, *see* Propylene dichloride											
Dichloropropene	SK	1	4.5	–	–	SK	1	5	10	50	
2,2-Dichloropropionic acid		–	5	–	–		–	–	–	–	
Dichlorotetrafluoro-ethane, *see* Cryofluorane											
Dichlorvos	SK	–	0.90	–	–	SK	0.1	0.92	0.3	2.8	
Dicrotophos	SK	–	0.25	–	–		–	–	–	–	
Dicyclohexyl phthalate		–	–	–	–		–	5	–	–	
Dicyclopentadiene		5	27	–	–		5	27	–	–	0.0057
Dicyclopentadienyl iron, *see* Ferrocene											
Dieldrin	SK	–	0.25	–	–	SK	–	0.25	–	0.75	
Diethanolamine		–	2	–	–		3	13	–	–	0.27
Diethylamine		5	–	25	75		10	30	25	75	0.13
2-Diethylamino-ethanol	SK	2	–	–	–	SK	10	49	–	–	0.011
Diethylene glycol		–	–	–	–		23	100	–	–	

Table 5.12 Cont'd

Substance		*Threshold limit values (USA)* TWA (ppm)[a]	TWA (mg/m³)[b]	STEL (ppm)[a]	STEL (mg/m³)[b]		*Occupational exposure limits (UK)* TWA[c] (ppm)	TWA[c] (mg/m³)	STEL[c] (ppm)	STEL[c] (mg/m³)	*Air odour threshold* (ppm; v/v)
Diethylene triamine	SK	1	4.2	–	–	SK	1	4.3	–	–	
Diethyl ether, *see* Ethyl ether											
Di(2-ethylhexyl)phthalate, *see* Di-sec-octyl phthalate											
Diethyl ketone		200	716	300	1024		200	716	250	895	2.0
Diethyl phthalate		–	5	–	–		–	5	–	10	
Diethyl sulphate		–	–	–	–		0.05	0.32	–	–	
Difluorodibromomethane		100	858	–	–		100	860	150	1290	
Diglycidyl ether (DGE)		0.1	0.53	–	–		0.1	0.5	–	–	
Dihydroxybenzene, *see* Hydroquinone											
Diisobutyl ketone		25	145	–	–		25	148	–	–	0.11
Diisoctyl phthalate		–	–	–	–		–	5	–	–	
Diisodecyl phthalate		–	–	–	–		–	5	–	–	
Diisononyl phthalate		–	–	–	–		–	5	–	–	
Diisopropylamine	SK	5	21	–	–		5	21	–	–	1.8
Diisopropyl ether, *see* Isopropyl ether											
Dimethoxymethane, *see* Methylal											
Dimethyl acetamide	SK	10	36	–	–	SK	10	36	20	72	47
Dimethylamine		5	9	15	27		2	3.8	6	11	0.34
Dimethylaminobenzene, *see* Xylidene											
Dimethylamino ethanol		–	–	–	–		2	7.4	6	22	
Dimethylaniline (*N,N*-Dimethylaniline)	SK	5	25	10	50	SK	5	25	10	50	0.013
Dimethylbenzene, *see* Xylene											
1,3-Dimethylbutyl acetate		–	–	–	–		50	300	100	600	
Dimethyl carbamoyl chloride		[e]	[e]	–	–		–	–	–	–	
Dimethyl-1,2-dibromo-2,2-dichloroethyl phosphate, *see* Naled											
Dimethyl ether		–	–	–	–		400	766	500	958	
N-N-Dimethylethylamine		–	–	–	–		10	30	15	45	
Dimethylethoxy silane		0.5	–	1.5	–		–	–	–	–	
Dimethylformamide	SK	10	30	–	–	SK	10	30	20	60	2.2
2,6-Dimethyl-4-heptanone, *see* Diisobutyl ketone											
1,1-Dimethyl hydrazine	SK	0.01[e]	–	–	–		–	carcinogen	–	–	1.7
Dimethylnitrosoamine, *see N*-Nitrosodimethylamine											
Dimethylphthalate		–	5	–	–		–	5	–	10	
Dimethyl sulphate	SK	0.1[e]	0.52[e]	–	–	SK	0.05	0.25	–	–	

Dinitolmide		–	5	–	–		–	–	–	–	
Dinitrobenzene (all isomers)	SK	0.15	1	–	–	SK	0.15	1	0.5	3	
Dinitro-*o*-cresol	SK	–	0.2	–	–	SK	–	0.2	–	0.6	
1,2-Dinitroethane, *see* Ethylene glycol dinitrate											
1,2-Dinitropropane		–	–	–	–	SK	–	–	–	–	
3,5-Dinitro-*o*-toluamide, *see* Dinitolmide											
Dinitrotoluene	SK	–	0.2	–	–	SK	–	carcinogen	–	–	
Di-nonyl phthalate		–	–	–	–		–	5	–	–	
Dioxane	SK	20	–	–	–	SK	25	90	100	366	24
Dioxathion	SK	–	0.2	–	–	SK	–	0.2	–	–	
Diphenyl, *see* Biphenyl											
Diphenylamine		–	10	–	–		–	10	–	20	
Diphenyl ether (vapour), *see* Phenyl Ether											
Diphenylmethane diisocyanate, *see* Isocyanates; Methylene bisphenyl isocyanate											
Diphosphorus pentoxide		–	–	–	–		–	–	–	2	
Dipotassium peroxydisulphate, as S_2O_8		–	–	–	–		–	1	–	–	
Dipropylene glycol methyl ether	SK	100	606	150	909		–	–	–	–	
Dipropyl ketone		50	233	–	–		–	–	–	–	
Diquat		–	0.5	–	–		–	0.5	–	1	
Di-sec-octyl phthalate		–	5	–	–		–	5	–	10	
Disodium disulphite		–	5	–	–		–	5	–	–	
Disodium peroxydisulphate, as S_2O_8		–	–	–	–		–	1	–	–	
Disodium tetraborate											
anhydrous		–	–	–	–		–	1	–	–	
decahydrate		–	–	–	–		–	5	–	–	
pentahydrate		–	–	–	–		–	1	–	–	
Disulfiram		–	2	–	–		–	–	–	–	
Disulfoton		–	0.1	–	–		–	0.1	–	0.3	
Disulphur dichloride		–	–	C1	–		–	–	1	5.6	
Diuron		–	10	–	–		–	10	–	–	
Divinyl benzene		10	53	–	–		10	54	–	–	
Dusts								(S)			
Emery		–	10[i]	–	–		–	10 (total inhalable dust) 4 (respirable dust)	–	–	
Endosulfan	SK	–	0.1	–	–	SK	–	0.1	–	0.3	
Endrin	SK	–	0.1	–	–	SK	–	0.1	–	0.3	
Enflurane		75	566	–	–		50	383	–	–	

Table 5.12 Cont'd

Substance		Threshold limit values (USA) TWA (ppm)[a]	TWA (mg/m³)[b]	STEL (ppm)[a]	STEL (mg/m³)[b]		Occupational exposure limits (UK) TWA[c] (ppm)	TWA[c] (mg/m³)	STEL[c] (ppm)	STEL[c] (mg/m³)	Air odour threshold (ppm; v/v)
Enzymes, *see* Subtilisins											
Epichlorohydrin	SK	0.5	1.9	–	–	SK MEL	0.5	1.9	1.5	5.8	0.93
EPN	SK	–	0.1	–	–		–	–	–	–	
1,2-Epoxypropane, *see* Propylene oxide											
2,3-Epoxy-1-propanol, *see* Glycidol											
2,3-Epoxypropyl isopropyl ether		50	240	75	360		50	240	75	360	
Ethane			(d)					(d)			120 000
Ethanethiol, *see* Ethyl mercaptan											
Ethanol, *see* Ethyl alcohol											
Ethanolamine		3	7.5	6	15		3	7.6	6	15	2.6
Ethion	SK	–	0.4	–	–		–	–	–	–	
2-Ethoxyethanol (EGEE)	SK	5	18	–	–	SK	10	37 MEL	–	–	2.7
2-Ethoxyethyl acetate (EGEEA)	SK	5	27	–	–	SK	10	54 MEL	–	–	0.056
Ethyl acetate		400	1440	–	–		400	1460	–	–	3.9
Ethyl acrylate		5[e]	20[e]	15[g]	61[g]	SK	5	20	15	60	0.0012
Ethyl alcohol		1000	1880	–	–		1000	1920	–	–	84
Ethylamine		5	9	–	–		2	3.8	6	11	0.95
Ethyl amyl ketone		25	131	–	–		25	133	–	–	6.0
Ethyl benzene		100	441	125	552		100	441	125	552	2.3
Ethyl bromide		5	–	–	–		200	906	250	1130	3.1
Ethyl tert-butyl ether		5	–	–	–		–	–	–	–	
Ethyl butyl ketone		50	234	75	351		50	237	100	475	
Ethyl chloride		100	–	–	–		1000	2700	1250	3380	4.2
Ethylchloroformate		–	–	–	–		1	4.4	–	–	
Ethyl cyanoacrylate		0.2	–	–	–		–	–	–	0.3	1.5
Ethylene			(d)								290
Ethylene chlorohydrin	SK	–	–	C1	–	SK	–	–	1	3.4	
Ethylenediamine		10	25	–	–		10	25	–	–	1.0
Ethylene dibromide	SK	(e)	(e)	–	–	SK	0.5	4 MEL	–	–	2.5
Ethylene dichloride		10	40	–	–		5	21	–	–	88
Ethylene glycol, vapour and mist		–	–	–	C100		–	10 particulate 60 vapour	–	– 125	
Ethylene glycol dinitrate	SK	0.05	0.31	–	–	SK	0.2	1.3	0.2	1.3	
Ethylene glycol methyl ether acetate, *see* 2-Methoxyethyl acetate											
Ethylene oxide		1[e]	1.8[e]	–	–	SK	5	9.2 MEL	–	–	430
Ethylene imine	SK	0.5	0.88	–	–	SK			carcinogen		

Ethyl ether		400	1210	500	1520		400	1230	500	1540	8.9
Ethyl formate		100	303	–	–		100	308	150	462	31
2-Ethylhexyl chloroformate		–	–	–	–		1	7.9	–	–	
Ethylidene chloride, *see* 1,1-Dichloroethane											
Ethylidene norbornene		–	–	C5	–						0.014
Ethyl mercaptan		0.5	1.3	–	–		0.5	1.3	2	5.2	0.00076
N-Ethylmorpholine	SK	5	24	–	–	SK	5	24	20	96	1.4
Ethyl silicate		10	85	–	–		10	87	30	260	17
Fenamiphos	SK	–	0.1	–	–		–	–	–	–	
Fenchlorphos, *see* Ronnel											
Fensulfothion		–	0.1	–	–		–	–	–	–	
Fenthion	SK	–	0.2	–	–		–	–	–	–	
Ferbam		–	10	–	–		–	10	–	20	
Ferrocene		–	10	–	–		–	10	–	20	
Ferrous foundry particulate											
total inhalable dust		–	–	–	–		–	10	–	–	
respirable dust		–	–	–	–		–	4	–	–	
Ferrovanadium dust		–	1	–	3						
Fibrous glass dust		–	10	–	–		–	–	–	–	
Flour dust		–	5	–	–		–	–	–	–	
Fluorides as F		–	2.5	–	–		–	2.5	–	–	
Fluorine		1	1.6	2	3.1		–	–	1	1.6	0.14
Fluorotrichloromethane, *see* Trichlorofluoromethane											
Fonofos	SK	–	0.1	–	–		–	–	–	–	
Formaldehyde		–	–	C0.3[e]	–		2	2.5 MEL	2	2.5 MEL	0.83
Formamide	SK	10	18	–	–		20	37	30	56	
Formic acid		5	9.4	10	19		5	9.6	–	–	49
Furfural	SK	2	7.9	–	–	SK	2	8	5	20	0.078
Furfuryl alcohol	SK	10	40	15	60	SK	5	20	15	60	8.0
Gasoline		300	890	500	1480		–	–	–	–	
Germanium tetrahydride		0.2	0.63	–	–		0.2	0.6	0.6	1.9	
Glass, fibrous or dust, *see* Fibrous glass dust											
Glutaraldehyde		–	–	C0.05	–	SEN	0.05	0.2	0.05	0.2	
Glycerin mist		–	10	–	–		–	10	–	–	
Glycidol		2	–	–	–		–	–	–	–	
Glycol monoethyl ether, *see* 2-Ethoxyethanol											
Glycerol trinitrate	SK	0.05	–	–	–	SK	0.2	1.9	0.2	1.9	
Grain dust (oat, wheat, barley)		–	4	–	–		–	–	–	–	
Graphite, all forms except graphite fibres		–	2	–	–		–	10 (total (inhalable dust) / 4 (respirable dust)	– / –	– / –	

Table 5.12 Cont'd

Substance		Threshold limit values (USA) TWA (ppm)[a]	TWA (mg/m³)[b]	STEL (ppm)[a]	STEL (mg/m³)[b]		Occupational exposure limits (UK) TWA[c] (ppm)	TWA[c] (mg/m³)	STEL[c] (ppm)	STEL[c] (mg/m³)	Air odour threshold (ppm; v/v)
Gypsum, *see* Calcium sulphate											
Hafnium		–	0.5	–	–		–	0.5	–	1.5	
Halothane		50	404 (f)	–	–		10	82 (f)	–	–	33
Helium											
Heptachlor	SK	–	0.5	–	–		–	–	–	–	
Heptane (*n*-Heptane)		400	1640	500	2050		–	–	–	–	150
2-Heptanone, *see* Methyl n-amyl ketone											
3-Heptanone, *see* Ethyl butyl ketone											
Hexachlorobenzene		–	0.002	–	–		–	–	–	–	
Hexachlorobutadiene	SK	0.02[e]	0.21[e]	–	–		–	–	–	–	
Hexachlorocyclohexane, *see* Lindane											
Hexachlorocyclo-pentadiene		0.01	0.11	–	–		–	–	–	–	0.030
Hexachloroethane		1	9.7	–	–		5	50 (vapour)	–	–	0.15
							–	10 (total inhalable dust)			
							–	4 (respirable dust)	–	–	
Hexachloronaphthalene	SK	–	0.2	–	–		–	–	–	–	
Hexafluoroacetone	SK	0.1	0.68	–	–		–	–	–	–	
Hexahydro-1,3,5-trinitro-1,3,5-triazine, *see* Cyclonite											
Hexamethylene diisocyanate		0.005 (e)	0.034 (e)	–	–		*see* Isocyanates				
Hexamethyl phosphoramide	SK			–	–		–	–	–	–	
Hexane (*n*-hexane)		50	176	–	–		20	72	–	–	130
Other isomers		500	1760	1000	3500						
Hexanediamine		0.5	–	–	–		–	–	–		
1,6-Hexanolactam, *see* Caprolactam											
2-Hexanone, *see* Methyl *n*-butyl ketone											
1-Hexene		30	–	–	–	–	–	–	–		
Hexone, *see* Methyl isobutyl ketone											
sec-Hexyl acetate		50	295	–	–	SK	50	205	75	300	
Hexylene glycol		–	–	C25	–		25	123	25	123	50
Hydrazine	SK	0.01	–	–	–	SK	0.02	0.03	0.1	0.13 MEL	3.7
Hydrazoic acid, vapour		–	–	–	–		–	–	0.1	0.18	
Hydrogen			(d)					(d)			
Hydrogenated terphenyls (non irradiated)		0.5	4.9	–	–		–	–	–	–	

Hydrogen bromide		–	–	C3	C9.9		–	–	3	10	2.0
Hydrogen chloride		–	–	C5	C7.5		1	2	5	8	0.77
Hydrogen cyanide and cyanide salts (excluding cyanogen and cyanogen chloride)											
Hydrogen cyanide		–	C4.7	–	–		–		MEL 10	11	
Calcium cyanide		–	–	–	C5						
Potassium cyanide		–	–	–	C5		–	5	–	–	
Sodium cyanide		–	–	–	C5						
Hydrogen fluoride as F		–	–	C3	–		–	–	3	2.5	0.042
Hydrogen peroxide		1	1.4	–	–		1	1.4	2	2.8	0.30
Hydrogen selenide as Se		0.05	0.16	–	–		0.05	0.17	–	–	0.0081
Hydrogen sulphide		10	14	15	21		10	14	15	21	
Hydroquinone		–	2	–	–		–	2	–	4	
4-Hydroxy-4-methyl-2-pentanone, *see* Diacetone alcohol											
2-Hydroxypropyl acrylate		0.5	2.8	–	–	SK	0.5	2.7	–	–	
Indene		10	48	–	–		10	48	15	72	0.015
Indium and compounds as In		–	0.1	–	–		–	0.1	–	0.3	
Iodine		–	–	C0.1	–		–	–	0.1	1	
Iodoform		0.6	10	–	–		0.6	9.8	1	16	0.005
Iron oxide dust + fume (Fe_2O_3) as Fe		(t)	5	–	–		–	5	–	10	
Iron pentacarbonyl as Fe		0.1	0.23	0.2	0.45		0.01	0.08	–	–	
Iron salts, soluble, as Fe		–	1	–	–		–	1	–	2	
Isoamyl acetate		50	266	100	532		50	270	100	541	0.025
Isoamyl alcohol		100	361	125	452		100	366	125	458	0.042
Isobutyl acetate		150	713	–	–		150	724	187	903	0.64
Isobutyl alcohol	SK	50	152	–	–		50	154	75	231	1.6
Isocyanates, all, as-NCO		–	–	–	–	SK	–	0.02 MEL	–	0.07 MEL	
Isofluorane		–	–	–	–		50	383	–	–	
Isooctyl alcohol	SK	50	266	–	–		50	270	–	–	
Isopentyl acetate		–	–	–	–		50	270	100	541	
Isophorone		–	–	C5	C28		–	–	5	29	0.20
Isophorone diisocyanate	SK	0.005	0.045	–	–		*see* Isocyanates				
Isopropoxyethanol		25	106	–	–		–	–	–	–	
Isopropyl acetate		250	1040	310	1290		–	–	200	849	2.7
Isopropyl alcohol		400	983	500	1230	SK	400	999	500	1250	22
Isopropylamine		5	12	10	24		–	–	–	–	1.2
N-Isopropylaniline	SK	2	11	–	–		–	–	–	–	
Isopropyl chloroformate		–	–	–	–		1	5	–	–	
Isopropyl ether		250	1040	310	1300		250	1060	310	1310	0.017
Isopropyl glycidyl ether (IGE), *see* 2,3-Epoxypropyl isopropyl ether											
Kaolin		–	2	–	–		–	2	–	–	
Ketene		0.5	0.86	1.5	2.6		0.5	0.87	1.5	2.6	

Table 5.12 Cont'd

Substance		Threshold limit values (USA) TWA (ppm)[a]	TWA (mg/m^3)[b]	STEL (ppm)[a]	STEL (mg/m^3)[b]		Occupational exposure limits (UK) TWA[c] (ppm)	TWA[c] (mg/m^3)	STEL[c] (ppm)	STEL[c] (mg/m^3)	Air odour threshold (ppm; v/v)
Lead, inorganic dusts and fumes,		–	0.05	–	–						
as Pb							[k]				
lead alkyls							[k]				
Lead arsenate as $PbHAsO_4$		–	0.15	–	–						
Lead chromate as Cr		–	{ 0.05 as Pb[e] 0.012 as Cr[e]								
Limestone, *see* Calcium carbonate											
Lindane	SK	–	0.5	–	–	SK	–	0.1	–	–	
Lithium hydride		–	0.025	–	–		–	0.025	–	–	
Lithium hydroxide		–	–	–	–		–	–	–	1	
LPG (Liquefied petroleum gas)		1000	1800	–	–		1000	1750	1250	2180	
Magnesite		–	10[l]	–	–	{	–	10 (total inhalable dust)	–	–	
							–	4 (respirable dust)	–	–	
Magnesium oxide fume		–	10	–	–	{		4 (fume, respirable dust)	–	10	
								10 (total inhalable dust)			
Malathion	SK	–	10	–	–	SK	–	10	–	–	
Maleic anhydride		0.1	–	–	–		–	1	–	3 MEL	0.32
Manganese as Mn											
Dust and compounds		–	–	–	–		–	5	–	–	
Fume		–	–	–	3		–	1	–	3	
Elemental and inorganic compounds		–	0.2	–	–		–	–	–	–	
Manganese cyclopenta-dienyl tricarbonyl as Mn	SK	–	0.1	–	–	SK	–	0.1	–	0.3	
Manganese methyl-pentadienyl tricarbonyl		–	–	–	–	SK	–	0.2	–	0.6	
Manganese tetroxide		–	–	–	–		–	1	–	–	
Man-made mineral fibre		–	–	–	–		–	5 MEL	–	–	
Marble, *see* Calcium carbonate											
Mercaptoacetic acid, *see* Thioglycolic acid											
Mequinol (INN)		–	5	–	–		–	5	–	–	
Mercury, as Hg											
Alkyl compounds		–	0.01	–	0.03		–	0.01	–	0.03	
Aryl compounds		–	0.1	–	–		–	–	–	–	

Inorganic forms including metallic mercury		–	0.025	–	–		–	0.025	–	–	
Mesityl oxide		15	60	25	100		15	61	25	102	0.45
Methacrylic acid		20	70	–	–		20	72	40	143	
Methacrylonitrile		–	–		–	SK	1	2.8	–	–	
Methane			(d)					(d)			
Methanethiol, *see* Methyl mercaptan											
Methanol, *see* Methyl alcohol											
Methomyl		–	2.5	–	–	SK	–	2.5	–	–	
Methoxychlor		–	10	–	–		–	10	–	–	
2-Methoxyethanol (EGME)	SK	5	16	–	–	SK	5	16 MEL	–	–	2.3
2-Methoxyethyl acetate (EGMEA)	SK	5	24	–	–	SK	5	24 MEL	–	–	–
2-Methoxymethylethoxy propanol		–	–	–	–		50	308	–	– SK	
4-Methoxyphenol, *see* Mequinol (INN)											
1-Methoxypropan-2-ol		100	375	150	560	SK	100	375	300	1120	
1-Methoxypropyl acetate		–	–	–	–		50	274	150	822	
Methyl acetate		200	606	250	757		200	616	250	770	4.6
Methyl acetylene		1000	1640	–	–		–	–	–	–	50
Methyl acetylene-propadiene mixture (MAPP)		1000	1640	1250	2050		–	–	–	–	
Methyl acrylate	SK	2	7	–	–		10	36	–	–	0.0048
Methylacrylonitrile	SK	1	2.7	–	–		–	–	–	–	7.0
Methylal		1000	3110	–	–		1000	3160	1250	3950	
Methyl alcohol	SK	200	262	250	328	SK	200	266	250	333	100
Methylamine		5	6.5	15	19.5		10	13	–	–	3.2
Methyl amyl alcohol, *see* Methyl isobutyl carbinol											
Methyl *n*-amyl ketone		50	233	–	–		50	237	100	475	0.35
N-Methyl aniline	SK	0.5	2.2	–	–	SK	0.5	2.2	–	–	1.7
Methyl bromide	SK	1	4	–	–	SK	5	20	15	60	
3-Methylbutan-1-ol		–	–	–	–		100	366	125	458	
1-Methylbutyl acetate		–	–	–	–		50	270	100	541	
Methyl-tert-butyl ether		40	–	–	–		25	92	75	275	
Methyl *n*-butyl ketone	SK	5	20	10	40	SK	5	20	–	–	0.076
Methyl chloride	SK	50	103	100	207		50	105	100	210	
Methyl chloroform		350	1910	450	2460		200	1110	400	2220	120
Methyl cyanoacrylate		0.2	1	4	18		–	–	0.3	1.4	2.2
Methylcyclohexane		400	1610	–	–		–	–	–	–	630
Methylcyclohexanol		50	234	–	–		50	237	75	356	500 (cis)
o-Methylcyclohexanone		50	229	75	344	SK	50	233	75	350	
Methyl demeton	SK	–	0.5	–	–			–	–	–	
2-Methyl-4,6-dinitrophenol		–	–	–	–	SK	–	0.2	–	0.6	
Methylene bisphenyl isocyanate (MDI)		0.005	0.051	–	–	*see* Isocyanates					
Methylene chloride, *see* Dichloromethane)											

Table 5.12 Cont'd

Substance		*Threshold limit values (USA)* TWA (ppm)[a]	TWA (mg/m³)[b]	STEL (ppm)[a]	STEL (mg/m³)[b]		*Occupational exposure limits (UK)* TWA[c] (ppm)	TWA[c] (mg/m³)	STEL[c] (ppm)	STEL[c] (mg/m³)	*Air odour threshold* (ppm; v/v)
4-4′-Methylene bis(2-chloroaniline)	SK	0.01[e]	0.11[e]	–	–		–	0.005	–	–	
Methylene bis(4-cyclo-hexylisocyanate)		0.005	0.054	–	–	*see* Isocyanates					
4,4′-Methylenedianiline	SK	0.1[e]	0.81[e]	–	–		0.01	0.08	–	–	
Methyl ethyl ketone (MEK)		200	590	300	885		200	600	300	899	5.4
Methyl ethyl ketone peroxide		–	–	C0.2	–		–	–	0.2	1.5	
Methyl formate		100	246	150	368		100	250	150	375	600
5-Methyl-3-heptanone, *see* Ethyl amyl ketone											
5-Methylhexan-2-one		–	–	–	–		50	237	100	475	
Methyl hydrazine	SK	0.1	–	–	–		–	–	–	–	1.7
Methyl iodide	SK	2[e]	12[e]	–	–	SK	2	12	–	–	
Methyl isoamyl ketone		50	234	–	–		50	237	100	475	0.012
Methyl isobutyl carbinol	SK	25	104	40	167	SK	25	106	40	170	0.070
Methyl isobutyl ketone		50	205	75	307	SK	50	208	100	416	0.68
Methyl isocyanate	SK	0.02	0.047	–	–	*see* Isocyanates					2.1
Methyl isopropyl ketone		200	705	–	–		–	–	–	–	1.9
Methyl mercaptan		0.5	0.98	–	–		0.5	1	–	–	0.0016
Methyl methacrylate		50	208	100	416		50	208	100	416	0.083
Methyl parathion	SK	–	0.2	–	–	SK	–	0.2	–	0.6	
2-Methylpentane-2,4-diol		–	–	–	–		25	125	25	125	
4-Methylpentan-2-ol		–	–	–	–	SK	25	100	40	170	
4-Methylpentan-2-one		–	–	–	–	SK	50	205	100	416	
4-Methylpent-3-en-2-one		–	–	–	–		15	60	25	100	
2-Methylpropan-1-ol		–	–	–	–		50	150	75	225	
Methyl propyl ketone		200	705	250	881		200	716	250	895	11
1-Methyl-2-pyrrolidone		–	–	–	–		25	103	75	309	
Methyl silicate		1	6.3	–	–		1	6.3	5	32	
Methyl styrene (all isomers except α-methyl styrene)		–	–	–	–		100	491	150	736	
α-Methyl styrene		50	242	100	483		–	–	100	491	0.29
N-Methyl-*N*,2,4,6-tetranitroaniline		–	–	–	–		–	1.5	–	3	
Methyl vinyl ketone		–	–	C0.2	–		–	–	–	–	
Metribuzin		–	5	–	–		–	–	–	–	
Mevinphos	SK	0.01	0.09	0.03	0.27	SK	0.01	0.1	0.03	0.28	
Mica		–	3 (respirable dust)	–	–	{	–	10 (total inhalable dust)	–	–	
							–	0.8 (respirable dust)	–	–	

Mineral wool fibre		–	10[i]				–	MEL	–	–	
Molybdenum as Mo											
Soluble compounds		–	5	–	–		–	5	–	10	
Metal and insoluble compounds		–	10	–	–		–	10	–	20	
Monochloroacetic acid		–	–	–	–	SK	0.3	1.2	–	–	
Monochlorobenzene, *see* Chlorobenzene											
Monocrotophos		–	0.25	–	–		–	–	–	–	
Morpholine	SK	20	71	–	–	SK	20	72	30	109	0.01
Naled	SK	–	3	–	–		–	3	–	6	
Naphthalene		10	52	15	79		10	53	15	80	0.084
β-Naphthylamine			(g)								
Neon			(d)					(f)			
Nickel											
Metal		–	1.5	–	–		–	0.1	–	–	
Insoluble compounds as Ni		–	0.2[g]	–	–		–	0.5	–	–	
Soluble compounds as Ni		–	0.1	–	–		–	0.1	–	–	
								(inorganic)			
Nickel carbonyl as Ni		0.05	0.35	–	–		–	–	0.1	0.24	0.30
Nickel organic compounds as Ni		–	–	–	–		–	1	–	3	
Nickel subsulphide		–	0.1[g]	–	–		–	–	–	–	
Nickel sulphide roasting, fume and		–	0.1	–	–		–	–	–	–	
dust, as Ni								carcinogen			
Nicotine	SK	–	0.5	–	–	SK	–	0.5	–	1.5	
Nitrapyrin		–	10	–	20		–	10	–	20	
Nitric acid		2	5.2	4	10		2	5	4	10	
Nitric oxide		25	31	–	–		25	30	35	45	
p-Nitroaniline	SK	–	3	–	–	SK	–	6	–	–	
Nitrobenzene	SK	1	5	–	–	SK	1	5	2	10	0.018
p-Nitrochlorobenzene	SK	0.1	0.64	–	–		–	–	–	–	
4-Nitrodiphenyl			(e)								
Nitroethane		100	307	–	–		100	310	–	–	2.1
Nitrogen			(d)					(d)			
Nitrogen dioxide		3	5.6	5	9.4		3	5.7	5	9.6	0.39
Nitrogen trifluoride		10	29	–	–		10	30	15	45	
Nitroglycerin (NG), *see* Glycerol trinitrate											
Nitromethane		20	50	–	–		100	250	150	381	3.5
1-Nitropropane		25	91	–	–		25	93	–	–	11.0
2-Nitropropane		10[e]	36[e]	–	–		5	19	–	–	70
N-Nitrosodimethylamine	SK	–	(e)	–	–		–	–	–	–	
Nitrotoluene	SK	2	11	–	–	SK	5	30	10	57	0.045
Nitrotrichloromethane, *see* Chloropicrin											
Nitrous oxide		50	90	–	–		100	183	–	–	
Nonane		200	1050	–	–		–	–	–	–	47
Nuisance particulates, *see* Particulates not otherwise classified (PNOC); Dusts											

Table 5.12 Cont'd

Substance		*Threshold limit values (USA)* TWA (ppm)[a]	TWA (mg/m³)[b]	STEL (ppm)[a]	STEL (mg/m³)[b]		*Occupational exposure limits (UK)* TWA[c] (ppm)	TWA[c] (mg/m³)	STEL[c] (ppm)	STEL[c] (mg/m³)	*Air odour threshold* (ppm; v/v)
Octachloronaphthalene	SK	–	0.1	–	0.3	SK	–	0.1	–	0.3	
Octane		300	1400	375	1750		300	1450	375	1800	48
Oil mist, mineral		–	5[w]	–	10		–	5	–	10	
Osmium tetroxide as Os		0.0002	0.0016	0.0006	0.0047		0.0002	0.002	0.0006	0.006	0.0019
Oxalic acid		–	1	–	2		–	1	–	2	
Oxydiethanol		–	–	–	–		23	100	–	–	
Oxygen difluoride		0.05C	0.11C	–	–		–	–	–	–	0.10
Ozone											
Heavy work		0.05	–	–	–						
Moderate work		0.08	–	–	–		–	–	0.2	0.4	
Light work		0.10	–	–	–						
Any work loads up to 2 hrs		0.20	–	–	–						
Paracetamol											
total inhalable dust		–	–	–	–		–	10	–	–	
Paraffin wax fume		–	2	–	–		–	2	–	6	
Paraquat											
total inhalable dust		–	0.5	–	–		–	–	–	–	
respirable		–	0.1	–	–		–	0.08	–	–	
Parathion	SK	–	0.1	–	–	SK	–	0.1	–	0.3	
Parathion-methyl		–	–	–	–	SK	–	0.2	–	0.6	
Particulate polycyclic aromatic hydrocarbons (PPAH), *see* Coal tar pitch volatiles											
Particulates not otherwise classified (PNOC)		–	10	–	–	*see* Dusts					
Pentaborane		0.005	0.013	0.015	0.039		–	–	–	–	0.96
Pentachloronaphthalene		–	0.5	–	–		–	–	–	–	
Pentachlorophenol	SK	–	0.5	–	–	SK	–	0.5	–	1.5	
Pentaerythritol		–	10	–	–		–	10 (total inhalable dust)	–	20	
							–	4 (respirable dust)	–	–	
Pentane		600	1770	750	2210						
2-Pentanone, *see* Methyl propyl ketone											
Pentyl acetate, *see* Amyl acetates											
Perchloroethylene		50	339	200	1357		50	345	100	690	27
Perchloromethyl mercaptan		0.1	0.76	–	–		–	–	–	–	
Perchloryl fluoride		3	13	6	25		3	13	6	26	
Precipitated silica, *see* Silica, amorphous											

Perlite		–	10[i]	–	–		–	–	–	–	
Petroleum distillates, *see* Gasoline; Stoddard solvent; VM&P naphtha											
Phenacyl chloride, *see* α-Chloroacetophenone											
Phenol	SK	5	19	–	–	SK	5	19	10	38	0.040
Phenothiazine	SK	–	5	–	–		–	–	–	–	
N-Phenyl-β-naphthylamine		[e]	[e]	–	–		–	–	–	–	
p-Phenylenediamine	SK	–	0.1	–	–	SK	–	0.1	–	–	
Phenyl-2,3-epoxypropyl ether		–	–	–	–		1	6	–	–	
Phenyl ether vapour		1	7	2	14		1	7	–	–	0.0012
Phenylethylene, *see* Styrene, monomer											
Phenyl glycidyl ether (PGE)		1	6.1	–	–		–	–	–	–	
Phenylhydrazine	SK	5[e]	22[e]	10[g]	44[g]						
Phenyl mercaptan		0.5	2.3	–	–		–	–	–	–	0.00094
Phenylphosphine		0.05C	0.23C	–	–		–	–	–	–	
Phorate (ISO)	SK	–	0.05	–	0.2	SK	–	0.05	–	0.02	
Phosdrin, *see* Mevinphos											
Phosgene		0.1	0.40	–	–		0.02	0.08	0.06	0.25	0.90
Phosphine		0.3	0.42	1	14		–	–	0.3	0.4	0.51
Phosphoric acid		–	1	–	3		–	–	–	2	
Phosphorus (yellow)		0.02	–	–	–		–	0.1	–	0.3	
Phosphorus oxychloride		0.1	0.63	–	–		0.2	1.3	0.6	3.8	
Phosphorus pentachloride		0.1	0.85	–	–		0.1	0.87	–	–	
Phosphorus pentasulphide		–	1	–	3		–	1	–	3	
Phosphorus trichloride		0.2	1.1	0.5	2.8		0.2	1.3	0.5	2.9	
Phthalic anhydride		1	6.1	–	–	SEN	–	4	–	12	0.053
m-Phthalodinitrile		–	5	–	–		–	–	–	–	
Picloram		–	10	–	–		–	10	–	20	
Picric acid		–	0.1	–	–	SK	–	0.1	–	0.3	
Pindone		–	0.1	–	–		–	–	–	–	
Piperazine dihydrochloride		–	5	–	–			5	–	–	
Piperidine		–	–	–	–	SK	1	3.5	–	–	
2-Pivalyl-1,3-indandione, *see* Pindone											
Plaster of Paris, *see* Calcium sulphate											
Platinum											
Metal		–	1	–	–		–	5	–	–	
Soluble salts (except certain halogeno-platinum compounds) as Pt		–	0.002	–	–	SEN	–	MEL 0.002	–	–	
Polychlorobiphenyls, *see* Chlorodiphenyls											
Polytetrafluoroethylene decomposition products			[m]								

Table 5.12 Cont'd

Substance		Threshold limit values (USA) TWA (ppm)[a]	TWA (mg/m^3)[b]	STEL (ppm)[a]	STEL (mg/m^3)[b]		Occupational exposure limits (UK) TWA[c] (ppm)	TWA[c] (mg/m^3)	STEL[c] (ppm)	STEL[c] (mg/m^3)	Air odour threshold (ppm; v/v)
Polyvinyl chloride											
Total inhalable dust		–	–	–	–		–	10	–	–	
Respirable dust		–	–	–	–		–	4	–	–	
Portland cement											
Total inhalable dust		–	–	–	–		–	10	–	–	
Respirable dust		–	–	–	–		–	4	–	–	
Potassium hydroxide		–	–	–	2C		–	–	–	2	
Propane		–	(d)	–	–		–	–	–	–	16 000
Propane-1,2,-diol											
Total (vapour + particulates)		–	–	–	–		150	474	–	–	
Particulates		–	–	–	–		–	10	–	–	
Propane sultone		(e)	(e)								
Propargyl calcohol, *see* Prop-2-yn-1-ol)											
β-Propiolactone		0.5[e]	1.5[e]	–	–		–	–	–	–	
Propionic acid		10	30	–	–		10	31	15	46	0.16
Propoxur		–	0.5	–	–		–	0.5	–	2	
Propranolol		–	–	–	–		–	2	–	6	
n-Propyl acetate		200	835	250	1040		200	849	250	1060	0.67
n-Propyl alcohol	SK	200	492	250	614	SK	200	500	250	625	2.6
isoPropyl alcohol		–	–	–	–		400	999	500	1250	
Propylene			(d)								76
Propylene dichloride		75	347	110	508		–	–	–	–	0.25
Propylene glycol dinitrate	SK	0.05	–	–	–	SK	0.2	1.4	0.2	1.4	
Propylene glycol monomethyl ether, *see* 1-Methoxypropan-2-ol											
Propylene imine	SK	(e)	4.7[e]	–	–		–	–	–	–	
Propylene oxide		20	48	–	–		5	12	–	–	44
n-Propyl nitrate		25	107	40	172		–	–	–	–	
Propyne, *see* Methyl acetylene											
Prop-2-yn-1-ol	SK	1	2.3	–	–	SK	1	2.3	3	7	
Pulverized fuel ash											
Total inhalable dust		–	–	–	–		–	10	–	–	
Respirable dust		–	–	–	–		–	4	–	–	
Pyrethrins		–	–	–	–		–	5	–	10	
Pyrethrum		–	5	–	–		–	–	–	–	
2-Pyridylamine, *see* 2-aminopyridine											
Pyridine		5	16	–	–		5	16	10	33	0.17
Pyrocatechol, *see* Catechol											
Quartz, *see* Silica, crystalline											

Quinone	0.1	0.44	–	–		0.1	0.4	0.3	1.3	0.084
RDX, *see* Cyclonite										
Resorcinol	10	45	20	90		10	46	20	92	
Rhodium										
Metal	–	1	–	–		–	0.1 (Fume and dust)	–	0.3	
Insoluble compounds as Rh	–	1	–	–		–	–	–	–	
Soluble compounds as Rh	–	0.01	–	–		–	0.001	–	0.003	
Ronnel	–	10	–	–		–	10	–	–	
Rosin core solder, pyrolysis products	–	0.1	–	–	SEN	–	0.05 MEL	–	0.15	
Rotenone (commercial)	–	5	–	–		–	5	–	10	
Rouge	–	$10^{(f)}$	–	–		–	10 (total inhalable dust)	–	–	
						–	4 (respirable dust)	–	–	
Rubber fume	–	–	–	–		–	0.6 MEL	–	–	
Rubber process dust	–	–	–	–		–	6 MEL	–	–	
Rubber solvent (naphtha)	400	1590	–	–		–	–	–	–	
Selenium and compounds as Se	–	0.2	–	–		0.1	(except hydrogen selenide)	–	–	
Selenium hexafluoride as Se	0.05	0.16	–	–		–	–	–	–	
Sesone	–	10	–	–		–	10	–	20	
Silane, *see* Silicon tetrahydride										
Silica, Amorphous										
Total inhalable dust	–	$10^{(f)}$	–	–		–	6	–	–	
Respirable dust	–	$3^{(f)}$	–	–		–	2.4	–	–	
Precipitated	–	10	–	–		–	–	–	–	
Silica fume (Respirable)	–	2	–	–		–	–	–	–	
Silica fused (Respirable)	–	0.1	–	–		–	0.08	–	–	
Silica, Crystalline (Respirable)										
Cristobalite	–	0.05	–	–						
Quartz	–	$0.05^{(e)}$	–	–		–	0.3	–	–	
Tridymite	–	0.05	–	–						
Tripoli	–	0.1	–	–						
Silicon	–	$10^{(f)}$	–	–		–	10 (total inhalable dust)	–	–	
						–	4 (respirable dust)	–	–	
Silicon carbide	–	$10^{(f)}$	–	–		–	10 (total inhalable dust)	–	–	
						–	4 (respirable dust)	–	–	
Silicon tetrahydride	5	6.6	–	–		0.5	0.67	1	1.3	

Table 5.12 Cont'd

Substance		Threshold limit values (USA) TWA (ppm)[a]	Threshold limit values (USA) TWA (mg/m³)[b]	Threshold limit values (USA) STEL (ppm)[a]	Threshold limit values (USA) STEL (mg/m³)[b]		Occupational exposure limits (UK) TWA[c] (ppm)	Occupational exposure limits (UK) TWA[c] (mg/m³)	Occupational exposure limits (UK) STEL[c] (ppm)	Occupational exposure limits (UK) STEL[c] (mg/m³)	Air odour threshold (ppm; v/v)
Silver											
Metal		–	0.1	–	–		–	0.1	–	–	
Compounds as Ag		–	0.01	–	–		–	0.01	–	–	
Soapstone											
Respirable dust		–	3	–	–		–	–	–	–	
Total dust		–	6[i]	–	–		–	–	–	–	
Sodium azide											
as sodium azide		–	–	–	C0.29		–	–	–	0.3	
as hydrazoic acid vapour		–	–	C0.11	–						
Sodium bisulphite		–	5	–	–		–	5	–	–	
Sodium 2,4-dichloro-phenoxyethyl sulphate, *see* Sesone											
Sodium fluoroacetate	SK	–	0.05	–	0.15	SK	–	0.05	–	0.15	
Sodium hydroxide		–	–	–	C2		–	–	–	2	
Sodium metabisulphite, *see* Disodium disulphite											
Starch		–	10	–	–		–	10 (total inhalable dust)	–	–	
							–	4 (respirable dust)	–	–	
Stearates		–	10	–	–		–	–	–	–	
Stibine, *see* Antimony hydride											
Stoddard solvent		100	525	–	–		–	–	–	–	
Strontium chromate		–	0.0005[e]	–	–						
Strychnine		–	0.15	–	–		–	0.15	–	0.45	
Styrene, monomer	SK	20	–	40	–		100	430 MEL	250	1080 MEL	0.32
Subtilisins (proteolitic enzymes as 100% pure crystalline enzyme)		–	–	–	C0.00006		–	0.00006	–	0.000 06	
Sucrose		–	10	–	–		–	10	–	20	
Sulfotep	SK	–	0.2	–	–	SK	–	0.2	–	–	
Sulfometuron methyl		–	5	–	–						
Sulphur dioxide		2	5.2	5	13		2	5	5	13	1.1
Sulphur hexafluoride		1000	5970	–	–		1000	6070	1250	7590	
Sulphuric acid		–	1[e]	–	3		–	1	–	–	
Sulphur monochloride, *see* Disulphur dichloride											
Sulphur pentafluoride		–	–	C0.01	–		0.025	0.25	0.075	0.79	
Sulphur tetrafluoride		–	–	C0.1	–		0.1	0.4	0.3	1.2	
Sulphuryl fluoride		5	21	10	42		5	21	10	42	
Sulprofos		–	1	–	–		–	–	–	–	
Systox, *see* Demeton											
2,4,5-T		–	10	–	–		–	10	–	20	

Synthetic vitreous fibres											
Continuous filament glass fibres											
Respirable		–	1f/cc(h)	–	–						
Inhalable		–	5mg/m^3								
Glass wool fibres		–	1f/cc(h)								
Rock wool fibres		–	1f/cc(h)								
Stag wool fibres		–	1f/cc(h)								
Special purpose glass fibres		–	1f/cc(h)								
Talc (containing no asbestos fibres)		–	2 (respirable dust)	–	–		–	1 (respirable dust)	–	–	
Talc (containing asbestos fibres)			Use asbestos TLV-TWA								
Tantalum metal and oxide dusts		–	5	–	–		–	5	–	10	
TEDP, *see* Sulfotep											
Tellurium and compounds, as Te		–	0.1	–	–		–	0.1 (except hydrogen telluride)	–	–	
Tellurium hexafluoride as Te		0.02	0.10	–	–		–	–	–	–	
Temephos		–	10	–	–		–	–	–	–	
TEPP	SK	0.004	0.047	–	–	SK	0.004	0.05	0.01	0.12	
Terephthalic acid		–	10								
Terphenyls		–	–	–	C5		–	–	0.5	5	
1,1,2,2-Tetrabromoethane		–	–	–	–	SK	0.5	7	–	–	
1,1,1,2-Tetrachloro-2,2-difluoroethane		500	4170	–	–		100	847	100	847	
1,1,2,2-Tetrachloro-1,2-difluoroethane		500	4170	–	–		100	847	100	847	
1,1,2,2-Tetrachloroethane	SK	1	6.9	–	–		–	–	–	–	1.5
Tetrachloroethylene, *see* Perchloroethylene											
Tetrachloromethane, *see* Carbon tetrachloride											
Tetrachloronaphthalene		–	2	–	–		–	2	–	4	
Tetraethyl lead as Pb	SK	–	0.1	–	–		*SEE* NOTE (K)				
Tetraethyl orthosilicate, *see* Ethyl silicate											
Tetrafluoroethylene		2	–	–	–						
Tetrahydrofuran		200	590	250	737	100	300	200	599	2.0	
Tetramethyl lead as Pb	SK	–	0.15	–	–	*SEE* NOTE (K)					
Tetramethyl orthosilicate, *see* Methyl silicate											
Tetramethyl succinonitrile	SK	0.5	2.8	–	–	SK	0.5	2.8	2	11	
Tetranitromethane		0.005	–	–	–		–	–	–	–	
Tetrasodium pyrophosphate											
anhydride		–	5	–	–						
decahydrate		–	5	–	–						

Table 5.12 Cont'd

Substance		*Threshold limit values (USA)* TWA (ppm)[a]	TWA (mg/m³)[b]	STEL (ppm)[a]	STEL (mg/m³)[b]		*Occupational exposure limits (UK)* TWA[c] (ppm)	TWA[c] (mg/m³)	STEL[c] (ppm)	STEL[c] (mg/m³)	*Air odour threshold* (ppm; v/v)
Tetryl		–	2	–	–	SK	–	1.5	–	3	
Thallium, soluble compounds, as Tl	SK	–	0.1	–	–	SK	–	0.1	–	–	
4,4′-Thiobis(6-tert-butyl-*m*-cresol)		–	10	–	–		–	10	–	–	
Thioglycolic acid	SK	1	3.8	–	–		1	3.8	–	–	
Thionyl chloride		–	–	C1	C4.9		–	–	1	5	
Thiram		–	1	–	–		–	5	–	10	
Tin											
Metal		–	2	–	–		–	–	–	–	
Oxide and inorganic compounds except SnH_4, as Sn		–	2	–	–		–	2	–	4	
Organic compounds as Sn	SK	–	0.1	–	0.2	SK	–	0.1 (except Cyhexin)	–	0.2	
Titanium dioxide		–	10[i]	–	–		–	10 (total inhalable dust)			
							–	4 (respirable dust)	–	–	
Toluene		50	191	150	565	SK	50	191	150	574	2.9
Toluene-2,4-diisocyanate (TDI)		0.005	0.036	0.02	0.14	*see* Isocyanates					
p-Toluenesulphonyl chloride		–	–	–	–		–	–	–	5	0.17
o-Toluidine	SK	2[e]	8.8[e]	–	–		0.2	0.9	–	–	0.25
m-Toluidine	SK	2	8.8	–	–		–	–	–	–	
p-Toluidine	SK	2[e]	8.8[e]	–	–		–	–	–	–	
Toluol, *see* Toluene											
Toxaphene, *see* Chlorinated camphene											
Tributyl phosphate		0.2	2.2	–	–		–	5	–	5	
Trichloroacetic acid		1	6.7	–	–						
1,2,4-Trichlorobenzene		–	–	C5	C37		1	7.6	–	–	1.4
1,1,1-Trichlorobis (chlorophenyl)ethane		–	–	–	–		–	1	–	3	
1,1,1-Trichloroethane, *see* Methyl chloroform											
1,1,2-Trichloroethane	SK	10	55	–	–						
Trichloroethylene		50	269	100	535	SK	100	550 MEL	150	820 MEL	28
Trichlorofluoromethane		–	–	C1000	C5710		1000	5710	1250	7140	5
Trichloromethane, *see* Chloroform											
Trichloronaphthalene	SK	–	5	–	–		–	–	–	–	
Trichloronitromethane, *see* Chloropicrin											
1,2,3-Trichloropropane	SK	10	60	–	–		50	306	75	460	
1,1,2-Trichloro-1,2,2-trifluoroethane		1000	7670	1250	9590		1000	7790	1250	9740	45

Tricyclohexyltin hydroxide, *see* Cyhexatin										
Tridymite, *see* Silica, crystalline										
Triethanolamine		–	5	–	–					
Triethylamine		1	3	15	62	10	42	15	63	0.48
Trifluorobromomethane, *see* Bromotrifluoromethane										
1,3,5-Triglycidyl-s-triazinetrione		–	0.05	–	–					
Triglycidyl iso-cyanurate		–	–	–	–	–	0.1	–	–	
Trimellitic anhydride, *see* Benzene-1,2,4-tricarboxylic acid 1,2-anhydride										
Trimethylamine		5	12	15	36	10	25	15	37	0.00044
Trimethyl benzene		25	123	–	–	25	125	–	–	0.55
3,5,5-Trimethylcyclohex-2-enone		–	–	–	–	–	–	5	29	
Trimethyl phosphite		2	10	–	–	2	10	–	–	0.00010
2,4,6-Trinitrophenol, *see* Picric acid										
2,4,6-Trinitrophenylmethylnitramine, *see* Tetryl										
2,4,6-Trinitrotoluene (TNT)	SK	–	0.1	–	–	–	0.5	–	–	
Triorthocresyl phosphate	SK	–	0.1	–	–	–	0.1	–	0.3	
Triphenyl amine		–	5	–	–	–	–	–	–	
Triphenyl phosphate		–	3	–	–	–	3	–	6	
Tripoli, *see* Silica, crystalline										
Tungsten as W										
Insoluble compounds		–	5	–	10	–	5	–	10	
Soluble compounds		–	1	–	3	–	1	–	3	
Turpentine		100	566	–	–	100	566	150	850	
Uranium (natural)										
Soluble and insoluble compounds as U		–	0.2[g]	–	0.6[g]	–	0.2	–	0.6	
n-Valeraldehyde		50	176	–	–	–	–	–	–	0.028
Vanadium pentoxide as V_2O_5										
Respirable dust and fume		–	0.05	–	–	–	0.5 (total inhalable dust)	–	–	
						–	0.04 (Fume and respirable dust) as V	–	–	
Vegetable oil mists		–	10	–	–	–	–	–	–	
Vinyl acetate		10	35	15	52	10	36	20	72	0.50
Vinyl benzene, *see* Styrene										
Vinyl bromide		0.5[e]	2.2[e]	–	–	5	20	–	–	
Vinyl chloride		1[g]	2.6[g]	–	–	7[l]	– MEL	–	–	3000
Vinyl cyanide, *see* Acrylonitrile										
Vinyl cyclohexene dioxide	SK	0.1	0.6	–	–	–	–	–	–	
Vinyl fluoride		1[e]	–	–	–					

Table 5.12 Cont'd

Substance		*Threshold limit values (USA)* TWA (ppm)[a]	(mg/m³)[b]	STEL (ppm)[a]	(mg/m³)[b]		*Occupational exposure limits (UK)* TWA[c] (ppm)	(mg/m³)	STEL[c] (ppm)	(mg/m³)	*Air odour threshold* (ppm; v/v)
Vinylidene chloride		5	20	20	79		10	40 MEL	–	–	190
Vinylidene fluoride		500	–	–	–						
Vinyl toluene		50	242	100	483		–	–	–	–	10
VM & P naphtha		300	1370	–	–		–	–	–	–	
Warfarin		–	0.1	–	–		–	0.1	–	0.3	
Welding fumes (NOC)		–	5[i]	–	–		–	5	–	–	
White spirit		–	–	–	–		*see* EH 40 for calculation				
Wood dust (certain hardwoods e.g. beech, oak)		–	1[g]	–	–	SEN	–	5 MEL	–	–	
Softwood		–	5	–	10		–	5 MEL	–	–	
Wool process dust		–	–	–	–		–	10	–	–	
Xylene											
o-,m-,p-isomers		100	434	150	651	SK	100	441	150	662	1.1 (meta)
m-Xylene α, α′-diamine	SK	–	–	–	C0.1		–	–	–	–	
Xylidine, mixed isomers	SK	0.5[e]	2.5[e]	–	–	SK	2	10	10	50	0.056 (2,4-Xylidine)
Yttrium, metal and compounds, as Y			–	–	–		–	1	–	3	
Zinc chloride fume		–	1	–	2		–	1	–	2	
Zinc chromates as Cr		–	0.01[g]	–	–		–	–	–	–	
Zinc distearate											
Total inhalable dust		–	–	–	–		–	10	–	20	
Respirable dust		–	–	–	–		–	4	–	–	
Zinc oxide											
Fume		–	5	–	10		–	5	–	10	
Dust		–	10	–	–		–	–	–	–	
Zirconium and compounds, as Zr		–	5	–	10		–	5	–	10	

C Ceiling limit
MEL Maximum exposure limit
NOC Not otherwise classified
SEN Capable of causing respiratory sensitization; skin sensitizers have not been given a separate notation
SK Can be absorbed through skin

This table is a useful guide but because standards are continually under review and many caveats apply reference should be made to the most recent edition of HSE EH 40 and the AGGIH TLV list for current values and their interpretation. Often carcinogens are not assigned a hygiene standard.

[a] Parts of vapour or gas per million parts of contaminated air by volume at 25°C and 760 torr (1.013 bar).
[b] Milligrams of substance per m³ air.
[c] Value shown is OES unless otherwise indicated as MEL.

(d) Simple asphyxiant. Some gases and vapours present at high concentrations act as asphyxiants by reducing the oxygen content of air. Many of these are odourless and colourless. Many also pose a fire or explosion risk, often at values below which asphyxiation can occur. (Although capable of asphyxiation, they are not considered to be substances hazardous to health under COSHH.)

(e) Suspected human carcinogens – *see* TLV Appendix A, Category A2 (below).

(f) The value is for total dust containing no asbestos and <1% crystalline silica.

(g) Confirmed human carcinogen – *see* TLV Appendix, A, Category A1 (below).

(h) Fibres longer than 5 μm and with an aspect ratio ≥3:1 as determined by the membrane filter method at 400–450X magnification (4 mm objective) phase contrast illumination.

(i) Welding fumes cannot be classified simply. The composition and quantity of both are dependent on the alloy being welded and the process and electrodes used. Reliable analysis of fumes cannot be made without considering the nature of the welding process and system being examined; reactive metals and alloys such as aluminium and titanium are arc-welded in a protective inert atmosphere such as argon. These arcs create relatively little fume, but they do create an intense radiation which can produce ozone. Similar processes are used to arc-weld steels, also creating a relatively low level of fumes. Ferrous alloys also are arc-welded in oxidizing environments that generate considerable fume and can produce carbon monoxide instead of ozone. Such fumes generally are composed of discrete particles of amorphous slags containing iron, manganese, silicon, and other metallic constituents depending on the alloy system involved. Chromium and nickel compounds are found in fumes when stainless steels are arc-welded. Some coated and flux-cored electrodes are formulated with fluorides and the fumes associated with them can contain significantly more fluorides than oxides. Because of the above factors, arc-welding fumes frequently must be tested for individual constituents that are likely to be present to determine whether specific TLVs are exceeded. Conclusions based on total fume concentration are generally adequate if no toxic elements are present in welding rod, metal, or metal coating and conditions are not conducive to the formation of toxic gases.

Most welding, even with primitive ventilation, does not produce exposures inside the welding helmet above 5 mg/m^3. That which does, should be controlled.

(j) UK control limits for asbestos:

Chrysotile	0.3 fibres/ml of air averaged over any continuous 4 hr period
	0.9 fibres/ml of air averaged over any continuous 10 min period
Any other form of asbestos,	0.2 fibres/ml of air averaged over any continuous 4 hr period
alone or in mixtures	0.6 fibres/ml of air averaged over any continuous 10 min period

Action levels for cumulative exposures within a 12 week period:

(*a*) for chrysotile, 72 fibre hours/ml of air

(*b*) for any other form of asbestos, alone or in mixtures, 48 fibre hours/ml of air

(*c*) for both types of exposure at different times within the period, a proportionate number of fibre hours/ml

The lack of limits should not be taken to imply an absence of hazard. In the absence of a specific OEL for a particular dust, exposure should be adequately controlled and where there is no indication of the need for a lower value, personal exposure should be kept below both 10 mg/m^3 8 hr TWA total inhalable dust and 4 mg/m^3 8 hr TWA respirable dust.

(k) UK limits for lead are 8 hr TWA concentrations as follows:

Lead other than lead alkyls	–	0.15 mg/m^3 of air
Lead alkyls	–	0.1 mg/m^3 of air

These are ceiling values that must not be exceeded when calculating 8 hr TWA. They should be read in conjunction with biological limits for lead.

(l) As measured by the vertical elutriator cotton-dust samples.

(m) Polytetrafluoroethylene decomposition products: thermal decomposition of the fluorocarbon chain in air leads to the formation of oxidized products containing carbon, fluorine and oxygen. Because these products decompose in part by hydrolysis in alkaline solution, they can be quantitatively determined in air as fluoride to provide an index of exposure. No TLV is recommended pending determination of the toxicity of the products, but air concentration should be minimal. (Trade names: Algoflon, Fluon, Teflon, Tetran.)

(n) In the UK vinyl chloride is also subject to an overriding annual maximum exposure limit of 3 ppm.

(o) As sampled by a method that does not collect vapour.

TLV Appendix A: Carcinogens (excerpts)

The Chemical Substances Threshold Limit Values Committee classifies certain substances found in the occupational environment as either confirmed or suspected human

Table 5.12 Cont'd

carcinogens. The present listing of substances that have been identified as carcinogens takes two forms: those for which a TLV has been assigned and those for which environmental and exposure conditions have not been sufficiently defined to assign a TLV. Where a TLV has been assigned, it does not necessarily imply the existence of a biological threshold; however, if exposures are controlled to this level, we would not expect to see a measurable increase in cancer incidence or mortality.

Two categories of carcinogens are designated:

A1 – *Confirmed Human Carcinogens.* Substances, or substances associated with industrial process, recognized to have carcinogenic potential.
A2 – *Suspected human carcinogens.* Chemical substances, or substances associated with industrial process, which are suspect of inducing cancer, based on their limited epidemiological evidence or demonstration of carcinogenesis in one or more animal species by appropriate methods.

Exposures to carcinogens must be kept to a minimum. Workers exposed to A1 carcinogens without a TLV should be properly equipped to eliminate to the fullest extent possible all exposure to the carcinogen. For A1 carcinogens with a TLV and for A2 carcinogens, worker exposure by all routes should be carefully controlled to levels *as low as reasonably achievable* (ALARA) below the TLV.

Threshold limit values (TLV)

These values represent airborne concentrations of substances to which it is believed that nearly all workers may be repeatedly exposed by inhalation day after day without adverse health effects. Because of the wide variation in individual susceptibility, however, a small percentage of workers may experience discomfort from some substances at concentrations below the TLV. A smaller percentage may experience aggravation of a pre-existing condition/illness. Age, genetic factors or personal habits may make some individuals hypersensitive. Physical factors, e.g. UV, ionizing radiation, humidity, abnormal atmospheric pressure (altitude), excessive temperatures, or overtime working may add stress to the body so that effects from exposure at a TLV may be altered. Therefore best occupational hygiene practice is to maintain levels of all airborne contaminants as low as is reasonably practicable.

There are three categories of TLV:

1 *Time-weighted average (TWA) TLV* – the time-weighted average concentration for a normal 8 hr work day and a 40 hr work week, to which it is believed that nearly all workers may be repeatedly exposed, day after day, without untoward effects. TWA TLVs permit excursions above the TLV provided that they are compensated for by equivalent excursions below the TLV during the work day. The excursion above the TLV is a rule of thumb, as explained in the source reference.
2 *Short-term exposure limit (STEL) TLV* – the concentration to which it is believed that workers can be exposed continuously for a short period of time without suffering from irritation, chronic or irreversible tissue damage, or narcosis of sufficient degree to increase the likelihood of accidental injury, impair self-rescue or materially reduce work efficiency, and provided that the daily TWA limit is not exceeded. A STEL is a 15 min TWA exposure which should not be exceeded at any time during the work day even if the TWA is within the TLV. It should not occur more than four times per day or without at least one hour between successive exposures.
3 Ceiling TLV(C) – the concentration that should not be exceeded during any part of the working exposure. If instantaneous monitoring is not possible then the TLV(C) can be assessed over a 15 min sampling period, except for those substances that may cause immediate irritation when exposures are short.

Occupational exposure limits (OEL)

Occupational exposure limits for airborne contaminants are reviewed annually in the UK by the Health and Safety Executive. They are published as Guidance Note EH 40/-. The two types of exposure limit are:

- *Long-term exposure limits.* These are concerned with the total intake of a contaminant (or contaminants) over a long period. As such they are appropriate for protecting workers against the effects of long-term exposure, or reducing the risks to an insignificant level.
- *Short-term exposure limits.* These are concerned primarily with the avoidance of acute effects, or reducing the risk of the occurrence.

Long-term and short-term limits are expressed as time-weighted average concentrations. For a long-term limit the normal period is eight hours (although for vinyl chloride it is 1 year); for a short-term limit the normal period is 15 minutes.

Specific short-term exposure limits are listed by the HSE for those chemicals which pose a risk of acute effects such as eye irritation from brief exposures. For other chemicals a recommended guideline for controlling short-term excursions is to restrict them to 3 × long-term exposure limit averaged over a 15 min period.

Percutaneous absorption

For most chemicals, inhalation is the main route of entry into the body. Penetration via damaged skin (e.g. cuts, abrasions) should, however, be avoided. Certain chemicals (e.g. phenol, aniline, certain pesticides) can penetrate intact skin and so become absorbed into the body. This may occur through local contamination, e.g. from a liquid splash, or through exposure to high vapour concentrations. Special precautions to avoid skin contact are required with these chemicals and potential exposure via skin absorption has to be taken into account when assessing the adequacy of control measures.

Some chemicals able to penetrate intact skin are listed in Table 5.2.

MEL/OES

In the UK (under the Control of Substances Hazardous to Health Regulations 1999) there are maximum exposure limits (MEL) and occupational exposure standards (OES):

- *Maximum exposure limit* (MEL) – the maximum concentration of an airborne substance, averaged over a reference period, to which employees may be exposed by inhalation under any circumstance. Thus, exposure to a chemical assigned an MEL must be as low as is reasonably practicable and, in any case, below the MEL.
- *Occupational exposure standard* (OES) – the concentration of an airborne substance, averaged over a reference period, at which, according to current knowledge, there is no evidence that it is likely to be injurious to employees if they are exposed by inhalation, day after day, to that concentration. Exposure to a chemical assigned an OES must be at no greater than that concentration but there is, under the UK Control of Substances Hazardous to Health Regulations, no duty to reduce it further. (Also under COSHH, if the OES is in fact exceeded but an employer has identified the reasons, and is taking appropriate action to remedy it as soon as reasonably practicable, control may be regarded as being adequate.) However,
  (a) exposure should preferably be reduced below the OES to ensure that the exposure of all personnel does not exceed it;
  (b) further application of good occupational hygiene principles to reduce exposure below the OES should not be discouraged.

The application of standards

Caution is necessary in the application of control limits because of:

- The effects of mixtures of chemicals, e.g. additive or synergistic.
- The effects of extended working hours, exertion etc.
- The variation between workers in susceptibility to the effects of chemical exposures, e.g. inherently or due to a pre-existing medical condition.
- Changes in limits with increasing knowledge of toxicology.
- With some asthmagens, respiratory sensitization may occur at atmospheric concentrations below published limits.

In any event, hygiene standards:

(a) should not be used as an index of relative hazard or toxicity;
(b) are not well-defined dividing lines between ‘safe’ and ‘dangerous’ airborne concentrations;
(c) cannot be readily extrapolated to assess long-term, non-occupational exposures, e.g. to environmental pollutants.

Calculation of exposure

(a) *Single substances.* Hygiene standards are quoted for pure substances. The 8 hr TWA is best assessed by personal dosimetry (Chapter 10) in which exposure is continuously monitored throughout the work day wherever the operator goes. When data are available on the individual fluctuations in exposure, e.g. from a variety of tasks, the 8 hr TWA exposure can be calculated as in the following example:

Working period	*Exposure* (mg/m^3)
8.00–10.30	0.16
10.30–10.45	0.00
10.45–12.45	0.07
12.45–13.30	0.00
13.30–15.30	0.42
15.30–15.45	0.00
15.45–17.15	0.21

$$\text{8 hr TWA exposure} = \frac{0.16 \times 2.5 + 0.07 \times 2 + 0.42 \times 2 + 0.21 \times 1.5 + 0 \times 1.25}{8}$$

$$= \frac{0.40 + 0.14 + 0.84 + 0.32}{8}$$

$$= 0.21\ \text{mg/m}^3$$

(b) *Mixtures.* Often working practices result in exposures to mixtures of chemicals. The individual components of the mixture can act on the body independently, additively, synergistically or antagonistically. The ACGIH publication on TLVs and a Chemicals Industries Association booklet (*see* Bibliography) provide elementary advice on calculating hygiene standards for mixtures.

For compounds behaving additively, the relationship for compliance with the TLV of the mixture is given by

$$\frac{C_1}{T_1} + \frac{C_2}{T_2} + \frac{C_3}{T_3} + \ldots = 1$$

where C_1, C_2 and C_3 are the concentrations of the different components and T_1, T_2 and T_3 are the TLVs for each pure component.

Example. Air contains 200 ppm acetone (TLV = 750), 300 ppm sec-butyl acetate (TLV = 200) and 200 ppm of methyl ethyl ketone (TLV = 200):

$$\frac{\text{concentration}}{\text{TLV of mixture}} = \frac{200}{750} + \frac{300}{200} + \frac{200}{200} = 0.26 + 1.5 + 1 = 2.76$$

i.e. the TLV has been exceeded.

For independent effects, the relationship for compliance with the TLV of the mixture is given by:

$$\frac{C_1}{T_1} = 1, \quad \frac{C_2}{T_2} = 1, \quad \frac{C_3}{T_3} = 1 \text{ etc.}$$

Example. Air contains 0.10 mg/m^3 lead (TLV 0.15 mg/m^3) and 0.9 mg/m^3 sulphuric acid (TLV 1 mg/m^3):

$$\frac{0.10}{0.15} = 0.7 \quad \frac{0.9}{1.0} = 0.9$$

i.e. the TLV has not been exceeded.

For chemicals behaving antagonistically or synergistically, expert advice from a hygienist and toxicologist should be sought.

Biological exposure indices

Exposure levels to certain chemicals can be monitored by determination of levels of the substance or its metabolites in exhaled breath or in body tissues or fluids such as urine, blood, hair, nails etc. For example, blood lead levels have long been used to determine operator exposure to this chemical. Alternatively, exposure to some chemicals can be gauged by measurement of modifications to critical biochemical constituents, e.g. changes in activity of a key enzyme, or physiological changes, e.g. vitalograph measurement. The advantage of biological monitoring over environmental analysis is that the former measures the actual amount of substance absorbed into the body and reflects the worker's individual responses and overall exposure. Shortcomings, however, include wide variation in individual responses to a given chemical; the unpopularity of invasive techniques; and, most importantly, the fact that it reflects a reaction to an exposure that has already occurred.

Biological exposure indices (BEI) published by the ACGIH are given in Table 5.13. BEIs represent the levels of determinant which are most likely to be observed in specimens collected from a healthy worker who has been exposed to chemicals to the same extent as a worker with inhalation exposure to the TLV. Due to biological variability it is possible for an individual's measurements to exceed the BEI without incurring increased health risk. If, however, levels in specimens obtained from a worker on different occasions persistently exceed the BEI, or if the majority of levels in specimens obtained from a group of workers at the same workplace exceed the BEI, the cause of the excessive values must be investigated and proper action taken to reduce the exposure.

BEIs apply to 8 hr exposures, five days a week. However, BEIs for altered working schedules can be extrapolated on pharmacokinetic and pharmacodynamic bases. BEIs should not be applied, either directly or through a conversion factor, to the determination of safe levels for non-occupational exposure to air and water pollutants, or food contaminants. The BEIs are not intended for use as a measure of adverse effects or for diagnosis of occupational illness.

There is also a framework for the use of biological monitoring in the UK. A biological monitoring guidance value has been set where it is likely to be of practical value, suitable monitoring methods exist and sufficient data are available.

Two types of value are quoted in Table 5.14:

- A Health Guidance Value, set at a level at which there is no indication from the scientific evidence available that the substance is likely to be injurious to health.
- A Benchmark Guidance value, set as a hygiene-based guidance value – actually a level which 90% of available validated data are below and therefore achievable by the great majority of industry with good workplace practices.

Table 5.13 Adopted biological exposure determinants

Chemical determinant	*Sampling time*	*Biological exposure indices*	*Notation*
ACETONE			
Acetone in urine	End of shift	50 mg/l	Ns
ACETYLCHOLINESTERASE INHIBITING PESTICIDES			
Cholinesterase activity in red blood cells	Discretionary	70% of individuals baseline	Ns
ANILINE			
Total *p*-aminophenol in urine	End of shift	50 mg/g creatinine	Ns
Methemoglobin in blood	During or end of shift	1.5% of haemoglobin	B, Ns, Sq
ARSENIC, ELEMENTAL AND SOLUBLE INORGANIC COMPOUNDS			
Inorganic arsenic plus methylated metabolites in urine	End of workweek	35 μg/g As/l	B
BENZENE			
S-Phenylmercapturic acid in urine	End of shift	25 μg/g creatinine	B
t,t-Muconic acid in urine	End of shift	500 μg/g creatinine	B
CADMIUM AND INORGANIC COMPOUND			
Cadmium in urine	Not critical	5 μg/g creatinine	B
Cadmium in blood	Not critical	5 μg/l	B
CARBON DISULPHIDE			
2-Thiothiazolidine-4-carboxylic acid (TTCA) in urine	End of shift	5 mg/g creatinine	
CARBON MONOXIDE			
Carboxyhaemoglobin in blood	End of shift	3.5% of haemoglobin	B, Ns
Carbon monoxide in end-exhaled air	End of shift	20 ppm	B, Ns
CHLOROBENZENE			
Total 4-chlorocatechol in urine	End of shift	150 mg/g creatinine	Ns
Total *p*-chlorophenol in urine	End of shift	25 mg/g creatinine	Ns
CHROMIUM (VI), water-soluble fume	Increase during shift	10 μg/g creatinine	B
Total chromium in urine	End of shift at end of workweek	30 μg/g creatinine	B
COBALT			
Cobalt in urine	End of shift at end of workweek	15 μg/l	B
Cobalt in blood	End of shift at end of workweek	1 μg/l	B, Sq
N,N-DIMETHYLACETAMIDE			
N-Methylacetamide in urine	End of shift at end of workweek	30 mg/g creatinine	
N,N-DEMETHYLFORMAMIDE (DMF)			
N-Methylformamide in urine	End of shift	15 mg/l	
N-Acetyl-*S*-(*N*-methyl-carbamoyl) cysteine in urine	Prior to last shift of workweek	40 mg/l	Sq

Table 5.13 Cont'd

Chemical determinant	*Sampling time*	*Biological exposure indices*	*Notation*
2-ETHOXYETHANOL (EGEE) and 2-ETHOXYETHYL ACETATE (EGEEA)			
2-Ethoxyacetic acid in urine	End of shift at end of workweek	100 mg/g creatinine	
ETHYL BENZENE			
Mandelic acid in urine	End of shift at end of workweek	1.5 g/g creatinine	Ns
Ethyl benzene in end-exhaled air			Sq
FLUORIDES			
Fluorides in urine	Prior to shift	3 mg/g creatinine	B, Ns
	End of shift	10 mg/g creatinine	B, Ns
FURFURAL			
Total furoic acid in urine	End of shift	200 mg/g creatinine	B, Ns
n-HEXANE			
2,5-Hexanedione in urine	End of shift	5 mg/g creatinine	Ns
n-Hexane in end-exhaled air			Sq
LEAD (*see* note below)			
Lead in blood	Not critical	30 μg/100 ml	
MERCURY			
Total inorganic mercury in urine	Preshift	35 μg/g creatinine	B
Total inorganic mercury in blood	End of shift at end of workweek	15 μg/l	B
METHANOL			
Methanol in urine	End of shift	5 mg/l	B, Ns
METHEMOGLOBIN INDUCERS			
Methemoglobin in blood	During or end of shift	1.5% of haemoglobin	B, Ns, Sq
2-METHOXYETHANOL (EGME) and 2-METHOXYETHYL ACETATE (EGMEA)			
2-Methoxyacetic acid in urine	End of shift at end of workweek		Nq
METHYL CHLOROFORM			
Methyl chloroform in end-exhaled air	Prior to last shift of workweek	40 ppm	
Trichloroacetic acid in urine	End of workweek	10 mg/l	Ns, Sq
Total trichloroethanol in urine	End of shift at end of workweek	30 mg/l	Ns, Sq
Total trichloroethanol in blood	End of shift at end of workweek	1 mg/l	Ns
4,4′–METHYLENE BIS(2-CHLOROANILINE) (MBOCA)			
Total MBOCA in urine	End of shift		Nq

Table 5.13 Cont'd

Chemical determinant	*Sampling time*	*Biological exposure indices*	*Notation*
METHYL ETHYL KETONE (MEK)			
MEK in urine	End of shift	2 mg/l	
METHYL ISOBUTYL KETONE (MIBK)			
MIBK in urine	End of shift	2 mg/l	
NITROBENZENE			
Total *p*-nitrophenol in urine	End of shift at end of workweek	5 mg/g creatinine	Ns
Methemoglobin in blood	End of shift	1.5% of haemoglobin	B, Ns, Sq
PARATHION			
Total *p*-nitrophenol in urine	End of shift	0.5 mg/g creatinine	Ns
Cholinesterase activity in red cells	Discretionary	70% of individual's baseline	B, Ns, Sq
PENTACHLOROPHENOL (PCP)			
Total PCP in urine	Prior to last shift of workweek	2 mg/g creatinine	B
Free PCP in plasma	End of shift	5 mg/l	B
PERCHLOROETHYLENE			
Perchloroethylene in end-exhaled air	Prior to last shift of workweek	5 ppm	
Perchloroethylene in blood	Prior to last shift of workweek	0.5 mg/l	
Trichloroacetic acid in urine	End of shift at end of workweek	3.5 mg/l	Ns, Sq
PHENOL			
Total phenol in urine	End of shift	250 mg/g creatinine	B, Ns
STYRENE			
Manelic acid in urine	End of shift	800 mg/g creatinine	Ns
	Prior to next shift	300 mg/g creatinine	Ns
Phenylglyoxlic acid in urine	End of shift	240 mg/g creatinine	Ns
	Prior to next shift	100 mg/g creatinine	
Styrene in blood	End of shift	0.55 mg/l	Sq
	Prior to next shift	0.02 mg/l	Sq
TETRAHYDROFURAN			
Tetrahydrofuran in urine	End of shift	8 mg/l	
TOLUENE			
o-Cresol in urine	End of shift	0.5 mg/l	B
Hippuric acid in urine	End of shift	1.6 g/g creatinine	B, Ns
Toluene in blood	Prior to last shift of workweek	0.05 mg/l	
TRICHLOROETHYLENE			
Trichloroacetic acid in urine	End of workweek	100 mg/g creatinine	Ns
Trichloroacetic acid and trichlorethanol in urine	End of shift at end of workweek	300 mg/g creatinine	Ns
Free trichloroethanol in blood	End of shift at end of workweek	4 mg/l	Ns

Table 5.13 Cont'd

Chemical determinant	*Sampling time*	*Biological exposure indices*	*Notation*
Trichloroethylene in blood			Sq
Trichloroethylene in end-exhaled air			Sq
VANADIUM PENTOXIDE			
Vanadium in urine	End of shift at end of workweek	50 μg/g creatinine	Sq
XYLENES (technical-grade)			
Methylhippuric acids in urine	End of shift	1.5 g/g creatinine	

[1]Women of child-bearing potential, whose blood Pb exceeds 10 μg/dl are at risk of delivering a child with a blood Pb over the current Centres for Disease Control guidelines of 10 g/dl. If the blood Pb of such children remains elevated, they may be at increased risk of cognitive deficits. The blood Pb of these children should be closely monitored and appropriate steps should be taken to minimize the child's exposure to environmental lead. (CDC Preventing Lead Poisoning in Young Children, October 1991; *see* BEI and TLV documentation for lead.)

Notations

'B' = background

The determinant may be present in biological specimens collected from subjects who have not been occupationally exposed, at a concentration which could affect interpretation of the result. Such background concentrations are incorporated in the BEI value.

'Nq' = non-quantitative

Biological monitoring should be considered for this compound based on the review; however, a specific BEI could not be determined due to insufficient data.

'Ns' = non-specific

The determinant is non-specific, since it is also observed after exposure to other chemicals.

'Sq' = semi-quantitative

The biological determinant is an indicator of exposure to the chemical, but the quantitative interpretation of the measurement is ambiguous. These determinants should be used as a screening test if a quantitative test is not practical or as a confirmatory test if the quantitative test is not specific and the origin of the determinant is in question.

It is essential to consult the specific BEI documentation before designing biological monitoring protocols and interpreting BEIs.

(Biological limits are also in force for lead and its compounds under the Control of Lead at Work Regulations 1998; different blood lead action, and suspension from work, levels apply to women of reproductive capacity, young persons and other employees.

Biological monitoring of cadmium workers is also recommended; guidance on interpretation of results is given in EH1 – Cadmium: health and safety precautions.)

Odour thresholds

Some materials possess low odour thresholds: their smell gives warning of impending danger. Others possess odour thresholds well in excess of the hygiene standard. Examples are included in Table 5.12.

Reliance on the nose as an indicator, however, can be hazardous since:

- Untrained exposees may not understand the significance of an odour.
- Some materials with low odour thresholds may paralyse the olfactory nerves and cause the sense of smell to be lost within minutes (e.g. hydrogen sulphide).
- Some materials are odourless (e.g. nitrogen).
- Some materials, such as arsine, phosphine, toluene di-isocyanate and stibine, may be present in concentrations in excess of their hygiene standards yet undetectable by smell.
- Published odour threshold values vary widely from source to source.

Table 5.14 UK biological monitoring guidance values

Substance	*Biological monitoring guidance values*	*Monitoring schedule*
Butan-2-one	70 µmol butan-2-one/l in urine (HGV)	Post shift
2-Butoxyethanol	240 mmol butoxyacetic acid/mol creatinine in urine (HGV)	Post shift
N,N-Dimethylacetamide	100 mmol *N*-methyl acetamide/mol creatinine in urine (HGV)	Post shift
Carbon monoxide	30 ppm carbon monoxide in end-tidal breath (HGV)	Post shift
Dichloromethane	30 ppm carbon monoxide in end-tidal breath (HGV)	Post shift
Lindane (γ BHC (ISO))	35 nmol/l (10 ug/l) of Lindane in whole blood (equivalent to 70 nmol/l) of Lindane in plasma (BGV)	Random
MbOCA (2.2′-dichloro-4,4′-methylene dianiline)	15 µmol total MbOCA/mol creatinine in urine (BGV)	Post shift
Mercury	20 µmol mercury/mol creatinine in urine (HGV)	Random
4,4′-Methylenedianiline (MDA)	50 µmol total MDA/mol creatinine in urine (BGV)	Post shift for inhalation and pre-shift next day for dermal exposure
4-Methylpentan-2-one	20 µmol 4-methylpentan-2-one/l in urine (HGV)	Post shift

HGV = Health Guidance Value
BGV = Benchmark Guidance Value

- Workers may become acclimatized to a commonly-occurring odour, or be suffering temporarily from an impaired sense of smell, e.g. due to a cold.
- The odour of a toxic chemical may be masked by the odour from another substance, or a mixture.

Risk assessment of carcinogens

Arguably, risk assessment from exposure to carcinogens merits special consideration because of the low levels of exposure capable of producing an adverse response in certain individuals coupled with the often long time-lag (latency period) between exposure and onset of disease.

There are several formal lists of carcinogens. Thus, in the UK under the Control of Substances Hazardous to Health Regulations 1999 (see later) carcinogens are defined as:

- Any substance or preparation which if classified in accordance with the classification provided for by regulation 5 of the Chemicals (Hazard Information and Packaging for Supply) Regulations 1994 ('CHIPS') as amended would be in the category of danger, carcinogenic (category 1) or carcinogenic (category 2).
- Substances listed in Table 5.15.

Table 5.15 Substances and processes to which the definition of 'carcinogen' relates

Aflatoxins
Arsenic
Auramine manufacture
Calcining, sintering or smelting of nickel copper matte or acid leaching or electrorefining of roasted matte
Coal soots, coal tar, pitch and coal tar fumes
Hardwood dusts
Isopropyl alcohol manufacture (strong acid process)
Leather dust in boot and shoe manufacture, arising during preparation and finishing
Magenta manufacture
Mustard gas (β, β′-dichlorodiethyl sulphide)
Rubber manufacturing and processing giving rise to rubber process dust and rubber fume
Used engine oils

Substances or preparations requiring to be labelled with the risk phrase 'R45' (may cause cancer) or 'R49' (may cause cancer by inhalation) under 'CHIPS' are listed in Table 5.16 after the 5th edition of the Approved Supply List (Information approved for the classification and labelling of substances and preparations dangerous for supply). This list excludes certain coal and oil-based substances which attract the phrase 'R45' only when they contain a certain percentage of a marker substance (e.g. benzene).

A summary of the evaluation of carcinogenic risks to humans by the International Agency for Research on Cancer is given in Table 5.17 together with the IARC reference. IARC classify carcinogens into four groups thus:

Group 1: The agent (mixture) is carcinogenic to humans. The exposure circumstance entails exposures that are carcinogenic to humans.
Group 2: (two classifications):
Group 2A: The agent (mixture) is probably carcinogenic to humans. The exposure circumstance entails exposures that are probably carcinogenic to humans.
Group 2B: The agent (mixture) is possibly carcinogenic to humans. The exposure circumstance entails exposures that are possibly carcinogenic to humans.
Group 3: The agent (mixture, or exposure circumstance) is unclassifiable as to carcinogenicity in humans.
Group 4: The agent (mixture, or exposure circumstance) is probably not carcinogenic to humans.

Table 5.17 lists only Groups 1 and 2.

Risk control

Exposures to chemicals, resulting in toxic effects or oxygen-deficient atmospheres, may arise in a variety of industrial situations. A summary of common sources is given in Table 5.18: clearly this is not exhaustive since exposure may result whenever materials are mixed, machined, heated, dispersed or otherwise processed or used.

The precautions naturally vary in each case. For example, to avoid improper admixture of chemicals will require:

- Adequate training, instruction and supervision of workers.
- Identification of chemicals by name and code number.
- Segregated storage of incompatible substances.

Table 5.16 Substances assigned 'R45' or 'R49' risk phrases

Substances with the risk phrase 'R45' (may cause cancer)
Acrylamide
Acrylonitrile
5-Allyl-1,3-benzodioxide
4-Aminoazobenzene
4-Aminobiphenyl* (4-Aminodiphenyl*)
Salts of 4-Aminobiphenyl* (salts of 4-Aminodiphenyl*)
4-Amino-3-fluorophenol
Arsenic acid and its salts
Arsenic pentoxide
Arsenic trioxide
Asbestos (all types)
Azobenzene
Benzene
Benzidine*
Salts of benzidine*
Benzidine azo dyes (except those specified elsewhere in the Approved Supply List)
Benzo-(a)-anthracene
Benzo-(a)-pyrene
Benzo-(e)-pyrene
Benzo-(b)-fluoranthene
Benzo-(j)-fluoranthene
Benzo-(k)-fluoranthene
Beryllium compounds (except aluminium beryllium silicates)
Bis(chloromethyl)ether (BCME)
Butane [1], isobutane [2], containing ≥0.1% butadiene
Buta-1,3-diene
Cadmium chloride
Cadmium fluoride
Cadmium sulphate
Calcium chromate
Captafol (ISO)
Carbadox (INN)
Chloroaniline
2-Chloroallyl diethyldithiocarbamate (Sulfallate ISO)
Chlorodimethyl ether
1-Chloro-2,3-epoxypropane (epichlorohydrin)
Chromium III chromate (chromic chromate)
Chryene
CI Basic Red 9
CI Direct Black 38
CI Direct Blue 6
CI Direct Red 28
CI Disperse Blue 1
Clarified oils (petroleum), catalytic cracked
Clarified oils (petroleum), hydrodesulphurized catalytic cracked
Cobalt dichloride
Cobalt sulphate
Coke (coal tar), high temperature pitch
Coke (coal tar), mixed coal–high temperature pitch
Coke (coal tar), low temperature, high temperature pitch
Diaminotoluene
o-Dianisidine
Salts of *o*-dianisidine
o-Dianisidine-based azodyes
Diarsenic trioxide
Diazomethane
Dibenz(*a*,*h*)anthracene
1,2-Dibromo-3-chloropropane

Table 5.16 Cont'd

1,2-Dibromoethane (ethylene dibromide)
3,3′-Dichlorobenzidine
Salts of 3,3′-dichlorobenzidine
1,4-Dichlorobut-2-ene
1,2-Dichloroethane (ethylene dichloride)
2,2′-Dichloro-4,4′-methylenedianiline (MbOCA)
Salts of 2,2-dichloro-4,4′-methylenedianiline
1,3-Dichloro-2-propanol
1,2:3,4 Diepoxybutane
Diethyl sulphate
3,3′-Dimethylbenzidine (*o*-tolidine)
Salts of 3,3′-dimethylbenzidine (salts of *o*-tolidine)
Dimethylcarbamoyl chloride
1,2-Dimethylhydrazine
N,N-Dimethylhydrazine
Dimethylnitrosamine (*N*-nitroso dimethylamine)
Dimethylsulphamoyl chloride
Dimethyl sulphate
2,6-dinitrotoluene
1,2-diphenyl hydrazine
Disodium [5-((4′-((2.6-hydroxy-3-((2-hydroxy-5-sulphophenyl)azo)phenyl)azo)(1,1′-biphenyl)-4-yl)azo]salicylato(4)cuprate(2-)
Distillates (petroleum), intermediate vacuum
Distillates (petroleum), petroleum residues vacuum
Distillates (petroleum), chemically neutralized heavy paraffinic
Distillates (petroleum), hydrodesulphurized light catalytic cracked
Distillates (petroleum), hydrodesulphurized full-range middle
Distillates (petroleum), light paraffinic
Distillates (petroleum), light vacuum
Distillates (petroleum), vacuum
Distillates (petroleum), hydrodesulphurized middle coker
Distillates (petroleum), heavy naphthenic
Distillates (petroleum), heavy steam cracked
Distillates (petroleum), acid-treated light naphthenic
Distillates (petroleum), acid-treated light paraffinic
Distillates (petroleum), chemically neutralized light paraffinic
Distillates (petroleum), chemically neutralized heavy naphthenic
Distillates (petroleum), chemically neutralized light naphthenic
Distillates (petroleum), light catalytic cracked
Distillates (petroleum), intermediate catalytic cracked
Distillates (petroleum), light thermal cracked
Distillates (petroleum), light steam-cracked naphtha
Distillates (petroleum), cracked steam-cracked petroleum distillate
Distillates (petroleum), hydrodesulphurized thermal cracked middle
Distillates (petroleum), acid-treated heavy paraffinic
Distillates (petroleum), light catalytic cracked, thermally degraded
Distillates (petroleum), light naphthenic
Distillates (coal tar), benzole fraction
Distillates (coal tar), heavy oils
Distillates (petroleum), intermediate catalytic cracked, thermally degraded
Distillates (petroleum), acid treated heavy, naphthenic
Distillates (petroleum), heavy catalytic cracked
Distillates (petroleum), heavy thermal cracked
Distillates (petroleum), heavy paraffinic
Distillates (petroleum), hydrodesulphurized intermediate catalytic cracked
Distillates (petroleum), hydrodesulphurized heavy catalytic cracked
1,2-Epoxypropane
2,3 Epoxypropan-1-ol
Erionite

Table 5.16 Cont'd

Ethyleneimine
Ethylene oxide
Extracts (petroleum), heavy naphthenic distillate solvent
Extracts (petroleum), heavy paraffinic distillate solvent
Extracts (petroleum), light naphthenic distillate solvent
Extracts (petroleum), light paraffinic distillate solvent
Extracts (petroleum), light vacuum gas oil solvent
Fuel oil no. 6
Fuel oil, heavy, high sulphur
Fuel oil, residues–straight-run gas oils, high sulphur
Fuel oil, residual
Gas oils (petroleum), thermal cracked, hydrodesulphurized
Gas oils (petroleum), heavy atmospheric
Gas oils (petroleum), hydrodesulphurized coker heavy vacuum
Gas oils (petroleum), steam cracked
Gas oils (petroleum), hydrodesulphurized heavy vacuum
Gas oils (petroleum), heavy vacuum
Gas oils (petroleum), light vacuum, thermal-cracked hydrodesulphurized
Gas oils (petroleum), hydrotreated vacuum
Gas oils (petroleum), catalytic-cracked naphtha depropanizer overhead, C_3-rich acid-free
Gasoline, coal solvent extn, hydrocracked naphtha
Hexachlorobenzene
Hexamethylphosphoric triamide
Hydrazine
Salts of hydrazine
Hydrazine bis (3-carboxy-4-hydroxybenzene sulphonate)
Hydrazobenzene
Hydrocarbons C_{26-55}, aromatic-rich
Lead hydrogen arsenate
2-Methylaziridine
4,4′-Methylene dianiline (4,4′-diaminodiphenylmethane)
4,4′-Methylenedi-*o*-toluidine
Methyl acrylamidomethoxyacetate (containing ≥0.1% acrylamide)
Methyl acrylamidoglycolate (containing ≥0.1% acrylamide)
2-Methoxyaniline
4-Methyl-phenylenediamine (2,-4-toluenediamine)
1-Methyl-3-nitro-1-nitrosoguanidine
Methyl-onn-azoxymethyl acetate (Methyl azoxy methyl acetate)
2-Naphthylamine*
Salts of 2-naphthylamine*
5-Nitroacenaphthene
2-Nitroanisole
4-Nitrobiphenyl* (4-Nitrodiphenyl*)
Nitrofen (ISO)
2-Nitronaphthalene
2-Nitropropane
N-Nitrosodipropylamine
2,2′-(Nitrosoimino) bis ethanol
Petroleum gases, liquefied (various)
Potassium bromate
1,3-Propanesultone
3-Propanolide (Propiolactone)
Residual oils (petroleum)
Residues (petroleum), coker scrubber, condensed-ring-arom-containing
Residues (petroleum), hydrogenated steam-cracked naphtha
Residues (petroleum), atm tower
Residues (petroleum), vacuum, light
Residues (petroleum), steam-cracked naphtha distn
Residues (petroleum), steam cracked

Table 5.16 Cont'd

Residues (petroleum), heavy coker and light vacuum
Residues (petroleum), catalytic reformer fractionator
Residues (petroleum), hydrodesulphurized atmospheric tower
Residues (petroleum), topping plant, low sulphur
Residues (petroleum), heavy coker gas oil and vacuum gas oil
Residues (petroleum), thermal cracked
Residues (petroleum), catalytic reformer fractionator residue distillation
Residues (petroleum), catalytic cracking
Residues (petroleum), steam-cracked light
Residues (petroleum), hydrocracked
Residues (petroleum), light vacuum
Residues (petroleum), steam-cracked heat-soaked naphtha
Residues (petroleum), steam cracked
Residues (petroleum), steam cracked, distillates
Residues (petroleum), atmospheric
Residues, steam cracked, thermally treated
Strontium chromate
Styrene oxide
Tar, brown-coal
Tar, brown-coal, low temperature
Tar, coal, low temperature
Tar, coal, high temperature
Tar, coal
1,4,5,8-tetraamino-anthraquinone
Toluene-2,4-diammonium sulphate (4-methyl-*m*-phenylenediamine sulphate)
o-Toluidine
o-Toluidine-based azodyes
Thioacetamide
4-*o*-Tolyazo-*o*-toluidine
a,a,a-Trichlorotoluene
Urethane (INN)
Vinyl chloride (Chloroethylene)
Zinc chromates (including zinc potassium chromate)

Substances with the risk phrase 'R49' (may cause cancer by inhalation)
Beryllium
Beryllium compounds with the exception of aluminium beryllium silicates
Cadmium oxide
Cadmium sulphate
Chromium (VI) compounds (with the exception of barium chromate and of compounds specified elsewhere in the Approved Supply List)
Chromium trioxide
Chromyl chloride (chromic oxychloride)
Dinickel trioxide
Nickel dioxide
Nickel monoxide
Nickel subsulphide
Nickel sulphide
Potassium chromate
Potassium dichromate
Refactory ceramic fibres or special purpose fibres, with the exception of those specified elsewhere in the Approved Supply List (man-made vitreous (silicate) fibres with random orientation with alkaline oxide and alkali earth oxide (Na_2O + K_2O + CaO + MgO + BaO) content less than or equal to 18% by weight)
The classification as a carcinogen need not apply to fibres with a length weighted geometric mean diameter less two standard errors greater than 6 μm
Sodium dichromate
Sodium dichromate dihydrate

* The manufacture and use of these substances, or any substance containing them, in concentrations equal to or greater than 0.1% by weight, is prohibited in the UK (COSHH Reg. 4(1)).

Table 5.17 Evaluation of carcinogenic risks to humans by the IARC, as at 2000

Group 1: Carcinogenic to humans (78)

Agents and groups of agents
Aflatoxins, naturally occurring [1402-68-2] (Vol. 56; 1993)
4-Aminobiphenyl [92-67-1] (Vol. 1, Suppl. 7; 1987)
Arsenic [7440-38-2] and arsenic compounds (Vol. 23, Suppl. 7; 1987)
(NB: This evaluation applies to the group of compounds as a whole and not necessarily to all individual compounds within the group)
Asbestos [1332-21-4] (Vol. 14, Suppl. 7; 1987)
Azathioprine [446-86-6] (Vol. 26, Suppl. 7; 1987)
Benzene [71-43-2] (Vol. 29, Suppl. 7; 1987)
Benzidine [92-87-5] (Vol. 29, Suppl. 7; 1987)
Beryllium [7440-41-7] and beryllium compounds (Vol. 58; 1993)
(NB: Evaluated as a group)
N,N-Bis(2-chloroethyl)-2-naphthylamine (Chlornaphazine) [494-03-1] (Vol. 4, Suppl. 7; 1987)
Bis(chloromethyl)ether [542-88-1] and chloromethyl methyl ether [107-30-2] (technical-grade)
(Vol. 4, Suppl. 7; 1987)
1,4-Butanediol dimethanesulfonate (Busulphan; Myleran) [55-98-1] (Vol. 4, Suppl. 7; 1987)
Cadmium [7440-43-9] and cadmium compounds (Vol. 58; 1993)
(NB: Evaluated as a group)
Chlorambucil [305-03-3] (Vol. 26, Suppl. 7; 1987)
1-(2-Chloroethyl)-3-(4-methylcyclohexyl)-1-nitrosourea (Methyl-CCNU; Semustine) [13909-09-6] (Suppl. 7; 1987)
Chromium[VI] compounds (Vol. 49; 1990)
(NB: Evaluated as a group)
Cyclosporin [79217-60-0] (Vol. 50; 1990)
Cyclophosphamide [50-18-0] [6055-19-2] (Vol. 26, Suppl. 7; 1987)
Diethylstilboestrol [56-53-1] (Vol. 21, Suppl. 7; 1987)
Epstein–Barr virus (Vol. 70; 1997)
Erionite [66733-21-9] (Vol. 42, Suppl. 7; 1987)
Ethylene oxide [75-21-8] (Vol. 60; 1994)
(NB: Overall evaluation upgraded from 2A to 1 with supporting evidence from other data relevant to the evaluation of carcinogenicity and its mechanisms)
Etoposide [33419-42-0] in combination with cisplatin and bleomycin (Vol. 76; 2000)
[Gamma Radiation: *see* X- and Gamma (g)-Radiation]
Helicobacter pylori (infection with) (Vol. 61; 1994)
Hepatitis B virus (chronic infection with) (Vol. 59; 1994)
Hepatitis C virus (chronic infection with) (Vol. 59; 1994)
Human immunodeficiency virus type 1 (infection with) (Vol. 67; 1996)
Human papillomavirus type 16 (Vol. 64; 1995)
Human papillomavirus type 18 (Vol. 64; 1995)
Human T-cell lymphotropic virus type I (Vol. 67; 1996)
Melphalan [148-82-3] (Vol. 9, Suppl. 7; 1987)
8-Methoxypsoralen (Methoxsalen) [298-81-7] plus ultraviolet A radiation (Vol. 24, Suppl. 7; 1987)
MOPP and other combined chemotherapy including alkylating agents (Suppl. 7; 1987)
Mustard gas (Sulphur mustard) [505-60-2] (Vol. 9, Suppl. 7; 1987)
2-Naphthylamine [91-59-8] (Vol. 4, Suppl. 7; 1987)
Neutrons (Vol. 75; 2000)
Nickel compounds (Vol. 49; 1990)
(NB: Evaluated as a group)
Oestrogen therapy, postmenopausal (Vol. 72; 1999)
Oestrogens, nonsteroidal (Suppl. 7; 1987)
(NB: This evaluation applies to the group of compounds as a whole and not necessarily to all individual compounds within the group)
Oestrogens, steroidal (Suppl. 7; 1987)
(NB: This evaluation applies to the group of compounds as a whole and not necessarily to all individual compounds within the group)
Opisthorchis viverrini (infection with) (Vol. 61; 1994)
Oral contraceptives, combined (Vol. 72; 1999)
(NB: There is also conclusive evidence that these agents have a protective effect against cancers of the ovary and endometrium)
Oral contraceptives, sequential (Suppl. 7; 1987)

Table 5.17 Cont'd

Radon [10043-92-2] and its decay products (Vol. 43; 1988)
Schistosoma haematobium (infection with) (Vol. 61; 1994)
Silica [14808-60-7], crystalline (inhaled in the form of quartz or cristobalite from occupational sources) (Vol. 68; 1997)
Solar radiation (Vol. 55; 1992)
Talc containing asbestiform fibres (Vol. 42, Suppl. 7; 1987)
Tamoxifen [10540-29-1] (Vol. 66; 1996)
(NB: There is also conclusive evidence that this agent (tamoxifen) reduces the risk of contralateral breast cancer)
2,3,7,8-Tetrachlorodibenzo-para-dioxin [1746-01-6] (Vol. 69; 1997)
(NB: Overall evaluation upgraded from 2A to 1 with supporting evidence from other data relevant to the evaluation of carcinogenicity and its mechanisms)
Thiotepa [52-24-4] (Vol. 50; 1990)
Treosulfan [299-75-2] (Vol. 26, Suppl. 7; 1987)
Vinyl chloride [75-01-4] (Vol. 19, Suppl. 7; 1987)
X- and Gamma (g)-Radiation (Vol. 75; 2000)

Mixtures
Alcoholic beverages (Vol. 44; 1988)
Analgesic mixtures containing phenacetin (Suppl. 7; 1987)
Betel quid with tobacco (Vol. 37, Suppl. 7; 1987)
Coal-tar pitches [65996-93-2] (Vol. 35, Suppl. 7; 1987)
Coal-tars [8007-45-2] (Vol. 35, Suppl. 7; 1987)
Mineral oils, untreated and mildly treated (Vol. 33, Suppl. 7; 1987)
Salted fish (Chinese-style) (Vol. 56; 1993)
Shale-oils [68308-34-9] (Vol. 35, Suppl. 7; 1987)
Soots (Vol. 35, Suppl. 7; 1987)
Tobacco products, smokeless (Vol. 37, Suppl. 7; 1987)
Tobacco smoke (Vol. 38, Suppl. 7; 1987)
Wood dust (Vol. 62; 1995)

Exposure circumstances
Aluminium production (Vol. 34, Suppl. 7; 1987)
Auramine, manufacture of (Suppl. 7; 1987)
Boot and shoe manufacture and repair (Vol. 25, Suppl. 7; 1987)
Coal gasification (Vol. 34, Suppl. 7; 1987)
Coke production (Vol. 34, Suppl. 7; 1987)
Furniture and cabinet making (Vol. 25, Suppl. 7; 1987)
Haematite mining (underground) with exposure to radon (Vol. 1, Suppl. 7; 1987)
Iron and steel founding (Vol. 34, Suppl. 7; 1987)
Isopropanol manufacture (strong-acid process) (Suppl. 7; 1987)
Magenta, manufacture of (Vol. 57; 1993)
Painter (occupational exposure as a) (Vol. 47; 1989)
Rubber industry (Vol. 28, Suppl. 7; 1987)
Strong-inorganic-acid mists containing sulphuric acid (occupational exposure to) (Vol. 54; 1992)

Group 2A: Probably carcinogenic to humans (63)

Agents and groups of agents
Acrylamide [79-06-1] (Vol. 60; 1994)
(NB: Overall evaluation upgraded from 2B to 2A with supporting evidence from other data relevant to the evaluation of carcinogenicity and its mechanisms)
Adriamycin [23214-92-8] (Vol. 10, Suppl. 7; 1987)
(NB: Overall evaluation upgraded from 2B to 2A with supporting evidence from other data relevant to the evaluation of carcinogenicity and its mechanisms)
Androgenic (anabolic) steroids (Suppl. 7; 1987)
Azacitidine [320-67-2] (Vol. 50; 1990)
(NB: Overall evaluation upgraded from 2B to 2A with supporting evidence from other data relevant to the evaluation of carcinogenicity and its mechanisms)
Benz[a]anthracene [56-55-3] (Vol. 32, Suppl. 7; 1987)
(NB: Overall evaluation upgraded from 2B to 2A with supporting evidence from other data relevant to the evaluation of carcinogenicity and its mechanisms)
Benzidine-based dyes (Suppl. 7; 1987)

Table 5.17 Cont'd

(NB: Overall evaluation upgraded from 2B to 2A with supporting evidence from other data relevant to the evaluation of carcinogenicity and its mechanisms)
Benzo[a]pyrene [50-32-8] (Vol. 32, Suppl. 7; 1987)
(NB: Overall evaluation upgraded from 2B to 2A with supporting evidence from other data relevant to the evaluation of carcinogenicity and its mechanisms)
Bischloroethyl nitrosourea (BCNU) [154-93-8] (Vol. 26, Suppl. 7; 1987)
1,3-Butadiene [106-99-0] (Vol. 71; 1999)
Captafol [2425-06-1] (Vol. 53; 1991)
(NB: Overall evaluation upgraded from 2B to 2A with supporting evidence from other data relevant to the evaluation of carcinogenicity and its mechanisms)
Chloramphenicol [56-75-7] (Vol. 50; 1990)
(NB: Overall evaluation upgraded from 2B to 2A with supporting evidence from other data relevant to the evaluation of carcinogenicity and its mechanisms)
a-Chlorinated toluenes (benzal chloride [98-87-3], benzotrichloride [98-07-7], benzyl chloride [100-44-7]) and benzoyl chloride [98-88-4] (combined exposures) (Vol. 29, Suppl. 7, Vol. 71; 1999)
1-(2-Chloroethyl)-3-cyclohexyl-1-nitrosourea (CCNU) [13010-47-4] (Vol. 26, Suppl. 7; 1987)
(NB: Overall evaluation upgraded from 2B to 2A with supporting evidence from other data relevant to the evaluation of carcinogenicity and its mechanisms)
para-Chloro-ortho-toluidine [95-69-2] and its strong acid salts (Vol. 48, Vol. 77; 2000)
(NB: Evaluated as a group)
Chlorozotocin [54749-90-5] (Vol. 50; 1990)
(NB: Overall evaluation upgraded from 2B to 2A with supporting evidence from other data relevant to the evaluation of carcinogenicity and its mechanisms)
Cisplatin [15663-27-1] (Vol. 26, Suppl. 7; 1987)
(NB: Overall evaluation upgraded from 2B to 2A with supporting evidence from other data relevant to the evaluation of carcinogenicity and its mechanisms)
Clonorchis sinensis (infection with) (Vol. 61; 1994)
(NB: Overall evaluation upgraded from 2B to 2A with supporting evidence from other data relevant to the evaluation of carcinogenicity and its mechanisms)
Dibenz[a,h]anthracene [53-70-3] (Vol. 32, Suppl. 7; 1987)
(NB: Overall evaluation upgraded from 2B to 2A with supporting evidence from other data relevant to the evaluation of carcinogenicity and its mechanisms)
Diethyl sulphate [64-67-5] (Vol. 54, Vol. 71; 1999)
(NB: Overall evaluation upgraded from 2B to 2A with supporting evidence from other data relevant to the evaluation of carcinogenicity and its mechanisms)
Dimethylcarbamoyl chloride [79-44-7] (Vol. 12, Suppl. 7, Vol. 71; 1999)
(NB: Overall evaluation upgraded from 2B to 2A with supporting evidence from other data relevant to the evaluation of carcinogenicity and its mechanisms)
1,2-Dimethylhydrazine [540-73-8] (Vol. 4, Suppl. 7, Vol. 71; 1999)
(NB: Overall evaluation upgraded from 2B to 2A with supporting evidence from other data relevant to the evaluation of carcinogenicity and its mechanisms)
Dimethyl sulphate [77-78-1] (Vol. 4, Suppl. 7, Vol. 71; 1999)
(NB: Overall evaluation upgraded from 2B to 2A with supporting evidence from other data relevant to the evaluation of carcinogenicity and its mechanisms)
Epichlorohydrin [106-89-8] (Vol. 11, Suppl. 7, Vol. 71; 1999)
(NB: Overall evaluation upgraded from 2B to 2A with supporting evidence from other data relevant to the evaluation of carcinogenicity and its mechanisms)
Ethylene dibromide [106-93-4] (Vol. 15, Suppl. 7, Vol. 71; 1999)
(NB: Overall evaluation upgraded from 2B to 2A with supporting evidence from other data relevant to the evaluation of carcinogenicity and its mechanisms)
N-Ethyl-*N*-nitrosourea [759-73-9] (Vol. 17, Suppl. 7; 1987)
(NB: Overall evaluation upgraded from 2B to 2A with supporting evidence from other data relevant to the evaluation of carcinogenicity and its mechanisms)
Etoposide [33419-42-0] (Vol. 76; 2000)
(NB: Other relevant data taken into consideration in making the overall evaluation)
Formaldehyde [50-00-0] (Vol. 62; 1995)
Glycidol [556-52-5] (Vol. 77; 2000)
(NB: Other relevant data taken into consideration in making the overall evaluation)
Human papillomavirus type 31 (Vol. 64; 1995)
Human papillomavirus type 33 (Vol. 64; 1995)

Table 5.17 Cont'd

IQ (2-Amino-3-methylimidazo[4,5-f]quinoline) [76180-96-6] (Vol. 56; 1993)
(NB: Overall evaluation upgraded from 2B to 2A with supporting evidence from other data relevant to the evaluation of carcinogenicity and its mechanisms)
Kaposi's sarcoma herpesvirus/human herpesvirus 8 (Vol. 70; 1997)
5-Methoxypsoralen [484-20-8] (Vol. 40, Suppl. 7; 1987)
(NB: Overall evaluation upgraded from 2B to 2A with supporting evidence from other data relevant to the evaluation of carcinogenicity and its mechanisms)
4,4′-Methylene bis(2-chloroaniline) (MbOCA) [101-14-4] (Vol. 57; 1993)
(NB: Overall evaluation upgraded from 2B to 2A with supporting evidence from other data relevant to the evaluation of carcinogenicity and its mechanisms)
Methyl methanesulphonate [66-27-3] (Vol. 7, Suppl. 7, Vol. 71; 1999)
(NB: Overall evaluation upgraded from 2B to 2A with supporting evidence from other data relevant to the evaluation of carcinogenicity and its mechanisms)
N-Methyl-*N*′-nitro-*N*-nitrosoguanidine (MNNG) [70-25-7] (Vol. 4, Suppl. 7; 1987)
(NB: Overall evaluation upgraded from 2B to 2A with supporting evidence from other data relevant to the evaluation of carcinogenicity and its mechanisms)
N-Methyl-*N*-nitrosourea [684-93-5] (Vol. 17, Suppl. 7; 1987)
(NB: Overall evaluation upgraded from 2B to 2A with supporting evidence from other data relevant to the evaluation of carcinogenicity and its mechanisms)
Nitrogen mustard [51-75-2] (Vol. 9, Suppl. 7; 1987)
N-Nitrosodiethylamine [55-18-5] (Vol. 17, Suppl. 7; 1987)
(NB: Overall evaluation upgraded from 2B to 2A with supporting evidence from other data relevant to the evaluation of carcinogenicity and its mechanisms)
N-Nitrosodimethylamine [62-75-9] (Vol. 17, Suppl. 7; 1987)
(NB: Overall evaluation upgraded from 2B to 2A with supporting evidence from other data relevant to the evaluation of carcinogenicity and its mechanisms)
Phenacetin [62-44-2] (Vol. 24, Suppl. 7; 1987)
Procarbazine hydrochloride [366-70-1] (Vol. 26, Suppl. 7; 1987)
(NB: Overall evaluation upgraded from 2B to 2A with supporting evidence from other data relevant to the evaluation of carcinogenicity and its mechanisms)
Styrene-7,8-oxide [96-09-3] (Vol. 60; 1994)
(NB: Overall evaluation upgraded from 2B to 2A with supporting evidence from other data relevant to the evaluation of carcinogenicity and its mechanisms)
Teniposide [29767-20-2] (Vol. 76; 2000)
(NB: Other relevant data taken into consideration in making the overall evaluation)
Tetrachloroethylene [127-18-4] (Vol. 63; 1995)
ortho-Toluidine [95-53-4] (Vol. 27, Suppl. 7, Vol. 77; 2000)
Trichloroethylene [79-01-6] (Vol. 63; 1995)
1,2,3-Trichloropropane [96-18-4] (Vol. 63; 1995)
Tris(2,3-dibromopropyl) phosphate [126-72-7] (Vol. 20, Suppl. 7, Vol. 71; 1999)
(NB: Overall evaluation upgraded from 2B to 2A with supporting evidence from other data relevant to the evaluation of carcinogenicity and its mechanisms)
Ultraviolet radiation A (Vol. 55; 1992)
(NB: Overall evaluation upgraded from 2B to 2A with supporting evidence from other data relevant to the evaluation of carcinogenicity and its mechanisms)
Ultraviolet radiation B (Vol. 55; 1992)
(NB: Overall evaluation upgraded from 2B to 2A with supporting evidence from other data relevant to the evaluation of carcinogenicity and its mechanisms)
Ultraviolet radiation C (Vol. 55; 1992)
(NB: Overall evaluation upgraded from 2B to 2A with supporting evidence from other data relevant to the evaluation of carcinogenicity and its mechanisms)
Vinyl bromide [593-60-2] (Vol. 39, Suppl. 7, Vol. 71; 1999)
(NB: Overall evaluation upgraded from 2B to 2A with supporting evidence from other data relevant to the evaluation of carcinogenicity and its mechanisms)
Vinyl fluoride [75-02-5] (Vol. 63; 1995)

Mixtures
Creosotes [8001-58-9] (Vol. 35, Suppl. 7; 1987)
Diesel engine exhaust (Vol. 46; 1989)
Hot mate (Vol. 51; 1991)

Table 5.17 Cont'd

Non-arsenical insecticides (occupational exposures in spraying and application of) (Vol. 53; 1991)
Polychlorinated biphenyls [1336-36-3] (Vol. 18, Suppl. 7; 1987)

Exposure circumstances
Art glass, glass containers and pressed ware (manufacture of) (Vol. 58; 1993)
Hairdresser or barber (occupational exposure as a) (Vol. 57; 1993)
Petroleum refining (occupational exposures in) (Vol. 45; 1989)
Sunlamps and sunbeds (use of) (Vol. 55; 1992)

Group 2B: Possibly carcinogenic to humans (235)

Agents and groups of agents
A-a-C (2-Amino-9H-pyrido[2,3-b]indole) [26148-68-5] (Vol. 40, Suppl. 7; 1987)
Acetaldehyde [75-07-0] (Vol. 36, Suppl. 7, Vol. 71; 1999)
Acetamide [60-35-5] (Vol. 7, Suppl. 7, Vol. 71; 1999)
Acrylonitrile [107-13-1] (Vol. 71; 1999)
AF-2 [2-(2-Furyl)-3-(5-nitro-2-furyl)acrylamide] [3688-53-7] (Vol. 31, Suppl. 7; 1987)
Aflatoxin M1 [6795-23-9] (Vol. 56; 1993)
para-Aminoazobenzene [60-09-3] (Vol. 8, Suppl. 7; 1987)
ortho-Aminoazotoluene [97-56-3] (Vol. 8, Suppl. 7; 1987)
2-Amino-5-(5-nitro-2-furyl)-1,3,4-thiadiazole [712-68-5] (Vol. 7, Suppl. 7; 1987)
Amitrole [61-82-5] (Vol. 41, Suppl. 7; 1987)
Amsacrine [51264-14-3] (Vol. 76; 2000)
ortho-Anisidine [90-04-0] (Vol. 73; 1999)
Antimony trioxide [1309-64-4] (Vol. 47; 1989)
Aramite® [140-57-8] (Vol. 5, Suppl. 7; 1987)
Auramine [492-80-8] (technical-grade) (Vol. 1, Suppl. 7; 1987)
Azaserine [115-02-6] (Vol. 10, Suppl. 7; 1987)
Aziridine [151-56-4] (Vol. 9, Suppl. 7, Vol. 71; 1999)
(NB: Overall evaluation upgraded from 3 to 2B with supporting evidence from other data relevant to the evaluation of carcinogenicity and its mechanisms)
Benzo[b]fluoranthene [205-99-2] (Vol. 32, Suppl. 7; 1987)
Benzo[j]fluoranthene [205-82-3] (Vol. 32, Suppl. 7; 1987)
Benzo[k]fluoranthene [207-08-9] (Vol. 32, Suppl. 7; 1987)
Benzofuran [271-89-6] (Vol. 63; 1995)
Benzyl violet 4B [1694-09-3] (Vol. 16, Suppl. 7; 1987)
2,2-Bis(bromomethyl)propane-1,3-diol [3296-90-0] (Vol. 77; 2000)
Bleomycins [11056-06-7] (Vol. 26, Suppl. 7; 1987)
(NB: Overall evaluation upgraded from 3 to 2B with supporting evidence from other data relevant to the evaluation of carcinogenicity and its mechanisms)
Bracken fern (Vol. 40, Suppl. 7; 1987)
Bromodichloromethane [75-27-4] (Vol. 52, Vol. 71; 1999)
Butylated hydroxyanisole (BHA) [25013-16-5] (Vol. 40, Suppl. 7; 1987)
b-Butyrolactone [3068-88-0] (Vol. 11, Suppl. 7, Vol. 71; 1999)
Caffeic acid [331-39-5] (Vol. 56; 1993)
Carbon black [1333-86-4] (Vol. 65; 1996)
Carbon tetrachloride [56-23-5] (Vol. 20, Suppl. 7, Vol. 71; 1999)
Catechol [120-80-9] (Vol. 15, Suppl. 7, Vol. 71; 1999)
Ceramic fibres (Vol. 43; 1988)
Chlordane [57-74-9] (Vol. 53; 1991)
Chlordecone (Kepone) [143-50-0] (Vol. 20, Suppl. 7; 1987)
Chlorendic acid [115-28-6] (Vol. 48; 1990)
para-Chloroaniline [106-47-8] (Vol. 57; 1993)
Chloroform [67-66-3] (Vol. 73; 1999)
1-Chloro-2-methylpropene [513-37-1] (Vol. 63; 1995)
Chlorophenoxy herbicides (Vol. 41, Suppl. 7; 1987)
4-Chloro-ortho-phenylenediamine [95-83-0] (Vol. 27, Suppl. 7; 1987)
Chloroprene [126-99-8] (Vol. 71; 1999)
Chlorothalonil [1897-45-6] (Vol. 73; 1999)
CI Acid Red 114 [6459-94-5] (Vol. 57; 1993)

Table 5.17 Cont'd

CI Basic Red 9 [569-61-9] (Vol. 57; 1993)
CI Direct Blue 15 [2429-74-5] (Vol. 57; 1993)
Citrus Red No. 2 [6358-53-8] (Vol. 8, Suppl. 7; 1987)
Cobalt [7440-48-4] and cobalt compounds (Vol. 52; 1991)
(NB: Evaluated as a group)
para-Cresidine [120-71-8] (Vol. 27, Suppl. 7; 1987)
Cycasin [14901-08-7] (Vol. 10, Suppl. 7; 1987)
Dacarbazine [4342-03-4] (Vol. 26, Suppl. 7; 1987)
Dantron (Chrysazin; 1,8-Dihydroxyanthraquinone) [117-10-2] (Vol. 50; 1990)
Daunomycin [20830-81-3] (Vol. 10, Suppl. 7; 1987)
DDT [*p,p'*-DDT, 50-29-3] (Vol. 53; 1991)
N,N'-Diacetylbenzidine [613-35-4] (Vol. 16, Suppl. 7; 1987)
2,4-Diaminoanisole [615-05-4] (Vol. 27, Suppl. 7; 1987)
4,4'-Diaminodiphenyl ether [101-80-4] (Vol. 29, Suppl. 7; 1987)
2,4-Diaminotoluene [95-80-7] (Vol. 16, Suppl. 7; 1987)
Dibenz[a,h]acridine [226-36-8] (Vol. 32, Suppl. 7; 1987)
Dibenz[a,j]acridine [224-42-0] (Vol. 32, Suppl. 7; 1987)
7H-Dibenzo[c,g]carbazole [194-59-2] (Vol. 32, Suppl. 7; 1987)
Dibenzo[a,e]pyrene [192-65-4] (Vol. 32, Suppl. 7; 1987)
Dibenzo[a,h]pyrene [189-64-0] (Vol. 32, Suppl. 7; 1987)
Dibenzo[a,i]pyrene [189-55-9] (Vol. 32, Suppl. 7; 1987)
Dibenzo[a,l]pyrene [191-30-0] (Vol. 32, Suppl. 7; 1987)
1,2-Dibromo-3-chloropropane [96-12-8] (Vol. 20, Suppl. 7, Vol. 71; 1999)
2,3-Dibromopropan-1-ol [96-13-9] (Vol. 77; 2000)
para-Dichlorobenzene [106-46-7] (Vol. 73; 1999)
3,3'-Dichlorobenzidine [91-94-1] (Vol. 29, Suppl. 7; 1987)
3,3'-Dichloro-4,4'-diaminodiphenyl ether [28434-86-8] (Vol. 16, Suppl. 7; 1987)
1,2-Dichloroethane [107-06-2] (Vol. 20, Suppl. 7, Vol. 71; 1999)
Dichloromethane (methylene chloride) [75-09-2] (Vol. 71; 1999)
1,3-Dichloropropene [542-75-6] (technical-grade) (Vol. 41, Suppl. 7, Vol. 71; 1999)
Dichlorvos [62-73-7] (Vol. 53; 1991)
1,2-Diethylhydrazine [1615-80-1] (Vol. 4, Suppl. 7, Vol. 71; 1999)
Diglycidyl resorcinol ether [101-90-6] (Vol. 36, Suppl. 7, Vol. 71; 1999)
Dihydrosafrole [94-58-6] (Vol. 10, Suppl. 7; 1987)
Diisopropyl sulphate [2973-10-6] (Vol. 54, Vol. 71; 1999)
3,3'-Dimethoxybenzidine (ortho-Dianisidine) [119-90-4] (Vol. 4, Suppl. 7; 1987)
para-Dimethylaminoazobenzene [60-11-7] (Vol. 8, Suppl. 7; 1987)
trans-2-[(Dimethylamino)methylimino]-5-[2-(5-nitro-2-furyl)-vinyl]-1,3,4-oxadiazole [25962-77-0] (Vol. 7, Suppl. 7; 1987)
2,6-Dimethylaniline (2,6-Xylidine) [87-62-7] (Vol. 57; 1993)
3,3'-Dimethylbenzidine (ortho-Tolidine) [119-93-7] (Vol. 1, Suppl. 7; 1987)
1,1-Dimethylhydrazine [57-14-7] (Vol. 4, Suppl. 7, Vol. 71; 1999)
3,7-Dinitrofluoranthene [105735-71-5] (Vol. 65; 1996)
3,9-Dinitrofluoranthene [22506-53-2] (Vol. 65; 1996)
1,6-Dinitropyrene [42397-64-8] (Vol. 46; 1989)
1,8-Dinitropyrene [42397-65-9] (Vol. 46; 1989)
2,4-Dinitrotoluene [121-14-2] (Vol. 65; 1996)
2,6-Dinitrotoluene [606-20-2] (Vol. 65; 1996)
1,4-Dioxane [123-91-1] (Vol. 11, Suppl. 7, Vol. 71; 1999)
Disperse Blue 1 [2475-45-8] (Vol. 48; 1990)
1,2-Epoxybutane [106-88-7] (Vol. 47, Vol. 71; 1999)
(NB: Overall evaluation upgraded from 3 to 2B with supporting evidence from other data relevant to the evaluation of carcinogenicity and its mechanisms)
Ethyl acrylate [140-88-5] (Vol. 39, Suppl. 7, Vol. 71; 1999)
Ethylbenzene [100-41-4] (Vol. 77; 2000)
Ethylene thiourea [96-45-7] (Vol. 7, Suppl. 7; 1987)
Ethyl methanesulfonate [62-50-0] (Vol. 7, Suppl. 7; 1987)
Foreign bodies, implanted in tissues (Vol. 74; 1999)
  Polymeric, prepared as thin smooth films (with the exception of poly(glycolic acid))
  Metallic, prepared as thin smooth films
  Metallic cobalt, metallic nickel and an alloy powder containing 66–67% nickel, 13–16% chromium and 7% iron

Table 5.17 Cont'd

2-(2-Formylhydrazino)-4-(5-nitro-2-furyl)thiazole [3570-75-0] (Vol. 7, Suppl. 7; 1987)
Furan [110-00-9] (Vol. 63; 1995)
Glasswool (Vol. 43; 1988)
Glu-P-1 (2-Amino-6-methyldipyrido[1,2-a:3',2'-d]imidazole) [67730-11-4] (Vol. 40, Suppl. 7; 1987)
Glu-P-2 (2-Aminodipyrido[1,2-a:3',2'-d]imidazole) [67730-10-3] (Vol. 40, Suppl. 7; 1987)
Glycidaldehyde [765-34-4] (Vol. 11, Suppl. 7, Vol. 71; 1999)
Griseofulvin [126-07-8] (Vol. 10, Suppl. 7; 1987)
HC Blue No. 1 [2784-94-3] (Vol. 57; 1993)
Heptachlor [76-44-8] (Vol. 53; 1991)
Hexachlorobenzene [118-74-1] (Vol. 20, Suppl. 7; 1987)
Hexachloroethane [67-72-1] (Vol. 73; 1999)
Hexachlorocyclohexanes (Vol. 20, Suppl. 7; 1987)
Hexamethylphosphoramide [680-31-9] (Vol. 15, Suppl. 7, Vol. 71; 1999)
Human immunodeficiency virus type 2 (infection with) (Vol. 67; 1996)
Human papillomaviruses: some types other than 16, 18, 31 and 33 (Vol. 64; 1995)
Hydrazine [302-01-2] (Vol. 4, Suppl. 7, Vol. 71; 1999)
Indeno[1,2,3-cd]pyrene [193-39-5] (Vol. 32, Suppl. 7; 1987)
Iron-dextran complex [9004-66-4] (Vol. 2, Suppl. 7; 1987)
Isoprene [78-79-5] (Vol. 60, Vol. 71; 1999)
Lasiocarpine [303-34-4] (Vol. 10, Suppl. 7; 1987)
Lead [7439-92-1] and lead compounds, inorganic (Vol. 23, Suppl. 7; 1987)
(NB: Evaluated as a group)
Magenta [632-99-5] (containing CI Basic Red 9) (Vol. 57; 1993)
MeA-a-C (2-Amino-3-methyl-9H-pyrido[2,3-b]indole) [68006-83-7] (Vol. 40, Suppl. 7; 1987)
Medroxyprogesterone acetate [71-58-9] (Vol. 21, Suppl. 7; 1987)
MeIQ (2-Amino-3,4-dimethylimidazo[4,5-f]quinoline) [77094-11-2] (Vol. 56; 1993)
MeIQx (2-Amino-3,8-dimethylimidazo[4,5-f]quinoxaline) [77500-04-0] (Vol. 56; 1993)
Merphalan [531-76-0] (Vol. 9, Suppl. 7; 1987)
2-Methylaziridine (Propyleneimine) [75-55-8] (Vol. 9, Suppl. 7, Vol. 71; 1999)
Methylazoxymethanol acetate [592-62-1] (Vol. 10, Suppl. 7; 1987)
5-Methylchrysene [3697-24-3] (Vol. 32, Suppl. 7; 1987)
4,4'-Methylene bis(2-methylaniline) [838-88-0] (Vol. 4, Suppl. 7; 1987)
4,4'-Methylenedianiline [101-77-9] (Vol. 39, Suppl. 7; 1987)
Methylmercury compounds (Vol. 58; 1993)
(NB: Evaluated as a group)
2-Methyl-1-nitroanthraquinone [129-15-7] (uncertain purity) (Vol. 27, Suppl. 7; 1987)
N-Methyl-*N*-nitrosourethane [615-53-2] (Vol. 4, Suppl. 7; 1987)
Methylthiouracil [56-04-2] (Vol. 7, Suppl. 7; 1987)
Metronidazole [443-48-1] (Vol. 13, Suppl. 7; 1987)
Mirex [2385-85-5] (Vol. 20, Suppl. 7; 1987)
Mitomycin C [50-07-7] (Vol. 10, Suppl. 7; 1987)
Mitoxantrone [65271-80-9] (Vol. 76; 2000)
Monocrotaline [315-22-0] (Vol. 10, Suppl. 7; 1987)
5-(Morpholinomethyl)-3-[(5-nitrofurfurylidene)amino]-2-oxazolidinone [3795-88-8] (Vol. 7, Suppl. 7; 1987)
Nafenopin [3771-19-5] (Vol. 24, Suppl. 7; 1987)
Nickel, metallic [7440-02-0] and alloys (Vol. 49; 1990)
Niridazole [61-57-4] (Vol. 13, Suppl. 7; 1987)
Nitrilotriacetic acid [139-13-9] and its salts (Vol. 73; 1999)
(NB: Evaluated as a group)
5-Nitroacenaphthene [602-87-9] (Vol. 16, Suppl. 7; 1987)
2-Nitroanisole [91-23-6] (Vol. 65; 1996)
Nitrobenzene [98-95-3] (Vol. 65; 1996)
6-Nitrochrysene [7496-02-8] (Vol. 46; 1989)
Nitrofen [1836-75-5] (technical-grade) (Vol. 30, Suppl. 7; 1987)
2-Nitrofluorene [607-57-8] (Vol. 46; 1989)
1-[(5-Nitrofurfurylidene)amino]-2-imidazolidinone [555-84-0] (Vol. 7, Suppl. 7; 1987)
N-[4-(5-Nitro-2-furyl)-2-thiazolyl]acetamide [531-82-8] (Vol. 7, Suppl. 7; 1987)
Nitrogen mustard N-oxide [126-85-2] (Vol. 9, Suppl. 7; 1987)
Nitromethane [75-52-5] (Vol. 77; 2000)
2-Nitropropane [79-46-9] (Vol. 29, Suppl. 7, Vol. 71; 1999)

Table 5.17 Cont'd

1-Nitropyrene [5522-43-0] (Vol. 46; 1989)
4-Nitropyrene [57835-92-4] (Vol. 46; 1989)
N-Nitrosodi-*n*-butylamine [924-16-3] (Vol. 17, Suppl. 7; 1987)
N-Nitrosodiethanolamine [1116-54-7] (Vol. 17, Suppl. 7, Vol. 77; 2000)
N-Nitrosodi-*n*-propylamine [621-64-7] (Vol. 17, Suppl. 7; 1987)
3-(*N*-Nitrosomethylamino)propionitrile [60153-49-3] (Vol. 37, Suppl. 7; 1987)
4-(*N*-Nitrosomethylamino)-1-(3-pyridyl)-1-butanone (NNK) [64091-91-4] (Vol. 37, Suppl. 7; 1987)
N-Nitrosomethylethylamine [10595-95-6] (Vol. 17, Suppl. 7; 1987)
N-Nitrosomethylvinylamine [4549-40-0] (Vol. 17, Suppl. 7; 1987)
N-Nitrosomorpholine [59-89-2] (Vol. 17, Suppl. 7; 1987)
N'-Nitrosonornicotine [16543-55-8] (Vol. 37, Suppl. 7; 1987)
N-Nitrosopiperidine [100-75-4] (Vol. 17, Suppl. 7; 1987)
N-Nitrosopyrrolidine [930-55-2] (Vol. 17, Suppl. 7; 1987)
N-Nitrososarcosine [13256-22-9] (Vol. 17, Suppl. 7; 1987)
Ochratoxin A [303-47-9] (Vol. 56; 1993)
Oestrogen–progestogen therapy, postmenopausal (Vol. 72; 1999)
Oil Orange SS [2646-17-5] (Vol. 8, Suppl. 7; 1987)
Oxazepam [604-75-1] (Vol. 66; 1996)
Palygorskite (attapulgite) [12174-11-7] (long fibres, >5 micrometres) (Vol. 68; 1997)
Panfuran S [794-93-4] (containing dihydroxymethylfuratrizine)
(Vol. 24, Suppl. 7; 1987)
Phenazopyridine hydrochloride [136-40-3] (Vol. 24, Suppl. 7; 1987)
Phenobarbital [50-06-6] (Vol. 13, Suppl. 7; 1987)
Phenolphthalein [77-09-8] (Vol. 76; 2000)
Phenoxybenzamine hydrochloride [63-92-3] (Vol. 24, Suppl. 7; 1987)
Phenyl glycidyl ether [122-60-1] (Vol. 47, Vol. 71; 1999)
Phenytoin [57-41-0] (Vol. 66; 1996)
PhIP (2-Amino-1-methyl-6-phenylimidazo[4,5-b]pyridine) [105650-23-5] (Vol. 56; 1993)
Polychlorophenols and their sodium salts (mixed exposures) (Vol. 41, Suppl. 7, Vol. 53, Vol. 71; 1999)
Ponceau MX [3761-53-3] (Vol. 8, Suppl. 7; 1987)
Ponceau 3R [3564-09-8] (Vol. 8, Suppl. 7; 1987)
Potassium bromate [7758-01-2] (Vol. 73; 1999)
Progestins (Suppl. 7; 1987)
Progestogen-only contraceptives (Vol. 72; 1999)
1,3-Propane sultone [1120-71-4] (Vol. 4, Suppl. 7, Vol. 71; 1999)
b-Propiolactone [57-57-8] (Vol. 4, Suppl. 7, Vol. 71; 1999)
Propylene oxide [75-56-9] (Vol. 60; 1994)
Propylthiouracil [51-52-5] (Vol. 7, Suppl. 7; 1987)
Rockwool (Vol. 43; 1988)
Safrole [94-59-7] (Vol. 10, Suppl. 7; 1987)
Schistosoma japonicum (infection with) (Vol. 61; 1994)
Slagwool (Vol. 43; 1988)
Sodium ortho-phenylphenate [132-27-4] (Vol. 73; 1999)
Sterigmatocystin [10048-13-2] (Vol. 10, Suppl. 7; 1987)
Streptozotocin [18883-66-4] (Vol. 17, Suppl. 7; 1987)
Styrene [100-42-5] (Vol. 60; 1994)
(NB: Overall evaluation upgraded from 3 to 2B with supporting evidence from other data relevant to the evaluation of carcinogenicity and its mechanisms)
Sulfallate [95-06-7] (Vol. 30, Suppl. 7; 1987)
Tetrafluoroethylene [116-14-3] (Vol. 19, Suppl. 7, Vol. 71; 1999)
Tetranitromethane [509-14-8] (Vol. 65; 1996)
Thioacetamide [62-55-5] (Vol. 7, Suppl. 7; 1987)
4,4′-Thiodianiline [139-65-1] (Vol. 27, Suppl. 7; 1987)
Thiourea [62-56-6] (Vol. 7, Suppl. 7; 1987)
Toluene diisocyanates [26471-62-5] (Vol. 39, Suppl. 7, Vol. 71; 1999)
Toxins derived from *Fusarium moniliforme* (Vol. 56; 1993)
Trichlormethine (Trimustine hydrochloride) [817-09-4] (Vol. 50; 1990)
Trp-P-1 (3-Amino-1,4-dimethyl-5H-pyrido[4,3-b]indole) [62450-06-0] (Vol. 31, Suppl. 7; 1987)
Trp-P-2 (3-Amino-1-methyl-5H-pyrido[4,3-b]indole) [62450-07-1] (Vol. 31, Suppl. 7; 1987)
Trypan blue [72-57-1] (Vol. 8, Suppl. 7; 1987)

Table 5.17 Cont'd

Uracil mustard [66-75-1] (Vol. 9, Suppl. 7; 1987)
Urethane [51-79-6] (Vol. 7, Suppl. 7; 1987)
Vinyl acetate [108-05-4] (Vol. 63; 1995)
4-Vinylcyclohexene [100-40-3] (Vol. 60; 1994)
4-Vinylcyclohexene diepoxide [106-87-6] (Vol. 60; 1994)
Zalcitabine [7481-89-2] (Vol. 76; 2000)
Zidovudine (AZT) [30516-87-1] (Vol. 76; 2000)

Mixtures
Bitumens [8052-42-4], extracts of steam-refined and air-refined (Vol. 35, Suppl. 7; 1987)
Carrageenan [9000-07-1], degraded (Vol. 31, Suppl. 7; 1987)
Chlorinated paraffins of average carbon chain length C12 and average degree of chlorination approximately 60% (Vol. 48; 1990)
Coffee (urinary bladder) (Vol. 51; 1991)
(NB: There is some evidence of an inverse relationship between coffee drinking and cancer of the large bowel; coffee drinking could not be classified as to its carcinogenicity to other organs)
Diesel fuel, marine (Vol. 45; 1989)
(NB: Overall evaluation upgraded from 3 to 2B with supporting evidence from other data relevant to the evaluation of carcinogenicity and its mechanisms)
Engine exhaust, gasoline (Vol. 46; 1989)
Fuel oils, residual (heavy) (Vol. 45; 1989)
Gasoline (Vol. 45; 1989)
(NB: Overall evaluation upgraded from 3 to 2B with supporting evidence from other data relevant to the evaluation of carcinogenicity and its mechanisms)
Pickled vegetables (traditional in Asia) (Vol. 56; 1993)
Polybrominated biphenyls [Firemaster BP-6, 59536-65-1] (Vol. 41, Suppl. 7; 1987)
Toxaphene (Polychlorinated camphenes) [8001-35-2] (Vol. 20, Suppl. 7; 1987)
Welding fumes (Vol. 49; 1990)

Exposure circumstances
Carpentry and joinery (Vol. 25, Suppl. 7; 1987)
Dry cleaning (occupational exposures in) (Vol. 63; 1995)
Printing processes (occupational exposures in) (Vol. 65; 1996)
Textile manufacturing industry (work in) (Vol. 48; 1990)

- Identification by labels, numbers, colour coding etc. of vessels, transfer lines and valves.
- Clear, unambiguous operating instructions (updated and freely available).
- Segregated disposal of residues, 'empty' sacks, containers, liquid effluents, solid wastes, floor-washes etc.

Control strategies in general

Strategies for reducing the risk from a toxic chemical depends upon its nature (i.e. toxic, corrosive, dermatitic) and extent. A combination of the following measures may be appropriate.

Substitution

Hazardous chemicals or mixtures may be replaceable by safer materials. These may be less toxic *per se*, or less easily dispersed (e.g. less volatile or dusty). Substitution is also applicable to synthesis routes to avoid the use of toxic reactants/solvents or the production, either intentionally or accidentally, of toxic intermediates, by-products or wastes.

Minimization of inventory

As a general rule, it is preferable to minimize the amounts of toxic chemicals in storage and in process. There may be an advantage in handling chemicals in the most dilute practicable concentration.

Table 5.18 Common sources of toxic atmospheres

Source	*Examples*
Improper storage, handling, use or disposal of specific chemicals	Leakages[1] Improper venting or draining[1] Open handling[1] Incorrect notification on disposal Use of wrong material
Accidental release, spillage	Transport incidents Overfilling of containers Equipment failure Unexpected reactions Runaway reactions
Admixture of chemicals	By mistake, e.g. wrongly identified In wrong proportions In wrong circumstances[1] In wrong sequence
Fires	Pyrolysis products Combustion products[1] Vaporization Through domino effects
Operation in confined spaces	Improper isolation From residues Oxygen deficiency (inherent, from purging or from rusting)
Maintenance or cleaning of equipment	Residues Loss of containment (breaking lines) Stripping insulation Burning-off paint, flame heating components Reaction or vaporization of cleaning products
Wastes	Anaerobic breakdown Admixture of effluents Open handling of effluents or 'wastes' Atmospheric venting Solid wastes Uncontrolled incineration
Fabrication, manufacturing or machining operations etc.	Welding fumes[1] Spray painting, curing of paints[1] Use of adhesives, curing of adhesives[1] Cutting/grinding/fettling/shotblasting[1] Electroplating[1] Degreasing/cleaning/etching/pickling[1] Plastics forming or overheating[1]

[1] May result in long-term exposure (throughout operation or in workplace).

Mechanical handling

Release of, and exposure of personnel to, toxic chemicals can be reduced by appropriate mechanical handling and enclosed transfer, including:

- In-plant transfer via pipelines.
- In-plant transfer in specially designed containers, e.g. tote bins, corrosion-resistant containers, with provision for mechanical lifting.
- Use of enclosed transfer by pressurization or vacuum, with appropriate balancing or venting,

(e.g. with knock-out and/or scrubbing, filtration or incineration for vapours/gases), or recovery (e.g. adsorption or vapour recompression) provisions.
- Use of enclosed belt conveyors, chutes or pneumatic conveyors for solids.
- Enclosed transfer of solids using screw feeders.
- Transfer of solids in sealed containers or plastic sacks in preset batch weights, so avoiding emptying for reweighing.

Vessel or equipment cleaning can also be automated, e.g. using high-pressure liquid sprays. This is essential, whenever practicable, to avoid entry under the Confined Spaces Regulations 1997.

Process change

A modification to the chemical process or manufacturing operation can reduce risk, for example:

- Purification of raw materials.
- Use of solutions, slurries, pellets, granules or 'dust-free' (i.e. partially-wetted) powders instead of dry powders.
- Centralized make-up of toxic chemicals in master batches for transfer in sealed containers or impermeable bags.
- Transfer of an active chemical agent in an inherently safer form (e.g. sulphur dioxide as sodium metabisulphite, chlorine as sodium hypochlorite). Generation of an active agent in this manner clearly reduces the inventory in use.

Suppression

Release of liquids (as mists or sprays), of vapour or of dusts may be reduced in some cases by suppression methods. Such practices include:

- Pre-wetting of powders or fibrous solids. This extends to wet sweeping (if vacuum cleaning is impractical).
- Use of a cover on open-topped tanks, vats or portable containers when not in use, or other methods to reduce the exposed liquid surface.
- Use of floating roof tanks.
- Lowering the operating temperature of process liquids.
- Provision of a partial seal on exposed liquid interfaces, e.g. a foam blanket or a layer of floating inert spheres.
- Provision of a vapour recovery system on storage tanks.

Monitoring of equipment operation and of process parameters

The use of appropriate instruments to monitor equipment operation and relevant process variables will detect, and provide warning of, undesirable excursions. Otherwise these can result in equipment failure or escape of chemicals, e.g. due to atmospheric venting, leakage or spillage. Instruments may facilitate automatic control, emergency action such as coolant or pressure relief or emergency shutdown, or the operation of water deluge systems.

Parameters which may be monitored include:

- Electrical power drawn by prime movers.
- Equipment vibration.

- Coolant flow and temperatures (in and out); low flow and high temperature.
- Composition of process and effluent streams.
- Pressure or vacuum; high or low pressure.
- Process temperature; high or low temperature.
- Flow rates of process fluids; high or low flow.
- Pressure drop.
- Oxygen concentration.
- pH; high or low pH.
- Liquid (or particulate solid) level; high or low level.
- Atmospheric concentration of specific pollutants.

Segregation

Segregation is a common means of controlling toxic risks, or restricting the working area exposed to them. Segregation may be by any, or a combination, of:

- Distance, e.g. spacing of equipment, operating stations, storage, buildings.
- Physical barriers, e.g. splashguards, screens, or use of separate rooms. In special cases equipment enclosures and/or complete rooms may be maintained under a slight negative pressure.
- Time, e.g. performance of cleaning, demolition or stripping operations 'out-of-hours'. Rotation of jobs, and limitation of exposure, e.g. in confined spaces, are also examples of partial segregation.

Containment

Contamination of the working environment can be prevented by complete containment, i.e. complete enclosure as in glove boxes in a laboratory or operation in sealed equipment. For materials transfer, balancing is preferred to venting. However, additional precautions are necessary for cleaning, emergency venting, sampling – foreseeable events which could result in unplanned leakage or spillage. Minimize pumping or blowing, sample points, pipe joints and equipment requiring maintenance, and avoid hoses unless reinforced and properly secured, e.g. armoured hose. Consider tank draining/recovery arrangements. Provide adequate instrumentation on storage tanks, e.g. temperature indicator, high temperature alarm, pressure indicator, high pressure alarm and level gauge, together with relief if necessary (safety valve backed up by rupture disk to prevent seepage). Provide bunds to contain spillages and protect drainage systems and sewers.

Local exhaust ventilation

Because complete containment is physically impracticable in many cases, local exhaust ventilation is often applied to remove contaminants. The objective is to extract pollutant as near as practicable to its source and before it enters, or passes through, a worker's breathing zone. Vents should lead away from personnel to a safe location, with scrubbing/filtering as appropriate. Common examples are:

- Laboratory fume cupboards, operated with the front sash at the correct setting and providing the requisite linear air velocity at the gap.
- Lip extraction at anticipated points of leakage (e.g. around open-topped tanks, sampling, drumming/packing points).
- Open-fronted extraction booths for spraying operations, or adhesive application operations.

Typical minimum transport velocities are given in Table 12.13 and capture velocities for various applications in Table 12.12.

General ventilation, which relies upon dilution by a combination of fresh air make-up and removal, is a secondary measure.

Personal protection

The provision and use of properly selected personal protective equipment is normally regarded as back-up for the previous measures. Refer to Chapter 13. In some situations it is the only reasonably practicable measure to ensure personal safety and its use may be a legal requirement. Examples are:

- For entry into a confined space which may contain toxic chemicals or be oxygen deficient. Refer to Figure 13.1.
- During certain maintenance operations.
- During fire-fighting or emergency rescue operations.
- As a standby for emergency use in case of accidental release of toxic materials, e.g. during tanker unloading, or disconnection of temporary pipelines or when dealing with spillages generally, or if other protective measures, e.g. local exhaust ventilation, fail in service.
- As protection against chemicals to which no exposure is permissible or desirable.

Personal hygiene

A good standard of personal hygiene is required to minimize exposure by ingestion or skin absorption of chemicals. The measures include:

- Adequate washing facilities with hot and cold running water, soap or hand cleanser, and drying provisions all conveniently located.
- Supply of an appropriate barrier cream (i.e. for 'wet' or 'dry' work) and afterwork cream.
- Adequate showering/bathing facilities.
- Avoidance of use of solvents, abrasive powders or process chemicals for skin cleaning.
- Provision of overalls of an appropriate type and their frequent laundering either in-house or by an approved contractor, with a prohibition on their unauthorized removal from the workplace or use in e.g. canteens. Disposable overalls are appropriate in some situations.
- Prompt attention to, and covering of, damaged or perforated areas of skin.
- Avoidance of eating, drinking, the application of cosmetics, or smoking in the work area.

Medical supervision and biological monitoring may be appropriate.

Training for all staff, covering both normal operation and emergency situations, is essential.

The combination of measures used will depend upon the degree of hazard, and the scale and nature of the processes. For example, dust and fume control measures in the rubber industry are summarized in Table 5.19.

Control of substances hazardous to health

In Great Britain the COSHH Regulations cover virtually all substances hazardous to health. (Excluded are asbestos, lead, materials dangerous solely due to their radioactive, explosive, or flammable properties, or solely because of high or low temperatures or pressures, or where risk

Table 5.19 Combination of measures for dust and fume control in the rubber industry

Factory process	*Health hazard*	*Control measures*
Drug room	Dust from 'small drugs' (complex organic compounds)	Substitution Master batches Preweighed, sealed bags Dust-suppressed chemicals Local exhaust ventilation Care in handling
	Dust from bulk fillers and whitings	Local exhaust ventilation Care in handling
	Dust from carbon black	Master batches Local exhaust ventilation Totally enclosed systems *Not* by 'careful handling' alone
	Skin contact with process oils	Direct metering into mixer Care in handling and protective clothing.
Compounding	Dust	Local exhaust ventilation Master batches Preweighed, sealed bags Dust-suppressed chemicals Care in handling
	Fume	Local exhaust ventilation Removal of hot product from workroom – cool before handling
	Skin contact with process oils	Direct metering Care in handling and protective clothing
Moulding	Fume	Local exhaust ventilation Removal of hot product from workroom – cool before handling Deflection by shields
Calendering and extruding	Fume	Local exhaust ventilation Water cooling of extrudate
	Dust from release agents (chalk stearate or talc)	Substitution of wet methods Enclosure and local exhaust ventilation
Curing	Fume	Local exhaust ventilation at autoclave door and storage racks Allow autoclave to cool before opening
Spreading	Fume	Local exhaust ventilation Care in handling mixes

to health arises due to administration in the course of medical treatment, and substances below ground in mines, which have their own legislation.)

Substances 'hazardous to health' include substances labelled as dangerous (i.e. very toxic, toxic, harmful, irritant or corrosive) under any other statutory requirements, agricultural pesticides and other chemicals used on farms, and substances with occupational exposure limits. They include harmful micro-organisms and substantial quantities of dust. Indeed any material, mixture or compound used at work, or arising from work activities, which can harm people's health is apparently covered.

The regulations set out essential measures that employers (and sometimes employees) have to take:

- Prohibit use of substances listed in Table 5.20.
- Assess the risk to health arising from work, and what precautions are needed (see Figure 5.3).

Table 5.20 Prohibition of certain substances hazardous to health for certain purposes

Substance	*Purpose for which substance is prohibited*
2-Naphthylamine; benzidene; 4-aminodiphenyl; 4-nitrodiphenyl; their salts and any substance containing any of those compounds, in a total concentration equal to or greater than 0.1% by mass	Manufacture and use for all purposes including any manufacturing process in which the substance is formed
Sand or other substance containing free silica	Use as an abrasive for blasting articles in any blasting apparatus
A substance: (a) containing compounds of silicon calculated as silica to the extent of more than 3% by weight of dry material, other than natural sand, zirconium silicate (zircon), calcined china clay, calcined aluminous fireclay, sillimanite, calcined or fused alumina, olivine; or (b) composed of or containing dust or other matter deposited from a fettling or blasting process	Use as a parting material in connection with the making of metal castings
Carbon disulphide	Use in cold-cure process of vulcanizing in the proofing of cloth with rubber
Oils other than white oil, or of entirely animal or vegetable origin or entirely of mixed animal and vegetable origin	Use in oiling the spindles of self-acting mules
Ground or powdered flint or quartz other than natural sand	Use in relation to the manufacture or decoration of pottery for the following purposes: (a) the placing of ware for the biscuit fire; (b) the polishing of ware; (c) as the ingredient of a wash for saggers, trucks, bats, cranks or other articles used for supporting ware during firing; and (d) as dusting or supporting powder in potters' shops
Ground or powdered flint or quartz other than: (a) natural sand; or (b) ground or powdered flint or quartz which forms part of a slop or paste	Use in relation to the manufacture or decoration of pottery for any purpose except: (a) use in a separate room or building for (i) the manufacture of powdered flint or quartz or (ii) the making of frits or glazes or the making of colours or coloured slips for the decoration of pottery; (b) use for the incorporation of the substance into the body of ware in an enclosure in which no person is employed and which is constructed and ventilated to prevent the escape of dust
Dust or powder of a refractory material containing not less than 80% of silica other than natural sand	Use for sprinkling the moulds of silica bricks, namely bricks or other articles composed of refractory material and containing not less than 80% of silica
White phosphorus	Use in the manufacture of matches
Hydrogen cyanide	Use in fumigation except when: (a) released from an inert material in which hydrogen cyanide is absorbed; (b) generated from a gassing powder; or (c) applied from a cylinder through suitable piping and applicators other than for fumigation in the open air to control or kill mammal pests
Benzene and any substance containing benzene in a concentration equal to or greater than 0.1% by mass other than: (a) motor fuels } covered by specific EEC (b) waste } Directives	Uses for all purposes except: (a) use in industrial processes; and (b) for the purposes of research and development or for the purpose of analysis

Table 5.20 Cont'd

Substance	*Purpose for which substance is prohibited*
The following substances: chloroform, carbon tetrachloride; 1,1,2-trichloroethane, 1,1,2,2-tetratchloroethane; 1,1,1,2-tetrachloroethane; pentachloroethane, vinylidene chloride; 1,1,1-trichloroethane and any substance containing one or more of those substances in a concentration equal to or greater than 0.1% by mass, other than: (a) medical products; (b) cosmetic products	Supply for use at work in diffusive applications such as in surface cleaning and the cleaning of fabrics except for the purposes of research and development or for the purpose of analysis

- Introduce appropriate measures to prevent or control the risk.
- Ensure that control measures are used and that equipment is properly maintained and procedures observed.
- Where necessary, monitor the exposure of the workers and carry out an appropriate form of surveillance of their health.
- Inform, instruct and train employees about the risks and the precautions to be taken.

Assessment

The HSE provide elegant guidance and checklists for conducting and recording risk assessments in *COSHH Essentials* (see Bibliography). This is supplemented by guidance sheets on ventilation, engineering controls and containment for a variety of unit operations including charging reactors; dipping; filling/emptying sacks/kegs/drums; mixing; sieving; weighing.

The basic steps in any assessment include a review of:

1 What substances are present? In what form?

(a) Substances brought into the workplace.
(b) Substances given off during any process or work activity.
(c) Substances produced at the end of any process or work activity (service activities included).

Substances 'hazardous to health' can be identified by:

- for brought-in substances, checking safety information on labels and that legally obtainable from the suppliers, e.g. on their Material Safety Data sheet: making sure it is the most up-to-date version;
- use of existing knowledge, e.g. past experience, knowledge of the process, understanding of relevant current best industrial practice, information on related industrial health problems;
- seeking advice from a trade association, others in a similar business, consultants;
- checking whether a substance is mentioned in any COSHH Regulations or Schedules, or listed in Guidance Note EH 40;
- examination of published trade data, HSE guidance information, literature or documentation;
- checking Part 1 of the approved supply list under the Chemicals (Hazard Information and Packaging for Supply) Regulations 1994. (Anything listed as very toxic, toxic, corrosive, harmful or irritant is covered by COSHH.)

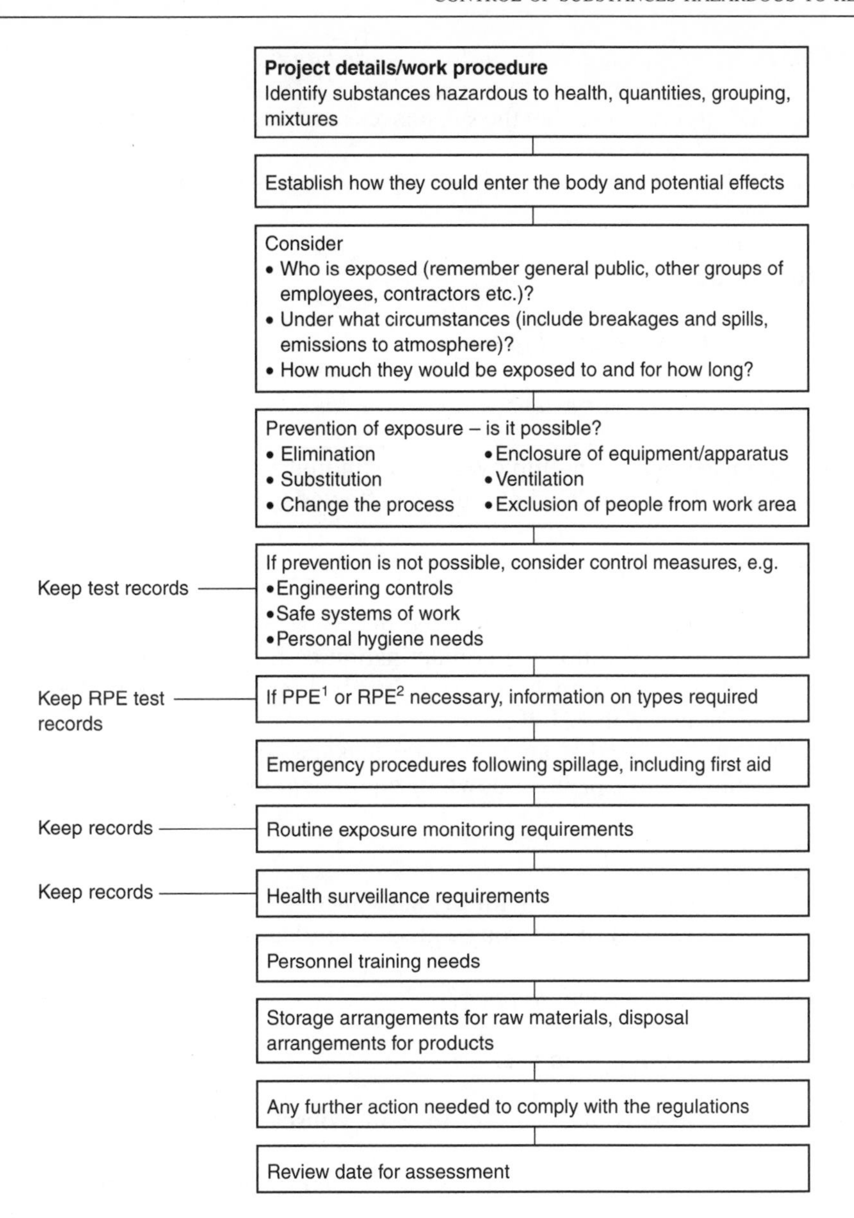

Figure 5.3 *COSHH assessment procedure*
[1] PPE = Personal Protective Equipment
[2] RPE = Respiratory Protective Equipment

2 What is the health hazard?

- if breathed in, on contact with the skin or eyes, or if ingested?
- quantity of material used, i.e. small (grams or millilitres), medium (kilograms or litres), or large (tonnes or cubic metres)?
- how dusty or volatile is the substance?

3 Where and how are the substances actually used or handled?

- Where and in what circumstances are the substances handled, used, generated, released, disposed of etc.?
- What happens to them in use (e.g. does their form change – such as from bulk solid to dust by machining)?
- Identify storage and use areas.
- Identify modes of transport.

4 What harmful substances are given off etc.?

5 Who could be affected, to what extent and for how long?

Identify both employees and non-employees – including cleaners, security staff, employees, contractors, members of the public who could be affected.

6 Under what circumstances?

- Is some of the substance likely to be breathed in?
- Is it likely to be swallowed following contamination of fingers, clothing etc.?
- Is it likely to cause skin contamination or be absorbed through the skin? (NB some materials have a definite Sk notation in EH 40.)
- Is it reasonably foreseeable that an accidental leakage, spill or discharge could occur (e.g. following an operating error or breakdown of equipment or failure of a control measure)?

Consider:

- How are people *normally* involved with the substance?
- How might they be involved (e.g. through misuse, spillage)?

7 How likely is it that exposure will happen?

Check control measures currently in use.

- Check on their effectiveness and whether they are conscientiously/continuously applied.

8 What precautions need to be taken to comply with the rest of the COSHH Regulations?

Having regard to

- who could be exposed,
- under what circumstances,
- the level and possible length of time,
- how likely exposure is,
- the environmental hazards,

together with knowledge about the hazards of the substance (i.e. its potential to cause harm), conclusions are reached about personal exposure.

The employer's duty is to ensure that the exposure of employees to a hazardous substance is prevented or, if this is not reasonably practicable, adequately controlled. Duties under the Regulations extend with certain exceptions to other persons, whether at work or not, who may be affected by the employers work.

Control

Prevention of exposure should be given priority, e.g. by:

- changing the process or method of work to eliminate the operation resulting in the exposure;
- process modification to avoid production of a hazardous product, by-product or waste product;
- substitution of a hazardous substance by a new, or different form of the same, substance which presents less risk to health.

If for a carcinogen prevention of exposure is not reasonably practicable by using an alternative substance or process there is a requirement to apply *all* the measures listed in Table 5.21. If these measures do not provide adequate control then suitable personal protective equipment as will adequately control exposure must be provided.

Table 5.21 Measures for the control of exposure to carcinogens

- Total enclosure of the process and handling systems unless not reasonably practicable.
- Use of plant, processes and systems of work which minimize the generation of, or suppress and contain, spills, leaks, dust, fumes and vapours.
- Limitation of quantities in the workplace.
- Keeping the number of persons who might be exposed to a minimum.
- Prohibition of eating, drinking and smoking in areas that may be contaminated.
- Provision of hygiene measures including adequate washing facilities and regular cleaning of walls and surfaces.
- Designation of those areas and installations which may be contaminated and the use of suitable and sufficient warning signs.
- Safe storage, handling and disposal, and use of closed and clearly labelled containers.

For hazardous substances not classified as carcinogens, where protection of exposure is not reasonably practicable, adequate control should be achieved by measures other than personal protection, so far as is reasonably practicable. This is subject to the degree of exposure, circumstances of use of the substance, informed knowledge about the hazards and current technical developments. Any combination of the measures listed in Table 5.22 are applicable.

Table 5.22 Measures for the control of exposure to hazardous substances not classified as carcinogens

- Totally enclosed process and handling systems.
- Plant or processes or systems of work which minimize generation of, or suppress or contain, the hazardous dust, fume, biological agent etc. and limit the area of contamination in the event of spills and leaks.
- Partial enclosure with local exhaust ventilation.
- Local extract ventilation.
- Sufficient general ventilation.
- Reduction of number of employees exposed.
- Exclusion of non-essential access.
- Reduction in the period of exposure for employees.
- Regular cleaning of contamination from, or disinfection of, walls, surfaces etc.
- Provision of means for safe storage and disposal.
- Prohibition of eating, drinking, smoking, application of cosmetics etc. in contaminated areas.
- Provision of adequate facilities for washing, changing and storage of clothing, with arrangements for laundering contaminated clothing.

Again when the measures in Table 5.22 do not prevent, or provide adequate control of exposure there is a requirement to provide suitable personal protective equipment to accomplish it. This includes respiratory protection, protective clothing generally, footwear and eye protection which, in the UK, complies with the Personal Protective Equipment Regulations 1992. All routes of exposure, e.g. inhalation, ingestion, absorption through the skin or contact with the skin, must be considered.

Control measures in existing work situations should be reviewed, extended or replaced as necessary to achieve and sustain adequate control.

If leaks, spills or uncontrolled releases of a hazardous substance could occur, means are required for limiting the extent of health risks and for regaining adequate control as soon as possible. Where appropriate means should include:

- establish emergency procedures;
- safe disposal of the substance;
- sufficient suitable personal protective equipment to enable the source of the release to be safely identified and repairs to be made;
- exclusion of all persons not concerned with the emergency action from the area of contamination;
- in the case of carcinogens, ensuring that employees and other persons who may be affected by an escape into the workplace are kept informed of the failure forthwith.

Exposure limits

Exposures require control such that nearly all people would not suffer any adverse health effects even if exposed to a specific substance (or mixture of substances) day after day. For certain substances there are set occupational exposure limits: refer to page 78.

As noted earlier, routes other than inhalation must also be considered. Thus exposure to a substance which can be hazardous upon ingestion, absorption through the skin or mucous membranes, or contact with skin or mucous membranes needs control to a standard such that nearly all the population could be exposed repeatedly without any adverse health effect. (Note that this will not necessarily protect those who are atopic or with a relevant pre-existing condition, e.g. dermatitis.)

Maintenance, examination and testing of control measures

An employer has specific obligations to ensure all control measures are kept in an efficient state, efficient working order and good repair. Engineering controls should be examined and tested at suitable intervals, e.g. local exhaust ventilation equipment must be tested at least once every fourteen months, and more often for processes specified in Table 5.23, and a record kept. Respirators and breathing apparatus must also be examined frequently and the checks recorded.

Monitoring

The exposure of workers should be monitored in certain cases, e.g.

- substances or processes listed in Table 5.24;
- where it is not certain that particular control measures are working properly;
- where it is not possible to be sure that exposure limits are not being exceeded;
- where there could be serious risks to health if control measures were to fail or deteriorate.

Table 5.23 Frequency of thorough examination and test of local exhaust ventilation plant used in certain processes (Schedule 4, Reg. 9(2)(a))

Process	*Minimum frequency*
Blasting in, or incidental to cleaning of metal castings, in connection with their manufacture	1 month
Processes, other than wet processes, in which metal articles (other than gold, platinum or iridium) are ground, abraded or polished using mechanical power, in any room for more than 12 hours per week	6 months
Processes giving off dust or fume in which non-ferrous metal castings are produced	6 months
Jute cloth manufacture	1 month

Table 5.24 Specific substances and processes for which monitoring is required (Schedule 5, Reg. 10(2))

Substance or process	*Minimum frequency*
Vinyl chloride monomer	Continuous or in accordance with a procedure approved by the Health and Safety Executive
Spray liberated from vessels at which an electrolytic chromium process is carried on, except trivalent chromium	Every 14 days while the process is being carried on

A record should be kept of any monitoring for at least 5 years, unless it is representative of personal exposure of identifiable employees when records must be retained for at least 40 years.

Personal/workplace air monitoring

Sampling strategies may include measurement of the hazardous substance:

- in the breathing zone of a worker (personal dosimetry); and/or
- in the workplace air (see Chapter 10).

Biological monitoring

For a few substances exposure may be assessed using biological monitoring (see page 114). Depending upon the substance the sampling strategy varies from post shift, random, or pre-shift the day after exposure.

Health surveillance

If a known adverse health effect can reasonably be anticipated under the circumstances of work – and could readily be observed – some form of health surveillance is appropriate. This may involve a doctor or trained nurse. It may include the checking of employees' skin for dermatitis or asking questions relevant to any asthmatic condition where work is with recognized causative agents (e.g. epoxy resin curing agents).

In the UK health surveillance is a statutory requirement for the agents, operations and processes summarized in Table 5.25. Advice on health surveillance is also given for the agents listed in Table 5.26.

Table 5.25 UK health surveillance requirements

Medical surveillance is required unless exposure is insignificant (Schedule 6 to COSHH Reg. 11(2)a and 5)	
Substance	*Process*
Vinyl chloride monomer (VCM)	In manufacture, production, reclamation, storage, discharge, transport, use or polymerization
Nitro or amino derivatives of phenol and of benzene or its homologues	In the manufacture of nitro or amino derivatives of phenol and of benzene or its homologues and the making of explosives with the use of any of these substances
Potassium or sodium chromate or dichromate	In manufacture
Orthotolidine and its salts Dianisidine and its salts Dichlorobenzidine and its salts	In manufacture, formation or use of these substances
Auramine Magenta	In manufacture
Carbon disulphide Disulphur dichloride Benzene, including benzol Carbon tetrachloride Trichloroethylene	Processes in which these substances are used, or given off as vapour, in the manufacture of indiarubber or of articles or goods made wholly or partially of indiarubber
Pitch	In manufacture of blocks of fuel consisting of coal, coal dust, coke or slurry with pitch as a binding substance
Health surveillance is appropriate unless exposure is insignificant (Control of Carcinogenic Substances ACOP, 15–18)	
	Selected relevant legislation
Asbestos	Control of Asbestos at Work Regulations 1987 and subsequent amendments
Compressed air (other than diving operations)	Work in Compressed Air Special Regulations 1958
Diving operations	Diving Operations at Work Regulations 1981 and subsequent amendments
Flint, quartz, transfers, colours, frits, glazes, dusts	Approved Code of Practice. Control of substances hazardous to health in the production of pottery. Control of substances hazardous to health regulations.
Ionizing radiations	Ionizing Radiations Regulations 1999
Lead	Control of lead at Work Regulations 1998
Pesticides	MAFF/HSC Code of Practice for the safe use of pesticides on farms and holdings. Approved Code of Practice. Safe use of pesticides for non-agricultural purposes. Control of substances hazardous to health regulations.

Health records must be kept of the health surveillance carried out for at least 40 years after the last entry. Appropriate action should be taken based upon the results, i.e. it should be established how and when workers should be referred for further examination and how the results will be used to improve the management of health risks.

Table 5.26 Agents for which health surveillance is advised

Agent	*UK HSE Guidance Note*
Agents liable to cause skin disease	EH 26
Antimony	EH 19
Arsenic	EH 8
Beryllium	EH 13
Cotton dust	MS 9
Agents causing genetic modification	ACGM/HSE Note 4
Isocyanates	EH 16 MS 8
Mineral wool	EH 46
Platinum	MS 22
Talc dust	EH 32
Biological monitoring	
Cadmium	EH 1
Mercury	EH 17 MS 12
Trichloroethylene	EH 5
Biological effect monitoring	
Organophosphorus pesticides	MS 17

Information supply

There is requirement to train and inform employees of:

- the risks arising from their work
- the precautions to be taken
- the results of any monitoring carried out
- the collective (anonymous) results of any health surveillance carried out.

Specific precautions

Ways in which these principles are applied in practice are illustrated in the following section using common potentially hazardous operations or substances:

- Everyday operations such as painting and welding.
- Toxic dusts such as asbestos and catalysts.
- Hyperpoisons such as cyanides.
- Insecticides.
- Primary irritants and corrosives.
- Common gases such as oxides of carbon and nitrogen, hydrogen sulphide, and inert gases.
- Liquids which pose a health hazard due to volatilization, e.g. mercury and degreasing with chlorinated solvent, i.e. dry cleaning with perchloroethylene or metal cleaning with trichloroethylene.
- Liquids posing problems because of the presence of impurities, e.g. mineral oils.
- Use of a strong disinfectant/biocide, i.e. glutaraldehyde.
- Machining operations on metals involving cooling by fluids.
- Application of synthetic resins, e.g. epoxy resins.
- Gases present in buildings, e.g. offices.

(i) Asbestos

This ubiquitous material previously found use in construction materials, lagging, brake linings etc. If inhaled, asbestos dust may result in serious respiratory disease (e.g. asbestosis, lung cancer, mesothelioma of the pleura). Therefore strict control must be exercised over all work with asbestos products which may give rise to dust. Within the UK, the Control of Asbestos at Work Regulations 1987 as amended by the Control of Asbestos at Work (Amendment) Regulations 1992 and 1998, and Approved Codes of Practice apply to all such work, including manufacturing, processing, repairing, maintenance, construction, demolition, removal and disposal. Because of their wider relevance their requirements are summarized in Table 5.27.

(ii) Catalysts

Catalysts are often used to increase the rate of reactions (Chapter 3). Like many chemicals they can pose health risks to workers (Table 5.28) unless handled with care. They can be either:

- homogeneous catalysts dispersed with reactants so that reaction takes place in a single phase. The catalyst is added to the reactor with other process ingredients and removed by the normal finishing separation processes. Worker exposure is similar to those for other process materials;
- heterogeneous catalysts where the catalysis occurs at a solid interface, often used in the form of fixed beds. These must be regenerated or replaced periodically posing significant exposure risks.

Heterogeneous catalysts are often located at the top of a reactor and manipulated with temporary handling equipment. To avoid exposure to toxic dust, local ventilation should be installed; if this is impracticable, scrupulous use of personal protective equipment and rigid compliance with systems-of-work are essential. Respiratory equipment may include self-contained or line-fed breathing apparatus.

Skin protection may necessitate use of full protective suits. When catalysts are dumped from reactors at the end of a process they may prove to be extremely dusty as a result of reduction in particle size during the reaction process. Again, depending upon the nature of the hazard, ventilation, personal protection, and use of temporary enclosures to prevent contamination of the general work area should be considered. Some catalysts are pyrophoric and some catalyst beds are inerted with the added possibility of fire, or release of inerting gas into the workplace which may cause asphyxiation.

Aluminium oxide may induce respiratory irritation upon inhalation of high concentrations resulting in emphysema and flu-like symptoms. Some catalysts are sensitive to exposure to moist air. Aluminium alkyls may be pyrophoric and personal protection must be worn to prevent skin burns. Aluminium chloride reacts with moisture in air to produce steam and irritant hydrogen chloride and with moisture in the eyes, mucous membranes or skin. It is on the basis that 3 moles of hydrogen chloride with a ceiling TLV of 5 ppm hydrolyse from one mole of $AlCl_3$, that an 8 hr TWA TLV of 2 mg/m^3 for $AlCl_3$ as Al has been set to offer the same degree of freedom from irritation that is provided by the TLV for HCl. The material should therefore be stored in a cool, dry, well-ventilated place and the bulk stocks must be waterproof and segregated from combustibles. Pressure build-up due to evolution of hydrogen chloride should be safely vented. Depending upon scale of operation, goggles, face-shield, gloves, shoes and overalls of acid-resistant materials should be worn. Transfer should be in dry air or under a nitrogen blanket. Process fumes/dust should be collected via a scrubber. Spillages should be collected before washing the area with copious volumes of water.

Table 5.27 Summary of precautions for work involving asbestos

Assessment

Before starting any work which is liable to expose employees to asbestos dust, an assessment of the work is required to help decide the measures necessary to control exposure. This should:

- Identify the type of asbestos (or assume that it is crocidolite or amosite, to which stricter controls are applicable than to chrysotile).
- Determine the nature and degree of exposure.
- Set out steps to be taken to prevent that exposure, or reduce it to the lowest level reasonably practicable.

The assessment should be in writing except if the work involves low level exposure and is simple, so that the assessment can be easily repeated and explained.

Control limits

(Fibres per millimetre)	4 hrs	10 min
chrysotile	0.3	0.9
any other form of asbestos alone or in mixtures	0.2	0.6

Employees should never breathe air containing a level of asbestos which exceeds these limits. Moreover the level should always be reduced so far as it reasonably can be. Use should be made of:

- Suitable systems of work.
- Exhaust ventilation equipment.
- Other technical measures.
- All of these techniques if reasonably practicable.

If the dust level is, or could be, above the control limit an employer must:

- Provide suitable respiratory protective equipment and ensure that it is used properly.
- Post warning notices that the area is a 'respirator zone'.

Action levels

Action levels are a measure of the total amount of asbestos to which a person is exposed within a 12 week period. These are set in fibres/hr per millilitre:

	over 12 weeks
where exposure is solely to chrysotile	72
where exposure is to any other form of asbestos, alone or in mixtures	48
where both types of exposure occur in the 12 week period at different times	a proportionate number

When these are, or may be, exceeded the employer must ensure that the enforcing authority has been notified, maintain a health record of exposed workers and make sure that they receive regular medical examinations, and identify work areas where the action level is liable to be exceeded as 'asbestos areas'.

Other provisions

There are also requirements for an employer to:

- Monitor the exposure of employees to asbestos where appropriate.
- Ensure that employees liable to be exposed to asbestos receive adequate information, instruction and training – so that they are aware of the risks and the precautions which should be observed.
- Provide protective clothing for workers when a significant quantity of asbestos is liable to be deposited on their clothes.
- Check that the plant or premises where work with asbestos is carried out is kept clean.
- Make sure that there are adequate washing and changing facilities.
- Provide separate storage areas for any protective clothing and respiratory protective equipment required, and for personal clothing.
- Make sure that all asbestos articles, substances and products for use at work are specially labelled.
- Keep raw asbestos and asbestos waste sealed and labelled.

Table 5.28 Health effects of catalysts

Catalyst	*Uses*	*Health effect*
Aluminium oxide	Hydrotreating petroleum feedstocks Fluid cracking Autoexhausts	Nuisance
Aluminium chloride	Resin manufacture by polymerization of low molecular-weight hydrocarbons Friedel–Crafts reactions to manufacture detergent alkylate, agrochemicals, drugs	Irritation due to formation of HCl with moisture
Aluminium alkyls	Alkylations/Grignard reactions	Acute thermal burns, lung damage
Chromic oxide		Cr^{3+} may be converted to the more toxic and carcinogenic Cr^{6+}
Colbalt	Hydrogenations of solid fuels and fuel oils Manufacture of terephthalic acid High pressure production of aldehydes	Lung irritation (hard metal disease); respiratory sensitization
Ferric oxide	Oxidations	Siderosis
Molybdenum compounds	Hydrodesulphurization and hydrotreating of petroleum Oxidation of methanol to formaldehdye Epoxidation of olefins Decomposition of alkali metal nitrides	Irritation of eyes and respiratory tract Pneumoconiosis
Nickel compounds	Hydrogenations (e.g. Raney nickel) Conversion of synthesis gas to methane Reduction of organo nitro compounds to amines	Carcinogenic (nickel subsulphide). Skin sensitization
Nickel carbonyl	Carbonylation of acetylene and alcohols to produce acrylic and methacrylic acids	Acute respiratory failure; carcinogenic
Platinum compounds	Hydrosilation cross-linking of silicone polymers Hydrogenation, isomerization and hydroformylation of alkenes Automobile exhaust catalyst	Sensitization dermatitis
Vanadium	Pollution control, e.g. removal of hydrogen sulphide and in manufacture of sulphuric acid	Respiratory irritation; green–black tongue (transient)

Chromium oxide and chromium supported on other oxides such as aluminium oxide are important catalysts for a wide range of reactions. Chromium forms several oxides, the most important of which are Cr_2O_3, CrO_2 and CrO_3. None are without problems and whilst it is often thought that trivalent Cr compounds are of low toxicity, dermatitis and pulmonary disease may result from exposure. The hexavalent compounds such as CrO_3 are more toxic with potential to cause irritant and allergic contact dermatitis, skin ulcers (including 'chrome holes'), nasal irritation and kidney damage. Some water-insoluble compounds have been associated with an increased risk of lung cancer. An 8 hr TWA MEL has been set at 0.05 mg/m^3 for Cr VI compounds. CrO_3 can react with reducing agents including organic compounds (e.g. acetic acid, aniline, quinoline, alcohol, acetone, thinners, and grease) vigorously to cause fires and explosions. On heating to 250°C it liberates oxygen to further support combustion. Containment, or use of ventilation, and personal protective equipment such as rubber gloves, respirators, overalls, rubber aprons, rubber boots may be necessary depending upon the risk and nature of exposure. If the process is routine, atmospheric analysis and biological monitoring backed up with health surveillance may also be required. Stocks should be protected from physical damage, stored in a dry place away from combustible materials and easily oxidizable substances. Avoid storage on wooden floors.

Long-term exposure to ferric oxide dust can cause changes to the lungs which are detectable by X-rays. For this reason an 8 hr TWA TLV of 5 mg/m^3 has been set. Good ventilation is important for processes involving this compound. For regular use routine medical examination and exclusion of staff with pulmonary disease may be necessary.

Some nickel compounds may be irritant to skin and eyes and dermal contact with nickel can result in allergic contact dermatitis. Nickel carbonyl is extremely toxic by inhalation and should be handled in totally enclosed systems or with extremely efficient ventilation. Air monitors linked to alarms may be required to detect leaks. Respiratory equipment must be available for dealing with leaks. Biological checks (e.g. nickel in urine) should be considered for routine operations involving nickel catalysts.

Platinum is used as a catalyst for nitric and sulphuric acid production, in petroleum refining and in catalytic mufflers to control air pollution. Platinum salts can cause respiratory complaints, asthma, and 'platinosis', an allergic response. Allergic dermatitis may also result from exposure to soluble platinum salts and once subjects have been sensitized it generally precludes continued occupational exposure at any level. The 8 hr TWA OEL for platinum metal is 5 mg/m^3 but for soluble platinum salts it is only 0.002 mg/m^3. Handling precautions must include containment where possible, ventilation, personal protection, and the screening out of individuals who have become sensitized.

Vanadium as the pentoxide is used as a catalyst in the oxidation of sulphur dioxide, oxides of nitrogen, and other substances. Vanadium is poisonous by any route in any but small doses and the pentavalent state, such as V_2O_5, is the most hazardous. Upon inhalation, the main effects are on the respiratory passages causing tracheitis, bronchitis, emphysema, pulmonary edema, or bronchial pneumonia. Symptoms of acute exposure may include nausea, vomiting, high temperature, diarrhoea, nervous malfunction and frequent coughs whilst those of chronic exposure are pale skin, anaemia, vertigo, cough, high blood pressure, green discoloration of tongue, tremor of fingers and nervous malfunction. In animal studies exposure to 70 mg/m^3 V_2O_5 dust was fatal within a few hours. An 8 hr TWA TLV of just 0.05 mg/m^3 has been set in the USA by the ACGIH. Clearly, processes must be designed such that dust formation is prevented. Where exposure is possible ventilation, personal protection including respiratory protection, medical surveillance, atmospheric monitoring and high standards of personal hygiene should be considered to ensure exposure is controlled.

(iii) Common gases (see also *Chapter 9)*

(a) Carbon dioxide

Carbon dioxide gas can act as an asphyxiant due to displacement of air, resulting in oxygen deficiency (page 262). Sources include:

- Fires, because it is inevitably a product of combustion from any carbon-based fuel.
- Use as an inert gas.
- Discharge of carbon dioxide extinguishers.
- Use of solid 'cardice' as a cryogen (page 261).
- Natural processes, e.g. fermentation.
- Water from certain underground strata, due to de-gassing (page 46).
- The neutralization of acids with carbonates or bicarbonates.
- As a byproduct of the synthesis of ammonia, hydrogen.

The hazard is particularly acute in confined spaces.

The gas is also toxic as exemplified by Table 5.29. Furthermore, the increased respiratory rate may cause increased amounts of other toxic gases, e.g. carbon monoxide in fires, to be inhaled.

Table 5.29 Typical reactions of persons to carbon dioxide in air

Carbon dioxide concentration (ppm)	(%)	*Effect*
5000	0.5	TLV/OEL-TWA: can be tolerated for 8 hr exposure with no symptoms and no permanent damage
15 000	1.5	OEL-STEL: 10 min
20 000	2.0	Breathing rate increased by 50%
30 000	3.0	TLV-STEL: breathing rate increased by 100%
50 000	5.0	Vomiting, dizziness, disorientation, breathing difficulties after 30 min
80 000	8.0	Headache, vomiting, dizziness, disorientation, breathing difficulties after short exposure
100 000	10.0	Headache, vomiting, dizziness, disorientation, unconsciousness, death after a few minutes

The special precautions appropriate for entry into confined spaces are summarized in Chapter 13. In fires, evacuation of burning buildings, prohibition on re-entry and the use of self-contained breathing apparatus by fire-fighters are key precautions.

(b) Carbon monoxide

Carbon monoxide is a colourless, odourless gas and – without chemical analysis – its presence is undetectable. It is produced by steam reforming or incomplete combustion of carbonaceous fuels; typical carbon monoxide concentrations in common gases are given in Table 5.30.

Table 5.30 Typical carbon monoxide concentrations in gases

Gas	*Typical carbon monoxide concentration* (%)
Blast furnace gas	20–25
Coal and coke oven gas	7–16
Natural gas, LPG (unburnt)	nil
Petrol or LPG engine exhaust gas	1–10
Diesel engine exhaust gas	0.1–0.5

Carbon monoxide is extremely toxic by inhalation since it reduces the oxygen-carrying capacity of the blood. In sufficient concentration it will result in unconsciousness and death. Typical reactions to carbon monoxide in air are summarized in Table 5.31.

Table 5.31 Typical reactions of persons to carbon monoxide in air

Carbon monoxide (ppm)	*Effect*
30	Recommended exposure limit (8 hr time-weighted average concentration)
200	Headache after about 7 hr if resting or after 2 hr exertion
400	Headache with discomfort with possibility of collapse after 2 hr at rest or 45 min exertion
1200	Palpitation after 30 min at rest or 10 min exertion
2000	Unconscious after 30 min at rest or 10 min exertion

The STEL is 200 ppm but extended periods of exposure around this, particularly without interruption, raise concern for adverse health effects and should be avoided.

If a potential carbon monoxide hazard is identified, or confirmed by atmospheric monitoring, the range of control techniques summarized on page 280 must be applied.

(c) Hydrogen sulphide

Hydrogen sulphide occurs naturally, e.g. in gases from volcanoes, undersea vents, swamps and stagnant water. It is also a byproduct of many industrial processes, e.g. coking and hydro-desulphurization of crude oil or coal. It is a highly toxic gas. Although readily detectable by odour at low concentrations, at high concentrations it paralyses the sense of smell and the nervous system controlling the lungs and hence acts as a chemical asphyxiant. Typical effects at different concentrations in air are summarized in Table 5.32.

Table 5.32 Typical effects or hydrogen sulphide concentrations in air

Concentration (ppm)	*Response*
0.2	Detectable odour
20–150	Conjunctivitis
150	Olfactory nerve paralysis
250	Prolonged exposure may cause pulmonary oedema
500	Systemic symptoms may occur in 0.5 to 1 hr
1000	Rapid collapse, respiratory paralysis imminent
5000	Immediately fatal

A hazard of hydrogen sulphide may be present in petroleum refining and recovery involving sour crudes, due to chemical breakdown of sulphides (e.g. by acids), or from anaerobic decomposition of sulphur-containing materials, e.g. in wells, sewers or underground pumping stations. Further properties and cylinder-handling precautions are given in Chapter 9. If hydrogen sulphide exposure is possible environmental levels should be monitored and if necessary ventilation provided and respiratory protection worn.

(d) Inert gases

Most toxicologically inert gases, e.g. nitrogen, argon, helium (and indeed common flammable gases, e.g. hydrogen, methane, propane, butane, acetylene) can generate oxygen-deficient atmospheres. These occur most often within confined spaces but may also be present near vents or open manways. The gases have no colour, smell or taste. Responses at given depleted oxygen levels are summarized in Table 5.7: to reduce the oxygen content to a fatal level requires a simple added asphyxiant gas concentration of approximately 50%.

Oxygen deficiency may arise through, for example:

- Use of nitrogen or argon to exclude air from vessels.
- Use of carbon dioxide fire extinguishers in a confined space.
- Excessive generation of e.g. nitrogen or helium gas from cryogenic liquids.
- Leakage of argon from an argon arc welding set in an unventilated enclosure.
- Formation of rust inside a closed steel tank (oxygen is removed from the atmosphere by the oxidation of iron).
- Neutralizing vessel contents with carbonate or bicarbonate, displacing the air with carbon dioxide.

Entry into a confined space requires strict control (page 417). Whenever oxygen deficiency may be encountered air quality checks should be made and appropriate breathing apparatus used.

(e) Oxides of nitrogen

Oxides of nitrogen comprise nitrous oxide (N_2O), nitric oxide (NO), nitrogen dioxide (NO_2), dinitrogen tetroxide (N_2O_4) and dinitrogen pentoxide (N_2O_5). N_2O_5 is a low-melting solid rapidly decomposing in air to NO_2/N_2O_4 and nitrogen hexoxide (NO_3). (The last is stable only below –142°C above which it decomposes into oxygen and nitrogen dioxide.)

Nitrous oxide is a colourless, non-flammable, non-corrosive gas with sweetish odour and taste. It is generally considered to be non-toxic and non-irritating but one of its main applications is use, in combination with air or oxygen, as a weak anaesthetic in medicine and dentistry. At low concentrations it produces hysteria (hence the term 'laughing gas'). A higher than expected incidence of spontaneous abortions among female workers exposed directly to anaesthetic gases has been reported but the current 8 hr TWA OES (page 99) of 100 ppm is believed sufficiently low to prevent embryofetal toxicity in humans. At high concentrations in the absence of air it is a simple asphyxiant. It is also used as a dispersing agent in whipping cream. It is oxidized in air to the dioxide. 'Nitrous fume' exposure in the main involves the inhalation of airborne NO_2/N_2O_4 mixtures – usually in an equilibrium ratio of approximately 3:7 – which at high concentrations exist as a reddish-brown gas. Sources of fume include:

- Fuming nitric acid.
- Chemical reactions with nitrogen-based chemicals, including the firing of explosives.
- Electric arc welding, flame-cutting using oxy-acetylene, propane or butane flames, or such flames burning in air.
- Forage tower silos.
- The exhaust of metal-cleaning processes.
- Fires, e.g. involving ammonium nitrate.
- Exhausts from diesel vehicles.

The effects of this mixture of gases are insidious: several hours may elapse before lung irritation develops. It is feebly irritant to the upper respiratory tract due to its relatively low solubility.

Effects of given concentrations of nitrogen oxides are listed in Table 5.33: the margin between concentrations that provoke mild symptoms and those proving to be fatal is small. A person with a normal respiratory function may be affected by exposure to as low as 5 ppm; diseases such as bronchitis may be aggravated by such exposures. The current 8 hr TWA OES is 3 ppm with an STEL (page 99) of 5 ppm.

Table 5.33 Effects of nitrogen oxides

Concentration (ppm in air)	*Effect*
<60	No warning effect (although the odour threshold is <0.5 ppm)
60–150	Can cause irritation and burning in nose and throat
100–150	Dangerous in 30–60 minutes
200–700	Fatal on short exposure (≤1 hour)

First-aid measures for people exposed to nitrogen dioxide are mentioned in Chapter 9. In any event, containment, ventilation and/or appropriate respiratory protection should be considered depending upon scale of operation and level of exposure.

(iv) Cyanides

As a group, the cyanides are among the most toxic and fast-acting poisons. (This is due to the cyanide ion which interferes with cellular oxidation.)

Hydrogen cyanide (prussic acid) is a liquid with a boiling point of 26°C. Its vapour is flammable and extremely toxic. The effects of acute exposure are given in Table 5.34. This material is a basic building block for the manufacture of a range of chemical products such as sodium, iron or potassium cyanide, methyl methacrylate, adiponitrile, triazines, chelates.

Table 5.34 Toxic effects of hydrogen cyanide

Concentration in air (ppm)	*Effect*
2–5	Odour detectable by trained individual
10	(UK MEL 10 mg/m^3 STEL (SK))
18–36	Slight symptoms after several hours
45–54	Tolerated for 3–60 min without immediate or late effects
100	Toxic amount of vapours can be absorbed through skin
110–135	Fatal after 30–60 min, or dangerous to life
135	Fatal after 30 min
181	Fatal after 10 min
270	Immediately fatal

Although organocyanides (alkyl cyanides, nitriles or carbonitriles), in which the cyanide group is covalently bonded, tend as a class to be less toxic than hydrogen cyanide, many are toxic in their own right by inhalation, ingestion or skin absorption. Some generate hydrogen cyanide under certain conditions, e.g. on thermal degradation.

The properties of selected cyanides of industrial importance are summarized in Table 5.35.

Depending upon scale of operation, precautions for cyanides include:

- techniques to contain substances and avoid dust formation (solid cyanides), aerosol formation (aqueous solutions), and leakages (gas);
- gloves, face and hand protection;
- high standards of personal hygiene;
- ventilation and respiratory protection (dust or gaseous forms);
- environmental monitoring for routine processes;
- health surveillance.

(v) Glutaraldehyde

Glutaraldehyde (1,3-diformyl propane) is a powerful, cold disinfectant. It is used principally in aqueous solution as a biocide and chemical disinfectant. It has been widely used in the health services, e.g. in operating theatres, endoscopy units, dental units and X-ray film processing.

The hazards with glutaraldehyde are those of irritation to the skin, eyes, throat, and lungs. It can cause dermal and respiratory sensitization, resulting in rhinitis and conjunctivitis or asthma. In the UK the Maximum Exposure Limit is just 0.05 ppm (8 hr TWA limit) and 0.05 ppm (15 min STEL) with a 'Sen' notation (p. 93).

Wherever practicable it is advisable for glutaraldehyde to be replaced by a less hazardous chemical, e.g. it should not be used as a general wipe-down disinfectant.

Table 5.35 Selected cyano compounds

Chemical	*Toxicity*	*Properties*
Acetone cyanohydrin (Oxyisobutyric nitrile) $(CH_3)_2C(OH)CN$	Highly toxic by inhalation or ingestion Irritating and moderately toxic upon skin contact Readily decomposes to HCN and acetone at 120°C, or at lower temperatures when exposed to alkaline conditions	Colourless combustible liquid Flash point 73°C Ignition temperature 68.7°C Completely soluble in water
Acetonitrile (Methyl cyanide) CH_3CN	Highly toxic by ingestion, inhalation or skin absorption Insufficient warning properties. Lethal amounts can be absorbed without great discomfort High concentrations rapidly fatal Possibility of severe delayed reactions	Colourless liquid with ether odour and sweet burning taste Flash point 73°C Ignition temperature 52.3°C Flammable limits 4.4%–16%
Acrylonitrile (Vinyl cyanide) CH_2CHCN	Closely resembles HCN in toxic action Poisonous by inhalation, ingestion or skin absorption Emits cyanides when heated or contacted by acids or acid fumes Symptoms: flushed face, irritation of eyes and nose, nausea etc.	Colourless flammable liquid with mild, faintly pungent odour Flash point 0°C. Dilute water solutions also have low flash points
Adiponitrile (Tetramethylene cyanide) $CN(CH_2)_4CN$	Can behave as a cyanide when ingested or otherwise absorbed into the body Combustion products may contain HCN	Water-white, practically odourless liquid Flash point 93°C Specific gravity 0.97 Vapour density 3.7
Calcium cyanide $Ca(CN)_2$	Reacts with air moisture to release HCN. If finely ground and the relative humidity of the air is >35%, this can occur fairly rapidly Releases HCN slowly on contact with water or CO_2, or rapidly with acids Do not handle with bare hands	Nonflammable white powder or crystals
Cyanogen (Ethane dinitrile, Prussite) $(CN)_2$	Highly poisonous gas similar to HCN	Colourless flammable gas with a pungent almondlike odour, becoming acrid in higher concentrations Water soluble Vapour density 1.8
Cyanogen bromide (Bromine cyanide) CNBr	Extremely irritating and toxic vapours Contact with acids, acid fumes, water or steam can produce toxic and corrosive fumes	Transparent crystals with a penetrating odour Melting point 52°C Boiling point 61°C Vapour density 3.6 Water soluble
Cyanogen chloride (Chlorine cyanide) CNCl	Poisonous liquid or gas Vapour highly irritating and very toxic	Colourless liquid with a strong irritating smell Boiling point 13°C Vapour density 2.1
Potassium cyanide KCN	On exposure to air, gradually decomposes to release HCN Poisonous by ingestion, inhalation or skin absorption Do not handle with bare hands. Strong solutions may be corrosive to the skin	Nonflammable white lumps or crystals Faint odour of bitter almonds Completely water soluble
Sodium cyanide NaCN	Poisonous by inhalation, ingestion or skin absorption Do not handle with bare hands Releases HCN slowly with water, more rapidly with acids	Nonflammable white granules, fused pieces or 'eggs' Odourless when dry; slight almond odour in damp air Completely water soluble

Basic precautions include those in Table 5.36.

Table 5.36 Basic precautions for handling gluteraldehyde

- Use with proper local extract ventilation or, as a minimum, in a well-ventilated area
- Replace lids on buckets, waste bins and troughs
- Use in a manner which enables splashes, skin contact and exposure to airborne droplets or fumes to be avoided
- Use appropriate personal protective equipment, e.g. gloves, apron, visor or goggles
- Automation of disinfection procedures, e.g. use of automatic machines, still with a good standard of general ventilation, for disinfecting endoscopes
- Establishment of a procedure to deal safely with any spillages

(vi) Insecticides

Insecticides may be in the form of liquid concentrates, requiring dilution in water or solvents; solutions, wettable powders, granules or pastes; or pressurized or liquefied gases. Application may be as fumigants or fogs, sprays, dust or granules. Obviously all such chemicals are toxic to varying degrees so that exposure via inhalation or ingestion, and in many cases via skin absorption, should be minimized.

The variation in toxicity of common organophosphate insecticides is exemplified in Table 5.37. The range of chlorinated hydrocarbon insecticides (Table 5.38) have, with the exception of Endrin and Isodrin, somewhat lower oral and dermal toxicities. The toxicities of a range of other insecticides, fungicides, herbicides and rodenticides are summarized in Table 5.39.

Essential precautions with insecticides are listed in Table 5.41.

(vii) Irritants and corrosives

As a class, primary irritants are the most widely encountered chemicals in industry and include inorganic acids and alkalis, halogens and halogen salts, chlorosilanes, detergents, organic solvents and organic acids and many derivatives, e.g. acid chlorides and anhydrides. In extreme cases, many are also corrosive (Table 5.4) and, in the case of organic compounds, possibly flammable. The skin, eyes and mucous membranes are at greatest risk although the respiratory tract is affected if the materials become airborne as dusts or aerosols, or if gaseous or volatile, e.g. halogens and inorganic anhydrous acids (Tables 5.42 and 5.43). Table 5.44 lists the properties of selected organic acids.

Typical precautions for work with irritant and corrosive chemicals are listed in Table 5.45.

(viii) Mercury

Mercury is used in the manufacture of thermometers, barometers and switchgear, and in the production of amalgams with copper, tin, silver and gold, and of solders. A major use in the chemical industry is in the production of a host of mercury compounds and in mercury cells for the generation of chlorine. Mercury has a significant vapour pressure at ambient temperature and is a cumulative poison.

The liquid attacks many metals, including aluminium, gold, copper and brass. Splashes break up into very small, mobile droplets, making clean-up of spillages difficult.

Mercury should not be left exposed in a laboratory. Reservoirs etc. should be covered with a layer of water or oil and, if practicable, the neck of the vessel plugged. The risk is increased by

Table 5.37 Organophosphate insecticides (see also Table 5.12)

Insecticide	*Oral LD_{50}* (mg/kg)	*Dermal LD_{50}* (mg/kg)	*UK OES 8 hr TWA value* (mg/m^3)
Abate	8600–13 000	4000	—
Azinphosmethyl (Guthion)	11–13	220	0.2 SK
Azodrin	17.5–20	112–126	—
Bidrin	22	225	—
Carbophenothion	10–30	27–54	—
Chlorthion	890–980	4100–4500	—
Ciodrin	125	385	—
Coumaphos (Co-Ral)	15.5–41	860	—
Demeton (Systox)	2.5–6.2	8.2–14	—
Diazinon	76–108	455–900	—
Dicapthon	330–400	790–1250	—
Dimethyldichlorovinyl Phosphate (DDVP)	56–80	75–107	—
Dimethoate	215	400–610	—
Dioxathion (Delnav)	23–43	63–235	0.2 SK
Disulfoton (Di-Syston)	2.3–6.8	6–15	0.1
O-ethyl-*O*-*p*-nitrophenyl Phenyl Phosphonothioate (EPN)	7.7–36	25–230	—
Ethion	27–65	62–245	8 SK
Fenthion (Baytex)	215–245	330	—
Malathion	1000–1375	4444	10 SK
Methyl Parathion	14–24	67	0.2 SK
Methyl Trithion	98–120	190–215	—
Naled	250	800	3 SK
Nemacide	270	—	—
NPD	—	1800–2100	—
Octamethyl Pyrophosphoramide (Schradan)	9.1–42	15–44	—
Parathion	3.6–13	6.8–21	0.1 SK
Phorate (Thimet)	1.1–2.3	2.5–6.2	0.05 SK
Phosdrin (Mevinphos)	3.7–6.1	4.2–4.7	0.1 SK
Phosphamidon	23.5	107–143	—
Ronnel (Korlan)	1250–2630	5000	10
Ruelene	460–635	—	—
Sulfotep, Tetraethyl Dithiopyrophosphate (TEDP)	—	—	0.2 SK
Tetraethyl Pyrophosphate (TEPP)	1.05	2.4	0.05 SK
Trichlorfon (Dipterex)	560–630	2000	—

SK Can be absorbed through skin.
The LD_{50} varies according to species of animal, sex, age and health.

heating, e.g. due to spillage on a hot surface; no glass blowing should therefore be done on mercury-contaminated glass.

Care is essential to avoid spillages. A fine capillary tube connected to a filter flask and filter pump should be used immediately to collect any spillage. Surfaces, e.g. floors, contaminated by minute mercury droplets should be treated with sulphur or zinc dust, or by use of a commercial clean-up kit.

Rooms in which mercury is regularly exposed should be subjected to routine atmospheric monitoring. Personnel in such rooms should receive periodic medical examinations.

For routine laboratory precautions refer to Table 5.40.

Table 5.38 Chlorinated hydrocarbon insecticides (see also Table 5.12)

Insecticide	*Oral LD_{50}* (mg/kg)	*Dermal LD_{50}* (mg/kg)	*UK OES 8 hr TWA* value (mg/m^3)
Aldrin	39–60	98	0.25 SK
Benzene Hexachloride (BHC)	1250	—	—
Chlordane	335–430	690–840	0.5 SK
Chlorobenzilate	1040–1220	<5000	—
DDT	113–118	2510	1.0
Dichloropropane-Dichloropropene	140	2100	—
Dicofol (Kelthane)	1000–1100	1000–1230	—
Dieldrin	46	60–90	0.25 SK
Dilan	—	5900–6900	—
Endosulfan (Thiodan)	18–43	74–130	—
Endrin	7.5–17.8	15–18	0.1 SK
Ethylene Dibromide	117–146	300	—
Ethylene Dichloride	770	3890	—
Heptachlor	100–162	195–250	0.5 SK
Isodrin	7.0–15.5	23–35	—
Kepone	125	<2000	—
Lindane	88–91	900–1000	0.5 SK
Methoxychlor	5000	—	10
Mirex	600–740	2000	—
Para-Dichlorobenzene	1000	—	—
Perthane	4000	—	—
Strobane	200	5000	—
TDE	4000	4000	—
Telone	200–500	—	—
Toxaphene	80–90	780–1075	—

SK can be absorbed through skin.
The LD_{50} varies according to species of animal, sex, age and health.

(ix) Mineral oil lubricants

Mineral oils, i.e. oils derived from petroleum, are widely used as lubricants, cutting oils, soluble oil coolants etc.

They have very low acute toxicities, i.e. oral LD_{50} values of around 10 g/kg. They are not absorbed via the skin and are insufficiently volatile to produce harmful vapours at room temperature. Additives are used in small quantities for specific properties but these do not normally affect the health and safety characteristics. Dermatitis may be caused by repeated or prolonged contact of mineral oils with the skin. Such contact with higher boiling fractions over many years can result in warty growths which may become malignant, e.g. on the scrotum following contamination of the front of overalls, possibly from oily rags in trouser pockets. Carcinogenic activity is reduced by solvent refining of the base stocks but can increase with use. Oil mists at concentrations normally encountered are primarily a nuisance, but very high concentrations could, on inhalation, cause irritation of the lungs leading to pneumonia. Because of the carcinogenic potential the atmospheric concentration should be controlled below 5 mg/m^3 as an 8 hr TWA concentration and 10 mg/m^3 as a 10 min STEL concentration. General recommendations for precautions with mineral oils are summarized in Table 5.46.

Table 5.39 Insecticides, rodenticides, fungicides and herbicides (see also Table 5.12)

Substance	*Oral LD_{50}* (mg/kg)	*Dermal LD_{50}* (mg/kg)	*UK OES* 8*hr TWA* value (mg/m^3)
Insecticides			
Binapacryl	58–63	720–810	—
Calcium Arsenate	298	2400	—
Carbaryl	500–850	4000	5
Cryolite	200	—	—
DN-111	330	1000	—
Lead Arsenate	—	2400	0.15
Metaldehyde	1000	—	—
Morestan	1800	—	—
Naphthalene	2400	2500	50
Nicotine Sulphate	83	285	—
Ovex	2050	—	—
Paris Green	100	2400	—
Pyrethrum	1500	1800	—
Rotenone	50–75	940	5
Ryania	1200	4000	—
Tetradifon	14 700	10 000	—
Zineb	5200	—	—
Rodenticides			
Sodium Fluoroacetate			0.05 SK
Strychnine			0.15
Thallium Sulphate			0.1 as Tl, SK
Warfarin			0.1
Fungicides			
Ferbam	17 000	—	10
Formaldehyde	—	—	2.5 MEL
Organic Mercurials	—	—	0.01
Maneb	7500	—	—
Nabam	395	—	—
Pentachlorophenol	—	—	0.5 SK
Ziram	1400	—	—
Herbicides			
2,4-D	375–700 (different acids, salts and esters)	—	10
2,4,5,-T	481–500 (different acids and esters)	—	10
Dinitrocresol (DNOC)	30	—	0.2 SK

SK Can be absorbed through skin.
The LD_{50} varies according to species of animal, sex, age and health.

(x) Metalworking fluids

Metalworking fluids contain mineral oils (refer to p. 80) or synthetic lubricants; they are used neat or in admixture with water. They may contain small amounts of biocides, stabilizers, emulsifiers, corrosion inhibitors, fragrances and extreme pressure additives. The formulations render them suitable for application to metal being worked, generally from a recirculatory system, to provide lubrication, corrosion protection, swarf removal and cooling of the tool and machined surface.

Table 5.40 Routine laboratory precautions with mercury

Avoid the use of mercury, if possible
Store in airtight containers or under water or liquid paraffin
Handle over a suitable tray near to the apparatus in use
Avoid wearing rings. Wear gloves. Wash the hands and gloves after handling mercury
Use catchpots under apparatus containing mercury
Use only apparatus strong enough to withstand the considerable force which may arise due to movements of mercury (e.g. rigid pvc or polythene)
Clean up all spillages immediately and check for pockets (e.g. in cracks and crevices) by monitoring
Decontaminate equipment such as vacuum pumps and glassware prior to service/maintenance

Table 5.41 Guidance on safety with pesticides

Approval	In the UK only approved pesticides may be supplied and used Each product has an approval number and conditions of use
Storage requirements	Suitable siting Adequate capacity and construction Designed to hold spillage Properly lit and ventilated Fire- and frost-resistant Designed so that containers can be safely stacked and moved Clearly marked Kept locked except when in use
Competence	Every user must be competent (Certificate of Competence required in UK)
Information	Workers must be supplied with sufficient information and guidance
Evaluation of possible problems	Product selection How to comply with the conditions of approval Selection of protective clothing How to avoid spray drift How to avoid environmental damage Need to warn neighbours and others who may be affected etc.
Training requirements	Pesticide legislation Decisions on whether a pesticide has to be used Selection of appropriate pesticide Interpretation of labels and codes of practice Hazards and risks to human health/the environment Selection and use of engineering controls and protective clothing Calibration and safe operation of application equipment Safe storage and disposal of pesticides Emergency action in case of poisoning or contamination How to contain and deal with accidental spillage Constraints imposed by weather or other factors Appropriate record keeping Need for exposure monitoring/health surveillance
Exposure control	Use engineering/technical means, e.g. Low-level filling bowls Suction probes Closed handling systems Soluble packs In-cab electronic sprayer controls Hydraulic boom-folding (These measures should be used in preference to protective clothing)
Disposal	Minimize disposal requirements by careful estimation of needs and correct measurement Dispose of dilute pesticides by using as a spray, in accordance with 'approval', in a safe/approved area Concentrated, unused pesticides should be stored, returned or disposed of as toxic waste

Table 5.42 Halogens

Halogen	*Melting point* (°C)	*Boiling point* (°C)	*Vapour density* (air = 1.0)	*Threshold limit value (ppm)*	*Reactivity and oxidizing strength*	*Appearance and state at 21°C*	*Colour of gas/vapour*
Fluorine (F_2)	–217	–188	1.3	0.1	Extremely active	Pale yellow gas	Pale yellow
Chlorine (Cl_2)	–101	–34	2.5	1.0	Very active	Greenish-yellow gas. Amber liquid at 5.8 bar pressure	Greenish-yellow
Bromine (Br_2)	–6.6	59	5.5	0.1	Active	Dark red liquid	Dark red to reddish-brown
Iodine (I_2)	113	185[1]	8.6	0.1	Least active	Bluish-black lustrous solid	Violet

[1] Readily sublimes at lower temperatures.

Table 5.43 Common anhydrous acids

Chemical	*Melting point* (°C)	*Boiling point* (°C)	*Typical cylinder pressure at 21°C* (bar)	*Vapour density* (air = 1.0)	*Properties*
Hydrogen bromide (Anhydrous hydrobromic acid) HBr	–86	–69	22	2.8	Colourless, corrosive nonflammable gas with an acrid odour Highly irritating to eyes, skin and mucous membranes Fumes in moist air, producing clouds with a sour taste Freely soluble in water
Hydrogen chloride (Anhydrous hydrochloric acid) HCl	–111	–83	42.4	1.3	Colourless, corrosive nonflammable gas with a pungent odour Considered somewhat more dehydrating and more corrosive than the mists and vapours of hydrochloric acid Fumes in air Very water soluble
Hydrogen fluoride (Anhydrous hydrofluoric acid) HF	–83	19	0.069	0.7	Colourless, corrosive nonflammable liquid or gas with a penetrating odour Highly irritating and poisonous Very soluble in water. Liquid liberates heat as it dissolves in water. The entrapment of water in an anhydrous hydrogen fluoride cylinder can cause rapid generation of heat and pressure which can lead to an explosion. Containers should never be heated to >52°C. A liquid hydrogen fluoride spill area should not be entered unless protective clothing (impervious to the compound) and a self-contained gas mask are worn Fumes in air
Hydrogen iodide (Anhydrous hydriodic acid) HI	–51	–35			Colourless, corrosive nonflammable gas with an acrid odour Highly irritating to eyes, skin and mucous membranes Attacks natural rubber Decomposed by light Extremely soluble in water Fumes in moist air

Hand dispensing is also used but on most modern machines application is by a continuous jet, spray or mist.

Skin contact with metalworking fluids may cause skin irritation or a contact irritant dermatitis. Contact with neat oils may cause folliculitis (oil acne). Contact with some aqueous-mix fluids may, depending upon the additives, e.g. biocides, cause an allergic contact dermatitis. Formerly the use of unrefined mineral oils posed a risk of skin cancer.

The fumes and mist from metalworking fluids may cause irritation of the eyes, nose and throat.

Table 5.44 Common organic acids

Acid	*Flash point (°C)*	*Ignition temperature (°C)*	*Specific gravity (water = 1.0)*	*Vapour density (air = 1.0)*	*Boiling point (°C)*	*Properties*
Acetic acid (glacial) (Ethanoic acid) CH_3COOH	40	426	1.05	2.07	118	Clear, colourless water-soluble mobile liquid Strong vinegar odour Lower explosive limit 4% Glacial acetic is a concentration of acetic acid >99% Expands on freezing into an icelike solid at 16.6°C; can break container
Butyric acid (Butanoic acid) $CH_3(CH_2)_2COOH$	72	452	0.96	3.04	163	Colourless water-soluble oily liquid with strong odour
Formic acid HCOOH	69	601	1.22	1.59	100	Colourless water-soluble fuming liquid with pungent penetrating odour Glacial acid freezes at 8°C Decomposes slowly in storage, liberating carbon monoxide Sufficient gas pressure can accumulate in tightly sealed tank to cause rupture or leakage Common concentrations all >90%
Propionic acid (Methylacetic acid) CH_3CH_2COOH	54	512	0.99	2.56	141	Colourless liquid with slightly pungent and rancid odour

Breathing difficulties, i.e. bronchitis or asthma, arising from sensitization to bacterial contamination or additive chemicals, have been reported.

The formation of nitrosamines, e.g. *n*-nitrosodiethanolamine, which are possible human carcinogens, can occur in synthetic or semi-synthetic fluids which contain a nitrite salt and diethanolamine or triethanolamine.

Basic precautions include those in Table 5.47.

(xi) Painting

Industrial painters may suffer adverse health effects from over exposure to paint by skin contact or accidental ingestion, from excessive inhalation of paint aerosol, solvent vapour, or of dust in the case of electrostatically-applied powder coatings (e.g. polyesters containing triglycidyl isocyanurate), or from exposure to thermal degradation products from heated paint or plastic coatings (Table 5.48).

Precautions for paintwork are summarized in Table 5.49.

Table 5.45 Precautions in handling primary irritants/corrosives

Avoid contact	Use in most dilute form practicable Minimize opportunity for contact and time of contact Select appropriate constructional materials (flexible hoses) Allow for clearance of blockage (e.g. steaming) *Containment* Avoid open tanks, even if only loose lids practicable Provide barriers Use handling technique that avoids airborne contamination *Mechanical handling* In-plant transfer preferably by pipeline Alternatively use special vessels, e.g. lined drums, or continuous arrangements for enclosed transfer Drums should be emptied using pumps/forklift trucks etc. Avoid manual tipping Provide warning notices to identify containers and areas where corrosive chemicals are in use, and instructions regarding necessary protection, particularly eye protection areas Identify vessels, pumps and pipelines (e.g. colour coding, numbering) *Spillage* Retain with bunds etc. Neutralize or mop up immediately Maintain supplies of neutralizing chemicals, e.g. soda ash Provide hosing-down facilities where appropriate Minimize joints, particularly drain and sample points, valves/pipe joints, flanges over access ways Avoid flexible hoses where possible; otherwise secure, shield and maintain Shield glass sections Provide separate FULL and EMPTY storage areas Use road or rail tanker for bulk transfer; if small containers are used, they should be of correct design (free space, pressure, corrosion) Label containers according to hazard, precautions, first aid Segregate incompatible materials Maintain good housekeeping
Personal protection	Depending on scale of operation, use impervious rubber gloves, eye protection (glasses/goggles/face shield), rubber aprons, boots, armlets, protective suits Provide respiratory protection against gases/dusts/fumes Provide shower and eyewash facilities Use protective/barrier creams and skin reconditioning creams Maintain high standard of personal hygiene
First aid measures	Investigate all complaints In the case of injury, obtain medical attention rapidly
Emergency measures	Copiously flush eyes with water for up to 15 min, and skin with water and soap – *except* in the case of substances such as quicklime whose reaction with water is exothermic (1 g generates >18 kcal), titanium or tin tetrachloride, both of which rapidly hydrolize to form hydrochloric acid Therapeutic measures for specific chemicals include: *White phosphorus.* This element burns in air and can produce severe thermal and chemical burns. It may reignite on drying. After washing, rapid but brief treatment with copper sulphate (to avoid systemic absorption and copper poisoning) is used to convert the phosphorus to copper phosphide which is then removed *Hydrogen fluoride.* This can form painful but delayed necrosis. Treat with calcium gluconate locally and monitoring of serum calcium levels, with administration of calcium where necessary

Table 5.46 General precautions with mineral oils

Avoid all unnecessary contact with mineral and synthetic oils, e.g. by carefully designed work practices
Avoid extreme exposure to oil mist or vapours
When used as a machine coolant:

- substitute by oil-in-water emulsions (suds) or by nitrite-free synthetic coolants;
- use of an adequate flow of cooling to minimize 'fuming';
- provide l.e.v.;
- provide and use enclosures and splash guards wherever practicable;
- regularly change coolant and machine system cleaning.
- with solutions or emulsions control concentration in use.

For machine operation, use of protective clothing, e.g. impervious aprons with detachable absorbent fronts (see also below).
Protective gloves may be helpful if they can be kept clean inside (porous gloves may prolong exposure)
Impervious elasticated armlets may be appropriate
Provide a readily available supply of disposable rags
Do not carry used rags in overall or trouser pockets
Wear goggles if eye contact is likely

Wear clean work clothes
Consider short-sleeved overalls for workers using metal cutting fluids (avoids skin friction from cuffs saturated with oil and holding particles of swarf)
Dry-clean oily overalls
Change underclothes that become wet with oils, and wash thoroughly

Wash skin thoroughly to remove all traces of oil
Avoid strong soaps, detergents, abrasive skin cleansers
Do not use paraffin (kerosene), petrol (gasoline), chlorinated hydrocarbons or proprietary solvents to cleanse skin
Use barrier cream before work and after washing hands (different barrier creams protect against different oils – a cream intended for soluble oil does not protect against straight oils)
Use skin reconditioning cream after washing hands at end of shift
See that all cuts and scratches receive prompt medical attention
Seek medical advice as soon as an irritation or other skin abnormality appears

Maintain a high standard of housekeeping – a clean workplace encourages clean work practices
Encourage self-checks and provide the necessary information – e.g. leaflet MS(B)5 available free from HSE
Use warning notices, placards etc. to promote good personal hygiene and good work practices

Table 5.47 Basic precautions for metalworking fluids

- Design, maintain and operate machine tools to minimize fumes and mist generation, splashing and skin contamination
- Provide local exhaust ventilation as appropriate
- Control fluid quality during use, involving checks on correct dilution and make-up, concentration and freedom from contamination in service, regular cleaning and fluid changing
- Provide and use appropriate personal protective equipment
- A high standard of personal hygiene
- Regular skin inspections, e.g. at a frequency of once per month; monitoring of any respiratory problems by simultaneous enquiries and completion of an annual questionnaire
- Prohibit air lines for blowing clean components
- Prohibit eating, drinking or smoking near machines
- Remove swarf and fines from, and prevent hydraulic oil leakage into, recirculated fluid
- Avoid the use of water-mix synthetic fluids containing nitrites if there is a technologically effective alternative.

Table 5.48 Potential pollutants from heated paints or plastic coatings

Elements in resin	*Chemical classification of resin*	*Possible products of pyrolysis*
Carbon, hydrogen and possibly oxygen	Resin and derivatives Natural drying oils Cellulose derivatives Alkyd resins Epoxy resins (uncured) Phenol-formaldehyde resins Polystyrene Acrylic resins Natural and synthetic rubbers	Carbon monoxide Aldehydes (particularly formaldehyde, acrolein and unsaturated aldehydes) Carboxylic acids Phenols Unsaturated hydrocarbons Monomers, e.g. from polystyrene and acrylic resins
Carbon, hydrogen, nitrogen and possibly oxygen	Amine-cured epoxy resins Melamine resins Urea-formaldehyde resins Polyvinyl pyridine or pyrrolidine Polyamides Isocyanate (polyurethanes) Nitrocellulose derivatives	As above, but also various nitrogen-containing compounds, including nitrogen oxides, hydrogen cyanide, isocyanates
Carbon, hydrogen and possibly halogens, sulphur and nitrogen	Polyvinyl halides Halogenated rubbers PTFE and other fluorinated polymers Thiourea derivatives Sulphonamide resins Sulphochlorinated compounds	As above, but also halogenated compounds. These may be particularly toxic when fluorine is present Hydrogen halides Carbonyl chloride (phosgene) Hydrogen sulphide Sulphur dioxide

Table 5.49 Precautions in preparation and paintwork

Information (i.e. at least a safety data sheet and comprehensive container label) and training related to the hazards in the handling and use of the range of chemicals.
Use where practicable of less harmful chemicals, e.g. water-based paints.
Provision and use of appropriate health surveillance, e.g. for signs of dermatitis, asthma, effects of specific solvent exposures.
Full use of any spray booth, enclosure, exhaust ventilation or dilution systems, and automatic handling equipment. (The efficiency of all local exhaust ventilation and other control systems should be maintained, and checked by testing.)
Where appropriate, atmospheric monitoring of airborne pollution levels.
Full use, where appropriate, of ventilation, e.g. by opening doors, windows.
Prompt attention to any damaged or malfunctioning equipment, e.g. general ventilation.
Replacement of lids on containers.
Correct disposal of paint, thinner, impregnated rags.
Use, where appropriate, of a properly-fitting respirator with correct filter or air-fed equipment.
Use of a vacuum cleaner or damping techniques to minimize dust generation.
Avoidance of the use of unauthorised thinners for paint dilution, surface preparation or cleaning of spray guns/brushes/rollers.
Avoidance of skin contact and ingestion of chemicals by:

- Use of protective clothing and eye protection.
- Use of barrier cream and skin conditioning cream.
- Removal of jewellery etc. which can trap chemicals in contact with the skin.
- Avoidance of excessive skin contact with solvents, e.g. when cleaning brush, spray guns; not washing hands in solvents.
- Avoidance of eating, drinking or smoking while painting.
- A good standard of personal hygiene, i.e. washing hands before eating, and showering or bathing at the end of work.
- Maintaining overalls and respiratory protection in a clean state.
- Leaving protective clothing at work.

(xii) Perchloroethylene (tetrachloroethene, tetrachloroethylene)

Perchloroethylene is a clear, dense, non-flammable volatile chlorinated solvent. It is widely used for dry cleaning; small quantities are used in adhesives and cleaning agents. It is miscible with organic solvents but only slightly soluble in water. Relevant physical properties are given in Table 5.50.

Table 5.50 Physical properties of perchloroethylene

Boiling point	121°C
Saturated vapour concentration	1.5×10^4 ppm at 20°C
Specific gravity	1.623 at 20°C
Vapour density (air = 1)	5.83 at 74°C
Solubility in water	0.015% w/w at 25°C

This solvent has an ethereal odour with a detection threshold between 5 and 80 ppm. It can be absorbed orally, by inhalation and through the skin, and small amounts of the compound can be detected in the breath of humans several days after exposure. Overexposure by inhalation can cause central nervous system (CNS) depression – characterized by dizziness, light-headedness, inebriation and difficulty in walking – and liver damage. Minimal CNS effects occur at exposures of about 100 ppm. Chronic exposure has resulted in peripheral neuropathy. Ingestion of large doses can cause internal irritation, nausea, vomiting and diarrhoea; it can cause drowsiness or unconsciousness. Contact with the skin will result in degreasing, resulting in mild irritation possibly leading to cracking and secondary infection; in extreme cases dermatitis may occur.

Eye contact with vapour above 100 ppm may cause irritation. Liquid splashes produce irritation.

Studies on the carcinogenicity are inconclusive but IARC classify tetrachloroethylene as a probable human carcinogen (Class 2A – see Table 5.17).

Thermal degradation in contact with flame or red hot surfaces will produce highly-toxic gases, e.g. acid chlorides and phosgene. Reaction with freshly-galvanized surfaces may produce dichloroacetylene, which is also highly toxic.

The long-term OES is 50 ppm (8 hr TWA), set to protect against CNS effects, which will also protect against liver or kidney damage and irritation. The short-term OES is 1000 ppm (15 minute reference period) to minimize exposures at irritant levels.

Tetratchloroethylene has been detected in the food chain as a contaminant: its volatility prevents significant bioaccumulation but some transfer to aquatic sediments is possible. At low concentrations it is slowly degraded under anaerobic conditions.

The precautions for the use of perchloroethylene correspond with those for trichloroethylene (Table 5.52). The dry cleaning process, and its safety measures comprise:

- cleaning in hot solvent in sealed machines;
- drying with hot air (after centrifugal removal of liquid solvent) which passes over a lint filter then cooled to condense solvent;
- deodorization to remove last traces of solvent with fresh air with venting to atmosphere;
- solvent recovery by distillation. Normal measures for solvent control use activated charcoal absorption filters often as disposable cartridges. Still residues are usually removed manually from each machine at least once per week, or in some machines they are pumped directly to a waste storage vessel, often a lidded metal container located outside the building.

Table 5.51 Physical properties of trichloroethylene

Boiling point	87°C
Saturated vapour concentration	7.9×10^4 ppm at 20°C
Specific gravity	1.464 at 20°C
Vapour density (air = 1)	4.54
Solubility in water	0.11% w/w at 25°C

Table 5.52 General safety precautions with trichloroethylene and tetrachloroethylene

Do not
Store liquid in buckets or other open storage vessels.
Lean into any vessel containing liquid or vapour.
Use in any location which is not well ventilated, but avoid extraneous draughts.
Enter any confined space, e.g. any tanks or pits, except in accordance with a permit-to-work system.

Do
Store in the open air if possible, or at least in a well-ventilated area which is not below ground level.
Provide and maintain efficient local exhaust ventilation, if enclosed plant cannot be used.
Avoid smoking in proximity to any open system.
Replace lids on any open-topped tanks.
Wear eye protection if there is any risk of splashing.
Avoid contact with the skin. If hand contact is likely, wear PVC gloves.
Avoid contact of liquid or vapour with naked flames or red-hot surfaces.
In vapour degreaser operation, do:
- control vapour at the specified level within the tank;
- employ effective l.e.v. and prevent draughts in the workroom, e.g. by suitable screening;
- avoid overheating of solvent and uncontrolled vaporization, including prevention by the use of appropriate thermostats;
- lower and raise work loads slowly, to minimize vapour displacement;
- carefully position, or drain, hollow articles.

When cleaning plant, if plant is within a pit, do:
- ensure ventilation draws air from the bottom of the pit;
- distil off vapour in accordance with specific operating instructions;
- allow any liquid to cool to ambient temperature before drawing off into a suitable receptacle;
- remove sludge by raking through sump door, after allowing several hours for ventilation;
- restrict any entry, in accordance with the Confined Space Regulations 1997.

(xiii) Trichloroethylene (trichloroethene)

Trichloroethylene is a colourless non-flammable chlorinated hydrocarbon liquid. It is mainly used for degreasing of metals in the engineering and electrical appliance industries; other outlets are as a solvent in inks, in dry-cleaning, in varnishes and adhesives, and as a solvent in the extraction of fats and oils. Relevant physical properties are given in Table 5.51.

The solvent has a sweetish odour similar to chloroform detectable at about 30 ppm. It is primarily a depressant of the central nervous system (CNS). Significant impairment of performance in behavioural tests and some CNS effects have occurred at 1000 ppm but not at 300 ppm; prenarcotic symptoms have occurred at mean levels of 200 ppm to 300 ppm.

Direct eye contact with liquid produces injury, generally transient, to the corneal epithelium. The liquid is mildly irritating to the skin due to the degreasing effect; repeated contact may cause dermatitis. Ingestion of substantial quantities of liquid can damage the mucous membranes, and produce acute effects ranging from mild discomfort to profound anaesthesia.

Evidence to indicate that exposure to trichloroethylene is associated with an increased incidence of cancer in man or in adverse effects in the offspring of women workers exposed to the compound is confused but IARC classify trichlorethylene as a probable human carcinogen, Class 2A (see Table 5.17).

Decomposition of trichloroethylene can occur upon contact with naked flames, red-hot surfaces, hot elements of electric heaters, or intense UV light with the generation of acidic and highly-toxic products. The presence of reactive contaminants, e.g. acids, strong alkalis, highly-reactive metals, may also result in decomposition to similar products.

The Occupational Exposure Standards imposed for trichloroethylene are Maximum Exposure Limits of 100 ppm (8 hr TWA) and 150 ppm (15 minute reference period). A skin notation 'Sk' is applicable because of the potential for skin absorption. Because of its volatility, trichloroethylene is not recommended for cold cleaning; it is normally used in partially enclosed vapour degreasing equipment provided with local exhaust ventilation.

Typical precautions with trichloroethylene are summarized in Table 5.52. An important factor is that the vapours are much heavier than air; they will therefore spread and may accumulate at low levels, particularly in undisturbed areas. Because of its volatility, releases to the environment usually reach the atmosphere. Here it reacts with hydroxyl or other radicals (estimated half-life for reaction with hydroxyl radicals is less than a week) and is not therefore expected to diffuse to the stratosphere to any significant extent. There is some evidence for both aerobic and anaerobic biodegradation of trichloroethylene.

(xiv) Sick building syndrome

When working in certain buildings some workers suffer temporarily from a group of symptoms including:

- lethargy/tiredness;
- irritability;
- lack of concentration/mental fatigue;
- headaches;
- nausea/dizziness;
- sore throats;
- dry eyes and skin;
- skin rash;
- asthma;
- blocked/runny nose.

The condition is usually non-specific and seldom traced to a single cause. This has been termed sick building syndrome. Despite much research, little has been proven but the building features associated with the condition are:

- hermetically sealed, airtight shell;
- mechanical heating, ventilation and air-conditioning;
- use of materials and equipment that emit a variety of irritating and sensitizing toxic fumes and/or dust;
- fluorescent lights;
- application of energy conservation measures;
- lack of individual control over environmental conditions;
- landscape plants;
- VDUs;
- draughts.

Whilst the causative agent(s) have not been established it is thought to be multifunctional and possibilities include physical factors (humidity, temperature, lighting), static electricity, electromagnetic radiation, air ion concentrations, fungi, noise, psychological stress, and chemicals. Chemicals which are not those involved in the normal work processes can become trapped within the building, albeit at concentrations below those known to cause ill-health effects, if:

- liberated from materials of construction or furnishings; or
- they could enter from outside.

Examples of common pollutants are given in Table 5.53.

Table 5.53 Common pollutants that may be found in buildings

Chemical	*Source*
Ammonia	Cleaning solutions
	Printers
	Cigarette smoke
Asbestos	Pipe lagging
	Air duct linings
	Ceiling and roof tiles
	Asbestos cement sheeting
Benzene	Synthetic fibres
	Plastics
	Cleaning solutions
	Tobacco smoke
Biocides	Air-conditioning systems
	Water treatment
	Humidifiers
	Disinfectants
Carbon dioxide	Exhaled breath
	Vehicle exhausts
	Smoking chimneys
	Portable heaters
Carbon monoxide	Tobacco smoke
	Gas cookers
	Gas and oil heaters
	Vehicle exhausts from loading bays or diesel trucks
Detergent dust	Carpet cleaners
Ethyl alcohol	Duplicating fluids
Fibre glass	Insulation
Formaldehyde	Insulation material
	Ceiling tiles
	Particle board
	Plywood
	Office furniture
	Carpet glues
	Various plastics
	Synthetic fibres and rugs
	Upholstery and other textiles
	Pesticides
	Paint
	Paper
	Tobacco smoke
Hydrocarbons	Paints, adhesives
	Solvents
	Synthetic materials

Table 5.53 Cont'd

Chemical	*Source*
	Floor and furniture polishes
	Vehicle exhausts
Hydrochloric acid	Electric stencil cutting machines
Methyl alcohol	Spirit duplicating machines
Micro-organisms	Duct work
	Humidifiers
Oxides of nitrogen	Vehicle exhausts
	Tobacco smoke
	Gas heaters
Ozone	Electrical discharges from equipment such as photocopiers, electrostatic precipitators, fluorescent lights
Paint fumes	Paint
PCBs	Transformers
	Ageing VDUs
	Fluorescent lights
Pesticides	Plant, timber and fabric treatments
	Air conditioning/ventilation units
Photochemical smog	Reaction between chemicals and ozone
Radon	Soil
	Concrete
	Stone
	Water supply
Solvents	Typist correction fluids
	Adhesives, glues
	Cleaning fluids
	Paint
	Felt-tip pens
	Stencil machines
	Inks
Sterilant gases	Humidifiers
	Air-conditioning units
Sulphur dioxide	Coal fires
	Power stations
	Chimneys
	Vehicle exhausts
Vinyl chloride	Plastic PVC pipes
	Light fittings
	Upholstery
	Carpets

Because the cause is unknown precautions are difficult to specify but general guidance includes that in Table 5.54.

Temporary problems of building pollution may occur during construction and engineering activities, refurbishment, painting and decorating, and cleaning in internal, or sometimes external, areas. The sources are, generally, more easily traced.

(xv) Synthetic resins

Synthetic resins are extensively used, e.g., in surface finishes, in the fabrication and repair of boat and motor vehicle bodies, in the manufacture of laminated boards, for electrical components, in pattern making and in paints and varnishes. Non-rubber adhesives made from fish glues and from cotton derivatives (e.g. cellulose acetate) tend not to be sensitizing but, depending upon composition and the manner of use, many other types may pose significant dermatitic and fume hazards.

Table 5.54 Precautions to avoid 'sick buildings'

- Avoid nuisance noise
- Provide adequate ventilation
- Avoid high uniform temperatures
- Avoid lack of air movement
- Provide adequate humidity
- Avoid uniform dull lighting and décor
- Provide natural lighting: avoid tinted glass
- Ensure staff are motivated
- Consider possible cause at building design, construction and commissioning stages
- 'Bake out' buildings to drive out pollutants
- Investigate complaints, especially if there are pockets of complaints, or epidemics with related symptoms.

The monomers, catalysts or hardeners, or plasticizers can include chemicals with the potential to irritate the skin, mucous membranes or respiratory tract. Some can promote skin or respiratory sensitization. The range of chemicals in use is extremely wide, so that reference should be made to the Materials Safety Data Sheet for each specific formulation or variation of it identifiable by reference to the supplier's proprietary name and code number. Some common resin types are summarized in Table 5.55.

Table 5.55 Synthetic resin types

Type	*Examples*
Acrylic	Polymethyl methacrylate
Amino	Melamine formaldehyde, urea formaldehyde, furfuryl alcohol – urea formaldehyde
Epoxy	Epichlorohydrin with bisphenol A. The curing agents may pose significant health hazards, e.g. amines (triethylamine, *p*-phenylenediamine, diethylenetriamine) or acid anhydrides (pyromellitic dianhydride)
Phenolic	Phenol formaldehyde. Formaldehyde is a respiratory irritant but is not classified as asthmagen.
Polyester	Powder coatings containing triglycidyl isocyanurate are possible asthmagens (unclassified)
Polyurethane	An organic isocyanate (MDI or a pre-polymer) with a hydroxy compound. The isocyanates are potent respiratory sensitizers, the risk increasing with volatility
Vinyl	Polyvinyl chloride, polyvinyl acetate.

With the exception of epoxy resins, when a resin is fully polymerized it loses any irritant properties. However, associated materials, e.g. glass fibre used as a filler, or the dust from plywood or veneers, may promote irritation. Partially-cured resins will retain some irritant properties. Traces of cutaneous or respiratory sensitizers liberated, e.g. by heating or machinery, may be problematic.

Epoxy resins are often mixtures of the epoxy resin, a curing agent (hardener), solvents, reactive diluents, antioxidants and filler. Examples are given in Table 5.56. Curing agents usually comprise peroxides, amines or anhydrides. These may cure at room temperature or require elevated temperatures. They may be irritant or damage skin, eyes or lungs; certain amines and peroxides are sensitizers. The solvents such as acetone, methyl ethyl ketone, toluene, xylene, glycol ethers, and alcohols pose both health and fire hazards. Glycidyl ethers are used as reactive diluents while

Table 5.56 Typical exposure effects associated with epoxy resin systems

Resin type	*Examples*	*Skin contact*	*Inhalation*	*Ingestion*
Liquid epoxy resin	based on the reaction product of epichlorohydrin and bisphenol A or bisphenol F	• mild to moderate irritants • mild to moderate sensitizers	• low volatility, exposure unlikely unless heated, sprayed, or spread over large unventilated surface	low toxicity
Solid epoxy resins	based on the reaction product of epichlorohydrin and bisphenol A or bisphenol F	• mild to moderate irritants and mild sensitizers • not readily absorbed through skin	• low volatility, exposure unlikely unless crushed or ground	low toxicity
Modified liquid epoxy resins	liquid epoxy resins with added reactive diluents or solvents	• mild to moderate irritants • moderate to strong sensitizers	• low volatility, exposure unlikely unless heated, sprayed, or spread over large unventilated surfaces	low toxicity
Aliphatic and cycloaliphatic amine curing agents	–	• irritants, sensitizers, corrosive, absorbed through skin	• respiratory irritants	high toxicity
Aromatic amine curing agents	–	• sensitizers, long-term health effects, absorbed through skin	• respiratory irritants • sensitizers	moderate to high toxicity
Anhydride curing agents	–	• corrosive, severe sensitizers	• dusts may be sensitizers	high toxicity
Reactive diluents	glycidyl ethers	• moderate to strong sensitizers	• moderate volatility, exposure possible	low toxicity
Solvents	acetone, methyl ethyl ketone (MEK), toluene, xylene, glycol, ethers, alcohol	• defats and dries skin • some may be absorbed • may carry other components through skin	• high volatility, exposure possible • irritation • central nervous system depression (e.g. dizziness, loss of coordination)	low to high toxicity, long-term effects
Fillers	fibreglass, silicas, calcium carbonate, powdered metal pigments	• some may be absorbed • potential primary irritant	• dust inhalation	low toxicity

phenols, phosphites, amines and thiobisphenols are employed as antioxidants in, e.g., polypropene, polyethylene and polystyrene resins. Typical precautions are summarized in Table 5.57.

Numerous alternatives to low-solid, solvent-based coatings traditionally used by, e.g., the wood furniture industry are shown in Table 5.58 with the estimated reduction in volatile organic compound emissions.

Table 5.57 General precautions with synthetic resins

Substitution using non-irritant resins and constituents, or less hazardous or less volatile constituents.
Limitation of skin contact, e.g. by covering benches, pre-mixing of resins, good housekeeping, mechanization.
Prevent or limit dust and vapour production.
Partial enclosure with local exhaust ventilation, or local exhaust ventilation, of working position: separate hazardous processes from other work, e.g. spray-painting with epoxy-containing sensitizing and carcinogenic compounds,
Enclose curing and mixing rooms.
Consider in advance procedures to be used for cleaning tools used in the process.
Provision and use of protective clothing, e.g. gloves, gauntlets, armlets, long-sleeved overalls; storage apart from ordinary clothing.
Provision and use of appropriate barrier cream.
Health surveillance on a regular basis covering skin inspections, enquiries/tests relating to respiratory function: consider the need to screen out staff with medical histories of allergic reactions.
Provision of first-aid treatment even for trivial injuries to the skin.
Provision of adequate washing facilities including resin-removal cream; enforce high standards of personal hygiene.
Limitation of the use of respiratory protection to those situations where control of exposure by inhalation cannot be controlled by other means.
Provision of a good standard of general ventilation.
Prohibition of eating, drinking, smoking or application of cosmetics in the workplace.
Plan and organize collection and disposal of waste.
Limit stocks, some ingredients such as MbOCA and MEK have potential for major accidents.

Table 5.58 Alternative wood glues

Coating type	*VOC reduction (%)*
Aqueous base	90–95
High solids (e.g. nitrocellulose)	17–40
UV cured	80–100
Polyester	80–100
Polyurethane	80–100

(xvi) Welding fume

Almost all welding, brazing, gas cutting, burning and similar processes produce polluting fume and gases which can be harmful. The composition of the fume varies and the quantity generated depends upon the type of process. A summary of the possible emissions is given in Figure 5.4. Generally the fume will comprise very fine particles of metals and their oxides. Gases such as carbon monoxide and nitrogen oxides may be generated and mix with any inert shielding gases used. The hazard may be increased if the metal is painted (e.g. if lead-based paint is present the fume will contain lead oxide) or coated (e.g. plastic coatings will emit thermal degradation products). (Examples are given in Table 5.48.) Metal coatings will also emit thermal degradation products, e.g. zinc oxide fumes from galvanized steel, lead oxide from steel painted with red lead for corrosion protection.

Measures to control welding fume include process modifications, engineering controls, system of work and administrative action as summarized in Figure 5.5.

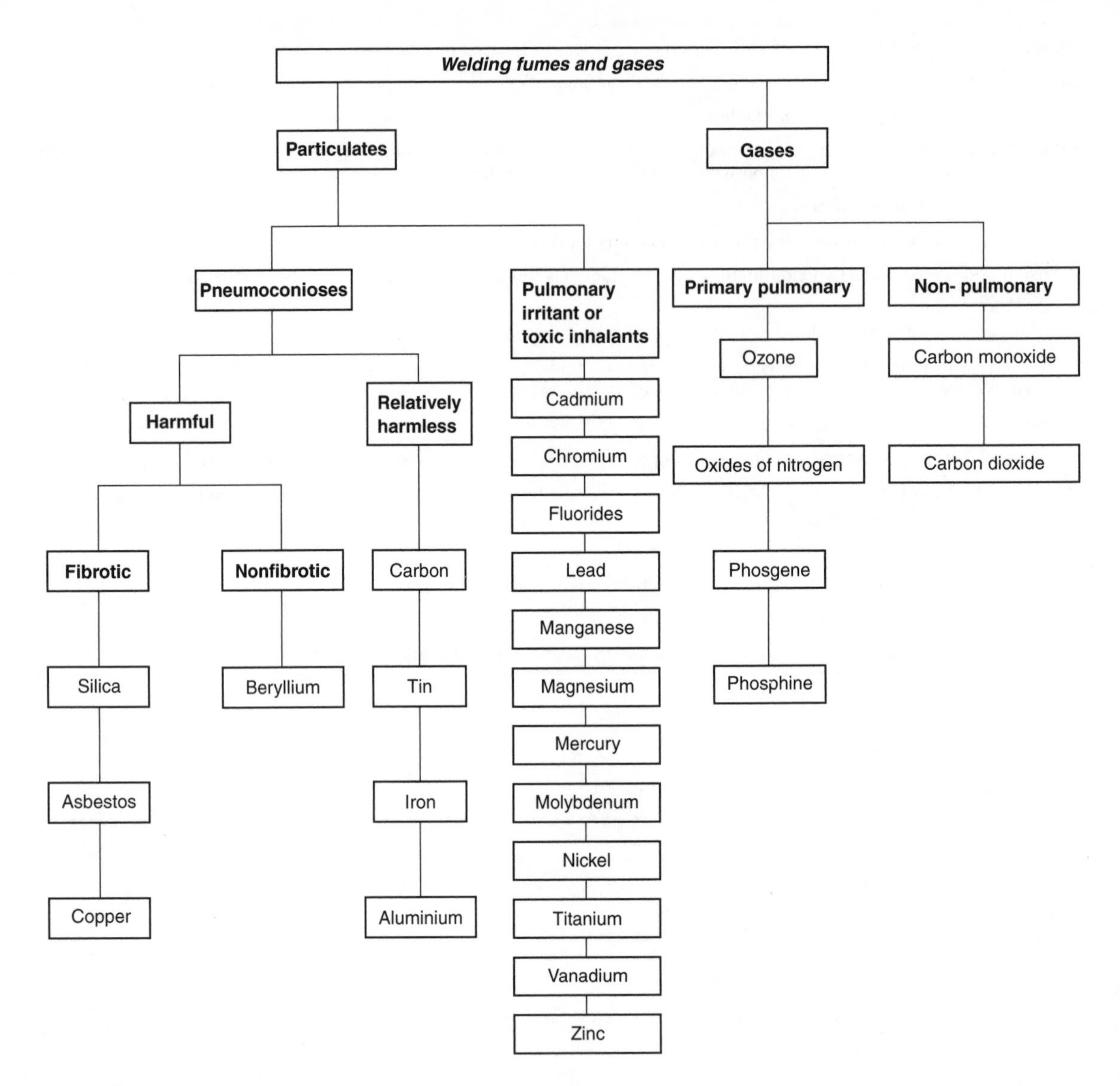

Figure 5.4 *Possible constituents in welding fumes and their effects*

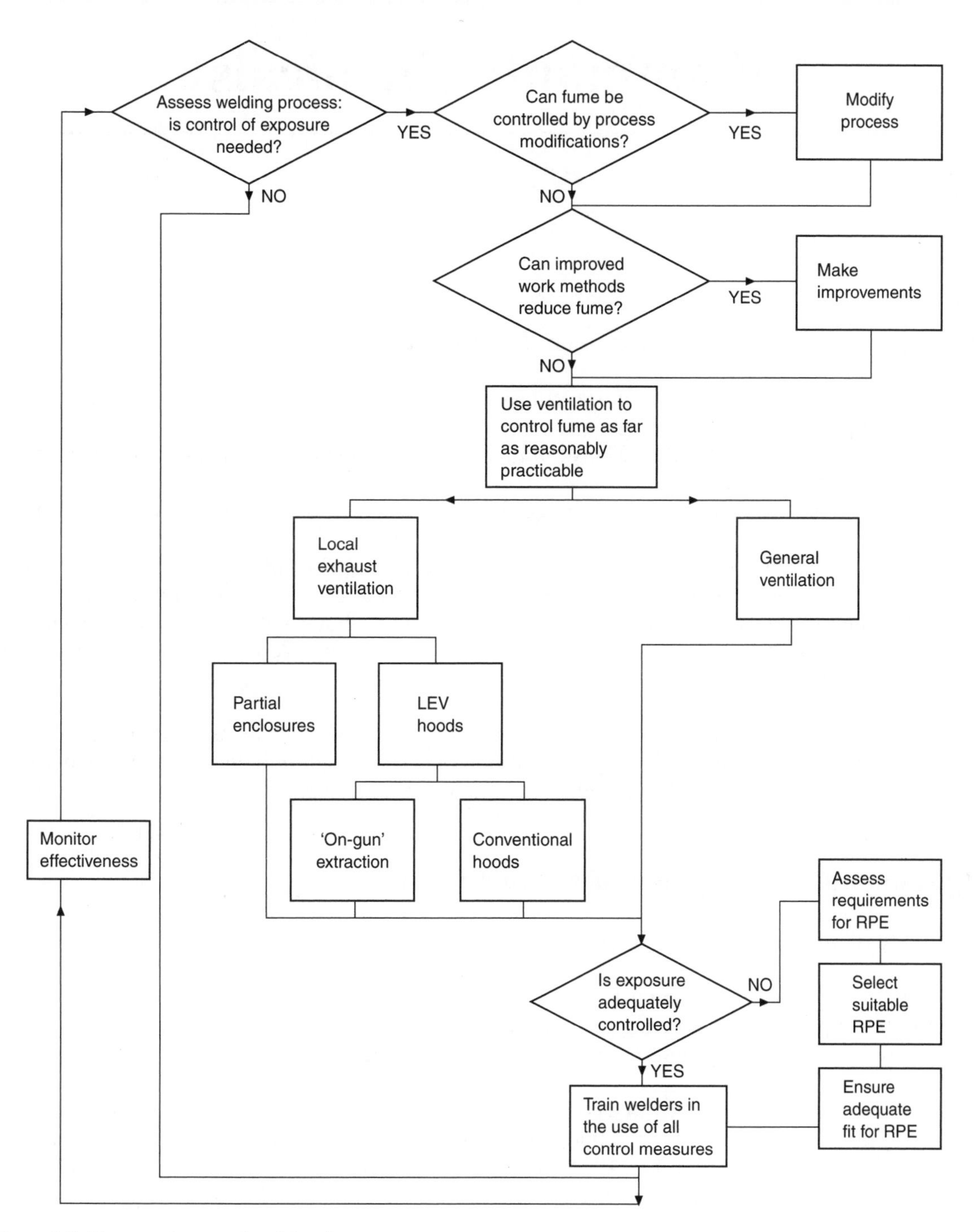

Figure 5.5 *Measures to control welding fume*

Flammable chemicals

Certain chemicals pose fire and explosion risks because:

- They ignite easily. Vapours often travel a considerable distance to an ignition source remote from the point of chemical escape.
- Considerable heat is generated. Many volatile substances liberate heat at a rate some ten times faster than burning wood.
- The fire spreads easily by, e.g., running liquid fire, a pool fire, a fire ball, heat radiation or thermal lift (convection).
- Explosion: a confined vapour cloud explosion (CVCE) can result from ignition of vapour within a building or equipment; a boiling liquid expanding vapour explosion (BLEVE) can result when unvented containers of flammable chemicals burst with explosive violence as a result of the build-up of internal pressure; unconfined vapour cloud explosion (UVCE) can result from ignition of a very large vapour or gas/air cloud.

Clearly, flammable chemicals also pose a health risk if the substance or its thermal degradation or combustion products are toxic, (e.g. carbon monoxide) or result in oxygen deficiency because oxygen is consumed. Hot smoke and other respiratory irritants, e.g. aldehydes, are also produced.

Ignition and propagation of a flame front

Normally flame propagation requires

(a) fuel, gas or vapour (or combustible dust) within certain concentration limits,
(b) oxygen supply (generally from air) above a certain minimum concentration, *and*
(c) ignition source of minimum temperature, energy and duration.

All three, represented by the three corners of a triangle (Figure 6.1), must generally be present. But no ignition source is needed if a material is above a specific temperature (see p. 214), and no additional oxygen is required if an oxidizing agent is present or in a few cases when oxygen is within the fuel molecule (e.g. ethylene oxide).

Fuel

Liquids and solids do not burn as such, but on exposure to heat vaporize or undergo thermal degradation to liberate flammable gases and vapours which burn. Some chemicals undergo spontaneous combustion (see page 214).

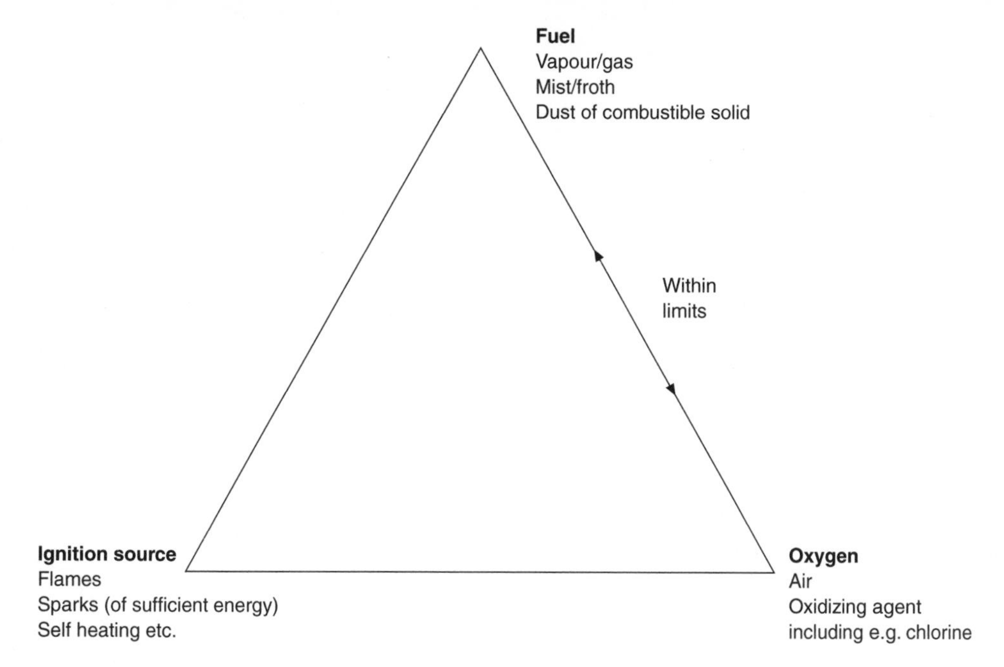

Figure 6.1 *Fire triangle*

Flammable limits

Flammable gases and volatile liquids are particularly hazardous because of the relative ease with which they produce mixtures with air within the flammable range. An increase in the surface area of any liquid facilitates vaporization. For each substance there is a minimum concentration of gas or vapour below which flame propagation will not occur (i.e. the mixture is too lean). There is also a concentration above which the mixture is too rich to ignite. The limits of flammability are influenced by temperature and pressure (e.g. the flammable range expands with increased temperature). Generally, the wider the flammable range the greater the fire risk. Flammability limits for a range of chemicals are summarized in Table 6.1.

The vapour pressure of a flammable substance also provides an indication of how easily the material will volatilize to produce flammable vapours; the higher the vapour pressure, the greater the risk. Lists of vapour pressures usually contain data obtained under differing conditions but inspection of boiling points (when the vapour pressure equals atmospheric pressure) gives a first approximation of the ease with which substances volatilize. Table 6.1 therefore includes both boiling point and vapour pressure data.

Flash point

The flash point represents the minimum temperature at which an ignitable mixture exists above a liquid surface. By definition, flash points are inapplicable to gases. Some solids, e.g. naphthalene and camphor, are easily volatilized on heating so that flammable mixtures develop above the solid surface and hence flash points can be determined. (However, although these substances can be ignited, they generally need to be heated above their flash points in order for combustion to be sustained: this is the 'fire point'.)

Flash point determinations may be made in 'closed' or 'open' containers, giving different values;

Table 6.1 Properties of flammable chemicals

Substance	Specific gravity	Vapour density (air = 1)	Flash point[1] (°C)	Ignition temp. (°C)	Flammable limits[1] (%)	Boiling point (°C)	Melting point (°C)	Solubility in water (g/100 g)	Vapour pressure (mm Hg/°C)
Acetal	0.83	4.08	–21	230	1.6–10.4	103	–100	21^{20}	10/8
Acetaldehyde	0.78	1.52	–38	185	4–57	21	–124	∞	–
Acetanilide	1.21	4.7	174	546	–	304	114	–	1/114
Acetic acid	1.05	2.1	43	426	4–16	118	17	∞	11.4/20
Acetic anhydride	1.08	3.5	54	385	3–10	140	–73	∞	10/36
Acetone	0.79	2.0	–18	538	3–13	56	–94	∞	400/40
Acetone cyanohydrin	0.93	2.9	74	688	–	82	–19	v. sol	–
Acetonitrile	0.79	1.4	6	524	4–16	80	–45	∞	100/27
Acetyl acetone, *see* 2,4-Pentanedione									
Acetyl chloride	1.1	2.7	4	390	–	51	–112	dec.	–
Acetylene	0.91	0.9	–	300	3–82	–83	Subl.	sl. sol.	40 atm/17
Acetylene dichloride, *see* 1,2-Dichloroethylene									
Acetyl peroxide	1.2	4.07	113oc	–	–	63exp.	30	sl. sol.	–
Acrolein	0.84	1.9	–26	278	3–31	53	–87	v. sol.	–
Acrolein dimer	1.1	–	48	–	–	151	–73	sol.	–
Acrylic acid	1.05	2.5	52oc	429	–	142	12	∞	10/39
Acrylonitrile	0.81	1.8	0oc	481	3–17	77	–83	sol.	100/23
Adipic acid	1.4	5.04	191	422	–	334	153	sl. sol.	1/160
Adiponitrile	0.97	3.73	93	–	–	295	2.3	sl. sol.	–
Aerozine 50, *see* Hydrazine									
Aldrin	–	–	66	–	–	–	104	insol.	–
Allyl acetate	0.93	3.4	21oc	374	–	103	–	sl. sol.	–
Allyl alcohol	0.85	2.0	21	378	3–18	97	–129	∞	10/10
Allyl amine	0.76	2.0	–29	374	2–22	55	–	∞	–
Allyl bromide	1.4	4.2	–1	295	4–7	70	–119	insol.	–
Allyl chloride	0.94	2.6	–32	392	3–11	45	–136	insol.	–
Allyl chloroformate	1.1	4.2	31	–	–	110	–	insol.	–
Allylene, *see* Propyne									
Allyl glycidyl ether	0.97	3.4	57	–	–	154	–100	sol.	–
2-Aminoethanol, *see* Ethanolamine									
Aminoethylethanol amine	1.03	3.6	129	368	–	244	–	v. sol	–
Ammonia, anhydrous	0.77	0.59	–	651	16–25	–33	–78	89.9°	10 atm/26
Ammonium nitrate	1.7	–	–	–	exp.	210/11 mm dec.	169	118°	–
n-Amyl acetate	0.88	4.5	25	379	1–7.5	148	–79	sl. sol.	–
iso-Amyl acetate	0.88	4.5	23	380	1–7.5	142	–	–	–
sec-Amyl acetate	0.86	4.5	32	–	1–7.5	121	–	–	–

n-Amyl alcohol	0.82	3.0	33	300	1–10	137	–79	sl. sol.	1/14
iso-Amyl alcohol	0.81	3.0	43	347	1–9	132	–	sl. sol.	–
tert-Amyl alcohol	0.81	3.0	19	437	1–9	102	–12	sl. sol.	–
Amylamine	0.8	3.0	7oc	–	–	103	–55	∞	–
n-Amyl bromide	1.22	5.2	32	–	–	130	–95	insol.	–
Amylene	0.66	2.4	–2	273	1.5–9	30	–124	insol.	–
n-Amyl ether	0.74	5.46	57	171	–	190	–70	insol.	–
iso-Amyl formate	0.89	4.0	26	–	–	131	–74	sl. sol	–
Amyl mercaptan	0.84	3.59	18	–	–	127	–76	insol.	13/25
iso-Amyl nitrate	0.99	–	52	–	–	152–7	–	sl.sol.	–
Amyl nitrite	0.85	4.0	10	209	–	104	–	sl. sol	–
Aniline	1.02	3.22	70	770	1.3	184	–6	sol.	1/35
o-Anisaldehyde	1.12	–	118	–	–	250	38	insol.	–
Anisole	1.0	3.72	52oc	–	–	154	–37	insol.	10/42
Anthracene	1.25	6.15	121	540	0.6–	340	217	insol.	1/145
Anthraquinone	1.44	7.16	185	–	–	380	286	insol.	1/190
Asphalt	1.1	–	204+	485	–	370–470	–	–	–
Aziridine	0.83	1.5	–11	322	3.6–46	56	–72	∞	–
Benzaldehyde	1.04	3.7	64	192	–	178	–26	sl. sol.	1/26
Benzene	0.88	2.8	–11	562	1.4–8	80	5.4	sl. sol	10/26
Benzene monochloride, *see* Chlorobenzene									
Benzoic acid	1.32	4.2	121	574	–	249	122	sl. sol.	1/96
p-Benzoquinone, *see* Quinone									
Benzonitrile	1.2	–	85oc	–	–	191	–13	–	1/28
Benzotrifluoride	1.19	5.04	12	–	–	101	–29	insol.	11/0
Benzoyl chloride	1.22	4.88	72	–	–	197	–0.5	dec.	1/32
Benzoyl peroxide	1.33	–	–	80	–	exp.	106	sl. sol.	–
Benzyl acetate	1.06	5.1	102	461	–	214	–51.5	sl. sol.	1/45
Benzyl alcohol	1.04	3.7	101	436	–	206	–15.3	4	1/58
Benzylamine	0.98	–	63	–	–	185		∞	–
Benzyl benzoate	1.11	7.3	148	481	–	323	21	insol.	–
Benzyl 'Cellosolve'	1.1	5.3	129	352	–	256	–75	–	–
Benzyl chloride	1.10	4.36	67	585	1.1–	179	–39	insol.	–
Benzylidene chloride, *see* Benzal chloride									
Benzyl mercaptan	1.06	4.3	70	–	–	194	–	insol.	–
Bicyclohexyl	0.9	5.7	74	244	1–5	240	2	–	–
Biphenyl	1.2	5.3	113	540	0.6–5.8	256	70	insol.	–
2-Biphenylamine	1.16	5.8	–	452	–	299	49	insol.	–
Borneol	1.01	5.31	66	–	–	–	212 subl	insol.	–
Boron hydrides, *see* Di-, Penta-, or Deca-Boranes									
Bromobenzene	1.50	5.4	51	566	–	155	–31	insol.	10/40
1-Bromobutane, *see* Butyl bromide									

Table 6.1 Cont'd

Substance	*Specific gravity*	*Vapour density* (air = 1)	*Flash point*[1] (°C)	*Ignition temp.* (°C)	*Flammable limits*[1] (%)	*Boiling point* (°C)	*Melting point* (°C)	*Solubility in water* (g/100 g)	*Vapour pressure* (mm Hg/°C)
Bromoethane	1.46	3.76	none	511	6.7–11.3	38	–119	sl. sol.	400/21
Bromomethane, *see* Methyl bromide									
Bromopentane, *see* Amyl bromide									
Bromopropane, *see* Propyl bromide									
3-Bromopropene, *see* Allyl bromide									
3-Bromopyne, *see* Propargyl bromide									
o-Bromotoluene	1.42	5.9	79	–	–	181	–27	insol.	–
1,3-Butadiene	0.62	1.97	<–7	429	2–11.5	–4.7	–109	insol.	1840/21
n-Butane	0.60	2.04	–60	405	1.9–8.5	–0.5	–138	v. sol.	2 atm/19
iso-Butane	0.56	2.01	–	462	1.8–8.4	–12	–160	sol.	
n-Butanol, *see* *n*-Butyl alcohol									
Butanone, *see* Methyl ethyl ketone									
1-Butene	0.60	1.9	–80	384	1.6–9.3	–6.1	–130	insol.	3480/21
2-Butene	0.62	1.9	–	324	1.8–9.0	1.1	–127	insol.	1410/21
2-Butene (trans)	0.6	2.0	–73	324	2–10	2.5	–106	–	1592/21
2-Butoxyethanol, *see* Butyl cellosolve									
n-Butyl acetate	0.88	4.0	27	399	1.4–7.6	125	–76	sl. sol.	15/25
iso-Butyl acetate	0.87	4.0	18	423	1.3–7.5	117	–99	sl. sol.	–
sec-Butyl acetate	0.86	4.0	31	–	1.7–	112	–	insol.	24
Butyl acetyl ricinoleate	0.9	13.7	110	385	–	220	–32	–	–
Butyl acrylate	0.9	4.4	49oc	–	–	69.50 mm	–65	–	10/36
n-Butyl alcohol	0.81	2.55	29	365	1.4–11	118	–89	sol.	6/20
iso-Butyl alcohol	0.81	2.55	28	427	1.7–10.9	107	–108	10^{15}	–
sec-Butyl alcohol	0.81	2.55	24	406	1.7–9.8	99.5	–115	12.5	10/20
tert-Butyl alcohol	0.78	2.55	10	478	2.4–8	83	25	∞	40/25
Butylamine	0.76	2.5	–12	312	1.7–9.8	78	–50	∞	72/20
tert-Butylamine	0.70	2.5	–	–	1.7–8.9	45	–67	∞	–
iso-Butylamine	0.73	2.5	–9	378	–	66	–104	∞	–
n-Butyl benzene	0.9	4.6	71	412	1–6	182	–81	–	1/23
iso-Butyl benzene	0.9	4.6	52	418	1–7	174	–83	–	1/19
n-Butyl bromide	1.28	4.7	18	265	2–6.6	101	–112	0.06	–
Butyl carbitol	1.0	5.6	78	228	–	231	–68	–	0.02/20
Butyl carbitol acetate	1.0	–	116	299	–	247	–32	–	<0.01/20

Butyl cellosolve	0.91	4.1	61	244	1.1–12.7	171	<–40	∞	0.6/20
n-Butyl chloride	0.88	3.2	7	471	1.9–10.1	78	–123	0.07	–
tert-Butyl chloride	0.85	3.2	<21	–	–	51	–27	sl. sol.	–
1,3-Butylene glycol	1.0	3.2	121	394	–	208	<–50	–	0.06/20
2,3-Butylene glycol	1.0	3.1	85	402	–	180	19	–	0.17/20
Butylene oxides	0.8	2.5	–15	–	1.5–18.3	63	–	–	–
Butyl ether, *see* Dibutyl ether									
n-Butyl formate	0.91	3.5	18	322	1.7–8	107	–90	sl. sol.	40/32
iso-Butyl formate	0.89	–	<21	–	–	98	–95	1.1^{22}	–
Butyl hydroperoxide	0.86	2.1	27	–	–	dec.	6	sol.	–
in heptane	0.68	–	–4	223	1.2–6.7	98	–	–	–
in hexane	0.69	–	–22	234	1.2–7.5	69	–	–	–
in pentane	0.70	–	–40	309	1.5–7.8	36	–	–	–
Butyl lactate	1.0	5.0	71	382	–	188	–43	–	0.4/20
Butyl mercaptan	0.84	3.1	2	–	–	98	–116	sl. sol.	–
Butyl methacrylate	0.89	4.8	52	294	2–8	163	–	insol.	5/20
Isobutyl methyl ketone		3.5	17	460	1.2–8	126	–57	–	–
tert-Butyl peracetate	0.93	–	27	–	–				
tert-Butyl perbenzoate	1.0	–	88	–	–	113dec	–	–	0.3/50
Butyl peroxypivalate (in 75% sol. of mineral spirits)		–	>68	–	–		–19		
Butyl peroxytrimethyl acetate, *see* Butyl peroxypivalate									
Butyl vinyl ether	0.77	3.4	–9	–	–	94	–92	insol.	–
n-Butyraldehyde	0.82	2.5	–6.7	230	2.5–	76	–99	4	–
iso-Butyraldehyde	0.79	2.5	–40	254	1.6–10.6	64	–66	4	–
n-Butyric acid	0.96	3.0	66	452	2–10	164	–7.9	∞	0.4/20
iso-Butyric acid	0.95	3.0	62	502	–	154	–47	20^{20}	–
n-Butyric anhydride	0.97	5.4	88	307	–	198	–73	dec.	–
2-Butyrolactone	1.05	3.0	98	–	–	206	–44	∞	–
n-Butyronitrile	0.8	–	26	–	–	117	–112	sl. sol.	–
Butyryl chloride	1.03	3.7	<21	–	–	107	–89	dec.	–
Camphor	0.99	5.24	66	466	0.6–3.5	204	180	sl. sol.	–
Caproic acid	0.93	4.0	102	–	–	205	–5.4	1.1^{20}	0.2/20
Capryl alcohol, *see* 2-Octanol									
Caprylaldehyde	0.8	4.5	52	–	–	168	–	–	
Caprylic acid, *see* Octanoic acid									
Caprylic alcohol, *see* 1-Octanol									
Carbitol, *see* Diethylene glycol monoethyl ether									
Carbolic acid, *see* Phenol									
Carbon disulphide	1.26	2.6	–30	100	1–44	46	–112	0.2°	400/28
Carbon monoxide	0.81	0.97	–	609	12.5–74	–192	–207	0.004°	–
Carbon oxysulphide, *see* Carbonyl sulphide									

Table 6.1 Cont'd

Substance	*Specific gravity*	*Vapour density* (air = 1)	*Flash point*[1] (°C)	*Ignition temp.* (°C)	*Flammable limits*[1] (%)	*Boiling point* (°C)	*Melting point* (°C)	*Solubility in water* (g/100 g)	*Vapour pressure* (mm Hg/°C)
Carbonyl sulphide	1.07	2.1	–	–	12–29	–50	–138	80[14]	
Carvene, *see* Dipentene									
Cellosolve	0.93	3.10	41	238	1.7–15.6	135	–	∞	3.8/20
Cellosolve acetate	0.97	4.7	55	382	1.2–12.7	156	–62	v. sol.	1.2/20
Cellulose nitrate	1.66	–	13	–	–	–	–	insol.	–
Chloral	1.51	5.1	none	–	–	98	–58	v. sol.	–
Chlordane	~1.6	–	56	–	–	175	–	–	–
Chlorine dioxide	3.09	2.3	–	100	–	9.9	–59	dec.	–
Chloroacetaldehyde	1.19	2.7	88	–	–	85	–16	–	100/45 (40% solution)
Chlorobenzene	1.11	3.9	29	638	1.3–7.1	132	–45	insol.	10/22
1-Chloro-1,1-difluoroethane	1.12	–	–	632	9–14.8	–9.2	–	–	–
1-Chloro-2,4-dinitrobenzene	1.7	–	194	432	2–22	315	43	insol.	–
Chlorodiphenyls	~1.4	–	176–80	–	–	340–75	–	–	30/200
1-Chloro-2,3-epoxypropane, *see* Epichlorohydrin									
Chloroethane, *see* Ethyl chloride									
Chloroethanol, *see* Ethylene chlorohydrin									
Chloromethane, *see* Methyl chloride									
Chloronaphthalene	1.19	5.6	132	>558	–	256	–20	insol.	–
Chloronitrobenzenes	1.37	–	127	–	–	242	32–46	insol.	–
1-Chloro-1-nitropropane	1.21	4.3	62	–	–	142	–	sl. sol.	–
o-Chlorophenol	1.24	–	64	–	–	175	7	v. sol.	1/12
p-Chlorophenol	1.24	–	121	–	–	220	43	sol.	1/49
o-Chlorophenyl diphenyl phosphate	1.3	12.5	>215	–	–	240–55/5 mm	<0	–	–
Chloroprene	0.95	3.0	–20	–	4–20	59	–	sl. sol.	–
Chlorotrifluoroethylene	1.31	–	–27	–	8.4–39	–28	158	–	–
Cinnamaldehyde	1.05	–	49	–	–	253	–7.5	sl. sol.	–
m-Cresol	1.03	3.7	94	559	1.06–1.35	203	12	sol.	1/52
o-Cresol	1.05	3.7	81	599	1.35–	191	31	sol. (hot)	1/38
p-Cresol	1.04	3.7	94	559	1.06–1.4	202	35	sol. (hot)	–
Creosote (mixed phenols)	1.07	–	74–82	336	–	200–250	–	–	–
Cresylic acid, *see* *o*-Cresol									
Crotonaldehyde	0.87	2.4	13	207	2.1–15.5	104	–76	v. sol.	–
Crotonitrile	0.83	2.3	<100	–	–	110–116	–52	–	–
Crude oil (petroleum)	0.78–0.97	–	(–7)–(+32)	–	–	–	<–46	–	–

Cumene	0.86	4.1	44	424	0.9–6.5	152	–96	insol.	10/38
Cumene hydroperoxide	1.05	–	79	–	–	153	–	–	–
Cyanogen	0.95	1.8	–	–	6–32	–21	–34	450[20]	–
Cyclohexane	0.77	2.91	–20	260	1.3–8.4	80	4.1	insol.	100/60
Cyclohexanol	0.96	3.45	68	300	–	161	25	3.6	1/21
Cyclohexanone	0.95	3.4	44	420	1.1–8.1	156	–45	sl. sol.	10/39
Cyclohexene	0.81	2.8	<–6	–	–	83	–104	insol.	160/38
Cyclohexylamine	0.87	3.4	32	293	–	135	–18	sol.	–
Cyclohexylbenzene	0.95	–	99	104	–	238	7.5	insol.	1/67
Cyclopentane	0.75	2.42	–7	–	–	49	–94	insol.	400/31
Cyclopentanone	0.95	2.3	26	–	–	131	–58	insol.	–
Cyclopropane	0.72	1.45	–	498	2.4–10.4	–33	–127	sl. sol.	–
p-Cymene	0.86	4.62	47	436	0.7–5.6	177	–68	insol.	1/17
Decaborane	0.94	–	80	149	–	213	100	sl. sol.	19/100
Decahydronaphthalene	0.87	4.76	57	250	0.7–4.9	186	–45	insol.	–
iso-Decaldehyde	0.8	5.4	85	–	–	197	–	–	–
Decalin, *see* Decahydronaphthalene									
n-Decane	0.73	4.90	46	208	0.8–5.4	174	–30	insol.	1/17
n-Decyl alcohol	0.83	5.3	82oc	–	–	229	7	insol.	1/170
Diacetone alcohol	0.93	4.0	64	603	1.8–6.9	168	–50	∞	1/20
Diaminoethane, *see* Ethylene Diamine									
Diazomethane	1.45	–	–	100exp.	–	–23	–145	dec.	–
Diborane	0.46	0.96	–90	145	0.9–98	–93	–165	dec.	–
Dibutylamine	0.50	4.46	52oc	–	–	159	–51	sol.	2/20
Dibutyldichlorotin	1.36	10.5	355	–	–	135 at 10 mm	43	dec.	–
Di-*n*-butyl ether	0.77	4.5	25	194	1.5–7.6	141	–98	insol.	–
Dibutyl oxalate	1.01	7.0	104	–	–	246	–30	insol.	–
Dibutyl peroxide	0.79	5.0	18	–	–	111	–40	insol.	20/20
Dibutyl phosphite	0.97	6.7	49	–	–	115 at 10 mm	–	–	–
Dibutyl phthalate	1.04	9.6	157	403	–	340	–35	insol.	–
Dibutyl tartrate	1.1	–	91	284	–	204/26 mm	21	–	–
Dibutyl tin dilaurate	1.05	21.8	235	–	–	–	27	insol.	–
Dichloroacetyl chloride	1.53	5.1	66	–	–	107	–	dec.	–
2,5-Dichloroaniline	–	5.6	166	–	–	251	50	sl. sol.	–
o-Dichlorobenzene	1.30	5.07	66	648	2–9	180	–18	insol.	–

Table 6.1 Cont'd

Substance	*Specific gravity*	*Vapour density* (air = 1)	*Flash point*[1] (°C)	*Ignition temp.* (°C)	*Flammable limits*[1] (%)	*Boiling point* (°C)	*Melting point* (°C)	*Solubility in water* (g/100 g)	*Vapour pressure* (mm Hg/°C)
1,4-Dichlorobutane	1.1	4.4	52	–	–	162	–39	insol.	4/20
1,3-Dichloro-2-butene	–	4.3	27	–	–	123	–	insol.	–
1,1-Dichloroethane	1.18	3.4	–6	458	5.6–11.4	57	–97	0.7°	–
1,2-Dichloroethane, *see* Ethylene dichloride									
1,1-Dichloroethylene, *see* Vinylidene chloride									
1,2-Dichloroethylene	1.3	3.3	2–4	–	9.7–12.8	48–60	–80	sl. sol.	–
2,2'-Dichloroethyl ether	1.22	4.9	55	369	–	178	–24	insol.	–
Dichloromethane	1.34	2.93	–	662	15.5–66 in O_2	40	–97	2^{20}	–
Dichloromonofluoromethane	1.48	3.8	–	552	–	9	–135	insol.	–
1,1-Dichloro-1-nitroethane	1.42	4.97	76	–	–	124	–	0.5^{20}	–
Dichloropentanes (mixed)	1.1	4.9	41	47	–	130	–	–	–
2,4-Dichlorophenol	1.38	5.6	114	–	–	210	45	sl. sol.	1/53
1,2-Dichloropropane	1.16	3.9	16	557	3.4–14.5	97	–100	sl. sol.	40/19
1,3-Dichloropropane	1.23	3.8	35	–	–	104	–	insol.	–
Dicyclohexylamine	0.93	6.3	99oc	–	–	256	–0.1	0.16^{28}	–
Dicyclopentadiene	0.93	4.55	35	–	–	170	33	–	10/48
Diethanolamine	1.09	3.6	152	662	–	270	28	v. sol.	5/188
1,1-Diethoxyethane, *see* Acetal									
Diethyladipate	1.01	–	–	–	–	240–5	–21	0.43^{30}	–
Diethylamine	0.71	2.5	<–26	312	1.8–10.1	56	–48	v. sol.	400/38
2-Diethyl-amino-ethanol	0.88	4.03	60	–	–	163	–	∞	–
N,N-Diethylaniline	0.94	5.15	85	332	–	216	–38	1.4^{12}	1/50
Diethylcarbonate	0.98	4.07	25	–	–	126	–43	insol.	10/24
Diethyl cellosolve	0.8	6.56	35oc	207	–	121	–74	–	9.4
Diethylene glycol	1.12	3.66	124	229	2–	245	–8	sol.	1/92
Diethylene glycol-monoethyl ether	1.11	4.6	96	204	1.2–	202	–10	∞	–
Diethylenetriamine	0.95	3.5	102	399	–	207	–39	∞	–
Diethyl ether, *see* Ethyl ether									
Diethyl ethyl phosphonate	1.03	5.7	105	–	–	83 at 11 mm	–	sl. sol.	–
Diethyl ketone	0.82	2.96	13	452	–	101	–42	4.7^{20}	–
Diethyl malonate	1.06	5.5	93	–	–	199	–50	2.1^{20}	1/40
o-Diethyl phthalate	1.1	7.7	163	–	–	302	–40	insol.	–
p-Diethyl phthalate	1.1	7.7	117	–	–	296	–5	insol.	–

Diethyl sulphate	1.18	5.3	104	436	–	208dec.	–25	insol.	1/47
Diglycol, *see* Diethylene glycol									
3,4-Dihydro-2H-pyran	0.92	2.9	–18	–	–	86	–	sol.	–
Diisobutyl ketone	0.81	4.9	60	–	0.8–6.2	168	–	insol.	–
Diisobutyl carbinol	0.8	5	74	–	0.8–6.1 (at 100°C)	173	–65	–	0.3/20
Diisopropyl amine	0.72	3.5	–1	–	–84	83	–61	sl. sol.	–
Diisopropyl ether	0.72	3.5	–28	443	1.4–7.9	69	–86	sl. sol.	150/25
Diisopropyl benzene	0.9	5.6	76oc	449	–	205	<–55	–	–
3,3′-Dimethoxybenzidine	–	8.5	206	–	–	–	137	insol.	–
Dimethoxyethane	0.85	3.1	40	–	–	65	–113	sol.	–
Dimethoxymethane	0.86	2.63	–18	237	–	46	–105	33	330/20
Dimethoxypropane	0.85	3.6	–7	–	–	95	–	–	–
N,N-Dimethyl acetamide	0.94	3.0	77	354	1.8–13.8	165	–20	∞	1.3/25
Dimethylamine	0.68	1.65	–50 (<–18)	402 (430)	2.8–14.4	7.4	–92	v. sol.	–
N,N-Dimethylaniline	0.95	4.17	63	371	–	193	2.5	sl. sol.	1/30
2,2-Dimethyl butane	0.65	3.00	–48	425	1.2–7.0	50	–98	insol.	400/31
2,3-Dimethyl butane	0.7	3.0	–29	420	1.2–7	58	–135	–	400/39
Dimethyl carbonate	1.1	3.1	19	–	–	90	0.5	insol.	–
Dimethyl chloracetal	1.1	4.3	43	232	–	130	–	–	–
Dimethyl ether, *see* Methyl ether									
Dimethylformamide	0.94	2.51	58	445	2.2–15.2	153	–61	∞	3.7/25
1,1-Dimethylhydrazine, *see* Unsymmetrical dimethylhydrazine									
2,6-Dimethyl-4-heptanone, *see* Diisobutyl ketone									
2,3-Dimethyl pentane	0.69	3.5	<–7	337	1–7	90	–135	–	40/14
Dimethyl phthalate	1.19	6.7	146	556	–	288	–	insol.	1/10
2,2-Dimethyl propane	0.61	2.48	<–7	450	1.4–7.5	9.5	–18	insol.	100/21
Dimethyl sulphate	1.33	4.35	83	188	–	189dec.	–31	sol.	–
Dimethyl sulphide	0.8	2.1	<–18	206	2.2–19.7	37	–83	insol.	–
Dimethyl sulphoxide	1.01	–	95	215oc	2.6–28.5	189	18	sol.	0.4/20
2,4-Dinitroaniline	1.62	6.3	244	–	–	–	188	insol.	–
m-Dinitrobenzene	1.58	–	150	severe explosion hazard when		291	90	0.3^{99}	–
o-Dinitrobenzene	1.31	5.8	150	exposed to shock or flame		319	118	sl. sol.	–
p-Dinitrobenzene	1.63	–	150	–	–	299	172	0.18^{38}	–
2,4-Dinitrotoluene	1.4	6.3	207	–	–	300	70	0.03^{20}	–
Di-*n*-Octyl phthalate	0.97	14.4	219	–	–	385	–30	–	–
Di-sec-Octyl phthalate	0.99	16	218	410	–	358	–55	–	1.2/200
1,4-Dioxane	1.04	3.0	12	180	2–22.2	101	10	∞	40/25
Dipentene	0.85	4.66	45	237	0.7–6.1	178	–97	insol.	1/14
Dipentene, dioxide *see* *p*-Mentha-1,8-diene									
Dipentene monoxide	0.93	4.45	67	–	–	75	<–6	–	–

Table 6.1 Cont'd

Substance	*Specific gravity*	*Vapour density* (air = 1)	*Flash point*[1] (°C)	*Ignition temp.* (°C)	*Flammable limits*[1] (%)	*Boiling point* (°C)	*Melting point* (°C)	*Solubility in water* (g/100 g)	*Vapour pressure* (mm Hg/°C)
Diphenylamine	1.16	5.8	153	634	–	302	53	insol.	1/108
Diphenylmethane	1.01	5.8	130	486	–	266	26	insol.	1/76
Dipropylene glycol methyl ether	0.95	5.1	85	–	–	189	–	–	–
n-Dodecane	0.75	5.96	74	204	0.6–	216	–12	insol.	1/48
Epichlorohydrin	1.18	3.29	41	–	–	117	–48	<5	10/17
1,2-Epoxy-propane	0.83	2.0	–37	–	2.1–21.5	34	–104	v. sol.	400/18
Ethane	0.57	1.04	–	515	3–12.5	–89	–183	insol.	–
Ethanediol, *see* Ethylene glycol									
Ethanethiol	0.84	2.14	<27	299	2.8–18	37	–144	1.5	–
Ethanoic acid, *see* Acetic acid									
Ethanol	0.79	1.59	12	423	3.3–19	79	–114	∞	40/19
Ethanolamine	1.02	2.11	85	–	–	170	11	∞	6/60
Ethoxy acetylene	0.79	2.4	<–7	100exp.	–	50	–	insol.	–
2-Ethoxy ethanol, *see* Cellosolve									
2-Ethoxy ethylacetate, *see* Cellosolve acetate									
Ethyl acetanilide	0.94	5.62	52	–	–	258	54	insol.	–
Ethyl acetate	0.90	3.04	–4.4	427	2.18–11.5	77	–84	7.5^{20}	100/27
Ethyl acetoacetate	1.03	4.48	84	–	–	181	–45	13^{17}	1/29
Ethyl acrylate	0.92	3.5	16	273	1.8–	100	<–72	sol.	29/20
Ethyl alcohol, *see* Ethanol									
Ethyl aldehyde, *see* Acetaldehyde									
Ethylamine	0.80	1.56	<–18	384	3.5–14	17	–81	∞	400/2
Ethyl amyl ketone	0.85	–	57	–	–	161	–	insol.	–
n-Ethyl aniline	0.96	4.18	85	–	–	205	–64	insol.	1/38
Ethyl benzene	0.9	3.7	15	432	1–6.7	136	–95	0.01^{15}	10/26
Ethyl benzoate	1.15	5.17	>96	–	–	213	–35	insol.	1/40
Ethyl bromide, *see* Bromoethane									
Ethyl bromoacetate	1.51	5.8	48	–	–	159	<–20	insol.	–
Ethyl butyl ketone	0.82	3.93	46	–	–	148	–37	insol.	–
2-Ethyl butyraldehyde	0.8	3.5	21oc	–	1–8	117	–90	–	14/20
Ethyl butyrate	0.88	4.0	26	463	–	121	–97	0.68^{25}	10/15
Ethyl chloride	0.92	2.2	–50	519	3.6–15.4	12	–139	0.45°	1000/20
Ethyl chloroacetate	1.26	4.3	66	–	–	144	– 27	insol.	10/38
Ethyl chloroformate	1.36	3.74	16	–	–	95	–81	dec.	–
Ethyl crotonate	0.92	3.93	2	–	–	143–7	45	insol.	–
Ethyl cyanoacetate	1.06	3.9	110	–	–	206	–23	2^{25}	1/68
Ethyl cyanide, *see* Propionitrile									
Ethylene	0.001	1.0	–	450	3.1–32	–104	–169	26^{0}	–

Ethylene chlorohydrin	1.21	2.78	60oc	425	4.9–15.9	128	–69	∞	10/30
Ethylene diamine	0.90	2.07	43	–	–	117	8.5	v. sol.	11/20
Ethylene dichloride	1.26	3.4	13	413	6.2–15.9	83	–36	sl. sol.	100/29
Ethylene glycol	1.11	2.14	111	413	3.2–	198	–13	∞	0.05/20
Ethylene glycol monobutyl ether, *see* Butyl cellosolve									
Ethylene glycol monoethyl ether, *see* Cellosolve									
Ethylene glycol monomethyl ether, *see* Methyl cellosolve									
Ethylenimine, *see* Aziridine									
Ethylene oxide	0.87	1.49	<–18	429	3–100	11	–111	sol.	1095/20
Ethyl ether	0.71	2.55	–45	180	1.85–48	34	–123	7.5^{20}	442/20
Ethyl ethynyl ether, *see* Ethoxyacetylene									
Ethyl formate	0.95	2.55	–20	455	2.7–13.5	54	–79	11^{18}	100/5
2-Ethyl hexanol	0.83	4.49	84	–	–	180–5	<–76	insol.	0.2/20
Ethyl lactate	1.04	4.07	46	400	1.5–30	154	–	∞	–
Ethyl malonate, *see* Diethyl malonate									
Ethyl mercaptan, *see* Ethanethiol									
Ethyl methyl ether	0.7	2.1	–37	190	2–10	11	–	–	–
N-Ethylmorpholine	0.92	4.00	32	–	–	138	–	∞	–
Ethyl nitrite	0.90	2.59	–35	90(explodes)	4.1–50	17	–	dec.	–
Ethyl oxalate	1.08	5.04	76	–	–	186	–41	sl. sol.	–
Ethyl propionate	0.9	3.5	12	477	2–11	99	–73	–	40/27
2-Ethyl-3-propyl acrolein	0.85	4.35	68oc	–	–	175	<100	insol.	1/20
Ethyl silicate	0.93	7.22	52	–	–	165	110subl.	dec.	1/20
Ethyl vinyl ether	0.75	2.46	<–46	202	1.7–28	36	–115	sl. sol.	–
Fluoroethylene	–	–	–	–	2.6–21.7	–51	–160	insol.	
Formaldehyde	0.82	1.08	–	430	7.0–73	–19	–92	sol.	–
Formalin (39% formaldehyde methanol free)	0.82	–	85	–	–	101	–	–	–
Formalin (37% formaldehyde-15% methanol)	–	–	50	–	–	101	–	–	–
Formamide	1.13	–	155oc	–	–	211dec.	2.6	∞	30/129
Form-dimethylamide, *see* Dimethylformamide									
Formic acid	1.22	1.59	69oc	601	18–57	101	8.2	∞	40/24
Furan	0.94	2.35	<0	–	2–14	32	–86	insol	–
Furfural	1.16	3.31	60	316	2.1–	162	–37	9.1^{13}	–
Furfuryl alcohol	1.13	3.37	75	491 (916)	1.8–16.3	171	–31	∞	1/32
Gasoline (petrol)	0.8	3.0–4.0	–43	280–456	1.4–7.6	38–204	–	–	–
Glycerol	1.26	3.17	160	354	–	290dec.	18	∞	1.26/20

Table 6.1 Cont'd

Substance	*Specific gravity*	*Vapour density* (air = 1)	*Flash point*[1] (°C)	*Ignition temp.* (°C)	*Flammable limits*[1] (%)	*Boiling point* (°C)	*Melting point* (°C)	*Solubility in water* (g/100 g)	*Vapour pressure* (mm Hg/°C)
Heptane	0.68	3.45	–4	223	1.2–6.7	98	–91	insol.	40/22
2-Heptanone, *see* Methyl amyl ketone									
3-Heptanone, *see* Ethyl butyl ketone									
n-Heptylamine	0.77	4.0	54oc	–	–	158	–23	sl. sol.	–
Hexachlorobenzene	1.57	9.8	242	–	–	322	230	insol.	–
2,4-Hexadienal	0.9	–	68oc	–	1–8	170	–	sl. sol.	–
1,4-Hexadiene	0.7	2.8	–21		2–6	64	–	insol.	–
Hexane	0.66	2.97	–22	261	1.1–7.5	68	–96	insol.	100/16
Hexanedioic acid, *see* Adipic acid									
Hexanoic acid, *see* Caproic acid									
n-Hexanol	0.81	3.52	60	293	–	158	–45	sl. sol	1/24
2-Hexanone, *see* Methyl butyl ketone									
1-Hexene	0.7	2.97	<–7	–	1.2–6.9	63	–139	insol.	310/38
2-Hexene	0.68	2.92	<–7	–	–	69	–146	insol.	–
Hexone, *see* iso-Butyl methyl ketone									
sec-Hexyl acetate	0.86	4.97	45	–	–	141	–64	insol.	4/20
Hexyl alcohol, *see* *n*-Hexanol									
Hexyl amine	0.76	3.49	29oc	–	–	129	–19	sl. sol.	–
Hydracrylic acid-β-lactone	1.15	2.5	74	–	2.9–	155	–33	dec.	–
Hydrazine	1.0	1.1	38	Varies with surface 23–264	4.7–100	113	1.4	v. sol.	14.4/25
Hydrocyanic acid	0.69	0.93	–18	538	6–41	26	–14	∞	400/10
Hydrogen	0.09	0.069	–	585	4–75	–253	–259	2.1°	–
Hydrogen sulphide	1.5	1.2	–	260	4.3–46	–60	–83	437°	20 atm/25
p-Hydroquinone	1.33	3.8	165	515	–	285	171	sol.	4/150
Hydroquinone monomethyl ether	1.55	–	131	421	–	246	54	sol.	–
Hydroxylamine	1.20	–	129exp.	–	–	56	34	sol.	–
Isoprene	0.68	2.35	–54	220	–	34	–147	insol.	–
Kerosene	0.81	4.5	38 dependent upon specific fraction	229	0.7–5	170–300	<–46	–	
Lactonitrile	0.99	2.45	77	–	–	182dec.	–40	∞	–
Linseed oil	0.9	–	22	343	–	–	–19	–	–
Maleic anhydride	0.9	3.4	102	477	1.4–7.1	202	58	dec.	1/44

p-Mentha-1,8-diene	3.84	7.4	45	237	0.7–6.1	170	<60	insol.	–
2-Mercaptoethanol	1.14	2.69	74oc	–	–	157	–	sol.	1/20
Mesityl oxide	0.86	3.5	31	344	–	130	–59	sol.	10/26
α-Methacrylic acid	1.02	–	77oc	–	–	158	16	sol.	1/25
Methane	0.42	0.6	–	537	5–15	–161	–183	sl. sol.	–
3-(3-Methoxypropoxy)-1-propanol, *see* Dipropylene glycol methyl ether									
Methyl acetate	0.97	2.55	–9	502	3.1–16	57	–99	v. sol.	100/9.4
Methyl acetylene, *see* Propyne									
Methyl acrylate	0.95	3.0	–3oc	–	2.8–25	80	–75	sl. sol.	100/28
Methylal, *see* Dimethoxymethane									
Methyl alcohol	0.79	1.11	12	464	6–36.5	65	–98	∞	100/21
Methyl amyl alcohol	0.80	3.5	41	–	1–5.5	130	<–90	sl. sol.	3/20
Methyl-*n*-amyl ketone	0.81	3.9	49oc	533	–	151	–35	sl. sol.	3/20
Methyl bromide	1.73	3.27	–	537	10–16	4	–95	insol.	–
2-Methyl-1-butene	0.66	2.4	<–7	–	–	39	–134	insol.	–
2-Methyl-2-butene	0.67	2.4	<–17	–	–	38	–123	sl. sol.	–
3-Methyl-1-butene	0.67	2.42	<–57	365	1.5–9.1	20	–168	insol.	–
N-Methylbutylamine	0.74	3.0	13oc	–	–	91	–	sol.	–
Methyl butyl ketone	0.81	3.45	35oc	533	1.2–8	126	–57	–	10/39
Methyl butyrate	0.90	3.53	14	–	–	102	<–97	sl. sol.	40/30
Methyl cellosolve	0.97	2.62	46oc	288	2.5–14	125	–87	∞	6/20
Methyl cellosolve acetate	1.01	4.07	56	394	1.7–8.2	145	–70	sol.	–
Methyl chloride	0.98	1.8	<0oc	632	10.7–17.4	–24	–98	sl. sol.	–
Methyl chloroform, *see* 1,-1,-1-Trichloroethane									
Methyl chloroformate	1.24	3.26	12	504	–	73	–	dec.	–
Methyl cyanide, *see* Acetonitrile									
Methyl cyclohexane	0.77	3.39	–4	285	1.2–	100	–126	insol.	40/22
σ-Methyl cyclohexanol	0.92	3.93	68	296	–	165	–20	sl. sol.	–
2-Methyl cyclohexanone	0.92	3.86	48	–	–	165	–	insol.	–
4-Methyl cyclohexene	0.80	3.34	–1oc	–	–	103	–116	insol.	10/38
Methylene chlorobromide, *see* Bromochloromethane									
Methyl ether	0.66	1.56	–41	350	3.4–18	–24	–139	sol.	–
Methyl ethyl ether	0.73	2.07	–37	190	2–10.1	11	–	sol.	–
Methyl ethyl ketone	0.81	2.5	–7	515	2–10	80	–87	v. sol.	–
Methyl formate	0.99	2.1	–19	456	5.9–20	32	–100	v. sol.	400/16
2-Methyl furan	0.92	2.8	–30	–	–	63	–89	insol.	139/20
5-Methyl-3-heptanone, *see* Ethyl amyl ketone									
Methyl hydrazine	0.9	1.6	<27	–	–	87	<–80	sol.	–
Methyl isobutyl ketone	0.80	3.5	23	460	1.4–7.5	117	–85	sl. sol.	16/20

Table 6.1 Cont'd

Substance	*Specific gravity*	*Vapour density* (air = 1)	*Flash point*[1] (°C)	*Ignition temp.* (°C)	*Flammable limits*[1] (%)	*Boiling point* (°C)	*Melting point* (°C)	*Solubility in water* (g/100 g)	*Vapour pressure* (mm Hg/°C)
Methyl isobutyrate	0.86	3.5	13oc	482	–	92	–84	sl. sol.	–
Methyl mercaptan	0.87	1.66	<–18	–	3.9–21.8	7.6	–123	sl. sol.	–
Methyl methacrylate	0.94	3.6	10	421	2.1–12.5	100	–50	sl. sol.	40/25
1-Methylnaphthalene	1.03	–	–	528	–	240–3	–22	insol.	–
2-Methyl-2-propanethiol, *see* *t*-Butyl mercaptan									
Methyl-*n*-Propyl ketone	0.81	3.0	7	505	1.6–8.2	102	–78	sl. sol.	–
1-Methlpyrrole	0.91	2.8	16	–	–	115	–57	insol.	–
Methyl salicylate	1.18	5.24	101	454	–	223	–8.3	sl. sol.	1/54
a-Methyl styrene	0.92	4.08	54	574	1.9–6.1	167–70	–23	insol.	–
m,-p-Methyl styrene	0.89	4.08	57	494	0.9	170	–83	–	–
Methyl sulphate, *see* Dimethyl sulphate									
Methyl sulphide, *see* Dimethyl sulphide									
Methyl vinyl ether	0.77	2.0	–51	–	–	8	–122	sl. sol.	1052/20
Mixed acids	–	–	None	–	None	Varies	Varies	–	–
Monomethylamine	–	1.1	–10	430	4.9–20.8	–6.3	–94	v. sol	–
Morpholine	0.99	3.00	38oc	310	–	128	–4.9	∞	–
Naphtha (coal tar)	0.87	–	42	277	–	149–216	–	–	–
Naphtha (petroleum), *see* Petroleum ether									
Naphtha, varnish makers and painters, 50° flash	<1	4.1	10	232	0.9–6.7	116–43	–	–	–
Naphtha, varnish makers and painters, high flash	<1	4.3	29	232	1–6	139–77	–	–	–
Naphtha, varnish makers and painters, regular	<1	–	–2	232	0.9–6	100–60	–	–	–
Naphthalene	1.15	4.42	79	526	0.9–5.9	210	80	insol.	–
2-Naphthol	1.22	4.97	153	–	–	295	123	insol.	10/45
1-Naphthylamine	1.12	4.93	157	–	–	301	50	sl. sol.	–
Natural gas	–	–	–	482	3.8–17	–	–	–	–
Nickel carbonyl	1.32	~6	–	60exp.	2–	43	–25	sl. sol.	400/26
Nicotine	1.01	5.61	–	244	0.7–4.0	247	<–80	∞	1/62
o-Nitroaniline	1.44	–	168oc	521	–	284	71	sl. sol.	1/104
p-Nitroaniline	1.44	–	199	–	–	336	146	insol.	1/124
Nitrobenzene	1.20	4.24	88	482	1.8– at 93°C	211	5	sl. sol.	1/44
o-Nitrobiphenyl	1.44	6.9	143	180	–	330	35	insol.	2/140

Nitroethane	1.05	2.58	38	360–415	3.4–	114	–90	sol.	16/20
Nitromethane	1.14	2.11	35	418	7.3–	101	–28	sol.	28/20
α-Nitronaphthalene	1.14	5.96	164	–	–	304	60	insol.	–
1-Nitropropane	0.99	3.06	49oc	421	2.6–	131	–108	sl. sol.	8/20
2-Nitropropane	0.99	3.06	39oc	428	2.6–	120	–93	sl. sol.	10/16
m-Nitrotoluene	1.15	4.7	106	–	–	232	15	insol.	1/50
o-Nitrotoluene	1.16	4.72	106	–	–	220	–4.1	insol.	1/50
p-Nitrotoluene	1.29	4.72	106	–	–	238	52	insol.	1/54
Nonyl phenol	0.95	7.6	141	–	–	290–301	–	–	–
Octane	0.70	3.86	13	220	1.0–4.66	125	–57	insol.	10/19
Octanoic acid	0.91	5.0	132oc	–	–	240	16	sl. sol.	–
1-Octanol	0.83	4.5	81	–	–	194	–17	sol.	–
2-Octanol	0.82	4.48	88	–	–	178	–39	sl. sol.	–
Oil, lubricating	–	–	>149oc	260–371	–	360	<–46	–	–
Oil, mineral oil mist	0.81	–	193oc	–	–	360	–	–	–
Oil, olive	0.9	–	225	343	–	–	–6	–	–
Oil, peanut	0.9	–	282	470	–	–	3	–	–
Oil, soybean	0.9	–	282	445	–	–	22	–	–
Oil, vegetable	<1	–	321	–	–	–	–9–1	–	–
Oleic acid	0.89	–	189	363	–	360	14	insol.	1/77
Paraffin wax	0.9	–	199	245	–	>370	42–60	–	–
Paraformaldehyde	1.39	–	70	300	–	–	120–170	–	145/25
Paraldehyde	0.99	4.55	36oc	238	1.3–	128	12	v. sol.	–
Pentaborane	0.66	2.2	30	–	0.4–	58	–47	dec.	66/0
n-Pentane	0.63	2.48	–49	309	1.4–8	36	–130	v. sol.	400/19
iso-Pentane	0.62	2.48	–51	420	1.4–7.6	28	–161	insol.	–
1,5-Pentanediol	0.99	3.59	135	334	–	240	–16	sol.	<0.01/20
2,4-Pentanedione	0.98	3.45	41oc	–	–	136–40	–23	v. sol.	–
n-Pentanol, *see n*-Amyl alcohol									
2-Pentanol	0.81	3.03	39	347	1.2–9.0	119	–	v. sol.	–
2-Pentanone, *see* Methyl *n*-propyl ketone									
3-Pentanone, *see* Diethyl ketone									
Pentene, *see* Amylene									
n-Pentyl acetate, *see n*-Amyl acetate									
sec-Pentyl acetate, *see* sec-Amyl acetate									
Pentyl alcohol, *see n*-Amyl alcohol									
Pentyl amine, *see n*-Amylamine									
iso-Pentyl nitrite, *see* Amyl nitrite									
Peracetic acid (40% acetic acid solution)	1.23	–	41	110exp.	–	105	–30	v. sol.	–
Petroleum ethers	0.6	2.50	<–17	288	1–6	30–160	<–73	–	–

Table 6.1 Cont'd

Substance	*Specific gravity*	*Vapour density* (air = 1)	*Flash point*[1] (°C)	*Ignition temp.* (°C)	*Flammable limits*[1] (%)	*Boiling point* (°C)	*Melting point* (°C)	*Solubility in water* (g/100 g)	*Vapour pressure* (mm Hg/°C)
Phenol	1.07	3.24	79	715	1.5–	181	40	sol.	–
Phenyl acetate	1.09	4.7	80	–	–	196	–	sl. sol.	–
Phenylcyclohexane, *see* Cyclohexylbenzene									
p-Phenylenediamine	–	3.7	156	–	–	267	140	sol.	–
Phenylethanolamine	1.09	–	152	–	–	285	35	4.6^{20}	<0.01/20
Phenyl ether	1.09	5.86	96oc	646	–	258	27	insol.	0.021/25
Phenyl ether – Biphenyl mixture	1.06	–	124oc	610	–	257	12	–	–
Phenylhydrazine	1.09	3.7	89	174	–	243dec.	20	sol.	1/72
o-Phenyl phenol	1.21	–	124	–	–	286	57	insol.	1/100
Phorone	0.88	4.8	85	–	–	197	28	sl. sol.	–
iso-Phorone	0.93	4.77	96oc	462	0.8–3.8	215	–8	sl. sol.	–
Phosdrin	1.23	–	79oc	–	–	107 at 1 mm	–	–	–
Phosphorus (white and yellow)	1.82	4.42	–	30	spontaneous ignition in dry air	280	44	sl. sol.	1/77
Phosphorus (red)	2.34	4.77	–	260	–	280	590 at 43 atm	v. sol.	–
Phosphorus pentasulphide	2.03	–	–	142	–	514	276	insol.	–
Phosphorus tribromide	2.8	–	–	100	–	173	–40	insol.	10/48
Phthalic anhydride	1.53	5.10	151	584	1.7–10.4	284	131	sl. sol.	1/97
2-Picoline	0.95	3.2	39oc	538	–	129	–70	v. sol.	10/24
4-Picoline	0.96	3.21	57oc	–	–	143	4	∞	–
Picric acid	1.76	7.9	150	300	–	>300exp.	122	sol.	–
2-Pinene	0.86	4.7	33	–	–	156	–55	sl. sol.	10/37
Piperidine	0.86	3.0	16	–	–	106	–7	∞	40/29
Piperylene	0.68	2.4	–43	–	2–8.3	42	–141	insol.	–
Propane	0.58	1.56		468	2.2–9.5	–45	–187	insol.	–
1,3-Propanediamine	0.86	2.56	24oc	–	–	136	–24	v. sol.	–
1,2-Propanediol	1.04	2.62	99	371	2.6–12.5	189	–59	∞	–
n-Propanol, *see n*-Propyl alcohol									
Propargyl alcohol	0.96	1.93	36oc	–	3.4–	115	–17	sol.	12/20
Propargyl bromide	1.56	4.1	10	324	3.0–	90	–61	–	–
Propene, *see* Propylene									
iso-Propenyl acetate	0.91	3.45	16	–	1.9–	93	–93	sl. sol.	–
β-Propiolactone, *see* Hydracrylic acid-β-lactone									
Propionaldehyde	0.81	2.0	–9oc	207	3–16	48	–81	sol.	–
Propionic acid	0.99	2.56	54	513	2.9–	141	–22	∞	10/40

Propionitrile	0.77	1.9	2	–	3.1–	97	–93	v. sol.	–
Propionyl chloride	1.06	3.2	12	–	–	80	–94	dec.	–
iso-Propyl acetate	0.87	3.52	4	460	1.8–7.8	93	–73	sol.	–
n-Propyl acetate	0.89	3.5	14	450	2–8	102	–95	sl. sol.	40/29
n-Propyl alcohol	0.78	2.07	15	433	2.1–13.5	97	–127	v. sol.	10/15
iso-Propyl alcohol	0.79	2.07	12	399	2.3–12.7	82	–89	∞	–
Propylamine	0.72	2.0	–37	318	2.0–10.4	49	–83	sol.	248/20
iso-Propylamine	0.69	2.03	–37oc	402	2.3–10.4	32	–101	∞	–
Propyl benzene	0.86	4.14	30	450	0.8–6	159	–100	insol.	10/43
iso-Propyl benzene, *see* Cumene									
iso-Propyl benzoate	1.01	5.67	99	–	–	218	–26	insol.	–
Propyl bromide	1.35	4.3	–	490	–	71	–110	sl. sol.	–
Propyl chloride	0.89	2.71	<–18	520	2.6–11.1	47	–123	sl. sol.	–
n-Propyl cyanide, *see* Butyronitrile									
Propylene	0.51	1.5	–108	460	2–11.1	–48	–185	v. sol.	10 atm/20
Propylene carbonate	1.21	3.5	135oc	–	–	242	–49	v. sol.	0.03/20
Propylene dichloride, *see* 1,2-Dichloropropane									
iso-Propyl ether	0.72	3.5	–28	443	1.4–21	69	–60	sl. sol.	150/25
n-Propyl formate	0.91	3.03	–3	455	2.3–	81	–93	sl. sol.	100/30
iso-Propyl formate	0.88	3.0	–6	485	–	68	–	sl. sol.	–
iso-Propyl glycidyl ether	0.92	4.15	–	–	–	137	–	–	–
Propyl nitrate	1.06	–	20	177	2–100	111	<–100	sl. sol.	–
iso-Propyl toluene, *see* Cymene									
Propyne	0.68	1.38	–	–	1.7	–23	–105	sl. sol.	3876/20
Pyridine	0.99	2.7	20	482	1.8–12.4	115	–42	∞	10/13
Pyrrolidine	0.85	2.45	3	–	–	89	–63	∞	128/39
Pyruvic acid	1.23	–	3.0	–	–	165	14	∞	–
Quinoline	1.09	4.45	–	480	1.2–	238	–20	sol.	1/60
Quinone	1.32	–	293	–	–	–	116subl.	sl. sol.	–
Resorcinol	1.27	3.79	127	608	1.4– at 200°C	281	110	sol.	1/108
Salicylaldehyde	1.15	–	78	–	–	197	–10	sl. sol.	1/33
Salicylic acid	1.44	4.8	157	545	–	211 at 203 mm	159	sl. sol.	1/114
Silane	0.68	–	–	Spontaneously flammable in air	–112	–185	insol.	–	
Sodium	0.97	–	–	>115 spontaneous ignition in dry air		892	98	dec.	–
Sodium acetate	1.53	–	–	607	–	–	324	119°	–
Stearic acid	0.95	9.8	196	395	–	358–83	69	insol.	1/174
Stoddard solvent	1.0	–	38–43	227–60	0.8–5	220–300	–	–	–
Styrene	0.909	3.6	31	490	1.1–6.1	146	–33	insol.	–
Styrene monomer	0.905	1	31	490	1.1–6.1	145	–31	insol.	–
Succinonitrile	0.98	2.1	132	–	–	266	58	v. sol.	2/100

Table 6.1 Cont'd

Substance	*Specific gravity*	*Vapour density* (air = 1)	*Flash point*[1] (°C)	*Ignition temp.* (°C)	*Flammable limits*[1] (%)	*Boiling point* (°C)	*Melting point* (°C)	*Solubility in water* (g/100 g)	*Vapour pressure* (mm Hg/°C)
Sulphur	2.07	–	207	232	–	444	119	insol.	1/184
Sulphur monochloride	1.69	4.7	118	234	–	136	–80	dec.	10/28
Tallow	0.895	–	265	–	–	–	32	–	–
Tannic acid	–	–	199oc	527	–	–	210dec.	sol.	–
m-Terphenyl	1.16	–	135oc	–	–	365	87	insol.	–
o-Terphenyl	1.14	7.9	163oc	–	–	332	57	insol.	–
1,2,4,5-Tetrachlorobenzene	1.86	–	155	–	–	243	139	insol.	<0.1/25
Tetradecane	0.76	6.8	100	202	0.5–	254	6	insol.	1/76
Tetraethylenepentamine	0.99	–	163oc	–	–	333	–	–	–
Tetraethyl lead	1.66	8.6	93	–	–	170exp.	–137	insol.	1/38
Tetrahydrofuran	0.89	2.5	–14	321	2–11.8	65	–65	v. sol.	114/15
Tetrahydronaphthalene	0.97	4.55	71	384	0.8–5 at 150°C	207	–30	insol.	1/38
2,2′-Thiodiethanol	1.18	4.2	160oc	–	–	28	–11	∞	–
Thiophene	1.06	2.9	–1	–	–	84	–38	–	40/12
Toluene	0.87	3.1	4.4	536	1.4–6.7	111	–95	insol.	37/30
Toluene-2,4-diisocyanate	1.2	6.0	132	–	0.9–9.5	251	20	–	–
m-Toluidine	0.99	3.9	86	482	–	203	–31	sl. sol.	1/41
o-Toluidine	1.004	3.7	85	482	–	200	–16	sl. sol.	1/44
p-Toluidine	1.046	3.9	87	482	–	200	44	sl. sol.	1/42
Triamylamine	0.8	7.8	102	–	–	232	–	–	–
Tri-*n*-butyl amine	0.8	6.38	86	–	–	216	–70	sl. sol.	–
Tributyl phosphate	0.97	9.2	146	–	–	292	<–80	sol.	–
1,2,4-Trichlorobenzene	1.45	6.3	99	–	–	214	17	insol.	1/38
1,1,1-Trichloroethane	1.34	4.6	none	–	–	74	–38	insol.	100/20
Trichloroethylene	1.46	4.54	–	410	12–90	87	–73	sl. sol.	100/32
1,2,3-Trichloropropane	1.39	5.0	82	304	3.2–12.6	156	–15	sl. sol.	100/46
1,1,2-Trichloro-1,2,2-trifluoro ethane	1.56	–	–	680	–	48	–36	insol.	–
Tricresyl phosphate, *see* Tritolyl phosphate									
Tridecanol	0.82	6.9	121	–	–	274	31	insol.	–
Triethyl aluminium	0.84	–	<–53	<–53	–	194	–53	exp. H_2O	4/83
Triethyl amine	0.73	3.48	<–7	–	1.2–8.0	89	–115	sol.	–
Triethanolamine	1.13	–	179	–	–	360	20	∞	10/205
Triethylene glycol	1.13	5.17	177	371	0.9–9.2	276	–4	∞	1/114
Triethylene-tetramine	0.98	–	135	338	–	267	12	sol.	<0.01/20
Triethyl *o*-formate	0.89	5.1	30	–	–	146	–	dec.	10/40
Triisobutyl aluminium	0.79	–	<0	<4	–	114	4	–	–

Trimethyl amine	0.66	2.0	–7	190	2–11.6	4	–117	v. sol.	–
Trimethyl borate	0.92	3.6	<27	–	–	67	–29	dec.	–
3,5,5-Trimethyl-2-cyclohexenone, *see* iso-Phorone									
2,2,4-Trimethyl pentane	0.69	3.9	–12	418	1.1–6.0	99	–107	insol.	41/21
2,4,4-Trimethyl-2-pentene	0.72	3.9	2	–	–	112	–107	insol.	77/38
1,3,5-Trioxane	1.17	3.1	45	414	3.6–29	115	62	v. sol.	13/25
Triphenyl phosphate	1.21	–	220	–	–	245 at 11mm	49	insol.	–
Triphenyl phosphine	1.19	90	180	–	–	>360	80	insol.	–
Tripropylamine	0.75	4.9	41	–	–	156	–94	sl. sol.	–
Tritolyl phosphate	1.17	12.7	225	385	–	410	11	–	–
Turpentine	0.87	4.6	35–39	253	0.8–	153–75	–	–	–
Unsymmetrical dimethyl hydrazine	0.79	1.94	~–15	249	2–95	63	–58	v. sol.	–
Valeraldehyde	0.81	3.0	12	–	–	102	–92	sl. sol.	–
Valeric acid	0.94	3.5	96	–	–	186	–35	sol.	–
Vinyl acetate	0.94	3.0	–8	427	2.6–13.4	73	–100	insol.	100/21
Vinyl chloride	0.91	2.15	–78	472	4–22	–14	–154	sl. sol.	2600/25
Vinyl cyanide, *see* Acrylonitrile									
Vinyl ether	0.77	2.4	<–30	360	1.7–37	39	–	–	–
Vinylidene chloride	1.3	3.4	–15oc	458	5.6–11.4	32	–122	insol.	–
Vinyl toluene, *see* Methyl styrene									
m-Xylene	0.87	3.7	29	528	1.1–7.0	139	–48	insol.	10/28
o-Xylene	0.90	3.7	32	464	1.0–6.0	144	–26	insol.	10/32
p-Xylene	0.86	3.6	27	529	1.1–7.0	138	13	insol.	10/27
Xylidine	0.99	4.2	97	–	–	224	<–15	sl. sol.	–

dec. Decomposes
exp. Explodes on contact with water
insol. Insoluble
oc Open cup
sl. sol. Slightly soluble (<5 g/100 g)
sol. Soluble (5–50 g/100 g)
v. sol. Very soluble (>50 g/100 g)
∞ Infinitely soluble (soluble in all proportions)
Superscript indicates °C.
[1] Unless otherwise stated, flammable limits relate to ambient temperature and atmospheric pressure and flash points relate to closed cup measurements.

these are non-equilibrium methods. Alternatively equilibrium methods are available. Typical flash points are quoted in Table 6.1 and, unless otherwise stated, these relate to closed cup measurements. In general, the lower the flash point the greater the potential for fire: materials with flash points at or below ambient temperature are highly flammable and can inflame at ambient temperature on contact with ignition sources. Flash point is used to classify liquids under many legislative systems: in the UK liquids with flash points <32°C (and which, when heated under specific test conditions and exposed to an external source of flame applied in a standard manner, supports combustion) are defined as 'highly flammable' under the Highly Flammable Liquid and Liquefied Petroleum Gas Regulations.

Chemicals may ignite below their flash points if the substance:

- Is in the form of a mist (or froth).
- Covers a large surface area (e.g. when absorbed on porous media).
- Contains a small amount of a more volatile flammable liquid, e.g. due to deliberate or accidental contamination.

In addition

- Flash points are reduced by increases in ambient pressure. Thus the flash point of toluene at sea level (101.3 kPa) is 4.5°C whereas at 83.3 kPa, e.g. in the mountains at 1685 m, the value is 1°C.
- Materials with high flash points such as heavy oils and resins can produce flammable vapours due to thermal degradation on heating. Dangers therefore arise when welding, flame cutting empty drums/vessels once used to contain such materials due to the presence of residues.

Substances may be heated to their flash points by other substances with lower flash points burning in close proximity. Storage of flammable chemicals, therefore, needs careful consideration.

Vapour density

The density of a vapour or gas at constant pressure is proportional to its relative molecular mass and inversely proportional to temperature. Since most gases and vapours have relative molecular masses greater than air (exceptions include hydrogen, methane and ammonia), the vapours slump and spread or accumulate at low levels. The greater the vapour density, the greater the tendency for this to occur. Gases or vapours which are less dense than air can, however, spread at low level when cold (e.g. release of ammonia refrigerant). Table 6.1 includes vapour density values.

Dust explosions

Increasing the surface area of a combustible solid enhances the ease of ignition. Hence dust burns more rapidly than the corresponding bulk solid; combustion of dust layers can result in rapid flame spread by 'train firing'. Solid particles less than about 10 μm in diameter settle slowly in air and comprise 'float dust' (see p. 51 for settling velocities). Such particles behave, in some ways, similarly to gas and, if the solid is combustible, a flammable dust–air mixture can form within certain limits. Larger particles also take part, since there is a distribution of particle sizes, and ignition can result in a dust explosion.

Dust explosions are relatively rare but can involve an enormous energy release. A *primary* explosion, involving a limited quantity of material, can distribute accumulations of dust in the atmosphere which, on ignition, produces a severe *secondary* explosion.

Small particles are required, to provide a large surface-area-to-mass ratio and for the solid to remain in suspension. Surface absorption of air (oxygen) by the solid, or the evolution of combustible gas or vapour on heating, may be a predisposing factor. The presence of moisture reduces the tendency to ignite: it also favours agglomeration to produce larger particles. An increase in the proportion of inert solid in particles tends to reduce combustibility.

The explosive range of dusts in air can be very wide. The limits vary with the chemical composition and with the size of the particles. The lower limits are equivalent to a dense fog in appearance. The upper limits are ill defined but are not generally of practical significance. The important characteristics are the ease of ignition, lower explosive limits, the maximum explosion pressure and the rates of pressure rise. Organic or carbonaceous materials, or easily oxidizable metals (e.g. aluminium or magnesium) are more hazardous than nitrogenous organic materials. The least hazardous materials are those which contain an appreciable amount of mineral matter.

For a summary of data for a range of dusts refer to Table 6.2.

Oxygen requirements

Most substances require a supply of oxygen in order to burn. Air contains about 21% oxygen. Gases and vapours can produce flammable mixtures in air within certain limits. When the oxygen content of air is increased (e.g. by enrichment with pure oxygen from a leaking cylinder) the fire hazard is increased. Conversely, lowering the oxygen by, for instance, the presence of an inert gas such as nitrogen, argon, or carbon dioxide, reduces the fire risk. Some chemicals contain their own supply of oxygen (e.g. perchlorates) and can burn even in an oxygen-deficient atmosphere. Just as chemicals can react violently with oxygen to produce a fire, certain substances can inflame on reaction with other oxidizing agents (e.g. hydrocarbons with chlorine). Upper and lower flammable limits exist for such systems. Oxidizing agents generally assist combustion (see page 234).

There is a critical oxygen content below which ignition of combustible dusts or gases will not occur and this can provide a means for safe operation under an inert atmosphere, i.e. 'inerting'.

Ignition sources

Combustion is generally initiated by the introduction of a finite amount of energy to raise a finite volume of the material to its ignition temperature. Potential ignition sources for vapour–air mixtures are listed in Table 6.3, and temperatures in Table 6.4. Heat sources can be chemical energy (spontaneous combustion, chemical reaction), mechanical energy (e.g. friction), radiant energy, solar energy, static energy or electrical current. Thus heat is generated from electrical current by resistance, arcing or sparking. Resistance arises when the current flow exceeds the capacity of the wire. The result is often a blown fuse, tripped circuit breaker or heating of the circuit wire. Arcing occurs when electrical current jumps from one point to another, e.g. in a switch or connection box when wires separate from connections, or as a result of worn insulation between positive and neutral wires.

Common ignition sources include:

- Naked flames (e.g. Bunsen burners, welding torches, blow lamps, furnaces, pilot lights, matches, glowing cigarettes or embers).
- Sparks created by arcs in electrical switchgear, engines, motors, or by friction (e.g. lighter spark). Aluminium, magnesium, titanium and their alloys have an affinity for oxygen and in a thermite reaction with rust produce temperatures ≤3000°C. A thermite flash can result from the

Table 6.2 Dust explosion characteristics of combustible solids

Dust	*Minimum ignition temperature (°C)*		*Minimum explosible concentration* (g/l)	*Minimum ignition energy* (mJ)	*Maximum explosion pressure* (psi)[1]	*Maximum rate of pressure rise* .(psi/s)	*Maximum oxygen concentration to prevent ignition* (% by volume)	*Notes*
	cloud	*layer*						
Acetamide	560	–	–	–	–	–	–	Group (b) dust
Aceto acetanilide	560	–	0.030	20	90	4800	–	
Acetoacet-*p*-phenetedide	560	–	0.030	10	87	>10 000	–	
Acetoacet-*o*-toluidine	710	–	–	–	–	–	–	
2-Acetylamino-5-nitro thiazole	450	450	0.160	40	137	9000	–	
Acetyl-*p*-nitro-*o*-toluidine	450	–	–	–	–	–	–	
Adipic acid	550		0.035	60	95	4000	–	
Alfalfa	460	200	0.100	320	88	1100	–	
Almond shell	440	200	0.065	80	101	1400	–	
Aluminium, atomized	650	760	0.045	50	84	>20 000	–	
Aluminium, flake	610	320	0.045	10	127	>20 000	–	
Aluminium–cobalt alloy	950	570	0.180	100	92	11 000	–	
Aluminium–copper alloy	–	830	0.100	100	95	4000	–	
Aluminium–iron alloy	550	450	–	–	36	300	–	
Aluminium–lithium alloy	470	400	<0.1	140	96	6000	–	
Aluminium–magnesium alloy	430	480	0.020	80	86	10 000	–	
Aluminium–nickel alloy	950	540	0.190	80	96	10 000		
Aluminum–silicon alloy	670	–	0.040	60	85	7500	–	
Aluminium acetate	560	640	–	–	59	950	–	Guncotton ignition source in pressure test
Aluminium octoate	460	–	–	–	–	–	–	
Aluminium stearate	400	380	0.015	10	86	>10 000	–	
2-Amino-5-nitrothiazole	460	460	0.075	30	110	5600	–	
Anthracene	505	Melts	–	–	68	700	–	
Anthranilic acid	580	–	0.030	35	84	6500	–	
Anthraquinone	670	–	–	–	–	–	–	
Antimony	420	330	0.420	1920	28	300	–	
Antipyrin	405	Melts	–	–	53		–	
Asphalt	510	500	0.025	25	94	4800	–	
Aspirin	550	Melts	0.015	16	87	7700	–	
Azelaic acid	610	–	0.025	25	76	4700	–	
α, α′-Azo isobutyronitrile	430	350	0.015	25	134	8000	–	
Barley	370	–	–	–	–	–	–	
Benzethonium chloride	380	410	0.020	60	91	6700	–	
Benzoic acid	600	Melts	0.011	12	95	10 300	–	
Benzotriazole	440		0.030	30	103	9200	–	

Benzoyl peroxide	–	–	–	21	–			
Beryllium	910	540		–	Did not ignite		–	Contained 8% oxide
Beryllium acetate, basic	620	–	0.080	100	87	2200	15	Inert gas carbon dioxide
Bis(2-hydroxy-5-chlorophenyl)-methane	570	–	0.040	60	70	2000	13	Inert gas carbon dioxide
Bis(2-hydroxy-3,5,6-trichlorophenyl)-methane	Did not ignite	450						
Bone meal	490	230			11	100		Guncotton ignition source in pressure test
Boron	730	390	Did not ignite		41	200		Guncotton ignition source in pressure test
Bread	450							
Brunswick green	360							
t-butyl benzoic acid	560	–	0.020	25	88	6500	–	
Cadmium	570	250		4000	7	100		
Cadmium yellow	390							
Calcium carbide	555	325			13			
Calcium citrate	470							Group (b) dust
Calcium gluconate	550							Group (b) dust
Calcium DL pantothenate	520	–	0.050	80	105	4600	–	
Calcium propionate	530	–	–	–	90	1900	–	
Calcium silicide	540	540	0.060	150	86	20 000	–	
Calcium stearate	400		0.025	15	97	>10 000	–	
Caprolactam	430	–	0.07	60	79	1700	8	
Carbon, activated	660	270	0.100	–	92	1700	–	Guncotton ignition source in min. expl. conc. and max. expl. pressure tests
Carbon, black	510							
Carboxy methyl cellulose	460	310	0.060	140	130	5000		
Carboxy methyl hydroxy ethyl cellulose	380	–	0.200	960	83	800	–	
Carboxy polymethylene	520	–	0.115	640	76	1200	–	
Casein	460	–	–	–	89	1200	–	
Cellulose	410	300	0.045	40	117	8000	–	
Cellulose acetate	340	–	0.035	20	114	6500	5	Inert gas nitrogen
Cellulose acetate butyrate	370	–	0.025	30	81	2700	7	
Cellulose propionate	460	–	0.025	60	105	4700	–	
Cellulose triacetate	390		0.035	30	107	4300	–	
Cellulose tripropionate	460	–	0.025	45	88	4000	–	
Charcoal	530	180	0.140	20	100	1800	–	
Chloramine-T	540	150	–	–	7	150	–	Guncotton ignition source in pressure test

Table 6.2 Cont'd

Dust	Minimum ignition temperature (°C)		Minimum explosible concentration (g/l)	Minimum ignition energy (mJ)	Maximum explosion pressure (psi)[1]	Maximum rate of pressure rise .(psi/s)	Maximum oxygen concentration to prevent ignition (% by volume)	Notes
	cloud	layer						
o-Chlorobenzmalono nitrile			0.025		90	>10 000		
o-Chloroaceto acetanilide	640		0.035	30	94	3900	–	
p-Chloroaceto acetanilide	650	–	0.035	20	85	5500	–	
Chloro amino toluene sulphonic acid	650	–	–	–	–	–	–	
4-Chloro-2 nitro aniline	590	120	<0.750	140	123	3500	–	
p-Chloro-*o*-toluidine hydrochloride	650	–	–	–	–	–	–	
Chocolate crumb	340	–	–	–	–	–	–	
Chromium	580	400	–	140	56	5000		
Cinnamon	440	230	0.230	30	121	3900	–	
Citrus peel	500	330	0.060	100	51	1200	–	
Coal, brown	485	230	0.060			–	–	See also Lignite
Coal, 8% volatiles	730			–		–	–	
Coal, 12% volatiles	670	240	–			–		
Coal, 25% volatiles	605	210	0.120	120	62	400	–	
Coal, 37% volatiles	610	170	0.055	60	90	2300		Standard Pittsburgh coal
Coal, 43% volatiles	575	180	0.050	50	92	2000		
Cobalt	760	370	–	–		–	–	
Cocoa	500	200	0.065	120	69	1200	–	
Coconut	450	280	–	–		–	–	
Coconut shell	470	220	0.035	60	115	4200	–	
Coffee	360	270	0.085	160	38	150	10	Inert gas carbon dioxide
Coffee, extract	600	–	–	–	47	–	–	
Coffee, instant	410	350	0.280	Did not ignite	68	500	–	
Coke	>750	430	–	–	–	–	–	
Coke, petroleum, 13% volatiles	670		1.00		36	200	–	Guncotton ignition source in min. expl. conc. and max. expl. pressure tests
Colophony	325	Melts		–			–	
Copal	330	Melts		–	68	–		See also Gum manila
Copper	700	–	–	Did not ignite	Did not ignite	Did not ignite	–	

Copper–zinc, gold bronze	370	190	1.00	–	44	1300	–	
Cork	460	210	0.035	35	96	7500		
Corn cob	450	240	0.045	45	127	3700		
Corn dextrine	410	390	0.040	40	124	7000		
Cornflour	390	–	–	–	–	–	–	
Cornstarch	390		0.040	30	145	9500		
Cotton flock	470	–	0.050	25	94	6000		
Cotton linters	520		0.50	1920	73	400	5	
Cottonseed meal	530	200	0.055	80	89	2200	–	
Coumarone–indene resin	550	–	0.015	10	93	11 000	11	
Crystal violet	475	Melts	–	–	–	–	–	
Cyclohexanone peroxide	–	–	–	21	84	5600	–	
Dehydroacetic acid	430	–	0.030	15	87	8000	–	
Dextrin	410	440	0.050	40	99	9000		
Dextrose monohydrate	350	–	–		–	–	–	
Diallyl phthalate	480	–	0.030	20	90	8500	–	
Diamino stilbene disulphonic acid	550	–				–	–	Group (b) dust
Diazo aminobenzene	550	–	0.015	20	114	>10 000	–	
Di-*t*-butyl-*p*-cresol	420		0.015	15	79	13 000	9	
Dibutyl tin maleate	600	–						
Dibutyl tin oxide	530							
Dichlorophene	770				72	3000		
2,4-Dichlorophenoxy ethyl benzoate	540		0.045	60	84	2200		
Dicyclopentadiene dioxide	420		0.015	30	89	9500		
Dihydrostreptomycin sulphate	600	230	0.520	–	42	200	7	
3,3′-Dimethoxy 4,4′-diamino diphenyl	–	–	0.030		82	>10 000	–	
Dimethylacridan	540						–	
Dimethyl diphenyl urea	490							
Dimethyl isophthalate	580		0.025	15	84	8000		
Dimethyl terephthalate	570		0.030	20	105	12 000	6	
S-S′-Dimethyl xanthogene thylene bis dithiocarbamate	400		0.300	3200	84	1500		
Dinitro aniline	470			–				
3,5-Dinitrobenzamide	500	Melts	0.040	45	163	6500	–	
3,5-Dinitrobenzoic acid	460	–	0.050	45	139	4300		
Dinitrobenzoyl chloride	380	–	–	–	–	–	–	
Dinitrocresol	340	Melts	0.030	–				
4,4′-Dinitro-sym-diphenyl urea	550	–	0.095	60	102	2500	–	
Dinitro stilbene disulphonic acid	450	–	–	–	–	–	–	
Dinitrotoluamide	500	–	0.050	15	153	>10 000	–	
Diphenyl	630	–	0.015	20	82	3700	–	

Table 6.2 Cont'd

Dust	*Minimum ignition temperature (°C)*		*Minimum explosible concentration* (g/l)	*Minimum ignition energy* (mJ)	*Maximum explosion pressure* (psi)[1]	*Maximum rate of pressure rise* .(psi/s)	*Maximum oxygen concentration to prevent ignition* (% by volume)	*Notes*
	cloud	*layer*						
4,4′-Diphenyl di sulphonylazide	590	140	0.065	30	143	5500		
Diphenylol propane (Bisphenol-A)	570		0.012	11	81	11 800	5	Inert gas nitrogen
Egg white	610	–	0.14	640	58	500		
Epoxy resin	490	–	0.015	9	94	8500	–	
Esparto grass	–	–	–	–	94	7300	–	
Ethyl cellulose	340	330	0.025	15	112	7000		
Ethylene diamine tetra acetic acid	450	–	0.075	50	106	3000		
Ethyl hydroxyethyl cellulose	390	–	0.020	30	94	2200		
Ferric ammonium ferrocyanide	390	210	1.500		17	100		
Ferric dimethyl dithio carbamate	280	150	0.055	25	86	6300		
Ferric ferrocyanide	370				82	1000		
Ferrochromium	790	670	2.00					
Ferromanganese	450	290	0.130	80	62	5000		
Ferrosilicon (45% Si)	640							
Ferrosilicon (90% Si)	Did not ignite	980	0.240	1280	113	3500		
Ferrotitanium	370	400	0.140	80	55	9500		
Ferrous ferrocyanide	380	190	0.400					
Ferrovanadium	440	400	1.300	400				
Fish meal	485							
Fumaric acid	520		0.085	35	103	3000	–	
Garlic	360	–	0.10	240	57	1300	–	
Gelatin, dried	620	480	<0.5	–	78	1200	–	
Gilsonite	580	500	0.020	25	78	4500		
Graphite	730	580				–		
Grass					56	400		
Gum arabic	500	260	0.060	100	117	3000		
Gum Karaya	520	240	0.100	180	116	2500	–	
Gum manila (copal)	360	390	0.030	30	89	6000		
Gum tragacanth	490	260	0.040	45	123	5000	–	
Hexa methylene tetramine	410	–	0.015	10	98	11 000	11	
Horseradish	–	–	<0.100		96	1600		

Hydrazine acid tartrate	570		0.175	460	30	200	–	
p-Hydroxy benzoic acid	620	–	0.040	–	37	–	–	
Hydroxyethyl cellulose	410	–	0.025	40	106	2600	–	
Hydroxyethyl methyl cellulose	410						–	
Hydroxy propyl cellulose	400	–	0.020	30	96	2900	–	
Iron	430	240	–	–	–	–		
Iron, carbonyl	420	230	0.105	100	47	8000		
Iron pyrites	380	280	1.00	8200	5	100	–	
Isatoic anhydride	700		0.035	25	80	4900	–	
Isinglass	520	–	–		Nil	Nil		
Isophthalic acid	700	–	0.035	25	78	3100		
Kelp	570	220	Did not ignite		19	200	–	
Lactalbumin	570	240	0.040	50	97	3500	–	
Lampblack	730	–	–	–		–	–	
Lauryl peroxide				12	90	6400		
Lead	790	290		Did not ignite	3	100	–	Flame ignition source in pressure test
Leather	390			–				
Lignin	450		0.040	20	102	5000	7	
Lignite	450	200	0.030	30	94	8000	–	
Lycopodium	480	310	0.025	40	75	3100	9	
Magnesium	560	430	0.030	40	116	15 000		
Maize husk	430		–	–	75	700		
Maize starch	410	–	–	–	–	–	–	
Maleic anhydride	500	Melts	–	–	–	–	–	
Malt barley	400	250	0.055	35	95	4400		
Manganese	460	240	0.125	305	53	4900	–	
Manganese ethylene bis dithio carbamate	270	–	0.07	35		–		
Manioc	430					–		
Mannitol	460		0.065	40	97	2800		
Melamine formaldehyde resin	410		0.02	50	93	1800		
DL Methionine	370	360	0.025	35	119	5700	7	
1-Methylamino anthraquinone	830	Melts	0.055	50	71	3300		
Methyl cellulose	360	340	0.030	20	133	6000	–	
2,2-Methylene bis-4-ethyl-6-*t*-butyl phenol	310				76	7300	–	
Milk	440	–	–	–	–	–	–	
Milk, skimmed	490	200	0.050	50	95	2300	–	
Milk sugar	450	Melts	–		31		–	
Molybdenum	720	360	–		–	–	–	
Molydenum disulphide	570	290	–	–	–	–	–	
Monochloracetic acid	620	–	–	–	–	–	–	
Monosodium salt of trichloro ethyl phosphate	540	–	–	–	–	–	–	Group (b) dust

Table 6.2 Cont'd

Dust	*Minimum ignition temperature (°C)*		*Minimum explosible concentration* (g/l)	*Minimum ignition energy* (mJ)	*Maximum explosion pressure* (psi)[1]	*Maximum rate of pressure rise* .(psi/s)	*Maximum oxygen concentration to prevent ignition* (% by volume)	*Notes*
	cloud	*layer*						
Moss, Irish	530	230	Did not ignite		21	300	–	
Naphthalene	575	Melts	–	–	87	–	–	
β-Naphththalene-azo-dimethyl aniline	510	Melts	0.020	50	70	2300	–	
β-Naphthol	670	–	–	–	–	–	–	
Naphthol yellow	415	395	–					
Nigrosine hydrochloride	630	–	–			–	–	
p-Nitro-*o*-anisidene	400				–			
p-Nitro-benzene arsonic acid	360	280	0.195	480	77	900	–	
Nitrocellulose	–	–	–	30	>256	>20 900	–	
Nitro diphenylamine	480	–	–	–	–	–	–	
Nitro furfural semi carbazone	240	–	–	–	>143	8600	–	
Nitropyridone	430	Melts	0.045	35	111	>10 000	–	
p-Nitro-*o*-toluidine	470	–	–	–	–	–	–	
m-Nitro-*p*-toluidine	470	–	–	–	–		–	
Nylon	500	430	0.030	20	95	4000	6	
Oilcake meal	470	285	–	–	–	–		
Onion, dehydrated	410	–	0.130	Did not ignite	35	500	–	
Paper	440	270	0.055	60	96	3600		
Para formaldehyde	410	–	0.040	20	133	13 000		
Peanut hull	460	210	0.045	50	116	8000		
Peat	420	295						
Peat, sphagnum	460	240	0.045	50	104	2200		
Pectin	410	200	0.075	35	132	8000		
Penicillin, *N*-ethyl piperidine salt of	310							
Penta erythritol	450	–	0.030	10	90	9500	7	
Phenol formaldehyde	450		0.015	10	107	6500		
Phenol furfural resin	530	–	0.025	10	88	8500	–	
Phenothiazine	540	–	0.030		56	3000	–	
p-Phenylene diamine	620	–	0.025	30	94	11 000	–	
Phosphorus, red	360	305	–					
Phosphorus pentasulphide	280	270	0.050	15	64	>10 000	–	
Phthalic acid	650	Melts		–	62	–		
Phthalic anhydride	605	Melts	0.015	15	72	4200	11	

Phthalimide	630	–	0.030	50	89	4800		
Phthalodinitrile	>700	Melts	–	–	43	–	–	
Phytosterol	330	Melts	0.025	10	76	>10 000		
Piperazine	480	–			72	1400		
Pitch	710	–	0.035	20	88	6000		
Polyacetal	440	–	0.035	20	113	4100		
Polyacrylamide	410	240	0.040	30	85	2500		
Polyacrylonitrile	500	460	0.025	20	89	11 000		
Polycarbonate	710		0.025	25	96	4700	–	
Polyethylene	390	–	0.020	10	80	7500	–	
Polyethylene oxide	350		0.030	30	106	2100	5	
Polyethylene terephthalate	500		0.040	35	98	5500	–	
Poly isobutyl methacrylate	500	280	0.020	40	74	2800	–	
Poly methacrylic acid	450	290	0.045	100	97	1800		
Polymethyl methacrylate	440		0.020	15	101	1800	7	
Polymonochlorotrifluoro ethylene	600	720	Did	not	ignite	–		
Polypropylene	420		0.020	30	76	5500	–	
Polystyrene	500	500	0.020	15	100	7000		
Polytetrafluoro ethylene	670	570	Did	not	ignite			
Polyurethane foam	510	440	0.030	20	87	3700		
Polyurethane foam, fire retardant	550	390	0.025	15	96	3700		
Polyvinylacetate	450		0.040	160	69	1000	11	Inert gas carbon dioxide
Polyvinyl alcohol	450	Melts			78			
Polyvinyl butyral	390		0.020	10	84	2000	5	
Polyvinyl chloride	670		Did not ignite		38	500		Flame ignition source Group (b) dust
Polyvinylidene chloride	670							
Polyvinyl pyrrolidone	465	Melts			15			
Potassium hydrogen tartrate	520							
Potassium sorbate	380	180	0.120	60	79	9500		
Potato, dried	450				97	1000		
Potato starch	430							
Provender	370				93	1400		
Pyrethrum	460	210	0.100	80	95	1500	–	
Quillaia bark	450							
Rape seed meal	465							
Rayon, viscose	420							
Rayon, flock			0.03					
Rice	440	240	0.050	50	105	2700		
Rosin	390		0.015	10	87	12 000		
Rubber	380							
Rubber, crude, hard	350		0.025	50	80	3800	13	
Rubber, crumb	440				84	3300		

Table 6.2 Cont'd

Dust	*Minimum ignition temperature (°C)*		*Minimum explosible concentration* (g/l)	*Minimum ignition energy* (mJ)	*Maximum explosion pressure* (psi)[1]	*Maximum rate of pressure rise* .*(psi/s)*	*Maximum oxygen concentration to prevent ignition* (% by volume)	*Notes*
	cloud	*layer*						
Rubber, vulcanized	360	–	–	–	40	–		
Rye flour	415	325	–	–	35	–		
Saccharin	690							
Salicylanilide	610	Melts	0.040	20	73	4800	–	
Salicylic acid	590	–	0.025		84	6800		
Sawdust	430	–	–	–	97	2000	–	
Sebacic acid	–	–	–	–	74	400	–	
Senna	440		0.010	105	49	300		
Shellac	400	–	0.020	10	73	3600	9	
Silicon	Did not ignite	760	<0.10	80	94	13 000	–	
Soap	430	600	0.085	100	77	2800		
Sodium acetate	590		0.030	35	90	4600		
Sodium amatol	580	Melts	0.140	–	65	800		
Sodium benzoate	560	680	0.050	80	91	3700		
Sodium carboxymethyl cellulose	320	–	1.10	440	49	400	5	
Sodium 2-chloro-5-nitro-benzene sulphonate	550	440	–	–	–	–	–	
Sodium 2,2-dichloro propionate	500	–	0.260	220	68	500		
Sodium dihydroxy naphthalene disulphonate	510	–	–				–	Group (b) dust
Sodium glucaspaldrate	600							
Sodium glucoheptonate	600							
Sodium monochloracetate	550						–	
Sodium *m*-nitrobenzene sulphonate	–	–	–	–	92	400	–	
Sodium *m*-nitrobenzoate	–				87	2900	–	
Sodium pentachlorophenate	Did not ignite	360			Did not ignite		–	
Sodium propionate	479				70	700		
Sodium secobarbital	520	–	0.100	960	76	800		
Sodium sorbate	400	140	0.050	30	87	6500		
Sodium thiosulphate	510	330		–	11	<100		Guncotton ignition source in pressure test

Sodium toluene sulphonate	530	–	–	–			–	
Sodium xylene sulphonate	490	–	–	–	–	–	–	
Soot	>690	535	–	–	Did not ignite		–	
Sorbic acid	440	460	0.020	15	106	>10 000	5	Inert gas nitrogen
L-Sorbose	370	–	0.065	80	76	4700	–	
Soya flour	550	340	0.060	100	94	800	9	
Soya protein	540		0.050	60	98	6500	9	
Starch	470	–	–	–	–	–	–	
Starch, cold water	490	–	–	–	–	–	–	
Stearic acid	290			25	80	8500		
Steel	450				–		–	
Streptomycin sulphate	700	–	–	–	–	–	–	
Sucrose	420	Melts	0.045	40	86	5500	–	
Sugar	370	400	0.045	30	109	5000		
Sulphur	190	220	0.035	15	78	4700	–	
Tantalum	630	300	<0.20	120	55	4400	–	
Tartaric acid	350							
Tea	500	–			93	1700		
Tea, instant	580	340	Did not ignite		48	400		
Tellurium	550	340	–	–	–	–	–	
Terephthalic acid	680	–	0.050	20	84	8000	–	
Tetranitro carbazole	395	Melts						
Thiourea	420	Melts	–	–	29	100	–	
Thorium	270	280	0.075	5	79	5500	–	
Thorium hydride	260	20	0.080	3	81	12 000		
Tin	630	430	0.190	80	48	1700	–	
Titanium	375	290	0.045	15	85	11 000	Ignites in carbon dioxide	
Titanium hydride	480	540	0.070	60	121	12 000	3	
Tobacco	485	290	–	–	–	–	–	
Tobacco, dried	320	–	–	–	85	1000	–	
Tobacco, stem	420	230	Did not ignite		53	400	–	
Tribromosalicyl anilide	880	Melts	–	–	–	–	–	
Trinitrotoluene	–	–	0.070	75	63	2100		
s-Trioxane	480	–	0.143	–	85	600	–	
α,α′-Trithiobis (*N*, *N*-dimethyl-thioformamide)	280	230	0.060	35	96	6000		
Tung	540	240	0.070	240	74	1900		
Tungsten	730	470	–	–	Did not ignite			
Uranium	20	100	0.060	45	69	5000		
Uranium hydride	20	20	0.060	5	74	9000		
Urea	900		Did	not	ignite			Group (b) dust

Table 6.2 Cont'd

Dust	*Minimum ignition temperature (°C)*		*Minimum explosible concentration* (g/l)	*Minimum ignition energy* (mJ)	*Maximum explosion pressure* (psi)[(1)]	*Maximum rate of pressure rise* .*(psi/s)*	*Maximum oxygen concentration to prevent ignition* (% by volume)	*Notes*
	cloud	*layer*						
Urea formaldehyde moulding powder	460		0.085	80	89	3600	9	
Urea formaldehyde resin	430		0.02	34	110	1600		
Vanadium	500	490	0.220	60	57	1000	10	
Vitamin B1 mononitrate	380	190	0.035	35	120	9000	–	
Vitamin C	460	280	0.070	60	88	4800		
Walnut shell	420	210	0.035	60	121	5500		
Wax, accra	260							
Wax, carnauba	340	–						
Wax, paraffin	340							
Wheat, flour	380	360	0.050	50	109	3700		
Wheat, grain dust	420	290	–	–	43	–		
Wheat starch	430		0.045	25	100	6500		
Wood	360	–		–	90	5700	5	
Wood, bark	450	250	0.020	60	103	7500		
Wood, flour	430	–	0.050	20	94	8500	7	
Wood, hard	420	315	–	–	66			
Wood, soft	440	325	–	–	63	–		
Yeast	520	260	0.050	50	123	3500		
Zinc	680	460	0.500	960	70	1800		
Zinc ethylene dithio carbamate	480	180			45	300		
Zinc stearate	315	Melts	0.020	10	80	10 000		
Zirconium	20	220	0.045	5	75	11 000	Ignites in carbon dioxide	
Zirconium hydride	350	270	0.085	60	90	9500	3	

(1) 1 psi = 0.069 bar = 6.9 kN/m^2.

Table 6.3 Sources of ignition

Mechanical sources	
Friction	Metal to metal
	Metal to stone
	Rotary impact
	Abrasive wheel
	Buffing disc
	Tools, drill
	Boot studs
	Bearings
	Misaligned machine parts
	Broken machine parts
	Chocking or jamming of material
	Poor adjustment of power drives
	Poor adjustment of conveyors
Missiles	Hot missiles
	Missile friction
Metal fracture	Cracking of metal
Electrical sources	
Electrical current	Switch gear
	Cable break
	Vehicle starter
	Broken light
	Electric motor
Electrostatic	Liquid velocity
	Surface charge
	Personal charge
	Rubbing of plastic or rubber
	Liquid spray generation
	Mist formation
	Water jetting
	Powder flow
	Water settling
Lightning	Direct strike
	Hot spot
	Induced voltage
Stray currents	Railway lines
	Cable break
	Arc welding
Radio frequency	Aerial connection
	Intermittent contact
Thermal sources	
Hot surface	Hot spot
	Catalyst hot spots
	Incandescent particles from incinerators, flarestacks, chimneys
	Vehicle exhaust
	Steam pipes
	Refractory lining, hot slag
	Foreign metal in crushing and grinding equipment
	Electric heater
	Smoking
	Glowing embers, brands
	Drying equipment
	Molten metal or glass
	Heat transfer salt
	Hot oil/salt transfer lines

Table 6.3 Cont'd

	Boiler ducts or flues Electric lamps Hot process equipment Welding metal Induction brazing Hot plates Soldering irons
Self-heating	Oxidation Reaction Activated carbon
Flames	Pilot light Primary fire involving liquid (running or pool), solid or gas Matches, cigarette lighters Cutting, welding Portable gas heaters Stoves; natural gas, LPG, oil or solid fuel – fired Burners Arson Blow torches Brazing
Compression	Pressure change Piston
Engines	Exhaust Engine overrun Hydraulic spray into air intake
Diffusion	High pressure change
Chemical sources	
Peroxides	Oxygen release Unstable Decomposition
Polymerization	Exothermic reaction Catalyst Lack of inhibitor Crystallization
Spontaneous	Pyrophoric deposit Deposits Water reactive Sulphides Oily rags, oil impregnation of lagging Heat transfer salt
Reaction with other substances	
Thermite reaction	Rust Exothermic reactions with aluminium, aluminium alloys
Unstable substances	Acetylides
Decomposition	Initiator Temperature Catalyst

striking of a smear or thin coating of alloy on rusty steel with a hammer. The glancing impact of stainless steel, mild steel, brass, copper–beryllium bronze, aluminium copper and zinc onto aluminium smears on rusty steel can initiate a thermite reaction of sufficient thermal energy to ignite flammable gas/vapour–air atmosphere or dust clouds.

Table 6.4 Approximate temperatures of common ignition sources

Flame/spark sources	*Ignition temperature* (°C)
Candles	640–940
Matches	870
Manufactured gas flame	900–1340
Propane flame	2000
Light bulb element	2483
Methane flame	3042
Electrical short circuit or arc	3870
Non-flame sources	
Steam pipes at normal pressure	100
Steam pipes at 10 psi (0.7 bar)	115
Light bulb, normal	120
Steam pipes at 15 psi (1 bar)	121
Steam pipes at 30 psi (2 bar)	135
Steam pipes at 50 psi (3.5 bar)	148
Steam pipes at 75 psi (5 bar)	160
Steam pipes at 100 psi (7 bar)	170
Steam pipes at 150 psi (10.5 bar)	185
Steam pipes at 200 psi (14 bar)	198
Steam pipes at 300 psi (21 bar)	217
Steam pipes at 500 psi (35 bar)	243
Steam pipes at 1000 psi (70 bar)	285
Cigarette, normal	299
Soldering iron	315–432
Cigarette, insulated	510
Light bulb, insulated	515

Petroleum vapour is unlikely to be ignited by impact of steel on steel produced by hand. Power operation can however produce incendive sparks. Hydrogen and perhaps ethylene, acetylene or carbon disulphide can be ignited by the impact of steel on steel using hand tools. If non-sparking tools are used, care must be taken to avoid embedded grit particles since impact of steel on 'rock' poses a greater hazard. Impact on flint or grit can produce incendive sparks irrespective of striking material. Friction in bearings is a common ignition source.

- Radiant heat sources include furnaces, vats, cooking stoves and other hot surfaces. Solar heat from the direct rays of the sun may directly, or if magnified by, e.g., glass bottles or flasks, provide sufficient energy to raise the temperature of chemicals to their flash-point.
- Vehicular petrol engines are potential ignition sources by means of the spark-ignition system, dynamo or battery, or hot exhaust pipe. Non-flameproof diesel engines are potential ignition sources due to a hot exhaust pipe or carbonaceous particles or flames from the exhaust.
- Spark due to static electricity associated with the separation of two dissimilar materials (Table 6.5). The charges may be transported/conducted some distance after separation before there is sufficient accumulation to produce a spark, e.g. in the flow of liquids or powders. The size of the charge is generally small but the potential difference may be very high such that a spark is of sufficient energy for ignition.

  Electrostatic charge generated by a liquid flow through a pipe depends on the electrical conductivity of the liquid. With a liquid of high electrical conductivity, the charge is easily generated but quickly dissipated. Hazardous liquids are generally those with conductivities in the range 0.1 to 1000 ps/m. The rate of charge generation increases with increase in flowrate and constrictions in the pipeline.
- Friction resulting from two surfaces rubbing together, e.g. drive belts in contact with their

Table 6.5 Operations which may result in static charge generation

Solid–solid	Persons walking Grit blasting Conveying of powders Belts and pulleys Fluidized beds
Solid–liquid	Flow of liquids in pipelines/filters Settling of particles in liquid (e.g. rust and sludge)
Gas–liquid	Released gas (air) bubbles rising in a large tank Mist formation from LPG evaporation Splash filling Cleaning with wet steam Mist formation from high pressure water jets
Liquid–liquid	Settling of water drops in oil
Solid–gas	Mixing of immiscible liquids Pneumatic conveying of solids Fluidized beds

housing or guard, or metal surfaces rubbing against one another. Usually friction arises from poor maintenance (e.g. loose guard, inadequate lubrication).
- Lightning. Protection is generally provided by earthing with low resistance, e.g. 7 Ω, which should be short and direct. The recommended value for protection of plant is ≤10 Ω.

A material that is above its autoignition temperature will ignite spontaneously on contact with air in the correct proportions (see Table 6.1 for minimum temperature of ignition source).

Ignition of a flammable dust–air mixture is more difficult than with flammable vapour–air mixtures. A larger source of heat is required, and a larger volume of fuel must be heated to the ignition point. The same range of potential ignition sources is applicable as for air–vapour mixtures.

At certain temperatures compounds will explode without application of a flame, as illustrated by the selection in Table 6.6.

Spontaneous combustion

Certain materials which are generally considered to be stable at ordinary temperatures can inflame even in the absence of normal ignition sources. Such spontaneous combustion results from exothermic autoxidation when the heat liberated exceeds that dissipated by the system. Materials prone to self-heating are listed in Table 6.7. In most cases, such fires involve relatively large, enclosed or thermally-insulated masses, and spontaneous combustion usually occurs after prolonged storage.

Pyrophoric chemicals

Pyrophoric chemicals are so reactive that on contact with air they undergo vigorous reaction with atmospheric oxygen (under ambient conditions or at elevated temperatures), or with water (Table 6.9). Examples include:

- Certain metals/alloys – the alkali metals (lithium, potassium, sodium) and even some metals/alloys which undergo slow oxidation or are rendered passive in bulk form but which, in the finely divided state, inflame immediately when exposed to oxygen (e.g. aluminium, magnesium, zirconium).

Table 6.6 Approximate temperatures at which selected substances will explode, without the application of a flame

Solids	*Temperature*[1] (°C)	
Gun cotton (loose)	137	139[2]
Cellulose dynamite	169	230
Blasting gelatine (with camphor)	174	
Mercury fulminate	175	
Gun cotton (compressed)	186	201
Dynamite	197	200
Blasting gelatine	203	209
Nitroglycerin	257	
Gunpowder	270	300
Gases		
Propylene	497	511
Acetylene	500	515
Propane	545	548
Hydrogen	555	
Ethylene	577	599
Ethane	605	622
Carbon monoxide	636	814
Manufactured gas	647	649
Methane	656	678

[1] The value quoted is that at which the substance itself explodes, not the temperature at which its container ruptures with the possible subsequent ignition of the contents.
[2] The higher temperature is applicable when the heat rise is very rapid, i.e. if the rate of rise is slow then the explosion will occur at the lower temperature.

- Phosphorus.
- Certain phosphides, hydrides and silanes (e.g. hydrogen phosphide, silane).
- Substances which react with water to liberate flammable gas, e.g. carbides (liberate acetylene), alkali metals (hydrogen), organometallics (hydrocarbons – see Table 6.8), and where the heat of reaction is sufficient to ignite the gas. Thus metals which are less electronegative than hydrogen (see Table 6.10) will displace this element from water or alcohols, albeit at different rates.

Explosions

Fires sometimes initiate, or are followed by, explosions resulting in blast damage, missiles etc. These may trigger secondary events, e.g. fires, toxic releases or further explosions.

Types of explosion

- Confined vapour cloud explosion: gas or vapour burns in a confined volume and rapid expansion of the combustion products is restrained until failure of the container or building occurs.
- Boiling liquid expanding vapour explosion: follows failure of a pressurized container of flammable liquid, e.g. LPG, or a sealed vessel containing volatile flammable liquids, under fire conditions. Ignition results in a fireball and missiles.
- Dust explosion (refer to page 220).
- Explosion due to thermal deflagration or detonation of a solid or liquid.
- Unconfined vapour cloud explosion: a large flammable gas or vapour–air cloud burns in free space with sufficient rapidity to generate pressure waves, which propagate through the cloud and into the surrounding atmosphere. Such events are extremely rare.

Table 6.7 Materials liable to self-heat

Liquid materials susceptible to self-heating when dispersed on a solid	
Bone oil	moderate
Castor oil	very slight
Coconut oil	very slight
Cod liver oil	high
Corn oil	moderate
Cotton seed oil (refined)	high
Fish oil	high
Lard	high
Linseed oil (raw)	very high
Menhaden oil	high
Neatsfoot oil	slight
Oleic acid	very slight
Oleo oil	very slight
Olive oil	slight
Palm oil	moderate
Peanut oil	moderate
Perilla oil	high
Pine oil	moderate
Rape seed oil	high
Rosin oil	high
Soya bean oil	moderate
Sperm oil	moderate
Tallow	moderate
Tallow oil	moderate
Tung oil	moderate
Turpentine	slight
Whale oil	high

Solid materials susceptible to self-heating in air	
Activated charcoal	Jaggery soap
Animal feedstuffs	Jute
Beans	Lagging contaminated with oils etc.
Bone meal, bone black	Lamp-black
Brewing grains, spent	Leather scrap
Carbon	Maize
Celluloid	Manure
Colophony powder material	Milk products
Copper powder	Monomers for polymerization
Copra	Palm kernels
Cork	Paper waste
Cotton	Peat
Cotton waste	Plastic, powdered (various)
Cottonseed	Rags, impregnated
Distillers dried grain	Rapeseed
Fats	Rice bran
Fertilizers	Rubber scrap
Fishmeal	Sawdust
Flax	Seedcake
Foam and plastic	Seeds
Grains	Silage
Grass	Sisal
Gum rosin	Soap powder
Hay	Soya beans
Hemp	Straw
Hides	Sulphur
Iron filings/wool/borings	Varnished fabric
Iron pyrites	Wood chips
Ixtle	Wood fibreboard

Table 6.7 Cont'd

Wood flour	Zinc powder
Wool waste	

Fibrous materials are subject to self-heating when impregnated with the following vegetable/animal oils (in decreasing order of tendency)

Cod liver oil
Linseed oil
Menhaden oil
Perilla oil
Corn oil
Cottonseed oil
Olive oil
Pine oil
Red oil
Soya bean oil
Tung oil
Whale oil
Castor oil
Lard oil
Black mustard oil
Oleo oil
Palm oil
Peanut oil

Other materials subject to self-heating (depending upon composition, method of drying, temperature, moisture content)

Desiccated leather
Leather scraps
Dried blood
Household refuse
Leather meal

NB This list is not exclusive

Table 6.8 Characteristics of some organometallic compounds in common use

Alkyl magnesium halides (Grignard reagents)	Usually prepared and handled in organic solvent For alkyl groups of ≤4 carbon atoms, the compounds react vigorously with water and the resulting alkane ignites
Butyl lithium	Pale yellow, caustic, extremely flammable liquid May ignite if exposed to air Reacts violently with water
Diethyl aluminium chloride	Colourless corrosive liquid Ignites immediately upon contact with air Reacts violently with water
Diethyl zinc Dimethyl zinc	Colourless malodorous liquids that are spontaneously flammable in air and react violently with water
Dimethyl arsine	Colourless poisonous liquid Ignites in air
Nickel carbonyl	Yellowish, volatile, toxic liquid, oxidizes in air and explodes at ~60°C Confirmed carcinogen
Sodium methylate	White powder, sensitive to air and decomposed by water
Triethyl aluminium Triethyl aluminium ethereate Trimethyl aluminium	Colourless liquids which ignite in air and decompose explosively in cold water

Table 6.9 Pyrophoric chemicals in common use

Pyrophoric alkyl metals and derivatives

Groups

- Alkyl lithiums
- Dialkylzincs
- Diplumbanes
- Trialkylaluminiums
- Trialkylbismuths

Compounds

- Bis-dimethylstibinyl oxide
- Bis(dimethylthallium) acetylide
- Butyllithium
- Diethylberyllium
- Diethylcadmium
- Diethylmagnesium
- Diethylzinc
- Diisopropylberyllium
- Dimethylberyllium
- Dimethylbismuth chloride
- Dimethylcadmium
- Dimethylmagnesium
- Dimethylmercury
- Dimethyl-phenylethynylthallium
- Dimethyl-l-propynlthallium
- Dimethylzinc
- Ethoxydiethylaluminium
- Methylbismuth oxide
- Methylcopper
- Methyllithium
- Methylpotassium
- Methylsilver
- Methylsodium
- Poly(methylenemagnesium)
- Propylcopper
- Tetramethyldistibine
- Tetramethyllead
- Tetramethylplatinum
- Tetramethyltin
- Tetravinyllead
- Triethylantimony
- Triethylbismuth
- Triethylgallium
- Trimethylantimony
- Trimethylgallium
- Trimethylthalium
- Trivinylbismuth
- Vinyllithium

Pyrophoric carbonyl metals

- Carbonylithium
- Carbonylpotassium
- Carbonylsodium
- Dodecacarbonyldivanadium
- Dodecacarbonyltetracobalt
- Dodecacarbonyltriiron
- Hexacarbonylchromium
- Hexacarbonylmolybdenum
- Hexacarbonyltungsten
- Nonacarbonyldiiron
- Octacarbonyldicobalt
- Pentacarbonyliron
- Tetracarbonylnickel

Pyrophoric metals
(in finely-divided state)

- Caesium
- Calcium
- Cerium
- Chromium
- Cobalt
- Hafnium
- Iridium
- Iron
- Lead
- Lithium
- Manganese
- Nickel
- Palladium
- Platinum
- Plutonium
- Potassium
- Rubidium
- Sodium
- Tantalum
- Thorium
- Titanium
- Uranium
- Zirconium

Alloys

- Aluminium–mercury
- Bismuth–plutonium
- Copper–zirconium
- Nickel–titanium

Pyrophoric non-metals and metal carbides

- Phosphorus
- Silane
- Phosphine
- Calcium carbide
- Uranium carbide

Pyrophoric metal sulphides

- (Ammonium sulphide)
- Barium sulphide
- Calcium sulphide
- Chromium (II) sulphide
- Copper (II) sulphide
- Diantimony trisulphide
- Dibismuth trisulphide
- Dicaesium selenide
- Dicerium trisulphide
- Digold trisulphide
- Europium (II) sulphide
- Germanium (II) sulphide
- Iron disulphide
- Iron (II) sulphide
- Manganese (II) sulphide
- Mercury (II) sulphide
- Molybdenum (IV) sulphide
- Potassium sulphide
- Rhenium (VII) sulphide
- Silver sulphide
- Sodium disulphide
- Sodium polysulphide
- Sodium sulphide
- Tin (II) sulphide
- Tin (IV) sulphide
- Titanium (IV) sulphide
- Uranium (IV) sulphide

Pyrophoric alkyl non-metals

- Bis(dibutylborino) acetylene
- Bis-dimethylarsinyl oxide
- Bis-dimethylarsinyl sulphide
- Bis-trimethylsilyl oxide
- Dibutyl-3-methyl-3-buten-1-ynlborane
- Diethoxydimethylsilane
- Diethylmethylphosphine
- Ethyldimethylphosphine
- Tetraethyldiarsine
- Tetramethyldiarsine
- Tetramethylsilane
- Tribenzylarsine
- mixo-Tributylborane
- Tributylphosphine
- Triethylarsine
- Triethylborane
- Triethylphosphine
- Triisopropylphosphine
- Trimethylarsine
- Trimethylborane
- Trimethylphosphine

Pyrophoric alkyl non-metal halides

- Butyldichloroborane
- Dichlorodiethylsilane
- Dichlorodimethylsilane
- Dichlor(ethyl)silane
- Dichloro(methyl)silane
- Iododimethylarsine
- Trichloro(ethyl)silane
- Trichloro(methyl)silane
- Trichloro(vinyl)silane

Pyrophoric alkyl non-metals

Hydrides

- Diethylarsine
- Diethylphosphine
- Dimethylarsine
- 1,1-Dimethyldiborane
- 1,2-Dimethyldiborane
- Dimethylphosphine
- Ethylphosphine
- Methylphosphine
- Methylsilane

Table 6.10 Electrochemical series

Metal	*Symbol*	*Electro-negativity*	*Occurrence*	*Reactivity with water*
Lithium	Li	0.97	Never found uncombined	React with cold water to yield hydrogen
Caesium	Cs	0.86		
Potassium	K	0.91		
Barium	Ba	0.97		
Strontium	Sr	0.99		
Calcium	Ca	1.04		
Sodium	Na	1.01		
Magnesium	Mg	1.23	Rarely found uncombined	Burning metals decompose water and hot metals decompose steam
Aluminium	Al	1.47		
Manganese	Mn	1.60		
Zinc	Zn	1.66		
Chromium	Cr	1.56		
Iron	Fe	1.64		
Cadmium	Cd	1.46		
Cobalt	Co	1.70		Very little reaction unless at white heat
Nickel	Ni	1.75		
Tin	Sn	1.72		
Lead	Pb	1.55		
Hydrogen	H	2.20		
Phosphorus	P	2.06		
Oxygen	O	3.50		
Bismuth	Bi	1.67	Sometimes found uncombined	Inactive with water or steam
Copper	Cu	1.75		
Mercury	Hg	1.44		
Silver	Ag	1.42		
Platinum	Pt	1.44	Found uncombined with other elements	
Gold	Au	1.42		

Other types of explosion involve,

- Pressure rupture, due to rapid release of high pressure. Blast is generated by rapid expansion of gas down to atmospheric pressure and rupture of the container generates missiles.
- Steam explosion: rapid vaporization of water within molten metal, molten salts or hot oil or through them contacting surface or adsorbed moisture (refer to page 47).

The last two types do not involve a combustion reaction but the damage they cause can similarly be related to the overpressure generated at a given distance from the event.

Control measures

Strategies for handling flammable materials

- Minimize *at the design stage* the risk of fire/explosion, e.g. by substitution with a less volatile chemical or operation at lower temperature, avoidance of air ingress or use of inerting. Design to minimize leakages and avoidance of potential ignition sources.

- Minimize the risk by *appropriate systems of work.*
- Mitigate *the effects of fire or explosion*, e.g. by detection provision, spacing, appropriate construction materials, shielding, venting, extinguishment, provision for evacuation of personnel.

Fire prevention

Theoretically, if one corner of the 'fire triangle' is eliminated a fire or explosion is impossible. However, in practice, if flammable gases or vapours are mixed with air in flammable concentrations, sooner or later the mixture is likely to catch fire or explode because of the difficulty of eliminating every source of ignition. For reliable control of flammable materials, including combustible dusts, the aim is to remove *two* corners from the fire triangle. This can include some combination of:

- Prevention of a mixture forming within the flammable range.
- Elimination of ignition sources (see Table 6.3).

Fire control

Fire detection and suppression form the basis of fire control, with emergency back-up procedures to mitigate the consequences. Selected key tactics for working with flammable chemicals are summarized in Table 6.11.

Refer also to 'Fire extinguishment' (page 221).

Dust explosions

The avoidance, and mitigation of the effects, of a dust explosion may involve some combination of:

- Elimination of ignition sources, which is inherently difficult to ensure.
- Atmosphere control, e.g. controlling dust concentrations or inerting.
- Containment of explosion overpressure, i.e. by designing plant capable of withstanding in excess of the maximum explosion overpressure, or safe venting of forces, e.g. via blow-off panels, doors, membranes.
- Limitation of inventory.
- Restriction of spread by means of baffles, chokes or by advance inerting.
- Use of water sprays or very rapid injection of suppressant gas or powder.
- Good housekeeping, particularly to avoid a devastating secondary explosion, following redispersion of any accumulations of combustible dust.

A similar logic is applicable to the control of explosions involving gas or vapour, but other measures, e.g. dispersion by steam or containment by water curtains, may be applicable to vapour clouds in the open air. Containment or diversion of a blast (e.g. by blast walls) and reducing its effect by appropriate spacing of equipment, buildings etc. are also applicable.

Pyrophorics

Control measures to reduce the risk from handling pyrophorics include:

- Handling and storing the minimum quantities necessary at any time.

Table 6.11 Control measures for working with flammable chemicals

Substitute with less volatile/flammable material where possible (i.e. higher flash point/autoignition temperature, lower vapour pressure)
Check on legal requirements and relevant standards/codes etc.
Minimize quantities in use/in store
Keep below LEL, e.g. chill to lower airborne concentration, use exhaust ventilation, inerting, keep air out.
Design plant/equipment so as to contain the material and provide adequate dilution or exhaust ventilation as appropriate
Provide means to contain spillages, e.g. bund walls, kerbs
Eliminate ignition sources
Eliminate static
Consider need for inerting, flame arresters, pressure relief valves, explosion vents (venting to safe location)
Consider need for checks on oxygen levels or loss of inert medium
Apply appropriate zoning criteria, e.g. with respect to standards of electrical equipment (refer to Table 12.6)
Set up procedures to prevent inadvertent introduction of other ignition sources and to avoid oxygen enrichment: Physical segregation, e.g. fences Warning signs to indicate flammable hazard, no smoking etc. Permits–to-work (including hot work permits) Safe systems of work to control plant modifications etc
Keep flammable chemicals apart from oxidizing agents
Design layout to avoid domino effects/fire spread
Segregate 'empty' and 'full' containers
Check for plant integrity/flammable leaks periodically or continuously on-line, as appropriate
Install appropriate fire/smoke detection, audible alarms
Provide adequate fire suppression systems
Deal with mishaps such as spillage immediately
Train staff in hazards and precautions, and practise emergency evacuation drills
Remember that flammable chemicals can also be *toxic* or *asphyxiant*

- Segregation of the material from other chemicals, particularly 'fuels', i.e. solvents, paper, cloth etc.
- Handling in dry, chemically-inert atmospheres or beneath other appropriate media, e.g. dry oil or inert gas.
- Handling in solution (e.g. aluminium alkyls in petroleum solvents).
- Immediate destruction and removal of spilled materials.
- Careful selection and provision of appropriate fire extinguishers in advance.
- Provision and use of appropriate eye/face protection, overalls and gloves.

Fire extinguishment

Detection

If a flammable gas or vapour is present, a pre-fire condition may be identified by a flammable-gas detector. This will actuate an alarm at a fraction of the LFL. Banks of detectors may be installed at high or low level depending, in part, upon the gas density. Fire detection may be by:

- Personnel, e.g. operating, maintenance or security staff, or neighbours, or passers-by.
- Heat sensing, as actual temperature or rate-of-temperature rise, and depending upon melting of a metal (fusion); expansion of a solid, liquid or gas; electrical sensing.
- Smoke detection depending upon absorption of ionizing radiation by smoke particles; light scattering by smoke particles; light obscuration.
- Flame detection by ultraviolet radiation or infra-red radiation sensing.

A combination of detectors may be appropriate. They may activate an alarm only, or actuate a combined alarm/extinguishment system. With a bank of detectors a voting system may be used to increase reliability and reduce the frequency of spurious alarms. Detection/alarm systems may also be interlinked with, e.g., fire-check doors held back on electromagnetic catches such that the doors close automatically upon activation of the detection system.

Extinguishment

Removal of one of the corners of the fire triangle normally results in extinguishment of a fire. Propagation of a flame can also be stopped by inhibition of the chain reactions, e.g. using dry powders or organo–halogen vaporizing liquids.

Classification of fires

Class	*Type*
A	Fire involving solid materials, generally organic materials, in which combustion normally takes place with the formation of glowing embers.
B	Fire involving a liquid or liquefiable solid (the miscibility or otherwise with water is an important characteristic).
C	Fire involving a gas.
D	Fire involving a burning metal, e.g. magnesium, aluminium, sodium, calcium or zirconium.

An additional class not currently included in British Standard EN2 is Class F fires including cooking oils or fats. Electrical fires are not classified since any fire involving, or initiated by, electrical equipment will fall within Class A, B or C.

Fire-extinguishing materials

The penetration and cooling action of *water* is required with Class A fires, e.g. those involving paper, wood, textiles, refuse. Water is applied in the form of a jet or spray; foam or multi-purpose powder extinguishers are alternatives. Extinguishment of a Class B fire can be achieved by the smothering action of dry chemical, carbon dioxide or foam. Most flammable liquids will float on water (refer to Table 6.1 under 'Specific gravity'), so that water as a jet is unsuitable: a mist may, however, be effective. Water is also widely used to protect equipment exposed to heat. Dry powders are effective on flammable liquid or electrical fires.

Foam is a proportioned mixture of water and foam concentrate aspirated with air to cause expansion, e.g. from 6 to 10 times the volume (low expansion foam) up to >100 times (high expansion foam). It transports water to the surface of flammable liquids and enables it to float and extinguish the fire. An effective system depends upon:

- The type of flammable liquid – determines the type of foam, e.g. standard or alcohol-resistant grade. Aqueous film-forming foam may be used for rapid 'knock-down'.

- The type of hazard – determines the method and rate of application, e.g. by fixed pourers, mobile monitors, portable foam-towers or fixed semi-subsurface systems.
- The size of the hazard – determines the requirements for foam concentrate and water.

Carbon dioxide is useful where the minimum damage should be caused to the materials at risk, on fires in liquid, solids or electrical fires but not where there is a high risk of reignition. It is likely to be ineffective outdoors due to rapid dispersal. It is unsuitable for reactive metals, metal hydrides or materials with their own oxygen supply, e.g. cellulose nitrate.

Dry powders are effective on flammable liquid or electrical fires. Special powders are available for use on metals. Dry powder extinguishers may be used on Class C fires, including gases and liquefied gases in the form of a liquid spillage or a liquid or gas leak. This must be accompanied by other actions, e.g. stopping the leak; this is necessary to avoid accumulation of an unburned flammable gas–air mixture which could subsequently result in an explosion. Activation may be automatic by a detection system, or manual.

Vaporizing liquid halogen agents are electrically non-conductive and are effective on a wide range of combustibles, particularly flammable liquids and electrical fires. A 'lock-off' system is required on fixed installations to protect personnel, the normal extinguishing concentration being 5% by volume. The use of such liquids is being phased out; except for defined essential uses they will be banned from 31 December 2003.

Portable extinguishers and *fire blankets* are normally provided at strategic points in the work area. The range of application of portable extinguishers is summarized in Table 6.12. British Standard EN3: Part 5 requires all new extinguisher bodies to be red. A zone of colour above, or within, the section used to provide operating instructions may be used to identify the type of extinguisher. The colours used are:

Standard dry powder or multi-purpose dry powder	blue
AFFF (aqueous film-forming foam)	cream
Water	red
Vaporizing liquid, including Halon	green
Carbon dioxide	black

Fixed installations for fire-fighting may be either:

- Manually operated for general protection, e.g. hose reels, hydrants and foam installations, or
- Automatically operated for general protection, e.g. sprinklers, or for special-risk protection, e.g. carbon dioxide installations.

The general requirements of such an installation are summarized in Table 6.13.

Fire precautions

A range of precautions are based on the principles summarized earlier. However, general precautions, applicable to the majority of work situations, are listed in Table 6.14, many of which are included in legal requirements. In the UK duties under the Fire Precautions (Workplace) (Amendment) Regulations 1999 are to:

- carry out a fire risk assessment of the workplace;
- identify significant findings and details of anyone specifically at risk;

Table 6.12 Portable fire extinguishers

Extinguisher type	*Water*	*Carbon dioxide*	*Dry powder*	*Foam*	*Vaporizing liquid*[3]	*Fire blanket*	*Sand*
Class A fire Wood, cloth, paper or similar combustible material *Cooling by water most effective*	Most suitable	Small fires only	Small fires only	Yes	Small fires only[1]	No, except for personal clothing on fire	No
Class B fire Flammable liquids, petrol, oils, greases, fats *Blanketing/ smothering most effective*	Dangerous	Most suitable	Most suitable	Most suitable[2]	Small fires only[1]	Most suitable Small fires only	Small fires only
Electrical plant, electrical installations *Non-conductivity of extinguishing agent most important*	Dangerous	Most suitable	Most suitable	No	Yes[1]	No	No

[1] Toxic products may be produced: care must be exercised after use in confined spaces.
[2] Special foam required for water-miscible liquids.
[3] Subject to replacement.

Table 6.13 General requirements for fixed fire-extinguishing systems (Activation may be automatic by a detection system, or manual)

- Capability to control and extinguish the anticipated fire condition without recourse to outside assistance (unless planned for).
- Reliability, allowing for environmental features likely to be detrimental to operation, e.g. dust, corrosion, tar.
- Agents must be compatible with the process, with each other, and with any other installed systems.
- Consideration of the potential toxicity of the agent, any thermal degradation products, or products generated on contact with chemicals present will dictate safety measures.
- If manual fire-fighting is also anticipated following agent discharge, visibility in the fire zone requires consideration.

Table 6.14 General fire precautions

Area designation	Zoning for electrics Control portable heaters etc. No smoking Restricted areas
Electrical equipment	Regular inspection and maintenance by qualified electricians Prohibition of makeshift installations
Waste disposal	Prevention of combustible waste accumulation in corners, passageways or other convenient 'storage' areas
Storage	Segregated storage Uncongested storage of combustibles: gangways/adequate breaks Material stacked in the open should be away from windows Flammable liquids in properly designed storerooms: bulk quantities in fixed, bunded, adequately spaced tanks
Contractors	Clearance Certificate control of contractors/temporary workers Close control of temporary heating, lighting, cooking etc.
Escape/access	Escape doors and routes must be kept free of obstructions Access for emergency services must be maintained
Fire equipment	Fire alarm and fire-fighting equipment must be regularly inspected, maintained and tested Portable extinguishers to have designated locations/be of correct type. Instructions must be provided as to where and how to use them. Practice is necessary
Flues	Passages for services or other ducts must be adequately fire-stopped to prevent their acting as flues for fire/smoke transmission
Sprinklers	Maintain sprinkler systems Institute alterations if building is modified, use changes etc. Observe use specifications, e.g. for stack heights, fire loading
Prevention of arson	Control access at all times Screen employees and casual labour Lock away flammable substances and keep combustibles away from doors, windows, fences Provide regular fire safety patrols, even where automatic systems are provided Secure particularly storage and unmanned areas
Fire and smoke stop doors	Ensure that fireproof doors and shutters are self-closing Keep *all* doors free from obstruction Ensure that fire check doors are kept closed

- provide and maintain fire precautions to safeguard those at the workplace;
- provide relevant information, instruction and training.

In the UK the Building Regulations impose fire safety requirements on:

- structural stability;
- compartmentalization to restrict fire spread;
- fire resistance of elements and structures;
- reduction of spread of flame over surfaces of walls and ceilings;
- space separation between buildings to reduce the risk of fire spread from one building to another;
- means of escape in case of fire;
- access for fire appliances and assistance to the fire brigade.

An action plan for risk assessment is given in Figure 6.2.

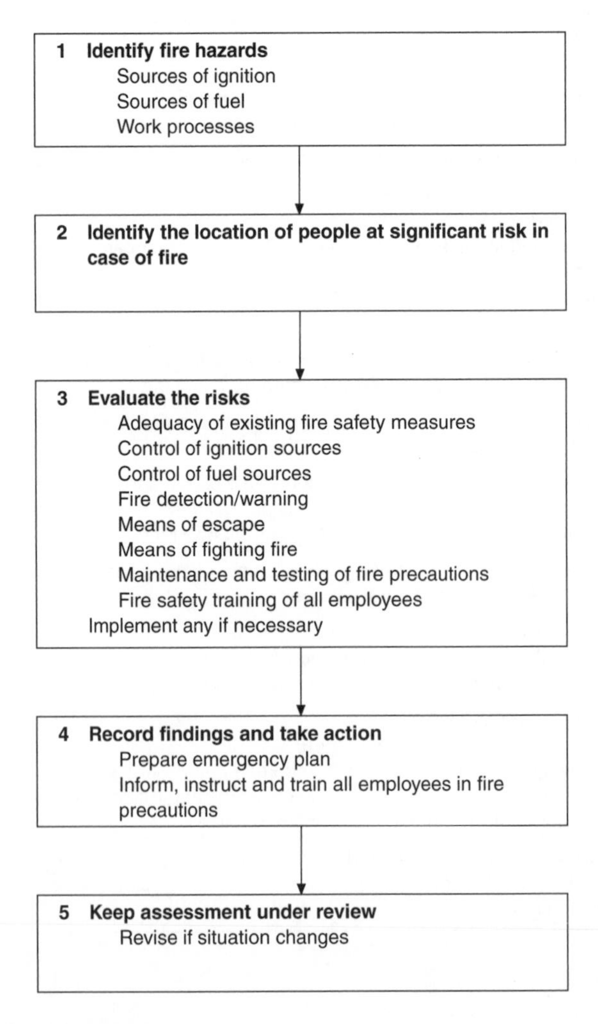

Figure 6.2 *Action plan for fire risk assessment*

The minimum fire instruction and training needs are summarized in Table 6.15.

Table 6.15 Essentials for fire instruction and training

Action to be taken upon discovering a fire
Action to be taken upon hearing the fire alarm
Raising the alarm, including the location of alarm call points, internal fire alarm telephones and alarm indicator panels
Correct method of calling the fire service
The location and use of fire-fighting equipment
Knowledge of the escape routes
The importance of fire doors and the need to close all doors at the time of a fire and on hearing the fire alarm
Stopping machines and processes and isolating power supplies where appropriate
Evacuation of the building:
• procedures for alerting visitors, members of the public etc., including if necessary directing/escorting them to exits;
• familiarity with how to open all escape doors and the use of any emergency fastenings;
• understanding of the reason for not using lifts except those designated, and of special design, for evacuation of disabled persons

7

Reactive chemicals

From Chapter 3 all chemical reactions involve energy changes as a combination of activation energy and reaction energy. Whilst in some cases the heat of reaction is absorbed into the products and the reaction cools (endothermic reactions), most reactions evolve energy as heat (exothermic reactions), and sometimes as light and sound. When the heat liberated or absorbed cannot be accommodated by the surrounding environment hazards are presented which can result in material damage. This can arise because either the *amount* of energy (thermodynamics) or the *rate* of energy release due to the speed of reaction (reaction kinetics) is excessive. The aim therefore is to establish the thermal stability of the system and then to control the extent and rate of heat release so as to minimize risk from energetic hazards. Thousands of reactive hazards have been documented by Bretherick (see Bibliography). The present chapter provides an insight into the hazards and the basic precautions to control the risk.

Reactive chemical hazards may arise from the inherent properties of the chemicals handled, used or disposed of and/or from their admixture or processing.

A hazard may arise with a chemical because of its tendency to decompose spontaneously or to react violently on contact with other common chemicals, as illustrated in Figure 7.1. The case of pyrophoric chemicals is summarized in Chapter 6; some dangerous reactions of compressed gases are mentioned in Chapter 9; other cases are summarized here.

However, extreme caution is necessary with mixed chemical systems since many which are thermodynamically unstable exhibit considerable kinetic stability. The kinetic barrier to stability may be overcome if traces of catalyst are present, and result in a violent reaction. The most common catalysts derive from metals, or their compounds, and the unpredictable behaviour of many reactions arises from the unwitting presence of impurities. Other catalysts include acids, bases, organic free-radical precursors, etc. Hence any system must be treated with care which

(a) is thermodynamically unstable *or*
(b) may contain a catalyst, or impurities which could serve as a catalyst.

Water-sensitive chemicals

Some chemicals are 'water-sensitive': in contact with water they can generate flammable or toxic gases and/or undergo a vigorous reaction. Refer to Table 7.1. Such reactions can cause overpressure in sealed equipment or pipework. Selected water-sensitive chlorine compounds are given in Table 7.2. With flammable gas generation the heat of reaction may cause ignition, depending upon the compound in question, as illustrated by the list of hydrides in Table 7.3.

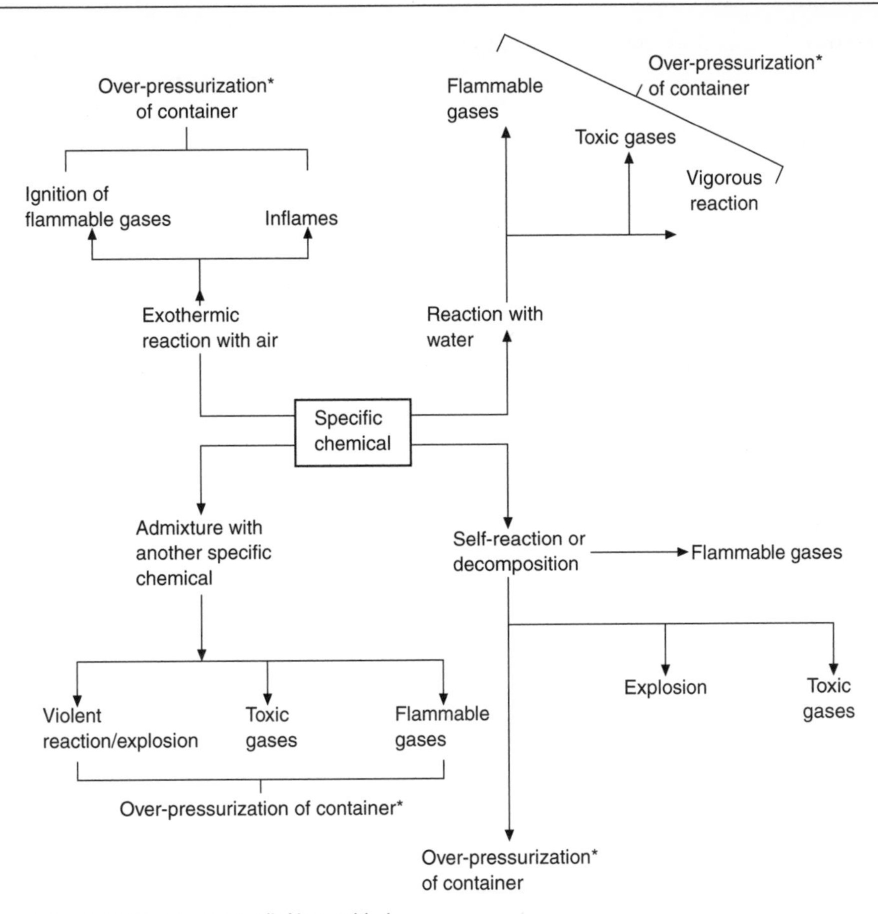

Figure 7.1 *Possible reactive chemical hazards (consequences are not mutually exclusive)*

Precautions for safe handling

- Store and use in such a way that accidental ingress of water, or contact with it, is avoided (roofs of storage areas should be regularly maintained to minimize leaks).
- Provide covered storage, off the ground, away from sprinkler systems, safety showers, overhead water lines or condensate lines.
- Keep away from water taps or sinks.
- Store under a chemically-inert medium when appropriate (stocks should be checked regularly to ensure that an adequate level of inert medium is maintained).
- Segregate from other flammable materials, e.g. solvents and combustibles.
- Use appropriate eye/face protection, overalls and gloves.

Toxic hazards from mixtures

Undesirable emissions of toxic gases may occur as a result of mixing relatively common chemicals. Refer to Table 7.4. Chemicals which are incompatible in this way must be brought into contact only under strictly controlled conditions.

Table 7.1 Water-sensitive chemicals: consequences of water contact

	F	T	V
Acetyl bromide		T	
Acetyl chloride		T	V
Acetylcholine bromide		T	
Aluminium (powder)	F		
Aluminium alkyls	F		V
Aluminium isopropoxide	F		
Aluminium lithium hydride	F		
Aluminium selenide		T	
Aluminium phosphide	F	T	
Boron tribromide		T	
Calcium (granules)	F		
Calcium carbide	F		
Calcium hydride	F		
Calcium phosphide	F	T	
Chlorosulphonic acid		T	V
Disulphur dichloride		T	V
Ethoxides, alkaline			V
Lithium (metal)	F		V
Lithium aluminium deuteride	F		
Lithium aluminium dihydride	F		
Lithium borohydride	F		
Lithium hydride	F		
Lithium methoxide	F		
Magnesium (powder)	F		
Magnesium alkyls	F		
Magnesium phosphide	F	T	
Methoxides, alkaline	F		V
Nickel sulphide		T	
Phosphorus pentasulphide	F	T	
Phosphorus sesquisulphide	F	T	
Phosphorus pentachloride		T	
Phosphorus pentabromide		T	
Potassium (metal)	F		V
Potassium borohydride	F		
Potassium methoxide	F		
Silicon tetrachloride		T	V
Sodium (metal)	F		V
Sodium aluminium hydride	F		
Sodium borohydride	F	T	
Sodium hydride	F		
Sulphur dichloride		T	V
Sulphuric acid, fuming (Oleum)		T	V
Sulphur tetrachloride		T	V
Sulphuryl chloride		T	V
Thionyl chloride		T	V
Titanium tetrachloride		T	V
Trichlorophenylsilane		T	
Trichlorosilane	F		
Zinc (powder)	F		
Zinc alkyls		T	V
Zirconium (powder)	F		

F Flammable gases
T Toxic products
V Vigorous reaction

Table 7.2 Examples of reactive chlorine compounds

Compound	*Description*	*Reactivity*
Acetyl chloride CH_3COCl	Colourless, fuming, corrosive liquid Flash point 4°C When heated, emits phosgene	Decomposes violently with water to produce heat and toxic fumes: HCl
Aluminium chloride (anhydrous) $AlCl_3$	Orange, yellow, grey or white powder which is a severe respiratory irritant and can cause skin/eye burns	Reacts with air moisture to form corrosive HCl gas Violent reaction when a stream of water hits a large amount Do not use water in vicinity
Benzoyl chloride C_6H_5COCl	Colourless, fuming, corrosive liquid with a strong odour Combustible: flash point 72°C Generates phosgene gas when heated	Reacts strongly with water or water vapour, producing heat and toxic/corrosive fumes Use of water must be considered carefully
Calcium hypochlorite $Ca(ClO)_2$	Water soluble white crystals or powder with strong chlorine odour Non-flammable but can evolve Cl_2 and O_2 May undergo decomposition	Water spray may be used but evolves Cl_2 gas freely at ordinary temperatures with moisture
Sulphur monochloride S_2Cl_2	Yellowish-red oily fuming liquid with a strong odour Combustible: flash point 118°C Ignition temp. 233°C Liquid and vapours are irritating	Decomposes when contacted by water, to produce heat and toxic/corrosive fumes Do not allow water to enter containers: reaction can be violent Wash down spills with flooding amounts of water
Titanium tetrachloride $TiCl_4$	Colourless to light yellow fuming corrosive liquid Vapour is irritating	Reacts vigorously with water, liberating heat and corrosive HCl gas Reacts more strongly with hot water Use water spray to keep exposed containers cool in fire

Precautions

- Provision of adequate training, instruction, and supervision.
- Prohibition on unauthorized mixing, e.g. of cleaning agents or 'wastes'.
- Correct labels.
- Segregated storage, to avoid accidental mixing.
- Dispose of wastes and 'empty' containers by different routes.
- Specific precautions against the inherent hazards of each individual chemical.

Reactive hazards from mixtures

Many chemicals are 'incompatible' because a violent reaction may occur on mixing. This can, in some conditions, result in an explosion. Refer to Table 7.5.

An appraisal is needed of all chemicals which may be present, even if unintentionally (e.g. as intermediates, byproducts or wastes) and how they can react under the most extreme conditions

Table 7.3 Variation in reactivity of hydrides with humid air or water

Substance	*Reaction (ambient temperature)*	
	Humid air	*Water*
Aluminium borohydride ($Al(BH_4)_3$)	Explosive	Explosive
Aluminium hydride (AlH_3)	Slow	Moderate
Antimony hydride (Stibine) (SbH_3)	Rapid	Very slow
Arsenic hydride (Arsine) (AsH_3)	Moderate	Very slow
Barium hydride (BaH_2)	Rapid	Rapid
Beryllium borohydride ($Be(BH_4)_2$)	Explosive	Explosive
Beryllium hydride (BeH_2)	Slow	Slow
Calcium hydride (CaH_2)	Moderately fast	Rapid
Cerium hydride (CeH_3)	Pyrophoric	Slow
Caesium hydride (CsH)	Inflames	Violent
Copper hydride (CuH)	Rapid	Slow
Diborane (B_2H_6)	Explosive	Moderate
Lead hydride (PbH_4)	Instant (unstable gas)	—
Lithium aluminium hydride ($LiAH_4$)	Rapid	Violent
Lithium borohydride ($LiBH_4$)	Rapid	Vigorous
Lithium hydride (LiH)	Can ignite	Rapid
Magnesium aluminium hydride ($Mg(AlH_4)_2$)	Vigorous	Vigorous
Magnesium borohydride ($Mg(BH_4)_2$)	Very slow	Violent
Magnesium hydride (MgH_2)	Known to ignite	Rapid
Pentaborane (B_5H_9)	Ignites	Rapid
Phosphorus hydride (Phosphine) (PH_3)	Pyrophoric	Very slow
Potassium borohydride (KBH_4)	Very slow	Very slow
Potassium hydride (KH)	Inflames	Vigorous
Rubidium hydride (RbH)	Inflames	Violent
Silicon hydride (Silane) (SiH_4)	Explosive	Rapid
Sodium aluminium hydride ($NaAlH_4$)	Rapid	Ignites, may explode
Sodium borohydride ($NaBH_4$)	Slow	Slow
Sodium hydride (NaH)	Ignites	Violent
Uranium hydride (UH_3)	Pyrophoric	Moderate

Table 7.4 Toxic hazards from incompatible chemical mixtures
Substances in column 1 must be stored/handled so that they cannot accidentally contact corresponding substances in column 2 because toxic materials (column 3) would be produced.

Column 1	*Column 2*	*Column 3*
Arsenical materials	Any reducing agent	Arsine
Azides	Acids	Hydrogen azide
Cyanides	Acids	Hydrogen cyanide
Hypochlorites	Acids	Chlorine or hypochlorous acid
Nitrates	Sulphuric acid	Nitrogen dioxide
Nitric acid	Copper, brass any heavy metals	Nitrogen dioxide (nitrous fumes)
Nitrites	Acids	Nitrous fumes
Phosphorus	Caustic alkalis or reducing agents	Phosphine
Selenides	Reducing agents	Hydrogen selenide
Sulphides	Acids	Hydrogen sulphide
Tellurides	Reducing agents	Hydrogen telluride

Table 7.5 Reactive hazards of incompatible chemicals
Substances in column 1 must be stored/handled so that they cannot contact corresponding substances in column 2 under uncontrolled conditions, or violent reactions may occur.

Column 1	*Column 2*
Acetic acid	Chromic acid, nitric acid, hydroxyl-containing compounds, ethylene glycol, perchloric acid, peroxides, or permanganates
Acetone	Concentrated nitric and sulphuric acid mixtures
Acetylene	Chlorine, bromine, copper, silver, fluorine or mercury
Alkali and alkaline earth metals, e.g. sodium, potassium lithium, magnesium, calcium, powdered aluminium	Carbon dioxide, carbon tetrachloride, or other chlorinated hydrocarbons. (Also prohibit, water, foam and dry chemical on fires involving these metals – dry sand should be available)
Anhydrous ammonia	Mercury, chlorine, calcium hypochlorite, iodine, bromine or hydrogen fluoride
Ammonium nitrate	Acids, metal powders, flammable liquids, chlorates, nitrites, sulphur, finely-divided organics or combustibles
Aniline	Nitric acid, hydrogen peroxide
Bromine	Ammonia, acetylene, butadiene, butane or other petroleum gases, sodium carbide, turpentine, benzene, or finely-divided metals
Calcium oxide	Water
Carbon, activated	Calcium hypochlorite
Chlorates	Ammonium salts, acids, metal powders, sulphur, finely-divided organics or combustibles
Chromic acid and chromium trioxide	Acetic acid, naphthalene, camphor, glycerol, turpentine, alcohol or other flammable liquids
Chlorine	Ammonia, acetylene, butadiene, butane or other petroleum gases, hydrogen, sodium carbide, turpentine, benzene or finely-divided metals
Chlorine dioxide	Ammonia, methane, phosphine or hydrogen sulphide
Copper	Acetylene, hydrogen peroxide
Fluorine	Isolate from everything
Hydrazine	Hydrogen peroxide, nitric acid, or any other oxidant
Hydrocarbons (benzene, butane, propane, gasoline, turpentine, etc.)	Fluorine, chlorine, bromine, chromic acid, peroxides
Hydrocyanic acid	Nitric acid, alkalis
Hydrofluoric acid, anhydrous (hydrogen fluoride)	Ammonia, aqueous or anhydrous
Hydrogen peroxide	Copper, chromium, iron, most metals or their salts, any flammable liquid, combustible materials, aniline, nitromethane
Hydrogen sulphide	Fuming nitric acid, oxidizing gases
Iodine	Acetylene, ammonia (anhydrous or aqueous)
Mercury	Acetylene, fulminic acid (produced in ethanol – nitric acid mixtures), ammonia
Nitric acid (conc)	Acetic acid, acetone, alcohol, aniline, chromic acid, hydrocyanic acid, hydrogen sulphide, flammable liquids, flammable gases, or nitratable substances, paper, cardboard or rags
Nitroparaffins	Inorganic bases, amines
Oxalic acid	Silver, mercury
Oxygen	Oils, grease, hydrogen, flammable liquids, solids or gases
Perchloric acid	Acetic anhydride, bismuth and its alloys, alcohol, paper, wood, grease, oils
Peroxides, organic	Acids (organic or mineral), avoid friction, store cold
Phosphorus (white)	Air, oxygen
Potassium chlorate	Acids (*see also* chlorates)
Potassium perchlorate	Acids (*see also* perchloric acid)
Potassium permanganate	Glycerol, ethylene glycol, benzaldehyde, sulphuric acid
Silver	Acetylene, oxalic acid, tartaric acid, fulminic acid (produced in ethanol–nitric acid mixtures), ammonium compounds
Sodium	See alkali metals (above)
Sodium nitrite	Ammonium nitrate and other ammonium salts
Sodium peroxide	Any oxidizable substance, such as ethanol, methanol, glacial acetic acid, acetic anhydride, benzaldehyde, carbon disulphide, glycerol, ethylene glycol, ethyl acetate, methyl acetate or furfural
Sulphuric acid	Chlorates, perchlorates, permanganates

(e.g. concentration, agitation, temperature, pressure) likely to arise. Sometimes special testing is required.

For reactions with air or water, refer to pyrophoric chemicals (Chapter 6).

In acid-base reactions, the heat of neutralization of aqueous acids and bases can be sufficient to cause 'spitting' from containers when the concentrated reagents interact. This is also encountered when concentrated sulphuric acid is diluted (refer to Table 7.1); the acid should always be added cautiously to water and not vice versa. Eye and skin protection is obligatory when using such reagents.

Oxidizing agents

Oxidizing agents, although not normally spontaneously flammable, often represent a source of oxygen that can support combustion. They will react readily in contact with reducing reagents. Hence an oxidizing agent will invariably accelerate the rate of burning of a combustible material. In finely divided state such mixtures may react explosively.

Some common oxidizing agents are classified according to stability in Table 7.6.

Table 7.6 Common oxidizing agents classified according to stability

Classification	*Example*
Relatively stable • Increase the burning rate of combustible materials • Form highly flammable or explosive mixtures with finely divided combustible materials	Aluminium nitrate Ammonium persulphate Barium nitrate/peroxide Calcium nitrate/peroxide Cupric nitrate Hydrogen peroxide solutions (8–27.5% by weight) Lead nitrate Lithium peroxide/hypochlorite Magnesium nitrate/perchlorate Nickel nitrate Nitric acid (concentrations ≤70%) Potassium dichromate/nitrate/persulphate Silver nitrate Sodium dichromate/nitrate/nitrite/perborate/persulphate/chlorite (≤40% by weight) Strontium nitrate/peroxide Zinc peroxide
Moderately unstable/reactive • Undergo vigorous decomposition on heating • Explode when heated in a sealed container • Cause spontaneous heating of combustible materials	Ammonium dichromate Barium chlorate Calcium chlorate/hypochlorite Chromium trioxide (chromic acid) Hydrogen peroxide solutions (27.5–91% by weight) Nitric acid (concentrations >70%) Potassium bromide/chlorate/permanganate/peroxide Sodium chlorate/permanganate/peroxide/chlorite (>40% by weight) Strontium chlorate
Unstable • Explode when catalysed or exposed to heat, shock or friction • Liberate oxygen at room temperatures	Ammonium chlorate/perchlorate/permanganate Benzoyl peroxide Guanidine nitrate Mercury chlorate Methyl ethyl ketone peroxide Potassium superoxide

Safe handling

- Handle and store the minimum quantities practicable for the process or experiments in progress.
- Segregate the materials from other chemicals, particularly reducing agents, paper, straw, cloth or materials of low flash point.
- Handle in the most dilute form possible in clearly designated areas, away from potential ignition sources.
- Provide and use appropriate eye/face protection, overalls and gloves.

Hazards arising from the oxidation of organic compounds are greater when the reactants are volatile, or present as a dust or an aerosol. Liquid oxygen and various concentrated acids, e.g. nitric, sulphuric or perchloric acid, and chromic acid are strong oxidizing agents. The use of perchloric acid or perchlorates has resulted in numerous explosions; their use should be avoided when possible (refer to Table 7.5).

Explosive chemicals

Explosions involving flammable gases, vapours and dusts are discussed in Chapter 6. In addition, certain chemicals may explode as a result of violent self-reaction or decomposition when subjected to mechanical shock, friction, heat, light or catalytic contaminants. Substances containing the atomic groupings listed in Table 7.7 are thermodynamically unstable, or explosive. They include acetylides and acetylenic compounds, particular nitrogen compounds, e.g. azides and fulminates, peroxy compounds and vinyl compounds. These unstable moieties can be classified further as in Table 7.8 for peroxides. Table 7.9 lists a selection of potentially-explosive compounds.

More specific definitions of 'explosives' appear in legislation, e.g. in the UK under the Explosives Act 1875 as amended, which covers:

- High explosives which detonate to produce shock waves. Materials which are easily detonated by mechanical or electrical stimuli are termed 'primary explosives'. Those requiring an impinging shock wave to initiate them are 'secondary explosives'.
- Pyrotechnics which burn to produce heat, smoke, light and/or noise.
- Propellants which burn to produce heat and gas as a means of pressurizing pistons, start engines, propel projectiles and rockets.

The precautions with any particular explosive depends on the hazard. In the UK explosives are classified as: 1 – Gunpowder; 2 – Nitrate mixture; 3 – Nitro compound; 4 – Chlorate mixture; 5 – Fulminate; 6 – Ammunition and 7 – Fireworks.

For the purposes of safety distances in connection with the issue of licences for factories and magazines, explosives have been categorized as: X – fire or slight explosion risks or both, with only local effect; Y – mass fire risks or moderate explosion risk, but not mass explosion risk; Z – mass explosion risk with serious missile effect; ZZ – mass explosion risk with minor missile effect.

Hazards can be illustrated by reference to Table 7.10 (showing the explosive effects of small quantities of high explosives in a 6 m × 6 m single-storey building) and to Figure 7.2 (relating the size of fireball to quantity of burning pyrotechnic, high explosive or propellant). With pyrotechnics the hazard is related to the violence with which the chemical burns. One scheme used to classify pyrotechnics is given in Table 7.11. This is used to restrict quantities in use/storage and for

Table 7.7 Atomic groupings characterizing explosive compounds

Bond groupings	*Class*
$-C\equiv C-$	Acetylenic Compounds
$-C\equiv C-$Metal	Metal Acetylides
$-C\equiv C-X$	Haloacetylene derivatives
N=N ring with C (three-membered ring, $>C<$)	Diazirines
$>CN_2$	Diazo Compounds
$\geq C-N=O$	Nitroso Compounds
$\geq C-NO_2$	Nitroalkanes, C–Nitro and Polynitroaryl compounds
$>C(NO_2)_2$	Polynitroalkyl compounds
$\geq C-O-N=O$	Acyl or alkyl nitrites
$\geq C-O-NO_2$	Acyl or alkyl nitrates
$>C-C<$ bridged by O	1,2–Epoxides
$>C=N-O-$Metal	Metal Fulminates or aci – nitro salts
$-C(NO_2)_2-F$	Fluorodinitromethyl compounds
$>N-$Metal	N–Metal Derivatives
$>N-N=O$	N–Nitroso Compounds
$>N-NO_2$	N–Nitro Compounds
$\geq C-N=N-C\leq$	Azo Compounds
$\geq C-N=N-O-C\leq$	Arenediazoates
$\geq C-N=N-S-C\leq$	Arenediazo aryl sulphides
$\geq C-N=N-O-N=N-C\leq$	Bis–arenediazo oxides
$\geq C-N=N-S-N=N-C\leq$	Bis–arenediazo sulphides
$\geq C-N=N-N(R)-C\leq$	Trizazenes (R=H, –CN, –OH, –NO)
$-N=N-N=N-$	High–nitrogen compounds tetrazoles
$\geq C-O-O-H$	Alkylhydroperoxides
$\geq C-CO-OOH$	Peroxyacids
$\geq C-O-O-C\leq$	Peroxides (cyclic, diacyl, dialkyl)
$\geq C-CO-OOR$	Peroxyesters
$-O-O-$Metal	Metal peroxides, peroxoacid salts
$-O-O-$Non–metal	Peroxoacids
$N\rightarrow Cr-O_2$	Aminechromium peroxo–complexes
$-N_3$	Azides (acyl, halogen, non–metal, organic)
$>C-N_2^+O^-$ (bridged)	Arenediazoniumolates
$\geq C-N_2^+S^-$	Diazonium sulphides and derivatives, 'xanthates'
N^+-HZ^-	Hydrazinium salts, oxosalts of nitrogenous bases
$-N^+-OH\ Z^-$	Hydroxylammonium salts

Table 7.7 Cont'd

Bond groupings	*Class*
$\rangle C-N_2^+Z^-$	Diazonium carboxylates or salts
$(N\text{–}Metal)^+Z^-$	Aminemetal oxosalts
Ar–Metal–X X–Ar–Metal	Halo–Arylmetals
N–X	Halogen Azides, N–Halogen compounds, N–Haloimides
$-NF_2$	Difluoroamino compounds
–O–X	Alkyl perchlorates, Chlorite salts, Halogen oxides, Hypohalites, Perchloric acid, Perchloryl Compounds

Table 7.8 Classification of organic peroxides

Peroxide class	*General structures or characteristic group*
Hydroperoxides	ROOH $R_mQ(OOH)_n$ (Q = metal or metalloid)
α-Oxy- and α-peroxy-hydroperoxides and peroxides	contain the grouping: 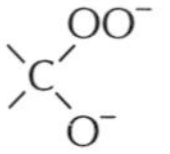 $\rangle C(OO^-)(O^-)$
Peroxides	ROOR′ $R_mQ(OOR)_n$ R_mQOOQR_n
Peroxyacids	$R(CO_3H)_n$ RSO_2OOH
Diacyl peroxides	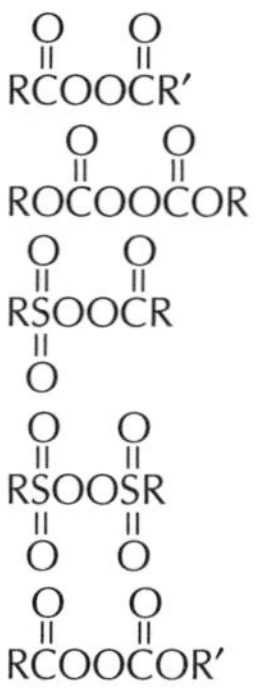 RC(=O)OOC(=O)R′ ROC(=O)OOC(=O)OR RS(=O)(=O)OOC(=O)R RS(=O)(=O)OOS(=O)(=O)R RC(=O)OOC(=O)OR′
Peroxyesters	$R(CO_3R')_n$ $R'(O_3CR)_2$ 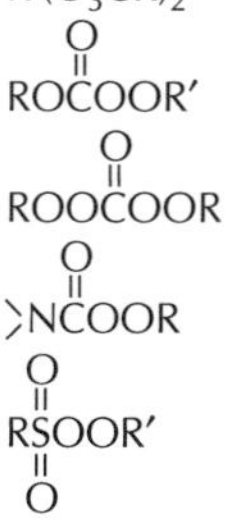 ROC(=O)OOR′ ROOC(=O)OOR $\rangle$NC(=O)OOR RS(=O)(=O)OOR′

Table 7.9 Selected potentially explosive compounds

(a) Peroxy compounds

(i) *Organic peroxy compounds*

Acetyl cyclohexane-sulphonyl peroxide (70%)
Acetyl cyclohexane-sulphonyl peroxide (28% phthalate solution)
o-Azidobenzoyl peroxide
t-Butyl mono permaleate (95% dry)
t-Butyl peracetate (70%)
t-Butyl peroctanoate
t-Butyl perpivalate (75% hydrocarbon solution)
t-Butyl peroxy isobutyrate
Bis-hexahydrobenzoyl peroxide
Bis-monofluorocarbonyl peroxide
Bis-benzenesulphonyl peroxide
Bis-hydroxymethyl peroxide
Bis (1-hydroxycyclohexyl) peroxide
2,2-Bis (t-butylperoxy) butane
2,2-Bis-hydroperoxy diisopropylidene peroxide
Barium methyl peroxide
Benzene triozonide
Cyclohexanone peroxide (95% dry)
Diacetyl peroxide
Di-*n*-butyl perdicarbonate (25% hydrocarbon solution)
2:4-Dichlorobenzoyl peroxide (50% phthalate solution)
Dicaproyl peroxide
Dicyclohexyl perdicarbonate
Di-2-ethylhexyl perdicarbonate (40% hydrocarbon solution)
Dimethyl peroxide
Diethyl peroxide
Di-t-butyl-di-peroxyphthalate
Difuroyl peroxide
Dibenzoyl peroxide
Dimeric ethylidene peroxide
Dimeric acetone peroxide
Dimeric cyclohexanone peroxide
Diozonide of phorone
Dimethyl ketone peroxide
Ethyl hydroperoxide
Ethylene ozonide
Hydroxymethyl methyl peroxide
Hydroxymethyl hydroperoxide
1-Hydroxyethyl ethyl peroxide
1-Hydroperoxy-1-acetoxycyclodecan-6-one
Isopropyl percarbonate
Isopropyl hydroperoxide
Methyl ethyl ketone peroxide
Methyl hydroperoxide
Methyl ethyl peroxide
Monoperoxy succinic acid
Nonanoyl peroxide (75% hydrocarbon solution)
1-Naphthoyl peroxide
Oxalic acid ester of t-butyl hydroperoxide
Ozonide of maleic anhydride
Phenylhydrazone hydroperoxide
Polymeric butadiene peroxide
Polymeric isoprene peroxide
Polymeric dimethylbutadiene peroxide
Polymeric peroxides of methacrylic acid esters and styrene
Polymeric peroxide of asymmetrical diphenylethylene
Peroxyformic acid
Peroxyacetic acid
Peroxybenzoic acid
Peroxycaproic acid
Polymeric ethylidene peroxide
Sodium peracetate
Succinic acid peroxide (95% dry)
Trimeric acetone peroxide
Trimeric propylidene peroxide
Tetraacetate of 1,1,6,6-tetrahydroperoxycyclodecane

(ii) *Inorganic peroxy compounds*

Peroxides

Hydrogen peroxide (>30%)
Mercury peroxide

Peroxyacids

Peroxydisulphuric acid
Peroxynitric acid
Peroxy ditungstic acid

Peroxyacid salts

Sodium peroxyborate (anhydrous)
Sodium triperoxychromate
Sodium peroxymolybdate
Sodium peroxynickelate
Sodium diperoxytungstate
Potassium peroxyferrate
Potassium peroxynickelate
Potassium hyperoxytungstate
Potassium peroxy pyrovanadate
Calcium diperoxysulphate
Calcium peroxychromate
Zinc tetraaminoperoxydisulphate
Ammonium peroxyborate
Ammonium peroxymanganate
Ammonium peroxychromate

Superoxides

Potassium superoxide
Ozone (liquid >30%)
Potassium ozonide
Caesium ozonide
Ammonium ozonide

Inorganic peracids and their salts (common examples which are particularly hazardous)

Ammonium perchlorate
Ammonium persulphate
Ammonium pernitrate
Perchloric acid (>73%)
Performic acid
Silver perchlorate
Tropylium perchlorate

Table 7.9 Cont'd

(b) Halo-acetylenes and acetylides

Lithium bromoacetylide
Dibromoacetylene
Lithium chloroacetylide
Sodium chloroacetylide
Dichloroacetylene
Bromoacetylene
Chloroacetylene
Fluoroacetylene
Diiodoacetylene
Silver trifluoromethylacetylide
Chlorocyanoacetylene
Lithium trifluoromethylacetylide
3,3,3-Trifluoropropyne
1-Bromo-2-propyne
1-Chloro-2-propyne
1-Iodo-1,3-butadiyne
1,4-Dichloro-2-butyne
1-Iodo-3-penten-1-yne
1,6-Dichloro-2,4-hexadiyne
2,4-Hexadiynylene bischlorosulphite
Tetra (chloroethynyl) silane
2,4-Hexadiynylene bischloroformate
1-Iodo-3-Phenyl-2-propyne
1-Bromo-1,2-cyclotridecadien-4,8,10-triyne

(c) Metal acetylides

Disilver acetylide
Silver acetylide-silver nitrate
Digold(I) acetylide
Barium acetylide
Calcium acetylide (carbide)
Dicaesium acetylide
Copper(II) acetylide
Dicopper(I) acetylide
Silver acetylide
Caesium acetylide
Potassium acetylide
Lithium acetylide
Sodium acetylide
Rubidium acetylide
Lithium acetylide-ammonia
Dipotassium acetylide
Dilithium acetylide
Disodium acetylide
Dirubidium acetylide
Strontium acetylide
Silver trifluoromethylacetylide
Sodium methoxyacetylide
Sodium ethoxyacetylide
1,3-Pentadiyn-1-ysilver
1,3-Pentadiyn-1-ylcopper
Dimethyl-1-propynylthallium
Triethynylaluminium
Triethynylantimony
Bis(Dimethylthallium) acetylide
Tetraethynylgermanium
Tetraethynyltin
Sodium phenylacetylide
3-Buten-1-ynyldiethylaluminium
Dimethyl-phenylethynylthallium
3-Buten-1-ynyltriethyllead
3-Methyl-3-buten-1-ynyltriethyllead
3-Buten-1-ynyldiisobutylaluminium
Bis(Triethyltin) acetylene

(d) Metal azides

Aluminium triazide
Barium diazide
Boron triazide
Cadmium diazide
Calcium diazide
Chromyl azide
Copper(I) azide
Copper(II) azide
Lead(II) azide
Lead(IV) azide
Lithium azide
Lithium boroazide
Mercury(I) azide
Mercury(II) azide
Potassium azide
Silicon tetraazide
Silver azide

(e) Metal azide halides

Chromyl azide chloride
Molybdenum azide pentachloride
Molybdenum azide tetrachloride
Silver azide chloride
Tin azide trichloride
Titanium azide trichloride
Tungsten azide pentabromide
Uranium azide pentachloride
Vanadium azide dichloride
Vanadyl azide tetrachloride

(f) Diazo compounds

1,1 Benzoylphenyldiazomethane
2 Butan-1-yl diazoacetate
t-Butyl diazoacetate
t-Butyl-2-diazoacetoacetate
Diazoacetonitrile
2-Diazocyclohexanone
Diazocyclopentadiene
1-Diazoindine
Diazomethane (The precursor of this compound (N-Methyl-N-nitriso-toluene-4-sulphonamide) is available commercially)
Diazomethyllithium
Diazomethylsodium
Dicyanodiazomethane
Dinitrodiazomethane
Isodiazomethane
Methyl diazoacetate

(g) Metal fulminates

Cadmium fulminate

Table 7.9 Cont'd

Copper fulminate
Dimethylthallium fulminate
Diphenylthallium fulminate
Mercury(II) methylnitrolate
Mercury(II) formhydroxamate
Mercury(II) fulminate
Silver fulminate
Sodium fulminate
Thallium fulminate

(h) Nitro compounds
(i) *C-Nitro compounds*
4-Chloro-2,6,-dinitroaniline
2-Chloro-3,5-dinitropyridine
Chloronitromethane
1-Chloro-2,4,6-trinitrobenzene (picryl chloride)
Dinitroacetonitrile
2,4-Dinitroaniline
Dinitroazomethane
1,2-Dinitrobenzene
1,3-Dinitrobenzene
1,4-Dinitrobenzene
2,4-Dinitrobenzenesulphenyl chloride
3,5-Dinitrobenzoic acid
3,5-Dinitrobenzoyl chloride
2,6-Dinitrobenzyl bromide
1,1-Dinitro-3-butene
2,3-Dinitro-2-butene
3,5-Dinitrochlorobenzene
2,4-Dinitro-1-fluorobenzene
2,6-Dinitro-4-perchlorylphenol
2,5-Dinitrophenol
2,4-Dinitrophenylacetyl chloride
2,4-Dinotrophenylhydrazine
2,4-Dinitrophenylhydrazinium perchlorate
2,7-Dinitro-9-phenylphenanthridine
2,4-Dinitrotoluene
1-Fluoro-2,4-dinitrobenzene
4-Hydroxy-3,5-dinitrobenzene arsonic acid
1-Nitrobutane
2-Nitrobutane
1-Nitro-3-butene
Nitrocellulose
1-Nitro-3 (2,4-dinitrophenyl) urea
Nitroethane
2-Nitroethanol
Nitroglycerine
Nitromethane
1-Nitropropane
2-Nitropropane
5-Nitrotetrazole
Picric acid (2,4,6-trinitrophenol)
Potassium-4,6-dinitrobenzofuroxan hydroxide complex
Potassium-3,5-dinitro-2(1-tetrazenyl) phenolate
Potassium trinitromethanide ('Nitroform' salt)
Sodium 5-dinitromethyltetrazolide
Tetranitromethane
Trichloronitromethane (chloropicrin)
2,2,4-Trimethyldecahydroquinoline picrate
Trinitroacetonitrile
1,3,5-Trinitrobenzene
2,4,6-Trinitrobenzenesulphonic acid (picryl sulphonic acid)
Trinitrobenzoic acid
2,2,2-Trinitroethanol
Trinitromethane
2,4,6-Trinitrophenol (picric acid)
2,4,6-Trinitroresorcinol
2,4,6-Trinitrotoluene (TNT)
2,4,6-Trinitro-*m*-xylene

(ii) *N-Nitro compounds*
1-Amino-3-nitroguanidine
Azo-N-nitroformamidine
1,2-Bis(difluoroamino)N-nitroethylamine
N,N' Diacetyl-N,N'-dinitro-1,2-diaminoethane
N,N' Dinitro-1,2-diaminoethane
N,N' Dinitro-N-methyl-1,2-diaminoethane
1-methyl-3-nitro-1-nitrosoguanidine
Nitric amide (nitramide)
1-Nitro-3(2,4-dinitrophenyl) urea
Nitroguanidine
N-Nitromethylamine
Nitrourea
N,2,4,6-Tetranitro-N-methylaniline (tetryl)
1,3,5,7-Tetranitroperhydro-1,3,5,7-tetrazocine

(iii) *Ammonium nitrate*

(i) Reactive vinyl monomers
Acrylic acid
Acrylonitrile
n-Butyl acrylate
n-Butyl methacrylate
4-Chlorostyrene
Divinyl benzene
Dodecyl methacrylate
Ethyl acrylate
Ethylene dimethacrylate
2-Hydroxypropyl methacrylate
Methyl acrylate
Methyl methacrylate
α-Methyl styrene
Methyl vinyl ether
Styrene
Vinyl acetate
Vinyl bromide

Table 7.10 Explosive effects of small quantities of high explosive in a 6 m × 6 m room

Quantity of explosive	*Effect*
1 g	Serious injury to a person holding the explosive
10 g	Very serious injury to a person close to the explosive
	1% of persons at a distance of 1.5 m are also liable to ear-drum rupture
100 g	50% of windows in room likely to be blown out
	1% ear-drum rupture at 3.5 m
	50% ear-drum rupture at 1.5 m
	Almost certain death of persons in very close proximity (e.g. holding the explosive)
500 g	Complete structural collapse of brick-built building probable
	Probable survival of steel of concrete-framed building
	Almost certain death of persons very close to blast
	Persons close to blast seriously injured by lung and hearing damage, fragmentation effects, and from being thrown bodily
	Ear-drum rupture of almost all persons within the room

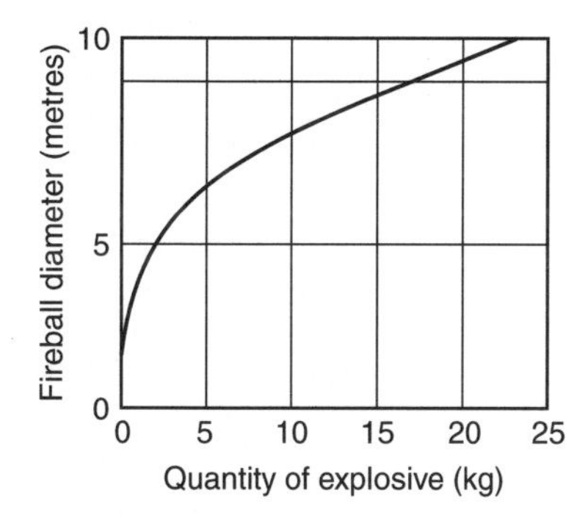

Figure 7.2 *Diameter of fireball versus quantity of explosive*

selection of the appropriate safety precautions. Propellant hazards are akin to those for pyrotechnics except that confinement can lead to detonation.

Precautions

Expert advice is required before handling any of these materials, many of which are governed by legislation regulating, e.g., licensing of premises, use, storage, transportation, import/export, sale, labelling, disposal. Some general considerations are given in Table 7.12. Table 7.13 summarizes basic precautions for work on a laboratory scale.

Disposal of explosive waste and the repair or dismantling of contaminated plant need extreme care. Table 7.14 provides guidance on techniques for disposal of the more commonly encountered explosives by experts at proper disposal sites. Collection of the waste should be in well labelled, distinctive, specially designed containers. Cleaning and decontamination of plant comprises removal of gross contamination under wet or solvent conditions using tools made of soft material, final cleaning with solvent or chemical reagent, and finally 'proving' of the equipment by heating to temperatures exceeding those for decomposition of the contaminant. Repair work should be the subject of a permit-to-work system (Chapter 13); it should be assumed that explosives may have penetrated threads, joints and other crevices and bolts, flanges etc. These should be thoroughly decontaminated prior to dismantling. Operatives should be suitably trained and protected.

Table 7.11 Classification of pyrotechnics

Composition	*Behaviour*	*Artefacts*	*Characteristics*
Group 1			
Chlorate and metal perchlorate report or whistling compositions Dry non-gelatinized cellulose nitrates Barium peroxide/zirconium compositions	Burn very violently	Flash shells (maroons) Casings containing flash compositions Sealed hail-preventing rockets	Mass explosion risk
Group 2			
Nitrate/metal/sulphur compositions Compositions with >65% chlorate Black powder Nitrate/boron compositions	Burn violently	Large firework shells Fuse unprotected signal flares Non-pressed report bullets (bird scarers) Report cartridges (unpacked) Black matches (uncovered)	Accelerating single-item explosions
Group 3			
Nitrate/metal compositions without sulphur Compositions with <35–65% chlorate Compositions with black powder Lead oxide/silicon with >60% lead oxides Perchlorate/metal compositions other than report	Burn fast	Large firework shells Fuse protected signal flares Pressed report cartridges in primary packagings Quickmatches in transport packagings Waterfalls; Silver wheels; Volcanoes Black powder delays	Burn very violently with single-item explosions
Group 4			
Coloured smoke compositions White smoke compositions (except those in Group 5) Compositions with <35% chlorate Thermite compositions Aluminium/phosphorus pesticide compositions	Low/medium speed burning	Large firework shells without flash compositions in transport packagings Signal ammunition without flash compositions, ≤40 g of composition Small fireworks, fuse protected (except volcanoes and silver wheels)	Single-item ignitions/explosions
Group 5			
Slow burning/heating compositions White smoke compositions based on hexachloroethane with zinc, zinc oxide and <5% aluminium or <10% calcium silicon	Burn slowly	Small fireworks in primary packagings Signal ammunition in transport packaging Delays without black powder Coloured smoke devices Sealed table bombs White smoke devices unpacked (see Group 5 composition)	Slow single-item ignitions/explosions

Table 7.12 General considerations for work with explosive chemicals

Consult with experts on the hazards and on technical, administrative and legal requirements.
The chance of accidental initiation is related to the energy imparted to the substance and the sensitivity of the compound. Hence the sensitivity of compounds should be established (e.g. Table 7.15) before devising appropriate control measures. (Many sensitive explosives have ignition energies of 1–45 mJ, while some very sensitive materials have ignition energies <1 mJ.)
Depending on scale, specially designed facilities may be required, remote from other buildings, accessways or populated areas. Remote handling procedures may be required (Figure 7.3), possibly with closed-circuit TV monitoring utilizing concrete outbuildings or bunkers.
Consider fire protection, detection and suppression requirements (Chapter 6), and means of escape, alarms, etc.
Minimize stocks and segregate from other chemicals and work areas. Where appropriate, keep samples dilute or damp and avoid formation of large crystals when practicable. Add stabilizers if possible, e.g. to vinyl monomers. Store in specially-designed, well-labelled containers in 'No Smoking' areas, preferably in several small containers rather than one large container. Where relevant, store in dark and under chilled conditions, except where this causes pure material to separate from stabilizer (e.g. acrylic acid).
Do not decant in store.
Consider need for high/low temperature alarms for refrigerated storage; these should be inspected and tested regularly.
Consider need for mitigatory measures (fire, blast, fragment-resistant barricades/screens), electrical and electrostatic safeguards, personal protection, disposal etc.
Stores and work areas should be designated 'No Smoking' areas and access controlled. Depending upon scale, explosion-proof electrics and static elimination may be required.
Deal with spillages immediately and make provisions for first aid.
All staff should be adequately trained, and written procedures provided.

Table 7.15 gives selected methods for testing explosives.

There is also a range of chemicals, sometimes termed 'blowing agents' (e.g. hydrazides) which decompose at low temperature producing large volumes of gas such as nitrogen and steam. Examples are listed in Table 7.16. Equipment for such products requires special design and a knowledge is required of activators, decomposition rates and temperatures.

General principles for storage

The general principles for storage of chemicals, which follow from previous summaries, are listed in Table 7.17. Those for compressed gases are given in Chapter 9.

Hazards arising in chemicals processing

Some factors contributing to chemical process hazards are summarized in Chapter 6: the roles of individual chemicals can be assessed from the preceding part of this chapter.

Chemical reaction hazards

Examples of hazardous reactions are given in Table 7.18. Table 7.19 gives basic precautions in monomer storage; Table 7.20 lists properties of common monomers.

Reaction characteristics

- Reaction in gas, liquid (neat or in solution suspension/emulsion) or solid phase.
- Catalytic or non-catalytic.

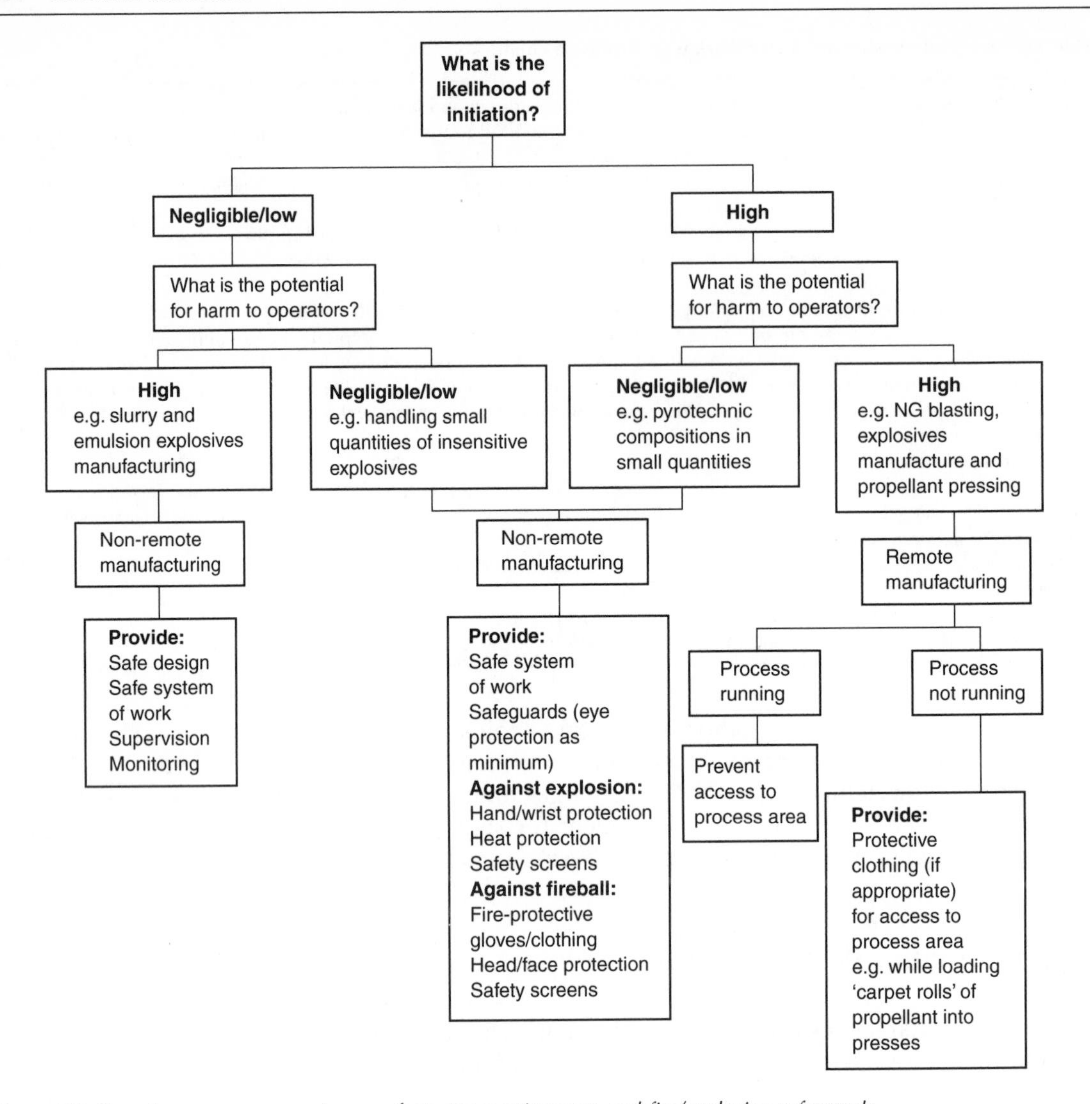

Figure 7.3 *Remote versus non-remote manufacturing requirements and fire/explosion safeguards*

- Exothermic, endothermic, or negligible heat loss/gain.
- Reversible or irreversible.
- First or second order or complex kinetics.

Reactors may be operated batchwise or continuously, e.g. in tubular, tubes in shell (with or without internal catalyst beds), continuous stirred tank or fluidized bed reactors. Continuous reactors generally offer the advantage of low materials inventory and reduced variation of operating parameters. Recycle of reactants, products or of diluent is often used with continuous reactors, possibly in conjunction with an external heat exchanger.

Adequate heat removal facilities are generally important when controlling the progress of exothermic chemical reactions. Common causes of thermal runaway in reactors or storage tanks are shown in Figure 7.4. A runaway reaction is most likely to occur if all the reactants are initially mixed together with any catalyst in a batch reactor where heat is supplied to start the reaction.

Table 7.13 Precautions for handling explosives in the laboratory

Storage

The quantities of potentially explosive materials in store and in use should be strictly limited.

Stores should be specially designed, constructed of non-combustible material, and located away from other hazards (e.g. brick 'coal bunkers' are suitable for small samples, but purpose-built constructions with explosion-proof lights etc. are required for larger quantities). They should be designated 'No Smoking' areas and be well labelled.

Stores should be used exclusively for these materials. Other combustible material such as fabric, paper, organic solvents should not be stored there.

Generally the substances in this class are unstable when heated or exposed to light; they should be stored cool and in the dark. However, for liquids with added stabilizer cooling may cause separation of the material from the stabilizer. Similarly, precipitation of a potentially explosive compound from a diluent may occur on cooling. In both cases this can represent a hazardous situation.

Stores should be ventilated and sound, e.g. no cracks in floors, no rusty window frames, no water seepages, etc.

Stores should be clean, tidy and locked. Contamination must be avoided and a high standard of housekeeping maintained.

Heat sources should not be permitted nearby.

Material should be purchased in several small containers rather than one large container and always stored in original containers. Integrity of the labels should be checked.

Use

Use must be restricted to experienced workers, aware of the hazards and the necessary precautions.

Records of usage should be kept and stock rotated. Old material should be disposed of.

Work should be on a scale of <0.5 g for novel but potentially explosive material until the hazards have been fully evaluated and <5 g for established, commercially available, substances such as peroxide free-radical initiators.

For the above scales, eye protection should be worn and work should be undertaken in a standard fume-cupboard behind a well-anchored polycarbonate screen. It is advisable to wear a protective apron and hand protection; whether leather gauntlets or tongs should be used will be dictated by circumstances. Such measures are recommended but it should be ensured that they do not precipitate a hazard as a result of loss of tactile sensitivity (e.g. dropping a flask, overtightening clamps, exerting excessive pressure when assembling apparatus). The material of gloves needs consideration. (PVC but *not* rubber is suitable for tert-butyl peroxide.)

For large-scale work, armour-plated fume cupboards are likely to be required.

Skin contact, inhalation and ingestion must be avoided. Splashes in eyes or on skin should be washed away immediately with copious quantities of water. Medical attention should be sought. If material is swallowed, medical aid is required immediately.

Glass apparatus should be pickled (e.g. in nitric acid) and thoroughly rinsed after use.

Sources of ignition such as hot surfaces, naked flames etc. must be avoided and smoking prohibited where explosives are used. Accidental application of mechanical energy should be avoided (e.g. material should not be trapped in ground-glass joints): seized stoppers, taps etc. must not be freed by the application of force. To minimize risk of static electricity, laboratory coats of natural fibre rather than synthetic fabrics are preferred. It is important to neutralize any spillage on the coat immediately, since delay could result in the impregnated garment becoming a fire hazard.

To prevent glass fragments from flying in the event of an explosion, use should be made of metal gauzes to screen reaction flasks etc., or cages, e.g. for desiccators. Vessels of awkward size/shape may be covered with cling film.

Whenever possible a stabilizer or diluent should be used and separation of the pure material should be avoided.

Any waste material (and contaminated cloths, tissues, clothing etc.) must be rendered safe by chemical means or by controlled incineration of dilute solution where practical prior to disposal.

In the event of fire, the area should be evacuated, the alarm raised and the fire brigade summoned. Only if it is clearly safe to do so should the fire be tackled with an *appropriate* extinguisher.

Chemical engineering operations

Many chemical engineering unit operations may be linked together in chemical processing. The commonest are:

Fluid mechanics

Pumping of liquids. Compression of gases
Mixing (solids, liquids, gases; possibly multiphase)
Atomization, dispersion

Table 7.14 Suggested methods of disposal for commonly encountered explosives

	Burn	*Detone*	*Dissolve*	*Chemical*	*Drowning*
NG-based blasting explosive	yes	yes	no	no	yes
ANFO	yes	yes	yes	no	yes
Pyrotechnic compositions	yes	no	yes	no	yes
Black powder	yes	no	yes	no	yes
Detonators	yes	yes	no	no	no
Detonating cord	yes	yes	no	no	no
Nitroglycerine	yes	no	no	yes	no
Slurry explosive	yes	yes	yes	no	yes
Contaminated paper waste	yes	no	no	no	no
Fireworks (finished)	yes	no	no	no	no
Initiatory explosives	no	yes	no	yes	no
Propellants	yes	no	no	no	no
Non-NG-based blasting explosives	yes	yes	no	no	yes
Shotgun cartridges	yes	no	no	no	no
Sporting/small arms ammunition	yes	no	no	no	no

Table 7.15 Methods for testing explosives

Flame test	A few crystals (or drop of solution) are heated in the non-luminous flame of a Bunsen burner. Melting with quiet burning is at one end of the spectrum; cracking, flashing-off or flaring are considered hazardous.
Impact	Impact sensitivity can be gauged by striking a few crystals of the compound on a metal last with the ball of a ball-pein hammer. Ignition, smoking, cracking or other sign of decomposition are considered hazardous.
Differential scanning calorimetry **Differential thermal analysis** **Hot stage microscope**	If the decomposition reaction follows the general rate law, the activation energy, heat of decomposition, rate constant and half-life for any given temperature can be obtained on a few milligrams using the ASTM method. Hazard indicators include heats of decomposition in excess of 0.3 kcal/g, short half-lives, low activation energies and low exotherm onset temperatures, especially if heat of decomposition is considerable. The DTA or hot-stage microscope can be used under ignition conditions to obtain an ignition temperature. The nature of the decomposition can also be observed at a range of temperatures. Observations such as decomposition with evolution of gases prior to ignition are regarded as potentially hazardous.
Bomb calorimetry	Use of oxygen and an inert gas enables the heat of combustion and the heat of decomposition to be evaluated respectively.
Deflagration	A melting point test has been described for diazo compounds. The first 1 mm of a melting-point tube filled with c. 10 mg of test compound is inserted in a melting-point apparatus heated at 270°C. Once decomposition starts, the tube is removed. The decomposition rapidly propagates through the entire mass for unstable diazo compounds; no such propagation is reported for stable versions.

Heat transfer

Convective heat exchange, natural or forced
Radiant heat transfer, e.g. furnaces
Evaporation, e.g. in evaporators
Condensation, e.g. in shell and tube heat exchanges
Heat transfer to boiling liquids, e.g. in vaporizers, boilers, re-boilers

Table 7.16 Substances with a high rate of decomposition

These substances decompose rapidly to produce large volumes of gas. They are substances not classified as deflagrating or detonating explosives but exhibit violent decomposition when subject to heat.

Material	*Trauzel lead block value* (cm^3/g)	*Combustion properties*
1:8, Bis (dinitrophenoxy)4,5-dinitro anthraquinone	18.5	Combustion propagates fully and fast with flame
100% Dinitrosopentamethylene tetramine	18.5	
2,4-Dinitroaniline	17.5	
2-Amino-3,5-dinitrothiophene	13–17.5	
1:5 Bis (dinitrophenoxyl) 4:8-dinitro anthraquinone	10.5	
2-formylamino-3,5-dinitrothiophene	8	
2-acetyl-amino-3,5-dinitrothiophene	7	
2-anisidine nitrate	6	
80% DNPT	2.5	
6-nitro-1-diazo-2-naphthol-sulphonic acid	5	Combustion propagates fully and fast by smouldering
Ammonium nitrate	23	Local decomposition – no propagation of the decomposition
2-bromo-4,6-dinitroaniline	17.5	
6-bromo-2,4-dinitroaniline	17	
2-chloro-4,6-dinitroaniline	16.5	

This substance exhibits a high rate of decomposition without combustion when exposed to heat and certain initiators.

Material	*Decomposition temperature*	*Property*
P,P′-oxybis (benzenesulphonyl hydrazide)	150°C	Decomposes and propagates

Mass transfer operations (in which a material is transferred across a phase boundary or interface)

Distillation, either batchwise or continuous
Liquid–liquid extraction (solvent extraction)
Solid–liquid extraction (leaching)
Gas absorption, scrubbing; desorption, stripping
Humidification and water cooling
Dehumidification and air conditioning
Drying of solids, solutions/slurries
Adsorption (in which a gas or liquid is taken up on a solid, e.g. activated charcoal, molecular sieves)
Crystallization
Less common means of separation, e.g. dialysis, ion-exchange, osmosis, chromatography, electrophoresis.

Non-mass transfer operations

Filtration of suspensions, e.g. in filter presses, rotary vacuum filters
Sedimentation, gravity settling
Centrifugation, for immiscible liquid or solid separation and recovery
Classification by sieving, screening
Size reduction, e.g. milling, crushing, grinding

Table 7.17 General principles for chemical storage

- Store minimum quantities
- Control stock, i.e. first-in/first-out, move redundant stock
- Segregate chemicals, e.g. from water, air, incompatible chemicals, sources of heat, ignition sources
- Segregate 'empties', e.g. cylinders, sacks, drums, bottles
- Monitor stock, e.g. temperature, pressure, reaction, inhibitor content, degradation of substance, deterioration of packaging or containers/corrosion, leakages, condition of label, expiry date, undesirable by-products (e.g. peroxides in ethers)
- Spillage control; bund, spray, blanket, containment. Drain to collection pit
- Decontamination and first-aid provisions, e.g. neutralize/destroy, fire-fighting
- Contain/vent pressure generated to a safe area
- Store in 'safest' form, e.g. as pre-polymers, as chemical for generation of requirements (e.g. hypochlorites for chlorine) in dilute form
- Handle solids as prills or pellets rather than powders to minimize the possibility of dust formation
- Split-up stocks into manageable lots, e.g. with reference to fire loading/spillage control. Limit stack heights; generally chemicals should be stored off the ground (e.g. to facilitate cleaning, to keep above any ingress of water in the event of flooding)
- Select correct materials of construction; allow for reduction in resistance due to dilution/concentration, presence of impurities, catalytic effects
- Transport infrequently to minimize stocks for both safety and to reduce costs and environmental hazards arising from the need to dispose of surplus or expired material
- Ensure appropriate levels of security, hazard warning notices, fences, patrols. Control access including vehicles
- Segregate/seal drains
- Appropriate gas/vapour/fume/pressure venting, e.g. flame arrestors, scrubbers, absorbers, stacks
- Ensure adequate natural or forced general ventilation of the storage area
- Provide adequate, safe lighting
- Label (name and number); identify loading/unloading/transfer couplings
- Facilitate sampling (for quality assurance and stock monitoring)
- Provide appropriate fire protection (sprinkler, dry powder, gas)
- Consider spacings from buildings, road, fence
- Ensure adequate access for both normal and emergency purposes with alternative routes
- Protect from vehicle impact, e.g. by bollards
- Assign responsibility for administration, maintenance, cleaning and general housekeeping

Gas cleaning by filtration, demisting, electrostatic precipitation, wet collection of particulates, cyclonic separation.

Dependent upon the chemicals in-process, each of these may introduce a range of hazards, e.g. chemical, flammable or mechanical. These must be checked in every case. Safety features which may be required are summarized in Table 7.21.

Common reaction rate v. temperature characteristics for reactions are illustrated in Figure 7.5.

To avoid runaway conditions (Fig. 7.5a) or an explosion (Figure 7.5c), control may involve:

- Use of dilute solutions, emulsions, or suspensions.
- Feeding one reactant in at a controlled rate depending upon reactant's temperature.
- Refluxing of solvent from a condenser.
- Imposing a limit on reactor size, to ensure adequate heat transfer area per unit volume.
- Careful selection of reactant and coolant temperature.
- Provision of efficient agitation.

Mitigation of a runaway reaction may involve:

- Emergency cooling.
- Dumping of reactants into an empty vessel or one containing a compatible quenching liquid.

Table 7.18 Characteristics of some different types of reaction

Oxidation	Feedstocks generally hydrocarbons Hazard of fire/explosion arises from contact of flammable material with oxygen Reactions highly exothermic: equilibrium favours complete reaction
Polymerization	Exothermic reaction which, unless carefully controlled, can run-away and create a thermal explosion or vessel overpressurization Refer to Table 7.20 for common monomers Certain processes require polymerization of feedstock at high pressure, with associated hazards Many vinyl monomers (e.g. vinyl chloride, acrylonitrile) pose a chronic toxicity hazard Refer to Table 7.19 for basic precautions
Halogenation	The commercially important halogens are chlorine, bromine, fluorine, iodine. Refer to Table 5.19 for properties All are highly toxic Reactions are highly exothermic and chain reactions can occur, which may result in detonation
Hydroprocesses	Hydrogen is chemically stable and relatively unreactive at ordinary temperatures; most processes utilizing it require a catalyst. Above 500°C it reacts readily with oxygen and confined flammable mixtures explode violently if ignited Main hazards: fire, explosion, metallurgical problems arising from hydrogen attack
Nitration	Hazards arise from the strong oxidizing nature of the nitrating agents used (e.g. mixture of nitric and sulphuric acids) and from the explosive characteristics of some end products Reactions and side reactions involving oxidation are highly exothermic and may occur rapidly Sensitive temperature control is essential to avoid run-away
Alkylation	Hazards arise from the alkylating agents, e.g. dimethyl sulphate (suspected human carcinogen), hydrogen fluoride (highly toxic irritant gas) Thermal alkylation processes require higher temperatures and pressures, with associated problems
High pressure reactions	High inventories of stored pressure (e.g. in pressurized reactors or associated plant) can result in catastrophic failure of the pressure shell

Table 7.19 Basic precautions in monomer storage

Indoors
Cool, well-ventilated area
Non-combustible construction
Segregated from other flammables/reactants

Outdoors
Well-spaced tanks, possibly with water cooling, refrigeration, buried
Some monomers (e.g. acrylic acid) may require provisions to avoid freezing
Provision for inhibitor/stabilizer addition
Provision for atmosphere inerting may be required

- Venting to a knock-out vessel, to remove non-gaseous substances. (This may be followed by a scrubbing unit for gases or a flare-stack.)

The kinetics and thermodynamics of the reaction, and of possible side reactions, need to be understood. The explosive potential of chemicals liable to exothermic reaction should be carefully appraised.

A thorough assessment should be made before undertaking:

Table 7.20 Properties of common monomers

	Flash point (°C)	*Ignition temp.* (°C)	*Flammable limits* (% by vol. in air)	*Specific gravity* (Water = 1.0)	*Vapour density* (Air = 1.0)	*Boiling point* (°C)	*Properties*
Acetaldehyde (Acetic aldehyde, ethanal) CH_3CHO	–38	185	4.0–55.0	0.8	1.5	21	Colourless fuming liquid Pungent odour Irritant Water soluble Can polymerize exothermically, form explosive peroxides, or react violently with other chemicals
Acrolein (Allyl aldehyde, propenal) CH_2:CHCHO	–26	278	2.8–31.0	0.8	1.9	53	Colourless/yellow liquid Pungent unpleasant odour Water soluble Irritant Can polymerize exothermically with strong alkalis, heat or light Can form peroxides
Acrylic acid (Propenoic acid, propene acid) CH_2:CHCOOH	54	–	–	1.1	2.5	140	Colourless, water soluble liquid Freezing point 14°C Polymerizes readily with oxygen Must be inhibited
Acrylonitrile (Vinyl cyanide, propenenitrile) CH_2:CHCN	0	481	3.0–17.0	0.8	1.8	77	Colourless, partially water soluble liquid Experimental carcinogen Polymerizes violently with organic peroxides or concentrated caustic alkalis Highly toxic Usually inhibited
1,3-Butadiene (Butadiene, vinylethylene) CH_2:CHCH:CH_2	–76	450	2.0–11.5	0.6	1.9	–4	Colourless, odourless liquefiable gas Polymerizes readily, particularly if O_2 or traces of catalyst present Can form explosive peroxides Normally contains inhibitor (liquid phase) and antioxidant
Epichlorhydrin (Chloropropylene oxide) CH_2:$OCHCH_2Cl$	32	–	–	1.2	3.3	115	Colourless, partly water soluble liquid Highly toxic Polymerizes exothermically with acids, bases, certain salts and catalysts Can react with water
Ethyl acrylate CH_2:$CHCOOC_2H_5$	15	–	1.8––	1.2	–	100	Colourless liquid Acrid odour Polymerizes readily, accelerated by heat, light, organic peroxides Irritant

Table 7.20 Cont'd

	Flash point (°C)	*Ignition temp.* (°C)	*Flammable limits* (% by vol. in air)	*Specific gravity* (Water = 1.0)	*Vapour density* (Air = 1.0)	*Boiling point* (°C)	*Properties*
Ethylene oxide $CH_2{:}CH_2$ (with O bridging the two carbons)	<–18	429	3.0–100	0.9	1.5	11	Colourless gas at room temperature Irritant to eyes and respiratory tract, and an experimental carcinogen Polymerizes uncontrollably with immense explosive force on contact with certain chemicals (e.g. ammonia)
Formaldehyde (Oxymethylene) HCHO	gas	430	7.0–73.0	–	1.1	–21	Colourless Water soluble gas producing formalin solutions Suffocating odour Polymerizes readily Highly toxic Respiratory sensitizer
Methacrylic acid $CH_2{:}C(CH_3)COOH$	77	–	–	–	–	158	Colourless, water soluble liquid Polymerizes readily unless inhibited or stored <15°C Irritant
Methyl acrylate $CH_2{:}CHCOOCH_3$	–3	–	2.8–25.0	1.0	3.0	80	Colourless liquid Acrid odour Extremely irritating to respiratory system, skin and mucous membranes
Methyl methacrylate $CH_2{:}C(CH_3)COOCH_3$	29	–	2.1–12.5	0.9	3.4	101	Colourless liquid Acrid odour
Styrene (Vinyl benzene) $C_6H_5CH{:}CH_2$	32	490	1.1–6.1	0.9	3.6	145	Colourless/oily yellow liquid Penetrating odour Polymerizes slowly in air or light, accelerated by heat or catalysts Ignition/explosion possible Usually inhibited Store <21°C
Vinyl acetate $CH_3COOCH{:}CH_2$	–8	427	2.6–13.4	1.1	3.0	72	Colourless, partially water soluble liquid Faint odour Polymerizes with heat or organic peroxides
Vinyl chloride (Chloroethene) $CH_2{:}CHCl$	–78	472	4.0–22.0	1.0	2.1	–14	Colourless, sweet smelling liquefiable gas Polymerizes with light, heat, air or catalysts Normally inhibited Human carcinogen
Vinylidene chloride (Dichloroethylene-1,1) $CH_2{:}CCl_2$	–10	458	5.6–11.4	1.3	3.3	37	Colourless volatile liquid Polymerizes unless inhibited Decomposes at 457°C

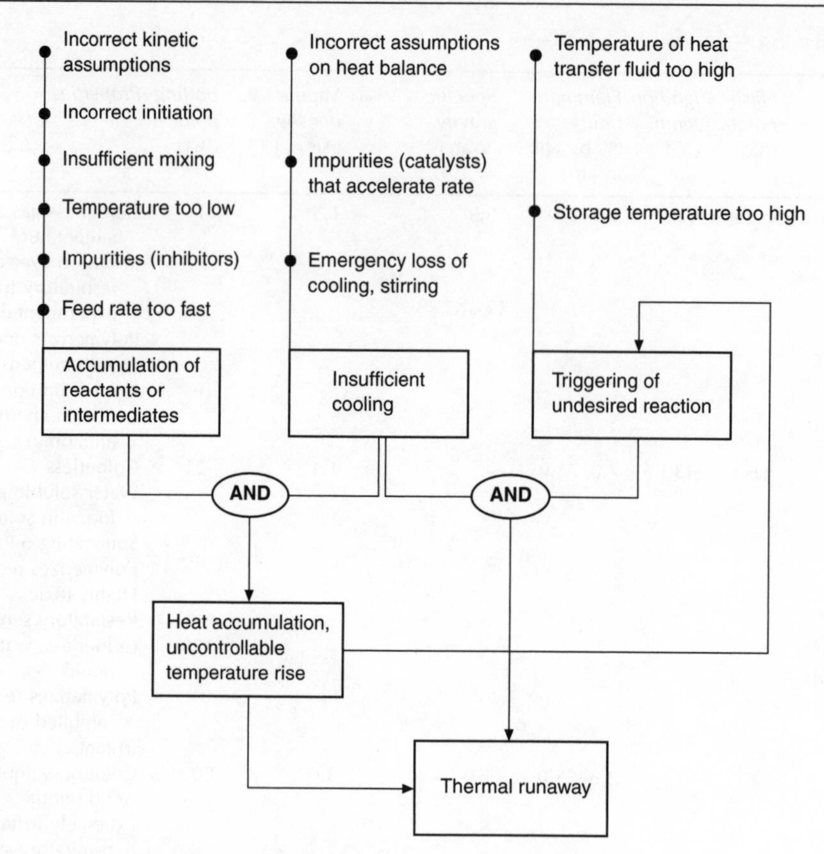

Figure 7.4 *Common causes of thermal runaway in reactors or storage tanks*

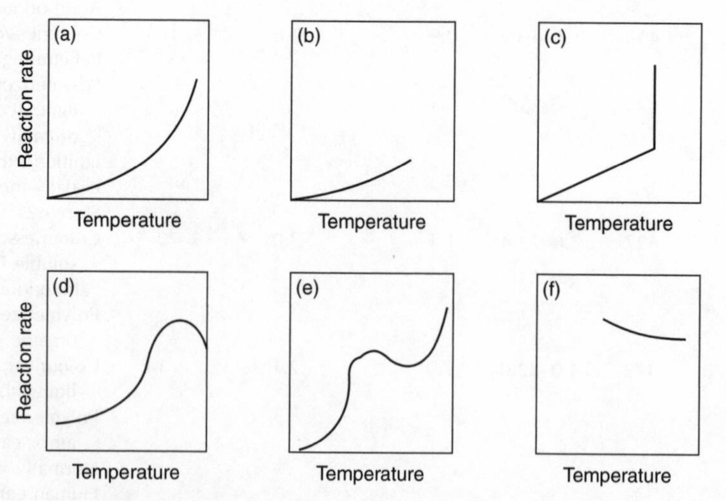

Figure 7.5 *Types of reaction rate/temperature curve*

a Rapid increase with temperature – normal characteristic
b Slow increase in rate with temperature – characteristic of some heterogeneous reactions
c Very rapid increase at one point – the ignition point in an explosion
d Decrease in rate at higher temperature – characteristic of catalytic reactions
e Decrease in rate at intermediate temperatures, followed by an increase
f Slow decrease in rate with temperature

Table 7.21 Safety features in chemical engineering operations

Inventory	Reduce inventory of chemicals: Continuous operation may be preferable to batch Low residence time contacting equipment may be better than cheaper alternatives etc.
Monitoring	Monitor temperature, pressure flow, composition, freedom from contamination and other appropriate properties of all streams where relevant. Consider automatic control
Isolation	Provide for isolation from upstream and downstream operations. Consider provision of automatic and/or remotely operated isolation. Consider isolation for cleaning needs
Contaminants	Provide measures to remove unacceptable contaminants from feed materials, process streams and services, e.g. entrained liquid, tramp metal, unwanted particulate solids
Pressure/temperature	Operate at moderate temperature and pressure where possible. Avoid superheated liquids, which will flash-off, if practicable Allow for effects of over-/under-temperature, over-/under-pressure. Following assessment (e.g. by HAZOP)
Continuous flow	With continuous flow operations consider (e.g. using a HAZOP procedure) the effects of: No flow Reduced flow Reverse flow Increased flow Contaminated flow Flow of a substituted material, etc.
Start-up/shutdown	Provide for safe start-up, including purging if necessary Provide for safe shutdown: Normal By a trip On standby In various emergency situations, etc.
Instrumentation	Provide safety instrumentation in addition to process instrumentation Consider high–high and low–low alarms. Consider automatic activation of emergency responses, e.g. venting, emergency cooling, recycling, discharge of liquid streams, shutdown High/low temperature pressure flow level etc. linked to trips for automatic operation where appropriate
Protective features	Provide protective as well as control features, e.g. pressure and vacuum relief, explosion suppression relief, advance inerting, containment
Waste streams	Cater for routine and emergency, safe discharge of all waste streams, e.g. atmospheric venting, possibly after treatment, discharge of liquid effluents including out-of-specification streams, discharges of particulate or bulk solids
Common-mode failure	Avoid common-mode failure possibilities with services, control systems, safety systems etc.

- Scale-up (since it may result in a decrease in heat transfer capacity per unit mass of reactant).
- Modifications in reactor geometry, agitation and control (e.g. instrumentation, cooling, venting).
- Changes in reaction materials, (e.g. source, purity, concentration), diluents, catalysts charging procedures.
- Changes in operating conditions.

The characteristics of some potentially hazardous reactions are summarized in Tables 7.18 and 7.22.

Many processes require equipment designed to rigid specifications together with automatic control and safety devices. Consideration should be given to the control, and limitation of the effects, of equipment malfunction or maloperation including:

Table 7.22 Hazard rating of chemical reactions

Reaction	*Degree of hazard*
Reduction	
Clemmensen	D
Sodium-amalgam	D
Zinc–acetic acid	E
Zinc–hydrochloric acid	E
Zinc–sodium hydroxide	E
Ferrous ammonium sulphate	E
Lead tetraacetate	E
Meerwein–Pondorff	D
Lithium aluminium hydride	B
Dialkyl aluminium hydride	B
Rosenmund	A
Catalytic high pressure	A
Catalytic low pressure	B
Oxidation	
Hydrogen peroxide – dilute aqueous	E
Air or I_2 (mercaptan to disulphide)	D
Oppenauer	D
Selenium dioxide	D
Aqueous solution nitric acid, permanganate, manganic dioxide, chromic acid, dichromate	E
Electrolytic	B
Chromyl chloride	C
Ozonolysis	A
Nitrous acids	A
Peracids – low molecular weight or two or more positive groups	A
Peracids – high molecular weight	B
t-Butyl hypochlorite	C
Chlorine	C
Alkylation	
Carbon–carbon	
Alkali metal	C
Alkali metal alcoholate	D
Alkali metal amides and hydrides	C
Reformatsky	E
Michael	E
Grignard	B
Organometallics, such as dialkyl zinc or cadmium–alkyl or aryl lithium	B
Alkali acetylides	A
Diels–Alder	D
Arndt–Eistert	A
Diazoalkane and aldehyde	A
Aldehydes or ketones and hydrogen cyanide	C
Carbon–oxygen	
Williamson	D
Formaldehyde – hydrochloric acid	E
Ethylene oxide	C
Dialkyl sulphate	D
Diazoalkane	A
Carbon–nitrogen	
Cyanomethylation	C
Chloromethylation	D
Ethylenimine	C
Ethylene oxide	C
Quaternization	D
Condensation	
Erlenmeyer	D
Perkin	D
Acetoacetic ester	D
Aldol	D
Claisen	D
Knoevenagel	D
Condensations using catalysts such as phosphoric acid; $AlCl_3$; $KHSO_4$; $SnCl_4$; H_2SO_4; $ZnCl_2$ $NaHSO_2$; $POCl_2$; HCl; $FeCl_2$	E
Acyloin	C
Diketones with hydrogen sulphide	C
Diketones with diamines→quinazolines	D
Diketones with NH_2OH→isoxazolines	D
Diketones with NH_2NH_2→pyrazoles	C
Diketones with semicarbazide → pyrazoles	D
Diketones with ammonia→pyrazoles	D
Carbon disuphide with aminoacetamide→thiazolone	A
Nitriles and ethylene diamines → imidazolines	D
Amination	
Liquid ammonia	B
Aqueous ammonia	E
Alkali amides	C
Esterification	
Inorganic	E
Alkoxy magnesium halides	B
Organic:	
Alcohol and acids or acid chloride or acid anhydride	D
Alkyl halide and silver salts of acids	E
Alkyl sulphate and alkali metal salt of acid	D
Alkyl chlorosulphates and	

Table 7.22 Hazard rating of chemical reactions

Reaction	*Degree of hazard*
alkali salts of carboxylic acid	D
Ester-exchange	D
Carboxylic acid and diazomethane	A
Acetylene and carboxylic acid-vinyl ester	A
Hydrolysis, aqueous nitriles, esters	E
Simple metathetical replacement	D
Preparation and reaction of peroxides and peracids	
Concentrated	A
Dilute	D
Pyrolysis	
Atmospheric pressure	D
Pressure	B
Schmidt	B
Mannich	D
Halogenation	
SO_2X_2, SOX_2, SX, POX_2, PX_5	D
HX	D
Cl_2, Br_2	C
Nitration	
Dilute	D
Concentrated	B

Hazardous
A Highly flammable
Develops high pressure instantly
Highly toxic

Special
B Flammable, perhaps explosive, mixtures form
C Flammable, or generates toxic ubstances

Conventional
D Slightly flammable
Generates or uses mildly toxic substances
E Non-flammable
Does not use or generate toxic substances

Named reactions

Aldol reaction is the condensation of an aldehyde to produce longer-chain hydroxy aldehydes

$$2CH_3CHO \xrightarrow{NaOH} CH_3CHOH \cdot CH_2CHO$$

Arndt-Eistert reaction is used to convert an acid compound into the next higher homologue by reaction with diazomethane

$$RCOCl + 2CH_2N_2 \longrightarrow R \cdot COCH_2CO_2H + 2N_2 + CH_3Cl$$

Claisen reaction is the condensation of benzaldehyde with aliphatic aldehydes and ketones containing α-hydrogen

$$C_6H_5CHO + CH_3CHO \xrightarrow{NaOH} C_6H_5CH{=}CH \cdot CHO + H_2O$$

$$C_6H_5CHO + CH_3COCH_3 \longrightarrow C_6H_5CH{=}CH \cdot CO \cdot CH_3 + H_2O$$

Clemmensen reaction is the reduction of carbonyl compounds with amalgamated zinc and concentrated hydrochloric acid

$$R \cdot CO \cdot R \xrightarrow{e;H^+} RCH_2R$$

Diels-Alder is the preparation of cyclic olefins from dienes and a dienophile

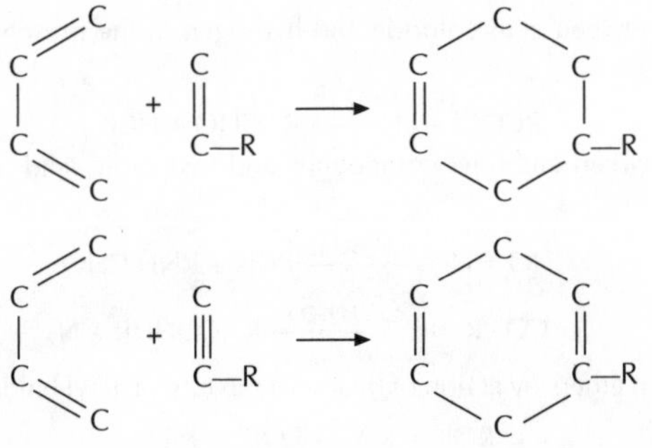

Erlenmeyer reaction is the condensation of aromatic aldehydes with hippuric acid to form azlactones (important intermediates in the preparation of amino- and keto-acids)

$$C_6H_5CHO + \underset{\displaystyle |\atop NHCOC_6H_5}{CH_2CO_2H} \longrightarrow C_6H_5CH{=}C{-}C{=}O \;\; (\text{ring: } C{-}N{=}C(C_6H_5){-}O{-}C{=}O)$$

Grignard reaction is the use of alkyl magnesium halides to form a host of products by reaction with a variety of chemicals

$$RMgX + H_2O \longrightarrow RH + Mg(OH) \cdot X$$

$$RMgX + R'OH \longrightarrow RH + Mg(OR') \cdot X$$

$$RMgX + NH_3 \longrightarrow RH + Mg(NH_2) \cdot X$$

$$RMgX + RN{=}C \longrightarrow R{-}N(MgX){=}CR + R{-}N{=}C(R)MgX$$

Knoevenagel reaction is the synthesis of α, β-unsaturated acids by reaction of aldehydes and compounds with active methylene groups in the presence of an organic base

$$RCHO + CH_2(CO_2C_2H_5) \xrightarrow{\text{base}} R \cdot CH{=}C(CO_2C_2H_5)_2 + H_2O$$

$$\downarrow KOH$$

$$R \cdot CH{=}CH \cdot CO_2H \xleftarrow{150\text{–}200°} R \cdot CH{=}C(CO_2H)_2$$

Mannich reaction is the condensation between formaldehyde, ammonia, or a primary or secondary amine (preferably as the hydrochloride), and a compound containing at least one active hydrogen atom

$$R_2CH \cdot NO + HCHO + NH_4Cl \longrightarrow R_2\overset{NO_2}{\overset{|}{C}} \cdot CH_2NH_2HCl + H_2O$$

Meerwein-Pondorff reduction is the synthesis of alcohols by heating carbonyl compounds with aluminium isopropoxide in isopropanol and distilling off acetone by-product

$$R_2C{=}O + (CH_3)_2CHOAl_3 \longrightarrow R_2CHO\ Al_3 + CH_3CO\ CH_3 \longrightarrow R_2CHOH$$

Michael condensation is the addition of a compound with an active methylene group to an α, β-unsaturated keto-compound

$$CH_2(CO_2C_2H_5)_2 + (CH_3)_2C{=}CH\ CO\ CH_3 \xrightarrow{C_2H_5O^-} (CH_3)_2\underset{\displaystyle |\atop CH(CO_2C_2H_5)}{C}CH_2COCH_3 + C_2H_5O^-$$

Oppenauer reaction is oxidation of secondary alcohols to ketones using aluminium t-butoxide

$$RCH(OH)R' + CH_3COCH_3 \xrightarrow{\text{catalyst}} RCOR' + CH_3CH(OH)CH_3$$

Perkin condensation is the reaction between aromatic aldehydes and aliphatic acid anhydrides (in the presence of the sodium salt) to form β-arylacrylic acid

$$C_6H_5CHO + (CH_3CO)_2O \xrightarrow{CH_3CO_2Na} C_6H_5CH{=}CH\ CO_2H$$

Reformatsky reaction is the formation of β-hydroxyesters by reaction of α-bromoacid ester and a carbonyl compound, usually in the presence of zinc

$$R'_2CO + RCH(Br)CO_2C_2H_5 \xrightarrow{Zn} R'C(OH)CHR\ CO_2C_2H_5$$

Rosenmund reaction is the action between acid chloride and hydrogen in the presence of palladium catalyst to produce aldehydes

$$RCOCl + H_2 \xrightarrow{Pd} R \cdot CHO + HCl$$

Schmidt reaction is the reaction between carbonyl compounds and hydrazoic acid in the presence of e.g. concentrated sulphuric acid

$$RCHO + HN_3 \xrightarrow{H_2SO_4} RCN + RNH\ CHO$$

$$R \cdot CO \cdot R{=}HN_3 \xrightarrow{H_2SO_4} R \cdot CONHR + N_2$$

Williamson reaction is the synthesis of ethers by action of heat on a mixture of alkyl haldie and sodium or potassium alkoxide

$$ROK + R'X \rightarrow ROR' + KX$$

- Stirrer failure, mechanical or electrical.
- Attainment of abnormal reaction conditions, e.g. overpressure, over-temperature, segregation of reactants, excessive reaction rate, initiation of side reactions.
- Power failure, affecting agitator, pumps, instruments.
- Error in valve, switch or associated equipment operation.
- Failure to actuate agitation at the proper time.
- Instrument failure, pressure, flow, temperature, level or a reaction parameter, e.g. concentration.
- Failure of instrument air or electricity.
- Loss of inert gas blanket.
- Failure of relief devices, e.g. pressure relief valves or rupture discs.
- Failure of coolant, refrigerant, or other utilities.
- Failure of high or low pressure alarms or cut-outs.
- Addition of wrong material or wrong quantities.
- Addition of materials in incorrect sequence.
- Failure to add material, e.g. short-stop or inhibitor, at correct stage.
- Spillage of material.
- Improper venting to atmosphere, i.e. other than via vents with flame arresters or scrubbers, or via a knockout drum, or to the correct flare systems.
- Restricted or blocked vent.
- Restricted material flows in or out.
- Leakage of *materials out*, e.g. due to a gasket failure, or *air in*.

8

Cryogens

Cryogenics, or low-temperature technology, is the science of producing and maintaining very low temperatures usually below 120 K, as distinct from traditional refrigeration which covers the temperature range 120 to 273.1 K. At or below 120 K, the permanent gases including argon, helium, hydrogen, methane, oxygen and nitrogen can be liquefied at ambient pressure as exemplified by Table 8.1. Any object may be cooled to low temperatures by placing it in thermal contact with a suitable liquefied gas held at constant pressure. Applications can be found in food processing, rocket propulsion, microbiology, electronics, medicine, metal working and general laboratory operations. Cryogenic technology has also been used to produce low-cost, high-purity gases through fractional condensation and distillation. Cryogens are used to enhance the speed of computers and in magnetic resonance imaging to cool high conductivity magnets for non-intrusive body diagnostics. Low-temperature infrared detectors are used in astronomical telescopes.

Table 8.1 Properties of common cryogens

Gas	*Boiling point (°C)*	*Volume of gas produced on evaporation of 1 litre of liquid (litres)*
Helium	– 269	757
Hydrogen	– 253	851
Neon	– 246	1438
Nitrogen	– 196	696
Fluorine	– 187	888
Argon	– 186	847
Oxygen	– 183	860
Methane	– 161	578
Krypton	– 151	700
Xenon	– 109	573
Chlorotrifluoromethane	– 81	–
Carbon dioxide	– 78.5	553

Every gas has a critical temperature above which it cannot be liquefied by application of pressure alone (Chapter 4). As a result, gases used, e.g., as an inert medium to reduce oxygen content of atmospheres containing flammable gas or vapour (Chapter 6) are often shipped and stored as cryogenic liquid for convenience and economy.

In the laboratory, a range of 'slush baths' may be used for speciality work. These are prepared by cooling organic liquids to their melting points by the addition of liquid nitrogen. Common examples are given in Table 8.2. Unless strict handling precautions are instituted, it is advisable to replace the more toxic and flammable solvents by safer alternatives.

Table 8.2 Working temperatures of cryogenic slush baths

Bath liquid	*Temperature (°C)*
Carbon tetrachloride	− 23
Chlorobenzene	− 45
Solid carbon dioxide	− 63
in acetone or methylated spirits[1]	− 78
Toluene	− 95
Carbon disulphide	− 112
Diethyl ether	− 120
Petroleum ether	− 140

[1]Liquid nitrogen is omitted from this mixture and the solvent is used to improve the heat transfer characteristics of cardice.

Typical insulating materials include purged rockwool or perlite, rigid foam such as foam-glass or urethane, or vacuum. However, because perfect insulation is not possible heat leakage occurs and the liquefied gas eventually boils away. Uncontrolled release of a cryogen from storage or during handling must be carefully considered at the design stage. The main hazards with cryogens stem from:

- The low temperature which, if the materials come into contact with the body, can cause severe tissue burns. Flesh may stick fast to cold uninsulated pipes or vessels and tear on attempting to withdraw it. The low temperatures may also cause failure of service materials due to embrittlement; metals can become sensitive to fracture by shock.
- Asphyxiation (except with oxygen) if the cryogen evaporates in a confined space.
- The very large vapour-to-liquid ratios (Table 8.1) so that a large cloud, with fog, results from loss of liquid.
- Catastrophic failure of containers as cryogen evaporates to cause pressure build-up within the vessel beyond its safe working pressure (e.g. pressures ≤280 000 kPa or 40 600 psi can develop when liquid nitrogen is heated to ambient temperature in a confined space).
- Flammability (e.g. hydrogen, acetylene, methane), toxicity (e.g. carbon dioxide, fluorine), or chemical reactivity (fluorine, oxygen).
- Trace impurities in the feed streams can lead to combination of an oxidant with a flammable material (e.g. acetylene in liquid oxygen, solid oxygen in liquid hydrogen) and precautions must be taken to eliminate them.
- Several materials react with pure oxygen so care in selection of materials in contact with oxygen including cleaning agents is crucial.

Key precautions are given in Table 8.3.

The cryogens encountered in greatest volume include oxygen, nitrogen, argon and carbon dioxide. Their physical properties are summarized in Table 8.4.

Liquid oxygen

Liquid oxygen is pale blue, slightly heavier than water, magnetic, non-flammable and does not produce toxic or irritating vapours. On contact with reducing agents, liquid oxygen can cause explosions.

Table 8.3 General precautions with cryogenic materials

Obtain authoritative advice from the supplier.

Select storage/service materials and joints with care, allowing for the reduction in ductility at cryogenic temperatures.
Provide special relief devices as appropriate.

Materials of construction must be scrupulously clean, free of grease etc.
Use only labelled, insulated containers designed for cryogens, i.e. capable of withstanding rapid changes and extreme differences in temperature, and fill them slowly to minimize thermal shock.
Keep capped when not in use and check venting.
Glass Dewar flasks for small-scale storage should be in metal containers, and any exposed glass taped to prevent glass fragments flying in the event of fracture/implosion.
Large-scale storage containers are usually of metal and equipped with pressure-relief systems.

In the event of faults developing (as indicated by high boil-off rates or external frost), cease using the equipment.

Provide a high level of general ventilation taking note of density and volume of gas likely to develop: initially gases will slump, while those less dense than air (e.g. hydrogen, helium) will eventually rise.
Do not dispose of liquid in a confined area.
Prevent contamination of fuel by oxidant gases/liquids.

With flammable gases, eliminate all ignition sources (refer to Chapter 6). Possibly provide additional high/low level ventilation; background gas detectors to alarm, e.g. at 40% of the LEL. With toxic gases, possibly provide additional local ventilation; monitors connected to alarms; appropriate air-fed respirators. (The flammable/toxic gas detectors may be linked to automatic shutdown instrumentation.)

Limit access to storage areas to authorized staff knowledgeable in the hazards, position of valves and switches.
Display emergency procedures.

Wear face shields and impervious dry gloves, preferably insulated and of loose fit.
Wear protective clothing which avoids the possibility of cryogenic liquid becoming trapped near the skin: avoid turnups and pockets and wear trousers over boots, not tucked in.
Remove bracelets, rings, watches etc. to avoid potential traps of cryogen against skin.

Prior to entry into large tanks containing inert medium, ensure that pipes to the tank from cryogen storage are blanked off or positively closed off: purge with air and check oxygen levels.
If in doubt, provide air-fed respirators and follow the requirements for entry into confined spaces (Chapter 13).
First aid measures include:
  Move casualties becoming dizzy or losing consciousness into fresh air and provide artificial respiration if breathing stops.
  Obtain medical attention (Chapter 13).
  In the event of 'frost-bite' do not rub the affected area but immerse rapidly in warm water and maintain general body warmth.
  Seek medical aid.
Ensure that staff are trained in the hazards and precautions for both normal operation and emergencies.

Gaseous oxygen is colourless, odourless and tasteless. It does not burn but supports combustion of most elements. Thus upon vaporization liquid oxygen can produce an atmosphere which enhances fire risk; flammability limits of flammable gases and vapours are widened and fires burn with greater vigour. It may cause certain substances normally considered to be non-combustible, e.g. carbon steel, to inflame. In addition to the general precautions set out in Table 8.3, the following are also relevant to the prevention of fires and explosions:

- Prohibit smoking or other means of ignition in the area.
- Avoid contact with flammable materials (including solvents, paper, oil, grease, wood, clothing) and reducing agents. Thus oil or grease must not be used on oxygen equipment.
- Purge oxygen equipment with oil-free nitrogen or oil-free air prior to repairs.
- Post warning signs.
- In the event of fire, evacuate the area and if possible shut off oxygen supply. Extinguish with

Table 8.4 Physical properties of selected cryogenic liquids

Property of liquid		*Oxygen*	*Nitrogen*	*Argon*	*Carbon dioxide*
Molecular weight		32	28	40	44
Boiling point (at atmospheric pressure)	°C	−183	−196	−186	−78
Freezing point	°C	−219	−210	−190	—
Critical temperature	K	154.8	126.1	150.7	—
Density of liquid (at atmospheric pressure)	kg/m³	1141	807	1394	1562 (solid)
Density of vapour (at NBP)	kg/m³	4.43	4.59	5.70	2.90
Density of dry gas at 15°C and at atmospheric pressure	kg/m³	1.34	1.17	1.67	1.86
Latent heat of vaporization at NBP and atmospheric pressure	kJ/kg	214	199	163	151
Expansion ratio (liquid to gas at 15°C and atmospheric pressure)		842	682	822	538[1]
Volume per cent in dry air	%	20.95	78.09	0.93	0.03

NBP Normal boiling point
[1]From liquid CO_2 at 21 bar −18°C.

water spray unless electrical equipment is involved, when carbon dioxide extinguishers should be used.

Liquid nitrogen and argon

Liquid nitrogen is colourless and odourless, slightly lighter than water and non-magnetic. It does not produce toxic or irritating vapours. Liquid argon is also colourless and odourless but significantly heavier than water. Gaseous nitrogen is colourless, odourless and tasteless, slightly soluble in water and a poor conductor of heat. It does not burn or support combustion, nor readily react with other elements. It does, however, combine with some of the more active metals, e.g. calcium, sodium and magnesium, to form nitrides. Gaseous argon is also colourless, odourless and tasteless, very inert and does not support combustion.

The main hazard from using these gases stems from their asphyxiant nature. In confined, unventilated spaces small leakages of liquid can generate sufficient volumes of gas to deplete the oxygen content to below life-supporting concentrations: personnel can become unconscious without warning symptoms (Chapter 5). Gas build-up can occur when a room is closed overnight.

Also, because the boiling points of these cryogenic liquids are lower than that of oxygen, if exposed to air they can cause oxygen to condense preferentially, resulting in hazards similar to those of liquid oxygen.

Liquid carbon dioxide

Liquid carbon dioxide is usually stored under 20 bar pressure at −18°C. Compression and cooling of the gas between the temperature limits at the 'triple point'and the 'critical point' will cause it to liquefy. The triple point is the pressure temperature combination at which carbon dioxide can exist simultaneously as gas, liquid and solid. Above the critical temperature point of 31°C it is impossible to liquefy the gas by increasing the pressure above the critical pressure of 73 bar. Reduction in the temperature and pressure of liquid below the triple point causes the liquid to disappear, leaving only gas and solid. (Solid carbon dioxide is also available for cryogenic work and at −78°C the solid sublimes at atmospheric pressure.)

Liquid carbon dioxide produces a colourless, dense, non-flammable vapour with a slightly pungent odour and characteristic acid 'taste'. Physical properties are given in Table 8.5 (see also page 277). Figure 8.1 demonstrates the effect of temperature on vapour pressure.

Table 8.5 Physical properties of carbon dioxide

Molecular weight	44.01
Vapour pressure at 21°C	57.23 bar
Specific volume at 21°C, 1 atm	547 ml/g
Sublimation point at 1 atm	– 78.5°C
Triple point at 5.11 atm	– 56.6°C
Density, gas at 0°C, 1 atm	1.977 g/l
Specific gravity, gas at 0°C, 1 bar (air = 1)	1.521
Critical temperature	31°C
Critical pressure	73.9 bar
Critical density	0.468 g/ml
Latent heat of vaporization	
at triple point	83.2 cal/g
at 0°C	56.2 cal/g
Specific heat, gas at 25°C, 1 atm	
C_p	0.205 cal/g °C
C_v	0.1565 cal/g °C
ratio C_p/C_v	1.310
Thermal conductivity at 0°C	3.5×10^{-5} cal/s cm² °C/cm
at 100°C	5.5×10^{-5} cal/s cm² °C/cm
Viscosity, gas at 21°C, 1 atm	0.0148 cP
Entropy, gas at 25°C, 1 atm	1.160 cal/g °C
Heat of formation, gas at 25°C	– 2137.1 cal/g
Solubility in water at 25°C, 1 atm	0.759 vol/vol water

Inhalation of carbon dioxide causes the breathing rate to increase (Table 8.6): 10% CO_2 in air can only be endured for a few minutes; at 25% death can result after a few hours exposure.

The 8 hr TWA hygiene standard (see Chapter 5) for carbon dioxide is 0.5%; at higher levels life may be threatened by extended exposure. The following considerations therefore supplement those listed in Table 8.3:

- Ensure that operator exposure is below the hygiene standard. (Note: For environmental monitoring, because of its toxicity, a CO_2 analyser must be used as distinct from simply relying on checks of oxygen levels.)
- When arranging ventilation, remember that the density of carbon dioxide gas is greater than that of air.
- Ensure that pipework and control systems are adequate to cope with the pressures associated with storage and conveyance of carbon dioxide, which are higher than those encountered with most other cryogenic liquids.

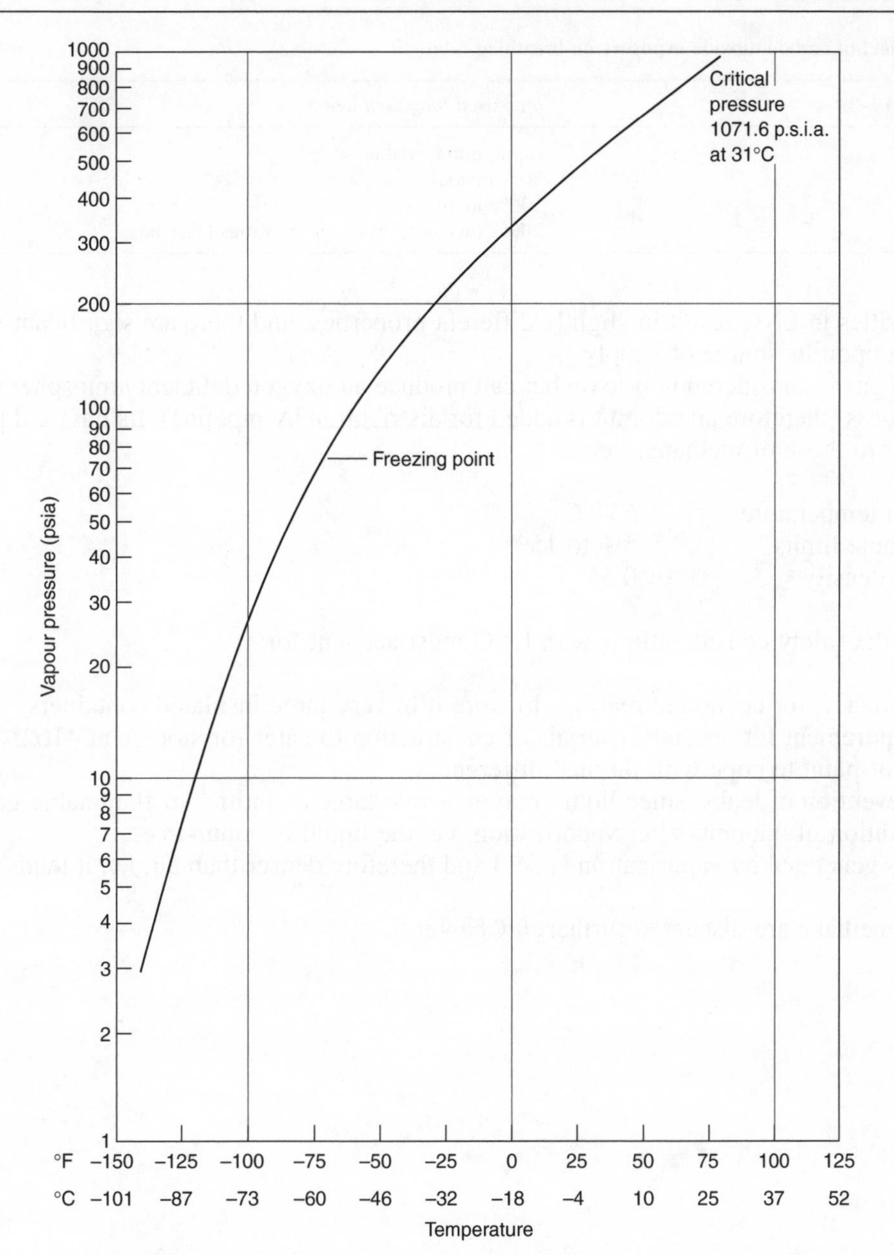

Figure 8.1 *Carbon dioxide vapour pressure versus temperature*

Liquefied natural gas

Liquefied natural gas is predominantly methane. The cryogenic properties of methane are:

Boiling point	–162°C
Critical temperature	–82°C
Critical pressure	45.7 atm
Liquid-to-gas ratio by volume	1 to 637

Table 8.6 Effect of carbon dioxide exposure on breathing rates

CO_2 in air (vol. %)	*Increased lung ventilation*
0.1–1	slight, unnoticeable
2	50% increase
3	100% increase
5	300% increase; breathing becomes laborious

The impurities in LNG result in slightly different properties, and there are significant variations depending upon its source of supply.

Natural gas is considered non-toxic but can produce an oxygen deficient atmosphere (p. 153). It is odourless (therefore an odorant is added for distribution by pipeline). Its physical properties are similar to those of methane, i.e.:

Ignition temperature	537°C
Flammable limits	5% to 15%
Vapour density	0.55

However, the safety considerations with LNG must account for:

- The tendency, for economic reasons, to store it in very large insulated containers.
- The requirement for special materials of construction to cater for storage at –162°C, and for design of plant to cope with thermal differences.
- The prevention of leaks, since liquid may generate large quantities of flammable gas.
- The addition of odorants after vaporization, i.e. the liquid is odour-free.
- The gas generated by vaporization is cold and therefore denser than air, i.e. it tends to slump.

LPG and methane are discussed further in Chapter 9.

9

Compressed gases

Whilst gases are sometimes prepared *in situ* for cost and safety reasons (e.g. to remove the risk associated with their transport, storage and piping to point of use) they are more often stored on an industrial scale at low pressure, either under refrigerated conditions, e.g. cryogens (Chapter 8), or at ambient temperature in 'gasholders', which 'telescope' according to the quantity of gas and are fitted with water or oil seals to prevent gas escape. Smaller quantities of gas at high pressure are usually stored in bottle-shaped gas cylinders. They find widespread use in welding, fuel for gas burners, hospitals, laboratories etc. The construction of compressed gas cylinders ensures that, when first put into service, they are safe for their designated use. Serious accidents can, however, result from ignorance of the properties of the gases, or from misuse or abuse. Great care is needed during the transportation, handling, storage and disposal of such cylinders.

Compressed gases can often be more dangerous than chemicals in liquid or solid form because of the potential source of high energy, low boiling-point of some liquid contents resulting in the potential for flashing (page 50), ease of diffusion of escaping gas, low flashpoint of some highly flammable liquids, and the absence of visual and/or odour detection of some leaking materials. The containers also tend to be heavy and bulky.

Compressed gases, therefore, present a unique hazard from their potential physical and chemical dangers. Unless cylinders are secured they may topple over, cause injury to operators, become damaged themselves and cause contents to leak. If the regulator shears off, the cylinder may rocket like a projectile or 'torpedo' dangerously around the workplace. Other physical hazards stem from the high pressure of a cylinder's contents, e.g. accidental application of a compressed gas/air hose or jet into eyes or onto an open cut or wound, whereby the gas can enter the tissue or bloodstream, is particularly dangerous.

A further hazard exists when compressed air jets are used to clean machine components in workplaces: flying particles have caused injury and blindness. Cylinders may fail if over-pressurized or weakened by the application of heat. Liquefied gases, e.g. butane or propane, respond more rapidly to heat than the permanent gases such as nitrogen or oxygen. Cylinders are normally protected by pressure relief valves, fusible plugs or bursting discs.

Low boiling-point materials can cause frostbite on contact with living tissue. While this is an obvious hazard with cryogenics, e.g. liquid nitrogen or oxygen, cylinders of other liquefied gases also become extremely cold and covered in 'frost' as the contents are discharged (page 47).

Precautions also have to be instituted to protect against the inherent properties of the cylinder contents, e.g. toxic, corrosive, flammable (refer to Table 9.1). Most gases are denser than air; common exceptions include acetylene, ammonia, helium, hydrogen and methane. Even these may on escape be much cooler than ambient air and therefore slump initially. Eventually the gas will rise and accumulate at high levels unless ventilated. Hydrogen and acetylene, which both have very wide flammable limits (Table 6.1), can form explosive atmospheres in this way.

More dense gases will on discharge accumulate at low levels and may, if flammable, travel a considerable distance to a remote ignition source.

Table 9.1 Compressed gases: hazards and construction materials for services

Gas	*Hazard*[(1)]	*Materials of construction for ancillary services*[(2)] *Compatible*	*Incompatible*
Acetylene	F	Stainless steel, aluminium, wrought iron	Unalloyed copper, alloys containing >70% copper, silver, mercury, and cast iron
Air	O	Any common metal or plastic	
Allene	F	Mild steel, aluminium, brass, or stainless steel	Copper, silver and their alloys, PVC and neoprene
Ammonia	C F T	Iron and steel	Copper, zinc, tin and their alloys (e.g. brass), and mercury
Argon		Any common metal	
Arsine	F T	Stainless steel and iron	
Boron trichloride	C T	Any common metal for dry gas Copper, Monel, Hastelloy B, PVC, polythene and PTFE if *moist* gas is used	Any metal incompatible with hydrochloric acid when moist gas is used
Boron trifluoride	C T	Stainless steel, copper, nickel, Monel, brass, aluminium for *dry* gas ≤200°C. Borosilicate glass for low pressures. For *moist* gas, copper and polyvinylidene chloride plastics	Rubber, nylon, phenolic resins, cellulose and commercial PVC
Bromine pentafluoride	C T O	Monel and nickel	
Bromine trifluoride	C T O	Monel and nickel	
Bromotrifluoroethylene	F T	Most common metals so long as gas is dry	Magnesium alloys and aluminium containing >2% magnesium
Bromotrifluoromethane		Most common metals	
1,3-Butadiene	F T	Mild steel, aluminium, brass, copper or stainless steel	PVC and Neoprene plastic
Butane	F	Any common metal	
1, Butene	F	Any common metal	
Carbon dioxide	T	Iron, steel, copper, brass, plastic for dry gas. For moist gas use stainless steel or certain plastics	For moist gas avoid materials attacked by acids
Carbon monoxide	F T	Copper-lined metals for pressures <34 bar. Certain highly alloyed chrome steels	Iron, nickel and certain other metals at high pressures
Carbon tetrafluoride		Any common metal	
Carbonyl fluoride	C F T	Steel, stainless steel, copper or brass for dry gas. Monel, copper or nickel for moist gas	
Carbonyl sulphide	F T	Aluminium and stainless steel	
Chlorine	C T O	Extra heavy black iron or steel for dry gas. Drop forged steel, PTFE tape. Moist gas requires glass, stoneware (for low pressures) and noble metals. High silica, iron, Monel and Hastelloy show some resistance	Rubber (e.g. gaskets)

Table 9.1 Cont'd

Gas	*Hazard*[1]	*Materials of construction for ancillary services*[2] *Compatible*	*Incompatible*
Chlorine trifluoride	C T O	Monel and nickel, PTFE and Kel-F, soft copper, 2S aluminium and lead are suitable for gaskets	
Chlorodifluoromethane		Steel, cast iron, brass, copper, tin, lead, aluminium at normal conditions Neoprene or chloroprene rubber and pressed fabrics are suitable for gaskets	Silver, brass, aluminium, steel, copper, nickel can cause decomposition at elevated temperatures. Magnesium alloys and aluminium containing >2% magnesium. Natural rubber
Chloropentafluoroethane		Neoprene or chloroprene rubber and pressed fabrics are suitable for gaskets	Silver, brass, aluminium, steel, copper, nickel can cause decomposition at elevated temperatures. Magnesium alloys and aluminium containing >2% magnesium. Natural rubber
Chlorotrifluoroethane	F T	Most common metals	
Chlorotrifluoromethane		As for chlorodifluoromethane	
Cyanogen	F T	Stainless steel, Monel and Inconel ≤65°C. Glass-lined equipment. Iron and steel at ordinary temperatures	
Cyanogen chloride	C T	Common metals for dry gas. Monel, tantalum. Glass for moist gas	
Cyclobutane	F	Most common metals	
Cyclopropane	F	Most common metals	
Deuterium	F	Most common metals	
Diborane	F T	Most common metals. Polyvinylidene chloride, polyethylene, Kel-F PTFE graphite and silicone vacuum grease	Rubber and certain hydrocarbon lubricants
Dibromodifluoromethane		Copper or stainless steel	Aluminium for wet gas
1,2-Dibromotetra-fluoroethane	C	Most common metals for dry gas. Stainless steel, titanium and nickel for moist gas	Zinc
Dichlorodifluoromethane		As for chlorodifluoromethane	
Dichlorofluoromethane		As for chlorodifluoromethane	
Dichlorosilane	C F T	Nickel and nickel steels and PTFE	Stainless steel for moist gas
1,2-Dichlorotetra-fluoroethane		As for chlorodifluoromethane	
1,1-Difluoro-1-chloroethane	F	Most common metals under normal conditions	Hot metals can cause degradation to toxic corrosive products
1,1-Difluoroethane	F	Most common metals under normal conditions	Hot metals can cause degradation to toxic corrosive products
1,1-Difluoroethylene	F	Most common metals	
Dimethylamine	C F T	Iron and steel	Copper, tin, zinc, and their alloys
Dimethyl ether	F T	Most common metals	
2,2-Dimethyl propane	F	Most common metals	
Ethane	F	Most common metals	
Ethyl acetylene	F	Steel and stainless steel	Copper, other metals capable of forming explosive acetylides

Table 9.1 Cont'd

Gas	*Hazard*[1]	*Materials of construction for ancillary services*[2] *Compatible*	*Incompatible*
Ethyl chloride	F T	Most materials for dry gases	
Ethylene	F	Any common metal	
Ethylamine	C F T	Iron and steel. Reinforced neoprene hose	Copper, tin, zinc and their alloys
Ethylene oxide	F T	Properly grounded steel	Copper, silver, magnesium and their alloys
Fluorine	C T O	Brass, iron, aluminium, magnesium and copper at normal temperatures. Nickel and Monel at higher temperatures	
Fluoroform		Any common metal	
Germane	F T	Iron and steel	
Helium		Any common metal	
Hexafluoroacetone	C T	For dry gas Monel, nickel, Inconel, stainless steel, copper and glass – Hastelloy C-line equipment	
Hexafluoroethane		Any common metal for normal temperatures. Copper, stainless steel and aluminium ≤150°C	
Hexafluoropropylene		Any common metal for dry gas	
Hydrogen	F	Most common metals for normal use	At elevated temperature and pressure hydrogen embrittlement can result
Hydrogen bromide	C T	Most common metals when dry. Silver, platinum and tantalum for moist gas. Heavy black iron for high-pressure work. High-pressure steel, Monel or aluminium pipe.	Most metals when gas is moist. Galvanized pipe or brass or bronze fittings
Hydrogen chloride	C T	Stainless steel, mild steel for normal conditions of temperature and pressure. When moist use silver, platinum or tantalum. Moist or dry gas use backed carbon, graphite. High pressure work in heavy black iron pipework. High pressure Monel or aluminium iron bronze valves	Galvanized pipes or brass or bronze fittings
Hydrogen cyanide	F T	Low-carbon steel at normal temp. and stainless steel for higher temperatures	
Hydrogen fluoride	C T	Steel in the absence of sulphur dioxide contaminants in the gas and at temperatures <65°C. Monel, Inconel, nickel and copper for liquid or gas at elevated temperature	Cast iron or malleable fittings
Hydrogen iodide	C T	Stainless steel, mild steel under normal temperature and pressure	Moist gas corrodes most metals

Table 9.1 Cont'd

Gas	*Hazard*[1]	*Materials of construction for ancillary services*[2] *Compatible*	*Incompatible*
		Silver, platinum and tantalum, carbon, graphite for wet gas. At higher pressures use extra heavy black iron pipe. High-pressure steel, Monel or aluminium-iron-bronze valves	Galvanized pipe or brass or bronze fittings
Hydrogen selenide	F T	Aluminium and stainless steel are preferred but iron, steel or brass are acceptable	
Hydrogen sulphide	F T	Aluminium preferred. Iron and steel are satisfactory. Brass, though tarnished, is acceptable	Many metals in the presence of moist gas
Isobutane	F	Most common metals	
Isobutylene	F	Most common metals	
Krypton		Most common metals	
Methane	F	Most common metals	
Methyl acetylene	F	Most common metals	Copper, silver, mercury and their alloys
Methylamine	C F T	Iron and steel	Copper, tin, zinc and their alloys. Avoid mercury
Methyl bromide	C F T	Most common metals when dry	Aluminium and its alloys
3-Methyl-1-butane	F	Most common metals	
Methyl chloride	F T	Most common metals when dry	Zinc, magnesium rubber and neoprene particularly when moist. Aluminium is forbidden
Methyl fluoride	F T	Most common metals	
Methyl mercaptan	F T	Stainless steel and copper-free steel alloys and aluminium. Iron and steel for dry gas	
Methyl vinyl ether	F	Most common metals	Copper and its alloys
Neon		Most common metals	
Nickel carbonyl	F T	Most common metals for pure gas. Copper or glass-lined equipment for carbonyl in the presence of carbon monoxide	
Nitric oxide	T O	Most common metals for dry gas. For moist gas use 18:8 stainless steel, PTFE	
Nitrogen		Any common metal	
Nitrogen dioxide	C T O	Most common metals for dry gas. For moist gas use 18:8 stainless steel	
Nitrogen trifluoride	T O	Nickel and Monel are preferred. Steel, copper and glass are acceptable at ordinary temperatures	Plastics
Nitrogen trioxide	C T O	Steel for *dry* gas otherwise use 18:8 stainless steel	
Nitrosyl chloride	C T O	Nickel, Monel and Inconel. For moist gas tantalum is suitable	
Nitrous oxide	O	Most common metals	
Octofluorocyclobutane	T	Cast iron and stainless steel <120°C, steel ≤175°C, Inconel, nickel and platinum ≤400°C	Avoid the metals opposite >500°C
Oxygen	O	Most common metals	On grease or combustible

Table 9.1 Cont'd

Gas	*Hazard*[1]	*Materials of construction for ancillary services*[2] *Compatible*	*Incompatible*
Oxygen difluoride	T O	Glass, stainless steel, copper, Monel or nickel ≤200°C. At higher temperatures only nickel and Monel are recommended	materials
Ozone	F T O	Glass, stainless steel, Teflon, Hypalon, aluminium, Tygon, PVC and polythene	Copper and its alloys, rubber or any composition thereof, oil, grease or readily combustible material
Perchloryl fluoride	T	Most metals and glass for dry gas at ordinary temperatures	Many gasket materials are embrittled At higher temperatures many organic materials and some metals can be ignited Some metals such as titanium show deflagration in contact with the gas under severe shock
Perfluorobutane		Most common materials	
Perfluorobutene	T	Most common materials when dry	
Perfluoropropane		Most common metals	
Phosgene	C T	Common metals for dry gas. Monel, tantalum or glass lined equipment for moist gas	
Phosphine	F T	Iron or steel	
Phosphorus pentafluoride	F T	Steel, nickel, Monel and Pyrex for dry gas. For moist gas hard rubber and paraffin wax	
Phosphorus trifluoride		Steel, nickel, Monel and the more noble metals and Pyrex for dry gas	
Propane	F	Most common metals	
Propylene	F	Most common metals	
Propylene oxide	F T	Steel or stainless steel preferred though copper and brass are suitable for acetylene-free gas. PTFE gaskets	Rubber
Silane	F T	Iron, steel, copper, brass	
Silicone tetrafluoride	C T	Most common metals for the dry gas. Steel, Monel and copper for moist gas	
Sulphur dioxide	C T O	Most common metals for dry gas. Lead, carbon, aluminium and stainless steel for moist gas	Zinc
Sulphur hexafluoride		Most common metals. Copper, stainless steel and aluminium are resistant to the decomposition products at 150°C	
Sulphur tetrafluoride	C T	Stainless steel or 'Hastelloy C' lined containers. Glass suitable for short exposures if dry. 'Tygon' for low-pressure connections	Glass for moist gas
Sulphuryl fluoride	T	Any common metal at normal temperatures and pressures	Some metals at elevated temperatures
Tetrafluoroethylene	F	Most common metals	

Table 9.1 Cont'd

Gas	*Hazard*[(1)]	*Materials of construction for ancillary services*[(2)] *Compatible*	*Incompatible*
Tetrafluorohydrazine	T O	Glass, stainless steel, copper or nickel to temperatures of 200°C. For higher temperatures use nickel and Monel	
Trichlorofluoromethane	T	Steel, cast iron, brass, copper, tin, lead, aluminium under normal, dry conditions	Some of the opposite at high temperatures magnesium alloys and aluminium coating >2% magnesium. Natural rubber
1,1,2-Trichloro-1,2,2-trifluoroethane		As above	As above
Trimethylamine	C F T	Iron, steel, stainless steel and Monel. Rigid steel piping	Copper, tin, zinc and most of their alloys
Vinyl bromide	F T	Steel	Copper and its alloys
Vinyl chloride	C F T	Steel	Copper and its alloys
Vinyl fluoride	F	Steel	Copper and its alloys
Xenon		Most common materials	

C Corrosive
F Flammable
T Toxic
O Oxidizing
(1) Even non-toxic gases are potentially hazardous owing to asphyxiation (oxygen deficiency). Irrespective of material, all equipment must be adequately designed to withstand process pressures.
(2) This is a guide and is no substitute for detailed literature.

To prevent interchange of fittings between cylinders of combustible and non-combustible gases, the valve outlets are screwed left-hand and right-hand thread, respectively (Table 9.2). Primary identification is by means of labelling with the name and chemical formula on the shoulder of the cylinder. Secondary identification is by use of ground colours on the cylinder body. Unless specified in Table 9.2, gas and gas mixtures shall be identified by a colour classification indicating gas properties in accordance with the risk diamond on the cylinder label e.g.

Toxic and/or corrosive	Yellow
Flammable	Red
Oxidizing	Light blue
Inert (non-toxic, non-corrosive, non-flammable, non-oxidizing	Bright green

The full scheme is given in BS EN 1089–3: 1997.

This should be consulted for the colour coding of gas mixtures used for inhalation e.g. medical and breathing apparatus mixtures containing oxygen.

Table 9.2 Specific colour codes for selected compressed gases

Gas	*Colour*
Acetylene	Maroon
Argon	Dark green
Carbon dioxide	Grey
Helium	Brown
Oxygen	White
Nitrogen	Black
Nitrous oxide	Blue

Table 9.3 provides general guidance for handling compressed gases.

Table 9.3 General precautions for handling compressed gases

Consult the supplier for data on the specification, properties, handling advice and on suitable service materials for individual gases.

Storage

Segregate according to hazard.

Stores should be adequately ventilated and, ideally, located outside and protected from the weather.

Store away from sources of heat and ignition.

Cylinders within workplaces should be restricted to those gases in use. Specially designed compartments with partitions may be required to protect people in the event of explosion. Take into account emergency exits, steam or hot water systems, the proximity of other processes etc. Consider the possibility of dense gases accumulating in drains, basements, cable ducts, lift shafts etc.

Where necessary, provide fireproof partitions/barriers to separate/protect cylinders.

Protect from mechanical damage.

All cylinders must be properly labelled and colour coded (BS 349).

Store full and empty cylinders separately.

Use in rotation: first in, first out.

Restrict access to the stores to authorized staff.

Display 'No smoking' and other relevant warning signs.

Ensure that all staff are fully conversant with the correct procedures when using pressure regulators. (For cylinders without handwheel valves, the correct cylinder valve keys should be kept readily available, e.g. on the valve. Only use such keys. Do not extend handles or keys to permit greater leverage; do not use excessive force, e.g. hammering, when opening/closing valves or connecting/disconnecting fittings. The pressure regulator must be fully closed before opening the cylinder valve. This valve can then be opened slowly until the regulator gauge indicates the cylinder pressure but should not be opened wider than necessary. The pressure regulator can then be opened to give the required delivery pressure. When a cylinder is not in use, or is being moved, the cylinder valve must be shut. When a cylinder has been connected, the valve should be opened with the regulator closed; joints should then be tested with soap/detergent solution.)

Clearly and permanently mark pressure gauges for use on oxygen. Do not contaminate them with oil or grease or use them for other duties.

Cylinders that cannot be properly identified should not be used; do not rely on colour code alone.

Never try to refill cylinders.

Never use compressed gas to blow away dust or dirt.

Provide permanent brazed or welded pipelines from the cylinders to near the points of gas use. Select pipe materials suitable for the gas and its application. Any flexible piping used should be protected against physical damage. Never use rubber or plastic connections from cylinders containing toxic gases.

On acetylene service, use only approved fittings and regulators. Avoid any possibility of it coming into contact with copper, copper-rich alloys or silver-rich alloys. (In the UK use at a pressure greater than 600 mbar g must be notified to HM Explosives Inspectorate for advice on appropriate standards.)

On carbon dioxide service, rapid withdrawal of gas may result in plugging by solid CO_2. Close the valve, if possible, to allow the metal to warm up; this will prevent a sudden gas discharge.

Replace the correct caps or guards on cylinder valves when not in use and for return to the supplier.

Test and inspect cylinders and pressure regulators regularly in accordance with current legislation.

Design and manage cylinder stores in accordance with suppliers' recommendations.

Wear appropriate personal protection when entering any store.

Inspect condition of cylinders regularly, especially those containing hazardous gases (e.g. corrosive).

Use

Transport gases in specially designed trolleys and use eye protection, stout gloves (preferably textile or leather) and protective footwear.

Do not roll or drop cylinders off the backs of wagons; never lift cylinders by the cap.

Ideally, depending on the length of pipe run, locate cylinders outside (for hazardous gases, valves installed within the workplace can be used for remote control of the main supply from the cylinder in the event of an emergency). Site cylinders so that they cannot become part of an electrical circuit.

Securely clamp, or otherwise firmly hold in position, cylinders on installation. (Unless otherwise specified, cylinders containing liquefied or dissolved gases must be used upright.)

Avoid subjecting cylinders containing liquid to excessive heat.

Table 9.3 Cont'd

Fit approved cylinder pressure regulators, selected to give a maximum pressure on the reduced side commensurate with the required delivery pressure. (The regulator and all fittings upstream of it must be able to withstand at least the maximum cylinder pressure.)
Fit in-line flame arresters for flammable gases and eliminate ignition sources.
Use compatible pipe fittings. (Flammable gas cylinders have valves with left-hand threads; cylinders for oxygen and non-flammable gases, except occasionally helium, have valves with right-hand threads. Certain liquefied gas cylinders have two supply lines, one for gas and one for liquid, dependent on cylinder position.)
Do not use oil, grease or joining compounds on any fittings for compressed gas cylinders.
Fit an excess flow valve to the outlet of a regulator, selected to allow the maximum required gas flow.
Use respirators and face protection etc. when changing regulators on cylinders of toxic gases.
Turn off gas supply at the cylinder at the end of each day's use.
Consider the need for gas detection/alarms, e.g. for hazardous gases left in use out of normal hours.
Periodic checks:
Ensure no gas discharge when gauge reading is zero
Ensure reading on gauge does not increase as the regulator valve is closed
Check for 'crawl' due to wear on the regulator valve and seat assembly
Ensure no leak between cylinder and regulator
Overhaul regulators on a 3–6 month basis for corrosive gases, annually for others
Train staff in hazards and correct handling procedures.

The hazards and safety precautions for selected common compressed gases are discussed below to illustrate the general approach. More details should be sought from suppliers. Some methods for their preparation *in situ* are noted; full experimental details must be obtained from the literature.

Acetylene

Acetylene is manufactured by the controlled reaction between water and calcium carbide:

$$CaC_2 + 2H_2O \rightarrow Ca\,(OH)_2 + C_2H_2$$

Alternatively it is obtained from cracking low molecular-weight aliphatic hydrocarbons, or by the partial oxidation of natural gas.

Because of its high chemical reactivity, acetylene has found wide use in synthesis of vinyl chloride, vinyl acetate, acrylonitrile, vinyl ethers, vinyl acetylene, trichloro- and tetrachloro-ethylene etc., in oxyacetylene cutting and welding, and as a fuel for atomic absorption instruments.

Acetylene is a simple asphyxiant and anaesthetic. Pure acetylene is a colourless, highly flammable gas with an ethereal odour. Material of commercial purity has an odour of garlic due to the presence of impurities such as phosphine. Its physical properties are shown in Table 9.4. Acetylene, which condenses to a white solid subliming at –83°C, is soluble in its own volume of water but highly soluble in acetone.

Under certain conditions acetylene can explode when mixed with air, hydrogen or ethylene.

Accidental heating of a small area of cylinder wall to 185°C or above may promote an extremely dangerous condition. Violent reactions have occurred between acetylene and oxidants such as oxides of nitrogen (see later), nitric acid, calcium hypochlorite, ozone and halogens. In the free state acetylene can decompose violently, e.g. above 9 psig (0.62 bar) undissolved (free) acetylene will begin to dissociate and revert to its constituent elements. This is an exothermic process which can result in explosions of great violence. For this reason acetylene is transported in acetone contained in a porous material inside the cylinder. Voids in the porous substance can result from settling, e.g. if the cylinder is stored horizontally or through damage to the cylinder in the form of denting. Voids may enable acetylene to decompose, e.g. on initiation by mechanical shock if the cylinder is dropped.

Table 9.4 Physical properties of acetylene

Molecular weight	26.038
Vapour pressure of pure liquid at 21°C (not cylinder pressure)	43.8 bar
Specific volume at 15.6°C, 1 atm	902.9 ml/g
Boiling point at 1.22 atm	–75°C
Sublimation point at 1 atm	–84.0°C
Triple point at saturation pressure	–80.8°C
Specific gravity, gas at 15.6°C, 1 atm (air = 1)	0.9057
Density, gas at 0°C, 1 atm	1.1709 g/l
Critical temperature	36.3°C
Critical pressure	62.4 bar
Critical density	0.231 g/ml
Latent heat of sublimation at –84°C	193.46 cal/g
Latent heat of fusion at triple point	23.04 cal/g
Flammable limits in air	2.5–81.0% by volume
Auto-ignition temperature	335°C
Gross heat of combustion at 15.6°C, 1 atm	13.2 cal/cc
Specific heat, gas at 25°C, 1 atm	
C_p	0.4047 cal/g°C
C_v	0.3212 cal/g °C
ratio C_p/C_v	1.26
Thermal conductivity, gas at 0°C	4.8×10^{-5} cal/s cm^2 °C/cm
Viscosity, gas at 25°C, 1 atm	0.00943 cP
Entropy, gas at 25°C, 1 atm	1.843 cal/g °C
Solubility in water at 0°C, 1 atm	1.7 vol/vol H_2O

Figure 9.1 illustrates the rise in cylinder pressure with temperature. Normally, acetylene cylinders are fitted with a fusible metal plug which melts at about 100°C.

Acetylene can form metal acetylides, such as copper or silver acetylide, which on drying become highly explosive: service materials require careful selection.

In addition to the general precautions for compressed gases in Table 9.3, the following control measures should be considered for acetylene:

- Never use free acetylene at pressures above 9 psig (0.62 bar) unless special safety features are employed.
- Store and use cylinders only in an upright position.
- Store reserves separate from oxygen cylinders.
- Ensure that no means of accidental ignition are in the area and provide adequate ventilation.
- Consult local regulations for use of this gas.
- Ensure that 'empty' cylinders have the valve closed to prevent evaporation of acetone.
- Close cylinder valve before shutting off regulator, to permit gas to bleed from regulator.
- When used e.g. for welding, avoid the careless use of flame which could fuse the metal safety plug in the cylinder.
- In the event of fire issuing from the cylinder, close the gas supply if it is safe to do so and evacuate the area.
- Consider the need for detection/alarm systems and in any event check periodically for leaks with e.g. soap solution, *never* with a naked flame.

Air

The physical properties of air are given in Table 9.5. Air is a mixture of nitrogen, oxygen, argon,

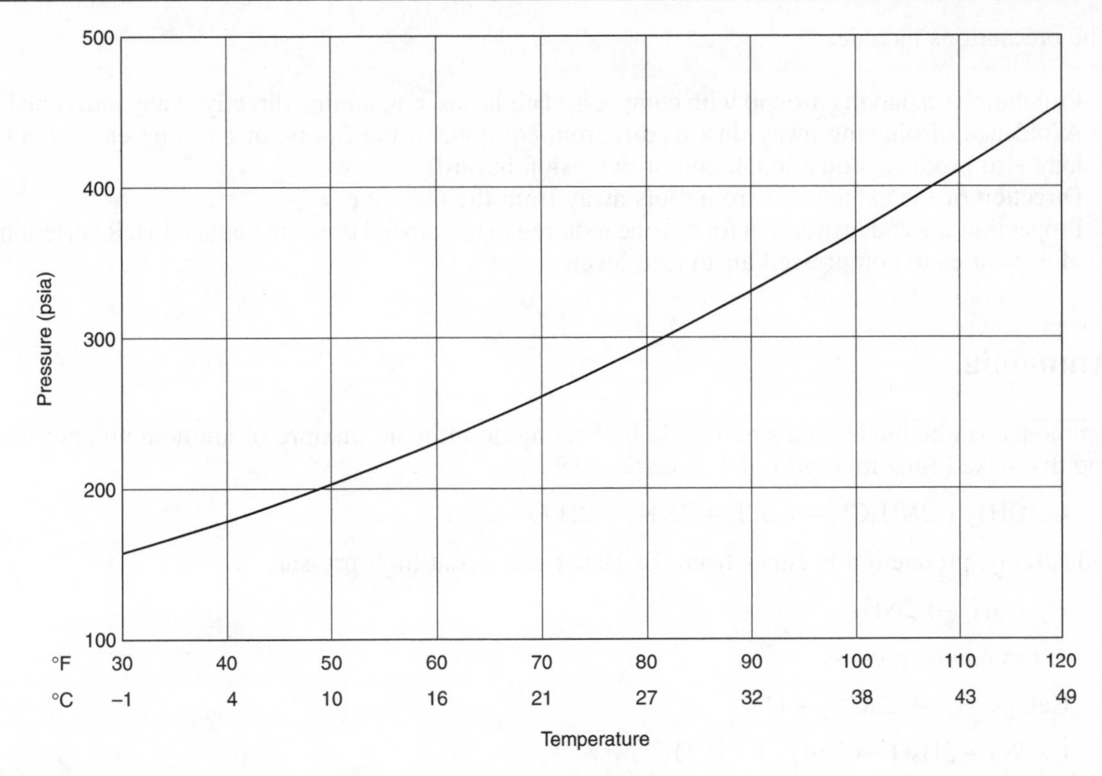

Figure 9.1 *Acetylene (in acetone): full cylinder pressure versus temperature*

carbon dioxide, water vapour, rare gases and trace quantities of ozone, oxides of nitrogen, acetylene, methane and other hydrocarbons. Its composition varies with altitude. Dry air is inert in its effect on metals and plastics. The hazards associated with compressed air, in addition to those associated with any pressure system (i.e. the potential for rupture of equipment or pipework), are:

- From inhalation at pressures above atmospheric, used in tunnelling or diving, or from breathing apparatus or resuscitation equipment, if the pressure is too high or exposure is prolonged. This may cause symptoms from pain to dyspnoea, disorientation and unconsciousness; it may be fatal.
- From particulate matter blown from orifices or surfaces, e.g. into the eyes.
- From entry into any of the body orifices, which can result in serious internal damage.
- From penetration of unbroken skin, or cuts. Foreign matter, e.g. grease, metal, concrete, may also be injected into subcutaneous tissues.
- From whipping of an unsecured hose on rapid gas release.

Table 9.5 Physical properties of air

Density @ 20°C, 1 atm	0.0012046 g/l
Critical temperature	–140.6°C
Critical pressure	546.8 psia (37.2 atm) (38.4 kg/cm^2 absolute)
Critical density	0.313 g/ml
Viscosity @ 0°C, 1 atm	170.9 micropoises

The precautions include:

- Prohibition on playing around with compressed air hoses, e.g. aiming directly at any individual.
- Avoidance of blowing away dust or dirt from equipment, the floors, or clothing etc. (which may also produce a dust inhalation or explosion hazard).
- Direction of the exhaust air from tools away from the operator.
- Proper training and instruction for anyone required to use air-fed breathing apparatus. Restriction of exposures to compressed air to safe levels.

Ammonia

Ammonia can be made on a small scale by heating an intimate mixture of ammonium chloride and dry slaked lime in a ratio of 1:3, respectively:

$$Ca(OH)_2 + 2NH_4Cl \rightarrow CaCl_2 + 2NH_3 + 2H_2O$$

Industrially, production is either from the Haber process at high pressure:

$$N_2 + 3H_2 \rightarrow 2NH_3$$

or the cyanamide process

$$CaC_2 + N_2 \rightarrow CaCN_2 + C$$

$$CaCN_2 + 3H_2O \rightarrow 2NH_3 + CaCO_3$$

At room temperature and atmospheric pressure ammonia is a colourless, alkaline gas with a pungent smell. It dissolves readily in water. Physical properties are summarized in Table 9.6. The effect of temperature on vapour pressure of anhydrous ammonia is shown in Figure 9.2.

Ammonia is shipped as a liquefied gas under its own vapour pressure of 114 psig (7.9 bar) at 21°C. Uses are to be found in refrigeration, fertilizer production, metal industries, the petroleum, chemical and rubber industries, domestic cleaning agents and water purification. Aqueous solutions of ammonia are common alkaline laboratory reagents; ca 0.88 solution is the strongest available. Ammonia gas is expelled on warming.

Ammonia gas is irritating to the eyes, mucous membranes and respiratory tract. Because of its odour few individuals are likely to be unwittingly over-exposed for prolonged periods. Table 9.7 summarizes the physiological effects of human exposure. Clearly at high concentrations the gas becomes corrosive and capable of causing extensive injuries. Thus 1% in air is mildly irritating, 2% has a more pronounced effect and 3% produces stinging sensations.

On contact with the skin, liquid ammonia produces severe burns compounded by frostbite due to the freezing effect from rapid evaporation from the skin.

Moist ammonia attacks copper, tin, zinc and their alloys. Ammonia is also flammable with flammability limits of 15–28%.

Ammonia can also react violently with a large selection of chemicals including ethylene oxide, halogens, heavy metals, and oxidants such as chromium trioxide, dichlorine oxide, dinitrogen tetroxide, hydrogen peroxide, nitric acid, liquid oxygen, and potassium chlorate.

Besides the control measures given in Table 9.3, the following precautions are appropriate:

- Wear rubber gloves, chemical goggles and, depending upon scale, a rubber apron or full chemical suit.
- Never heat ammonia cylinders directly with steam or flames to speed up gas discharge.

Table 9.6 Physical properties of ammonia

Molecular weight	17.031
Vapour pressure at 21°C (cylinder pressure)	7.87 bar
Specific volume at 21°C, 1 atm	1.411 ml/g
Boiling point at 1 atm	– 33.35°C
Triple point at 1 atm	– 77.7°C
Triple point pressure	1.33 mbar
Specific gravity, gas at 0°C, 1 atm (air = 1)	0.5970
Density, gas at boiling point	0.000 89 g/ml
Density, liquid at boiling point	0.674 g/ml
Critical temperature	132.44°C
Critical pressure	113 bar
Critical density	0.235 g/ml
Flammable limits in air	15–28% by volume
Latent heat of vaporization at boiling point	327.4 cal/g
Specific heat, liquid at –20°C	1.126 cal/g K
Specific heat, gas at 25°C, 1 atm	
C_p	0.5160 cal/g°C
C_v	0.4065 cal/g °C
ratio, C_p/C_v	1.269
Thermal conductivity, gas at 25°C, 1 atm	5.22×10^{-5} cal/s cm^2 °C/cm
Entropy, gas at 25°C, 1 atm	2.7 cal/g °C
Heat of formation, gas at 25°C	–648.3 cal/g
Solubility at 0°C, 1 atm	
in water	42.8% by weight
in methanol, absolute	29.3% by weight
in ethanol, absolute	20. 95% by weight
Viscosity, gas at 0°C, 1 atm	0.009 18 cP
Viscosity, liquid at –33.5°C	0.266 cP

- Use under well-ventilated conditions and provide convenient safety showers and eye-wash facilities.
- Ensure that gas cannot be accidentally ignited.
- Check for leaks, e.g. with moist litmus paper or concentrated hydrochloric acid (which forms dense white fumes of ammonium chloride).
- In the event of accident, administer first aid (see Table 9.9).

Carbon dioxide

Carbon dioxide is present in air and is a constituent of natural gas escaping from mineral springs and fissures in the earth's surface. It is also the ultimate product of combustion of carbon and its compounds. Laboratory scale preparation usually entails reaction between dilute hydrochloric acid and marble (calcium carbonate):

$$2HCl + CaCO_3 \rightarrow CaCl_2 + H_2O + CO_2$$

Industrially, it is obtained as a by-product of fermentation of sugar to alcohol:

$$C_6H_{12}O_2 \rightarrow 2C_2H_5OH + 2CO_2$$

or by burning coke/limestone mixture in a kiln:

$$CaCO_3 \rightarrow CO_2 + CaO$$

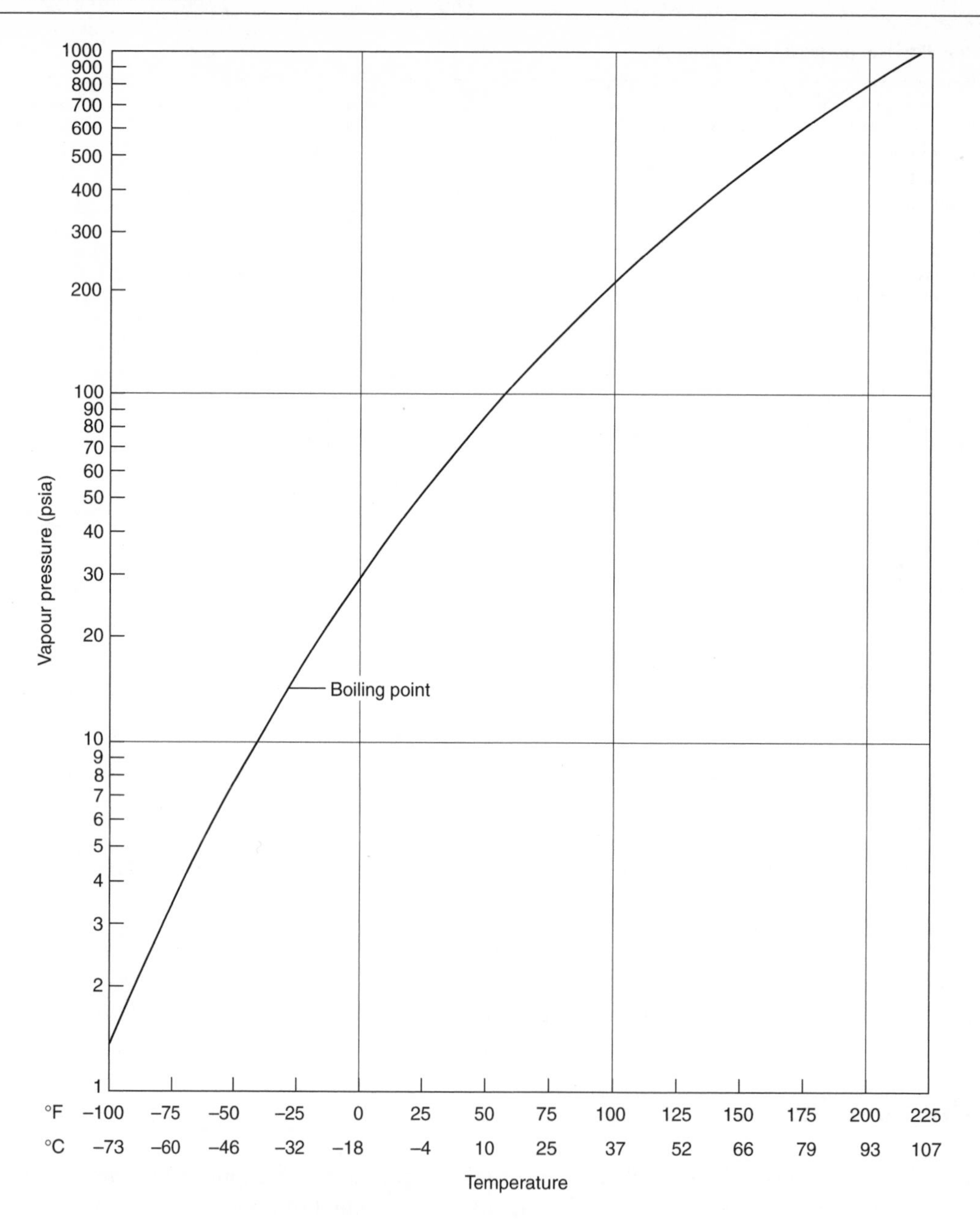

Figure 9.2 *Ammonia vapour pressure versus temperature*

Liquid carbon dioxide is discussed on page 261. Carbon dioxide gas is commonly used for carbonating drinks, in fire extinguishers, for gas-shielding of welding and in shell moulding in foundries. Its physical and toxicological properties are summarized in Tables 8.5, 8.6 and 5.29.

The gas is non-flammable, and is used for inert gas purging. Because it is 1.5 times heavier than air it may accumulate at low level. The general handling precautions are those in Table 9.3.

Table 9.7 Physiological effects of ammonia

Atmospheric concentration (ppm)	*Effects*
20	First perceptible odour
40	A few individuals may suffer slight eye irritation
100	Noticeable irritation of eyes and nasal passages after few minutes' exposure
400	Severe irritation of the throat, nasal passages and upper respiratory tract
700	Severe eye irritation
	No permanent effect if exposure <30 min
1700	Serious coughing, bronchial spasms, <30 min exposure may be fatal
5000	Serious oedema, strangulation, asphyxia
	Fatal almost immediately

Carbon monoxide

Carbon monoxide is produced by incomplete combustion of carbon and its compounds. In the laboratory it can be prepared by careful dehydration of formic or oxalic acid with sulphuric acid:

$$HCO_2H \rightarrow CO + H_2O$$

Traditionally, pure CO is not used industrially; water gas or producer gas are used instead. However, pure CO is made by thermal decomposition of nickel carbonyl:

$$Ni(CO)_4 \rightarrow Ni + 4CO$$

Carbon monoxide is a toxic, flammable, colourless and odourless gas which is slightly lighter than air and slightly soluble in water. Some physical constants are given in Table 9.8. Corrosion by pure carbon monoxide is considered negligible. It is shipped as a non-liquefied gas in high-pressure steel containers. Its main uses include fuel gas mixtures with hydrogen and for reduction of ores. Its main hazards are its extreme toxic effects which stem from its ability to complex with haemoglobin (with which it has an affinity 300 times that of oxygen) resulting in chemical asphyxiation (see Table 5.31). It burns in air with a characteristic blue flame. It combines directly with chlorine in sunlight to produce highly toxic phosgene:

Table 9.8 Physical properties of carbon monoxide

Molecular weight	28.01
Specific volume @ 21°C, 1 atm	13.8 cu.ft/lb (861.5 ml/g)
Boiling point @ 1 atm	–191.5°C
Triple point	–205.01°C
Specific gravity @ 21°C, 1 atm	115.14 mm Hg
Density (liquid) @ bp	0.9678
Latent heat of vaporization @ bp	1444 cal/mole
Latent heat of fusion @ tp	200.9 cal/mole
Flammable limits in air	12.5–74%
Autoignition temperature	650°C
Specific heat (gas) @ 25°C, 1 atm	
C_p	0.2491 cal/g°C
C_v	0.1774 cal/g°C
ratio C_p/C_v	1.4
Viscosity (gas) @ 0°C, 1 atm	0.0166 centipoise
Entropy (gas) @ 25°C	47.266 cal/mole°C
Heat of formation (gas) @ 25°C	–26.417 kcal/mole

$$CO + Cl_2 \rightarrow COCl_2$$

Liquid carbon monoxide in the presence of nitrous oxide poses blast hazards.

Precautions for handling carbon monoxide in compressed gas cylinders in addition to those given in Table 9.3 include:

- Handle in well-ventilated conditions.
- Consider the need for respiratory equipment.
- Use CO gas detection system if used indoors or in confined spaces.
- Check the system periodically for leaks.
- Avoid accidental contact with ignition sources.
- Segregate stocks from oxygen cylinders or other oxidizing or flammable substances.

Table 9.9 First aid measures following exposure to a compressed gas

Obtain medical help immediately	
Inhalation	Remove victim to uncontaminated area and carry out artificial respiration In the case of hydrogen sulphide, ensure that the patient remains rested and refrains from exercise for 24 hr For chlorine gassing, lay victim on stomach with head and shoulders slightly lowered; discourage from coughing
Skin contact	Use emergency shower, removing contaminated clothing and shoes at the same time
Eye contact	Wash promptly with copious amounts of water for ≥15 min

Chlorine

Chlorine can be made on a small scale by oxidation of hydrogen chloride with, e.g., manganese dioxide:

$$MnO_2 + 4HCl \rightarrow MnCl_2 + Cl_2 + 2H_2O$$

Industrially, chlorine is obtained as a by-product in the electrolytic conversion of salt to sodium hydroxide. Hazardous reactions have occurred between chlorine and a variety of chemicals including acetylene, alcohols, aluminium, ammonia, benzene, carbon disulphide, diethyl ether, diethyl zinc, fluorine, hydrocarbons, hydrogen, ferric chloride, metal hydrides, non-metals such as boron and phosphorus, rubber, and steel.

Chlorine is very reactive and finds wide use, e.g. in water purification, sanitation, as a bleaching agent, as a versatile raw material in synthetic chemistry etc. In liquid form, chlorine is a clear amber dense liquid. The gas is greenish-yellow, about 2.5 times as dense as air. Although non-flammable, it will support combustion. Liquid chlorine causes severe irritation and blistering of skin. The gas has a pungent suffocating odour and is irritant to the nose and throat. It is an extremely powerful blistering agent and respiratory irritant. Persons exposed to chlorine become restless, sneeze, develop sore throat and salivate copiously. Effects on the body are summarized in Table 9.10 and physical characteristics are given in Table 9.11.

Moist chlorine is corrosive to skin and to most common materials of construction. Wet chlorine at low pressure can be handled in chemical stonewear, glass or porcelain and in certain alloys and plastics.

The effect of temperature on vapour pressure is shown in Figure 9.3. Cylinders are normally protected from over-pressurization by a fusible metal plug melting at about 85°C.

Table 9.10 Physiological effects of chlorine

Atmospheric concentration (ppm)	*Effects*
1	Minimum concentration causing slight symptoms after several hours
3.5	Minimum concentration detectable by odour
4	Maximum concentration that can be breathed for 1 hr without damage
15	Minimum concentration causing throat irritation
30	Minimum concentration causing coughing
40–60	Concentration dangerous within 30 min
1000	Concentration likely to be fatal after a few deep breaths

Table 9.11 Physical properties of chlorine

Molecular weight	70.906
Vapour pressure at 21°C	5.88 bar
Specific volume at 21°C, 1 atm	337.1 ml/g
Boiling point at 1 atm	–34.05°C
Freezing point at 1 atm	–100.98°C
Specific gravity, gas at 0°C, 1 atm (air = 1)	2.49
Specific gravity, liquid at 20°C	1.41
Density, gas at 0°C, 1 atm	3.214 g/l
Density, liquid at 0°C, 3.65 atm	1.468 g/l
Critical temperature	144°C
Critical pressure	77.1 bar
Critical density	0.573 g/ml
Latent heat of vaporization at boiling point	68.8 cal/g
Heat of fusion at flash point	22.9 cal/g
Specific heat, liquid at 0–24°C	0.226 cal/g °C
Specific heat, gas at 15°C, 1 atm	
C_p	0.115 cal/g °C
C_v	0.085 cal/g °C
ratio C_p/C_v	1.355
Thermal conductivity, gas at 0°C	1.8×10^{-5} cal/s cm^2 °C/cm
Viscosity, gas at 20°C, 1 atm	0.0147 cP
Viscosity, liquid at 20°C	0.325 cP
Solubility in water at 20°C, 1 atm	7.30 g/l

The following safety measures supplement the general precautions listed in Table 9.3:

- Provide convenient showers, eye-wash facilities and appropriate respiratory protection for emergencies.
- Work in well-ventilated area wearing appropriate skin protection and respiratory equipment.
- Check for leaks (e.g. with aqueous ammonia) and consider the need for detection/alarm systems. Leaks should be dealt with immediately after evacuating the area.
- Never connect the cylinder directly to vessels of liquid since suck-back into the cylinder may result in violent reaction. Insert a trap in the line between the chlorine supply and the receiver of sufficient capacity to accommodate all the liquid.
- Never supply heat directly to the cylinder.
- Segregate stocks of chlorine from acetylene, hydrogen, ammonia and fuel gases and ensure no accidental contact with ethers, hydrocarbons and other organics and finely divided metals. Never mix chlorine with another gas in the cylinder.
- In the event of exposure, apply first aid as in Table 9.9 (refer also to Table 13.17).

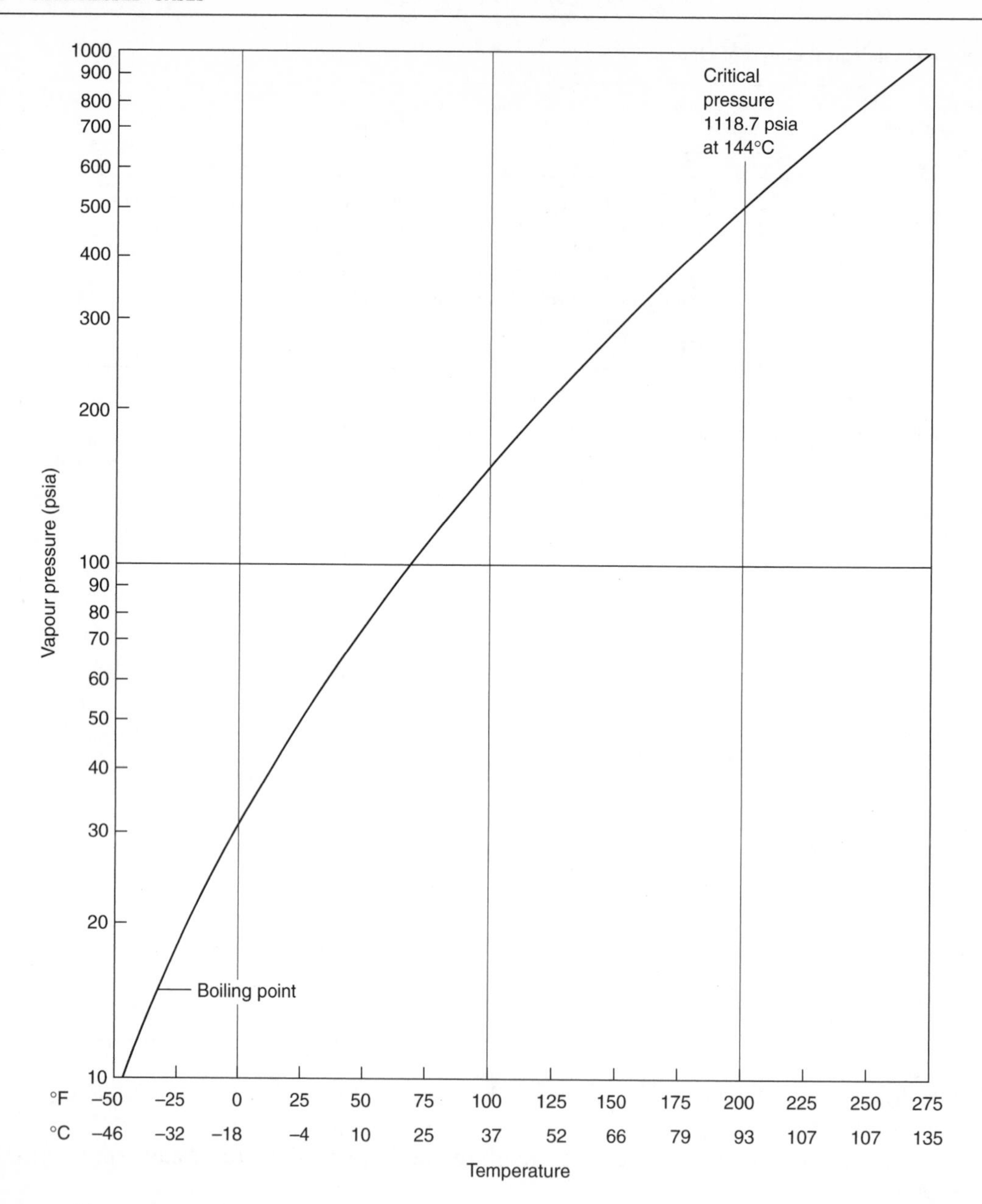

Figure 9.3 *Chlorine vapour pressure versus temperature*

Hydrogen

Hydrogen does not appear free in the atmosphere except at levels below 1 ppm, since rapid diffusivity enables molecules to escape the earth's gravitational field and it is continuously lost from the atmosphere. It is present in the earth's crust at about 0.87% in combination with oxygen in water and with carbon and other elements in organic substances. It is prepared commercially on a small scale by action of sulphuric acid on zinc:

$$Zn + H_2SO_4 \rightarrow ZnSO_4 + H_2$$

and industrially by electrolysis of sodium chloride, or sodium hydroxide, or by reduction of steam by carbon monoxide:

$$C + H_2O \rightarrow CO + H_2O \text{ (water gas)}$$

$$H_2 + CO + H_2O \rightarrow CO_2 + 2H_2$$

Hydrogen is used for the hydrogenation of oils and fats, in metallurgy, metal welding/cutting, ammonia synthesis and petroleum refining. It is the lightest gas known. It is colourless and odourless, only slightly soluble in water but readily soluble in hydrocarbons. Hydrogen is non-toxic but can act as an asphyxiant. It is usually shipped in containers at 2000 psig (137.9 bar) at 21°C, often protected by frangible discs backed up by a fusible metal plug melting at 100°C. Physical properties are given in Table 9.12.

Table 9.12 Physical properties of hydrogen

Molecular weight	2.016
Specific volume at 21°C, 1 atm	11 967 ml/g
Boiling point at 1 atm	–252.9°C
Triple point at 0.0695 atm	–259.3°C
Specific gravity, gas at 23.9°C, 1 atm (air = 1)	0.06952
Density, gas at 0°C, 1 atm	0.0899 g/l
Density, liquid at –253°C, 1 atm	0.0708 g/ml
Critical temperature	–240.2°C
Critical pressure	12.98 bar
Critical density	0.03136 g/ml
Latent heat of vaporization at boiling point	106.5 cal/g
Latent heat of fusion at triple point	13.875 cal/g
Flammable limits in air	4.0–75% by volume
Auto-ignition temperature	585°C
Specific heat, gas at 0–200°C, 1 atm	
C_p	3.44 cal/g °C
C_v	2.46 cal/g °C
ratio C_p/C_v	1.40
Thermal conductivity at 0°C	0.00040 cal/s cm^2 °C/cm
Viscosity, gas at 15°C, 1 atm	0.0087 cP
Solubility in water at 15.6°C, 1 atm	0.019 vol/vol H_2O

The main danger with hydrogen is of fire or explosion. Hydrogen burns in chlorine to yield hydrogen chloride. Although relatively inactive at ambient temperature it reacts with many elements either at high temperatures or in the presence of catalysts and can react dangerously with air, acetylene, aromatics, unsaturated organic matter, halogens, metals such as lithium, calcium, barium, strontium and potassium, and with oxidants such as chlorine dioxide, oxides of nitrogen and palladium oxides. The following precautions are important to supplement those in Table 9.3:

- Use only in well-ventilated conditions to avoid accumulation at high levels.
- Eliminate means of accidental ignition.
- Use only explosion-proof electrical equipment and spark-proof tools.
- Ground all equipment and lines used with hydrogen.
- Check for leaks with soapy water and consider the need for automatic detection/alarms.

Hydrogen chloride

Hydrogen chloride may be conveniently prepared by heating sodium chloride with sulphuric acid:

$$NaCl + H_2SO_4 \rightarrow NaHSO_4 + HCl$$

When this reaction has occurred accidentally sufficient hydrogen chloride has been liberated to explosively burst the vessel. The purest form of hydrogen chloride is made by the action of water on silicon tetrachloride:

$$2H_2O + SiCl_4 \rightarrow SiO_2 + 4HCl$$

Commercially, hydrogen chloride is obtained either as a by-product in the manufacture of salt cake from sodium chloride, or by allowing chlorine produced as a by-product in electrolytic processes to react with hydrogen in the presence of activated charcoal. It is also formed as a by-product in the manufacture of phenol.

Anhydrous hydrogen chloride is a colourless, pungent, heavy, corrosive, thermally-stable gas with a suffocating odour. It is heavier than air and fumes strongly in moist air and is highly soluble in water with evolution of much heat. Physical properties are given in Table 9.13 and its pressure vs temperature profile in Figure 9.4. It is shipped as a liquefied gas with a cylinder pressure of about 613 psig at 21°C and platinum coated frangible bursting discs and fusible metal plugs. Its main uses are as a chemical intermediate and in hydrochlorinations. Its toxicity results from its severe irritating effects to the upper respiratory tract and corrosivity towards skin, eyes and mucous membranes. Neutralization of alkalis in tissues can result in death from oedema or spasm of the larynx. At exposures of 50–100 ppm work is impossible, and difficult at 10–50 ppm; the TLV is a short-term exposure limit of 5 ppm ceiling.

Table 9.13 Physical properties of hydrogen chloride

Molecular weight	36.46
Vapour pressure @ 21°C	613 psig
Specific volume @ 21°C, 1 atm	661.7 ml/g
Boiling point @ 1 atm	–85.03°C
Freezing point @ 1 atm	–114.19°C
Specific gravity (gas) @ 0°C, 1 atm	1.268
Density (gas) @ 0°C, 1 atm	1.639 g/l
Density (liquid) @ –36°C	1.194 g/l
Critical temperature	51.4°C
Critical pressure	1198 psia (81.5 atm)
Critical density	0.42 g/ml
Latent heat of vaporization @ bp	103.12 cal/g
Specific heat (gas) @ 15°C, 1 atm	
C_p	0.1939 cal/g °C
C_v	0.1375 cal/g °C
ratio C_p/C_v	1.41
Viscosity (gas) @ 20°C, 1 atm	0.0156 cP
Solubility in water @ 0°C, 1 atm	82.31 g/100 g water

The gas is essentially inert to common materials of construction such as stainless steel under normal conditions of use. Platinum and gold are also not attacked by pure hydrogen chloride.

In the presence of moisture, however, most metals are corroded and it is advised that the proposed use and pressures are discussed with the supplier so that suitable construction materials are established prior to installation. Hydrogen chloride neither burns nor supports combustion,

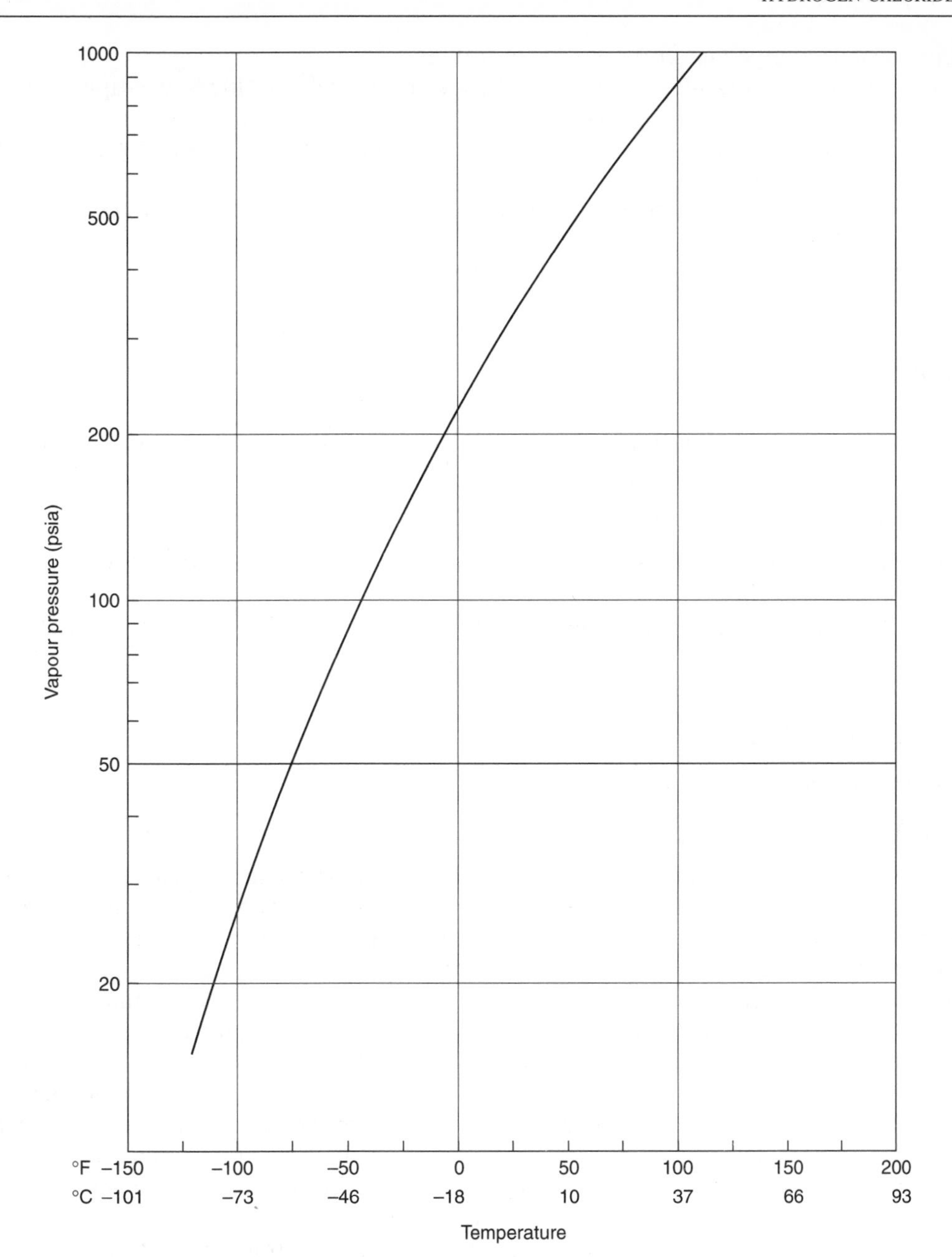

Figure 9.4 *Hydrogen chloride vapour pressure vs temperature*

although burning sodium will continue to burn in it forming hydrogen and sodium chloride. Hydrogen chloride gas has produced runaway reactions with dinitrotoluene, fluorine (with ignition), sodium, and alcoholic hydrogen cyanide.

Liquid hydrogen chloride does not conduct electricity and is without action on zinc, iron, magnesium, calcium oxide and certain carbonates. However, it does dissolve aluminium.

Some special precautions for use of compressed hydrogen chloride gas include:

- Store and use under ventilated conditions.

- Avoid galvanized pipe and brass or bronze fittings.
- Wear protective clothing such as rubber or plastic aprons, rubber gloves, gas-tight goggles and respiratory equipment as appropriate.
- Ensure fast-acting showers are available close to site of use/storage plus eye-wash fountains or similar facilities for eye irrigation.
- Prevent suck-back of foreign material into the cylinder by use of check valves or vacuum break traps.
- Switch off gas lines from use backwards to the cylinder.

Hydrogen sulphide

Hydrogen sulphide is usually prepared on a small scale by the action of hydrochloric acid on ferrous sulphide:

$$FeS + 2HCl \rightarrow FeCl_2 + H_2S$$

The purest form is obtained by passing a mixture of sulphur vapour and hydrogen over finely divided nickel at 450°C.

Hydrogen sulphide is used in the preparation of metal sulphides, oil additives etc., in the purification and separation of metals, as an analytical reagent and as raw material in organic synthesis. It burns in air with a blue flame:

$$2H_2S + 3O_2 \rightarrow 2SO_2 + 2H_2O$$

or if oxygen is depleted:

$$2H_2S + O_2 \rightarrow 2H_2O + S$$

Hydrogen sulphide occurs naturally, e.g. in natural gas and petroleum, volcanic gases, and from decaying organic matter. It may be present near oil wells and where petroleum is processed. Commercially it is obtained as a by-product from many chemical reactions including off-gas in the production of some synthetic polymers (e.g. rayon, nylon) from petroleum products, and by the action of dilute mineral acids on metal sulphides. Physical properties are summarized in Table 9.14 and effects of temperature on vapour pressure are shown in Figure 9.5.

Hydrogen sulphide is a dense, colourless, highly flammable water-soluble gas with an offensive odour of rotten eggs. It is highly toxic; its effects on the body are given in Table 5.32. Acute poisoning may result from exposures at or above 700 ppm due to systemic effects, including attack on the nervous system and respiratory collapse. Hydrogen sulphide may become rapidly oxidized on contact with a range of metal oxides and in certain cases may ignite or explode. It can also react dangerously with a host of oxidants, rust and soda lime.

Cylinders are typically protected from over-pressurization by frangible gold-plated discs and fusible plugs.

Important precautions include:

- Use in well-ventilated conditions and eliminate sources of ignition.
- Operators should work in pairs.
- Do not rely on the sense of smell to detect hydrogen sulphide leaks. Strips of wet lead acetate paper turn black on exposure to hydrogen sulphide and offer a simple indicator, as do colour indicator tubes. For plant-scale operations, instrumental multi-point detectors and alarms are likely to be more appropriate.

Table 9.14 Physical properties of hydrogen sulphide

Molecular weight	34.08
Vapour pressure at 21°C	17.4 bar
Specific volume at 21°C, 1 atm	701 ml/g
Boiling point at 1 atm	–60.33°C
Freezing point at 1 atm	–85.49°C
Specific gravity, gas at 15°C, 1 atm (air = 1)	1.1895
Density, gas at 0°C, 1 atm	1.5392 g/l
Density, liquid at boiling point	0.993 g/ml
Critical temperature	100.4°C
Critical pressure	90.23 bar
Critical density	0.349 g/ml
Latent heat of vaporization at boiling point	131 cal/g
Latent heat of fusion at melting point	16.7 cal/g
Specific heat, gas at 25°C, 1 atm	
C_p	0.240 cal/g °C
C_v	0.181 cal/g °C
ratio C_p/C_v	1.32
Thermal conductivity at 0°C	3.05×10^{-5} cal/s cm^2 °C/cm
Flammable limits in air	4.3–45% by volume
Auto-ignition temperature	260°C
Solubility in water at 20°C, 1 atm	0.672 g/100 ml water
Viscosity, gas at 0°C, 1 atm	0.01166 cP

- Segregate cylinders of hydrogen sulphide from oxygen or other highly-oxidizing or combustible materials.
- Ground all lines and equipment used with hydrogen sulphide.
- Insert traps in the line to prevent suck-back of liquid into the cylinder.
- Provide respiratory protection for emergencies.
- In the event of exposure, apply first aid as indicated in Table 9.9.

Liquefied petroleum gases (LPG)

LPG is a mixture of propane and *n*- and iso-butanes, plus small amounts of their olefinic counterparts. The main sources are natural gas wells, gas from crude oil wells and the cracking of crude oil. The requirements for commercial LPG are defined in national standards and a stenching agent is added for some uses.

The common LPGs in general use are commercial propane, comprising predominantly propane and/or propylene, and commercial butane. The physico-chemical properties of propane and the butanes are given in Table 9.15. These compounds are gaseous at normal ambient temperature and pressure but are readily liquefied by the application of moderate pressure. They are stored and distributed as liquids in low pressure cylinders or bulk containers at ambient temperature and allowed to revert to gas at the point of use. Large-scale storage and shipment by sea is in refrigerated vessels at close to atmospheric pressure.

Butane itself is considered to be insoluble in water. Exposures of up to 5% for 2 hours appear not to present problems. The TLV is 800 ppm. The relationship between pressure and temperature is given by Figure 9.6.

Propane has a characteristic natural gas odour and is basically insoluble in water. It is a simple asphyxiant but at high concentrations has an anaesthetic effect. The TLV is 2500 ppm. It is usually shipped in low-pressure cylinders as liquefied gas under its own vapour pressure of ca 109 psig at 21°C. Its pressure/temperature profile is given in Figure 9.7.

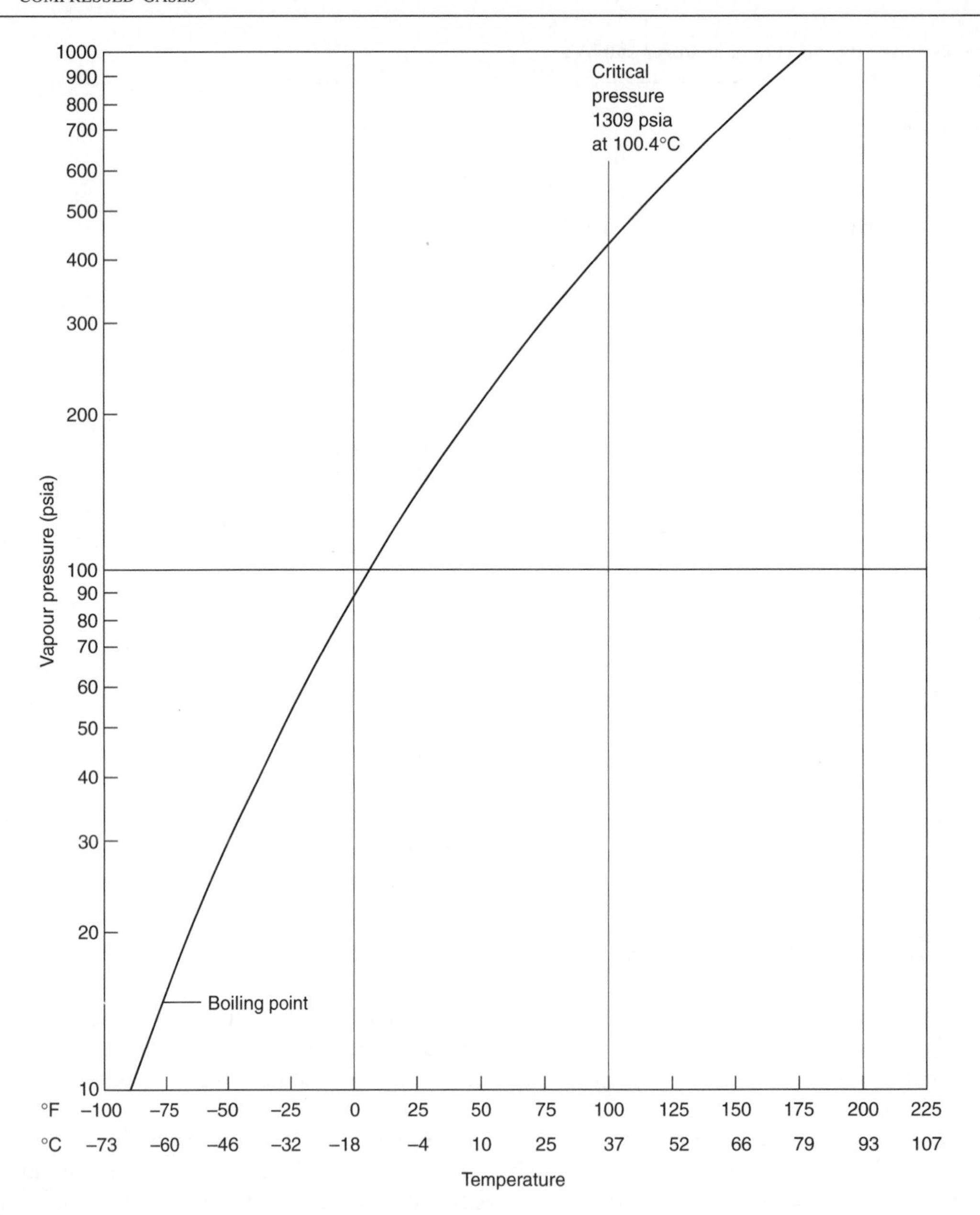

Figure 9.5 *Hydrogen sulphide vapour pressure vs temperature*

LPG is considered to be non-toxic with no chronic effects, but the vapour is slightly anaesthetic. In sufficiently high concentrations, resulting in oxygen deficiency, it will result in physical asphyxiation. The gases are colourless and odourless but an odorant or stenching agent (e.g. methyl mercaptan or dimethyl sulphide) is normally added to facilitate detection by smell down to approximately 0.4% by volume in air, i.e. one-fifth of the lower flammable limit. The odorant is not added for specific applications, e.g. cosmetic aerosol propellant.

The main danger with LPG arises from its flammability. Fire or explosion may be fuelled by gas escape from leaking cylinders, from an appliance which has not been turned off properly, or

Table 9.15 Properties of LPGs

	Propane	*Butane*	*isoButane*
Molecular weight	44.1	58.1	58.1
Vapour pressure at 21°C, i.e. cylinder pressure (kg/cm^2 gauge)	7.7	1.15	21.6
Specific volume at 21°C/l atm (ml/g)	530.6	399.5	405.8
Bp at 1 atm (°C)	–42.07	–0.5	–11.73
Mp at 1 atm (°C)	–187.69	–138.3	–159.6
SG, gas at 16°C/l atm (air = 1)	1.5503	2.076	2.01
Density, liquid at sat. pressure (g/ml)	0.5505 (20°C)	0.5788 (20°C)	0.563 (15°C)
Density, gas at 0°C/l atm (kg/m^3)	2.02	2.70	–
Critical temperature (°C)	96.8	152	135
Critical pressure (atm)	42	37.5	37.2
Critical density (g/ml)	0.220	0.225	0.221
Latent heat of vap. at bp (cal/g)	101.76	92.0	87.56
Latent heat of fusion at mp (cal/g)	19.10	19.17	18.67
Specific heat, liquid at 16°C (cal/g°C)	–	0.5636	0.5695
Specific heat, gas at 16°C:			
C_p (cal/g°C)	0.3885	0.3908	0.3872
C_v (cal/g°C)	0.3434	0.3566	0.3530
ratio C_p/C_v	1.13	1.1	1.1
Specific heat ratio at 16°C/l atm, C_p/C_v	1.131	1.096	1.097
Gross heat of combustion at 16°C/l atm (cal/ml)	22.8	30.0	29.8
Viscosity, gas at 1 atm (centipoise)	0.00803 (16°C)	0.0084 (15°C)	0.00755 (23°C)
Coefficient of cubical expansion at 15°C (per °C)	0.0016	0.0011	–
Surface tension (dynes/cm)	16.49 (–50°C)	16.02 (–10°C)	15.28 (–20°C)
Solubility in water at 1 atm (volumes/100 volumes water)	6.5 (18°C)	–	1.7 (17°C)
Flammable limits in air (% by volume)	2.2–9.5	1.9–8.5	1.8–8.4
Autoignition temp. (°C)	467.8	405	543
Max. explosion pressure (MPa)	0.86	0.86	–
Min. ignition energy (MJ)	0.25	0.25	–
Max. flame temperature (°C)	2155	2130	–
Max. burning velocity (m/s)	0.45	0.38	–
Necessary min. inert gas conc. for explosion prevention in case of emergent outflow of gas in closed volumes (% v/v):			
Nitrogen	45	41	–
Carbon dioxide	32	29	–

from an appliance in which the flame has been extinguished. Any fire near an LPG cylinder may cause it to overheat and catch fire, or result in a BLEVE with missiles projected over long distances.

The following safety measures supplement the general precautions listed in Table 9.3:

- Do not store or use a cylinder on its side, but upright with the valve uppermost.
- Do not store or leave full or empty cylinders below ground level.
- Never change a cylinder without first closing the cylinder valve and extinguishing naked lights in the vicinity.

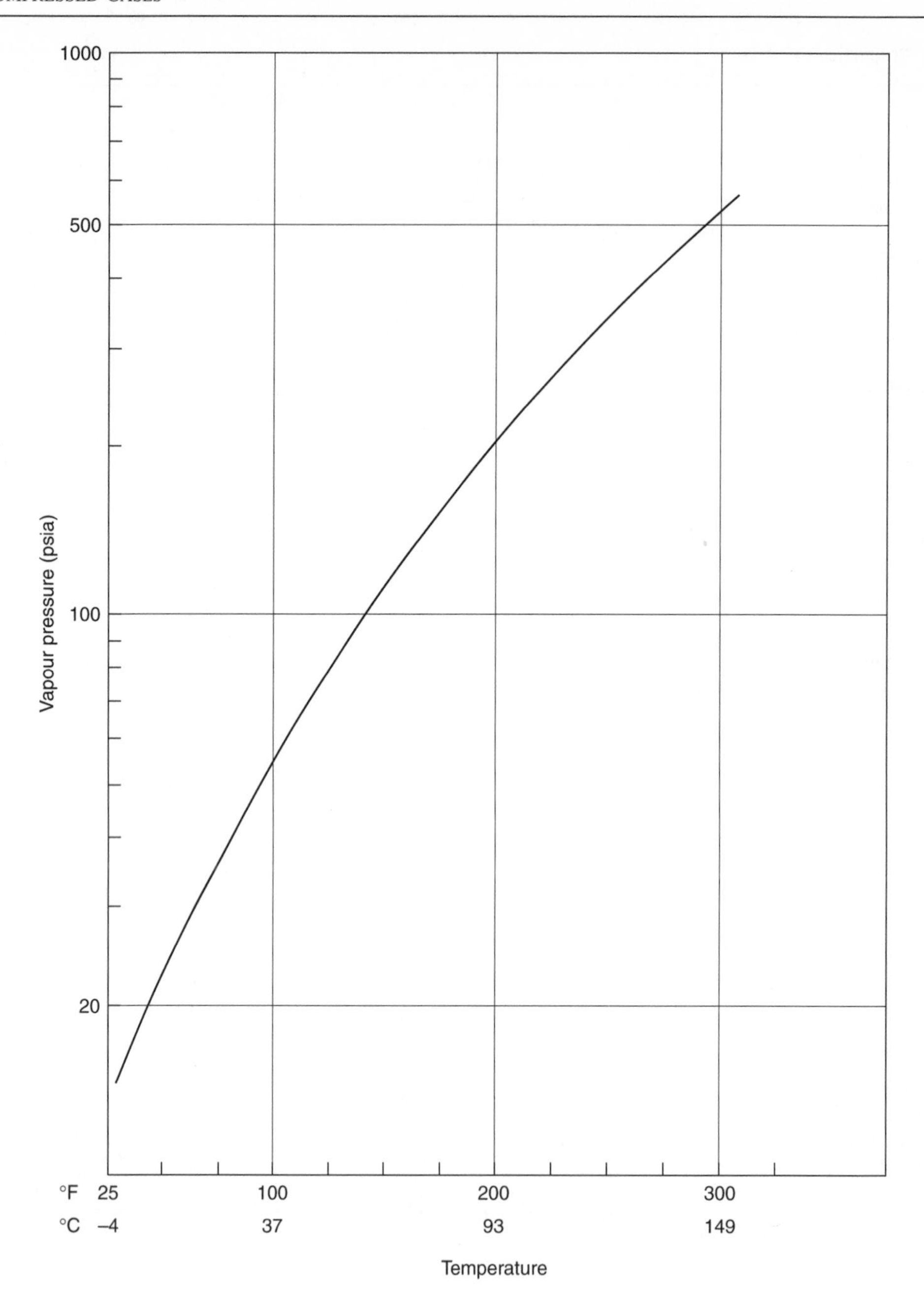

Figure 9.6 *n-Butane vapour pressure vs temperature*

- Store cylinders, both full and empty, in a cool location away from flammable, toxic, or corrosive materials and preferably at least 6.1 m from any source of ignition or heat.
- Replace any valve-protecting covers on empty cylinders, or those not in use. Handle cylinders with care.
- Unless the installation is designed to be permanent, always disconnect a cylinder after use.
- Always wear appropriate protective equipment, including gloves and goggles, when filling an LPG cylinder.

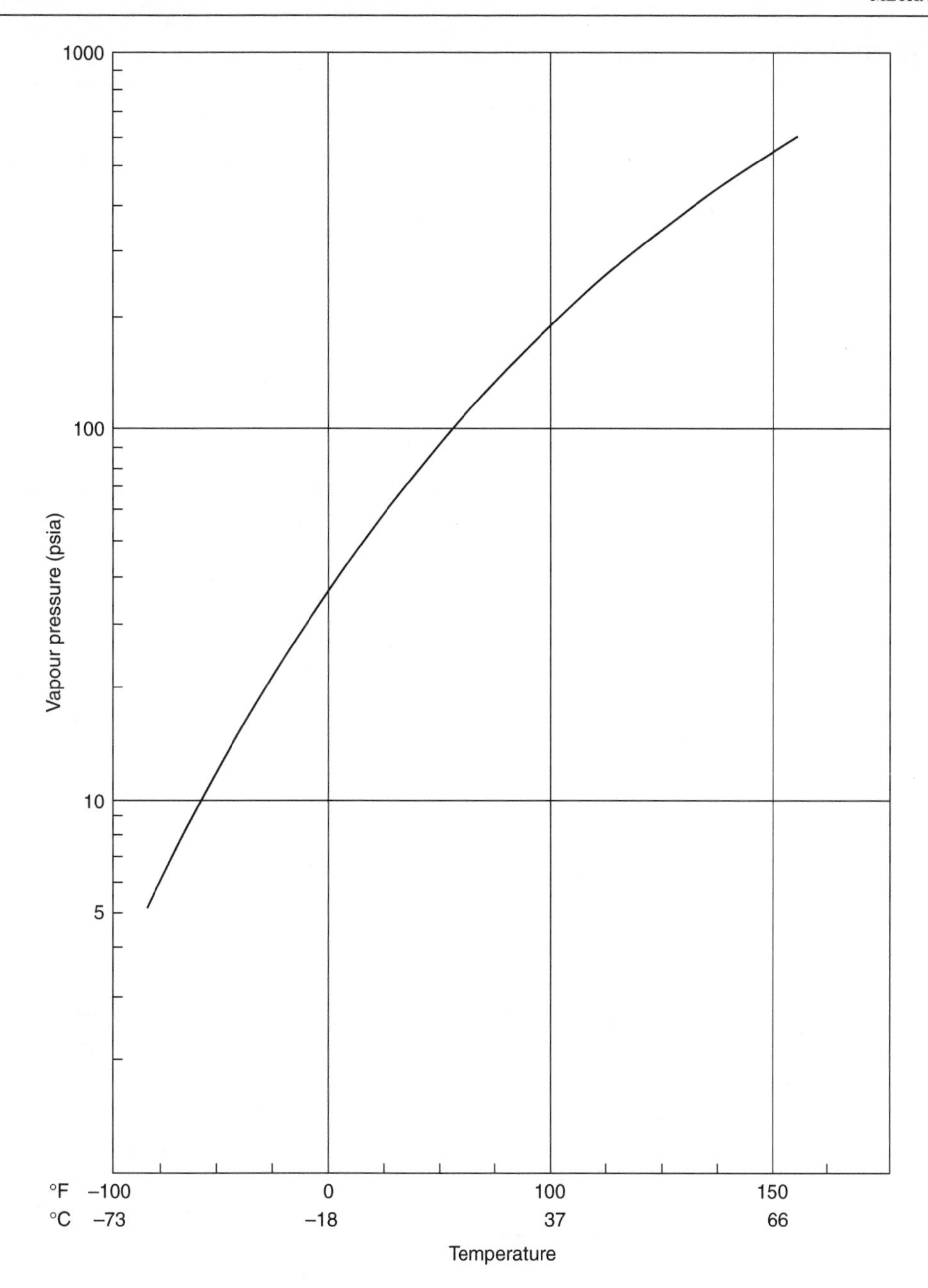

Figure 9.7 *Propane vapour pressure vs temperature*

The general requirements for the storage of LPG cylinders are given in Tables 9.16 to 9.18. Considerations in the transport of cylinders by road are given in Chapter 15.

Methane

Methane is obtained commercially from natural gas wells and from cracking of petroleum fractions.

Table 9.16 General requirements for storage of LPG cylinders

- Store in the open in a well-ventilated area at ground level.
- Avoid other materials stacked near restricting natural ventilation.
  Flammable liquid and combustible, corrosive, oxidizing or toxic substances and compressed gases should be kept separately.
- Avoid cylinders impeding or endangering means of escape from premises or from adjoining premises.
- The store floor should be level. A load-bearing surface (concrete, paved or compacted) is required where cylinders are to be stacked.
- Security arrangements should be such as to prevent tampering or vandals. Fencing should be of robust wire mesh which does not obstruct ventilation.
- The area should be at the separation distance from the property boundary, any building or fixed source of ignition quoted in Table 9.17. This may only be reduced if suitable fire-resisting separation is provided.
- To give additional protection from thermal radiation from any cylinder stack fire the minimum separation distance to any nearby building housing a vulnerable population should, for quantities >400 kg, be 8 m or as in Table 9.17, column 3, whichever is the greater. This may be reduced to those in Table 9.17, column 4, by the installation of fire-resisting separation or a fire wall.
- Prohibit all sources of ignition, including smoking, within a store or within the separation distance (exclude motor vehicles other than fork-lift trucks and those for delivery/collection from open-air stores).
- Avoid openings into buildings, cellars, or pits within 2 m or the separation distances, whichever is greater. (Any gully or drain unavoidably within 2 m should have the opening securely covered or fitted with a water seal to prevent vapour ingress.)
- Electrical equipment suitable for a Zone 2 area (e.g. to BS 5345) and constructed to a recognized standard should be installed within the store and separation distance. Zone 2 areas are summarized in Table 9.18.
- Clearly mark the area with notices indicating an LPG storage area, flammable contents, prohibition of ignition sources and procedures to follow in case of fire.
- Avoid accumulation of rubbish, small bushes, dry leaves etc. within the separation distance. Remove weeds and long grass within the separation distance and up to 3 m from cylinders.
- Cylinders should be stored upright with the valves closed with any protective cover, cap or plug in place. Cartridges without valves may be stored on their sides.
- Cylinders on a vehicle or trailer parked overnight rate as a single stack so that the separation distance in Table 9.17, column 3, is applicable.
- Cylinders received into store and taken out for delivery should be checked for damage or leakage. Stacks should be inspected daily for stability and that they contain no damaged/leaking cylinders.
- Cylinders should be handled carefully to avoid personal injury or damage to them.

The physical properties of methane are given in Table 9.19 and the relationship between pressure and temperature in Figure 9.8. It is considered to be non-toxic but is an asphyxiant at high concentrations. Coalminers have inhaled air containing up to 9% methane with no apparent untoward effect but higher concentrations of at least 10% in air result in a feeling of pressure on the eyes and forehead which will disappear after breathing fresh air. Characteristic symptoms of asphyxiation occur at higher concentrations (i.e. rapid respiration, fatigue, nausea, vomiting possibly leading to loss of consciousness and anoxia). At low concentrations the odour of methane is undetectable; at exposures of >5000 ppm there is a sweet oil-type odour. Selected grades contain an odorizer, e.g. natural gas for commercial and domestic full.

It is considered to be chemically inert at room temperature and atmospheric pressure but it does react under more forcing conditions and vigorous reactions have occurred with halogens, interhalogen compounds, and oxidizing agents such as liquid oxygen and dioxygen difluoride.

The main hazard is that of flammability. The following precautions supplement those in Table 9.3 for the storage of methane gas cylinders:

- Store in a purpose-built compound preferably in the open air and well ventilated.
- Locate free from fire risk and away from sources of heat and ignition.
- Maintain clear access, and restrict access to the compound. Mark with hazard warning signs.
- Prohibit the use of naked flames or smoking in the vicinity of, or inside, the compound.

Table 9.17 Recommended minimum separation distances for total storage of LPG in cylinders or size of maximum stack (the greater of the two distances is advised)

(1) *Total quantity LPG store* (kg)	(2) *Size of largest stack* (kg)	(3) *Minimum separation distance to boundary, building or fixed ignition sources from the nearest cylinder (where no fire wall provided)* (m)	(4) *Minimum separation distance to boundary, building or fixed ignition source from fire wall (where provided)* * *** (m)
From 15 to 400		1***	Nil
" 400 to 1000	Up to 1000	3	1
" 1000 to 4000		4	1
" 4000 to 6000	From 1000 to 3000	5	1.5
" 6000 to 12 000		6	2
" 12 000 to 20 000	" 3000 to 5000	7	2.5
" 20 000 to 30 000	" 5000 to 7000	8	3
" 30 000 to 50 000	" 7000 to 9000	9	3.5
" 50 000 to 60 000	" 9000 to 10 000	10	4
" 60 000 to 100 000		11	4.5
" 100 000 to 150 000	" 10 000 to 20 000	12	5
" 150 000 to 250 000	" 20 000 to 30 000	15	6
Above 250 000		20	7

*The distance from the nearest cylinder to a boundary, building etc. should be not less than in column 3 when measured around the fire wall.
**Minimum distance from nearest cylinder to fire wall should normally be 1.5 m except as qualified.
***No separation distance is required for these quantities where boundary walls and buildings are of suitable construction.

Table 9.18 Areas classified as requiring Zone 2 electrical equipment

Location	*Extent of classified area*
Storage in open air	In the storage area up to a height of 1.5 m above the top of the stack or beneath any roof over the storage place. Outside the storage area or the space covered by any roof up to 1.5 m above ground level and within the distance set out for a fixed source of ignition in Table 9.17, column 3.
Storage within a specially designed building or in a specially designed storage area within a building	The entire space within the building or storage area and outside any doorway, low level ventilator or other opening into the store within the separation distance set out in Table 9.17, column 3, up to a height of 1.5 m above ground level.

- Store separately from oxygen and oxidants, using a fire-resistant partition where necessary.
- Store the minimum quantity practicable.
- Check periodically for general condition and leakage.

Nitrogen

Nitrogen is an odourless, colourless gas which comprises ca 79% by volume of air and is an essential constituent of all living organisms, e.g. as protein. It is made in the laboratory by gently

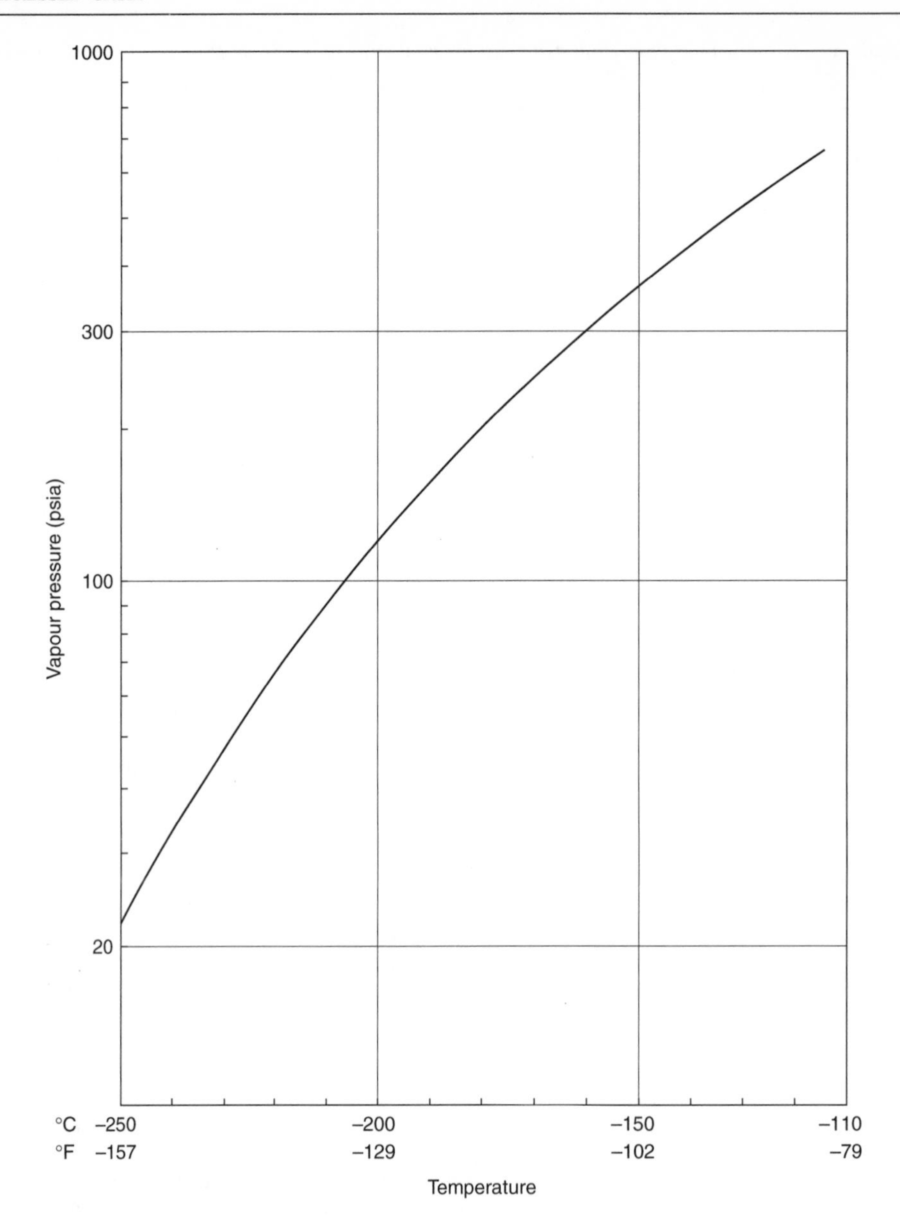

Figure 9.8 *Methane vapour pressure vs temperature*

warming a concentrated solution of equimolar proportions of ammonium chloride and sodium nitrite:

$$NH_4Cl + NaNO_2 \rightarrow NaCl + NH_4NO_2 \rightarrow N_2 + 2H_2O$$

On the industrial scale nitrogen is obtained by the fractional distillation of liquid air (see page 261). It will neither burn nor support combustion. Nitrogen is distributed as a pressurized gas in grey cylinders with black shoulders. (Oxygen-free nitrogen has opposed white spots on the cylinder shoulder.) It finds wide use in its inert capacity, e.g. in electrical equipment and in the chemical and food industries. Large volumes of atmospheric nitrogen are converted into ammonium sulphate for fertilizer and into nitric acid.

Table 9.19 Physical properties of methane

Molecular weight	16.04
Specific volume @ 21°C	1479.5 ml/g
Boiling point @ 1 atm	–161.5°C
Freezing point @ 1 atm	–182.6°C
Triple point	–182.5°C
Triple point pressure	0.115 atm
Specific gravity (gas) @ 16°C, 1 atm	0.5549
Density (gas) @ 0°C, 1 atm	0.72 g/l
Density (liquid) @ bp	0.4256 g/l
Critical temperature	–82.1°C
Critical pressure	673.3 psia (45.8 atm)
Critical density	0.162 g/ml
Latent heat of vaporization @ bp	121.54 cal/g
Specific heat (gas) @ 16°C, 1 atm C_p/C_v	1.307
Flammable limits in air	5.3–14%
Autoignition temperature	540°C
Minimum ignition energy	0.28 mJ
Flame temperature	1880°C
Limiting oxygen index	11.5%
Viscosity (gas) @ 1 atm, 4.4°C	0.0106 cP

Since it is chemically inert no special materials of construction are required. Selected physical properties are listed in Table 9.20.

Nitrogen is non-toxic but will cause asphyxiation through oxygen depletion of air and is the most common cause of gassing accidents in industry. There are no warning signs before unconsciousness occurs, and at high concentrations almost instant unconsciousness may occur and prove fatal (page 77). Table 5.7 summarizes the health effects of oxygen-deficient atmospheres.

Nitrogen oxides

The various oxides of nitrogen are given in Chapter 5.

Nitrous oxide is a colourless, non-flammable, non-corrosive gas with a sweetish odour and taste. It is prepared both in the laboratory and on a commercial scale by heating ammonium nitrate:

$$NH_4NO_3 \rightarrow N_2O + 2H_2O$$

Heating must be terminated when two-thirds of the nitrate has decomposed since explosive nitrogen trichloride may be formed from traces of ammonium chloride impurity.

Nitrous oxide may also be obtained by the controlled reduction of nitrates or nitrites, decomposition of hyponitrites, or thermal decomposition of hydroxylamine.

It is generally considered not to be toxic and non-irritating. One of its main applications is, in combination with air or oxygen, as a weak anaesthetic in medicine and dentistry. At low concentrations it produces hysteria (hence the term laughing gas). At high concentrations in the absence of air it is a simple asphyxiant. It is also used as a dispersing agent in whipping cream.

It does not react with oxygen, ozone, hydrogen, chlorine, potassium, phosphine or aqua regia. However, whilst stable at ordinary temperatures, it decomposes readily to oxygen and nitrogen at 600°C and therefore supports combustion of burning substances. When nitrous oxide is transferred from a stock cylinder to smaller cylinders the gas expansion results in cooling and a reduction in pressure. On heating the stock container to increase pressure and thereby facilitate transfer the gas

Table 9.20 Physical properties of nitrogen

Molecular weight	28.0134
Specific volume @ 21°C	861.5 ml/g
Boiling point @ 1 atm	–195.8°C
Triple point	–210.0°C
Triple point pressure	94.24 mm Hg
Density (gas) @ 20°C	1.250 g/l
Density (liquid) @ bp	0.8064 g/ml
Critical temperature	–147.1°C
Critical pressure	33.5 atm
Critical density	0.311 g/ml
Latent heat of vaporization @ bp	47.51 cal/g
Specific heat (gas) @ 16°C, 1 atm	
C_p	0.2477 cal/g °C
C_v	0.1765 cal/g °C
ratio C_p/C_v	1.4
Viscosity (gas) @ 15°C, 1 atm	0.01744 centipoises
Dielectric constant (liquid) @ bp	1.433
Solubility in water @ 0°C	2.3 ml/100 ml water

decomposes causing explosive rupture of the stock cylinder. This emphasizes the importance of keeping cylinders away from heat sources, e.g. oxyacetylene welding operations. Explosive reaction has also occurred between nitrous oxide and a variety of chemicals including amorphous boron, ammonia, carbon monoxide, hydrogen, hydrogen sulphide, and phosphine. Because the oxygen content of 36.4% is higher than the 21% in air, combustion or oxidation in nitrous oxide is much faster than in air.

It is transported in high pressure steel cylinders equipped with brass valves.

Physical properties are summarized in Table 9.21 and the vapour pressure/temperature relationship is depicted by Figure 9.9. The general precautions in Table 9.3 apply.

Nitric oxide is made commercially by oxidation of ammonia above 500°C in the presence of platinum, or by reduction of nitrous acid with ferrous sulphate or ferrous halides. The physical

Table 9.21 Physical properties of nitrous oxide

Molecular weight	44.013
Vapour pressure @ 21°C	745 psig
Specific volume @ 21°C, 1 atm	543 ml/g
Boiling point @ 1 atm	–89.5°C
Freezing point @ 1 atm	–90.84°C
Specific gravity (gas) @ 15°C, 1 atm	1.530
Density (gas) @ 0°C, 1 atm	1.907 g/l
Density (liquid) @ bp	1.266 g/l
Critical temperature	36.5°C
Critical pressure	1054 psia (71.7 atm)
Critical density	0.457 g/l
Latent heat of vaporization @ bp	89.9 cal/g
Specific heat gas @ 25°C, 1 atm	
C_p	0.2098 cal/g °C
C_v	0.1610 cal/g °C
ratio C_p/C_v	1.3
Viscosity (gas) @ 0°C, 1 atm	0.01362 centipoise
Solubility in water @ 0°C, 1 atm	1.3 volumes/volume of water

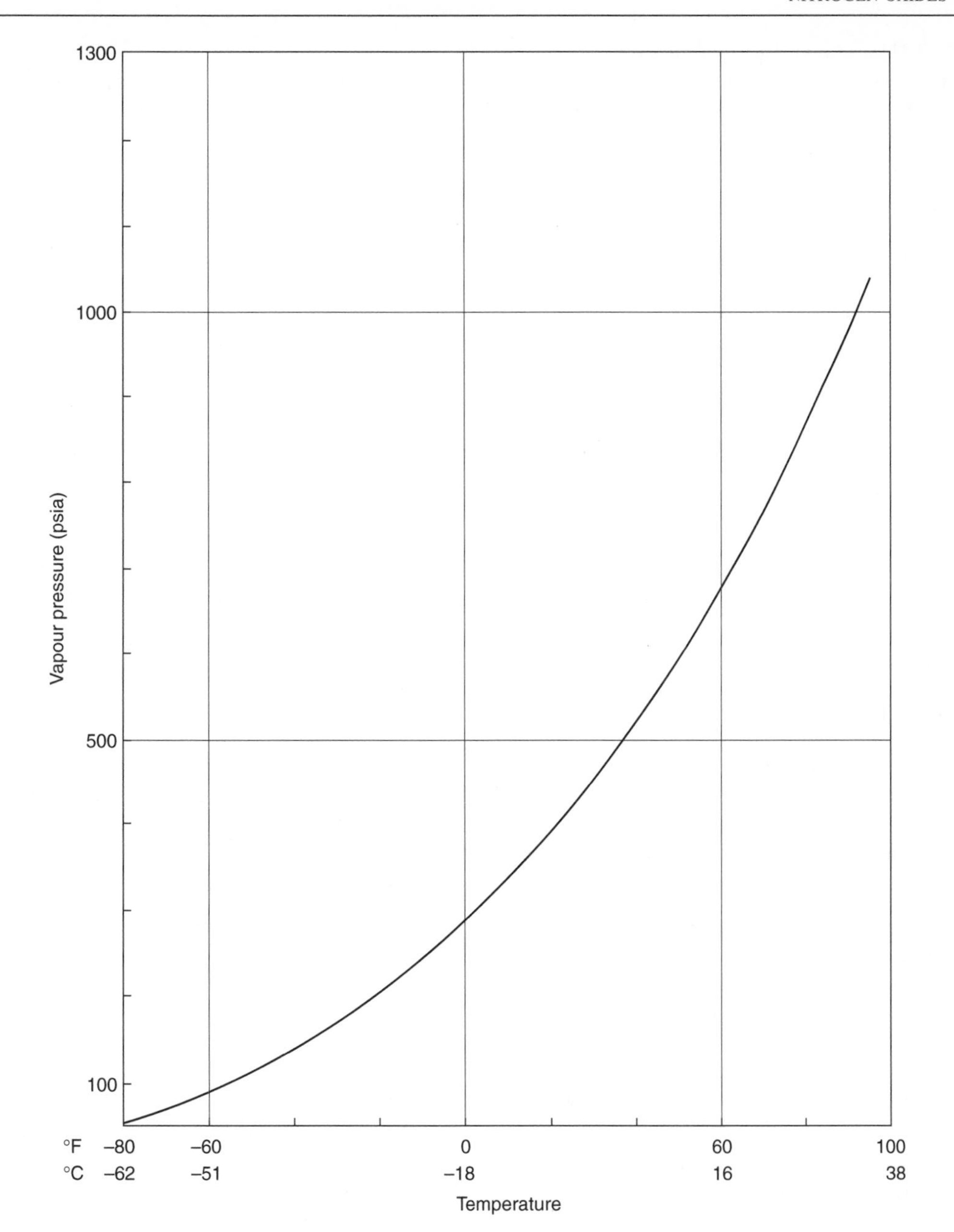

Figure 9.9 *Nitrous oxide vapour pressure vs temperature*

properties of nitric oxide are given in Table 9.22. This is also a colourless, non-flammable gas which is basically non-corrosive to standard materials of construction. Liquid nitric oxide like other cryogenic oxidizers (e.g. ozone) is very sensitive to detonation in the absence of fuel. It is the simplest molecule that is capable of detonation in all three phases: the liquid oxide may explode during distillation. Nitric oxide is more toxic than nitrous oxide. It decomposes into its elements only when heated to 1000°C and therefore it is not a ready supporter of combustion. Vigorously burning phosphorus continues to burn in the gas but burning sulphur or charcoal is extinguished.

Table 9.22 Physical properties of nitric oxide

Molecular weight	30.006
Specific volume @ 21°C, 1 atm	811 ml/g
Boiling point @ 1 atm	–151.7°C
Freezing point @ 1 atm	–163.6°C
Density (gas) @ 0°C, 1 atm	1.3402 g/l
Density (liquid) @ bp	1.269 g/l
Critical temperature	–93°C
Critical pressure	940.8 psia (64 atm)
Critical density	0.52 g/ml
Latent heat of vaporization @ bp	110.2 cal/g
Specific heat (gas) @ 15°C, 1 atm	
C_p	0.2328 cal/g°C
C_v	0.1664 cal/g°C
ratio C_p/C_v	1.4
Viscosity (gas) @ 15°C 1 atm	0.0178 cP
Solubility in water @ 0°C, 1 atm	7.34 ml/100 g water

Nitric oxide combines readily with atmospheric oxygen at ambient temperature to produce brown fumes of pungent nitrogen dioxide, and in the presence of charcoal with chlorine to form nitrosyl chloride:

$$2NO + O_2 \rightarrow NO_2$$

$$2NO + Cl_2 \rightarrow 2NOCl$$

It is shipped in cylinders as a non-liquefied gas at a pressure of about 500 psig at 21°C and is primarily used as a chemical intermediate (e.g. in production of nitric acid). It is somewhat soluble in water producing nitric acid, making the gas slightly irritating to the lower reaches of the respiratory system and mucous membranes causing congestion to the throat, bronchi and oedema of the lungs with little warning. The acids neutralize alkali in tissue with subsequent adverse effect on blood pressure, producing headaches and dizziness. The OEL is 25 ppm (8 hr TWA) and 35 ppm (15 min STEL), but generation of the more toxic dioxide on release to air is always a major consideration.

Some special safety measures include:

- Handle only in well-ventilated areas, preferable with a hood equipped with forced ventilation.
- Provide adequate number of exits from the work area.
- Use approved personal protection as appropriate, including self-contained breathing apparatus and associated training.
- Provide suitable first-aid cover.
- Prevent air contamination in high-pressure reactions since the nitrogen dioxide which could form may pose ignition and detonation hazards.

Gaseous nitrogen dioxide is a brown, paramagnetic, non-flammable, toxic, strongly oxidizing, corrosive substance shipped in approved, low-pressure steel cylinders. It is prepared *in situ* by heating lead nitrate:

$$2Pb(NO_3)_2 \rightarrow 2PbO + 4NO_2 + O_2$$

It is available commercially from several routes including as a product from the manufacture of sodium nitrate from sodium chloride and nitric acid, and from a process involving the passage of ammonia and air over heated platinum and treating the nitric oxide so formed with oxygen.

The brown nitrogen dioxide gas condenses to a yellow liquid which freezes to colourless crystals of dinitrogen tetroxide. Below 150°C the gas consists of molecules of dinitrogen tetroxide and nitrogen dioxide in equilibrium and the proportion of dinitrogen tetroxide increases as the temperature falls. Above 150°C nitrogen dioxide dissociates into nitric oxide and oxygen.

Nitrogen dioxide is an oxidizing agent; it gives up all, or part, of its oxygen to reducing agents, leaving a residue of nitrogen and nitric oxide. It reacts with potassium, hydrogen sulphide, mercury, burning phosphorus or carbon, heated iron and copper. Explosions have been reported between nitrogen dioxide and a host of materials including alcohols (to produce alkyl nitrates), boron compounds, carbonyl metals, propyl nitrite, nitroaniline dust, sodium amide, triethylamine, and vinyl chloride.

Nitrogen dioxide reacts with water, giving first a mixture of nitrous and nitric acids, and ultimately nitric acid and nitric oxide:

$$H_2O + 2NO_2 \rightarrow HNO_3 + HNO_2$$

$$H_2O + 3NO_2 \rightarrow 2HNO_3 + NO$$

When dry the gas is not corrosive to mild steel at normal temperatures and pressures. Metals and alloys such as carbon steel, stainless steel, aluminium, nickel and Inconel are satisfactory. For wet usage stainless steels resistant to 60% nitric acid are suitable. Important uses include use as a bleaching agent, an oxidation catalyst, polymerization inhibitor, a nitrating agent, oxidizing agent, rocket fuel, and in explosives manufacture. Its physical properties are summarized in Table 9.23 and its vapour pressure/temperature relationship is shown in Figure 9.10.

Table 9.23 Physical properties of nitrogen dioxide

Molecular weight	46.005 (or 92.01 for the tetroxide)
Vapour pressure @ 21°C	14.7 psia
Specific volume @ 21°C, 1 atm	293.4 ml/g
Boiling point @ 1 atm	21.25°C
Freezing point @ 1 atm	–9.3°C
Specific gravity (gas) @ 20°C, 1 atm	1.58
Density (gas) @ 21°C, 1 atm	3.3 g/l
Density (liquid) @ 20°C	1.448 g/ml
Critical temperature	158.0°C
Critical pressure	1470 psia (100 atm)
Critical density	0.56 g/ml
Latent heat of vaporization @ bp	99.0 cal/g
Specific heat (gas) @ 25°C, 1 atm C_p	0.1986 cal/g °C
Viscosity (liquid) @ 20°C	4.275 millipoises

A key feature of its toxicity (page 154) at low concentrations is the delay between exposure and onset of symptoms. The OES is 3 ppm (8 hr TWA) and 5 ppm (15 min STEL). Effects of exposure are summarized in Table 5.33. Chronic exposures to low concentrations may cause chronic irritation of the respiratory tract with cough, headache, loss of weight, loss of appetite, dyspepsia, corrosion of the teeth and gradual loss of strength. Concentrations above 60 ppm produce immediate irritation of the nose and throat with coughing, choking, headache, shortness of breath and restlessness. Even brief exposures above 200 ppm may prove fatal. Liquid nitrogen dioxide (dinitrogen tetroxide) is corrosive to skin.

First-aid measures include removal from the contaminated atmosphere, rest and administration of pure oxygen. Skin or eyes in contact with liquid should be thoroughly irrigated. Medical attention should be sought. Other special precautions include:

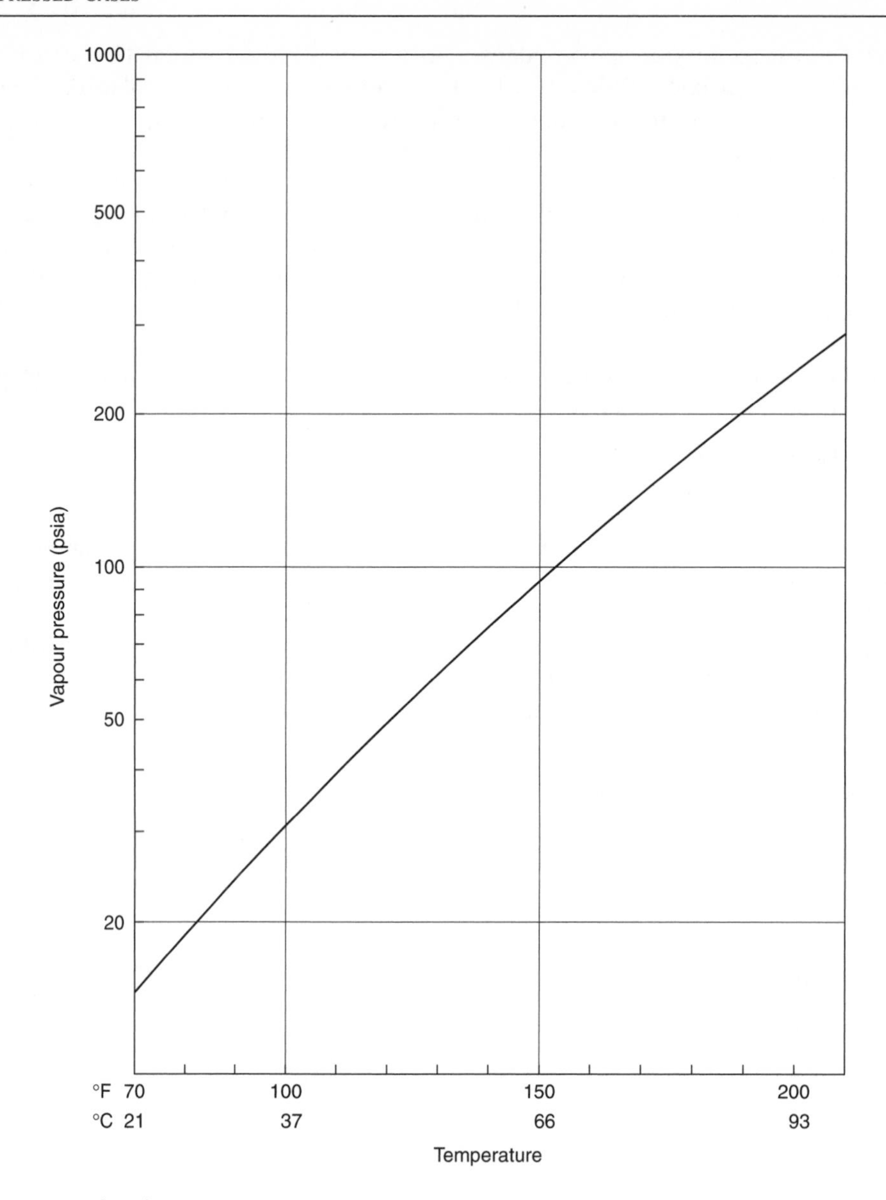

Figure 9.10 *Nitrogen dioxide vapour pressure vs temperature*

- Handle in well-ventilated area, preferably with hood equipped with forced extract ventilation.
- Ensure adequate number of emergency exits.
- Consider the need for personal protection such as skin, eye (goggles/face shield) and respiratory protection including self-contained breathing apparatus.
- Provide instant-acting safety showers and eye-wash facilities in the location of the work area.
- Train staff in appropriate first-aid measures including how to seek immediate medical assistance.

Oxygen

Oxygen occurs free in air in which it forms 21% by volume. It is also found combined with hydrogen in water and constitutes 86% of the oceans, and with other elements such as minerals constituting ca 50% of the earth's crust. In the laboratory it is usually prepared by the thermal decomposition of potassium chlorate in the presence of manganese dioxide catalyst:

$$2KClO_3 \rightarrow 2KCl + 3O_2$$

Industrially, it is manufactured either by fractional distillation of air, or by electrolysis of sodium hydroxide and it is distributed as a non-liquefied gas in pressurized black cylinders at ca 2200 psig at 21°C. Since it is non-corrosive no special materials of construction are required.

Selected physical properties of oxygen are included in Table 9.24. It is a colourless, odourless and tasteless gas which is essential for life and considered to be non-toxic at atmospheric pressure. It is somewhat soluble in water and is slightly heavier than air. Important uses are in the steel and glass industries, oxyacetylene welding, as a chemical intermediate, waste-water treatment, fuel cells, underwater operations and medical applications.

Table 9.24 Physical properties of oxygen

Molecular weight	32.00
Specific volume @ 21°C, 1 atm	755.4 ml/g
Boiling point @ 1 atm	–183.0°C
Triple point	–218.8°C
Density (gas) @ 0°C, 1 atm	1.4291 g/l
Density (liquid) @ bp	1.141 g/l
Critical temperature	1118.4°C
Critical pressure	737 psia (50.14 atm)
Critical density	0.427 g/l
Latent heat of vaporization @ bp	50.94 cal/g
Specific heat (gas) @ 15°C, 1 atm	
C_p	0.2200 cal/g°C
C_v	0.1554 cal/g°C
ratio C_p/C_v	1.42
Viscosity (gas) @ 25°C, 1 atm	0.02064 centipoise
Solubility in water @ 0°C, 1 atm	1 volume/21 volumes water

Explosive reactions can occur between oxygen and a wide range of chemicals including organic compounds (such as acetone, acetylene, secondary alcohols, hydrocarbons), alkali and alkaline earth metals, ammonia, biological specimens previously anaesthetized with ether, hydrogen and foam rubber.

Oxygen supports combustion and the hazard is increased if the concentration in air exceeds 21% (page 199) or at pressures above atmospheric pressure. Substances ignite more readily, burn at a faster rate, generate higher temperatures and may be extremely difficult to extinguish. Substances, e.g. plastics, clothing or metals, which may not normally burn easily can burn vigorously in oxygen-enriched air. Oxygen may become trapped within clothing; this can then be ignited and cause serious burn injuries. Because some substances, e.g. oil and grease, may react explosively with pressurized oxygen it is important never to use lubricants on oxygen equipment and to free pipelines from deposits of them.

Enrichment of the atmosphere in any workplace to about 25% oxygen can be hazardous; this is particularly so in a confined space. Inhalation of 100% oxygen at atmospheric pressure for

16 hr per day for several days poses no undue problems but longer periods of exposure to high pressures can adversely affect neuromuscular coordination and the power of concentration.

In addition to the control measures given in Table 9.3, the following precautions are appropriate when using industrial and medical oxygen, or mixtures of oxygen with other gases.
Never:

- Use oxygen to sweeten the air of a workplace.
- Use oxygen instead of fresh air to ventilate or cool a confined space.
- Use oxygen instead of compressed air or as a source of pressure, e.g. to clear blockages in pipelines or to power air-driven tools.
- Use oxygen to blow down clothing, benches or machinery.
- Use oxygen to inflate vehicle tyres, rubber boats etc.
- Use oxygen to start a diesel engine.
- Use oxygen to cool the person.
- Use oil or grease on oxygen equipment.
- Use jointing compound or tape to cure leaks in oxygen equipment, e.g. at cylinder connections.
- Cut off the supply of oxygen by nipping or kinking flexible hose when changing equipment.

Always:

- Only use materials and equipment which are suitable for oxygen service and to a recognized standard.
- Regularly check equipment for signs of leaks, e.g., at hose connections; replace damaged or worn items. Usually the system is pre-tested with, e.g., nitrogen.
- Ensure that equipment, e.g. pipework for oxygen service, is kept clean and free from oil, grease or dust.
- Ensure that the rated maximum inlet pressure of the regulator is not less than the cylinder supply pressure. (For cylinder pressures up to 200 bar, pressure regulators should comply with BS 5741. For higher cylinder pressures check with the manufacturer that the pressure regulator has been shown to be suitable by appropriate testing.)
- Ensure that the pressure adjusting screw of a pressure regulator is fully unwound, so that the regulator outlet valve is closed before opening the oxygen cylinder valve.
- Open cylinder valves slowly.
- Ensure that the cylinder valves are closed and piped supplies isolated whenever work is stopped.
- Ensure that flexible oxygen hose for welding etc. complies with BS 5120. The correct colour for oxygen is blue.
- Ensure that proper fresh air ventilation is provided in oxy-fuel gas welding and cutting operations.
- Ensure that high-pressure oxygen systems are designed, constructed, installed and commissioned by competent people with specialized knowledge of the subject.
- When working on ships, remove pipes or hoses when work stops, other than for short intervals.
- Store oxygen cylinders in a well-ventilated area or compound away from combustible materials and separated from cylinders of flammable gases.
- Handle oxygen cylinders carefully, preferably using a purpose-built trolley and keep secured to prevent cylinders from falling.
- Wear protective clothing appropriate to the work, e.g. leather gloves, fire-retardant overalls, safety shoes or boots and eye protection.
- If applicable, locate oxygen cylinders outside any confined space and in an area of good ventilation.

Ozone

Ozone is an allotrope of oxygen containing three oxygen atoms. It occurs naturally in the upper atmosphere and is formed in small quantities during electrical discharges from electrical machines or when white phosphorus smoulders in air. In the laboratory it is most conveniently obtained by subjecting air to electrical discharges in 'ozonizers' when some of the oxygen molecules dissociate into oxygen atoms which then combine with other oxygen molecules. However, the yield of ozone is only 10% even when pure oxygen is used. It may also be prepared by electrolysis of ice-cold dilute sulphuric acid using a high current density. Here the concentration of ozone liberated at the platinum-in-glass anode is about 14%.

Pure ozone is made by fractional distillation of the blue liquid resulting from the cooling of ozonized oxygen in liquid air. Commercially it is often supplied dissolved in chlorofluorocarbons in stainless steel cylinders at ca 475 psig cylinder pressure at 20°C often transported chilled with dry ice. These solutions can be handled safely at vapour concentrations of ca 20% by volume of ozone.

The physical properties of ozone are summarized in Table 9.25.

Table 9.25 Physical properties of ozone

Molecular weight	47.998
Boiling point @ 1 atm	–111.9°C
Freezing point @ 1 atm	–192°C
Density (gas) @ 0°C, 1 atm	2.143 g/l
Density (liquid) @ –183°C	1.571 g/l
Critical temperature	–12.1°C
Critical pressure	802.6 psia (54.6 atm)
Viscosity (liquid) @ –183°C	1.57 cP
Latent heat of vaporization @ bp	3410 cal/mole
Dielectric constant (liquid) @ –183°C	4.79
Dipole moment	0.55D
Solubility in water @ 0°C, 1 atm	0.494 volume/volume of water

Ozone is strongly exothermic in its reactions and neat solid or liquid phases are highly explosive.

Pure ozone is a toxic, slightly bluish, unstable, non-flammable but potentially explosive gas with a smell akin to that of much-diluted chlorine. It is used mainly because of its extreme oxidizing ability (second only to fluorine in oxidizing power) in chemical syntheses or because of its powerful germicidal activity on many bacterial organisms, e.g. as a water purification agent (e.g. swimming pools) although it may leave an unpleasant taste, as a bleach, in treatment of industrial waste, sterilization of air (e.g. in the ventilation of premises of underground railways with limited access to fresh air), deodorizing sewage and stack gases, and in food preservation. Indeed, algae and certain fungi resistant to chlorine are highly susceptible to ozone. It decomposes slowly at room temperature and rapidly at 200°C and is decomposed by many finely-divided metals. It reacts readily with unsaturated organic moieties to form ozonides:

$$R_1{-}CH{=}CH{-}R_2 + O_3 \rightarrow R_1{-}\overset{\displaystyle O\text{ (bridging)}}{CH\quad CH}{-}R_2 \;\; \text{(ring: CH–O–CH, CH–O–O–CH)}$$

(Ozonide)

where R_1 and R_2 are univalent organic groups.

Violent reactions have occurred between ozone and many chemicals, a small selection being acetylene, alkenes, dialkyl zincs, benzene/rubber solution, bromine, carbon monoxide and ethylene, diethyl ether, hydrogen bromide, and nitrogen oxide.

Ozone is an irritant to eyes and mucous membranes. Inhalation can cause pulmonary oedema and bleeding at 'high' concentrations, whilst at lower exposures symptoms include headache, shortness of breath, or drowsiness. Long-term effects may include chronic pulmonary effects, ageing, and possibly lung cancer. The ACGIH classify ozone as one of those 'Agents which cause concern that they could be carcinogenic for humans but which cannot be assessed conclusively because of the lack of data. In vitro or animal studies do not provide indications of carcinogenicity which are sufficient to classify the agent into one of the other categories.' The TLV is set as a sliding scale for exposures during different workloads, thus:

- 0.05 ppm for heavy work.
- 0.08 ppm for moderate work.
- 0.10 ppm for light work.
- 0.20 ppm for heavy, moderate, or light workloads for up to a maximum exposure of 2 hours.

Because of the odour threshold of ca 0.015 ppm, exposure to ozone below the TLV can usually be detected by smell.

Precautions in addition to those in Table 9.3 include:

- Keep cylinders chilled (to help prevent decomposition rather than as a safety measure).
- Avoid copper and copper alloys since these can catalyse the decomposition, and rubber components are unsuitable.
- Pre-test systems for leaks with inert gas.
- Prevent contact with grease, oil or other combustible material.
- Clean all equipment for oxygen service.
- Handle in a ventilated hood to protect surrounding atmosphere from leaks.
- Consider the need for appropriate personal protection including eye, skin and respiratory equipment.
- Consider the need for leak detection systems.
- Where possible avoid explosions by working with dilute solutions at low temperatures in suitable solvents.

Sulphur dioxide

Sulphur dioxide is used as a preservative for beer, wine and meats; in the production of sulphites and hydrosulphites; in solvent extraction of lubricating oils; as a general bleaching agent for oils and foods; in sulphite pulp manufacture; in the cellulose and paper industries; and for disinfection and fumigation.

It is a non-flammable colourless gas which is twice as dense as air, and slightly soluble in water forming sulphurous acid. It is readily liquefied as a gas under its own vapour pressure of about 35 psig (2.4 bar) at 21°C. Figure 9.11 depicts the effect of temperature on vapour pressure; Table 9.26 lists the physical properties. Cylinders tend to be protected against over–pressurization by metal plugs melting at about 85°C.

Gaseous sulphur dioxide is highly irritant and practically irrespirable. Effects on the body are summarized in Table 5.3. It can be detected at about 3.5 ppm and the irritating effects would

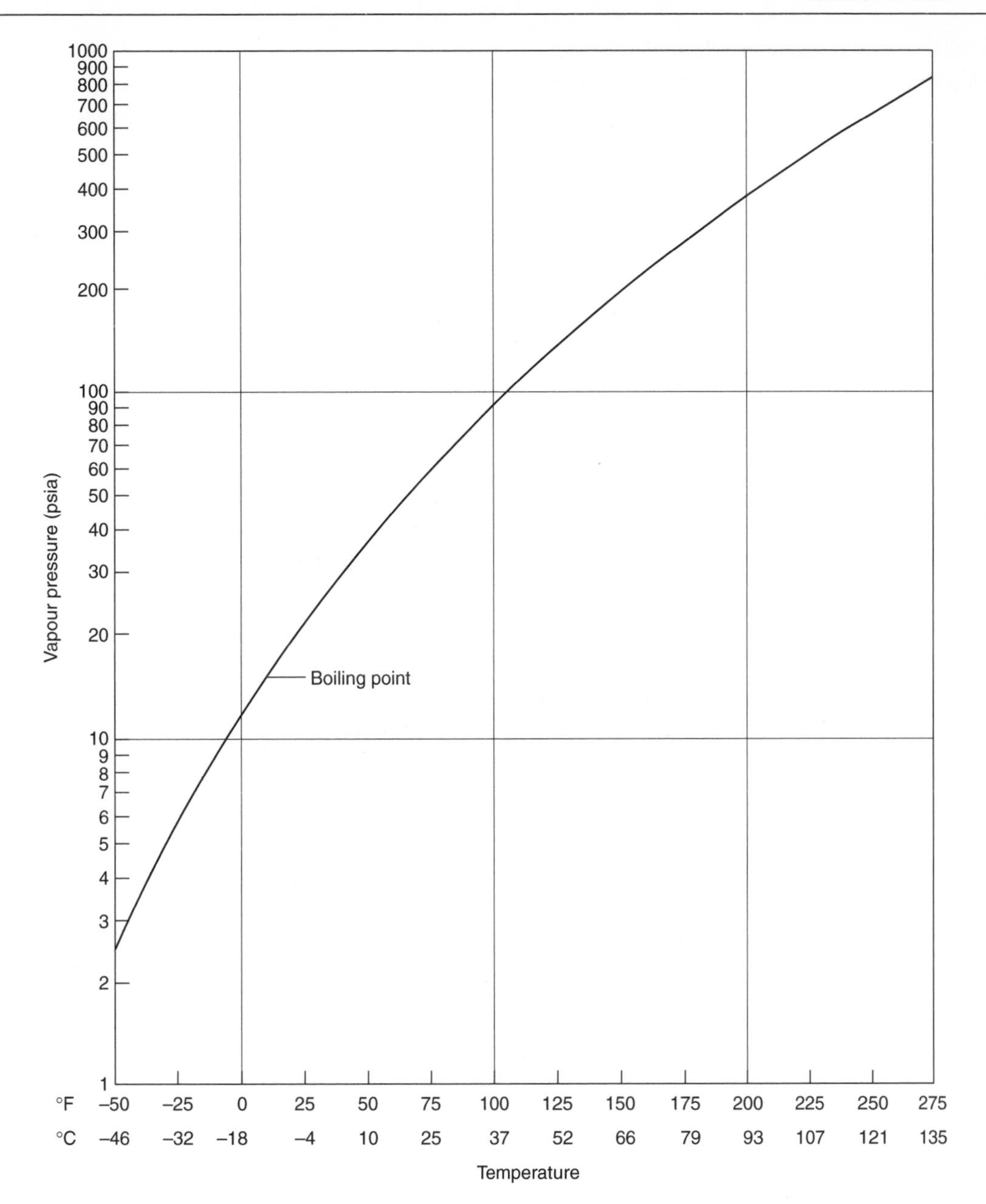

Figure 9.11 *Sulphur dioxide vapour pressure vs temperature*

preclude anyone from suffering prolonged exposure at high concentrations unless unconscious, or trapped.

Liquid sulphur dioxide may cause eye and skin burns resulting from the freezing effects upon evaporation. Dry sulphur dioxide is non-corrosive to common materials of construction except zinc. The presence of moisture renders the environment corrosive.

In addition to the precautions listed in Table 9.3, the following controls are appropriate:

- Use in well-ventilated areas.

Table 9.26 Physical properties of sulphur dioxide

Molecular weight	64.063
Vapour pressure at 21°C	2.37 bar
Specific volume at 21°C, 1 atm	368.3 ml/g
Boiling point at 1 atm	–10.0°C
Freezing point at 1 atm	–75.5°C
Specific gravity, gas at 0°C, 1 atm (air = 1)	2.264
Density, gas at 0°C, 1 atm	2.927 g/l
Density, liquid at –10°C	1.46 g/ml
Critical temperature	157.5°C
Critical pressure	78.8 bar
Critical density	0.524 g/ml
Latent heat of vaporization at boiling point	92.8 cal/g
Latent heat of fusion at melting point	27.6 cal/g
Specific heat, liquid at 0°C	0.318 cal/g °C
Specific heat, gas at 25°C, 1 atm	
C_p	0.1488 cal/g °C
C_v	0.1154 cal/g °C
ratio C_p/C_v	1.29
Thermal conductivity at 0°C	2.06×10^{-5} cal/s cm^2 °C/cm
Viscosity, gas at 18°C, 1 atm	124.2 mP
Solubility in water at 0°C, 1 atm	18.59% by weight
at 20°C, 1 atm	10.14% by weight

- Wear eye/face protection, approved footwear and rubber gloves.
- Showers and eye-wash facilities and respiratory protection should be conveniently located for emergencies.
- Insert traps in the line to avoid liquid suck-back into the cylinder.
- Check for leaks with soap solution, aqueous ammonia or colour indicator tubes.
- First aid measures include those in Table 9.9.

10
Monitoring techniques

As mentioned in Chapters 4, 5 and 16 chemicals can be a nuisance or pose safety, health and environmental risks, or become wasteful of expensive resources if allowed to escape excessively and uncontrollably into the general or workplace environment. Escapes can result from inadequate process control, errors in operation or maintenance, incomplete understanding of the process, etc. Such problems can arise from both: periodic emissions of chemicals due to the need to open, or enter, the 'system' occasionally (e.g. during sampling, cleaning, line-breaking) including both planned and unplanned releases (e.g. due to accidents, human error) and, continuous low-level fugitive emissions from normally-closed points, e.g. valve seals, flange gaskets, pump seals, drain valves.

The need to monitor the impact of activities involving chemicals on the environment may stem from sound management practice or to satisfy a host of specific legal requirements. Thus, in the UK under the Environmental Protection (Prescribed Substances and Processes) Regulations 1991, operators must apply BATNEEC to prevent or minimize the release of prescribed substances into the environment, or to render harmless any emissions. The prescribed substances for release into the air are given in Table 10.1. No prescribed process may be operated without an authorization from the Environment Agency and air pollutants which must be measured and the frequency of monitoring are set out in the authorization. Compliance with emission limits for municipal waste incineration plants (Table 10.2) also requires monitoring.

Table 10.1 Prescribed substances for release into the air

Oxides of sulphur and other sulphur compounds
Oxides of nitrogen and other nitrogen compounds
Oxides of carbon
Organic compounds and partial oxidation compounds
Metals, metalloids and their compounds
Asbestos, glass fibres and mineral fibres
Halogens and their compounds
Phosphorus and its compounds
Particulate matter

In addition to pollution episodes, risks may arise due to atmospheric oxygen concentrations fluctuating beyond its normal level of 21% posing health (page 72) or fire hazards. Fire and explosion dangers may also arise from the presence of flammable gases, vapours, or dusts in the atmosphere (Chapter 6).

Thus, as illustrated by Table 17.13 monitoring emissions of hazardous chemicals into the environment may be required for a variety of reasons such as:

Table 10.2 Selected emission limits for municipal waste incineration (units: mg/m³)

Country	*EU*	*EU*	*EU*	*UK*	*UK*
Plant capacity (tonne per hour)	3	1–3	<1	>1	<1
Particulates	30	100	200	30	200
CO	100	100	100	100	100
SO_2	300	300	–	300	300
Volatile organic compounds	20	20	20	20	20
HF	2	4	–	2	–
NO_x as NO_2	–	–	–	350	–
Cr, Cu, Mn, Pb	5	5	–		
Ni, As	1	1	–	1 (incl Sn)	5
Cd, Hg	0.2	0.2	–	0.1 each	
Dioxins	–	–	–	1	

- Assessing fire or explosion risks from atmospheres containing flammable gas, vapour or dust.
- Determining oxygen content of the working atmosphere.
- Determining sources of leaks of toxic, flammable, or nuisance pollutants.
- Identifying unknown pollutants.
- Assessing process efficiency and control.
- Assessing environmental risk from effluent discharges or for formal environmental impact assessment.
- Determining employee exposure to known toxic substances.
- Providing data for internal company environmental audits.
- Investigation of the causes of accidents.
- Investigation of the cause and nature of problems (e.g. local complaints of odour) or pollution incidents.

Selected general analytical techniques for monitoring environmental pollution

Stages in environmental monitoring include obtaining representative samples of the environment in question and subsequent analysis of physical, chemical or microbiological attributes. Monitoring techniques range from sophisticated in-line, continuous sampling and instantaneous analysis linked to audible/visual alarms or features to control the pollution; samplers running continuously, e.g. throughout a normal day for subsequent analysis; to grab samples (i.e. samples collected over a short time span of, e.g., a few minutes). Continuous monitoring is common in personal dosimetry studies where an appropriate collection device samples air wherever a worker is throughout a specified period, e.g. 15 minutes or 8 hours (page 111). A selection of common analytical techniques include the following.

Gases and vapours

Atomic absorption/emission spectrometry

Metal ions are most commonly measured using atomic absorption spectrometry. In this technique the test sample is aspirated into a flame where chemical reduction of metal ions to metal atoms occurs. Light is emitted at the discrete absorption wavelength for the metal. A disadvantage is the need for a separate lamp for each element. Emission spectrometry is therefore preferred. Here test material is heated and vaporized using a DC or inductively-coupled plasma generator, after which an optical emission spectrum as a function of wavelength is recorded. An advantage of this technique is that a range of metals or metalloids can be analysed simultaneously.

Chemiluminescence

Here a chemical reaction produces a molecule with electrons in an excited state. Upon decay to the ground state the liberated radiation is detected. One such example is the reaction between ozone and nitric oxide to form nitrogen dioxide emitting radiation in the near infra-red in the 0.5–3μ region. The technique finds use for measuring nitric oxide in ambient air or stack emissions.

Chromatography

This technique permits the separation of a mixture of compounds by their partition between two immiscible heterogeneous phases, one of which is stationary. It detects substances qualitatively and quantitatively. The chromatogram retention time is compound-specific, and peak-height indicates the concentration of pollutant in the sample. Detection systems include flame ionization, thermal conductivity and electron capture. With gas chromatography the mixtures to be separated are in the vapour phase under the operating conditions of the equipment. A gas is used as the mobile phase to carry the sample over a column of stationary phase. Flame ionization detection operates by ionisation of molecules in a hydrogen flame and detection of the current change using a pair of biased electrodes. The current signal is directly related to the number of carbon atoms in the sample. Thermal conductivity detectors measure the change in electrical resistance of a heated filament as gas flows over it. It is most suitable for gases with very high, or very low, conductivity. Traditionally gas chromatography is a laboratory analysis but portable versions are now available for field work.

In classical liquid chromatography a solution of solute percolates under gravity through a column packed with finely-divided solid when different compounds elute at different rates. In high-performance liquid chromatography (HPLC) the liquid is eluted from a packed column under high pressure using solvent. Detection systems include differential refractive index, diode array, electrochemical and ultra-violet-visible absorption. HPLC is used for analysis of less volatile compounds in liquid samples than those in gas chromatography.

Because of its sensitivity (<1 ppm), ion chromatography has become extremely popular for analysis of ions in solution. It is a column-based method for separating ions similar to HPLC but using ion exchange columns and either high- or low-conductivity eluent. The most common detectors are electrical conductivity and ultra-violet absorption. It finds wide use in air pollution monitoring of rain waters, impinger solutions and filter extracts for anions such as sulphate, nitrate, chloride, and cations, including ammonia and metals.

Colorimetry

Use is made of colour changes resulting from reaction of pollutant and chemical reagents: colour intensity indicates concentration of pollutant in the sample. Reaction can take place in solution or on solid supports in tubes or on paper strips, e.g. litmus or indicator paper. Quantitative assessment of colour formation can also be determined using visible spectroscopy. Instruments are calibrated such that colour intensity is directly related to concentration of contaminant in the test sample. Arguably the technique could also embrace simple acid-base titrations which utilize colour indicators to determine the end-point.

Electrochemical techniques

In electrochemical cells sample oxidation produces an electric current proportional to the concentration of test substance. Sometimes interferences by other contaminants can be problematic and in general the method is poorer than IR. Portable and static instruments based on this method are available for specific chemicals, e.g. carbon monoxide, chlorine, hydrogen sulphide.

Coulometry measures the amount of current flowing through a solution in an electrochemical oxidation or reduction reaction and is capable of measuring at ppm or even ppb levels of reactive gases. Thus a sample of ambient air is drawn through an electrolyte in a cell and the required amount of reactant is generated at the electrode. This technique tends to be non-specific, but selectivity can be enhanced by adjustment of pH and electrolyte composition, and by incorporation of filters to remove interfering species.

Ion-selective electrodes are a relatively cheap approach to analysis of many ions in solution. The emf of the selective electrode is measured relative to a reference electrode. The electrode potential varies with the logarithm of the activity of the ion. The electrodes are calibrated using standards of the ion under investigation. Application is limited to those ions not subject to the same interference as ion chromatography (the preferred technique), e.g. fluoride, hydrogen chloride (see Table 10.3).

Table 10.3 Examples of applications of ion-selective electrodes

Electrode	*Measurement range (ppm)*	*Major interference*
Ammonia	0.01–17 000	Ionic – none
Bromide	0.08–80 000	CN^-, I^-, S^{--}
Chloride	0.35–3500	CN^-, I^-, Br^-, S^{--}
Cyanide	0.003–2600	I^-, S^{--}
Fluoride	0.02–1900	OH^-
Iodide	0.013–127 000	CN^-, S^{--}
Nitrate	0.6–62 000	I^-, Br^-, SCN^-, ClO_4^-
Sulphide	0.32–32 100	Ag^+, Hg^+

Infra-red spectroscopy

The basis of this technique is absorption of IR radiation by molecules over a wide spectrum of wavelengths to give a characteristic 'fingerprint' spectrum providing both qualitative and quantitative data on the substance. This versatile technique owes its success in occupational hygiene to the development of a portable spectrometer of the non-dispersive type which focuses on specific parts of the spectrum in which the pollutant shows peak absorption as opposed to scanning the entire spectrum. Table 10.4 identifies principal absorption peaks for selected gases. One advantage of IR is that the detector does not 'react' with the gases and the major functional components are protected and easily removed for maintenance. Since IR detection is potentially sensitive to temperature, the instrument requires approximately 15 minutes to equilibrate prior to use. Water vapour can seriously affect performance.

Table 10.4 Main IR absorption peaks for selected gases

Chemical	*Wavelength (micron)*
Carbon dioxide	4.35
Carbon monoxide	4.60
Methane	3.30
Nitric oxide	5.30
Nitrogen dioxide	3.70
Nitrous oxide	4.50
Sulphur dioxide	4.00

Mass spectrometry

This technique relies on the formation of ions by various means in a high-vacuum chamber, their acceleration by an electrical field and subsequent separation by mass/charge ratio in a magnetic field and the detection of each species. It can be used for both inorganic and organic substances, be very sensitive, and be of value in examining mixtures of compounds especially if linked to glc. Usually this is a laboratory technique but portable or 'transportable' models are now available.

Ultra-violet spectrometry

Outermost valency electrons in atoms are excited by ultra-violet radiation. The excited electrons return to the ground state liberating energy by disassociation, re-emission, fluorescence, or phosphorescence. The level of UV radiation absorbed follows the Beer–Lambert law (page 312). The peak wavelength for selected gases is given in Table 10.5. Photo ionization detectors (PIDs) use ultraviolet light to ionize gas molecules such as volatile organic compounds; the free electrons collected at electrodes result in a current flow proportional to the gas concentration. The lamp requires constant cleaning and hence may have limited life expectancy.

Table 10.5 UV wavelengths for selected gases

Gas	*Wavelength (nm)*
Ammonia	200
Chlorine	280–380
Hydrogen sulphide	200–230
Nitrogen dioxide	380–420
Oxygen/ozone	170
Sulphur dioxide	285

Compound specific analysers

Several instruments are available that are designed to monitor a specific compound rather than a wide range of substances. The detection system varies according to the pollutant. A selection is given in Table 10.6.

Table 10.6 Selected examples of compound specific instruments

Compound	Detection system
Ammonia	Coulometry (e.g. Nessler method)
	Ion selective electrode
	Oxidation to NO_x and chemiluminescence
Carbon monoxide	Polarography
	Infra-red
Chlorine	Ultraviolet spectroscopy
Hydrogen chloride	Polarography
	Ion-selective electrode
Hydrogen fluoride	Polarography
	Colorimetry
Inorganic cyanides	Colorimetry
	Ion-selective electrode
Oxides of nitrogen	Chemiluminescence
	Coulometry
Oxygen	Polarography
	Paramagnetic susceptibility
	Fluorescence
Ozone	Ultraviolet spectroscopy
	Chemiluminescence
	Coulometry
Phosgene	Ultraviolet spectroscopy
Phosphorus and its compounds	Flame photometry
Toluene	Ion-mobility spectroscopy
Total hydrocarbons	Flame ionization
Sulphur compounds	Flame photometry
	Coulometry
	UV fluorescence

Particulates

Samples of particulate matter can be subjected to many of the above analytical techniques in chemical characterization. The following methods are, however, particularly applicable to analysis of physical characteristics of particulate matter isolated from air sampling.

Mass concentration

Simple gravimetry of the sample is likely to be an integral component of the determination of, e.g., the concentration of, or exposures to, airborne dust. Care is required to avoid errors arising from absorption of atmospheric moisture. This can be avoided by using blank filters, by conditioning the filters in an atmospherically-controlled room, or use of a desiccator.

Automatic aerosol mass concentration can be achieved directly by collecting particles on a surface followed by use of a piezoelectric or oscillation microbalance, or by β-attenuation sensing techniques, or indirectly using light scattering. The piezoelectric microbalance contains an electrostatic precipitator to deposit particles onto a vibrating silica crystal. The change in resonance frequency is converted into mass concentration using a microprocessor. Oscillating balances operate on the principle that air at 50°C (to avoid condensation) passes through a filter attached to the top of a tapered glass tube which vibrates at its natural frequency. As material is deposited on the filter the oscillation frequency changes directly in proportion to the increased mass. Beta gauges rely on the principle that when low-energy β particles pass through a material the intensity of the beam is attenuated according to Beer–Lambert law:

$$I = I_0^{-ux}$$

where I and I_0 are the initial and attenuated beam intensities, u is the mass absorption coefficient and x is the absorber thickness.

Light scattering

Optical particle counters provide information on the particles present in different size ranges. A beam of light is collimated and focused onto a measurement cell. Light impinging on a particle is scattered and reaches a photomultiplier tube and converted to an output proportional to particle size. Particle size distributions are computed by appropriate software.

Electrostatic sampling

Particles become positively charged by a corona discharge and travel out of the charging chamber and collect on a substrate such as a microscope slide. Thus, the method is useful for particles which are to be examined by optical or electron microscopy.

Optical microscopy

This technique is invaluable for measurement of particle size, for counting the number of particles and for identification of particles by:

- morphology, e.g. by comparison with standard particles, and
- refractive index using polarized light microscopy.

Electron microscopy

With a resolution of 0.01 μm this technique outperforms optical light microscopy (0.1 μm) and is used, e.g., to examine fine particles such as metal fume. When linked to other facilities such as dispersive X-ray analysis, quantitative data can be obtained.

X-ray techniques

Crystals produce different diffraction patterns when subjected to bombardment of monochromatic X-ray sources and thereby provide unequivocal identification of crystalline materials.

In X-ray fluorescence incident radiation induces electronic fluorescent emission in most atoms. The effects can be used both qualitatively and quantitatively for metals, alloys etc.

Monitoring water quality

Water is essential to man both directly and indirectly through agriculture and industry in which vast quantities are used for cooling, energy production, irrigation, refrigeration, washing, solvents etc. Risk of contamination can render water dangerous, unpleasant, or unusable. Point sources of water pollution include domestic and industrial waste whilst non-point sources include agricultural and urban run-offs. Analysis of water is important for estimating the nature and concentration of contaminants and hence fitness for use. Artificial contaminants are mainly of domestic and industrial origin, and are increasing in similarity because of the expanding domestic use of chemicals (cosmetics, detergents, paints, garden insecticides and fertilizers). Water quality can be assessed by direct analysis of chemical substances or by indirect effects, e.g. pH, colour, turbidity, odour, impact on dissolved oxygen content.

Chemical pollutants are classified as inorganic or organic. The former include metals (e.g. Mn, Fe, Cu, Zn, Hg, Cd, As, Cr), anions (e.g. Cl^-, SiO_3^{2-}, CN^-, F^-, NO_3^-, NO_2^-, PO_4^{---}, SO_3^{--}, SO_4^{--}, S^{--}), and gases (Cl_2, NH_3, O_2, O_3). Methods for the examination of waters and associated materials published by the UK Department of the Environment are listed in Table 10.7. Selected methods for metal analysis are summarized in Table 10.8. Sampling protocols are described in, e.g., BS 6068 and BS EN 2567. Examples of BS methods for analysis of chemical contaminants in water are illustrated by Table 10.9. Biological methods are also given in BS 6068.

Monitoring land pollution

Sources of land pollution include transport accidents, spillage during chemical handling, loss of containment from storage tanks, leakage and landfill of waste effluent. An appreciation of the processes governing retention, degradation and removal of pollutants and the behaviour of specific pollutants in soil are essential in devising correct sampling and analytical strategies for assessing land contamination. Even soil itself varies in dynamics and composition from one site to another. Constituents include solid phase materials (such as complex mixtures of clays, minerals, organic matter), liquid aqueous phase of solutions (e.g. natural minerals, fertilizers, pesticides and industrial wastes) and gaseous phase components (e.g. oxygen, nitrogen, carbon dioxide, oxides of nitrogen, ammonia, hydrogen sulphide). The determination of toxic elements and organic substances in soils is a requisite of some EC directives as a means of controlling environmental pollution. Analyses are important when certain types of waste are recycled, e.g. by spreading sludge from water purification units on land, composting from household refuse. The choice of analytical method will be dictated by accuracy, sensitivity etc. Some key techniques are summarized in Table 10.10 and selected BS methods for monitoring soil quality are listed in Table 10.11.

Monitoring air pollution

Sampling

Differences exist between the monitoring of pollution levels in ambient and workplace air. These reflect the differences in levels of contaminant, environmental standards, purposes for which data are used, etc. (see also Table 16.8). Thus, although similarities may exist in detection techniques, the sampling regimes, analytical details and hardware specifications may differ for assessment of the two environments. In general, atmospheric levels of contaminants are much lower in ambient air than those encountered in the workplace. As a result larger volumes of sample are often needed for ambient air analyses. This can be achieved using pumps of larger flow rate capacities, or by longer sampling times.

Atmospheric monitoring involves first obtaining samples of the air with subsequent analysis of the samples collected. Examples of sampling techniques for gases and vapours are given in Table 10.12. Air samples can be pumped into instruments for direct analysis and data readout. Alternatively, they are collected in air-tight bags, or absorbed in liquids, or onto solid sorbents, for subsequent laboratory analysis using techniques such as those described on page 308. Common solid sorbents

Table 10.7 Methods for the examination of water and associated materials published by the UK Department of the Environment

A review and methods for the use of epilithic diatoms for detecting and monitoring changes in river water quality, 1993

Chlorphenylid, Flucofuron and Sulcofuron Waters (Tentative Methods based on methylation and GC-ECD, ion-pair HPLC and hydrolysis of Sulcofuron to 4-chloro-3-trifluoromethylaniline by GC-ECD), 1993

Cyanide in Waters etc. (by Reflux Distillation followed by either Potentiometry using a Cyanide Selective Electrode or Colorimetry, or Continuous Flow Determination of Cyanide or Determination by Microdiffusion), 1988

Determination of Aldicarb and other *N*-methyl carbamates in Waters (by HPLC or Confirmation of total Aldicarb residues and other *N*-methyl carbamates in waters by GC), 1994

Determination of the pH Value of Sludge, Soil, Mud and Sediment; and the Lime Requirement of Soil (Second Edition) (by Determination of the pH Value of Sludge, Soil, Mud and Sediment or by Determination of the Lime Requirement of Soil), 1992

Flow Injection Analysis, An Essay Review and Analytical Methods

Information on Concentration and Determination Procedures in Atomic Spectrophotometry, 1992

Isolation and Identification of Giardia Cysts, Cryptosporidium Oocysts and Free Living Pathogenic Amoebae in Water etc., 1989

Kjeldahl Nitrogen in Waters [including Mercury Catalysed Method, Semi-automated Determination of Kjeldahl Nitrogen (Copper Catalysed, Multiple Tube, Block Digestion Method followed by Air Segmented Continuous Flow Colorimetry) Determination of Kjeldahl Nitrogen in Raw and Potable Water (Hydrogen Peroxide, Multiple tube, Block Digestion Method followed by Manual or Air-Segmented Continuous Flow Colorimetry) Semi-automated Determination of Kjeldahl Nitrogen (Copper/Titanium Catalysed, Multiple Tube, Block Digestion Method followed by Distillation and Air Segmented Continuous Flow Colorimetry), Air-segmented Continuous Flow Colorimetric Analysis of Digest Solutions for Ammonia], 1987

Linear Alkylbenzene Sulphonates (LAS) and Alkylphenol Ethoxylates (APE) in Waters, Wastewaters and Sludges by High Performance Liquid Chromatography, 1993

Phenylurea herbicides (urons), Dinocap, Dinoseb, Benomyl, Carbendazim and Metamitron in Waters [e.g. determination of phenylurea herbicides by reverse phase HPLC, phenylurea herbicides by dichloromethane extraction, determination by GC/NPD, phenylurea herbicides by thermospray LC-MS, Dinocap by HPLC, Dinoseb water by HPLC, Carbendazim and Benomyl (as Carbendazim) by HPLC], 1994

Phosphorus and Silicon in Waters, Effluents and Sludges [e.g. Phosphorus in Waters, Effluents and Sludges by Spectrophotometry-phosphomolybdenum blue method, Phosphorus in Waters and Acidic Digests by Spectrophotometry-phosphovanadomolybdate method, Ion Chromatographic Methods for the Determination of Phosphorus Compound, Pretreatment Methods for Phosphorus Determinations, Determination of silicon by Spectrophotometric Determination of Molybdate Reactive Silicon-1-amino-2-naphthol-4, sulphonic acid (ANSA) or Metol reduction methods or ascorbic acid reduction method, Pretreatment Methods to Convert Other Forms of Silicon to Soluble Molybdate Reactive Silicon, Determination of Phosphorus and Silicon Emission Spectrophotometry], 1992

Sulphate in Waters, Effluents and Solids (2nd Edition) [including Sulphate in Waters, Effluents and Some Solids by Barium Sulphate Gravimetry, Sulphate in waters and effluents by direct Barium Titrimetry, Sulphate in waters by Inductively Coupled Plasma Emission Spectrometry, Sulphate in waters and effluents by a Continuous Flow Indirect Spectrophotometric Method Using 2-Aminoperimidine, Sulphate in waters by Flow Injection Analysis Using a Turbidimetric Method, Sulphate in waters by Ion Chromatography, Sulphate in waters by Air-Segmented Continuous Flow Colorimetry using Methylthymol Blue], 1988

Temperature Measurement for Natural, Waste and Potable Waters and other items of interest in the Water and Sewage Disposal Industry, 1986

The Determination of 6 Specific Polynuclear Aromatic Hydrocarbons in Waters [Using High-Performance Liquid Chromatography,Thin-layer Chromatography], 1985

The Determination of Taste and Odour in Potable Waters, 1994

Use of Plants to Monitor Heavy Metals in Freshwaters [Methods based on Metal Accumulation and on Techniques other than Accumulation], 1991

Table 10.8 Methods for analysis of metal content of water

Method	*Comment*	*Sensitivity*
Spectrophotometry–Colorimetry	One of most useful and versatile methods but can be time consuming	10^{-5} to 10^{-7} M (10^{-9} with pre-concentration)
Kinetic analysis (metal ion acts as catalyst)	Sensitive, highly selective, only needs small samples	10^{-8} to 10^{-9} M
Atomic absorption spectrophotometry:		
• Flame	Simple, versatile; measures total metal content. Knowledge of interfering effects important.	10^{-6} to 10^{-7} M
• Flameless		10^{-8} to 10^{-9} M
Flame and spark emission spectroscopy	Not very accurate. Gives multi-element analyses	10^{-5} to 10^{-8} M
Neutron activation analysis	Specialized, expensive	10^{-9} to 10^{-10} M
Ion-selective electrodes	Highly selective but insensitive and imprecise	10^{-5} to 10^{-6} M
Polarography:	Restricted to electroactive metals	
• Conventional		10^{-5} to 10^{-6} M
• Modified		10^{-6} to 10^{-8} M
Anodic stripping voltametry	Applicable to few metals and dependent on metal speciation	10^{-8} to 10^{-10} M

for gases and vapours (Table 10.13) include silica gel, activated charcoal, or organic resins. Silica gel is most useful for polar compounds whereas charcoal finds wide use for non-polar substances. Subsequently, the pollutant is generally removed from the solid phase by thermal desorption or by solvent extraction. The advantages and disadvantages of the two desorption techniques are summarized in Table 10.14. Their versatility is illustrated by Table 10.15 for use of charcoal.

Pumps vary from large, stationary high-volume versions to pocket-size devices for use in personal dosimetry when operators wear sampling devices in the form of tubes or badges in lapels to collect air sampled in their breathing zone. Passive samplers are also available for monitoring gases and vapours in air. These are inexpensive devices which do not require a mechanical pump but rely on the concentration gradient between the air and sorbent material and the resultant molecular diffusion of the pollutant towards the sorbent according to Fick's law:

$$Q = D\mathrm{d}C/\mathrm{d}Z$$

where Q = molar flux (mol $cm^{-2}s^{-1}$),
D = diffusion coefficient (cm^2s^{-1}),
C = concentration (mol cm^{-3})
Z = diffusion path length (cm).

Because of the low rates of molecular diffusion, assessment of workplace air quality using passive samplers usually entails sampling for a working shift, and exposure periods of one to four weeks tend to be needed to measure concentrations in ambient air.

Gases and vapours

Analyses of gases and vapours tend to utilize the techniques described on page 308. Many of these methods were traditionally limited to laboratory analyses but some portable instruments are now available for, e.g., gas chromatography (Table 10.16) and non-dispersive infra-red spectrometry (Table 10.17).

Table 10.9 Selected British Standard methods for monitoring water quality

BS reference	*Substance*	*Method*
BS 1427	General guide to methods for analysis of water	field and on-site techniques
BS 2690	free EDTA	
	hydrazine	spectrophotometry
	silica	
	cyclohexylamine	spectrophotometry
	morpholine	spectrophotometry
	long-chain fatty acids	spectrophotometry
BS 6068	mercury	flameless atomic absorption
	ammonium	distillation and titration
	ammonium	potentiometric
	ammonium	manual spectometry
	phenol index	spectrometric
	cyanide	diffusion at pH6
	cyanogen chloride	
	ammonium	automated spectrometry
	nitrate	spectrometry using sulphosalicylic acid
	inorganically bound total fluoride	digestion and distillation
	pH	
	iron	photometric
	calcium and magnesium	EDTA titrimetric
	nonionic surfactants	Dragendorff reagent
	anionic surfactants	methylene blue titration
	arsenic, cadmium, cobalt, copper, lead, magnesium, nickel, zinc	atomic absorption spectrometry spectrometry
	arsenic (total)	spectrometry
	manganese	formaldoxime spectrometry
	chemical oxygen demand	
	sulphate	barium chloride gravimetry
	borate	spectrometry
	sodium and potassium	atomic absorption spectrometry and by flame emission spectrometry
	selenium	atomic absorption spectrometry
	aluminium	spectrometry
	herbicides	gas chromatography and mass spectrometry
	organochlorine compounds	gas chromatography
	dissolved anions (bromide, chloride, nitrate, nitrite, phosphate, sulphate, chromate, iodide, sulphite, thiocyanate, thiosulphate, chlorate, chlorite)	liquid chromatography
		liquid chromatography
	organic nitrogen and phosphorus	gas chromatography
	33 elements	inductively coupled plasma atomic emission spectrometry
	nitrogen	chemiluminescence
	mercury	enrichment by amalgamation
	free and total chlorine	titrimetric or colorimetric
	chlorophenols	gas chromatography
	organophosphates	gas chromatography
	dissolved oxygen	iodometry or electrochemical probe
	phosphorus	spectrometry

Table 10.9 Cont'd

BS reference	*Substance*	*Method*
	chromium	atomic absorption spectrometry
	hydrocarbon oil index	solvent extraction and gas chromatography
	halogenated hydrocarbons (volatile)	gas chromatography
	total organic carbon and dissolved organic carbon	
	organically bound halogen	
	alpha and beta activity	
	suspended solids	filtration through glass fibre
	electrical conductivity turbidity	
BS EN 11732	ammonium nitrogen	flow analysis and spectrometry
BS EN 12020	aluminium	atomic absorption spectrometry
BS EN 1483	mercury	
BS EN 1484	total organic carbon and dissolved organic carbon	
BS EN 12338	mercury	enrichment by amalgamation
BS EN 14911	dissolved sodium lithium ammonium, potassium, manganese, calcium, magnesium, strontium, barium	ion chromatography
BS EN 25663	nitrogen	Kjeldahl
BS EN 26777	nitrite	molecular absorption spectrometry
BS EN 7887	colour	
BS EN 1622	odour	
BS EN ISO 12020	aluminium	atomic absorption spectrometry
BS EN 13395	nitrate and nitrite nitrogen	flow analysis and spectrometry
BS EN ISO 9377	hydrocarbon oil index	solvent extraction/gravimetry

Table 10.10 Common instrumental techniques for soil analysis

Pollutant	*Method*
• Phosphorus or mineral nitrogen	Spectrophotometry
• Alkaline earth metals and transition metals	Flame atomic absorption spectrometry
• Alkali metals	Flame emission spectrometry
• Aluminium, boron, silicon	Inductively coupled plasma atomic emission spectrometry
• Lead in soil slurries	Electrothermal atomic absorption spectrometry
• Toxic organic compounds	High pressure liquid chromatography

One of the most commonly used portable, gas detection systems is based on colour indicator tubes employed, in the main, for grab sampling. Tubes are available to detect over 300 substances (Table 10.18) and, using a combination of tubes, a range of concentrations can be measured. The technique relies on a manually-operated bellows or piston pump to aspirate a fixed volume of atmosphere through a glass tube containing crystals (e.g. silica gel or alumina) impregnated with a reagent which undergoes a colour change upon reaction with a specific pollutant or class of pollutant. The length of stain that develops is proportional to the concentration of contaminant and the tube is generally calibrated to permit direct read-off in parts per million. Sample lines of several metres between the pump and tube allow atmospheres to be sampled, e.g., inside vessels. A smaller number of special tubes (Table 10.18(b)) are available for longer-term monitoring and these

Table 10.11 Selected British Standard methods for monitoring soil quality

British Standard reference	*Contaminant*	*Method*
BS 7755	dry matter	gravimetric
	pH	
	effective cation exchange capacity	barium chloride reagent
	phosphorus	spectrometry
	total nitrogen	modified Kjeldahl analysis
	organic and total carbon	elementary analysis
	trace elements	extraction with aqua regia
	carbonate content	volumetric
	water- and acid-soluble sulphate	barium chloride reagent
	potential cation exchange capacity	
	cadmium, chromium, cobalt, copper, lead, manganese, nickel, and zinc	extraction with aqua regia with flame and electrothermal atomic absorption spectrometry
	polynuclear aromatic hydrocarbons	high performance liquid chromatography
	redox potential	field method
	trace elements	extraction with DTPA or digestion with hydrofluoric and perchloric acids
BS 8855	coal tar derived phenolic compounds	

are used in conjunction with battery-powered pumps operating at a flowrate of 10–20 ml/s. Again the average concentrations can be read off directly after an exposure period of, e.g., 8 hours. Direct-reading colorimetric diffusion tubes requiring no pump are available for a small number of substances (Table 10.18(c)). Because of their simplicity, colour indicator tubes are widely used but their limitations must be appreciated; sources of inaccuracy are given in Table 10.19.

Portable or fixed multi-point colorimetric detectors are available which rely on paper tape impregnated with the reagent. A cassette of the treated paper is driven electrically at constant speed over a sampling orifice and the stain intensity measured by an internal reflectometer to provide direct read out of concentration of contaminant in sample. Such instruments are available for a range of chemicals including the selection given in Table 10.20.

Flammable gases

Flammable atmospheres can be assessed using portable gas chromatographs or, for selected compounds, by colour indicator tubes. More commonly, use is made of 'explosimeters' fitted with Pellistors (e.g. platinum wire encased in beads of refractory material). The beads are arranged in a Wheatstone bridge circuit. The flammable gas is oxidized on the heated catalytic element, causing the electrical resistance to alter relative to the reference. Instruments are calibrated for specific compounds in terms of 0–100% of their lower flammable limit. Recalibration or application of correction factors is required for different gases. These types of sensor are subject to 'poisoning' by certain chemicals such as silicone, sulphur, and halogen compounds. Points to consider are listed in Table 10.21.

Semiconductors resembling Pellistors are also available, e.g. a platinum/rhodium filament is used to heat a pellet of doped oxide which absorbs any flammable gas passing over it, causing the electrical conductivity to alter and subsequently the voltage in the electrodes attached to the pellet. Amplification of the signal registers a deflection on the meter. Again, errors can arise if the instrument is used for gases other than those for which it is calibrated. These sensors offer a long life expectancy; they are not restricted to flammable substances.

Table 10.12 Selected examples of sampling techniques for air contaminants

Contaminant	*Principle*	*Apparatus*	*Collecting agent*	*Sampling rate* (l/min)	*Collection efficiency* (%)	*Remarks*
Gases or vapours that are water soluble or miscible or that are only soluble or highly reactive in other agents	Absorption with simple washing	Petri bottle Wash bottle Drechsel bottle Impinger	Water, acid, alkali, or organic solvent	1–5	90–100	Two units in series can be used for highly reactive gases in other reagents
Gases or vapours of all types Mists and fumes	Absorption with multiple contact washing by dispersing gas into fine bubbles of large surface area	Fritted or sintered glass, alundum, stainless steel, or plastics	Water, acid, alkali, or organic solvents	1–15 (depends upon flask and sintered surface dimensions)	95–100	May plug if large particulates are present or precipitates form from reactions Fumes require very slow rates
Gases or vapours of all types Mists	Absorption with multiple contact provided by wetted surfaces	Packed columns of glass beads, glass spirals, or fibres Plastic packing also feasible in some cases Large surface wall area units also included	Water, acid, alkali, or organic solvents	1–5	90–100	May incorporate device for continually wetting column High loadings of particulates may plug unit with packing
Gases or vapours that are water soluble or miscible or that are only soluble or highly reactive in other agents	Absorption with multiple surface contact by atomizing liquid with spray nozzle or jet impaction	Crabtree ozone analyser or midget venturi scrubber	Water, acid, or alkali	5–25	60–100	Venturi scrubber satisfactory if dust is present Atomizer absorber will plug
Combustible gases or vapours that are not water soluble but are slowly reactive with absorbing agents	Combustion and absorption	Quartz or ceramic furnace with absorbers	Water, acid, or alkali	1–5	90–100	Can also be used for organic halogenated fumes
Gases or vapours that are not water soluble but are slowly reactive, with absorbing agent	Condensation	Freezing traps or low temperature condensers Glass or metal	Dry	1–5	90–100 (depends on vapour pressures at reduced temperature)	Can be packed with glass beads or other extended surfaces

Table 10.13 Common sorbent materials and applications

Sorbent material	*Application*
• Anasorb activated charcoal	Solvent desorption. Polar organic compounds
• Anasorb graphitized charcoal	Various grades for collection of substances of a range of volatilities
• Charcoal	Usually for solvent desorption especially of non-polar compounds
• Chromasorb	Hydrocarbons, high molecular-weight organic vapours, chlorinated organic compounds
• Florisil	Halogenated hydrocarbons including PCBs
• Hopcalite	Mercury vapour
• Molecular sieve	Nitrogen oxides
• Polyurethane foam	Halogenated compounds including PCBs, dioxins, furans and organophosphorus compounds
• Porous polymer	Various types usually for solvent desorption; suitable for a range of organic compounds including highly polar substances
• Silica gel	Aldehydes, amines, ammonia, chorophenols, esters, inorganic acid gases, methanol, nitro compounds
• Tenax	Various grades for collection of a range of boiling-point compounds. Suitable for low ambient concentrations

Table 10.14 Thermal desorption of sorbed gas from sample tubes

Advantages over solvent extraction
Elimination of sample preparation and handling of toxic solvents such as carbon disulphide
Absence of solvent simplifies chromatograph
Increased sensitivity
Sample tubes can be reused
Greater range of detection systems to which the desorbed gas can be subjected (e.g. chromatography, infra-red and ultraviolet spectroscopy, colorimetry)

Limitations
Certain resins undergo degradation even below 250°C
Test sample may be thermally unstable
Not all compounds readily desorb
The entire test sample is used with no opportunity for repeat analyses

Toxic particulates

Airborne particulates include dust, fume and aerosols. Many such particles are invisible to the naked eye under normal lighting but are rendered visible, by reflection, when illuminated with a strong beam of light. This is the 'Tyndall effect' and use of a dust lamp provides a simple technique for the rapid assessment of whether a dust is present, its flow pattern, leak sources, the effects of ventilation, etc. More sophisticated approaches are needed for quantitative data. Whether personal, spot or static sampling is adopted will depend upon the nature of the information required.

Air in the general atmosphere, or in the breathing zone of individuals, may be collected using a pump coupled to a means of isolating particulate matter for subsequent analysis or determination (Table 10.23). It is important to differentiate between 'total inhalable dust' i.e., the fraction of airborne material which enters the nose and mouth during breathing and is hence available for deposition in the respiratory tract, and 'respirable dust', i.e. the fraction which penetrates to the gas exchange region of the lung. For this purpose techniques for separating dust or aerosol

Table 10.15 Charcoal tube user guide

Substance	*Expected concentration range*[1] (ppm)	(mg/m^3)	*Recommended sampling*[2] (ml/min) 2 hr	4 hr	8 hr	*Recommended maximum tube load*[3] (mg)	*Approximate desorption efficiency*[4] (%)	*Eluent*[5]
Acetaldehyde	2–50	4–90	100	100	50			
	50–400	90–720	50	25	10			
Acetone	5–200	12–480	100	50	25	9	86 ± 10	CS_2
	200–2000	480–4800	10	5	1			
Acetic acid	0.1–1	0.25–2.5	–	200	200	10.4		11
	1–20	2.5–50	200	100	50			
Acetonitrile	1–10	1.8–17.5	100	100	50	2.7		
	10–80	17.5–140	50	25				
Acrolein	0.02–0.2	0.05–0.5	–	–	200			
Acrylonitrile	0.2–2	0.45–4.5	200	200	100	>2	<80	12
	2–40	4.5–90	100	50	25			
Allyl alcohol	0.1–0.5	0.24–1.20	–	200	200	<0.4	89 ± 5	2
	0.5–4	1.2–9.6	200	100	50			
Allyl chloride	0.05–2	0.15–6	–	–	200	0.75		13
n-Amyl acetate	1–25	5.3–131.3	100	100	50	15	86 ± 5	CS_2
	25–200	131.3–1050	50	25	10			
sec-Amyl acetate	1–25	5.3–131.3	100	100	50	15.5	91 ± 10	CS_2
	25–250	131–1300	50	25	10			
Isoamyl alcohol	1–25	3.6–90	100	100	50	10		
	25–200	90–720	50	25	10			
Benzene	0.1–1	0.31–3.1	–	200	200		96	CS_2
	1–20	3.1–62.6	100	100	50			
Benzyl chloride	0.05–2	0.25–10	–	200	200	>0.4	90 ± 5	CS_2
Bromoform	0.05–1	0.5–10	200	100	50	>0.25		CS_2
Butadiene	5–100	11–220	100	50	25	4		
	100–2000	220–4400	10	5	1			
2-Butoxyl ethanol	0.5–10	2.4–48	100	100	50		95 ± 5	6
	10–100	48–480	100	50	25			
n-butyl acetate	1–25	4.7–118	100	100	50	15	95	CS_2
	25–300	118–1420	50	25	10			
sec-Butyl acetate	1–50	4.8–237.5	100	100	50	15	91 ± 5	CS_2
	50–400	237.5–1900	50	25	10			
tert-Butyl acetate	1–50	4.8–237.5	100	100	50	12.5	94 ± 5	CS_2
	50–400	237.5–1900	50	25	10			
Butyl alcohol	1–25	3–75	200	100	50	10.5	88 ± 5	3
	25–200	75–600	100	50	25			
sec-Butyl alcohol	1–30	3–90	200	100	50	6	93 ± 5	3
	30–300	90–900	50	25	10			

tert-Butyl alcohol	1–25	3–75	200	100	50	5	90 ± 5	3
	25–200	75–600	50	25	10			
Butylamine	0.1–1	0.3–3	–	200	200			
	0.5–10	1.5–30	200	100	50			
Butyl glycidyl ether	0.5–10	1.4–54	200	100	50	11.5	85 ± 10	CS_2
	10–100	54–540	100	50	25			
p-tert Butyl toluene	0.1–1	0.6–6	200	200	100	2.5	100+	CS_2
	1–20	6–120	100	50	25			
Camphor	0.05–0.5	0.63–6.3	200	200	100	13.4	98 ± 5	1
	0.1–4	1.3–25	200	100	50			
Carbon disulphide	0.5–5	1.5–15	200	200	100		95	13
	5–40	15–120	200	100	50			
Carbon tetrachloride	0.2–2	1.3–13	200	200	100	7.5	97 ± 5	CS_2
	2–20	13–130	200	100	50			
Chlorobenzene	0.75–10	3.5–23.3	100	100	50	15.5	90 ± 5	CS_2
	10–150	23.3–700	50	25	10			
Chlorobromomethane	2–20	10.5–105	100	50	25	9.3	94 ± 5	CS_2
	20–400	105–2100	25	10	5			
Chloroform	0.5–10	2.4–48	200	100	50	11	96 ± 5	CS_2
	10–100	48–480	100	50	25			
1-Chloro-1-Nitropropane	0.2–5	1–25	200	100	50			
	5–40	25–200	100	50	25			
Chloroprene	0.5–5	1.8–18	200	100	50			
	5–50	18–180	100	50	25			
o-Chlorotoluene	1–10	5–50	100	50	25			CS_2
	10–100	50–500	50	25	10			
Cresol (all isomers)	0.1–1	0.44–4.4	200	200	100	>2.0		
	1–10	4.4–44	100	100	50			
Crotonaldehyde	0.1–1	0.3–3	–	200	200			
	0.5–4	1.5–12	100	100	50			
Cumene	0.5–10	2.5–49	100	100	50	11	100+	CS_2
	10–100	49–490	50	25	10			
Cyclohexane	3–50	10.5–175	100	50	25	6.3	100+	CS_2
	50–600	175–2100	25	10	5			
Cyclohexanol	0.5–10	2–40	200	100	50	10	99 ± 5	2
	10–100	40–400	100	50	25			
Cyclohexanone	0.5–10	2–40	200	100	50	13	78 ± 5	CS_2
	10–100	40–400	100	50	25			
Cyclohexene	3–50	10.2–169.2	100	50	25		100+	CS_2
	50–600	169.2–2030	25	10	5			
Diacetone alcohol	0.5–10	1.4–48	200	100	50	12	77 ± 10	2
	10–100	48–480	100	50	24			
o-Dichlorobenzene	0.5–10	3–60	200	100	50	15	85 ± 5	CS_2
	10–100	60–600	50	25	10			
p-Dichlorobenzene	1–25	6–150	100	100	50		85 ± 5	CS_2
	25–150	150–900	50	25	10			

Table 10.15 Cont'd

Substance	*Expected concentration range*[1] (ppm)	(mg/m^3)	*Recommended sampling*[2] (ml/min) 2 hr	4 hr	8 hr	*Recommended maximum tube load*[3] (mg)	*Approximate desorption efficiency*[4] (%)	*Eluent*[5]
Dichlorodifluoromethane	5–100	24.8–495	100	50	25			
(Freon 12)	100–2000	495–9900	10	5	1			
1,1-Dichloroethane	1–15	4–60	100	100	50	7.5	100+	CS_2
	15–200	60–300	50	25	10			
1,2-Dichloroethylene	2–25	7.9–99	100	50	25	5.1	100+	CS_2
	25–400	99–1580	25	10	5			
1,1-Dichloro-1-nitroethane	0.1–1.5	0.6–9	200	200	100	9		CS_2
	1.5–20	9–120	100	50	25			
Dichloroethyl ether	0.5–5	3–30	200	100	50			
	5–30	30–180	100	50	25			
Dichloromonofluoromethane	5–100	21–420	100	50	25			
(Freon 21)	100–2000	420–8400	10	5	1			
Dichlorotetrafluoroethane	5–100	35–700	50	25	10			
	100–2000	7000–14 000	5	1	1			
Difluorodibromomethane	1–15	8.6–129	100	50	25	15		CS_2
	15–200	129–1720	25	10	5			
Diisobutyl ketone	0.5–7.5	2.9–43.5	200	100	50	12.5		CS_2
	7.5–100	43.5–580	100	50	25			
Dimethylaniline	0.1–1	0.5–5	200	200	100	>1.1	>80	CS_2
	1–10	5–50	100	50	25			
Dimethylformamide	0.5–4	1.5–12	200	100	50			
	2–20	6–60	100	100	50			
p-Dioxane	1–15	3.6–54	200	100	50	13	91 ± 5	CS_2
	15–200	54–720	100	50	25			
Dipropylene glycol methyl ether	1–15	6–90	100	50	25		75 ± 15	CS_2
	15–200	90–1200	25	10	5			
Epichlorohydrin	0.1–1	0.4–4	–	200	100	>1	>80	CS_2
	1–10	4–40	100	50	25			
2-Ethoxyethanol	2–30	7.4–111	100	50	25			
	30–400	111–1480	50	25	10			
2-Ethoxyethyl acetate	1–15	5.4–81	200	100	50	19	74 ± 10	CS_2
	15–200	81–1080	50	25	10			
Ethyl acetate	5–75	17.5–263	100	50	25	12.5	90 ± 10	CS_2
	75–800	263–2800	25	10	5			
Ethyl acrylate	0.5–5	2–20	200	200	100	<5	95 ± 5	CS_2
	5–50	20–200	200	100	50			
Ethyl alcohol	5–100	9.4–188.5	100	50	25	2.6	77 ± 10	5
	100–2000	188.5–3770	5	1	1			

Ethyl benzene	1–15	4.4–65.3	200	100	50	16	100+	CS_2
	15–200	65.3–870	100	50	25			
Ethyl bromide	2–50	8.9–223	100	50	25	7.1	83 ± 5	4
	50–400	223–1780	25	10	5			
Ethyl butyl ketone	0.5–10	2.3–46	200	100	50	>5.5	93 ± 5	1
	10–100	46–460	50	25	10			
Ethyl chloride	10–150	26–390	100	50	25	9.7		CS_2
	150–2000	390–5200	10	5	1			
Ethyl ether	5–75	15–227	100	50	25	7.5	98 ± 5	7
	75–800	227–2420	10	5	1			
Ethyl formate	1–15	3–45	200	100	50	4.8	80 ± 10	CS_2
	15–200	45–600	50	25	10			
Ethylene chlorohydrin[6]	0.1–2	0.32–6.4	–	200	200	16.0	92 ± 5	10
	1–10	3.2–32	200	100	50			
Ethylene dibromide	0.2–5	1.6–38.88	200	200	100	>10.7	93 ± 5	CS_2
	2–40	15.5–310	100	50	25			
Ethylene dichloride	0.5–10	2–40.5	200	200	100	12	95 ± 5	CS_2
	10–100	40.5–405	100	50	25			
Ethylene oxide	0.5–10	0.9–18	200	100	50	1.1		CS_2
	10–100	18–180	100	50	25			
Fluorotrichloromethane	5–100	28–560	100	50	25			
(Freon 11)	100–2000	560–11 200	5	1	1			
Furfural	0.1–2	0.4–8	200	200	100		>80	CS_2
	1–10	4–40	200	100	50			
Furfuryl alcohol	0.5–20	2–80	200	100	50			
	10–100	40–400	50	25	10			
Glycidol	0.5–20	1.5–60	200	100	50	22.5	90 ± 5	8
	10–100	30–300	100	50	25			
Heptane	5–100	20–400	100	50	25	12.5	96 ± 5	CS_2
	100–1000	400–4000	10	5	1			
Hexachloroethane	0.05–2	0.5–20	200	200	100		98 ± 5	CS_2
Hexane	5–100	18–360	100	50	25	11	94 ± 5	CS_2
	100–1000	360–3600	10	5	1			
Isoamyl acetate	1–15	5.3–78.8	200	100	50	16.5	90 ± 5	CS_2
	15–200	78.8–1050	50	25	10			
Isoamyl alcohol	1–15	3.6–54	200	100	50	10	99 ± 5	2
	15–200	54–720	50	25	10			
Isobutyl acetate	1–20	4.7–93	200	100	50	14	92 ± 5	CS_2
	20–300	93–1400	50	25	10			
Isobutyl alcohol	1–15	3–46	200	100	50	10.5	84 ± 10	3
	15–200	46–610	50	25	10			
Isophorone	0.5–5	2.8–28	200	100	50	13	>80	CS_2
	5–50	28–280	50	25	10			
Isopropyl acetate	2–50	7.6–190	100	100	50	13	85 ± 5	CS_2
	50–500	190–1900	25	10	5			

Table 10.15 Cont'd

Substance	*Expected concentration range*[1] (ppm)	(mg/m³)	*Recommended sampling*[2] (ml/min) 2 hr	4 hr	8 hr	*Recommended maximum tube load*[3] (mg)	*Approximate desorption efficiency*[4] (%)	*Eluent*[5]
Isopropyl alcohol	5–75	12.3–185	100	50	25	5.6	94 ± 5	5
	75–800	185–1970	25	10	5			
Isopropyl ether	5–100	21–420	100	50	25			
	100–1000	420–4200	10	5	1			
Isopropyl glycidyl ether	0.5–10	2.4–48	200	100	50	10.5	80 ± 10	CS_2
	10–100	48–480	100	50	25			
Mesityl oxide	0.5–10	2–40	200	200	100	4.8	79 ± 5	1
	5–50	20–200	100	50	25			
Methyl acetate	1–50	3.1–152.5	200	100	50	7	88 ± 5	CS_2
	50–400	152.5–1220	25	10	5			
Methyl acrylate	0.1–2	0.35–7	–	200	100	>1.5	80 ± 10	CS_2
	1–20	3.5–70	200	100	50			
Methyl alcohol	Charcoal tube method not recommended. Use silica gel tubes.							
Methylal (dimethoxymethane)	5–100	15.5–311	100	50	25	11.5	78 ± 10	9
	100–2000	311–6220	10	5	1			
Methyl amyl ketone	1–15	4.7–70	200	100	50	7.5	80 ± 10	1
	15–200	70–930	50	25	10			
Methyl bromide	0.5–4	2–16	200	100	50	1.5		CS_2
	4–40	16–160	50	25	10			
Methyl butyl ketone (2-hexanone)	1–15	4.1–61.5	200	100	50	2.0	79 ± 10	CS_2
	15–200	61.5–820	50	25	10			
Methyl isobutyl ketone (hexanone)	1–15	4.1–61.4	200	100	50	10	81 ± 5	CS_2
	15–200	61.4–818	50	25	10			
Methyl cellosolve	0.3–10	1–32	200	200	100	10	97 ± 5	6
	5–50	16–160	100	50	25			
Methyl cellosolve acetate	0.3–10	1.5–48	200	100	50	5	76 ± 10	CS_2
	5–50	24–240	100	50	25			
Methyl chloride	1–15	2.1–31.5	200	100	50	1		CS_2
	15–200	31.5–420	25	10	5			
Methyl chloroform (1,1,1 trichloroethane)	1–75	5.4–407	200	100	50	18	98+	CS_2
	75–700	407–3800	25	10	5			
Methyl cyclohexane	5–100	20–400	100	50	25		95 ± 5	CS_2
	100–1000	400–4000	10	5	1			
Methyl ethyl ketone (2-butanone)	2–50	5.9–148	100	100	50	9.5	89 ± 10	CS_2
	50–400	148–1180	50	25	10			
Methyl formate	1–15	2.5–37.5	200	100	50			
	15–200	37.5–500	50	25	10			
5-Methyl-3-heptanone	0.25–5	1.3–26	200	100	50	>5		1
	5–50	26–260	100	50	25			

Methyl iodide	0.1–1	0.56–5.6	200	200	100	7		14
	1–10	5.6–56	100	50	25			
Methyl isobutyl carbinol	0.5–5	2.1–21	200	100	50	5.7	99 ± 5	2
	5–50	21–210	200	100	50			
Methyl isoamylacetate	1–10	6–60	200	100	50	14		CS_2
	10–100	60–600	100	50	25			
α-methyl styrene	1–15	4.8–72	200	200	100	21	91 ± 5	CS_2
	15–200	72–960	100	50	25			
Methylene chloride	5–75	17.4–261	100	100	50	9.3	95 ± 10	CS_2
	75–1000	261–3480	10	5	1			
Naphtha (coal tar)	1–15	4–60	200	100	50	14.8	88 ± 5	CS_2
	15–200	60–800	100	50	25			
Naphthalene	0.1–2	0.5–10	200	200	100			
	1–20	5–100	100	50	25			
Nitromethane	1–15	3.1–46.5	200	200	100			
	15–200	46.5–620	50	25	10			
1-Nitropropane	0.5–10	1.8–36	200	100	50			
	5–50	18–180	100	50	25			
2-Nitropropane	0.5–10	1.8–36	200	100	50			
	5–50	18–180	100	50	25			
Nonane	2–30	10.5–157.5	100	50	25			CS_2
	30–400	157.5–2100	25	10	5			
n-Octane	5–75	23.5–352	100	50	25	15	93 ± 5	CS_2
	75–1000	352–4700	10	5	1			
Pentane	5–100	15–295	200	100	50	9	96 ± 5	CS_2
	100–2000	295–5900	10	5	1			
2-Pentanone	1–75	3.5–263	200	100	50		88 ± 5	CS_2
	75–400	263–1400	25	10	5			
Perchloroethylene (tetrachloroethylene)	1–15	6.8–102	200	100	50	29	95 ± 5	CS_2
	15–200	102–1362	50	25	10			
Petroleum distillates	5–75	20–300	100	50	25	12.3	96 ± 5	CS_2
	75–1000	300–4000	10	5	1			
Phenol	0.1–2	0.38–7.6	–	200	100			
	1–10	3.8–38	200	100	50			
Phenyl ether (vapour)	0.05–2	0.35–14	–	200	200	0.6	90 ± 5	CS_2
Phenyl glycidyl ether	0.5–4	3–24	200	100	50	12.5	97 ± 5	CS_2
	2–20	12–120	100	50	25			
n-Propyl acetate	1–75	4.2–315	200	100	50	14.5	93 ± 5	CS_2
	75–400	315–1680	50	25	10			
n-Propyl alcohol	1–75	2.5–184	200	100	50	9	87 ± 5	3
	75–400	184–980	50	25	10			
Propylene dichloride	1–15	4.6–70	200	100	50	5	97 ± 5	CS_2
	15–150	70–700	50	25	10			
Propylene oxide	1–15	2.4–36	200	100	50	2	90 ± 5	CS_2
	15–200	36–480	25	10	5			

Table 10.15 Cont'd

Substance	*Expected concentration range*[1]		*Recommended sampling*[2] (ml/min)			*Recommended maximum tube load*[3] (mg)	*Approximate desorption efficiency*[4] (%)	*Eluent*[5]
	(ppm)	(mg/m^3)	2 hr	4 hr	8 hr			
n-Propyl nitrate	0.5–5	2.2–22	200	100	50	12		CS_2
	5–50	22–220	100	50	25			
Pyridine	0.5–2	1.5–6	200	200	100	>7.3	70 ± 10	CS_2
	1–10	3–30	200	100	50			
Stoddard solvent	5–75	29.5–443	100	50	25	13	96 ± 5	CS_2
	75–1000	443–5900	10	5	1			
Styrene (monomer)	1–15	4.3–64	200	100	50	18	87 ± 5	CS_2
	75–1000	64–850	100	50	25			
1,1,1,2-Tetrachloro-2,2-Difluoroethane	5–75	42–625	100	50	25	19.5	100+ CS_2	
	75–1000	625–8340	10	5	1			
1,1,2,2-Tetrachloro-1,2-Difluoroethane	5–75	42–625	100	50	25	26	96 ± 5	CS_2
	75–1000	625–8340	10	5	1			
1,1,2,2-Tetrachloroethane[6]	0.1–2	0.7–14	200	200	100	4.5	85 ± 5	CS_2
	1–10	7–70	100	100	50			
Tetrahydrofuran	1–75	2.5–184	200	100	50	7.5	92 ± 5	CS_2
	75–400	184–1180	25	10	5			
Tetramethyl succinonitrile	0.05–1	0.3–6	–	200	200	>0.8		CS_2
Toluene	1–15	3.8–56	200	100	50		92 ± 5	CS_2
	15–200	56–750	50	25	10			
1,1,2-trichloroethane	0.1–2	0.6–11	200	100	50	5	96 ± 5	CS_2
	2–20	11–110	100	50	25			
Trichloroethylene	1–15	5.4–80	200	100	50	21	96 ± 5	CS_2
	15–200	80–1070	100	50	25			
1,2,3-Trichloropropane	0.5–10	3–60	200	100	50	14		CS_2
	10–100	60–600	100	50	25			
1,1,2-Trichloro-1,2,2-Trifluoroethane (Freon 113)	10–150	77–1150	100	50	25	20	100+	CS_2
	150–2000	1150–15 300	10	5	1			
Trifluoromonobromomethane[7]	10–150	61–914	100	50	25	25		
	150–2000	914–12 180	10	5	1			
Turpentine	1–15	5.6–84	200	100	50	13	96 ± 5	CS_2
	15–200	84–1120	50	25	10			
Vinyl chloride[7]	0.05–1	0.13–2.5	–	–	200		85 ± 5	CS_2
	0.5–2	1.3–5.2	200	200	100			
Vinyl toluene	1–15	4.8–72	200	100	50	17	85 ± 10	CS_2
	15–200	72–960	100	50	25			
Xylene	1–15	4.4–65	200	100	50		95 ± 5	CS_2
	15–200	65–870	50	25	10			

(1) Two concentration ranges are given for most substances. The low range is approximately 1%–15% of the TLV and the high range 15%–200% TLV. The user should select from these two ranges the expected mean concentration.

(2) A sampling rate is recommended for each concentration range and for a 2 hr, 4 hr, or 8 hr sampling period. Each sampling rate is given in ml/min and has been calculated to provide a minimum tube loading of at least 0.01 mg at the minimum concentration shown and not to exceed the recommended tube loading at the highest concentration shown for that range. These figures are based on the use of 100 mg coconut-shell charcoal tubes to the NIOSH recommended design except where otherwise noted.

(3) A recovery of 5% of the total sample from the back-up section of charcoal in a sample tube was defined as the breakthrough point: 50% of this value is shown as the recommended maximum tube loading, to allow for high humidity or the presence of other substances which reduce the normal tube capacity.

(4) The figures given are not intended to be used as exact desorption efficiencies and are only given as a guide when carrying out system calibrations. Actual desorption efficiencies should always be determined at the time of analysis. However, these figures represent the best obtainable data from several sources, and any significant deviation should be regarded as a possible indication of a systematic error in the analytical technique. The figure given for desorption efficiency is an average figure. The desorption efficiency for a compound will vary with the amount of substance on the tube. With reduced tube loadings, in most cases, the desorption efficiency will be lower. Significant errors may be introduced when analysing small amounts of substance and an average desorption efficiency factor is used.

(5) The desorption efficiencies given are directly related to the eluent used. All data in the desorption efficiency column correspond to the specific eluent listed:

1 CS_2 + 1% methanol
2 CS_2 + 5% 2-propanol
3 CS_2 + 1% 2-propanol
4 Isopropanol
5 CS_2 + 1% butanol
6 Methylene chloride + 5% methanol
7 Ethyl acetate 0.5 ml
8 Tetrahydrofuran 0.5 ml
9 Hexane 0.5 ml
10 CS_2 + 5% isoamyl alcohol
11 Formic acid
12 Methanol
13 Benzene
14 Toluene

(6) 100 mg petroleum-based charcoal tubes based on the NIOSH recommended design should be used for sample collection of these compounds.

(7) These compounds migrate rapidly to the back-up section of the charcoal tube. A 400 mg tube should be used for sample collection with a second 100 mg tube in series behind the large tube to determine breakthrough.

Table 10.16 Chromatographic column guide for Century organic vapour analyser

Compound	*Relative response*[1] (%)	*Column temperature*[2] (°C)	*Column packing retention time* (min:secs)		
			B-8	*G-8*	*T-8*
Acetone	60	0	0:37	0:27	1:50
		40	0:20	0:15	0:30
Acetonitrile	70	0	0:40	0:30	5:24
		40	0:20	0:18	1:15
Acrylonitrile	70	0	0:35	0:30	3:45
		40	0:19	0:17	0:51
Allyl alcohol	30	0	1:15	0:44	11:45
		40	0:37	0:26	1:35
Allyl chloride	50	0	0:16	0:28	0:31
		40	0:08	0:16	0:15
Benzene	150	0	0:50	1:19	1:43
		40	0:22	0:25	0:32
2-Bromo-2-chloro-1,1,1-trifluoroethane	45	0	0:37	0:35	0:58
(Halothane)		40	0:17	0:14	0:19
Bromoethane	75	0	0:15	0:26	0:23
		40	0:05	0:15	0:14
1-Bromopropane	75	0	0:21	0:58	0:40
		40	0:08	0:22	0:18
2-Butane	60	0	0:15	0:15	0:10
		40	–	–	–
n-Butanol	50	0	4:10	2:20	15:07
		40	0:24	0:53	1:44
2-Butanol	65	0	1:45	0:55	6:07
		40	0:24	0:24	0:53
n-Butyl acetate	80	0	4:15	7:30	11:40
		40	0:50	1:14	1:31
n-Butyl acrylate	60	0	–	–	–
		40	2:30	2:15	2:40
2-Butyl acrylate	70	0	–	–	–
		40	1:22	1:45	1:46
n-Butyl formate	50	0	2:15	2:57	4:40
		40	0:28	0:34	0:49
2-Butyl formate	60	0	1:22	2:00	3:31
		40	0:20	0:26	0:37
n-Butyl methacrylate	60	0	–	–	–
		40	4:03	5:46	5:10
2-Butyl methacrylate	80	0	–	–	–
		40	2:46	3:40	3:34
Carbon tetrachloride	10	0	0:20	1:24	0:37
		40	0:10	0:25	0:17
Chlorobenzene	200	0	5:45	8:00	11:20
		40	1:08	1:24	1:35
Chlorodifluoromethane (Freon 22)	40	0	0:11	0:11	0:15
		40	–	–	–
Chloroform	65	0	0:55	0:57	2:00
		40	0:20	0:20	0:31
1-Chloropropane	75	0	0:16	0:31	0:23
		40	0:05	0:16	0:14
2-Chloropropane	90	0	0:15	0:23	0:18
		40	0:05	0:05	0:05
2-Chloro-1,1,2-trifluoroethyl difluoromethyl ether	150	0	0:36	0:26	1:22
(Ethrane)		40	0:13	0:12	0:19
Cumene	100	0	11:00	20:00	12:45
		40	2:20	3:03	2:19

Table 10.16 Cont'd

Compound	*Relative response*[1] (%)	*Column temperature*[2] (°C)	*Column packing retention time* (min:secs)		
			B-8	*G-8*	*T-8*
Cyclohexane	85	0	0:36	1:25	0:19
		40	0:18	0:26	0:14
Cyclohexanone	100	0	18:00	12:45	–
		40	3:00	2:00	–
n-Decane	75	0	–	–	–
		40	2:57	6:20	1:35
o-Dichlorobenzene	50	0	–	–	–
		40	8:06	10:00	11:20
Dichlorodifluoromethane (Freon 12)	15	0	0:10	0:11	0:12
		40	–	–	–
1,1-Dichloroethane	80	0	0:17	0:37	0:45
		40	0:08	0:17	0:18
1,2-Dichloroethane	80	0	1:14	1:08	3:50
		40	0:23	0:22	0:43
trans 1,2-Dichloroethylene	50	0	0:16	0:35	0:31
		40	0:05	0:18	0:16
Dichlorofluoromethane (Freon 21)	70	0	0:16	0:15	0:23
		40	–	–	–
Dichloromethane	100	0	0:27	0:29	1:08
		40	0:10	0:10	0:22
1,2-Dichloropropane	90	0	0:41	1:49	2:56
		40	0:18	0:29	0:36
1,3-Dichloropropane	80	0	1:32	4:12	4:24
		40	0:26	0:47	1:20
1,2-Dichloro 1,1,2,2,-tetrafluoroethane	110	0	0:12	0:12	0:14
(Freon 114)		40	–	–	–
Diethyl ether	50	0	0:20	0:26	0:19
		40	0:05	0:05	0:05
Diethyl ketone	80	0	2:00	2:01	3:16
		40	0:29	0:30	0:49
p-Dioxane	30	0	3:15	2:09	6:40
		40	0:44	0:34	1:19
Ethane	80	0	0:15	0:15	0:15
		40	–	–	–
Ethanethiol	30	0	0:18	0:24	0:26
		40	0:13	0:12	0:14
Ethanol	25	0	0:59	0:31	0:26
		40	0:26	0:22	0:43
Ethyl acetate	65	0	0:48	1:00	2:20
		40	0:20	0:20	0:31
Ethyl acrylate	40	0	1:40	2:10	4:08
		40	0:30	0:30	0:50
Ethyl benzene	100	0	6:10	9:31	7:44
		40	1:15	1:35	1:35
Ethyl butyrate	70	0	4:34	6:22	8:00
		40	0:48	1:07	1:24
Ethyl formate	40	0	0:25	0:29	1:05
		40	0:16	0:17	0:20
Ethyl methacrylate	70	0	4:13	6:13	6:20
		40	0:47	1:01	1:01
Ethyl propionate	65	0	1:40	2:48	4:10
		40	0:25	0:35	0:50
Ethylene dibromide	50	0	5:20	4:51	15:00
		40	1:00	0:55	2:43

Table 10.16 Cont'd

Compound	Relative response[1] (%)	Column temperature[2] (°C)	Column packing retention time (min:secs)		
			B-8	G-8	T-8
Ethylene dichloride	60	0	1:07	1:08	3:45
		40	0:20	0:21	0:50
Ethylene oxide	70	0	0:10	0:13	0:20
		40	–	–	–
Fluorotrichloromethane (Freon 11)	10	0	0:15	0:18	0:17
		40	–	–	–
Heptane	75	0	0:50	2:20	0:27
		40	0:20	0:30	0:16
Hexane	70	0	0:23	0:50	0:20
		40	0:13	0:20	0:13
Isoprene	50	0	0:10	0:23	0:20
		40	0:05	0:05	0:05
Methane	100	0	0:05	0:05	0:05
		40	–	–	–
Methyl alcohol	12	0	0:37	0:21	2:23
		40	0:22	0:14	0:45
Methyl acetate	41	0	0:30	0:30	1:30
		40	0:17	0:15	0:24
Methyl acrylate	40	0	0:50	0:58	2:30
		40	0:20	0:21	0:37
Methyl cyclohexane	100	0	0:54	2:37	0:25
		40	0:18	0:33	0:17
Methyl cyclopentane	80	0	0:22	1:02	0:17
		40	0:05	0:19	0:05
Methyl ethyl ketone	80	0	1:00	0:50	3:30
		40	0:22	0:20	0:43
Methyl isobutyl ketone	80	0	4:20	3:15	7:30
		40	0:42	0:40	1:25
Methyl methacrylate	50	0	1:41	2:22	4:08
		40	0:27	0:31	0:55
Methyl propyl ketone	70	0	2:20	2:05	6:14
		40	0:33	0:32	0:52
Nitromethane	35	0	0:51	0:40	3:00
		40	0:25	0:19	1:31
1-Nitropropane	60	0	4:50	2:50	25:00
		40	0:46	0:41	4:05
2-Nitropropane	70	0	2:53	1:51	10:00
		40	0:35	0:31	1:52
Nonane	90	0	7:26	8:00	2:32
		40	1:08	2:40	0:42
Octane	80	0	2:47	7:39	1:07
		40	0:39	1:09	0:27
Pentane	65	0	0:18	0:25	0:14
		40	0:12	0:12	0:12
Pentanol	40	0	12:00	6:00	20:00
		40	2:44	1:17	3:36
Propane	80	0	0:05	0:11	0:05
		40	–	–	–
n-Propanol	40	0	2:50	1:00	6:30
		40	0:35	0:30	1:05
2-Propanol	65	0	1:13	0:30	3:43
		40	0:25	0:20	0:38
n-Propyl acetate	75	0	2:04	2:52	5:52
		40	0:30	0:38	0:51

Table 10.16 Cont'd

Compound	*Relative response*[1] (%)	*Column temperature*[2] (°C)	*Column packing retention time* (min:secs) *B-8*	*G-8*	*T-8*
n-Propyl ether	65	0	1:37	2:06	1:15
		40	0:43	0:29	0:18
n-Propyl formate	50	0	0:46	1:07	2:10
		40	0:17	0:18	0:27
Pyridine	128	0	8:00	4:15	–
		40	–	1:26	–
Styrene	85	0	20:00	25:00	28:00
		40	2:06	2:26	2:39
1,1,1,2-Tetrachloroethane	100	0	8:25	7:48	10:00
		40	1:16	1:27	1:37
1,1,2,2-Tetrachloroethane	100	0	32:00	14:00	50:00
		40	4:09	2:37	7:51
Tetrachloroethylene	70	0	3:00	5:45	2:10
		40	0:41	1:06	0:33
Tetrahydrofuran	40	0	1:05	1:05	1:45
		40	0:23	0:23	0:30
Toluene	110	0	2:30	4:05	4:30
		40	0:38	0:47	0:53
1,1,1-Trichloroethane	105	0	0:31	1:10	0:47
		40	0:15	0:23	0:20
1,1,2-Trichloroethane	85	0	5:13	3:43	15:00
		40	0:40	0:45	2:30
Trichloroethylene	70	0	1:17	2:02	1:25
		40	0:23	0:28	0:28
Trichlorotrifluoroethane (Freon 113)	80	0	0:15	0:30	0:16
		40	0:13	0:14	0:13
Triethylamine	70	0	–	2:49	–
		40	–	–	–
Vinyl acetate	50	0	0:34	0:43	1:55
		40	0:15	0:17	0:30
Vinylidene chloride	40	0	0:20	0:25	0:22
		40	0:13	0:14	0:14
m-Xylene	111	0	2:39	12:03	8:31
		40	0:39	1:43	1:17
o-Xylene	116	0	3:29	15:07	8:40
		40	0:48	1:58	1:45
p-Xylene	116	0	2:46	12:25	8:23
		40	0:39	1:42	1:19

B 3% Diisodecyl phthalate on Chromosorb W, AW, 60/80 mesh
G 10% SP – 2100 on Supelcoport, 60/80 mesh
T 10% 1,2,3-Tris(2-cyanoethoxy) propane on Chromosorb P, AW, 60/80 mesh
[1] Century organic vapour analysers are factory calibrated to measure 'total organic' vapours according to a standard (methane). Since different organic vapours interact with the flame ionization detector (FID) to varying extents, it is vital that the instrument user be aware of the magnitude of the variation in order to obtain the most accurate data. Each user must determine relative responses for the individual instrument.
[2] For chromatographic work, the OVA can be used with a variety of column lengths and packing materials. For highest accuracy, temperature control for the column is mandatory. This is accomplished using the portable isothermal pack (PIP) kit which is supplied with three 8 in (203 mm) columns packed with B, G and T materials respectively. Isothermal control is accomplished non-electrically using an ice-water mixture for 0°C and a seeded eutectic mixture for 40°C. The data listed are for comparison purposes only since retention time for a compound can vary according to the condition of the column packing material, packing procedure and chemical interaction among the components of a vapour mix.
A blank in the table indicates that no data are available for the analysis.

Table 10.17 Compounds detectable by portable infrared analysis (MIRAN 1A)

Compound	*Analytical Wavelength* (μm)[1]	*Path length* (m)[2]	*Absorbence*[3]	*Minimum Detectable Concentration* (ppm)[4] (20 metre cell)
Acetaldehyde	9.0	20.25	0.23	0.8
Acetic acid	8.5	20.25	0.044	0.1
Acetic anhydride	8.9	20.25	0.18	0.02
Acetone	8.2	2.25	0.49	0.09
Acetonitrile	9.6	20.25	0.005	5.0
Acetophenone	7.9	20.25	0.5	0.7
Acetylene	3.03	6.75	0.19	0.4
Acetylene dichloride, *see* 1,2-Dichloroethylene				
Acetylene tetrabromide	8.85	20.25	0.0047	0.1
Acrolein	8.6	20.25	0.0002	0.3
Acrylonitrile	10.5	20.25	0.005	0.2
Allyl alcohol	9.8	20.25	0.0055	0.3
Allyl chloride	10.8	20.25	0.0032	0.2
Allylglycidylether	9.1	20.25	0.08	0.07
2-Aminoethanol, *see* Ethanolamine				
Ammonia	10.4	20.25	0.14	0.2
n-Amyl acetate	8.0	0.75	0.12	0.02
Aniline	9.3	20.25	0.011	0.3
Arsine	4.66	20.25	0.0003	0.1
Arylam, *see* Carbaryl				
Benzene	14.87	20.25	0.08	0.3
p-Benzoquinone, *see* Quinone				
Benzyl chloride	7.9	20.25	0.0012	0.4
Bisphenol A, *see* Diglycidyl ether				
Bis(chloromethyl) ether	8.99	20.25	0.0016	0.2
Boron trifluoride	6.8	20.25	0.006	0.5
Bromoform	8.7	20.25	0.005	0.05
Butadiene, (1,3-butadiene)	11.0	2.25	0.46	0.2
Butane	10.36	20.25	0.03	3.0
Butanethiol, *see* Butyl mercaptan				
2-Butanone (MEK)	8.5	20.25	0.55	0.15
2-Butoxyethanol (butyl cellosolve)	8.9	9.75	0.2	0.08
Butyl acetate (*n*-butyl acetate)	8.1	3.75	0.53	0.02
sec-Butyl acetate	9.9	9.75	0.48	0.15
tert-Butyl acetate	9.9	9.75	0.36	0.2
n-Butyl alcohol	9.6	8.25	0.15	0.17
sec-Butyl alcohol	10.1	9.75	0.15	0.35
tert-Butyl alcohol	8.2	20.75	0.47	0.05
Butylamine	13.0	20.25	0.025	0.3

Butyl ether	8.85	20.25	0.5	0.04
Butyl carbitol	8.9	20.25	0.32	0.08
n-Butyl glycidyl ether	8.9	20.25	0.55	0.05
Butyl mercaptan	3.37	20.25	0.04	0.08
p-tert-Butyltoluene	12.3	20.25	0.015	0.8
Carbon disulphide	4.54	20.25	0.015	0.5
Carbon dioxide	4.25	0.75	1.2	0.05
Carbon monoxide	4.61	20.25	0.045	0.2
Carbon tetrachloride	12.6	20.25	0.31	0.06
Carbonyl sulphide	4.85	20.25	0.5	0.02
Carbary[5]				
Chlorinated camphene[5]				
Chlorobenzene (monochlorobenzene)	9.2	20.25	0.26	0.2
Chlorobromomethane	8.1	9.75	0.29	0.2
2-Chloro-1,3-butadiene, *see* Chloroprene				
Chlorodifluoromethane (Freon 22)	9.0	5.25	0.17	0.02
1-Chloro-2,3,-Epoxypropane, *see* Epichlorohydrin				
2-Chloroethanol, *see* Ethylene chlorohydrin				
Chloroethylene, *see* Vinyl chloride				
Chloroform (trichloromethane)	13.0	5.25	0.45	0.06
1-Chloro-1-nitropropane	12.4	20.25	0.04	1.0
Chloropentafluoroethane (Genetron 115)	8.1	5.25	0.7	0.02
Chloropicrin (trichloronitromethane)	11.5	20.25	0.002	0.05
Chloroprene (2-chloro-1,3-butadiene)	11.4	20.25	0.08	0.4
Chlorotrifluoroethylene	9.24	20.25	0.04	0.06
Cresol (all isomers)	8.6	20.25	0.0066	0.3
Crotonaldehyde (trans-2-butenal)	8.7	20.25	0.01	0.1
Cumene (isopropyl benzene)	9.75	20.25	0.031	0.7
Cyanogen	4.7	20.25	0.016	3.0
Cyclohexane	3.4	2.25	0.56	0.03
Cyclohexanol	9.3	20.25	0.28	0.10
Cyclohexanone	8.3	20.25	0.054	0.5
Cyclohexene	8.8	20.25	0.12	0.3
Deuterium oxide	3.68	20.25	0.067	1.0
DDVP, see Dichlorvos				
Diacetone alcohol (4-hydroxy-4-methyl-2-pentanone)	8.5	20.25	0.335	0.08
1,2-Diaminoethane, *see* Ethylenediamine				
Diborane	3.9	20.25	0.0005	0.1
Dibromochloropropane	8.27	20.25	0.19	0.2
1,2-Dibromotetrafluoroethane	8.45	20.25	0.7	0.02
o-Dichlorobenzene	13.4	20.25	0.213	0.4
p-Dichlorobenzene	9.1	9.75	0.47	0.06
Dichlorodifluoromethane (Freon 12)	9.1	0.75	1.2	0.02
1,1-Dichloroethane	9.4	9.75	0.15	0.3
1,2-Dichloroethylene	12.1	0.75	0.1	0.07

Table 10.17 Cont'd

Compound	*Analytical Wavelength* (μm)(1)	*Path length* (m)(2)	*Absorbence*(3)	*Minimum Detectable Concentration* (ppm)(4) (20 metre cell)
Dichloroethyl ether	8.85	20.25	0.13	0.06
Dichloromethane, *see* Methylene chloride				
Dichloromonofluoromethane (Freon 21)	9.3	0.75	0.42	0.08
1,1-Dichloro-1-nitroethane	9.1	20.25	0.1	0.07
1,2-Dichloropropane, *see* Propylene dichloride				
Dichlorotetrafluoroethane (Freon 114)	8.4	0.75	1.43	0.02
Dichlorvos DDVP	9.4	20.25	0.0014	0.02
Diethylamine	8.8	20.25	0.1	0.2
Diethylamino ethanol	9.4	20.25	0.08	0.1
Diethylether, *see* Ethyl ether				
Diethyl ketone	9.0	20.25	0.2	0.1
Diethyl malonate	9.5	20.25	0.11	0.05
Difluorodibromomethane	9.2	2.25	0.36	0.02
Diglycidyl ether(6)				
Dihydroxybenzene, *see* Hydroquinone				
Diisobutyl ketone	8.6	20.25	0.12	0.2
Diisopropylamine	8.5	20.25	0.026	0.1
1,2-Dimethoxyethane	8.8	5.25	0.3	0.1
Dimethoxymethane, *see* Methylal				
N,N-Dimethylacetamide	9.9	20.25	0.02	0.3
Dimethylamine	8.7	20.25	0.02	0.5
Dimethylaminobenzene, *see* Xylidene				
Dimethylaniline (N, N dimethyl aniline)	8.6	20.25	0.015	0.2
Dimethylbenzene, *see* Xylene				
Dimethylformamide	9.2	20.25	0.063	0.1
2,6-Dimethylheptanone, *see* Diisobutyl ketone				
Dimethylsulphate	9.9	20.25	0.04	0.02
Dimethyl sulphoxide	9.0	20.25	0.5	0.2
Dioxane (diethylene dioxide)	8.9	5.25	0.38	0.05
Diphenylmethane diisocyanate, *see* Methylene bisphenyl isocyanate, MDI				
Enflurane	8.7	20.25	0.10	0.01
Epichlorohydrin	11.8	20.25	0.013	0.3
1,2-Epoxypropane, *see* Propylene oxide				
2,3-Epoxy-1-propanol, *see* Glycidol				
Ethanethiol, *see* Ethyl mercaptan				
Ethane	12.1	20.25	0.15	3.0
Ethanolamine	13.0	20.25	0.0017	1.2
2-Ethoxyethanol (cellosolve)	8.9	2.25	0.24	0.06

2-Ethoxyethyl acetate (cellosolve acetate)	8.8	8.25	0.58	0.03
Ethyl acetate	8.0	0.75	0.39	0.02
Ethyl acrylate	8.4	20.25	0.33	0.04
Ethyl alcohol (ethanol)	9.5	2.25	0.5	0.2
Ethylamine	3.4	20.25	0.015	0.2
Ethyl sec-amyl ketone (5-methyl-3-heptanone)	9.0	20.25	0.04	0.3
Ethylbenzene	9.7	20.25	0.06	1.0
Ethyl bromide	8.0	5.25	0.17	0.2
Ethyl butyl ketone (3-heptanone)	9.0	20.25	0.12	0.2
Ethyl chloride	10.4	20.25	0.62	0.8
Ethyl ether	8.8	2.25	0.58	0.03
Ethyl formate	8.5	2.25	0.1	0.06
2-Ethyl hexanol	9.7	20.25	0.02	0.2
Ethyl mercaptan	3.3	20.25	0.029	0.8
Ethyl silicate	9.1	5.25	0.93	0.02
Ethylene	10.6	20.25	0.12	0.5
Ethylene chlorohydrin	9.3	20.25	0.026	0.08
Ethylenediamine	13.0	20.25	0.035	0.4
Ethylene dibromide (1,2-dibromoethane)	8.45	20.25	0.07	0.1
Ethylene dichloride (1,2-dichloroethane)	8.1	20.25	0.09	0.3
Ethylene glycol monomethyl ether acetate, *see* Methyl cellosolve acetate				
Ethylene oxide	11.8	20.25	0.12	0.4
Ethylidine chloride, *see* 1,1-Dichloroethane				
Fluorobenzene	8.1	20.25	0.4	0.06
Fluorotrichloromethane (Freon 11)	11.9	0.75	1.7	0.01
Fluroxene	8.5	20.25	0.086	0.01
Formaldehyde	3.58	20.25	0.015	0.2
Formic acid	8.9	20.25	0.055	0.05
Furfural	13.3	20.25	0.05	0.1
Furfuryl alcohol	9.8	20.25	0.09	0.3
Glycidol (2,3-epoxy-1-propanol)	9.9	20.25	0.020	1.3
Glycol monoethyl ether, *see* 2-Ethoxyethanol				
Guthion(R), *see* Azinphosmethyl				
Halothane	12.3	20.25	0.027	0.08
Heptane (*n*-heptane)	3.4	0.75	0.29	0.01
1-Heptanol	9.5	20.25	0.2	0.1
Hexachloroethane	12.8	20.25	0.04	0.03
Hexafluoropropene	8.3	20.25	0.16	0.06
Hexane (*n*-hexane)	3.4	0.75	0.26	0.02
2-Hexanone	8.6	20.25	0.42	0.1
Hexone (methyl isobutyl ketone)	8.5	20.25	0.31	0.2
sec-Hexyl acetate	9.8	9.75	0.1	0.2
Hydrazine	10.55	20.25	0.001	0.25
Hydrogen chloride	3.4	20.25	0.002	1.0

Table 10.17 Cont'd

Compound	*Analytical Wavelength* (μm)[1]	*Path length* (m)[2]	*Absorbence*[3]	*Minimum Detectable Concentration* (ppm)[4] (20 metre cell)
Hydrogen cyanide	3.04	20.25	0.0083	0.4
Hydroquinone[5]				
Isoamyl acetate	9.4	9.75	0.27	0.14
Isoamyl alcohol	9.4	9.75	0.22	0.13
Isobutyl acetate	8.2	2.25	0.4	0.02
Isobutyl alcohol	9.6	8.25	0.30	0.08
Isodecanol	9.6	20.25	0.25	0.3
Isoflurane	8.25	20.25	0.07	0.02
Isophorone	3.4	20.25	0.12	0.4
Isoprene	11.2	5.25	0.078	0.2
Isopropyl acetate	8.0	2.25	0.63	0.02
Isopropyl alcohol	8.7	6.75	0.45	0.12
Isopropylamine	8.5	20.25	0.025	0.1
Isopropyl ether	8.9	0.75	0.17	0.07
LPG (liquefied petroleum gas)	3.4	0.75	0.24	0.4
Mesityl oxide	8.2	20.25	0.06	0.2
Methane	7.65	20.25	0.072	0.2
Methanethiol, *see* Methyl mercaptan				
Methoxyflurane	12.0	20.25	0.03	0.07
2-Methoxyethanol, *see* Methyl cellosolve				
Methyl acetate	9.5	5.25	0.31	0.2
Methyl acetylene (propyne)	7.9	20.25	0.77	1.0
Methyl acrylate	8.4	20.25	0.15	0.03
Methylal (dimethoxymethane)	9.5	0.75	0.43	0.05
Methyl alcohol (methanol)	9.5	5.25	0.3	0.1
Methylamine	3.4	20.25	0.02	0.1
Methyl amyl alcohol, *see* Methyl isobutyl carbinol				
Methyl *n*-amyl ketone (2-heptanone)	8.6	9.75	0.14	0.2
Methyl bromide	7.6	20.25	0.021	0.4
Methyl butyl ketone, *see* 2-Hexanone				
Methyl cellosolve	8.8	20.25	0.19	0.05
Methyl cellosolve acetate	8.0	20.25	0.39	0.02
Methyl chloride	13.4	20.25	0.14	1.5
Methyl chloroform	9.2	2.25	0.4	0.06
Methylcyclohexane	3.4	2.25	0.84	0.04
Methylcyclohexanol	9.5	20.25	0.24	0.2
o-Methylcyclohexanone	8.9	20.25	0.18	0.3
Methylene bisphenyl isocyanate, MDI[5]				
Methylene chloride	13.4	0.75	0.25	0.2

Methyl ethyl ketone (MEK), *see* 2-Butanone				
N-methyl formamide	8.4	5.25	0.065	0.2
Methyl formate	8.5	0.75	0.13	0.02
Methyl iodide	7.9	20.25	0.009	0.4
Methyl isoamyl ketone	8.6	20.25	0.6	0.04
Methyl isobutyl carbinol	8.7	20.25	0.053	0.25
Methyl isobutyl ketone, *see* Hexone				
Methyl isocyanate	11.6	20.25	0.00003	0.6
Methyl isopropyl ketone	8.8	20.25	0.1	0.3
Methyl mercaptan	3.38	20.25	0.008	0.4
Methyl methacrylate	8.5	2.25	0.21	0.03
Methyl propyl ketone, *see* 2-Pentanone				
α-Methyl styrene	11.1	20.25	0.3	0.3
Monomethylaniline	7.9	20.25	0.007	0.2
Morpholine	9.0	20.25	0.05	0.2
Nickel carbonyl	4.86	20.25	0.0003	0.005
Nitric oxide (NO)	5.3	20.25	0.015	2.0
Nitrobenzene	11.8	20.25	0.005	0.2
Nitroethane	9.0	20.25	0.08	0.6
Nitrogen dioxide (NO_2)	6.17	20.25	0.048	0.1
Nitrogen trifluoride	11.0	20.25	0.35	0.03
Nitromethane	9.3	20.25	0.056	0.9
Nitrotoluene	11.8	20.25	0.0076	0.7
Nitrotrichloromethane, *see* Chloropicrin				
Nitrous oxide	4.50	20.25	0.3	0.07
Octane	3.4	0.75	0.40	0.02
Pentane	3.4	0.75	0.42	0.02
2-Pentanone	8.5	5.25	0.23	0.1
Perchloroethylene	10.9	5.25	0.63	0.05
Petroleum distillates	3.4	2.25	0.65	0.02
Phenyl ether-biphenyl mixture (vapour)	8.1	20.25	0.011	0.04
Phenylethylene, *see* Styrene				
Phenylhydrazine	8.5	20.25	0.0029	0.9
Phosgene (carbonyl chloride)	11.8	20.25	0.0027	0.03
Phosphine	10.1	20.25	0.0003	1.0
Picric acid(5)				
Propane	3.35	2.25	0.79	0.03
n-Propyl acetate	8.1	0.75	0.18	0.02
Propyl alcohol	9.4	9.75	0.39	0.2
n-Propyl chloride	7.9	20.25	0.2	0.7
n-Propyl nitrate	10.4	20.25	0.11	0.1
Propylene dichloride	9.8	20.25	0.12	0.3
Propylene oxide	12.0	20.25	0.32	0.3
Propyne, *see* Methyl acetylene				
Pyridine	14.2	20.25	0.05	0.2

Table 10.17 Cont'd

Compound	*Analytical Wavelength* (μm)[1]	*Path length* (m)[2]	*Absorbence*[3]	*Minimum Detectable Concentration* (ppm)[4] (20 metre cell)
Quinone[5]				
Stoddard solvent	3.4	0.75	0.39	0.01
Styrene	11.0	12.75	0.16	0.2
Sulphur dioxide	8.6	20.25	0.003	0.5
Sulphur hexafluoride	10.7	0.75	0.7	0.02
Sulphuryl fluoride	11.5	20.25	0.10	0.04
Systox, *see* Demeton[R]				
1,1,2,2-Tetrachloro-1,2-difluoroethane (Freon 112)	9.7	0.75	0.19	0.06
1,1,2,2-Tetrachloroethane	8.3	20.25	0.014	0.2
1,1,1,2-Tetrachloroethane	10.4	20.25	0.4	0.2
Tetrachloroethylene, *see* Perchloroethylene				
Tetrachloromethane, *see* Carbon tetrachloride				
Tetrahydrofuran	9.2	9.75	0.47	0.2
Tetryl[5]				
Toluene	13.7	6.75	0.38	0.5
o-Toluidine	13.5	20.25	0.029	0.5
Toxaphene, *see* Chlorinated camphene				
Tributyl phosphate	10.7	20.25	0.020	0.3
1,1,1-Trichloroethane, *see* Methyl chloroform				
1,1,2-Trichloroethane	11.8	5.25	0.27	0.1
Trichloroethylene	12.4	20.25	0.16	0.4
Trichloromethane, *see* Chloroform				
1,2,3-Trichloropropane	8.4	0.75	0.46	0.02
1,1,2-Trichloro-1,2,2-Trifluoroethane (Freon 113)	9.3	20.25	0.077	0.3
Trifluoromonobromoethane (Freon 13B1)	8.3	0.75	1.5	0.03
2,4,6-Trinitrophenol, *see* Picric acid				
2,4,6-Trinitrophenylmethyl-nitramenine, *see* Tetryl				
Turpentine	3.4	2.25	0.13	0.02
Vinyl benzene, *see* Styrene				
Vinyl acetate	8.1	20.25	0.37	0.02
Vinyl bromide	10.85	20.25	0.096	0.4
Vinyl chloride	10.9	20.25	0.0017	0.3
Vinyl cyanide, *see* Acrylonitrile				
Vinylidene chloride	9.18	20.25	0.24	0.1
Vinyl toluene	11.1	20.25	0.2	0.4
Xylene (xylol)	12.6	20.25	0.33	0.6
Xylidene	7.2	20.25	0.056	0.2

(1) The analytical wavelength has usually been chosen as that of the strongest band in the spectrum which is free from interference due to atmospheric water and CO_2. If more than one infra-red absorbing material is present in the air in significant concentration, the use of another analytical wavelength may be necessary.
(2) Path lengths are chosen to optimize readings at the exposure limits. All measurements were made using a 1 mm slit.
(3) Equivalent to the OSHA limit. Absorbance less than the tabulated value indicates concentrations below the exposure limits regardless of the presence of interfering compounds.
(4) The concentration that would produce an absorbance equal to the peak-to-peak noise of the instrument.
(5) Solid or vapour pressure too low for analysis.
(6) Difficult to obtain commercially.

Table 10.18(a) Substances for which colour detector tubes are available from one supplier (Drager) – short-term tubes

Substance	*Measuring range*
Acetaldehyde	0.2–5 ppm
	10–150 ppm
	100–1000 ppm
Acetic acid	1–15 ppm
	20–320 ppm
Acetic anhydride	20–320 ppm
Acetone	100–12 000 ppm
Acetonitrile	40–1000 ppm
Acetylene	25–300 ppm
	100–2500 ppm
	500–3000 ppm
Acrolein	0.1–10 ppm
Acrylonitrile	0.5–20 ppm
	5–30 ppm
	25–250 ppm
Alcohols (general)	25–5000 ppm
	100–3000 ppm
Aliphatic hydrocarbons (bp 50–200°C)	2–23 mg/l
Allyl propyl disulphide	*
Amines (general)	*
Aminobutane	*
Aminoethanol	0.5–6 ppm
1-Aminopropane	2–30 ppm
	5–60 ppm
2-Aminopropane	2–30 ppm
	5–60 ppm
Ammonia	0.25–3 ppm
	2–30 ppm
	2.5–100 ppm
	5–700 ppm
	500–100 000 ppm
iso-Amyl acetate	200–1000 ppm
n-Amyl acetate	200–1000 ppm
Aniline	0.5–10 ppm
	1–20 ppm
o-Anisidine	1–20 ppm
Antimony hydride	0.05–3 ppm
Arsenic trioxide/pentoxide	0.2 mg/m^3
Arsine	0.05–60 ppm
Aziridine	0.25–3 ppm
Benzene	0.5–10 ppm
	2–60 ppm
	5–50 ppm
	15–420 ppm
Benzyl chloride	0.2–8 ppm
	1–15 ppm
Biphenyl ether	*
bis (2-Chloromethyl) ether	*
bis (Dimethylaminoethyl) ether	*
Bromine	0.2–30 ppm
	50–500 ppm
Bromobenzene	10–500 ppm
Bromochlorodifluoromethane	*
Bromochloromethane	30–350 ppm
1-Bromo-2-chloropropane	*
Bromoethane	20–200 ppm
	40–400 ppm

Table 10.18(a) Cont'd

Substance	*Measuring range*
Bromomethane	20–200 ppm
	40–400 ppm
	500–25 000 ppm
Butadiene	1–60 ppm
	50–1000 ppm
	100–1200 ppm
iso-Butane	100–1000 ppm
n-Butane	100–8000 ppm
Butane thiol	*
iso-Butanol	40–1000 ppm
n-Butanol	30–50 ppm
	100–5000 ppm
Butan-2-one	100–12 000 ppm
Buten-2-al	*
2-Butoxyethanol	*
2-Butoxyethyl acetate	*
n-Butyl acetate	200–3000 ppm
Butyl acrylate	*
n-Butylamine	5–60 ppm
	0.075–15 vol %
Butyl chloroformate	*
n-Butylethylamine	*
Butyl glycol	*
Butyl mercaptan	*
Butyl methacrylate	*
p-tert-Butyl toluene	*
iso-Butyraldehyde	*
n-Butyraldehyde	*
Calcium chromate	*
Carbonization untreated gases	*
Carbon dioxide	100–3000 ppm
	0.01–0.3 vol %
	0.1–6 vol %
	0.5–10 vol %
	1–20%
	5–60%
Carbon disulphide	3–95 ppm
	2.5–120 ppm
	0.1–60 mg/l
Carbon monoxide	2–300 ppm
	5–250 ppm
	5–700 ppm
	8–150 ppm
	10–3000 ppm
	0.001–0.3 vol %
	0.3–7 vol %
Carbonyl sulphide	*
Carbon tetrachloride	1–15 ppm
	5–50 ppm
Chlorine	0.2–30 ppm
	50–500 ppm
Chlorine anhydrate	*
Chlorine dioxide	0.1–1.5 ppm
Chloroacetic acid	2–30 ppm
Chloroacetyl chloride	*
Chlorobenzene	5–200 ppm
2-Chloro-1,3-butadiene	5–90 ppm

Table 10.18(a) Cont'd

Substance	*Measuring range*
1-Chlorobutane	10–60 ppm
2-Chlorobutane	10–60 ppm
Chlorobromomethane	*
4-Chlorobutryl chloride	0.5–4.5 ppm
Chlorodifluoromethane	*
1-Chloro-2,3-epoxypropane	*
Chloroethane	5–50 ppm
2-Chloroethanol	*
Chloroform	2–10 ppm
Chloroformates	0.2–10 ppm
o,m,p-Chlorophenol	*
Chloropicrin	1–15 ppm
Chloroprene	5–90 ppm
o,m,p-Chlorotoluene	5–50 ppm
	50–400 ppm
Chromates (alkali)	*
Coke oven gas	*
Cresol	0.8–16 ppm
Crotonaldehyde	*
Crude petroleum	*
Cumeme	30–400 ppm
	50–400 ppm
	100–1800 ppm
Cyanides (Na, K, Ca)	2–15 mg/m^3
Cyanogen chloride	0.25–5 ppm
Cyclohexane	100–1500 ppm
Cyclohexanol	*
Cyclohexanone	25–100 ppm
Cyclohexene	*
Cyclohexylamine	2–30 ppm
Cyclopentadiene	*
Cyclopentane	100–1500 ppm
Cyclopropane	100–1500 ppm
Decane	10–200 ppm
Decene	10–200 ppm
Dementon	*
Denmenton methyl	*
1,2-Diaminoethane	0.5–12 ppm
Diaminotoluene	*
Dibenzylamine	0.25–3 ppm
	5–60 ppm
1,2-Dibromo-3-chloropropane	5–50 ppm
Dibromofluoromethane	*
1,2-Dibromoethane	5–50 ppm
Dibutylamine	2.5–30 ppm
Dibutylether	100–4000 ppm
o,m,p-Dichlorobenzene	*
1,4-Dichlorobut-2-ene	*
Dichlorodiisopropyl ether	*
Dichlorodimethyl ether	*
1,2-Dichloroethane	10–100 ppm
1,1-Dichloroethylene	0.38–3 ppm
1,2-Dichloroethylene	7.5–75 ppm
Dichlorodifluoromethane	*
1,1-Dichloro-1-fluoroethane	*
Dichlorofluoromethane	*
Dichloromethane	10–2000

Table 10.18(a) Cont'd

Substance	*Measuring range*
3,3-Dichloro-1,1,2,2-pentachloropropane	*
1,3-Dichloro-1,1,2,2-pentachloropropane	*
Dichloroprene	*
1,2-Dichloropropane	50–100 ppm
1,3-Dichloropropene	5–50 ppm
1,2-Dichloro-1,1,2,2-tetrafluoroethane	*
2,2-Dichloro-1,1,1-trifluoroethane	*
1,1-Dichloro-2,2,2-trifluoroethane	*
Dichlorotoluene	*
Dicyclopentadiene	1.5–15 mg/m^3
Diesel	*
Diethanolamine	*
Diethylamine	5–60 ppm 0.05–10 vol %
2-Diethylaminoethanol	0.5–10 ppm
Diethylbenzene	*
Diethylcarbonate	*
Diethylene dioxide	12.5–175 ppm
Diethylene glycol	*
Diethylenediamine	5–60 ppm
Diethylenetriamine	*
Diethyl ether	10–4000 ppm
Diethyl sulphate	*
Diethyl sulphide	0.1–1.5 ppm
Difluorobromomethane	*
Diisobutylene	2.5–30 ppm
Diisobutyl ketone	*
Diisopropanolamine	*
Diisopropylamine	*
Dimethoxyethane	*
Dimethoxymethane	*
Dimethyl acetamide	100–400 ppm
Dimethylamine	0.25–6 ppm 5–60 ppm
Dimethylaminoacetonitrile	*
Dimethylaminopropylamine	5–60 ppm
N,N-Dimethylaniline	*
2,5-Dimethylanisole	*
N,N-Dimethylbenzylamine	*
1,3-Dimethyl butylacetate	*
Dimethylcyclohexylamine	5–60 ppm
Dimethylethanolamine	5–60 ppm
Dimethylether	*
Dimethylethylamine	0.2–2 ppm 5–60 ppm
Dimethylformamide	100–400 ppm
2,6-Dimethylheptane-4-one	*
2,4-Dimethyl hexane	*
3,3-Dimethyl hexane	*
1,1-Dimethyl hydrazine	0.25–3 ppm 5–70 ppm
Dimethylisopropylamine	15–70 ppm
N,N'-Dimethylpiperazine	*
Dimethyl sulphate	0.005–0.05 ppm
Dimethyl sulphide	1–15 ppm
4,7-Dimethylundecane	*
1,4-Dioxane	12.5–175 ppm

Table 10.18(a) Cont'd

Substance	*Measuring range*
1,3-Dioxolan	*
Dipentene	*
Dipropylamine	*
Dipropylether	100–4000 ppm
Dipropylglycol methyl ether	*
Di-sec octyl phthalate	10–20 mg/m^3
Disulphoton	*
Dodecane	50–400 ppm
Dodecene	*
Epichlorohydrin	5–50 ppm
2,3-Epoxy-1-chloropropane	*
1,2-Epoxypropane	*
Ethanethiol	0.1–20 ppm
Ethanol (ethyl alcohol)	25–2000 ppm
	100–3000 ppm
	0.5–5.2 vol %
Ethanolamine	0.5–6 ppm
Ether	100–4000 ppm
2-Ethoxyethanol	*
2-Ethoxyethyl acetate	50–700 ppm
Ethyl acetate	200–3000 ppm
Ethyl acrylate	5–300 ppm
Ethylamine	5–60 ppm
Ethylamyl ketone	*
Ethyl benzene	5–300 ppm
	30–600 ppm
Ethyl bromide	20–200 ppm
	40–400 ppm
Ethyl bromoacetate	*
Ethyl chloride	5–50 ppm
Ethylene	0.1–5 ppm
	50–2500 ppm
	0.5–2 vol %
Ethylene bromide	5–50 ppm
Ethylene chlorohydrin	*
Ethylene diamine	0.5–12 ppm
Ethylene dibromide	1–50 ppm
	5–50 ppm
Ethylene glycol	10–180 mg/m^3
Ethylene glycol dimethyl ether	50–500 ppm
Ethylene glycol dinitrate	*
Ethylene glycol monobutylether	*
Ethylene glycol monoethylether	*
Ethylene glycol monoethylether acetate	50–700 ppm
Ethylene glycol monomethylether	*
Ethyleneimine	*
Ethylene oxide	1–15 ppm
	25–500 ppm
Ethyl ether	100–4000 ppm
Ethyl formate	*
Ethyl glycol	*
Ethyl glycol acetate	50–700 ppm
2-Ethyl-1-hexanol	*
Ethyl hexylacrylate	*
2-Ethylhexylamine	*
Ethylidene dicycloheptene	*
Ethyl lactate	*

Table 10.18(a) Cont'd

Substance	*Measuring range*
Ethyl mercaptan	0.1–20 ppm
Ethyl methacrylate	*
Ethyl morpholine	*
Ethyl vinyl ether	20–300 ppm
Exxsol D40	*
Fenthion	*
Fluorine	0.1–2 ppm
Fluorodichloromethane	*
Formaldehyde	0.04–25 ppm
	2–40 ppm
Formic acid	1–15 ppm
Furfural	*
Furfuryl alcohol	*
Furfurylamine	*
Gasoline	*
Glycerol trinitrate	*
Glycidol	*
Glycol dinitrate	*
Heptane	7–2100 ppm
	200–600 ppm
	500–1000 ppm
2-Heptanone	*
Heptene	10–1000 ppm
1,2,3,4,4-hexachlor-1,3-butadiene	*
Hexachlorocyclopentadiene	1–50 ppm
Hexamethylene diamine	*
Hexane	5–150 ppm
	100–3000 ppm
Hexene	5–60 ppm
Hexone	1300–15 600 ppm
sec-Hexyl actetate	*
Hexylene glycol	*
Hydrazine	0.05–0.6 ppm
	0.1–10 ppm
Hydrobromic acid	0.5–10 ppm
Hydrocarbons (general)	0.1–1.3 vol %
	2–28 mg/l
Hydrochloric acid	1–20 ppm
	50–5000 ppm
Hydrocyanic acid	2–150 ppm
Hydrofluoric acid	1.5–60 ppm
Hydrogen	0.2–2 vol %
	0.5–3 vol %
Hydrogen bromide	*
Hydrogen peroxide	0.1–3 ppm
Hydrogen sulphide	0.02–6 ppm
	0.05–150 ppm
	0.2–5 ppm
	0.2–16 ppm
	1–60 ppm
	1–200 ppm
	5–600 ppm
	100–2000 ppm
	0.2–7 vol %
	2–40 vol %
Iron pentacarbonyl	*
Isoamyl alcohol	*

Table 10.18(a) Cont'd

Substance	*Measuring range*
Isobutyl acetate	*
Isobutyl acrylate	*
Isooctyl nitrate	*
Isopar G	150–1000 ppm
Isopentane	10–50 ppm
Isoprene	10–30 ppm
Isopropanol	15–40 ppm
	50–4000 ppm
	100–3000 ppm
Isopropenyl benzene	10–50 ppm
Isopropyl acetate	200–3000 ppm
Isopropylamine	2–30 ppm
	5–60 ppm
Isopropyl ether	100–4000 ppm
Isopropylnitrate	*
Kerosene	*
Limonene	*
Mercaptans (general)	0.5–5 ppm
	20–1000 ppm
Mercury	0.01–2 mg/m^3
Mesityl oxide	25–500 ppm
Mesitylene	10–100 ppm
Methacrylic acid	1–15 ppm
Methane thiol	0.1–20 ppm
Methane	*
Methanol	25–5000 ppm
	100–3000 ppm
2-Methoxyethanol	*
2-Methoxyethyl acetate	*
1-Methoxy-2-propanol	50–200 ppm
2-Methoxy-1-propanol	*
1-Methoxy-2-propylacetate	50–200 ppm
2-Methoxy-1-propylacetate	*
Methoxypropylamine	*
Methoxy propoxy propanol	*
Methyl acetate	200–3000 ppm
Methyl acetylene	*
Methyl acrylate	5–200 ppm
Methyl allyl chloride	*
Methylamine	5–60 ppm
2-Methylaminoethanol	*
Methyl-*n*-amyl ketone	*
Methylated spirits	*
Methyl bromide	3–100 ppm
	5–50 ppm
Methyl cellosolve	*
Methyl cellosolve acetate	*
Methyl chloroacetate	*
Methyl chloroform	10–600 ppm
Methyl cyclohexane	200–3000 ppm
2-Methyl cyclohexane-2-one	*
Methyl cyclopentane	*
N-Methyldiethanolamine	*
N-Methyldiethylamine	5–60 ppm
Methylene chloride	10–2000 ppm
Methyl ethyl ketone	100–12 000 ppm
Methyl ethyl sulphide	*

Table 10.18(a) Cont'd

Substance	*Measuring range*
Methyl formate	*
Methyl glycol	*
Methyl glycol acetate	*
Methyl heptanone	*
Methyl iodide	*
Methyl isobutylketone	130–15 600 ppm
Methyl mercaptan	0.1–20 ppm
	2–1000 ppm
Methyl methacrylate	15–700 ppm
n-Methyl morpholine	*
1-Methylpropanol	*
2-Methylpropan-2-ol	50–500 ppm
Methyl propyl ketone	130–15 600 ppm
N-Methylpyrolidone	*
α-Methyl styrene	10–50 ppm
Methyl styrene	10–100 ppm
Methyl-*tert*-butylether	*
Methyl vinyl ketone	*
Morpholine	*
Naphthalene	*
Natural gas	*
Nickel carbonyl	0.1–1 ppm
Nickel chloride	0.1–1 mg/m^3
Nickel tetracarbonyl	0.1–1 ppm
Nitric acid	1–50 ppm
Nitroglycerine	*
Nitrogen dioxide	0.5–25 ppm
	2–100 ppm
Nitroglycol	*
2-Nitropropane	*
Nitrous fumes	0.025–10 ppm
	2–150 ppm
	20–500 ppm
	50–2000 ppm
	100–5000 ppm
n-Nonane	12–360 ppm
	500–1000 ppm
iso-Octane	10–200 ppm
	10–1500 ppm
n-Octane	1–300 ppm
	10–2500 ppm
Octene	20–1000 ppm
Oil	1–10 mg/m^3
	2.5–10 mg/m^3
Oil (mist and vapour)	2.5–10 mg/m^3
Olefins (general)	1–55 mg/l
Organic arsenic compounds and arsine	*
Organic basic nitrogen compounds	*
Oxirane	1–15 ppm
	25–500 ppm
Oxygen	5–23 vol %
Ozone	0.05–1.4 ppm
	10–300 ppm
2,4-Pentadione	*
Pentane	100–1500 ppm
Pentan-2-one	130–15 600 ppm
Pentyl acetate	200–1000 ppm

Table 10.18(a) Cont'd

Substance	*Measuring range*
Perchloroethylene	0.1–4 ppm
	0.7–300 ppm
	10–500 ppm
Petroleum hydrocarbons (general)	10–300 ppm
	100–2500 ppm
Phenol	1–20 ppm
Phenylhydrazine	*
Phenyl mercaptan	2–100 ppm
Phosgene	0.02–1 ppm
	0.25–25 ppm
Phosphine	0.01–1 ppm
	0.1–40 ppm
	1–100 ppm
	25–10 000 ppm
	15–3000 ppm
Phosphoric acid ester	0.05 ppm
α-Pinene	10–200 ppm
Polytest	*
Potassium dichromate	*
Propanal	*
Propane	0.5–1.3 vol %
Propan-1-ol	100–3000 ppm
Propan-2-ol	15–40 ppm
	50–4000 ppm
	100–3000 ppm
Propargyl alcohol	*
Propionic acid	1–15 ppm
n-Propyl benzene	5–300 ppm
Propylene glycol	*
Propylene oxide	4–60 ppm
	50–2000 ppm
Pyridine	5 ppm
Pyrrolidine	*
Shellsol	*
Sodium chromate	0.1–0.5 mg/m^3
Strontium chromate	0.2–1 mg/m^3
Styrene	10–250 ppm
	50–400 ppm
Sulphur dioxide	0.1–3 ppm
	0.5–25 ppm
	1–25 ppm
	20–2000 ppm
	50–8000 ppm
Sulphuric acid	1–5 mg/m^3
Tar from brown coal	*
Tar vapours	*
Tetrabromoethane	*
1,1,2,2-Tetrabromoethane	0.5–3 ppm
1,1,2,2-Tetrachloro-1,2-difluoroethane	*
1,1,1,2-Tetrachloro-2,2-difluoroethane	*
1,1,2,2-Tetrachloroethane	*
Tetrafluoroethane	*
Tetrafluoromethane	*
Tetrahydrofuran	*
Tetrahydrothiophene	1–16 ppm
Thioether	*
Thionyl chloride	1–30 ppm

Table 10.18(a) Cont'd

Substance	*Measuring range*
Thiophene	*
Toluene	5–300 ppm
	50–400 ppm
	100–1800 ppm
Toluene diisocyanate	0.02–0.2 ppm
o-Tolidine	1–30 ppm
Tributylamine	2.5–50 ppm
1,2,4-Trichlorobenzene	*
Trichlorotoluene	*
Trichloroethane	50–600 ppm
Trichloroethylene	2–250 ppm
	50–500 ppm
Trichlorofluoromethane	100–1400 ppm
Trichloronitromethane	1–15 ppm
1,2,3-Trichloropropane	*
1,1,2-Trichloro-1,2,2-trifluoroethane	*
Triethylamine	5–60 ppm
Triethylene diamine	*
Triethylene tetratriamine	*
Trifluorobromomethane	*
Trimethylamine	0.25–3 ppm
	5–60 ppm
Trimethyl benzene	10–100 ppm
Trimethyl phosphate	*
Turpentine oil	*
Undecane	10–200 ppm
Vinyl acetate	*
Vinyl bromide	0.7–300 ppm
Vinyl chloride	0.5–30 ppm
	1–50 ppm
	100–3000 ppm
Vinyl ethyl ether	20–300 ppm
Vinylidene chloride	2.5–25 ppm
Vinyl trimethoxy silane	100–1000 ppm
Water vapour	0.05–1 mg/l
	0.5–18 mg/l
	1–40 mg/l
White spirit	30–200 ppm
	100–1600 ppm
Xylene (all isomers)	10–400 ppm
Zinc chromate	0.2–1 mg/m^3

* Consult with supplier.

Table 10.18(b) Dräger tubes for long-term measurements – with pump

Dräger tube	*Range of measurement for maximum period of use*
Acetic acid 5/a-L	1.25–40 ppm (4 hr)
Acetone 500/a-L	63–10 000 ppm (8 hr)
Ammonia 10/a-L	2.5–100 ppm (4 hr)
Benzene 20/a-L	5–200 ppm (4 hr)
Carbon dioxide 1000/a-L	250–6000 ppm (4 hr)
Carbon disulphide 10/a-L	1.3–100 ppm (8 hr)
Carbon monoxide 10/a-L	2.5–100 ppm (4 hr)
Carbon monoxide 50/a-L	6.3–500 ppm (8 hr)
Chlorine 1/a-L	0.13–20 ppm (8 hr)
Ethanol 500/a-L	63–8000 ppm (8 hr)
Hydrocarbon 100/a-L	25–3000 ppm (4 hr)
Hydrochloric acid 10/a-L	1.3–50 ppm (8 hr)
Hydrocyanic acid 10/a-L	1.3–120 ppm (8 hr)
Hydrogen sulphide 5/a-L	0.63–60 ppm (8 hr)
Methylene chloride 50/a-L	13–800 ppm (4 hr)
Nitrogen dioxide 10/a-L	1.3–100 ppm (8 hr)
Oxides of nitrogen 5/a-L ($NO + NO_2$)	1.3–50 ppm (4 hr)
Oxides of nitrogen 50/a-L ($NO + NO_2$)	13–350 ppm (2 hr)
Perchloroethylene 50/a-L	13–300 ppm (4 hr)
Sulphur dioxide 5/a-L	1.25–50 ppm (4 hr)
Sulphur dioxide 2/a-L	0.5–20 ppm (4 hr)
Toluene 200/a-L	25–4000 ppm (8 hr)
Trichloroethylene 10/a-L	2.5–200 ppm (4 hr)
Vinyl chloride 10/a-L	1–50 ppm (10 hr)

Table 10.18(c) Direct-indicating Dräger diffusion tubes – no pump required

Substance	*Measuring range*	*Max. operating time (hours)*
Acetic acid 10/a-D	10–200 ppm × hr	8
Ammonia 20/1-D	20–1500 ppm × hr	8
Butadiene 10/a-D	10–300 ppm × hr	8
Carbon dioxide 500/a-D	500–20 000 ppm × hr	8
(1%/a-1)	1–30 vol % × hr	360
Carbon monoxide 50/a-D	50–600 ppm × hr	360
Ethanol 1000/a-D	1000–25 000 ppm × hr	8
Ethyl acetate 500/a-D	500–10 000 ppm × hr	8
Hydrochloric acid 10/a-D	10–200 ppm × hr	8
Hydrocyanic acid 20/a-D	20–200 ppm × hr	8
Hydrogen sulphide 10/a-D	10–300 ppm × hr	72
Nitrogen dioxide 10/a-D	10–2000 ppm × hr	24
Olefines 100/a-D	100–2000 ppm × hr	8
Perchloroethylene 200/a-D	200–1500 ppm × hr	8
Sulphur dioxide 5/a-D	5–150 ppm × hr	10
Toluene 100/a-D	100–3000 ppm × hr	8
Trichloroethylene 200/a-D	200–1000 ppm × hr	8
Water vapour 5/a-D	5–100 mg/l × hr	8

Table 10.19 Selected sources of inaccuracy in use of colour detector tubes

Failure to break both ends of the sealed tube before insertion of the tube into the pump housing.

Insertion of the tube incorrectly into the pumphousing (the correct direction is indicated on the tube).

Reuse of previously used tubes. It is advisable not to reuse tubes even if previous use indicated zero.

Leaks in sample lines, or insufficient time allowed to lapse between pump strokes when extensions are used.

Use of tubes beyond expiry of the shelf-life. Tubes should be stored under refrigerated conditions but allowed to warm to ambient temperature prior to use.

Ill-defined stain format because it is irregular, diffuse or has failed, i.e. not at right angles to tube wall. (This can be caused by poor quality of granular support medium used by manufacturer.) It is advisable to read the maximum value indicated.

Use of tubes under conditions of temperature, pressure or humidity outside the range of calibration.

Blockages or faulty pumps. Pumps should be checked periodically as instructed by the manufacturer. They can be calibrated using rotameters or bubble flowmeters. Unless pumps possess a limiting orifice they should be calibrated with the air indicator tube in position.

Misuse of the pump, e.g. incomplete stroke or wrong number of strokes.

Mismatch of tubes with type of pump.

Interference due to the presence of other contaminants capable of reacting with the tube reagent. This can result in over- or under-estimation of concentrations. The former is the more likely and hence errs on the side of safety.

Tube blockage caused by airborne dusts, affecting the flow rate.

Table 10.20 Examples of chemicals for which paper-tape colorimetric instruments are available

Ammonia	Hydrogen selenide
Arsine	Hydrogen sulphide
Chlorine	Isocyanates
Diborane	Nitrogen dioxide
Germane	*p*-Phenylene diamine
Hydrazines	Phosgene
Hydrogen chloride	Phosphine
Hydrogen cyanide	Silane
Hydrogen fluoride	Sulphur dioxide

Table 10.21 Considerations when using instruments with catalytic detection

Portable instrument should be of explosion-proof design; fixed point systems may rely on remote sensing heads
For zero adjustment, place instrument in uncontaminated air or use activated charcoal filters to remove flammable vapours
Sources of error include: Inadequate calibration Drift due to age Design not fail-safe (i.e. no indication of component failure) Poisoning of Pellistor by, e.g., silicones, halocarbons, leaded petrol Too high a sampling rate (causing cooling of the elements) Sampling lines and couplings not airtight Condensation of high-boiling-point components in the line between sample head and sensor Hostile environment

particles of respirable dimensions from non-respirable fractions include horizontal elutriation and centrifugation. Equipment for personal monitoring comprises a lapel-mounted filter holder connected to a portable pump with a flow rate of about 3 litres/min. Respirable matter can be separated by use of a small cyclone. In order to ensure uniformity of fractionation, smooth and constant flow rates are essential. The dust collection and analytical stages are separate operations. For background monitoring, miniaturization is unimportant and as a consequence equipment incorporates pumps of higher flow rates, typically ≤100 l/min. This enables sampling times to be short and larger samples to be obtained (e.g. for laboratory analysis). Both direct-reading and absolute methods are available.

The main principles of instrument design are summarized in Table 10.23. In filtration, e.g. for gravimetric analysis, selection of filter material (Table 10.22) requires careful consideration in terms of application, strength, collection efficiency, compatibility with pump, water uptake, etc. Humidity-controlled balance rooms, microbalances and careful handling techniques may be required.

Table 10.22 Examples of filter material for collection of particulates

Filter Material	*Application*	*Characteristics*
• Cellulose	Washing of samples to determine water soluble fraction or for ashing to determine organic content	High flowrates, low pressure drop, low impurity levels
• Glass fibre	High-flow samplers for gravimetric assessment	High efficiencies, high flowrates, high wet strength, good temperature stability, low pressure drop
• Mixed celluloses, e.g. nitrate, ester	Microscopy (asbestos); metal content (by atomic absorption, atomic emission, fluorescence and infra-red spectrometry)	Low levels of metal impurities; oxidizable during digestion
• Polycarbonate	Optical microscopy; organic content	Transparent grades available, non-hygroscopic, low ash content, solvent resistant
• PTFE	Sampling for HPLC or UV analysis. PAHs	Inert, hydrophobic
• PVC	Gravimetric analysis, carbon black, quartz, silica	Acid and alkali resistant; low water pick-up
• Silver membrane	Crystalline materials for X-ray diffraction	Costly. High collection efficiency. Uniform pore size

Table 10.23 Particulates monitoring – principles of apparatus

Principle	*Examples*	*Collection*	*Sampling rate* (l/min)	*Collection efficiency* (%)	*Analysis*	*Advantages/disadvantages*
Impinger	Midget impinger	By bubbling through liquid phase			Microscopy	Aggregates broken up; only particles >1 μm collected. In wet impingers particles must be water insoluble.
Impactor	1. Konimeter	Impaction on gel-coated disc	1–37 (depending on type)	60–100	Built-in microscope	Underestimates small particles, overestimates large particles. Particles between 0.5 and 5.0 μm collected.
	2. Cascade impactor	Impaction on 4 stages on glass disc			Microscopy	
	3. Andersen sampler	Impaction in 8 stages onto glass or metal discs			Gravimetric or chemical	Instrument inflexible.
Electrostatic or thermal precipitation	Casella thermal precipitator	Deposition on glass slides or discs	1–85	90–100	Microscopy	Poor for large particles. Collection efficiency increases as particle size decreases.
Filtration	1. Fibrous filter		1–50	85–100 Depends on particle size values stated for those usually encountered	Gravimetric or chemical	Fibrous filter good for gravimetric analysis for a range of particle sizes (fast and relatively easy).
	2. Membrane filter				Microscopy, gravimetric or chemical	Membrane filter good for microscopy identification of particles and counting where required.
Respirable dust separation	1. Hexhlet (horizontal elutriator)	Fibrous filter	1–50	60–100	Gravimetric or chemical	Instrument must be kept horizontal for sampling; relatively large quantities of dust collected in short period.
	2. Casella cyclone	Fibrous or membrane filter			Gravimetric, microscopy or chemical	The only instrument for carrying out personal respirable dust sampling.
	3. Anderson sampler	Selective inlet 2 dust fractions (<10 μm and <2.5 μm)	16.7		Gravimetric	Can be used for unattended operations
Beta attenuation		Impaction on disc or filtration			Attenuation of beta radiation Direct reading	Provides short- or long-term TWA (e.g. up to 8 hr, depending on model) of dust or fume mass concentration. Suitable for unattended automated continuous methods.

Table 10.23 Cont'd

Principle	*Examples*	*Collection*	*Sampling rate (l/min)*	*Collection efficiency (%)*	*Analysis*	*Advantages/disadvantages*
Photometry	1. Number concentration, e.g. Royco				Light scattered on to a photomultiplier Direct reading	Gives automatic particle sizing but accuracy only guaranteed if calibrated for particulate of interest.
	2. Mass concentration e.g. Simslin				Light scattered onto a photomultiplier Direct reading	Very versatile – only accurate continuous long-term mass monitoring instrument; sample may also be collected on a filter. Suitable for automated operations.
Piezoelectric		Electrostatic frequency of crystal. Direct reading			Change in resonant frequency	Unsuitable for ambient air monitoring.

Official methods

Regulatory and advisory bodies publish methods for *ambient* air analysis such as those issued by the British Standards Institute and the US Environment Protection Agency (Tables 10.24 and 10.25, respectively). Methods for assessment of *workplace* air are published by the Health and Safety Executive. Some of these are generic methods (Table 10.26) whilst others are compound specific (Table 10.27). Examples of other official methods for monitoring workplace air quality are those published by the British Standards Institute (Table 10.28), and the US National Institute of Occupational Safety and Health (Table 10.29). Table 10.30 provides additional guidance on analytical techniques for a selection of substances.

Table 10.24 Selected British Standards relating to ambient air pollution measurements

British Standard	*Subject*	*Method*
BS 893	Particulate matter in ducts	
BS 1747	Deposit gauges	
	Particulate matter	
	Sulphur dioxide	Thorin spectrophotometry
	Directional dust gauges	
	Sampling equipment for determination of gaseous sulphur	
	Nitrogen dioxide in ambient air	Modified Griess–Salzman method
	Nitrogen oxides in ambient air	Chemiluminescence
	Sulphur dioxide in ambient air	Tetrachloromercurate/pararosaline
	Black smoke index in ambient air	
	Ozone	Chemiluminescence
	Particulate lead in aerosol	Collected on filter with atomic absorption spectrometry
BS 1756	Flue gases	General methods
BS 2811, 2742	Smoke monitoring	
BS 3048	Flue gases	Continuous and automatic
BS 3405	Particulate matter including grit and dust	
BS 3406	Particle size distribution	Various including microscopy
BS 4995	Gas mixtures	Preparation
BS 5243	Airborne radioactive gases and particulates	Various
BS 5343	Long-term gas detector tubes	
BS ISO 6768	Nitrogen dioxide	Modified Griess–Salzman
BS ISO 10312	Asbestos	Direct-transfer transmission Electron microscopy
BS ISO 10498	Sulphur dioxide	Ultra-violet fluorescence
BS ISO 10473	Particulate mass on filters	Beta-ray absorption
BS ISO 12884	Polycyclic aromatic hydrocarbons	Collection of filters with gas chromatography/mass spectrometry
BS ISO 13794	Asbestos	Indirect-transfer transmission Electron microscopy
BS ISO 13964	Ozone	Ultra-violet photometry
BS ISO 14965	Non-methane organics	Flame ionization
BS ISO 16000	Indoor formaldehyde and other carbonyl compounds	Active and diffusive sampling
BS EN 12341	Suspended matter	Reference and field methods
BS EN 13528	Gases and vapours	Diffusive samplers

Table 10.25 Selected EPA standard methods for air monitoring (Code of Federal register-protection of the environment section, 40)

Subject	*Method reference (see original reference)*
Part 60 Appendix A	
• Various measurement techniques for sample and velocity from stationary sources, stacks, ducts, pipes	1, 1A, 2, 2A, 2B, 2C
• CO_2 oxygen, excess air, dry molecular weight	3
• O_2 and CO_2	3A
• Moisture from stacks	4
• Particulate	5
– stationary sources	5A
– asphalt	5B
– non-sulphuric acid matter	5D
– from positive pressure fabric filters	5E
– wool fibre-glass	5F
– non-sulphate matter	5G
– wood heaters	5H
– in-stack filtration method	17
• SO_2 from stationary sources	6
– fossil fuel	6A, 6B
– stationary sources	6C
• NO_x from stationary sources	7 (various methods 7A–7E)
• Sulphuric acid mist and SO_2 from stationary sources	8
• Visual opacity	9
• CO from stationary sources	10, 10A, 10B
• H_2S (refineries)	11
• Inorganic lead	12
• Total fluoride	13A, 13B
• Fluoride (aluminium plants)	14
• H_2S, COS, CS_2	15
• Total reduced sulphur (sulphur recovery plants)	15A
• Sulphur	16
• Total reduced sulphur (stationary sources)	16A, 16B
• Gaseous organics	18
• SO_2, particulates and NO_x emission rates	19, 20
• Volatile organic leaks	21
• Visual assesssment of fugitive emissions from material sources and smoke from flares	22
• Chlorinated dioxins and dibenzofurans	23
• Volatile content, water, density, volume and weight of surface coatings	24, 24A
• Total gaseous organics	25, 25A, 25B
• HCl	26
• Vapour tightness of gasoline delivery tanks	27
Part 61 Appendix B (Hazardous air pollutants)	
• Particulate and gaseous Hg	101, 101A, 102
• Beryllium	103,104
• Vinyl choride	106
• Gaseous and particulate arsenic	108
• Benzene	110
• Polonium-210	111
• Radionuclide	114
• Radon-222	115

Table 10.26 HSE generic techniques for monitoring quality of workplace air

Subject	*MDHS number*
Generation of test atmospheres of organic vapours by the syringe injection technique	3
Generation of test atmospheres of organic vapours by the permeation tube method	4
On-site validation of sampling methods	5
General methods for sampling and gravimetric analysis of respirable and inhalable dust	14/3
Protocol for assessing the performance of a diffusive sampler	27
Sorbent tube standards preparation by the syringe injection technique	33/2
Protocol for assessing the performance of a pumped sampler for gases and vapours	54
General methods for sampling airborne gases and vapours	70
Analytical quality in workplace air monitoring	71
Measurement of air change rates in factories and offices	73
Dustiness of powders and materials	81
The dust lamp	82
Discrimination between fire types in samples of airborne dust on filters using microscopy	87

Sampling strategies

The results of environmental monitoring exercises will be influenced by a variety of variables including the objectives of the study, the sampling regime, the technical methods adopted, the calibre of staff involved, etc. Detailed advice about sampling protocols (e.g. where and when to sample, the volume and number of samples to collect, the use of replicates, controls, statistical interpretation of data, etc.) and of individual analytical techniques are beyond the scope of this book. Some basic considerations include the following, with examples of application for employee exposure and incident investigation.

Sampling

It is crucial to consider the sampling protocol, equipment, calibration, and validation. Tightly sealed sample containers of adequate strength, and generally protected from heat and light, are required. Extreme care must be taken with sample identification and labelling. Sample containers must not become contaminated with the substance under study or by any major interfering chemicals. Precautions must also prevent accidental loss of material collected awaiting analysis, e.g. during storage or transport. For example, water samples can become affected by evaporation, degassing, chemical degradation, photophysical degradation, precipitation, or damage of suspended matter.

Samples must be representative of the environment in relation to study objectives and to permit comparison of data with appropriate standards, i.e. average concentrations, time-weighted exposures, peak concentrations, etc. Replicate samples may be advisable.

Methods

Consideration must be given to equipment calibration and method suitability in terms of sensitivity, limits of detection, accuracy, precision, repeatability.

Table 10.27 HSE methods for measuring levels of specific airborne chemicals

Substance	*Method*	*MDHS number*
Acrylamide in air	Lab method with high performance liquid chromatography after collection in an impinger containing water	57
Acrylonitrile in air	Charcoal adsorption tube and gas chromatography	1
	Lab method using porous polymer adsorption tube and thermal desorption with gas chromatography	2
	Lab method using porous polymer diffusive samplers with thermal desorption and gas chromatography	55
Aromatic amines in air and on surfaces	Lab method using pumped acid-coated filters, desorption and liquid chromatography	75
Aromatic carboxylic acid anhydrides in air	Lab method using glass-fibre/Tenax tube sampling and high performance liquid chromatography	62
Aromatic isocyanates in air	Field method using acid hydrolysis, diazotization, coupling and spectrophotometry	49
Arsenic and inorganic compounds of arsenic (except arsine)	Lab method using continuous flow or flow injection analysis hydride generation and atomic absorption spectrometry	41/2
Arsine in air	Colorimetric field method using silver diethyldithiocarbamate in presence of excess silver nitrate	34
Asbestos fibres in air	Sampling and evaluation by phase-contrast microscopy	39/4
Asbestos in bulk material	Sampling and identification by polarized light microscopy	77
Azodicarbonamide in air	Lab method using high performance liquid chromatography	92
Benzene in air	Charcoal adsorbent tubes, solvent desorption, and gas chromatography	17
	Lab method using pumped porous polymer adsorbent tubes, thermal desorption and gas chromatography	22
	Lab method using porous polymer diffusive samplers, thermal desorption and gas chromatography	50
Beryllium and beryllium compounds in air	Lab method using flame atomic absorption spectrometry or electrothermal atomic absorption spectrometry	29/2
Butadiene in air	Lab method using pumped molecular sieve sorbent tubes, thermal desorption and gas chromatography	53
	Lab method using molecular sieve samplers, thermal desorption and gas chromatography	63
Cadmium and inorganic cadmium compounds in air	Flame atomic adsorption or electrothermal atomic absorption spectrometry	12/2
Carbon disulphide in air	Charcoal adsorbent tubes with solvent desorption and gas chromatography	15

Table 10.27 Cont'd

Substance	*Method*	*MDHS number*
Chlorinated hydrocarbon solvent vapours in air	Lab method using pumped charcoal adsorption tubes, solvent desorption and gas chromatography	28
Chromium and inorganic compounds of chromium in air	Flame atomic absorption spectrometry	12/2
Chromium (hexavalent) in chromium plating mists	Colorimetric field method using 1,5-diphenylcarbazide	52/3
Chromium (total and speciated chromium) in chromium plated mists	Colorimetric field method using 1,5-diphenyl carbazide after oxidation with silver (1)-catalysed peroxydisulphate	67
Coal tar pitch volatiles: measurement of particulates and cyclohexane soluble material in air	Lab method using filters and gravimetric analysis	68
Cobalt and cobalt compounds in air	Lab method using flame atomic absorption spectrometry	30/2
Cristobalite in respirable airborne dust	Lab method using X-ray diffraction (direct method)	76
Diethyl sulphate and dimethyl sulphate	Lab method using Tenax sorbent tube, thermal desorption and gas chromatography with mass spectrometry	89
Dioctyl phthalate	Lab method using Tenax sorbent tubes, solvent desorption and gas chromatography	32
Ethylene dibromide	Lab method with pumped Tenax absorbent tubes, solvent desorption and electron capture gas chromatography	45
Ethylene oxide	Lab method using charcoal absorbent tubers, solvent desorption and gas chromatography	26
Formaldehyde in air	Lab method using diffusive sampler, solvent desorption and high performance liquid chromatography	78
Glutaraldehyde in air	Lab method using high performance liquid chromatography	93
Glycol ether and glycol ether acetate vapours in air	Lab method using charcoal adsorbent tubes, solvent desorption and gas chromatography	21
Glycol ether and glycol acetate vapours in air	Lab method using Tenax sorbent tubes, thermal desorption and gas chromatography	23
n-Hexane in air	Lab method using charcoal diffusive samplers, solvent desorption and gas chromatography	74
Hydrazine in air	Lab method using sampling either onto acid-coated glass-fibre filters followed by solvent desorption or into specially constructed impingers. Final analysis by derivatization and high performance liquid chromatography	86
Hydrocarbons (mixed C_3–C_{10})	Lab method using pumped porous polymer and carbon sorbent tubes, thermal desorption and gas chromatography	60
Hydrocarbons (mixed C_5–C_{10})	Lab method using porous polymer diffusive samplers, thermal desorption and gas chromatography	66

Table 10.27 Cont'd

Substance	*Method*	*MDHS number*
Hydrogen cyanide in air	Lab method using an ion-selective electrode	56/2
Hydrogen fluoride and fluorides in air	Lab method using an ion-selective electrode or ion chromatography	35/2
Lead and inorganic compounds of lead in air	Lead method using flame or electrothermal atomic absorption spectrometry	6/3
	Lab method using X-ray fluorescence spectrometry	7
	Colorimetric field method	8
Mercury vapour in air	Diffusive samplers with qualitative on-site colorimetric analysis and quantitative cold vapour atomic absorption spectrometry in the laboratory	59
Metals and metalloids in workplace air	X-ray fluorescence spectrometry	91
Metalworking fluids (water based)	Elemental marker method using flame atomic absorption spectrometry or inductively coupled plasma emission spectrometry	95
Mineral oil mist from mineral oil-based metalworking fluids	Pumped filters with gravimetric determination	84
Newspaper print rooms: measurement of total particulate and cyclohexane-soluble material	Lab method using filters and gravimetric evaluation	48
Nickel and inorganic compounds of nickel in air (except nickel carbonyl)	Lab method using flame atomic absorption spectrometry or electrothermal atomic absorption spectrometry	42/2
Organic isocyanates in air	Lab method with sampling either onto coated glass-fibre filters followed by solvent desorption, or into impingers and analysis using high performance liquid chromatography	25/3
Peroxodisulphate salts in air	Lab method using mobile phase ion chromatography	79
Pesticides in air and on surfaces	Pumped filters/sorbent tubes with gas chromatography	94
Platinum metal and soluble platinum compounds in air	Lab method using electrothermal atomic absorption spectrometry or inductively coupled plasma mass spectrometry	46/2
Quartz in respirable airborne dusts	Lab method using infra-red spectroscopy (KBr disc technique)	38
Resin acids in rosin (colophony) solder flux fume	Lab method using gas chromatography	83
Rubber fume in air measured as total particulate and cyclohexane-soluble material	Lab method using filters and gravimetric estimation	47/2
Styrene in air	Lab method using charcoal adsorbent tubes, solvent desorption and gas chromatography	20
	Lab method using porous polymer adsorbent tubes, thermal desorption and gas chromatography	31

Table 10.27 Cont'd

Substance	*Method*	*MDHS number*
	Lab method using porous polymer diffusive samplers, thermal desorption and gas chromatography	43
	Lab method using charcoal diffusive samplers	44
Tetra-alkyl lead compounds in air	Personal monitoring with atomic absorption analysis or electrothermal atomization or X-ray fluorescence spectrometry or on-site colorimetry	9
Toluene in air	Lab method using pumped charcoal adsorption tubes, solvent desorption and gas chromatography	36
	Lab method using pumped porous polymer adsorption tubes, thermal desorption and gas chromatography	40
	Lab method using charcoal diffusive samplers, solvent desorption and gas chromatography (using Drager ORSA monitor)	64
	Lab method using charcoal diffusive samplers, solvent desorption and gas chromatography	69
Triglycidyl isocyanurate (and coating powders containing triglycidyl isocyanurate)	Lab method using pumped filter, desorption and liquid chromatography	85
Vinyl chloride in air	Lab method using charcoal adsorbent tubes, solvent desorption and gas chromatography	24
Volatile organic compounds in air	Lab method using pumped solid sorbent tubes, thermal desorption and gas chromatography	72
	Lab method using diffusive solid sorbent tubes, thermal desorption and gas chromatography	80
	Lab method using diffusive samplers, solvent desorption and gas chromatography	88
	Lab method using pumped solid sorbent tubes, solvent desorption and gas chromatography	96

Selected strategies for determining employees' exposure to airborne chemicals

The scheme in Figure 10.1 illustrates a general approach for devising a monitoring strategy. Where doubt exists about the level of exposure, a crude assessment can be made by determining levels under expected worst-case situations, paying attention to variations and possible errors. More detailed assessment may be required, depending upon the outcome. Sampling times should be long enough to overcome fluctuations but short enough for results to be meaningfully associated with specific activities and for corrective actions to be identified. For monitoring particulates, sampling times may be determined from the following equation:

$$\text{Minimum volume (m}^3) = \frac{10 \times \text{sensitivity of analytical method (mg)}}{\text{suitable hygiene standard (mg/m}^3)}$$

Table 10.28 Selected British Standards for analysis of workplace air quality

Standard	*Subject*	*Method*
BS 6069	Vinyl chloride	Charcoal tube and gas chromatography
	Sulphur dioxide	
	Particulate lead and lead compounds	Flame atomic absorption spectrometry
	Chlorinated hydrocarbon vapours	Charcoal tube, solvent desorption and gas chromatography
	Aromatic hydrocarbon vapours	Charcoal tube, solvent desorption and gas chromatography
	Asbestos	Fibre count
	Particulates	Gravimetry
BS 7384	Arc welding paticulates	Various
BS EN 1231	Short-term detector tubes	Colorimetry
BS EN 45544	Electrical apparatus used for the detection and direct concentration measurement to toxic gases. General requirements and test methods	Various
BS 50054	Electrical apparatus for the detection and measurement of combustible gases. General requirements and test methods	Various
BS EN 50104	Electrical apparatus for the detection and measurement of oxygen. Performance requirements and test methods	
BS EN 61779	Electrical apparatus for the detection of flammable gases. General requirements and test methods	
BS EN 10882	Welding fumes and gases	Personal dosimetry
BS ISO 11041	Particulate arsenic and arsenic compounds and arsenic trioxide	Hydride generation and atomic absorption spectrometry
BS ISO 11174	Particulate cadmium and cadmium compounds	Flame and electrothermal atomic absorption spectrometry
BS ISO 15202	Metals and metalloids in particulate matter	Inductively coupled plasma atomic emission spectrometry
BS ISO 16107	Evaluation of diffusive samplers	
BS EN 838	Diffusive samplers	Test methods
BS EN 1076	Pumped sorbent tubes	Test methods
BS EN 1232	Personal dosimetry pumps	Test methods
BS EN 12919	Sampling pumps of flowrate over 5 l/min	Test methods
BS EN 45544	Electrical apparatus for the direct detection and measurement of toxic gases and vapours	
BS ISO 15202	Metals and metalloids	Inductively coupled plasma atomic emission spectrometry
BS ISO 16200	Volatile organic compounds	Solvent desorption and gas chromatography

Table 10.29 Compounds for which there are analytical methods recommended by NIOSH

Acenaphthene
Acetaldehyde
Acetic acid
Acetic anhydride
Acetone
Acetone cyanohydrin
Acetonitrile (methyl cyanide)
Acetylene dichloride (1,2-dichloroethylene)
Acetylene tetrabromide (tetrabromoethane)
Acetylene tetrachloride
Acid mists
Acrolein
Acrylonitrile
Alachlor
ALAD (δ-aminolevulinic acid dehydratase)
Aldehydes (screening)
Aldecarb
Aldrin
Alkaline dusts
Allyl alcohol
Allyl chloride
Allyl glycidyl ether
Allyl trichloride
Alumina
Aluminium
Amines, alipathic
Amines, aromatic
Amino benzene
4-Aminobiphenyl
Aminoethanol compounds
bis (2-Aminoethyl) amine (diethylenetriamine)
p-Aminophenylarsonic acid
2-Aminopyridine
2-Aminotoluene
Ammonia
Ammonium sulphamate
Amorphous silica
n-Amyl acetate
sec-Amyl acetate (α-methylbutyl acetate)
Aniline
Anisidine
Anthanthrene
Anthracene
Antimony
ANTU (δ-naphthyl thiourea)
p-Arsanilic acid
Arsenic
Arsenic trioxide
Arsine
Asbestos
Aspartame
Asphalt fume
Atrazine
Azinphos methyl
Azelaic acid
Aziridine
Azo dyes

Barium

Benomyl
Benzaldehyde
Benz(c)acridine
1,3-benzenediol
Benz(a)anthracene
Benz(a,h)anthracene
Benz(a)anthrone
Benzene
Benzene, chlorinated
Benzene-solubles
Benzidine
Benzidine-based dyes
Benzo Azurine G
Benzo(a)fluoranthene
Benzo(b)fluoranthene
Benzo(j)fluoranthene
Benzo(k)fluoranthene
Benzo(g,h,i)perylene
Benzopurpurine 4B
Benzo(a)pyrene
Benzo(c)pyrene
Benzo(e)pyrene
Benzosulphonazole
Benzothiazole
Benzoyl peroxide
Benzyl chloride
Beryllium
Beryllium and compounds
Bibenzyl
Biphenyl (diphenyl)
Biphenyl-phenyl ether mixutre (phenyl ether-biphenyl vapour mixture)
Bis(chloromethyl)ether
2,2-Bis[4-(2,3-epoxypropoxy)phenyl] propane
Bismuth
Bisphenol A
2,2-Bis(*p*-chlorophenyl) 1,1,1-trichloroethane (DDT)
Boron carbide
Boron oxide
Bioaerosol
Bitumen fume
Bromotrifluoromethane
Bromoxynil
Bromoxynil octanoate
Butadiene (1,3-butadiene)
1-Butanethiol (*n*-butyl mercaptan)
2-Butanone (methy ethyl ketone or MEK)
2-Butoxy ethanol (butyl cellosolve)
sec-Butyl acetate
tert-Butyl acetate
Butyl acetate (*n*-Butyl acetate)
sec-Butyl alcohol
tert-Butyl alcohol
Butyl alcohol (*n*-butyl alcohol)
Butyl cellosolve (2-butoxy ethanol)
n-Butyl glycidyl ether
n-Butyl mercaptan
n-Butylamine
1,3-Butylene glycol

Table 10.29 Cont'd

Butyraldehyde
p-tert-Butyltoluene

Cadmium
Cadmium and compounds
Calcium
Calcium and compounds
Calcium arsenate
Calcium oxide
Camphor
Capsaicin
Captan
Carbaryl® (Sevin)
Carbendazim
Carbitol
Carbofuran
Carbon black
Carbon dioxide
Carbon disulphide
Carbon, elemental
Carbon monoxide
Carbon tetrachloride
Carbonyl chloride (phosgene)
3-Carene
Chlordane
Chloroacetic acid
Chloride
Chlorinated camphene (toxaphene)
Chlorinated diphenyl oxide
Chlorine
2-Chloro-1,3-butadiene (chloroprene)
Chlorodifluoroethane
1-Chloro-2,3-epoxypropane (epichlorohydrin)
4-Chloronitrobenzene
1-Chloro-1-nitropropane
Chloroacetaldehyde
α-Chloroacetophenone
Chlorobenzene (monochlorobenzene)
o-Chlorobenzylidine malononitrile
Chlorobromomethane
Chlorodiphenyl (42% chlorine)
Chlorodiphenyl (54% chlorine)
2-Chloroethanol (ethylene chlorohydrin)
Chloroform
Chloroform methyl ether
p-Chlorophenol
Chloroprene
Cloropropham
Chloryrfos
Chromic acid
Chromium
Chromium fume
Chromium, hexavalent
Chrysene
Chrysotile
Coal-tar naphtha (naphtha, coal tar)
Coal-tar pitch volatiles
Cobalt
Cobalt and compounds
Cobalt, metal, dust, and fume
Congo red
Copper
Copper dust and mists
Copper fume
Crag herbicide I
Cresol, all isomers
Cristobalite
Crotonaldehyde
Cryofluorane
Cumeme
Cyanide
Cyanuric acid
Cyanazine
Cyclohexane
Cyclohexanol
Cyclohexanone
Cyclohexene
Cyclohexylamine
Cyclopentadiene

DBPC
2,4 D
2,4-D acid
2,4-D 2-ethylhexyl ester
2,4-D 2-butoxyethyl ester
DDT (2,2-Bis(*p*-chlorophenyl)-1,1,1-trichloroethane)
DDVP
n-Decane
Demeton
Diacetone alcohol (4-hydroxy-4-methyl-2-pentanone)
1,2-Diaminoethane (ethylene diamine)
o-Dianisidene-based dyes
Diatomaceous earth
Diazomethane
Diazonium salts
Dibenz(a,h)anthracene
Diborane
Dibromodifluoromethane
1,2-Dibromoethane (ethylene dibromide)
2-Dibutylaminoethanol (aminoethanol compounds)
Dibutyl phosphate
Dibutylphthalate
Dibutyl tin bis (isooctylmercaptoacetate)
Dichloromethane
1,1-Dichloro-1-nitroethane
o-Dichlorobenzene
p-Dichlorobenzene
3,3′-Dichlorobenzidine
Dichlorodifluoromethane (Refrigerant 12)
1,1-Dichloroethane (ethylidene chloride)
1,2-Dichloroethane (ethylene dichloride)
Dichloroethyl ether
1,2-Dichloroethylene (acetylene dichloride)
Dichloromethane (methylene chloride)
Dichloromonofluoromethane (Refrigerant 21)
2,4-Dichlorophenoxyacetic acid and salts
1,2-Dichloropropane
Dichlorotetrafluoroethane (Refrigerant 114)

Table 10.29 Cont'd

Dichloro-5-triazine-2,4,6-trione, sodium salt
Dichlorovos
Dicrotophos
Dieldrin
Diesel exhaust
Diethanolamine
2-Diethylaminoethanol
Diethylamine
Diethylcarbamoyl chloride
Diethylene dioxide (dioxane)
Diethylene glycol
Diethylene glycol ether
Di-(2 ethyl hexyl) phthalate
Diethylenetriamine
Difluorodibromomethane
Difluorodichloromethane
Diglycidyl ether of Bisphenol A
Dihydrocapsaicin
Diisobutyl ketone (2,6-dimethyl-4-heptanone)
Diisopropylamine
Dimethoxymethane (methylal)
Dimethyl acetamide
Dimethyl benzene (xylene)
1,3-Dimethyl butyl acetate (sec-hexyl acetate)
Dimethyl formamide
2,6-Dimethyl-4-heptanone (diisobutyl ketone)
Dimethyl sulphate
N,N-Dimethyl-*p*-toluidine
Dimethylamine
4-Dimethylaminoazobenzene
2,4-Dimethylaminobenzene (xylidine)
N,N-dimethylaniline
Dimethylarsenic acid
1,1-Dimethylhydrazine
Dimethylnitrosamine
bis (Dimethylthiocarbamoyl) disulphide
Dinitrobenzene (all isomers)
Dinitro *o*-cresol
Dinitrotoluene
Dioxane
Diphenyl
4,4'-Methylenebisphenyl isocyanate (MDI)
Dipropylene glycol methyl ether
Direct Black 38
Direct Blue 6
Direct Blue 8
Direct Brown 95
Direct Red 2
Direct Red 28
Disulphotondiuron
2,6-Di-tert-butyl-*p*-cresol
Dowtherm A (phenyl etherbiphenyl vapour mixture)

Elemental carbon
Elements
Endrin
Epichlorohydrin (1-chloro-2,3-epoxypropane)
EPN (O-ethyl-O-*p*-nitrophenyl
  phenyl-phosphonothiolate)
2,3-Epoxy-1-propanol (glycidol)
1,2-Epoxypropane (propylene oxide)
2,2-bis[4-(2,3-Epoxypropoxy) phenyl] propane
Ethanol (ethyl alcohol)
Ethanolamine (aminoethanol compounds)
Ether
Ethion
Ethoprop
2-Ethoxyethanol
2-Ethoxyethylacetate
Ethyl acrylate
Ethyl alcohol (ethanol)
Ethyl benzene
Ethyl bromide
Ethyl butyl ketone (3-heptanone)
Ethyl chloride
Ethyl ether
Ethyl formate
Ethyl mercaptan
Ethyl sec-amyl ketone (5-methyl-3-heptanone)
Ethyl silicate
O-Ethyl-O-*p*-nitrophenyl phenyl-phosphonothiolate
  phosphonate (EPN)
Ethylamine
Ethylene chloride (ethylene dichloride)
Ethylene chlorohydrin (2-chloroethanol)
Ethylenediamine
Ethylene dibromide (1,2-dibromoethane)
Ethylene dichloride (1,2-dichloroethane)
Ethylene glycol
Ethylene glycol dinitrate
Ethylene oxide
Ethylene thiourea
Ethylenimine
di-2-Ethylhexylphthalate (di-sec-octyl phthalate)
Ethylidene chloride (1,1-dichloroethane)
N-Ethylmorpholine

Fenamiphos
Fibrous glass
Fluoranthene
Fluorene
Fluoride
Fluoroacetate, sodium
Fluorotrichloromethane (Refrigerant 11)
Fonofos
Formaldehyde
Formetanate.HCl
Formic acid
Furfural
Furfuryl alcohol

Galena
Gallium
Glutaraldehyde
Glycerin mist
Glycidol (2,3-epoxy-1-propanol)
Glycols

Hafnium
Heptachlor

Table 10.29 Cont'd

Heptane
Heptanal
3-Heptanone (ethyl butyl ketone)
2-Heptanone (methyl (*n*-amyl) ketone)
Herbicides
Hexachlorobutadiene
Hexachlorocyclopentadiene
Hexachloroethane
Hexachloronaphthalene
Hexamethylene diisocyanate
Hexamethylenetetramine
Hexanal
Hexane
2-Hexanone
Hexavalent chromium
2-Hexanone (methyl butyl ketone or MBK)
Hexone (methyl isobutyl ketone or MIBK)
sec-Hexyl acetate (1,3-dimethyl butyl acetate)
Hippuric acid
Hydrazine
Hydrogen bromide
Hydrogen chloride
Hydrogen cyanide
Hydrogen fluoride
Hydrogen sulphide
Hydroquinone
4-Hydroxy-4-methyl-2-pentanone (diacetone alcohol)

2-Imidazolidinethione (ethylene thiourea)
Indeno[1,2,3-cd]pyrene
Indium
Iodine
Iron
Iron oxide fume
Isoamyl acetate
Isoamyl alcohol
Isobutyl acetate
Isobutyl alcohol
Isobutyraldehyde
Isocyanates
Isophorone
Isopropanol (isopropyl alcohol)
Isopropyl acetate
Isopropyl alcohol (isopropanol)
Isopropyl benzene (cumeme)
Isopropyl glycidyl ether
Isopropylamine
4,4'-Isopropylidenediphenol
Isovaleraldehyde

Kepone
Kerosene
Ketene

Lead
Lead sulphide
Limonene
Lindane
Liquefied petroleum gas (LPG)
Lithium
Lithium hydroxide
LPG (liquefied petroleum gas)

MAPP (methyl acetylene/propadiene)
Magnesium
Magnesium oxide fume
Malathion
Maleic anhydride
Manganese
Manganese fume
MBK (2-hexanone)
MDI (4,4'-methylenebisphenyl isocyanate)
MEK (2-butanone)
Mercaptans
Mercury
Mesityl oxide
Metals in air
Methamidophos
Methanol (methyl alcohol)
Methiocarb
Methomyl
2-Methoxyethanol (methyl cellosolve)
2-Methoxethyl acetate
Methozychlor
Methyl(*n*-amyl)ketone (2-heptanone)
5-Methyl-3-heptanone
Methyl acetate
Methyl acetylene
Methyl acetylene propadiene mixture
Methyl acrylate
Methyl alcohol (methanol)
4-Methylbenzenesulphonic acid
Methylamyl ketone
Methyl arsonic acid
Methyl bromide
α-Methyl butyl acetate (sec-amyl acetate)
Methyl butyl ketone (2-hexanone)
Methyl cellosolve (2-methoxyethanol)
Methyl cellosolve acetate
Methyl chloride
Methyl chloroform (1,1,1-trichloroethane)
Methyl cyanide (acetonitrile)
Methyl ethyl ketone (2-butanone)
Methyl ethyl ketone peroxide
Methyl formate
Methyl iodide
Methyl isoamyl acetate
Methyl isobutyl carbinol
Methyl isobutyl ketone (hexone)
Methyl methacrylate
α-Methyl styrene
Methylal (dimethoxymethane)
Methlyamine
Methylcyclohexane
Methylcyclohexanol
Methylcyclohexanone
4,4'-Methylenebis (2-chloroaniline)
4,4'-Methylenebisphenyl isocyanate (MDI)
Methylene chloride (dichloromethane)
4',4'-Methylenedianiline

Table 10.29 Cont'd

5-Methyl-3-heptanone
Methylhydrazine
Metolachlor
Methyl mercaptan
Methyl parathion
Methyl phenol
N-Methyl-2-pyrrolidinone
Methyl tert-butyl ether
Mevinphos®
MIBK (hexone)
Mineral spirits
MOCA
Molybdenum
Molybdenum insoluble compounds
Molybdenum soluble compounds
Monochloroacetic acid
Monochlorobenzene (chlorobenzene)
Monocrotophos
Monomethyl aniline
Monomethyl hydrazine
Monomethylarsonic acid
Morpholine

Naphtha, coal tar (coal-tar naphtha)
Naphthalene
Naphthylamines
α-Naphthyl thiourea
Nickel
Nickel carbonyl
Nickel fume
Nicotine
Nitric acid
Nitric oxide
p-Nitroaniline
Nitrobenzene
Nitrobenzol
4-Nitrobiphenyl
p-Nitrochlorobenzene
Nitroethane
Nitrogen peroxide
2-Nitropropane
Nitrogen dioxide
Nitroglycerin
Nitroglycol
Nitromethane
Nitrosamines
N-Nitrosodimethylamine
Nitrous oxide
Nitrotoluene
Nuisance dusts

Octachloronaphthalene
Octamethylcyclotetrasiloxane
Octane
1-Octanethiol
di-sec-Octyl phthalate (di-2-ethylhexylphthalate)
Oil mist
Organic solvents
Organo(alkyl)mercury
Organoarsenicals
Organonitrogen pesticides
Organophosphorus pesticides
Oxamyl
Ozone
Oxygen

Palladium
Paraquat
Parathion
Particulates
Pentamidine
PCBs
Pentachlorobenzene
Pentachloroethane
Pentachloronaphthalene
Pentachlorophenol
Pentane
2-Pentanone
Perchloroethylene (tetrachloroethylene)
Perylene
Pesticides
Petroleum distillates (petroleum naphtha)
Petroleum naphtha (petroleum distillates)
Phenacylchloride
Phenanthene
Phenol
Phenyl ether
Phenyl ether-biphenyl vapor mixture (Dowtherm A)
Phenyl ethylene (styrene)
Phenyl glycidyl ether
Phenylhydrazine
Phenyloxirane
Phorate
Phosdrin
Phosgene (carbonyl chloride)
Phosphate
Phosphine
Phosphoric acid
Phosphorus (white, yellow)
Phosphorus pentachloride
Phosphorus trichloride
Phthalic anhydride
Picric acid
Pinene
Platinum, soluble salts
PNAs
Polyacrylate
Polychlorinated biphenyls (PCBs)
Polymethylsiloxane
Polynuclear aromatic hydrocarbons (PNAs)
Potassium
Potassium hydroxide
Propane
Propanol (Propyl alcohol)
Propham
Propionaldehyde
Propxur
n-Propyl acetate
Propyl alcohol (propanol)
n-Propyl nitrate

Table 10.29 Cont'd

Propylene dichloride
Propylene glycol
Propylene oxide (1,2-epoxypropane)
Propyne
Pyrene
Pyrethrum
Pyridine

Quartz (silica, crystalline)
Quinone

Refrigerant 11 (fluorotrichloromethane)
Refrigerant 113 (1,1,2-trichloro-1,2,2-fluoroethane)
Refrigerant 114 (dichlorotetrafluoroethane)
Refrigerant 12 (dichlorodifluoromethane)
Refrigerant 21 (dichloromonofluoromethane)
Resorcinol
Rhodium
Rhodium, metal fume and dust
Rhodium, soluble salts
Ribavirin
Ronnel
Rotenone
Rubber solvent
Rubidium

Selenium
Sevin (carbaryl)
Silica, amorphous
Silica, crystalline
Silica in coal mine dust
Silicon
Silicon dioxide
Silver
Silver, metal and soluble compounds
Simazine
Sodium
Sodium dichloroisocyanate dihydrate
Sodium fluoroacetate
Sodium-2,4-dichlorophenoxyethyl sulphate
Sodium hexafluoroaluminate
Sodium hydroxide
Stibine
Stoddard solvent
Strontium
Strychnine
Styrene (vinyl benzene)
Styrene oxide
Sulphate
Sulphite
Sulprofos
Sulphur dioxide
Sulphur hexafluoride
Sulphuric acid
Sulphuryl fluoride
Systox

2,4,5-T
TDI (Toluene 2,4-diisocyanate)
Tantalum
Tellurium
Tellurium hexafluoride
TEPP
Terbufos
Terpenes
Terphenyl
Tetrabromoethane (acetylene tetrabromide)
Tetrabutyltin
1,1,1,2-Tetrachloro-2,2-difluoroethane
1,1,2,2-Tetrachloro-1,2-difluoroethane (Refrigerant 112)
1,2,4,5-Tetrachlorobenzene
1,1,2,2-Tetrachloroethane
Tetrachloroethylene (perchloroethylene)
Tetrachloromethane (carbon tetrachloride)
Tetrachloronaphthalene
Tetraethyl lead
Tetraethylene glycol
Tetraethyl pyrophosphate
Tetrahydrofuran
Tetramethyl lead (as Pb)
Tetramethyl succinonitrile
Tetramethyl thiourea
Tetramethyl thiuram disulphide (thiram)
Tetranitromethane
Tetryl (2,4,6-Trinitrophenylmethyl-nitramine)
Thallium
Thiobencarb
Thiopene
Thiram (tetramethyl thiuram disulphide)
Tin
Tin, organic compounds
Tissue preparation
Titanium
Titanium diboride
Titanium dioxide
o-Tolidine based dyes
Toluene
2,4 and 2,6-Toluenediamine
Toluene-2,4-diisocyanate (TDI)
Toluene-2,6-diisocyanate
p-Toluene sulphonic acid
o-Toluidine
Toxaphene (chlorinated camphene)
Tribromomethane
Tributyltin chloride
Tributyl phosphate
1,1,2-Trichloro-1,2,2-trifluoroethane (Refrigerant 113)
1,2,4-Trichlorobenzene
1,1,1-Trichloroethane (Methyl chloroform)
1,1,2-Trichloroethane
Trichloroethylene
Trichloroisocyanuric acid
Trichloromethane
Trichloromonofluoromethane (Fluorotrichloromethane)
Trichloronaphthalene
2,4,5-Trichlorophenoxyacetic acid and salts
1,2,3-Trichloropropane
1,3,5-Trichloro-s-triazine-2,4,6-trione
Tricyclohexyltin hydroxide

Table 10.29 Cont'd

1,1,2-Trichloro-1,2,2-trifluoroethane
Tridymite
Triethanolamine
Triethylamine
Triethylene glycol
Triethylenetetramine
Trifluoromonobromomethane
Trimellitic anhydride
2,4,7-Trinitro-9-fluorenone
2,4,6-Trinitrophenylmethylnitramine (Tetryl)
Triorthocresyl phosphate
Triphenyl phosphate
Tungsten
Turpentine

Vanadium
Vanadium, V_2O_5 fume
Valeraldehyde
Vinyl acetate
Vinyl benzene (styrene)
Vinyl bromide
Vinyl chloride
Vinyl toluene
Vinylidene chloride
Volatile organic compounds

Warfarin
Wood alcohol

Xylene (xylol or dimethyl benzene)
Xylidene (2,4-dimethylamino-benzene)
Xylol (xylene)

Yttrium

Zinc
Zinc and compounds
Zinc fume
Zinc oxide
Zirconium compounds
Zirconium oxide

When compliance with hygiene standards is assessed using short-term sampling (e.g. 15 min), the number of samples obtainable within an 8 hr shift is 32. The average of these will indicate whether or not exposures have exceeded the TWA hygiene standard. When fewer than 32 results are available it is necessary to lower the acceptable upper limit for the average below the hygiene standard to compensate for the lack of data. The following defines the upper acceptable limit to the average, $\bar{x}_{max}$, thus:

$$\bar{x}_{max} = (\text{TWA hygiene standard}) - 1.6\left(\frac{1}{\sqrt{n}} - \frac{1}{32}\right)\sigma$$

where n = number of results , σ = standard deviation. If σ is unknown from previous results, an estimate can be made from a few samples and the maximum acceptance limit is hygiene standard ($k \times$ range), where k is obtained from Table 10.31. If the calculated limit is below the observed range, compliance can be assumed for that day, although the statistical significance over longer periods needs consideration.

For routine monitoring, frequency will be influenced by the level of exposure. The further the levels depart from the standard, the less the need for routine monitoring. A suggested guide is:

personal monitoring once per month if the TWA exposures are 1–2 × hygiene standard

once per quarter if 0.5–1 or 2–4 × hygiene standard

once per year if 0.1–0.5 or 4–10 × hygiene standard.

Routine monitoring becomes superfluous if exposures are very low (i.e. below 0.1 × hygiene standard) or very high (e.g. 10 × hygiene standard), although where standards are Control Limits routine sampling should be considered when exposures exceed 0.1 CL. (The frequencies refer to monitoring the entire shift period for every 10 employees.) Table 10.32 offers guidance for monitoring strategies for compounds assigned OELs (Chapter 5) through legislative requirements.

Table 10.30 Methods for sampling and analysis of a range of air pollutants

Substance	*Sampling method*[1]	*Min. sample size*[2] (l)	*Suggested max. sample rate*[3] (ml/min)	*Analytical technique*[4]
Abate	F (5 μm PVC)	250	2000	G
Acetaldehyde	I	10	2000	C
Acetic acid	I	100	2500	Titration
Acetic anhydride	I	100	1000	C
Acetone	CT	2	200	GLC
Acetonitrile	CT	10	200	GLC
Acetylene dichloride, *see* 1,2-Dichloroethylene				
Acetylene tetrabromide	ST	100	1000	GLC
Acrolein	I	10	1500	C
	AT	6	200	GLC
Acrylamide	I		1000	
	CT	10	200	GLC
Acrylonitrile	CT	20	200	GLC
Aldrin	F + I (GF)	180	1000	GLC
Allyl alcohol	CT	10	200	GLC
Allyl chloride	CT	100	1000	GLC
Allyl glycidyl ether (AGE)	CT	10	50	GLC
Allyl propyl disulphide	CT	10	1000	GLC
Alundum® (Al_2O_3)	F (5 μm PVC)	250	2000	G
4-Aminodiphenyl	F (0.8 μm MCEF)	540	4000	C
2-Aminoethanol, *see* Ethanolamine				
2-Aminopyridine				
Ammonia	I	5	1000	C
	OT	1	100	LDDT
	P			PC
Ammonium chloride – fume	F (0.8 μm MCEF)	100	2000	C
Ammonium sulphamate (ammate)	F (5 μm PVC)	250	2000	G
n-Amyl acetate	CT	10	200	GLC
sec-Amyl acetate	CT	10	200	GLC
Aniline	ST	20	200	GLC
Anisidine (*o*-, *p*-isomers)	ST	15	200	GLC
Antimony and compounds (as Sb)	F (0.8 μm MCEF)	360	1500	AAS
ANTU (α-naphthylthiourea)	F (0.8 μm MCEF)	300	1500	GLC
Arsenic and compounds (as As)	F (0.8 μm MCEF)	90	2000	AAS
Arsenic trioxide production (as As)		30	1700	C
Arsine	CT	10	200	AAS
Asbestos	F (0.8 μm MCEF)	100	1500	(microscopic fibre count)
Asphalt (petroleum) fumes	F (2 μm PVC)	250	2000	G
Azinophos methyl	I	100	1000	GLC

Barium (soluble compounds)	F (0.8 μm MCEF)	180	1500	AAS
Benzene	CT	12	200	GLC
	OT	2		LDDT
Benzidine production	F (0.8 μm MCEF)	480	4000	C
p-Benzoquinone, *see* Quinone				
Benzoyl peroxide	I	30	1000	C
Benzo(a)pyrene				
Benzyl chloride	CT	10	200	GLC
Beryllium	F (0.8 μm MCEF)	270	1500	AAS (30 min allowed at 0.025 mg/m^3
Biphenyl	CT	30	200	GLC
Bismuth telluride	F (0.8 μm MCEF)	100	1500	AAS
Boron oxide	F (2 μm PVC)	60	1000	G
Boron trifluoride	F + I	30	2500	C
Bromine	I	45	1000	C
Bromine pentafluoride	I	15	2500	ISE
Bromoform	CT	10	200	GLC
Butadiene (1,3-butadiene)	CT	1	50	GLC
Butanethiol, *see* Butyl mercaptan				
2-Butanone	CT	10	200	GLC
2-Butoxyethanol (butyl cellosolve)	CT	10	200	GLC
n-Butyl acetate	CT	10	200	GLC
sec-Butyl acetate	CT	10	200	GLC
tert-Butyl acetate	CT	10	200	GLC
n-Butyl alcohol	CT	10	200	GLC
sec-Butyl alcohol	CT	10	200	GLC
tert-Butyl alcohol	CT	10	200	GLC
Butylamine	I	15	1000	C
Butyl cellosolve, *see* 2-Butoxyethanol				
tert-Butyl chromate (as CrO_3)	F (0.8 μm MCEF)	23	1500	C
n-Butyl glycidyl ether	CT	10	200	GLC
n-Butyl lactate	F (0.8 μm MCEF)	30	1000	GLC
Butyl mercaptan	CT	10	1000	GLC
p-tert-Butyltoluene	CT	10	200	GLC
Cadmium, dust and salts (as Cd)	F (0.8 μm MCEF)	15	1000	AAS
Cadmium, fume (as Cd)				
Cadmium oxide fume (as Cd)	F (0.8 μm MCEF)	25	1500	AAS
Calcium arsenate (as As)	F (0.8 μm MCEF)	500	1500	AAS
				C
Calcium carbonate/marble	F			G (nuisance dust)
Calcium oxide	F (0.8 μm MCEF)	85	1500	AAS
Camphor, synthetic	CT	10	200	GLC
Caprolactam				
dust	F (0.8 μm MCEF)			GLC
vapour	I			GLC

Table 10.30 Cont'd

Substance	*Sampling method*[1]	*Min. sample size*[2] (l)	*Suggested max. sample rate*[3] (ml/min)	*Analytical technique*[4]
Carbaryl (Sevin®)	F (GF)	90	1500	C (do not use Tenite holder)
Carbon black	F (2 μm PVC)	200	1700	G (stainless steel support screen)
Carbon dioxide	(air bag)	4	50	GLC
Carbon disulphide	CT	6	200	GLC (used with drying tube)
Carbon monoxide	M		100	(dosimeter)
	OT	1	20	LDDT
Carbon tetrachloride	CT	5	1000	GLC (10 min allowed at 200 ppm)
Cellulose (paper fibre)	F			G (nuisance duet)
Chlordane	I	100	1000	GLC
Chlorinated camphene	F (0.8 μm MCEF)	15	1000	GLC
Chlorinated diphenyl oxide	F (0.8 μm MCEF)	90	1000	GLC
Chlorine	I	30	2000	C
Chlorine dioxide	I	40	1000	C
Chlorine trifluoride	I	15	1000	ISE
Chloroacetaldehyde	CT	10	1000	GLC
Chloroacetophenone (phenacyl chloride)	CT	100	1000	GLC
Chlorbenzene (monochlorobenzene)	CT	10	200	GLC
o-Chlorobenzylidene malononitrile (OCBM)	CT	10	1000	GLC
Chlorobromomethane/bromochloromethane	CT	5	200	GLC
2-Chloro-1,3-butadiene, *see* β-Chloroprene				
Chlorodiphenyl (42% chlorine)	F (0.8 μm MCEF)	100	1500	GLC
	OT	50	200	GLC
Chlorodiphenyl (54% chlorine)	F (0.8 μm MCEF)	100	1500	GLC
	OT	50	200	GLC
1-Chloro,2,3-epoxy-propane, *see* Epichlorhydrin				
2-Chloroethanol, *see* Ethylene chlorhydrin				
Chloroethylene, *see* Vinyl chloride				
Chloroform (trichloromethane)	CT	15	1000	GLC
bis-Chloromethyl ether	CT	30	200	GLC
1-Chloro-1-nitropropane	CT	10	200	GLC
Chloropicrin	I	100	1000	C
β-Chloroprene	CT	3	50	GLC
2-Chloro 6-trichloromethyl pyridine (N-Serve®)	CT	10	200	GLC
Chromates, certain insoluble forms	F (0.8 μm MCEF)	90	1500	AAS
Chromic acid and chromates (as Cr)	F (5 μm PVC)	22.5	1000	C

Chromium, soluble chromic and chromous salts (as Cr)	F (0.8 μm MCEF)	90	1500	AAS
Coal tar pitch volatiles, *see* Particulate polycyclic aromatic hydrocarbons (PPAH), as benzene solubles				
Cobalt metal, dust and fume (as Co)	F (0.8 μm MCEF)	720	1500	AAS
Copper				
fume	F (0.8 μm MCEF)	720	1500	AAS
dusts and mists (as Cu)	F (0.8 μm MCEF)	90	1500	AAS
Corundum (Al_2O_3)	F			G (nuisance dust)
Cotton dust (raw)	F (5 μm PVC)	540	1500	G
Crag® herbicide	F	100	2000	G
Cresol, all isomers	ST	20	200	GLC
Crotonaldehyde	CT	20	200	GLC
	I	100	2000	C
Cumene	CT	10	200	GLC
Cyanides (as CN)	F + 1 (0.8 μm MCEF)	90	1500	ISE
		1	20	LDDT
Cyanogen	I	15	500	C
Cyclohexane	CT	2.5	200	GLC
Cyclohexanol	CT	10	200	GLC
Cyclohexanone	CT	40	1000	GLC
Cyclohexene	CT	5	200	GLC
Cyclohexylamine	I	10	2000	C
Cyclopentadiene	CT	10	200	GLC
2,4-D (2,4-Diphenoxy-acetic acid)	I	100	1000	GLC
DDT (Dichlorodiphenyltrichloroethane)	F(GF)	90	1500	GLC
DDVP, *see* Dichlorvos				
Diacetone alcohol (4-hydroxy-4-methyl-2-pentanone)	CT	10	200	GLC
1,2-Diaminoethane, *see* Ethylenediamine				
Diazomethane	OT	10	200	GLC (XAD-2 resin tube)
Dibrom®	I	100	1000	GLC
1,2-Dibromoethane	CT	1	200	GLC (5 min allowed at 50 ppm)
2-*n*-Dibutylaminoethanol				
Dibutyl phosphate				
Dibutyl phthalate	F (0.8 μm MCEF)	30	1000	GLC
Dichloracetylene				
o-Dichlorobenzene	CT	10	200	GLC
p-Dichlorobenzene	CT	3	50	GLC
Dichlorobenzidine				
Dichlorodifluoromethane	CT	3	50	GLC
1,3-Dichloro-5,5-dimethyl hydantoin	I	100	1000	C

Table 10.30 Cont'd

Substance	*Sampling method*[(1)]	*Min. sample size*[(2)] (l)	*Suggested max. sample rate*[(3)] (ml/min)	*Analytical technique*[(4)]
1,1-Dichloroethane	CT	10	200	GLC
1,2-Dichloroethane	CT	3	200	GLC (12 min allowed at 200 ppm)
1,2-Dichloroethylene	CT	3	200	GLC
Dichloroethyl ether	CT	15	1000	GLC
Dichloromethane, *see* Methylene chloride				
Dichloromonofluoromethane	CT	3	50	GLC (use 2 large charcoal tubes back-to-back)
1,1-Dichloro-1-nitroethane	CT	15	1000	GLC (15 min sample)
1,2-Dichloropropane, *see* Propylene dichloride				
Dichlorotetrafluoroethane	CT	3	50	GLC
Dichlorvos (DDVP)	I	100	1000	GLC
Dieldrin	F (GF)	180	1500	GLC
Diethylamine	ST	50	1000	GLC
Diethylaminoethanol	I	30	1000	C
Diethylene triamine	I	100	1500	C
Diethyl ether, *see* Ethyl ether				
Diethyl phthalate	F (0.8 μm MCEF)	30	1000	GLC
Difluorodibromomethane	CT	10	200	GLC
Diglycidyl ether	CT	15	1000	GLC
Dimethoxyethane, *see* Methylal				
Diisobutyl ketone	CT	12	200	GLC
Diisopropylamine	I	15	200	C
Dimethyl acetamide	ST	50	1000	GLC
Dimethylamine	ST	50	200	GLC
Dimethylaminobenzene, *see* Xylidene				
Dimethylaniline (*N,N*-dimethylaniline)	CT	20	200	GLC
	ST	5	200	
Dimethylbenzene, *see* Xylene				
Dimethyl-1,2-dibromo-2-dichloroethyl phosphate, *see* Dibrom®				
Dimethylformamide	ST	50	1000	GLC
2,6-Dimethyl-4-heptanone, *see* Diisobutyl ketone				
1,1-Dimethylhydrazine	I	100	1000	C
Dimethylphthalate	CT	20	200	GLC
Dimethyl sulphate				
Dinitrobenzene (all isomers)	I	100	1000	C
Dinitro-*o*-cresol	CT	60	200	GLC

Dinitrotoluene	I	100	1000	C
Dioxane (tech. grade)	CT	10	200	GLC
Diphenyl, *see* biphenyl				
Diphenylmethane diisocyanate, *see* Methylene bisphenyl isocyanate (MDI)				
Dipropylene glycol methyl ether	CT	10	200	GLC
Di-sec, Octyl phthalate (di-2-ethylhexylphthalate)	CT	30	200	GLC
Dust, inert or nuisance				
respirable	F (5 µm PVC)	102	1700	G (use cyclone but no Tenite holders)
total				
Emery	F			G (nuisance dust)
Endrin	I	100	1000	GLC
Epichlorhydrin	CT	20	200	GLC
EPN	F (GF)	120	1500	GLC
1,2-Epoxypropane, *see* Propylene oxide				
2,3-Epoxy-1-Propanol, *see* Glycidol				
Ethanethiol, *see* Ethyl mercaptan				
Ethanolamine	CT	20	200	GLC
2-Ethoxyethanol	CT	10	200	GLC
2-Ethoxyethyl acetate (cellosolve acetate)	CT	10	200	GLC
Ethyl acetate	CT	6	200	GLC
Ethyl acrylate	CT	10	200	GLC
Ethyl alcohol (ethanol)	CT	1	50	GLC
Ethylamine	ST	30	200	GLC
Ethyl sec-amyl ketone (5-methyl-3-heptanone)	CT	10	200	GLC
Ethyl benzene	CT	10	200	GLC
Ethyl bromide	CT	4	200	GLC
Ethylbutyl ketone (3-heptanone)	CT	10	200	GLC
Ethyl chloride				
Ethyl ether	CT	3	200	GLC
Ethyl formate	CT	10	200	GLC
Ethyl silicate	OT	9	50	GLC
Ethylene chlorohydrin	CT	20	200	GLC
Ethylenediamine	I	30	1000	C
Ethylene dibromide, *see* 1,2-Dibromoethane				
Ethylene dichloride, *see* 1,2-Dichloroethane				
Ethylene glycol dinitrate and/or nitroglycerine	OT	15	1000	GLC (Tenax tube, 5 min sample)
Ethylene glycol monomethyl ether acetate (methyl cellosolve acetate)	CT	20	200	GLC
Ethylene oxide	CT	5	50	GLC (use 2 back-to-back large tubes)
Ethylidene chloride, *see* 1,1-Dichloroethane				
N-Ethylmorpholine	CT	10	200	GLC

Table 10.30 Cont'd

Substance	*Sampling method*[1]	*Min. sample size*[2] (l)	*Suggested max. sample rate*[3] (ml/min)	*Analytical technique*[4]
Ferbam	F (0.8 μm MCEF)	150	1500	C
Ferrovandium dust	F (0.8 μm MCEF)	100	1500	AAS
Fluoride (as F)	I	10	2500	ISE
	F			
Fluorotrichloromethane	CT	4	50	GLC
Formaldehyde	I	25	1000	C (10 min allowed at 10 ppm)
	AT	6	200	GLC
Formic acid	I	100	1000	GLC
				C
Furfural	CT	20	200	GLC
	AT	6	200	GLC
Furfuryl alcohol	CT	10	200	GLC
Glass, fibrous or dust	F			G (nuisance dust)
Glycerin mist	F			G (nuisance dust)
Glycidol (2,3-epoxy-1-propanol)	CT	50	1000	GLC
Glycol monoethyl ether, *see* 2-Ethoxyethanol				
Graphite (synthetic)	F			G (nuisance dust)
Guthion®, *see* Azinphos-methyl				
Gypsum	F			G (nuisance dust)
Hafnium	F	720	1500	AAS
Heptachlor	I	100	1000	GLC
Heptane (*n*-heptane)	CT	4	200	GLC
Hexachlorocyclopentadiene				
Hexachloroethane	CT	10	200	GLC
Hexachloronaphthalene	F (0.8 μm MCEF)	30	1000	GLC (1000 cc/min only)
Hexane (*n*-hexane)	CT	4	200	GLC
2-Hexanone (methyl butyl ketone)	CT	10	200	GLC
Hexone (methyl isobutyl ketone)	CT	10	200	GLC
sec-Hexyl acetate	CT	10	200	GLC
Hexylene glycol				
Hydrazine	I	100	1000	C
Hydrogen bromide	I	100	1000	ISE (1000 cc/min only)
Hydrogen chloride	I	15	1000	ISE (1000 cc/min only)
Hydrogen cyanide	F + I (0.5 μm MCEF)	10	2000	ISE
	OT	1	20	LDDT
Hydrogen fluoride	I	45	1500	ISE (1500 cc/min only)
Hydrogen selenide	I	100	1000	AAS
Hydrogen sulphide	I	30	2000	C
	OT	1	20	LDDT

Hydroquinone	I	75	1000	C
Indium and compounds (as In)	F (0.8 μm MCEF)	300	1500	C
Iodine	I	15	1000	C
Iron oxide fume (NOC)	F (0.8 μm MCEF)	30	1500	AAS
Iron pentacarbonyl	F (0.8 μm MCEF)	175	1500	AAS
Iron salts, soluble (as Fe)	F (0.8 μm MCEF)	4	1500	AAS
Isoamyl acetate	CT	10	200	GLC
Isoamyl alcohol	CT	10	200	GLC
Isobutyl acetate	CT	10	200	GLC
Isobutyl alcohol	CT	10	200	GLC
Isophorone	CT	12	200	GLC
Isopropyl acetate	CT	9	200	GLC
Isopropyl alcohol	CT	3	200	GLC
Isopropylamine	I	100	1000	GLC
Isopropyl ether	CT	3	50	GLC
Isopropyl glycidyl ether	CT	10	200	GLC
Kaolin	F			G (nuisance dust)
Ketene	I	50	1000	C
Lead, inorg., fumes and dusts (as Pb)	F (0.8 μm MCEF)	100	4000	AAS
Limestone	F			G (nuisance dust)
Lindane	F + 1	90	1500	GLC
Lithium hydride	F (0.8 μm MCEF)	720	1500	AAS
LPG (liquefied petroleum gas)	M	direct reading combustible gas meter		
Magnesite	F			G (nuisance dust)
Magnesium oxide fume (as Mg)	F (0.8 μm MCEF)	150	1500	AAS
Malathion	F (GF)	120	1000	GLC
	I			
Manganese and compounds (as Mn)	F (0.8 μm MCEF)	22.5	1500	AAS (1500 cc/min only)
Manganese cyclopentadienyl tricarbonyl (as Mn)	F (0.8 μm MCEF)	25	1500	AAS
Marble/calcium carbonate	F			G (nuisance dust)
Mercury (alkyl compounds) (as Hg)	OT	3	50	AAS (chromosorb)
Mesityl oxide	CT	10	200	GLC
Methanethiol, *see* Methyl mercaptan				
Methoxychlor	I	100	1000	GLC
2-Methoxyethanol (methyl cellosove)	CT	50	1000	GLC
Methyl acetate	CT	7	200	GLC
Methyl acetylene (propyne)	CT	2	50	GLC
Methyl acetylene propadiene mixture	CT	2	50	GLC
Methyl acrylate	CT	5	200	GLC
Methylal (dimethoxymethane)	CT	2	200	GLC
Methyl alcohol (methanol)	ST	3	50	GLC
Methylamine	I	50	2000	C
Methyl amyl alcohol, *see* Methyl isobutyl carbinol				
Methyl *n*-amyl ketone (2-heptanone)	CT	10	200	GLC

Table 10.30 Cont'd

Substance	*Sampling method*[1]	*Min. sample size*[2] (l)	*Suggested max. sample rate*[3] (ml/min)	*Analytical technique*[4]
Methyl bromide	CT	11	1000	GLC (2 large charcoal tubes in series)
Methyl butyl ketone, *see* 2-Hexanone				
Methyl cellosolve, *see* 2-Methoxyethanol				
Methyl cellosolve acetate, *see* Ethylene glycol monomethyl ether acetate				
Methyl chloroform (1,1,1-trichloroethane)	CT	6	200	GLC
Methylcyclohexane	CT	4	200	GLC (for methylcyclohexane)
Methylcyclohexanol	CT	10	200	GLC
o-Methylcyclohexanone	CT	10	200	GLC
Methylene bisphenyl isocyanate (MDI)	I	20	1000	C
Methylene chloride (dichloromethane)	CT	5	1000	GLC
4,4′-Methylene bis (2-chloraniline) (MOCA)	ST	3	500	GLC (use prefilter)
Methyl ethyl ketone (MEK), *see* 2-Butanone				
Methyl formate	CT	10	200	GLC
Methyl iodide	CT	50	1000	GLC
Methyl isoamyl ketone	CT	5	200	GLC
Methyl isobutyl carbinol	CT	10	200	GLC
Methyl isobutyl ketone, *see* Hexone				
Methyl mercaptan	CT	15	1000	GLC
	I	30	2000	C
Methyl methacrylate	CT	5	200	GLC
Methyl parathion	I	50	2800	GLC
Methyl propyl ketone, *see* 2-Pentanone				
α-Methyl styrene	CT	3	200	GLC
Molybdenum (as Mo)				
soluble compounds	F (0.8 μm MCEF)	90	1500	AAS
insoluble compounds	F (0.8 μm MCEF)	90	1500	AAS
Monomethyl aniline	I	100	1000	GLC
Monomethyl hydrazine	I	22.5	1500	C
Morpholine	ST	20	200	GLC
Naphtha (coal tar)	CT	10	200	GLC
Naphthalene	CT	200	1000	GLC
β-Naphthylamine				
Nickel carbonyl	I	200	2000	AAS
Nickel, metal				
Nickel, soluble compounds (as Ni)	F (0.8 μm MCEF)	90	1500	AAS
Nicotine	OT	100	1000	GLC (XAD-2 resin tube)
Nitric acid	I	200	2800	C (2800 cc/min only)

Nitric oxide	OT	1	20	C (LDDT)
		1	50	(MOL sieve tube)
p-Nitroaniline	ST	15	1000	GLC
Nitrobenzene	ST	50	1000	GLC
p-Nitrochlorobenzene	ST	50	1000	GLC
Nitroethane	CT	10	200	GLC
Nitrogen dioxide	OT	1	20	(LDDT)
	OT	1	50	(MOL sieve tube)
	I	10	4000	C
Nitrogen trifluoride	P			PC
Nitroglycerine	OT	15	1000	GLC (Tenax tube)
Nitromethane	CT	10	200	GLC
1-Nitropropane	CT	10	200	GLC
2-Nitropropane	CT	10	200	GLC
Nitrotoluene	ST	20	200	GLC
Nitrotrichloromethane, *see* Chloropicrin				
Octachloronaphthalene	F (0.8 μm MCEF)	30	1000	GLC
Octane	CT	4	200	GLC
Oil mist, mineral	F (0.8 μm MCEF)	100	1500	C (fluor. spect.)
Ozone	I	45	1000	C
Paraffin wax fume	F (SM)	200	2000	G
Paraquat, respirable sizes	F (0.8 μm MCEF)	150	1500	C
Parathion	I	25	1500	GLC
	F (GF)	120	1500	GLC
Particulate polycyclic aromatic hydrocarbons (PPAH), as benzene solubles	F (0.8 μm SM)	720	1500	G (use GF prefilter)
Pentachloronaphthalene	F + I (GF)	250	1500	GLC
Pentachlorophenol	F (0.8 μm MCEF)	150	1500	GLC
Pentaerythritol	F			G (nuisance dust)
Pentane	CT	2	50	GLC
2-Pentanone	CT	10	200	GLC
Perchloroethylene	CT	1	200	GLC (10 min allowed at 300 ppm)
Perchloromethyl mercaptan	CT	30	200	GLC
Perchloryl fluoride	I	30	1000	ISE
Petroleum distillates (naphtha)				
Phenol	I	100	1000	GLC
p-Phenylene diamine	I	101	1000	C
Phenyl ether (vapour)	CT	10	200	GLC
Phenyl ether–diphenyl mixture (vapour)	ST	10	200	GLC
Phenylethylene, *see* Styrene, monomer				
Phenyl glycidyl ether	CT	50	1000	GLC
Phenylhydrazine	I	100	1000	C
Phosdrin (Mevinphos®)	I	100	1000	GLC
Phosgene (carbonyl chloride)	I	25	1000	C

Table 10.30 Cont'd

Substance	*Sampling method*[1]	*Min. sample size*[2] (l)	*Suggested max. sample rate*[3] (ml/min)	*Analytical technique*[4]
Phosphine	F (0.8 μm MCEF)	5	50	C (GF prefilter)
Phosphoric acid	F (0.8 μm MCEF)	15	1500	C (four 2 hr samples)
Phosphorus (yellow)				
Phosphorus pentachloride	I	100	1000	C
				ISE
Phosphorus trichloride	I	100	1000	C
				ISE
Phthalic anhydride	F (0.8 μm MCEF)	100	1500	GLC
Picric acid				
Pival® (2-pivalyl-1,3-indandione)	I	100	1000	GLC
Plaster of Paris	F			G (nuisance dust)
Platinum, soluble salts (as Pt)	F (0.8 μm MCEF)	720	1500	AAS
Polychlorobiphenyls, *see* Chlorodiphenyls				
Polytetrafluoroethylene decomposition products				
Propane				*see* Fluoride procedure
Propargyl alcohol	CT	30	200	GLC
n-Propyl acetate	CT	10	200	GLC
Propyl alcohol	CT	10	200	GLC
n-Propyl nitrate	CT	70	1000	GLC
Propylene dichloride (1,2-dichloropropane)	CT	10	200	GLC
Propylene glycol monomethyl ether	CT	10	200	GLC
Propylene oxide	CT	5	200	GLC
Propyne, *see* Methyl acetylene				
Pyrethrum	I	100	1000	GLC
Pyridine	CT	100	1000	GLC
Quinone	I	100	1000	C
RDX	F (0.8 μm MCEF)	150	1500	UV
Rhodium (as Rh)				
metal fume and dusts	F (0.8 μm MCEF)	720	1500	AAS
soluble salts	F (0.8 μm MCEF)	370	1500	AAS
Ronnel	I	100	1000	GLC
Rosin core solder pyrolysis products (as formaldehyde)	I	25	1000	C
Rotenone (commercial)	I	60	1000	UV
Rouge	F			G (nuisance dust)
Selenium compounds (as Se)	F (0.8 μm MCEF)	360	1500	AAS
Selenium hexafluoride (as Se)	I	60	1000	AAS
Sevin® (*see* carbaryl)				
Silica				
respirable	F (5 μm PVC)	816	1700	(cyclone, X-ray diffraction)
total				

Silicon	F (SM, 1.2, 25)			G (nuisance dust)
Silicon carbide	F			G (nuisance dust)
Silver, metal and soluble compounds (as Ag)	F (0.8 μm MCEF)	45	1500	
Sodium fluoroacetate				
Sodium hydroxide	I	350	1000	C (for titration)
Starch	F			G (nuisance dust)
Stibine	CT	100	200	AAS
				C
Stoddard solvent	CT	3	200	GLC
Strychnine	F (0.8 μm MCEF)	360	1500	UV
Styrene, monomer (phenylethylene)	CT	1	200	GLC
Succinaldehyde, *see* Glutaraldehyde				
Sucrose	F			G (nuisance dust)
Sulphur dioxide	OT	1	20	LDDT
	F + I (0.8 μm MCEF)	100	2000	C (titration)
	P			PL
Sulphur hexafluoride	I	15	1000	UV
Sulphuric acid	F (0.8 μm MCEF)	10	1500	C (titration)
Sulphur monochloride	I	15	1000	ISE
Sulphuryl fluoride	I	15	1000	GLC
2,4,5-T	I	100	1000	GLC
Tantalum	F (0.8 μm MCEF)	720	1500	AAS
TEDP	I	100	1000	GLC
Teflon® decomposition products				*see* Fluoride procedure
Tellurium and compounds (as Te)	F (0.8 μm MCEF)	670	1500	AAS
Tellurium hexafluoride (as Te)	CT	360	1000	AAS (MCEF prefilter)
TEPP	I	100	1000	GLC
Terphenyls	F (0.8 μm MCEF)	15	1000	GLC
1,1,1,2-Tetrachloro 2,2-Difluoroethane	CT	2	35	GLC
1,1,2,2-Tetrachloro 1,2-Difluoroethane	CT	2	50	GLC
1,1,2,2-Tetrachloroethane	CT	10	200	GLC
Tetrachloroethylene, *see* Perchloroethylene				
Tetrachloromethane, *see* Carbon tetrachloride				
Tetrachloronaphthalene	F + 1 (GF)	100	1300	GLC
Tetraethyl lead (as Pb)	CT	60	1000	AAS
	F + I (0.8 μm MCEF)	60	2000	C
Tetrahydrofuran	CT	9	200	GLC
Tetramethyl lead (as Pb)	CT	60	1000	AAS
	F + I (0.8 μm MCEF)	60	2000	C
Tetramethyl succinonitrile	CT	50	1000	GLC
Tetranitromethane	I	250	1000	GLC
Tetryl (2,4,6-trinitrophenylmethylnitramine)	F (0.8 μm MCEF)	100	1500	C
	F (0.8 μm MCEF)	540	1500	AAS
Thallium, soluble compounds (as Tl)	F (0.8 μm MCEF)	180	1500	AAS
Thiram	F (0.8 μm MCEF)	180	1500	C

Table 10.30 Cont'd

Substance	*Sampling method*[(1)]	*Min. sample size*[(2)] (l)	*Suggested max. sample rate*[(3)] (ml/min)	*Analytical technique*[(4)]
Tin, inorganic compounds, except SnH_4 and SnO_2 (as Sn)	F (0.8 μm MCEF)	240	1000	AAS
Tin, organic compounds (as Sn)				
Tin oxide (as Sn)	F			G (nuisance dust)
Titanium dioxide (as Ti)	F (0.8 μm MCEF)	100	1500	AAS (nuisance dust)
Toluene (toluol)	CT	2	1000	GLC (10 min allowed at 500 ppm)
Toluene-2,4-diisocyanate (TDI)	I	20	2000	C
o-Toluidine	ST	50	1000	GLC
Toxaphene, *see* Chlorinated camphene				
Tributyl phosphate	F (0.8 μm MCEF)	100	1500	GLC
1,1,1-Trichloroethane, *see* Methyl chloroform				
1,1,2-Trichloroethane	CT	10	200	GLC
Trichloroethylene	CT	10	1000	GLC (10 min allowed at 300 ppm)
Trichloromethane, *see* Chloroform				
Trichloronaphthalene	F + I (GF)	100	1300	GLC
1,2,3-Trichloropropane	CT	10	200	GLC
1,1,2-Trichloro 1,2,2-Trifluoroethane	CT	1.5	50	GLC
Triethylamine	I	100	1000	GLC
Trifluoromonobromomethane	CT	1	50	GLC (2 charcoal tubes)
Trimethyl benzene	CT	5	200	GLC
2,4,6-Trinitrophenol, *see* Picric acid				
2,4,6-Trinitrophenyl-methylnitramine, *see* Tetryl				
2,4,6-Trinitrotoluene (TNT)	F (0.8 μm MCEF)	360	1500	C
Triorthocresyl phosphate	F (0.8 μm MCEF)	100	1500	GLC
Triphenyl phosphate	F (0.8 μm MCEF)	100	1500	GLC
Tungsten and compounds (as W)				
soluble				
insoluble				
Turpentine	CT	10	200	GLC
Uranium, natural (as U)				
soluble	F (0.8 μm MCEF)	45	1500	(fluorometric)
insoluble compounds	F (0.8 μm MCEF)	45	1500	(fluorometric)
Vanadium (V_2O_5) (as V)				
dust	F (0.8 μm MCEF)	25	1700	AAS
fume				
Vinyl acetate	CT	10	200	GLC
Vinyl benzene, *see* Styrene				
Vinyl chloride	CT	5	50	GLC (use 2 tubes)
	OT	1	20	LDDT

Vinyl cyanide, *see* Acrylonitrile				
Vinylidene chloride				
Vinyl toluene	CT	10	200	GLC
VM and P naphtha				
Warfarin				
Welding fumes (total particulate) (NOC)	F (0.8 μm MCEF)	720	1500	G
Xylene (*o*-, *m*-, *p*-isomers)	CT	12	1000	GLC
Xylidene	ST	20	200	GLC
Yttrium	F (0.8 μm MCEF)	500	1500	AAS
Zinc chloride fume	F (0.8 μm MCEF)	25	1500	AAS
Zinc oxide fume	F (0.8 μm MCEF)	360	1500	AAS (also X-ray diffraction)
	F (PVC)			
Zinc stearate	F			G (nuisance dust)
Zirconium compounds (as Zr)	F (0.8 μm MCEF)	720	1500	AAS

Based on *Sampling and Analytical Guide for Airborne Health Hazards*, Du Pont Company, Applied Technology Division.

(1) *Sampling methods*

AT	Alumina tube
CT	Charoal tube or PRO-TEK™ organic vapour
F	Filter
F + I	Impinger preceded by filter
GF	Glass fibre filter
I	Impinger or fritted bubbler
M	Miscellaneous sampler (dosimeter etc.)
MCEF	Mixed cellulose ester membranes filter
OT	Other tubes (long duration detector tubes, porous polymer tubes etc.)
P	PRO-TEK™ Colorimetric Air Monitoring Badge System
PVC	Polyvinyl chloride filter
SM	Silver membrane filter
ST	Silica gel tube

(2) The minimum air sample (litres) that will provide enough of the substance for the most accurate analysis at the TLV concentrations using the analytical procedures listed.

(3) The maximum flowrate recommended for the given collection method. Sampling may be done at lower rates as long as minimum sample size is met.

(4) *Analytical technique*

AAS	Atomic adsorption spectroscopy
C	Colorimetric
G	Gravimetric
GLC	Gas–liquid chromatography
ISE	Ion-specific electrode
LDDT	Long-duration detector tubes
PC	PRO-TEK™ PT_3 Colorimetric Readout
UV	Ultraviolet

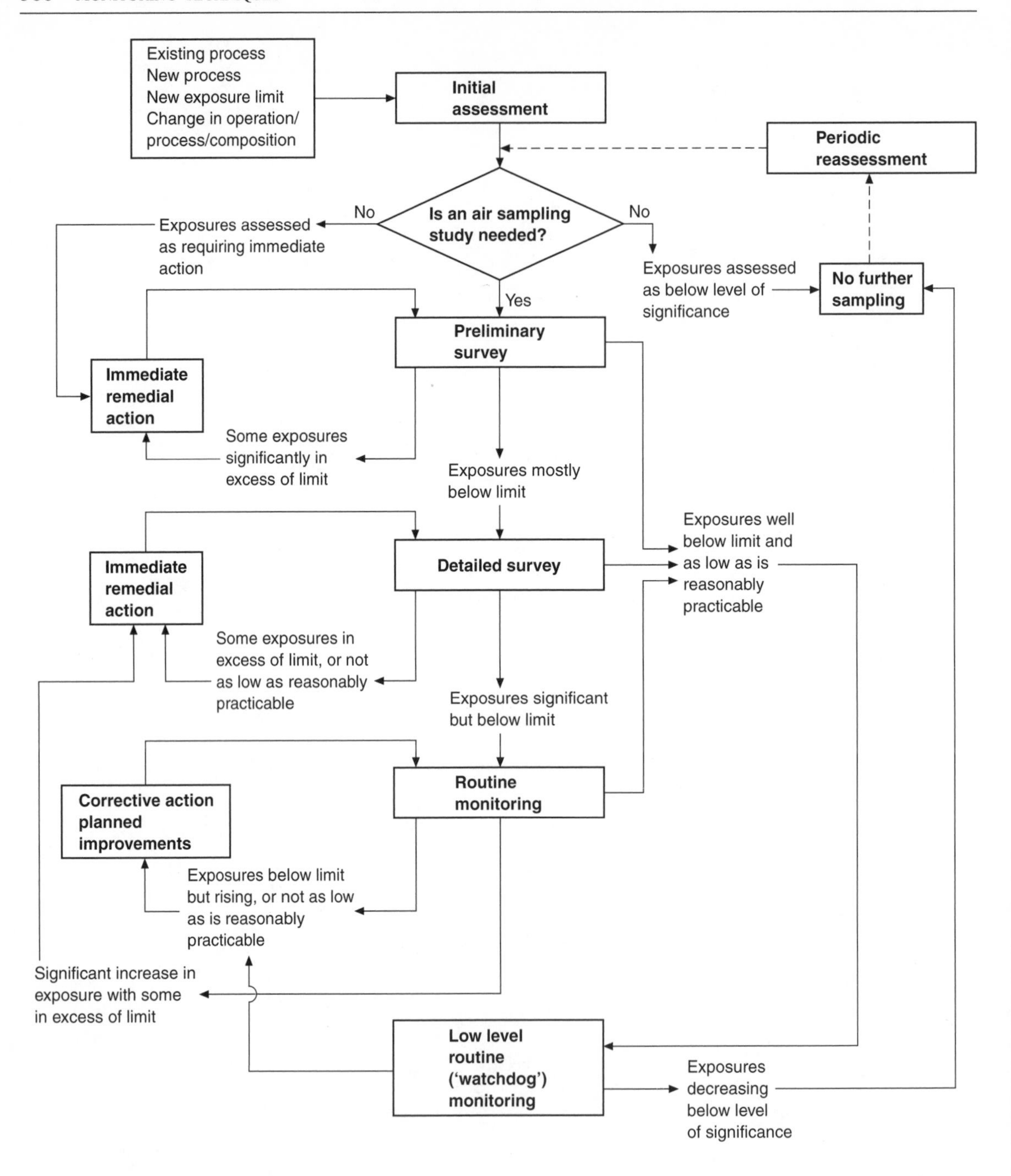

Figure 10.1 *Monitoring strategies flow diagram*

Table 10.31 Values for the range

Number of results	*k*
32	0
10–31	0.1
6–9	0.2
5	0.3
4	0.4
3	0.8
2	0.9

Table 10.32 Recommended guidance for routine monitoring

Survey	*Results*	*Action*
Preliminary	All <0.1 × OEL[1]	None if exposures are as low as reasonably practical.
	0.1–1.5 ×OEL	Investigate and carry out a more detailed survey.
	Some >1.5 ×OEL	Investigate, take remedial action and repeat survey.
Detailed	All <0.25 × OEL	None if exposure is as low as reasonably practical.
	≤1.25 × OEL	Investigate, take remedial action and repeat survey.
	Arithmetic mean <0.5 ×OEL	Consider routine monitoring and the appropriate frequency.
	Mean >0.5 × OEL (or with individual results scattered above the limit)	Investigate, assess control measures, improve where possible, repeat survey and consider routine monitoring.
Routine	All	Check values, mean, range, etc. Is there compliance with OEL and if so what is confidence?
	Differ significantly from previous survey	Investigate, consider remedial action detailed survey or change protocols.

[1] Higher values can be used if standard is based on nuisance or odour and there are no known effects of the maximum exposure concentration measured.

Pollution monitoring strategies in incident investigation

Industrial accidents involving hazardous chemicals include release of gas or vapour (including deposition to land), fire, explosion, spillages to land, and discharges to water courses (including surface waters, ground water, spring waters, saline waters, estuaries, potable waters, industrial waste waters). Fires in storage facilities housing even apparently non-toxic chemicals such as plastics can have an ecological impact because of plastic components or their pyrolytic breakdown products, e.g. PAHs, or combustion products (Table 5.48).

In such accidents it may be imperative to determine:

- the severity of the incident;
- those chemicals that may have entered the environment;
- whether long-term problems are posed.

Personal safety, e.g. during the collection of samples for accident investigation purposes is the joint responsibility of employers and employees, e.g. sampling at heights or in deep or fast-flowing waters or close to fires. For fires, the distance downwind for sampling will be dictated by the scale of the disaster as illustrated by Table 10.33 whereas for releases not involving a fire the sampling points will be dictated by the gas, i.e. whether lighter or heavier than air.

Table 10.33 Downwind distances to sample based on size of fire

Scale of fire	*Sample distances*
Major	100 m, 350 m, 700 m, 1.25 km, 2.30 km, 4.25 km,15 km, 50 km
Large	50 m, 180 m, 350 m, 600 m, 1 km, 2 km, 7.5 km, 25 km
Medium	20 m, 70 m, 250 m, 850 m, 3 km, 10 km
Small	5 m, 20 m, 60 m, 200 m, 750 m, 2.5 km
Minor	2 m, 10 m, 25 m, 85 m, 300 m, 1 km

For releases to land, influencing factors include:

- duration and extent of release;
- type of release (liquid or solid);
- hydrogeology;
- soil permeability;
- soil vulnerability;
- land use;
- screening techniques;
- levels and characteristics of historic contamination;
- depth to take sample.

The aims of sampling are to establish whether contaminants are present, their distribution and concentrations. Commonly-used sampling regimes include square grid, stratified random or simple random techniques. Evenly-spaced sampling points may be appropriate if the contamination is visible otherwise judgement is required based on whether the land slopes or is flat. Samples are also taken near to the point of release.

Table 10.34 Data to record during accident investigation

- Information about the accident, e.g. date, time, magnitude and nature of release.
- Speed of response/dates of actions.
- Sample numbers and history, e.g. date, time of collection.
- Ecosystems, structures and individuals potentially affected.
- Number of samples collected and sampling pattern.
- Sample quality and sample programme.
- Resources.
- Analytical techniques (including method, accuracy, precision, reproducibility, robustness, limits of detection).
- Sample package, storage, transport and security to ensure integrity.
- List of sample equipment and calibration record.
- Names and training records of staff involved in sampling and analysis, or details of contract analysts and any quality standards/accreditation schemes with which they comply.
- Physical and chemical properties of the chemical (*see* Chapter 12).
- The form in which it was released, and subsequent forms.
- Characteristics of the receiving ecosystem (e.g. solid type, ground water, drinking water, fish/shellfish water, bathing water) and how it migrates from one environment to another.
- Meteorology during and after the accident.
- Topography.
- Environmental, health, or safety properties of the pollutants including their acute and chronic effects and any synergistic effects.
- Visible environmental impact on site.
- Recorded quality of ecosystem prior to accident.
- Site plan with marked release points, sample points, detailed dimensions.
- Temperature of chemicals released.

Generally, sampling of waterways should be at the fastest flowing part of the stream/river, usually mid-depth unless the contaminant is less dense than water and could float, or is more dense and could accumulate near the river bed. For lakes representative samples should be taken near to the inflow, outflow and other locations. If two phases are present both may require sampling. Sample preservation by refrigeration, pH adjustment, elimination of light, filtration, and extraction may be important.

Table 10.34 summarizes the information to be recorded during an investigation. (A code of practice for the identification of potentially contaminated land and its investigation is given in British Standard DD 175/1988.) Consideration should be given to the application of appropriate quality control and quality assurance procedures such as those advocated by 'good laboratory practice' or ISO 17025 to ensure the sampling, the analysis and interpretation/reporting of data are robust since results will dictate action and may be subject to scrutiny by third parties. Duplicate samples may need to be retained for disclosure to third parties. There may be legal requirements to notify results to relevant authorities.

11

Radioactive chemicals

The main chemical elements are listed in Chapter 18. Each comprises a nucleus of positively-charged protons and neutral neutrons orbited by negative electrons. The mass number A is given by

$$A = Z + N$$

where Z is the number of protons, or atomic number
N is the number of neutrons.

Atoms with the same value of Z but different values of A are isotopes (Table 11.1). Many isotopes are stable but others are naturally or artificially radioactive, i.e. their atomic nuclei disintegrate, emitting particles or radiation. This changes the nuclear structure of the atom and often results in the production of a different element.

Table 11.1 Nuclear composition of selected istopes

Element	*Symbol*	*Atomic number*	*Protons*	*Neutrons*	*Total number of protons and neutrons*	*Atomic weight*
Hydrogen	$^{1}_{1}H$	1	1		1	1.0080
(Deuterium)	$^{2}_{1}H$	1	1	1	2	
Carbon	$^{12}_{6}C$	6	6	6	12	12.010
	$^{13}_{6}C$	6	6	7	13	
	$^{14}_{6}C$	6	6	8	14	
Nitrogen	$^{13}_{7}N$	7	7	6	13	14.008
	$^{14}_{7}N$	7	7	7	14	
	$^{15}_{7}N$	7	7	8	15	
Chlorine	$^{35}_{17}Cl$	17	17	18	35	35.457
	$^{37}_{17}Cl$	17	17	20	37	
Lead	$^{206}_{82}Pb$	82	82	124	206	207.21
	$^{207}_{82}Pb$	82	82	125	207	
	$^{208}_{82}Pb$	82	82	126	208	
Uranium	$^{234}_{92}U$	92	92	142	234	238.07
	$^{235}_{92}U$	92	92	143	235	
	$^{238}_{92}U$	92	92	146	238	

Natural sources of ionizing radiation include cosmic rays and nucleides such as potassium-40, carbon-14 and isotopes of thorium and uranium which are present in rocks, earth and building materials. Industrial sources of radiation include nuclear reactors, X-ray radiography, electron microscopy, X-ray diffractors, thickness gauges, smoke detectors, electron beam welding and certain processes including chemical analysis, polymer curing, chemical/biological tracing, food and medical sterilization, and mining. The radiation source can be sealed, when the radiation can be switched off, or unsealed. Examples of the former are smoke detectors and electrical devices for producing radiation.

Hazards

The chemistry, and hence hazards, of 'hot', or radioactive, elements parallel those of their 'cold' isotopes. However, the radiation poses additional toxicity hazards. A qualitative classification of selected isotopes in terms of toxicity is given in Table 11.2. The biological effects of ionizing radiation stem mainly from damage to individual cells following ionization of the water content. Oxidizing species, e.g. hydrogen peroxide, form together with ions and free radicals, all capable of chemical attack on important organic moieties within the cells, e.g. nucleic acids. Biological effects are influenced by the type of radiation, the dose, duration of exposure, exposed organ and route of entry. Effects on cells include death, mutation and delayed reproduction. Acute adverse effects of exposure are illustrated in Table 11.3.

Table 11.2 Classification of isotopes according to relative radiotoxicity per unit activity
The isotopes in each class are listed in order of increasing atomic number

Very high toxicity	Sr-90 + Y-90, *Pb-210 + Bi-210 (Ra D + E), Po-210, At-211, Ra-226 + 55 per cent *daughter products, Ac-227, *U-233, Pu-239, *Am-241, Cm-242.
High toxicity	Ca-45, *Fe-59, Sr-89, Y-91, Ru-106 + *Rh-106, *I-131, *Ba-140 + La-140, Ce-144 + *Pr-144, Sm-151, *Eu-154, *Tm-170, *Th-234 + *Pa-234, *natural uranium.
Moderate toxicity	*Na-22, *Na-24, P-32, S-35, Cl-36, *K-42, *Sc-46, Sc-47, *Sc-48, *V-48, *Mn-52, *Mn-54, *Mn-56, Fe-55, *Co-58, *Co-60, Ni-59, *Cu-64, *Zn-65, *Ga-72, *As-74, *As-76, *Br-82, *Rb-86, *Zr-95 + *Nb-95, *Nb-95, *Mo-99, Tc-98, *Rh-105, Pd-103 + Rh-103, *Ag-105, Ag-111, Cd-109 + *Ag-109, *Sn-113, *Te-127, *Te-129, *I-132, Cs-137 + *Ba-137, *La-140, Pr-143, Pm-147, *Ho-166, *Lu-177, *Ta-182, *W-181, *Re-183, *Ir-190, *Ir-192, Pt-191, *Pt-193, *Au-196, *Au-198, *Au-199, Tl-200, Tl-202, Tl-204, *Pb-203.
Slight toxicity	H-3, *Be-7, C-14, F-18, *Cr-51, Ge-71, *Tl-201.

*Gamma-emitter.

Types of radiation

The nature of the radioactive decay is characteristic of the element; it can be used to 'fingerprint' the substance. Decay continues until both the original element and its daughter isotopes are non-radioactive. The half-life, i.e. the time taken for half of an element's atoms to become non-radioactive, varies from millions of years for some elements to fractions of a second for others.

1. α-Particles (helium nuclei, i.e. 2 neutrons plus 2 protons): on emission the original isotope degrades into an element of two atomic numbers or less, e.g. uranium 238 produces thorium 234. Such transformations are usually accompanied by γ-radiation or X-radiation. α-Particles have a velocity about one-tenth that of light with a range in air of 3–9 cm. Because of their relatively

Table 11.3 Effects of acute exposures to X- and γ-radiation

Dose (Gy)	*Effects*
<1	No clinical effects but small depletions in normal white cells count and in platelets likely within 2 days.
1	About 15% of those exposed show symptoms of loss of appetite, nausea, vomiting, fatigue etc. 1 Gy delivered to whole body or 5 Gy delivered to bone marrow produces leukaemia.
2	Some fatalities occur.
3.5–4	LD_{50} (see Ch. 5), death occurring within 30 days. Erythema (reddening of skin) within 3 weeks.
7–10	LD_{100}, death occurring within 10 days.

large size and double positive charge they do not penetrate matter very readily and are stopped by paper, cellophane, aluminium foil and even skin. If inhaled or ingested, however, absorption of α-particles within tissues may cause intense local ionization.

2. β-Rays comprise electrons of velocity approaching that of light with a range of several metres and an energy of 0–4 MeV. β-Particles of <0.07 MeV do not penetrate the epidermis whereas those >2.5 MeV penetrate 1–2 cm of soft tissue. Thus β-emitters pose both an internal and an external radiation hazard: skin burns and malignancies can result. Once inside the body they are extremely hazardous, though less so than γ-rays. About 1 mm of aluminium is needed to stop these particles. Most β-emissions are accompanied by γ- or X-radiation and result in transformation into the element of one atomic number higher or lower but with the same atomic mass.

3. γ-Radiation is similar to, but shorter in wavelength than, X-rays and is associated with many α- or β-radiations. γ-Radiation does not transform isotopes/elements. Like X-rays, γ-rays are very penetrating; they are capable of penetrating the whole body and thus require heavy shielding, e.g. γ-rays from ^{60}Co penetrate 15 cm steel.

4. X-Radiation like γ-radiation is electromagnetic in nature. It can be emitted when β-particles react with atoms. More often it is electrically generated by accelerating electrons in a vacuum tube. The latter source can be switched off. X-rays are extremely penetrating and are merely attenuated by distance and shielding.

5. Neutron radiation is emitted in fission and generally not spontaneously, although a few heavy radionucleides, e.g. plutonium, undergo spontaneous fission. More often it results from bombarding beryllium atoms with an α-emitter. Neutron radiation decays into protons and electrons with a half-life of about 12 min and is extremely penetrating.

The same type of radiation emitted by different isotopes may differ significantly in energy, e.g. γ-radiation from potassium-42 has about four times the energy of γ-radiation from gold-198.

Units of radiation are the becquerel (Bq), the gray (Gy) and the sievert (Sv).

Control measures

The control of ionizing radiation is heavily regulated. Expert advice should be sought prior to introducing sources of radiation onto the premises. The general provisos for their control are that:

- All practices resulting in exposure shall be justified by the advantages produced.
- All exposures shall be as low as reasonably practicable.

- The dose received shall not exceed specified limits. As with most hygiene standards these limits vary slightly between nations: local values should be consulted. Limits set within the UK are summarized in Table 11.4.
- All regulatory requirements will be followed.
- All incidents will be investigated and reported.

Table 11.4 UK exposure limits (the Ionizing Radiations Regulations, 1999)

Dose (mSv in any calendar year)	*Employee aged 18 or over*	*Trainee under 18*	*All others*
Whole body	20*	6	1
Equivalent dose for the			
Lens of eye	150	50	15
Skin**	500	150	50
Arms, forearms, feet, ankles	500	150	50
Abdomen of women of child bearing capacity at work (in any consecutive period of 3 months)	13***		

* Where these limits are impracticable having regard to the nature of the work the employer may apply a dose limit of 100 mSv in any period of 5 consecutive months subject to a maximum effective dose of 50 mSv in any single calendar year, and to prior approval by the Radiation Protection Adviser, the affected employee(s), and the Health and Safety Executive.
** As applied to the dose averaged over an area of 1 cm^2 regardless of area exposed.
*** Once an employer has been informed that an employee is pregnant the equivalent dose to a foetus should not exceed 1 mSv during the remainder of the pregnancy and significant bodily contamination of breast-feeding employees must be prevented.

In the UK, annual doses of 20 mSv, or any case of suspected overexposure, must be investigated, reported and recorded. The HSE must also be notified of any spillages of radioactive substance beyond specified amounts. Companies are obliged to monitor exposures and investigate excursions beyond action limits, and to maintain records for specified periods. Checks on surface contamination are aimed at avoiding exposure, preventing spread of contaminant, detecting failures in containment or departures from good practice, and providing data for planning further monitoring programmes.

Air monitoring will be required, e.g., when volatiles are handled in quantity, where use of radioactive isotopes has led to unacceptable workplace contamination, when processing plutonium or other transuranic elements, when handling unsealed sources in hospitals in therapeutic amounts, and in the use of 'hot' cells/reactors and critical facilities. Routine monitoring of skin, notably the hands, may be required.

Monitoring for both external and internal radiation may be required, e.g. when handling large quantities of volatile 'hot' chemicals or in the commercial manufacture of radionucleides, in natural and enriched uranium processes, in the processing of plutonium or other transuranic elements, and in uranium milling and refining. The nature of biological monitoring is influenced by the isotope: e.g., faeces, urine and breath monitoring is used for α- and β-emitters and whole-body monitoring for γ-sources.

Exposure is minimized by choice of source, by duration of exposure, by distance from source (at 1 m the radiation level is reduced almost 10-fold), and by shielding. The greater the mass per unit area of shield material the greater the shielding efficiency. Whereas α- and β-particles pose few problems (the former can be absorbed by, e.g., paper and the latter by 1 cm Perspex) γ- and X-rays are not completely absorbed by shield material but attenuated exponentially such that radiation emerging from the shield is given by:

$$D_t = D_0 e^{-ut}$$

where D_0 is the dose without a shield
D_t is the dose rate emerging from a shield of thickness t
u is the linear absorption coefficient of shield material.
Half-thickness values (H-TV) i.e. thickness to reduce intensity to half the incidence value, for materials commonly used as shields for selected γ-rays are exemplified by Table 11.5.

Table 11.5 Approximate half-thickness values for a selection of shield materials and γ-emitters

Nucleide	*Half-life*	*γ-energy* (MeV)	*H-TV* (cm) *Concrete*	*Steel*	*Lead*
^{137}Cs	27 years	0.66	4.82	1.63	0.65
^{60}Co	5.24 years	1.17–1.33	6.68	2.08	1.20
^{198}Au	2.7 days	0.41	4.06	–	0.33
^{192}Ir	74 days	0.13–1.06	4.32	1.27	0.60
^{226}Ra	1622 years	0.047–2.4	6.86	2.24	1.66

Detailed precautions for handling radioactive substances will be dictated by the nature and quantity of isotope and the likely level of exposure. Thus for some materials laboratory coats and gloves may be adequate; for others a fully enclosed suit and respirator may be more appropriate. Some general precautions are listed in Table 11.6.

Table 11.6 General control measures for work with radioactive substances

Consult experts including competent authorities (in UK the HSE must be given 28 days prior notice of specified work with ionizing radiation).
Conduct a risk assessment to any employee and other persons to identify measures needed to restrict exposure to ionizing radiation and to assess magnitude of risk including identifiable accidents.
Conduct work in designated controlled areas (e.g. in UK these are areas in which instantaneous dose rates >7.5 μSv/hour occur, or where employees may exceed 6 mSv annual dose limit, or where air concentration or surface contamination exceeds specified levels).
Provide barriers for identification and display of appropriate warning notices, e.g. Trefoil symbol.
Control exposures by engineering techniques, e.g. containment, shielding, ventilation (consider need for in-duct filters to remove contamination prior to exhausting to atmosphere), backed up by systems of work and personal protection including approved respirators where necessary.
Use remote handling techniques where necessary.
Appoint a Radiation Protection Adviser: all staff involved with radioactive work should be adequately trained and instructed.
Limit access to designated areas to classified persons (e.g. in UK persons likely to receive doses in excess of 6 mSv per year or an equivalent dose which exceeds 30% of relevant hygiene standard). Access may need to be limited by trapped keys or interlocks for high dose rate enclosures.
Prepare as appropriate written rules for work in designated areas and appoint Radiation Protection Supervisors.
Check exposures routinely by personal dosimetry or following accidents (in the UK dosimetry services must be approved) and keep records (e.g. in UK for at least 50 years). Notify the relevant employees and authorities as appropriate. Monitor background contamination periodically using equipment that has been checked by qualified persons and keep records of levels (e.g. for 2 years).
Investigate accidents which may have led to persons receiving effective doses in excess of 6 mSv or an equivalent dose greater than 30% of any relevant dose limit. Investigate and report to the authorities loss of materials from accidental release to atmosphere, spillages, theft. The Regulations provide a comprehensive list of notifiable concentrations for each radionuclide isotope.
Provide mandatory medical surveillance for classified workers, e.g. medical examinations prior to commencement of radioactive work followed by check-ups annually or when overexposure may have occurred. (In the UK the surveillance must be undertaken by appointed doctors and records retained for at least 50 years.)
Maintain high standards of personal hygiene and housekeeping.

Table 11.6 Cont'd

Do not eat, drink, smoke, apply cosmetics or use mouth pipettes in controlled areas. Dress any wounds prior to entering the area.
Wherever practicable, store radioactive substances in sealed, properly labelled containers. Check for leaks periodically (e.g. 2-yearly intervals) and maintain records of stocks including sealed sources (e.g. for 2 years).
Carry out work over spill trays to contain leakages and use impervious work surfaces.
Decontaminate apparatus and prevent cross-contamination.
Collect waste for treatment or disposal and deal with spillages immediately.
Remove protective clothing in a changing area provided with wash basin, and lockers for clean and dirty clothing.

12

Safety by design

Plant and equipment design are regulated by substantial legislation. In the UK this includes: The Health and Safety at Work etc. Act 1974, the Provision and Use of Work Equipment Regulations 1998, the Control of Substances Hazardous to Health Regulations 1999, the Factories Act 1961, the Electricity at Work Regulations 1989, the Fire Precautions Act 1971, together with specific legislation, e.g. the Highly Flammable Liquids and LPG Regulations 1972.

Common matters, e.g. ventilation, temperature and lighting; floor, wall and ceiling surfaces; workspace allocation; workstation design and arrangement; floors and traffic routes; safeguards against falls or being struck by a falling object; glazing; doors and gates; travelators and escalators; sanitary and washing facilities; drinking water supply; accommodation for clothing; facilities for changing, resting and meals; are all covered by the Workplace (Health, Safety and Welfare Regulations) 1992.

Design procedures

To ensure safety consult flowsheets/engineering line diagrams and consider both the materials (raw materials storage, processing, product storage, disposal and transportation) and the process details (scale, batch vs continuous, temperature, pressure, materials of construction, monitoring, safety features, e.g. fail-safe or 'second chance' design). See Table 12.1. Subject the proposals to detailed scrutiny, as in Table 12.2 or using a HAZOP study, fault tree analysis, etc. for both the planned operation and anticipated major deviations from normal operation (Table 12.3).

A HAZOP (Hazard and Operability Study) involves a formal review of process and instrumentation diagrams by a specialist team using a structured technique, based upon key words. These comprise 'property words' and 'guide words', e.g. as in Table 12.4.

Where possible plants of intrinsically safe design are preferred, i.e. those which have been designed to be 'self-correcting' rather than those where equipment has been 'added on' to control hazards. Some characteristics of intrinsically safe plants are:

- Low inventory (small plant with less inherently hazardous materials on site).
- Substitution of hazardous materials with less dangerous chemicals.
- Attenuation of risk by using hazardous materials in the least dangerous form.
- Simplification of plant, instrumentation, operating procedures to reduce the chance of human error.
- Domino effects eliminated so that adverse events are self-terminated and do not initiate new events.
- Incorrect assembly of plant made impossible by equipment design.

Table 12.1 Some considerations in reaction process selection and design

Are unstable reactions and side reactions possible, e.g. spontaneous combustion or polymerization?
Could poor mixing or inefficient distribution of reactants and heat sources result in undesirable side reactions, hot spots, runaway reactions, fouling, etc?
Can hazards from the reaction be reduced by changing the relative concentration of reactants or other operating conditions?
Can side reactions produce toxic or explosive material, or cause dangerous fouling?
Will materials absorb moisture from the air and then swell, adhere to surfaces, form toxic or corrosive liquid or gas, etc?
What is the effect of impurities on chemical reactions and upon process mixture characteristics?
Are materials of construction compatible mutually and with process materials?
Can dangerous materials build up in the process, e.g. traces of combustible and non-condensible materials?
What are the effects of catalyst behaviour, e.g. aging, poisoning, disintegration, activation, regeneration?
Are inherently hazardous operations involved:
  Vaporization and diffusion of flammable/toxic liquids or gases?
  Dusting and dispersion of combustible/toxic solids?
  Spraying, misting or fogging of flammable/combustible materials or strong oxidizing agents?
  Mixing of flammable materials and combustible solids with strong oxidizing agents?
  Separation of hazardous chemicals from inerts or diluents?
  Increase in temperature and/or pressure of unstable liquids?

- The status of plant should be immediately obvious.
- The tolerance should be such that small mistakes do not lead to major problems.
- Leaks should be small.

Table 12.5 summarizes the application of these principles (after Kletz – see Bibliography). Where this approach is not feasible, external features of plant must ensure the minimization of unwanted consequences.

Layout

Factory layout has a significant bearing on safety. Relevant considerations include:

- Relative positions of storage and process areas; control room, laboratories and offices – i.e. areas of highest population density; switch-house; materials receipt and despatch areas; effluent treatment facilities. Spacing distances according to standard guidelines.
- Need for normal and emergency access (and escape).
- Security, e.g. fencing requirement, control of access.
- Topography.
- Tendency for flooding.
- Location of public roads.
- Prevailing wind direction.
- Zoning of electrical equipment (Table 12.6).
- Positions of neighbouring developments, housing, public roads, controlled water courses, etc.

Segregation is practised to allow for housekeeping, construction and maintenance requirements and to reduce the risk of an accident resulting in a 'domino effect', e.g. from a fire, explosion or toxic release. For very toxic substances, e.g. prussic acid (HCN) or tetraethyl lead, this may involve isolating the entire manufacturing operation in a separate unoccupied building or sealed-off area.

Table 12.2 Chemical process hazard identification

Materials and reaction	Identify all hazardous process materials, intermediates and wastes Produce material information sheets for each process material Check the toxicity of process materials, identify short and long term effects for various modes of entry into the body and different exposure tolerance Identify the relationship between odour and toxicity for all process materials Determine the means for industrial hygiene recognition, evaluation and control Determine relevant physical properties of process materials under all process conditions, check source and reliability of data Determine the quantities and physical states of material at each stage of production, handling and storage, relate these to the danger and second-degree hazards Identify any hazard the product might present to transporters and public while in transit Consult process material supplier regarding properties, characteristics, safety in storage, handling and use Identify *all* possible chemical reactions, both planned and unplanned Determine the inter-dependence of reaction rate and variables, establish the limiting values to prevent undesirable reactions, excessive heat development etc. Ensure that unstable chemicals are handled so as to minimize their exposure to heat, pressure, shock and friction Are the construction materials compatible with each other and with the chemical process materials, under all foreseeable conditions? Can hazardous materials build-up in the process, e.g. traces of combustible and noncondensible materials?
General process specification	Are the scale, type and integration of the process correct, bearing in mind the safety and health hazards? Identify the major safety hazards and eliminate them, if possible Locate critical areas on the flow diagrams and layout drawings Is selection of a specific process route, or other design option, more appropriate on safety grounds? Can the process sequence be changed to improve the safety of the process? Could less hazardous materials be used? Are emissions of material necessary? Are necessary emissions discharged safely and in accordance with good practice and legislation? Can any unit or item be eliminated and does this improve safety, e.g. by reducing inventory or improving reliability? Is the process design correct? Are normal conditions described adequately? Are all relevant parameters controlled? Are the operations and heat transfer facilities properly designed, instrumented and controlled? Has scale-up of the process been carried out correctly? Does the process fail safe in respect of heat, pressure, fire and explosion? Has second chance design been used?

Spacing distances may be reduced in the light of:

- explosion relief, blast-proofing, blast walls, earth banks;
- bunds, dykes;
- steam and water curtains; foam blanketing provisions;
- inter-positioning of sacrificial plant, e.g. cooling towers, unpopulated buildings;
- provision of refuges, e.g. for toxic release incidents.

Adequate distance frequently serves to mitigate the consequences of an accidental release of chemicals, e.g. a flammable liquid spillage or toxic gas escape.

Distances are recommended for zoning of electrical equipment, separation of storage from buildings etc. Distances are also proposed (on the basis of experience) to minimize the escalation

Table 12.3 Checklist for major deviations from normal operation

Start-up and shutdown;
What else, apart from normal operation, can happen?
Is suitable provision made for such events?
Can start-up and shutdown of plant, or placing of plant on hot standby, be expedited easily and safely?
Can the plant pressure or the inventory of process materials, or both, be reduced effectively and safely in a major emergency?
Are the operating parameter limits which require remedial action, known and measured, e.g. temperature, pressure, flow, concentration?
Should plant be shutdown for any deviation beyond the operating limits?
Does this require the installation of alarm, trip, or both, i.e. to what extern is manual intervention expected?
Does material change phase from its state in normal operation, during the start-up, and shutdown of plant? Is this acceptable, e.g. does it involve expansion or contraction, solidification, etc.?
Can effluent and relief systems cope with large or abnormal discharges, during start-up, shutdown, hot standby, commissioning and fire-fighting?
Are adequate supplies of utilities and miscellaneous chemicals available for all activities, e.g. absorbents for spillage control?
Is inert gas immediately available in all locations where it may be required urgently?
Is there a standby supply?
Is any material added during start-up and shutdown, which can create a hazard on contact with process or plant materials?
Is the means of lighting flames, e.g. on burners and flares, safe on every occasion?

Table 12.4 Key words for use in HAZOP studies

Typical property words

Common	*Application-dependent*
Flow	Viscosity
Temperature	Flash point
Pressure	Vapour pressure
Level	Heat transfer
Concentration	Separate
	Absorb
	etc.

Guide words for HAZOP studies

Guide words	*Meaning*	*Comment*
NO/NOT	The complete negation of these intentions	No part of the intentions is achieved but nothing else happens
MORE/LESS	Quantitative increase or decrease	Refer to quantities and properties such as flowrates and temperatures as well as activities like 'heat', 'react'
AS WELL AS	Qualitative increase	All the design and operating intentions are achieved together with some additional activity
PART OF	Qualitative decrease	Only some of the intentions are achieved: some are not
REVERSE	The logical opposite of the intention	Mostly applicable to activities, e.g. reverse flow, chemical reaction. Can also be applied to substances, e.g. 'poison' instead of 'antidote', 'D' instead of 'L' optical isomers
OTHER THAN	Complete substitution	No part of the original intention is achieved; something quite different happens

Table 12.5 Features of intrinsically safe plants

	Examples	
Characteristic	*Friendliness*	*Hostility*
Low inventory		
Heat transfer	Miniaturized	Large
Intermediate storage	Small or nil	Large
Reaction	Vapour phase	Liquid phase
	Tubular reactor	Pot reactor
Liquids separation	Centrifugal	Gravity settling
Substitution		
Heat transfer media	Non-flammable	Flammable
Solvents	Non-flammable/low toxicity	Flammable/high toxicity
Attenuation		
Liquefied gases	Refrigerated	Under pressure
Explosive powders	Slurried	Dry
Runaway reactants	Diluted	Neat
Any material	Vapour, superheated liquid	Liquid
Any process	Low pressure	High pressure/vacuum
	Moderate temperature	High or very low temperature
Simplification		
	Hazards avoided	Hazards controlled by added equipment
	Single stream	Multistream with many crossovers
	Dedicated plant	Multipurpose plant
Liquid or powder transfer	One big plant	Many small plants
	By gravity	Pressurized
Domino effects		
	Open construction	Closed buildings
	Fire breaks	No fire breaks
Tank roof	Weak seam	Strong seam
Horizontal cylinder	Pointing away from other equipment and buildings	Pointing at other equipment or buildings
Incorrect assembly impossible		
Compressor valves	Non-interchangeable	Interchangeable
Device for adding water to oil	Cannot point upstream	Can point upstream
Obvious state		
	Rising spindle valve or ball valve with fixed handle	Non-rising spindle valve
	Figure-8 plate	Spade
Tolerant of maloperation or poor maintenance	Continuous plant	Batch plant
	Spiral-wound gasket	Fibre gasket
	Expansion loop	Bellows
	Fixed pipe	Hose
	Articulated arm	Hose
	Bolted arm	Quick-release coupling
	Metal	Glass, plastic
	Self-sealing couplings	Standard couplings
Low leak rate	Spiral-wound gasket	Fibre gasket
	Tubular reactor	Pot reactor
	Vapour-phase reactor	Liquid-phase reactor

Table 12.6 Electrical zoning

Classification of hazard areas according to the probability of a flammable concentration of vapour occurring (to BS 5345 Part 1)

Zone 0	Area in which an explosive gas–air mixture is continuously present, or present for long periods
Zone 1	Area in which an explosive gas–air mixture is likely to occur in normal operation
Zone 2	Area in which an explosive gas–air mixture is not likely to occur in normal operation, and if it occurs will exist only for a short time
Safe area	By implication, an area that is not classified Zone 0, 1 or 2 is deemed to be a non-hazardous or safe area with respect to BS 5345

Examples of electrical area classification for various operations

Vapour source	*Extent of classified area*	*Area classification*
Vapour space inside storage tank	Within the vapour space of the tank	Zone 0
Storage tank outside buildings	Vertically from ground level up to 2 m above the tank connections and horizontally within 2 m from the tank connections or shell	Zone 2
Discharge from vent line	(a) Where liquids or vapours discharged from the vent may impinge	Fixed electrical equipment should not be installed
	(b) Within 2 m in all other directions from point of discharge	Zone 2
Tank vehicle loading	(a) Vertically from ground level up to 2 m above, and horizontally outwards for 2 m from any point where connections are regularly made or disconnected for product transfer	Zone 1
	(b) Vertically and horizontally between 2 m and 4 m from the points of connection or disconnection	Zone 2
Pumps and sample points in the open air	Within 2 m in all directions	Zone 2
Pump house building	Within the building	Zone 1

Notes
1. Where any area is classified under more than one factor, the higher classification should prevail.
2. Any bunded area, pit, trench, depression or drain falling within a Zone 1 or Zone 2 area should be treated as being a Zone 1 area throughout.
3. Pump seals should be properly maintained.

or effects on site of fire, explosion, toxic release or similar incident. Selected sources of information are summarized in Table 12.7.

Storage

Chemicals in packages

The design of any building or outside compound for the storage of chemicals in packages (e.g. drums, cylinders, sacks) will depend upon their hazardous characteristics (pages 228, 248 and 272).

For a storage building the considerations include:

- Siting to minimize risk to nearby premises on and off site in a fire.
- Access for delivery and transfer of chemicals and for emergency purposes.
- Fire-proof construction.

Table 12.7 Selected sources of spacing distances with hazardous chemicals (see Bibliography)

Preliminary minimum distances	
Liquid oxygen	*A Code of Practice for the Bulk Storage of Liquid Oxygen at Production Sites* (HSE, 1977)
Liquefied flammable gases	*Process Plant Layout* page 562 (Mecklenburgh, 1985)
Liquids stored at ambient temperature and pressure	*Process Plant Layout* page 564
Electrical area classification distances	*Process Plant Layout* pages 568–577
Distances for storage of explosives	*Explosive and Toxic Hazard Materials* page 370 (Meidl, 1970) *Safe Handling Requirements during Explosive, Propellant and Pyrotechnic Manufacture* (HSE, SIR 31)
General recommendations for spacing	*General Recommendations for Spacing in Refineries, Petrochemical Plants, Gasoline Plants, Terminals, Oil Pump Stations and Offshore Properties* (Oil Insurance Association, No. 361)
'Consultation' distances (in relation to major hazard and other sites, e.g. LPG, chlorine, for land use planning)	HSE assessments for consultation distances for major hazard installations, Chapter 5 in *Safety Cases* (Lees and Ang, 1989)
Minimum recommended spacing distances	
Flammable liquids	*Storage of Flammable Liquids in Tanks* (HSE, HSG 176) *Storage of Flammable Liquids in Containers* (HSE, HSG 51) *Storage of Highly Flammable Liquids* (HSE, CS2) *Highly Flammable Materials on Construction Sites* (HSE, HSG 3) *Storage of Flammable Liquids in Fixed Tanks, up to 10 000 m^3 Total Capacity* (HSE, HSG 50)
Liquefied petroleum gas	*Storage of LPG at Fixed Installations* (HSE, HSG 34) *Storage and Use of LPG at Metered Estates* (HSE, CS11) *Storage and Use of LPG on Construction Sites* (HSE, CS6) *Keeping of LPG in Cylinders and Similar Containers* (HSE, CS4)
Distances for storage and handling of dangerous chemicals	*Storage and Handling of Organic Peroxides* (HSE, CS21) *Chemical Warehousing Storage of Packaged Dangerous Substances* (HSE, HS(G) 71) *Safety Advice for Bulk Chlorine Installations* (HSE, HSG 28) *Storage and Handling of Industrial Nitrocellulose* (HSE, HSG 135) *Storage and Handling of Ammonium Nitrate* (HSE, CS18)

- Security, i.e. control of access by staff and means to prevent entry by vandals or trespassers.
- General ventilation provisions, particularly if volatile toxic or highly flammable liquids, or toxic or flammable pressurized gases are stored.
- Lighting, and emergency lighting, provisions.
- Correct selection of heating equipment and zoning of electrical equipment to reduce the chance of an ignition source arising.
- Provision of fire detection and extinguishment equipment.

The same factors arise for a dedicated compartment of a building. In either case identification and warning notices need to be provided.

Operation of the store (see Chapter 13) then needs to account for:

- A system for checking that all packages entering the store have identifiable labels which indicate their contents and any hazardous characteristics.
- A system for inspecting all packages received, and routinely in-store, for leaks, damage, external corrosion, etc.

- A system for the segregation of chemicals, following the principles explained on page 136 and 248.
- In-store handling of packages. Manual handling should be eliminated or reduced as far as possible. Fork-lift trucks should be regularly maintained, be provided with adequate access ways; driving should be restricted to fully-trained personnel.
- Safe stacking practice. This will include provision of suitable racking, limitations on stack size and height, e.g. having regard to the potential severity of a fire.
- Specification of a storage capacity and a procedure to avoid overfilling.
- Good housekeeping.
- Fire precautions including control of ignition sources, e.g. smoking, maintenance activities, vehicular access (see Chapter 6) and limitation of combustible materials, e.g. packaging.

Drum storage

Drums containing flammable liquids are preferably stored outside, so that any flammable vapour can readily disperse. Similar considerations may apply to the dispersion of any vapour/fumes from drums of toxic liquids or solids. In some cases weather protection is provided by a roof.

If outdoor storage is not reasonably practical a specially designed storeroom, preferably in a separate building, may be used.

A summary of recommendations for outdoor drum storage is given in Table 12.8.

Table 12.8 Outdoor drum storage recommendations

- If possible, situate area remote from buildings and plant. Avoid location beneath pipe bridges or electric cable runs.
- Secure the area to prevent tampering or trespassing.
- Limit permissible fire load to 79×10^5 kJ or 250 tons of hydrocarbon.
- Label each drum; affix appropriate warning, e.g. 'Highly Flammable', 'Corrosive'.
- Provide suitable absorbent materials to deal with spillages.
- Limit stack heights, e.g. 4.5 m for 200 litres drums., 5 m high for drums stored on end or 4 m high on pallets, 4 m high for drums on their sides.
- Segregate highly flammable liquids and mark the stack; classify the area as Zone 2. Place material >15 m from any working building, amenity building or plant and >7.5 m from plant boundary and boundary fence.
- Space combustible material stacks >7.5 m from buildings and 4 m from plant boundary fences.
- Limit areas for any stack of drums to 250 m^2 with generally a maximum length of 18.5 m. (A reduced area applies to particularly hazardous chemicals.)
- Restrict number of 180 litre drums to 1500.
- Ground should be impervious, and sloped and drained, with demarcation lines.
- Spillages from this area should be contained in a bund, or similar arrangement.
- Provide >5 m clearance between adjacent stacks with access on three sides for fire-fighting, etc.
- Provide an adequate number of fire hydrants for fixed monitors for drenching the stacks with water. Stack drums 2 m from any hydrants and leave clear access.
- Leave space for possible hose runs for a fire in any stack and provide dry powder extinguishers around the area.

LPG in cylinders

The risk with liquefied petroleum gas in cylinders is significantly greater than with a highly flammable liquid in drums because of the potential for rapid release of heavy flammable gas. In a fire around a cylinder there is a potential BLEVE hazard (see p. 178); ignition of a leak from a valve will cause a jet fire.

Therefore detailed recommendations relate to the keeping of liquefied petroleum gas in cylinders and similar containers. General recommendations for storage, other than on rooftops, are summarized in Tables 9.16–9.18.

Equipment design

The design and instrumentation of individual equipment items are beyond the scope of this text. It is important to note, however, that:

- All work equipment must be constructed or adapted to be suitable for its use, and its selection should have regard to the working conditions and risks to health and safety of persons where it is used and any additional risks it poses (Provision and Use of Work Equipment 1998 Reg. 2).
- In addition to guarding dangerous machinery (Reg. 11) measures are required to prevent or adequately control hazards from work equipment due to:
  - any article or substance falling or being ejected;
  - rupture or disintegration of parts;
  - it catching fire or overheating;
  - the unintended discharge or premature discharge of any gas, dust, liquid, vapour or other substance; consideration should also be given to fugitive emissions or leaks from valves, flanges, connections, seals, pumps, agitators, etc. by, for example, using average emission factors for common plant equipment such as those given in Table 12.9. (Such assessments may be important in reducing material loss or in estimating potential employee exposure or environmental pollution levels);
  - the unintended or premature explosion of it or any substance in it (Reg. 12).
- The selection of materials of construction is subject to the considerations of corrosion summarized on page 54.
- Consideration should be given to the safety features listed in Table 7.21.
- For any pressure system there are obligatory requirements summarized on page 423.

Table 12.9 Average emission factors for plant items

Equipment	*Service*	*Emission factors (kg/hr)**	
		Refineries	*Synthetic organic chemical manufacturing*
Valves	Gas	0.27	0.0056
	Light liquid	0.11	0.0071
	Heavy liquid	0.00022	0.00023
Pump seals	Light liquid	0.11	0.0494
	Heavy liquid	0.021	0.0214
Compressors	Gas/vapour	0.64	0.228
Pressure relief valves	Gas/vapour	0.16	0.104
Flanges	All	0.00025	0.00083
Open-ended lines	All	0.0023	0.0017
Sampling connections	All	0.015	0.015

*Multiplying the total number of items in a unit by the factor provides an indication of the emission rate.

Piping arrangements

Complex piping systems may be required for the transfer of chemicals, balancing of pressures, venting, drainage, supply of services, etc.

Safe operation is generally assisted by simplification of piping and valve arrangements, and by their identification, e.g. by colour coding or tags. Logical arrangement can also serve as a prompt in identification.

Design of piping systems is a special discipline but some guidance is given in Table 12.10.

Table 12.10 Design measures to reduce leaks from pipework

- Minimize the number of branches and deadlegs.
- Minimize the number of small drain lines.
- Design small diameter lines to the same codes, and test to the same standards, as main lines; reinforce junctions as necessary.
- Provide flexibility to allow for thermal expansion, or contraction, of pipework and connected equipment.
- Direct discharges from automatic drains to visible locations.
- Provide gaskets compatible with the internal fluid over the full range of temperatures and pressures.
- Provide removable plugs on valve sample points.
- Provide adequate pipe supports.
- Minimize the number of flanges on vacuum lines.
- Consider duplication, locking, temporary blanking of valves on 'open' discharge lines.
- Either avoid pipe runs along routes subject to possible mechanical damage, e.g. from fork-lift trucks or tankers, or provide shields and barriers.
- Provide walkways, ladders, etc. to avoid pipework being clambered over.
- Minimize the use of flexible hoses; if used:
  - ensure hose and couplings are to the appropriate standard;
  - provide for rapid isolation in any emergency;
  - use bolted, not jubilee, clips;
  - consider self-sealing couplings;
  - provide adequate support and protection whilst in use;
  - protect from crushing or contamination when not in use.

Services and utilities

Ensure the adequacy (in terms of quality, quantity and reliability) of services/utilities, e.g. steam, process/cooling water, electricity, compressed air, inert gas, fire suppression systems, ventilation. Stand-by or emergency services may be required. Some general safety design considerations are summarized in Table 12.11.

Ventilation can be provided as general dilution ventilation or local extraction ventilation.

Dilution ventilation (general ventilation)

Open construction is preferable wherever practicable for areas processing hazardous chemicals, to provide general ventilation and assist in the dispersion of leaking gas or vapour, to maximize explosion-venting area and to facilitate fire-fighting. Local dilution ventilation is provided to flush the workplace atmosphere with clean air and thus dilute the level of contaminant in ambient air to acceptable levels. The siting of the extract fans and the air inlets require careful consideration to minimize operator exposure (Figure 12.1). The general rules are:

- Site the exhaust fan near to the source of contaminant.
- Ensure that fresh air movement is from worker to leak source and not vice versa.
- Ensure that air inlet supply is not contaminated with exhaust effluent.
- Provide back-up for air inlet where necessary.

Table 12.11 Safety considerations for plant services design

- The need to duplicate power supplies to equipment.
- The need to duplicate or triplicate equipment.
- Selection between steam, or electricity, or a combination for pumps and compressors.
- Provision of automatic start for spare pumps.
- Provision of voltage protection for key equipment which must be kept on-line or be restarted quickly.
- Order for restarting equipment after a power failure.
- Extent of emergency power supplies for lighting, communication systems, and key items of equipment (e.g. cooling facilities, reactor agitators, exhaust ventilation) and instruments/alarms.
- Provision for cooling emergency equipment, e.g. diesel generators if cooling water is disrupted.
- Required positioning of control valves (i.e. shut, open or 'as is') upon power failure.
- Provision of compressed air, inert gas from reservoirs for a limited period.
- Provision for alternative fire water supply.
- Provision of efficient drift eliminators of water cooling towers; consideration of replacement by air cooling systems.
- Thermal insulation to protect personnel from contact with hot or cold surfaces; prevention of water supply disruption by freezing.
- Design of hot and cold water services to avoid water standing undisturbed for long periods. Use of covered tanks and cisterns with approved fittings and materials.
- Avoidance of cold water temperatures of 20°C–45°C, storage of hot water at 60°C and circulation at 50°C.

For continuous release of gas or vapour the steady-state dilution ventilation required to reduce the atmospheric pollutant to a level below its hygiene standard is given by

$$Q = \frac{3.34 \times 10 \times SG \times ER \times K}{MW \times HS}$$

where Q is the ventilation rate (m/min)
SG is the specific gravity of the evaporating liquid
ER is the rate of evaporation (l/hr)
MW is the molecular weight
K is the design factor
HS is the hygiene standard.

In general, dilution ventilation alone is inappropriate for highly-toxic substances, carcinogens, dusts or fumes or for widely fluctuating levels of pollutants. Since hygiene standards are often revised (usually downwards), specifications of existing systems may prove inadequate. An inherent problem is that contaminants may have to pass through a worker's breathing zone before dilution.

Local exhaust ventilation

Local exhaust ventilation serves to remove a contaminant near its source of emission into the atmosphere of a workplace. A system normally comprises a hood, ducting which conveys exhausted air and contaminants, a fan, equipment for contaminants collection/removal and a stack for dispersion of decontaminated air. Hoods normally comprise an enclosure, a booth, a captor hood or a receptor hood. Those relying on other than complete enclosure should be as close as practicable to the source of pollution to achieve maximum efficiency.

Total enclosure may be in the form of a room with grilles to facilitate air flow; this functions as a hood and operates under a slight negative pressure with controls located externally. Entry is restricted and usually entails use of comprehensive personal protective equipment. Ancillary requirements may include air filters/scrubbers, atmospheric monitoring, decontamination procedures and a permit-to-work system (see page 417).

Partial enclosure allows small openings for charging/removal of apparatus and chemicals. The requisite air velocity to prevent dust or fumes leaking out determines the air extraction rate, e.g.

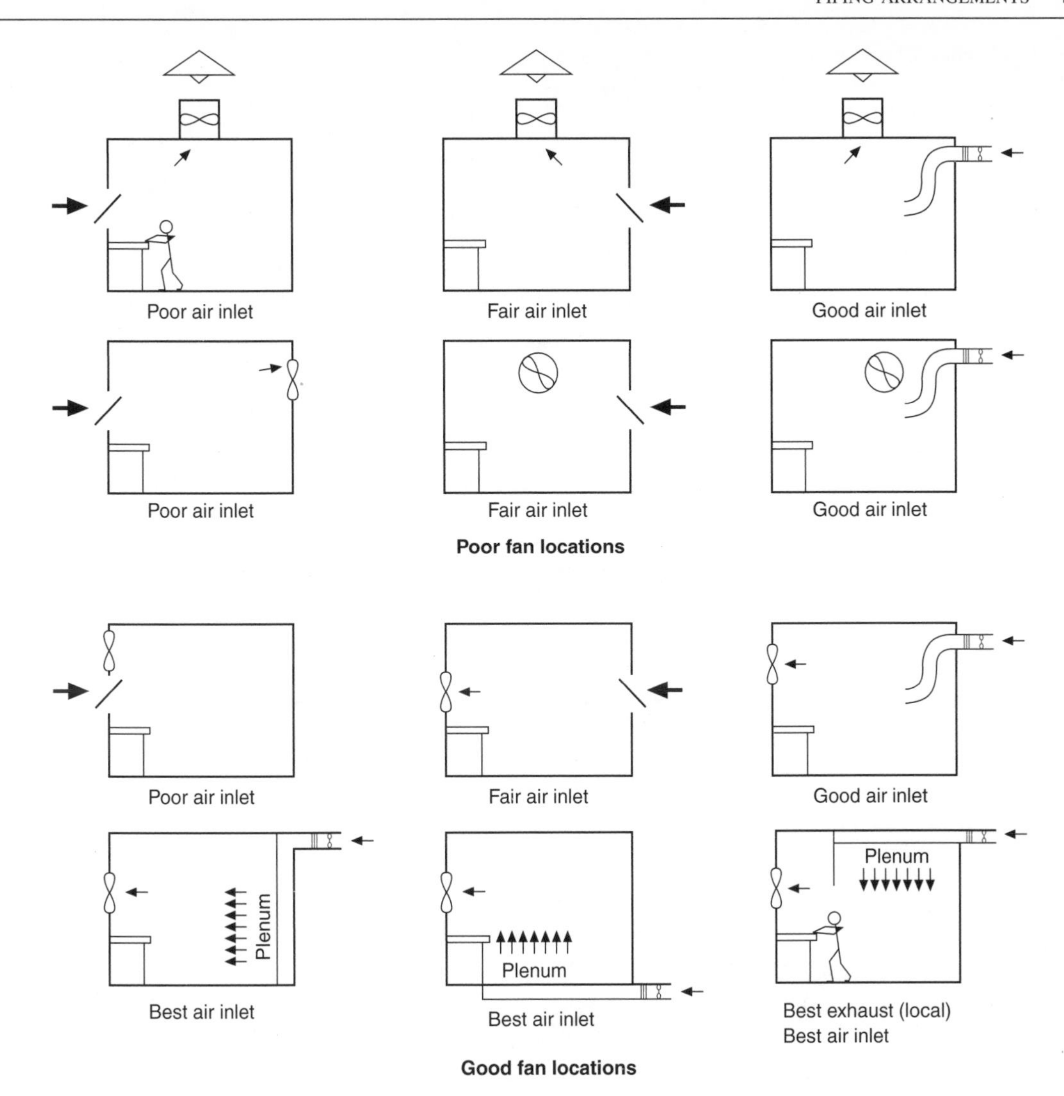

Figure 12.1 *Dilution ventilation: inlet air may require tempering according to outside conditions*

in the range 0.28 to 0.56 m/s. A higher velocity is required if there is significant dispersion inside the enclosure.

A *booth* should be of sufficient size to contain any naturally occurring emissions and so minimize escape via the open face. An air velocity of 0.56 m/s is required over the whole open face; a higher velocity is needed if there is significant air movement within the booth or to cope with convection currents. Booths should be deep enough to contain eddies at the rear corners; baffle plates or multiple offtakes may be necessary with shallow booths.

No operator should work between the source of pollutant and the rear of the booth (i.e. all work should be handled from the front open face, as illustrated by Figure 12.1).

If enclosure or use of a booth is impracticable, a *captor hood* is used. This is placed some distance from the source of pollution and the rate of air flow needs to be such as to capture contaminants at the furthermost point of origin.

Typical capture velocities are given in Table 12.12. Since velocity falls off rapidly with distance from the face of the hood, as shown in Figure 12.2, any source of dust should be within one hood

Table 12.12 Range of capture velocities

Condition of dispersion of contaminant	*Examples*	*Capture velocity* (m/s)
Released with practically no velocity into quiet air	Evaporation from tanks Degreasing vats etc.	0.25–0.51
Released at low velocity into moderately still air	Spray booths Intermittent container filling Low-speed conveyor transfers Welding Plating Pickling	0.51–1.02
Active generation into zone of rapid air motion	Spray painting in shallow booths Barrel filling Conveyor loading Crushers	1.02–2.54
Released at high initial velocity into zone of very	Grinding rapid air motion Tumbling	2.54–10.2 Abrasive blasting

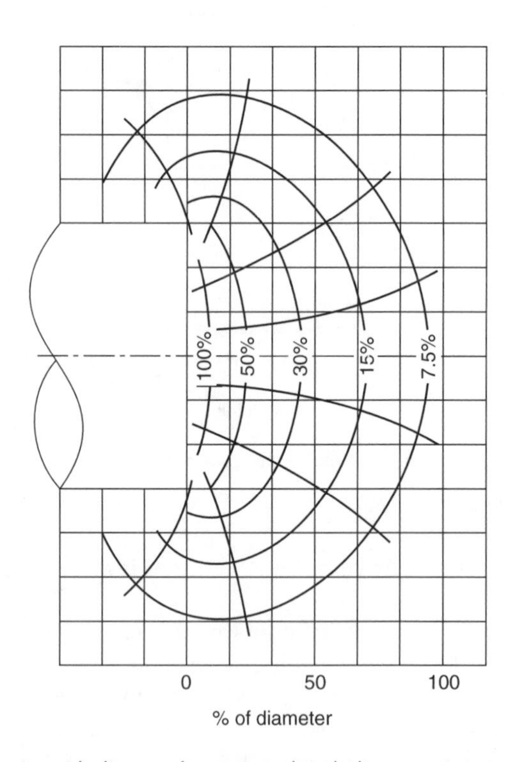

Figure 12.2 *Reduction in air velocity with distance from captor hood (distance given as % of hood diameter)*

diameter. Efficiency can be significantly improved by the use of flanges and by avoiding abrupt changes in direction of the ducting.

A *receptor hood* receives a contaminant driven into it by the source of generation. The flowrate needs to ensure that the hood is emptied more rapidly than the process fills it and to overcome draughts. No operator should work between the hood and the source of the contaminants.

Whatever the type of ventilation, air flow must be adequate to prevent particulate matter settling in the ducting; typical transport velocities are given in Table 12.13.

Table 12.13 Typical recommended transport velocities

Material, operation or industry	*Minimum transport velocity* (m/s)	(ft/min)
Abrasive blasting	17.8–20.3	3500–4000
Aluminium dust, coarse	20.3	4000
Asbestos carding	15.2	3000
Bakelite moulding powder dust	12.7	2500
Barrel filling or dumping	17.8–20.3	3500–4000
Belt conveyors	17.8	3500
Bins and hoppers	17.8	3500
Brass turnings	20.3	4000
Bucket elevators	17.8	3500
Buffing and polishing		
dry	15.2–17.8	3000–3500
sticky	17.8–22.9	3500–4500
Cast iron boring dust	20.3	4000
Ceramics		
glaze spraying	12.7	2500
brushing	17.8	3500
fettling	17.8	3500
dry pan mixing	17.8	3500
dry press	17.8	3500
sagger filling	17.8	3500
Clay dust	17.8	3500
Coal (powdered) dust	20.3	4000
Cocoa dust	15.2	3000
Cork (ground) dust	12.7	2500
Cotton dust	15.2	3000
Crushers	15.2	3000
Flour dust	12.7	2500
Foundry, general	17.8	3500
sand mixer	17.8–20.3	3500–4000
shakeout	17.8–20.3	3500–4000
swing grinding booth exhaust	15.2	3000
tumbling mills	20.3–25.4	4000–5000
Grain dust	12.7–15.2	2500–3000
Grinding, general	17.8–22.9	3500–4500
portable hand grinding	17.8	3500
Jute		
dust	12.7–15.2	2500–3000
lint	15.2	3000
dust shaker waste	16.3	3200
pickerstock	15.2	3000
Lead dust	20.3	4000
with small chips	25.4	5000
Leather dust	17.8	3500
Limestone dust	17.8	3500
Lint	10.2	2000
Magnesium dust, coarse	20.3	4000
Metal turnings	20.3–25.4	4000–5000
Packaging, weighing etc.	15.2	3000
downdraft grille	17.8	3500
Pharmaceutical coating pans	15.2	3000
Plastics dust (buffing)	19.0	3800
Plating	10.2	2000

Table 12.13 Cont'd

Material, operation or industry	*Minimum transport velocity*	
	(m/s)	(ft/min)
Rubber dust		
fine	12.7	2500
coarse	20.3	4000
Screens		
cylindrical	17.8	3500
flat deck	17.8	3500
Silica dust	17.8–22.9	3500–4500
Soap dust	15.2	3000
Soapstone dust	17.8	3500
Soldering and tinning	12.7	2500
Spray painting	10.2	2000
Starch dust	15.2	3000
Stone cutting and finishing	17.8	3500
Tobacco dust	17.8	3500
Woodworking		
wood flour, light dry sawdust and shavings	12.7	2500
heavy shavings, damp sawdust	17.8	3500
heavy wood chips, waste, green shavings	20.3	4000
hog waste	15.2	3000
Wool	15.2	3000
Zinc oxide fume	10.2	2000

Fire protection

The installation of fixed fire protection depends on analysis of potential fire characteristics (see Chapter 6). A summary of the factors is given in Table 12.14.

Table 12.14 Fire characteristics determining protection system design

- Types of combustibles, e.g. resulting in surface fires for which inerting or an inhibiting agent may be best, or deep-seated combustion requiring a cooling agent, e.g. water.
- Relative importance of speed of detection.
- Degree of risk to personnel, e.g. toxic combustion or fire-fighting agent decomposition products.
- Indoors, indicative of a total flooding system; outdoors requiring assessment of exposure of nearby hazards, involvement of other combustibles, wind effects and difficulty of extinguishment.
- Probability of fire spread, with a need for, e.g., fire walls, fire doors, fire stops.
- Requirement for detection and control, or detection and extinguishment.
- Any features which rule out specific extinguishing agents, or making other types preferable.
- Any advantages of a dual agent system, e.g. water and carbon dioxide.
- Inherent or possible explosion risks (e.g. static discharge associated with extinguishant discharge).
- Need for operation with personnel present.
- Need for rapid 'knock-down'.

Activation may be automatic, by a detection system, or manual. The general requirements are summarized in Table 12.15.

Table 12.15 General requirements for fixed fire extinguishing system

- Capability to control and extinguish the anticipated fire condition without outside assistance, unless planned for (dependent on requisite spacing of discharge nozzles to cover all areas at risk, application rates, discharge times, quantities of extinguishant available).
- Reliability, allowing for environmental features possibly detrimental to operation, e.g. corrosion, dust, tar.
- Agents must be compatible with the process, chemicals present (Chapter 7) and any other installed system.
- Potential toxicity of the agent, its thermal breakdown products, or products generated on contact with chemicals will determine safety measures necessary (e.g. in occupied areas a 'double-knock' system, allowing evacuation time, may be required.
- Visibility in the fire zone after extinguishant discharge, if manual fire-fighting is anticipated.

Installation and operation

Engineering specification and purchasing procedures are essential to ensure that all items are to the design specification and to comply with company or national standards. During installation the features to consider are foundations, selection of materials, fabrication, assembly, supports, pressure testing, etc.

13

Operating procedures

Safety is a management responsibility requiring an awareness of all relevant local legislation and a strategy that comprises:

- A clear, widely-published statement of company policy on safety.
- Clear lines of responsibility identified for each level of management and the workforce to enable the policy to be implemented.
- Company management systems and 'safe systems of work'.
- Emergency procedures.
- Procedures for monitoring compliance with the policy and the safety performance.

In the UK these duties are imposed by the Health and Safety at Work etc. Act. Employees also have a legal responsibility to cooperate with management by using designated protective devices and not interfering with such apparatus.

An excellent framework for the management of occupational health and safety is provided by OHSAS 18001:1999 which is compatible with ISO 9001: 1994. Key elements of the standard addressed include

- Policy.
- Legislation.
- Hazard identification and risk assessment and risk control.
- Structure and responsibilities.
- Training, awareness and competence.
- Consultation and communication.
- Documentation.
- Documentation and data control.
- Operational control.
- Emergency preparedness and response.
- Performance measurement and monitoring.
- Accidents, incidents, non-conformances and corrective and preventative action.
- Records and their management
- Audits.
- Management review and continuous improvement.

Plant must be properly designed, installed, commissioned, operated and maintained. Depending upon the risk, working procedures should be documented, e.g. as codes of practice. Consideration of safety at the design stage was discussed in Chapter 12.

Commissioning

Commissioning embraces those activities essential for bringing a newly installed plant into routine production. It includes mechanical completion and provisional acceptance, pre-commissioning, first start-up and post-commissioning.

Special efforts are needed to ensure safety at this stage of the development. Since start-up and shutdown procedures are responsible for many accidents, these procedures merit special attention.

Operation

Operation includes normal start-up, normal and emergency shutdown, and most activities performed by the production team. Whilst inherently safe plant design limits inventories of hazardous substances, inherently safe operation ensures the number of individuals at risk are minimized. Access to the plant for non-essential operational people such as maintenance engineers, post staff, administrators, quality control samplers, warehouse staff delivering raw material or plant items or collecting finished product, members of security, visitors etc., must be controlled.

Safe handling of chemicals demands a combination of 'hardware' and 'software' such as operating procedures, staff selection and training. Within the UK there is a duty under the Management of Health and Safety at Work Regulations 1992 for an employer to make appropriate arrangements for the effective planning, organization, control, monitoring and review of the preventive and protective measures. Systems of work will generally include:

- The selection and supervision of personnel with clearly-defined job descriptions.
- Planning and coordination of all activities – in particular those involving different sections, departments or tradesmen.
- Training and instruction of the workforce. Retraining. This should cover both normal and emergency situations.
- Provision of operating instructions and procedures. These should eliminate confusion and provide continuity on, e.g., shift changeover. Errors in identification of valves, pumps, pipes, storage tanks, and the sequence in which they are to be operated is a common cause of accidents, e.g. on staff changeovers.
- Provision of adequate supervision.
- Inspection, testing, maintenance and replacement of equipment as necessary.
- General conditions of work, e.g. maintenance of adequate general ventilation, local exhaust ventilation, access, lighting, fire precautions, occupational health measures, housekeeping.
- Auditing of health, safety and environmental matters.

A checklist is given in Table 13.1.

Maintenance

Maintenance embraces regular inspection, periodic examination by 'competent persons', and repairs. Accidents may arise as a result of the lack of maintenance, during maintenance, and as a consequence of faulty maintenance. Preventive maintenance can be a legal requirement for, e.g., pressure vessels/boilers, lifting gear, power presses. A system is required for monitoring compliance with scheduled preventive maintenance programmes, backed up by in-service inspection. Breakdown

Table 13.1 Safety systems in operation

Access	*System for* Control of access to site; maintenance of security Designation of restricted areas, e.g. containing flammable materials, 'eye protection' zones, 'hearing protection' zones, radiological hazards, microbiological hazards Ensuring freedom from obstruction of roads, stairs, gangways, escape routes Control of vehicles
Communication systems	*Maintenance of adequate* Written instructions, maintenance procedures etc. Log books, recipe sheets, batch sheets Identification of vessels, lines, valves etc. Intershift communication and records System for reporting and follow-up of Plant defects Process deviations Hazardous occurrences ('near misses') Accidents: minor/major injury and/or material loss Warnings, notices Notices relating to specific hazards Notices relating to temporary hazards (and their subsequent removal) (See also permit-to-work systems and operating procedures) Alarms; emergency communications
Control of contractors	Familiarization with plant hazards, rules and safety practices, security Clear delineation of work, responsibilities and handover (*See also* maintenance, permit-to-work systems, personal protection, site restrictions, access, modifications, personal hygiene etc.)
First aid provisions	*Maintenance of adequate provision for* Emergency showering Eye-washing Decontamination from chemicals (and antidotes if relevant) First aid (equipment, materials, trained personnel) and resuscitation
Housekeeping	Regular cleaning of floors, ledges, windows, by appropriate procedures (e.g. 'dustfree' methods) Removal of empty containers, scrap, waste Prompt treatment of spillages Redecoration
Inspection and testing (including cleaning/ maintenance where necessary)	*Implementation of a system for routine, regular inspection of* General ventilation provisions Local exhaust ventilation provisions (test and report at regular intervals) Fire protection Alarm systems (HTA, HPA, fire, evacuation etc.) Plant and equipment generally; portable tools; flexible hoses; lifting equipment; pressure systems Safety devices, e.g. PSVs, explosion reliefs, trips Monitoring and control instruments Protective clothing/equipment and emergency equipment Drains and floor drainage; bund walls Electrical equipment; earthing and bonding
Maintenance	*Systems include arrangements for* Issue, control, testing and repair of lifting gear, ladders, scaffolding etc. Scheduled maintenance of key plant items Use of job cards/log book to identify tasks precisely Permit-to-work systems Flame cutting/welding (or soldering, brazing etc.) Line-breaking Electrical work – certain classifications Equipment removal, e.g. to workshop Work on roof/at heights

Table 13.1 Cont'd

	Isolation of safety services Excavations Introduction of non-flameproof electrical equipment into restricted area Mechanical isolation, e.g. conveyors, lifts Confined space entry Radioactive areas Microbiological ventilation systems, fume cupboards etc. Other unusual/non-routine situations Provision for trained persons to be present/on call, e.g. rescue, first aid, fire-fighting Instruction and training of personnel Maintenance of guards on machinery, open vessels, handrails, screens at sampling/drumming-up points etc. and on glass equipment Use of sparkproof tools where appropriate Record-keeping Inspections, e.g. pressure systems, trips, relief devices, guards, lighting (*see* inspection and testing)
Materials	*Maintenance of adequate procedures for* In bulk, e.g. checks on tanker contents, earthing and bonding, identification of receiving vessel In small containers, e.g. carboy handling, cylinder unloading Despatch of materials Inspection of tankers, carboys, cylinders etc. prior to filling Identification and labelling of materials; driver instructions Use of earthing clips on portable containers of flammable liquids
Modifications to plant or process	*System for* Control of changes in raw materials Formal approval for plant/process changes Hazard and operability study prior to implementation (repeat when in use) Updating of all operating instructions, notices, procedures Removal/isolation of obsolete plant/lines Control of changes in personnel (*see* training)
Monitoring and follow-up	*Systems for monitoring* Working environments Toxic contaminants Temperature Noise Lighting Vibration Ionizing radiation Occupational health Pre-employment medical for selection and establishment of base levels Working conditions in specific areas Employees' routine medicals, exposure profiles Biological monitoring where appropriate
Operating procedures	*Systems include* Provision of clearly written operating instructions, accessible to operators, referring to numbered/identified plant items, covering Start-up and shutdown Normal operation, draining, purging, excursions etc. Operation at high or low rates Emergency shutdown, service failures etc. Procedures for non-routine operations, e.g. clearance of blockages, reprocessing of materials, temporary process alteration Pressure testing, sampling, chemical and atmospheric testing Established procedures for draining, purging, venting, isolating, testing and inspection prior to opening/entering/maintaining plant (see permit-to-work systems, maintenance)

Table 13.1 Cont'd

	Quality control on raw materials, materials in process, products and wastes Materials control to enable losses, over-use, under-use or accumulations to be detected; control of quantities in storage
Permit-to-work systems	*See* under 'Maintenance' Entry into confined spaces (e.g. vat, vessel, flue, sewer, boiler or similar) Use of non-flameproof electrical equipment where flammable liquids/vapours/dusts may arise
Personal hygiene	*Checks on* Adequacy of washing/showering facilities (location and provision) Skin cleansing, barrier and conditioning cream provision Double locker system
Personal protection	*Systems to ensure provision of* Overalls: special requirements (flame retardant, antistatic), frequency of laundering Protective clothing: suits, spats, armlets, helmets, gloves for specific applications, footwear (industrial and/or antistatic) Hearing protectors Eye protection: specific provision for various duties Respirators: specific types for different applications
Safety management	*Systems include* Safety audit system: hazard analysis Checks on the adequacy and location of fire-fighting, emergency rescue and alarm equipment Practice in emergency situations Emergency procedures for Fire/explosion Toxic release Serious accidents Spillage Pollution incidents (particularly when manning levels are low, e.g. weekends) Major emergency procedures Internal procedures (evacuation, communications, shutdown, damage control) External liaison (fire, police, hospitals, neighbours)
Site restrictions	*Systems include* Control over chemicals, containers etc. removal from site (*see also* waste disposal) Prohibition of eating/drinking except in designated areas Prohibition of smoking, carrying matches/lighters, except in designated areas; arrangements for enforcing this Restriction on employees permitted to drive works transport Control of lone working Control over equipment removal from site
Storage	*Procedures for ensuring* Identification of all materials Segregation of incompatible materials Good housekeeping Compliance with limits set for stocks of potentially hazardous chemicals Storage, segregation and handling of gas cylinders Display of appropriate warnings/notices
Training: management	*Provision for* Refresher course on fundamentals of safety, hazard recognition, procedures Participation in hazard and operability studies on existing operations and procedures
Training: operating personnel	*Provision for* Formal training programme; refresher courses Instruction in hazards associated with the work and safe procedures Introduction of new legislation Instruction and practice in emergency procedures

Table 13.1 Cont'd

	Encouragement/enforcement of use of personal protection Special training needs, e.g. first aid, emergency rescue/fire-fighting
Waste disposal	*Systems for* Collection of combustible waste in appropriate containers (e.g. oily rags/other material subject to spontaneous combustion) Identification of wastes; analytical control procedures; labelling of containers Segregation of wastes; identification of hazards due to inadvertent mixing of wastes Selection of appropriate disposal routes for liquids/solids to comply with acceptable discharge levels Disposal of toxic waste

maintenance is also required to correct faults and repair damage. All maintenance on equipment used with chemicals should be properly planned following a risk assessment, and recorded. Maintenance operations, particularly those which are non-routine, require a sound system of work with strict administrative procedures, e.g. permit-to-work, to avoid risks arising from modifications to established safe practices. A permit-to-work should be used wherever the method by which a job is done is likely to be critical to the safety of the workers involved, nearby workers, or the public; it is required whenever the safeguards provided in normal operations are no longer available. The system has written instructions to explain how it operates and to which jobs it is applicable. A form is used to control the work. This must be issued by a responsible person and the recipient must be trained in the system and how to perform the work. Different forms are applied to different tasks. Examples include:

- Entry into confined spaces where there is likely to be a danger from toxic or flammable gases, ingress or presence of liquids, free-flowing solids, or oxygen deficiency or enrichment, etc. (Table 13.2). Such work requires well-rehearsed procedures regulated by a permit. Figure 13.1 identifies the procedures for entry into confined spaces and Table 13.3 lists associated safety requirements with a specimen permit (Figure 13.2). Detailed requirements in the UK are specified in the Confined Spaces Regulations 1997.
- Hot work on plant, or in areas where flammable materials are handled, e.g. arc or gas welding (see also Figure 5.4), flame-cutting, use of blow torches, grinding chipping concrete, introduction of non-sparkproof vehicles.

Table 13.2 Control of oxygen gas

Ease of ignition, combustion rate and combustion temperature all increase when the oxygen content of an atmosphere >21% (Chapter 6). Enrichment of a room atmosphere >25% creates a hazardous situation: enrichment in a confined space is particularly dangerous.

Precautions include
Never use oxygen to 'sweeten' the atmosphere in a confined space
When flame-cutting metals:
    Use proper ventilation (all the oxygen supply to the torch may not be consumed)
    Do not nip/kink the hose to cut off the oxygen supply while changing torches – always use the isolation valve
Do not leave blowpipes or torches inside a confined space during breaks
Wherever possible locate the cylinder outside the confined space
Do not use oxygen gas as a source of pressure in place of compressed air
Keep regulators, valves and piping on compressed gas systems clean and free from oil and grease
Close the cylinder valve when not in use

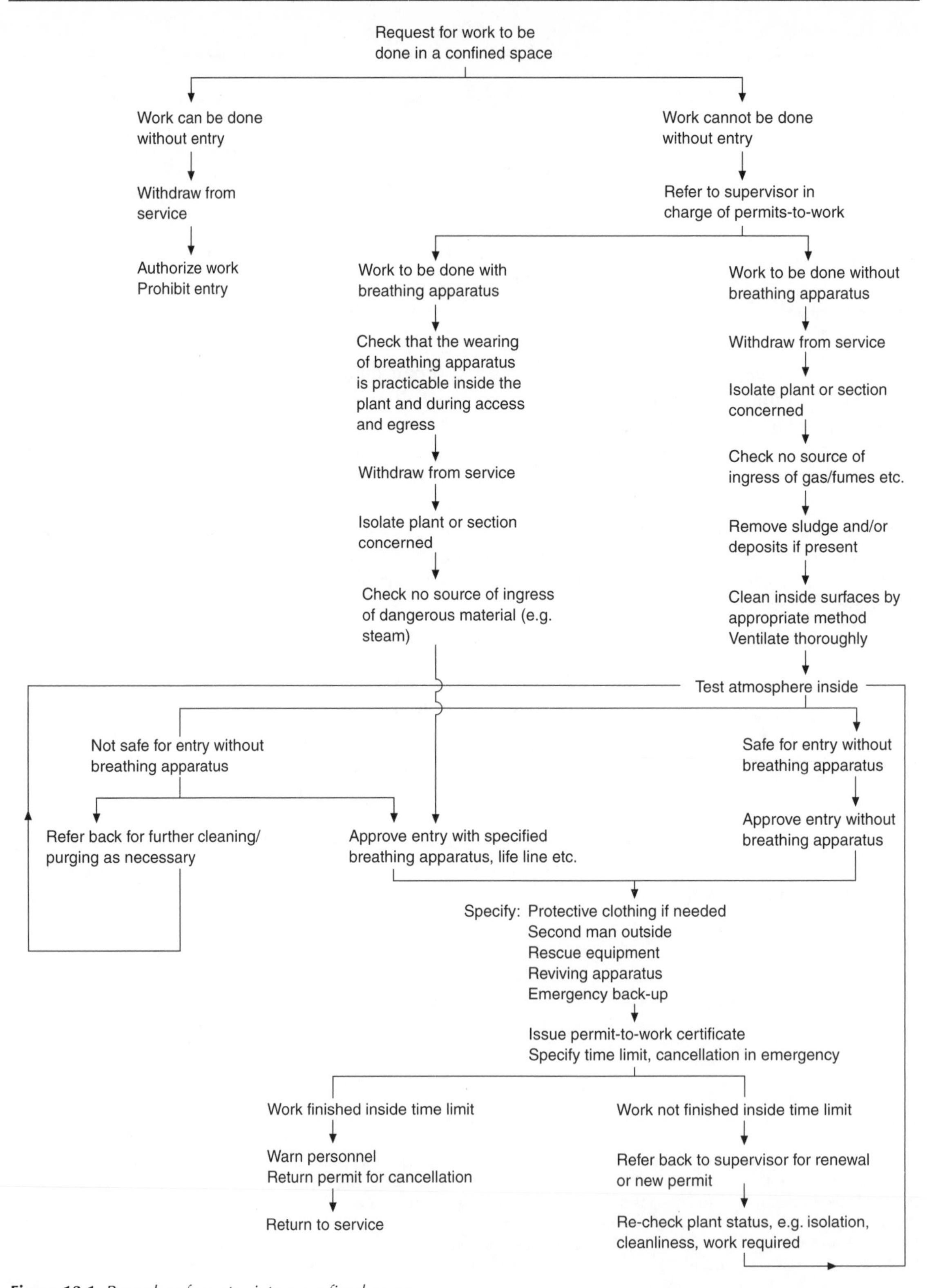

Figure 13.1 *Procedure for entry into a confined space*

Table 13.3 Entry into confined spaces (see Figure 13.1)

A permit-to-work system is essential to control entry into confined spaces, e.g. tanks, reactors or underground chambers.

Requirements

No entry without a specific permit-to-work authorizing the work; entry without a permit is forbidden except for rescue purposes

Strict compliance with all instructions on the permit, e.g. the wearing of breathing apparatus and other safety equipment; if a safety belt or harness is used, it must be attached to a lifeline, the free end of which is held by a person outside the confined space

If a person is appointed to hold the lifeline and to keep continual watch over the person within the confined space, it is forbidden to leave the post. It is necessary to be ready to summon help if required. (A specific emergency plan should be established)

Rescue of any person from a confined space where dangerous fumes, or an oxygen-deficient atmosphere, may be present is prohibited without wearing appropriate safety equipment

Application of a welding torch or burner to a tank or drum containing flammable material, either as solid, liquid or vapour or their residues, can cause an explosion. Such vessels, although apparently 'empty', may have residue in the bottom and/or in seams and crevices.

Heat should not be applied if cold methods, e.g. cold-cutting practices or cold-setting resins, can be used. Otherwise a permit-to-work system is required. Refer to Figure 13.3. The vessel must first be rendered safe, e.g. by thorough cleaning, using low-pressure steam or by boiling strong detergent solution, for ≥30 min. It must then be inspected internally; mirrors and torches can be used for this but any lights must be flameproof or intrinsically safe. An explosimeter can be used to check for vapour but will not detect solid residues or involatile liquid. Alternatively, the vessel may be filled with water or inert gas.

Do not:

- Attempt to drive off liquid using a naked flame;
- Assume that a vessel is clean without inspecting it;
- Rely upon blowing through with air or water washing;
- Use chlorinated solvents to remove residues.

The need to remove all flammable material from the area should be considered. Where this is impractical, checks on its containment and inerting are necessary, together with environmental monitoring to ensure that atmospheric levels are well below the lower explosive limit (Figure 13.3).

Other situations for which permits-to-work are required include:

- Maintenance of apparatus used for microbiology.
- Pipeline breaking, e.g. carrying chemicals, steam or those under pressure.
- Work on high-voltage electrical supply, radioactive equipment, roof work such as checking toxic exhaust from ventilation outlets.
- The isolation of certain mechanical equipment, e.g. conveyors, work on lifts, excavations, entry and positioning of cranes, isolation of various 'safety services', e.g. water or inert gas, stand-by power generation, water supply to sprinkler systems, compressed air for breathing apparatus.
- Removal of potentially contaminated equipment to workshop for maintenance, introduction of non-flameproof electrical equipment into a classified area.
- In general, any other work should be covered which the responsible manager considers non-routine and/or for which a special hazard exists.

PLANT DETAILS (Location identifying number, etc.)	
WORK TO BE DONE	
WITHDRAWAL FROM SERVICE	The above plant has been removed from service and persons under my supervision have been informed Signed Date Time
ISOLATION	The above plant has been isolated from all sources of ingress of dangerous fumes, liquids, solids, etc. Signed The above has been isolated from all sources of electrical and mechanical power Signed The above plant has been isolated from all sources of heat Signed Date Time
CLEANING AND PURGING	The above plant has been freed of dangerous materials Material(s): Method(s): Signed Date Time
TESTING	Contaminants tested Results Signed Date Time
I certify that I have personally examined the plant detailed above and satisfied myself that the above particulars are correct (1) The plant is safe for entry without breathing apparatus (2) Approved breathing apparatus must be worn Other precautions necessary: Time of expiry of certificate: Delete (1) or (2)	Signed Date Time
ACCEPTANCE OF CERTIFICATE	I have read and understood this certificate and will undertake to work in accordance with the conditions in it. Signed Date Time
COMPLETION OF WORK	The work has been completed and all persons under my supervision, materials and equipment withdrawn Signed Date Time
REQUEST FOR EXTENSION	The work has not been completed and permission to continue is requested Signed Date Time
EXTENSION	I have re-examined the plant detailed above and confirm that the certificate may be extended to expire at: Further precautions: Signed Date Time
THE PERMIT TO WORK IS NOW CANCELLED. A NEW PERMIT WILL BE REQUIRED IF WORK IS TO CONTINUE	Signed Date Time
RETURN TO SERVICE	I accept the above plant back into service Signed Date Time

Figure 13.2 *'Permit-to-work' certificate for entry into a confined space. A certificate number is added for identification, authenticity checking etc.*

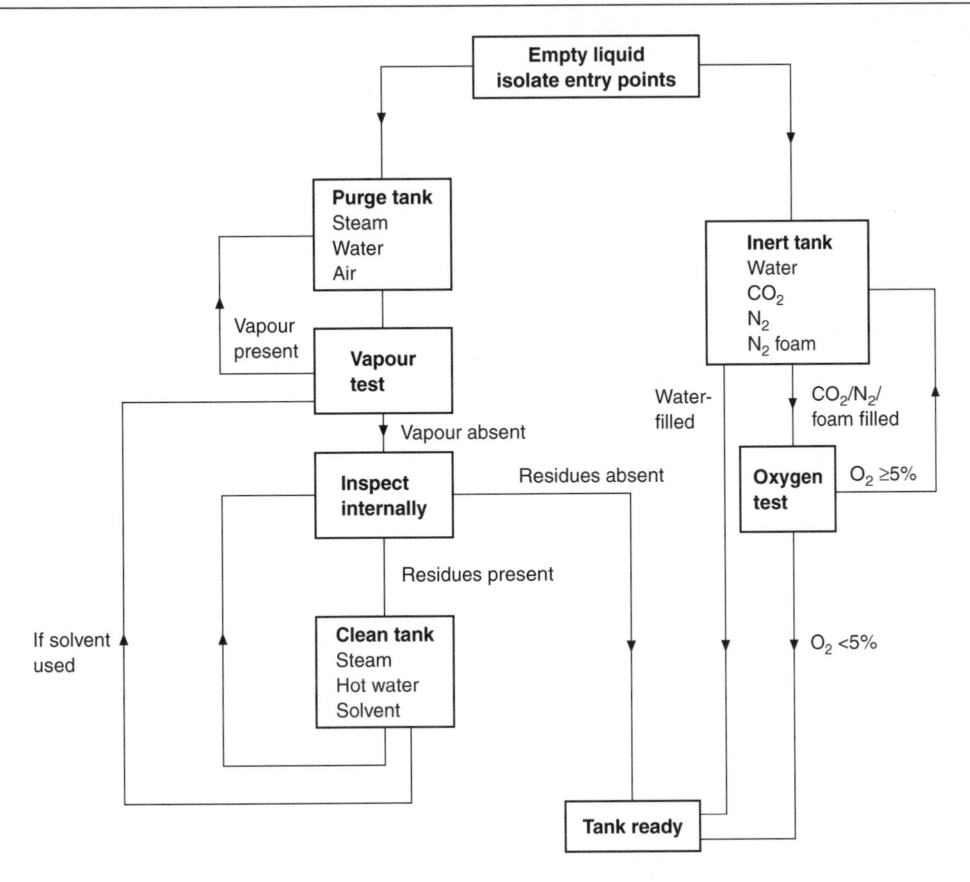

Figure 13.3 *Procedure for 'hot work' on equipment that contains, or has contained, a flammable chemical (general summary only: reference should be made to detailed procedures)*

A permit should document:

- The person issuing the permit and his/her position in the company.
- The date/time on which the permit becomes effective.
- The precise work to be done and specify unambiguously the plant items involved.
- Precautions which have been taken and those which need to be observed during the work, including which apparatus or plant items have been made safe.
- The associated outstanding risks with it.
- Precautions to be taken on completion of the work, e.g. a fire-watch following welding in an area containing combustible solids.
- The completion of the work.
- The handback of plant from maintenance to production staff.

Thus operating personnel are responsible for ensuring that an area or plant item is isolated and free of danger; at the end of the work, maintenance staff are responsible for ensuring that any hazard that may have been introduced during their work has been removed and the plant etc. checked and shown to be safe prior to handback to operating staff. A checklist for the review of permit-to-work systems is given in Table 13.4 and Figure 13.4 illustrates a typical permit-to-work. Safety can only be guaranteed if all participants follow the permit conscientiously and

CERTIFICATE OF CONDITIONS AND RESPONSIBILITY OTHER THAN FOR ENTRY INTO VESSELS OR CONFINED SPACES
Part 1

WORKS		Certificate No.
DEPARTMENT		Date of issue Date .... Time ....
SECTION OF PLANT		Period valid to ....
CIRCUMSTANCES		
NOTE: strike through where not applicable		
(1) The above item of plant is isolated from every (dangerous) source of steam, gas, fume, liquid, solids, motive power, heat and electricity. Details of isolation:		
(2) The above item of plant is not isolated from every (dangerous) source of steam, gas, fume, liquid, solids, motive power, heat and electricity. These special precautions must be taken in addition to (3) and (4) below.		
(3) Naked Flame or Other Source of Ignition	Not Permissible	Permissible: Type:
(4) Other hazards which may be encountered		Precautions
Hand over Issuing Dept .... (*Signature*) Date .... Time .... Receiving Dept. .... (*Signature*)		Hand Back Issuing Dept .... (*Signature*) Date .... Time .... Receiving Dept. .... (*Signature*)

AUTHORIZATION TO WORK OTHER THAN IN VESSELS OR CONFINED SPACES
Part 2

WORKS		Authorization No.
DEPARTMENT		Date of issue Date .... Time ....
SECTION OF PLANT		Period Valid to ....
Certificate No.		
(1) Work to be carried out.		
(2) The following precautions have been taken.		
(3) Naked Flame or Other Source of Ignition	Not Permissible	Permissible Type:
(4) Precautions to be taken in addition to (2) and (3) above		
Work may start .... (*Signature*)		Date .... Time ....
Work completed .... (*Signature*)		Date .... Time ....

Figure 13.4 *(facing page) 'Hot work' permit. Part 1 records conditions and steps required; Part 2 authorizes work and specifies precautions.*
This type of certificate may be designed to cover work on roofs, in storage bunkers and on plant containing gas, fume, steam or corrosive, poisonous or radioactive materials; welding and burning in potentially hazardous areas; work being carried out where there is a risk of fire, explosion, electric shock, flooding or high pressure. It is not considered appropriate for work on electrical equipment operating at >650 V.

without short cuts. Only one permit relating to one piece of equipment should be allowed at any one time. Best practice is for only one person to have authority to issue permits for a specific area. These permits should be cross-referenced and collected.

Table 13.4 Checklist for permit-to-work systems

Are staff trained in the use of the permit-to-work system and its significance?
Is it clearly laid down who should issue permits and to whom?
Is the system audited regularly?
Does each permit clearly identify the work to be done and the hazards associated with it?
Does each permit state the precautions which have been taken and those needed while the work is in progress (e.g. isolations, purging, personal protective equipment)?
Is there a system of clear cross-referencing when two or more jobs subject to permits may affect each other?
Does each permit have a hand-back procedure which incorporates statements that the maintenance work has finished and that the plant has been returned to production staff in a safe condition?
Are time limits included, and shift changeovers covered?
Is it clear what procedures to follow if work is suspended for any reason?
Is the permit form clearly laid out, avoiding statements or questions which could be misleading or ambiguous, but being sufficiently flexible for use in 'unusual' circumstances, however rare?
Does the system cover contractors?
Is a copy of the permit on clear display in the work area?
Have signatories of permits been trained?

Pressure systems

The safety of pressure systems in Great Britain is controlled by the Pressure System Safety Regulations 2000. These probably apply if

(a) the plant contains relevant fluid (e.g. steam, compressed air or liquefied gas, or a gas dissolved under pressure in a solvent at ambient temperature and which could be released from the solvent without application of heating) at a pressure >0.5 bar above atmospheric,
(b) there is a pressure vessel in the system *and*
(c) the plant is used at work by employees and/or the self-employed.

All steam systems with a pressure vessel used at work are covered, irrespective of pressure.

Basic requirements

- Establish the safe operating limits of the plant.
- Have a suitable written scheme drawn up or certified by a competent person for the examination at appropriate intervals of most pressure vessels and all safety devices, and any pipework which is potentially dangerous. Advice may be sought from any competent person when deciding what vessels and parts of the pipework need to be included.

- Have examinations carried out by a competent person at the intervals set down in the scheme.
- Provide adequate operating instructions to ensure that the plant is operated within its safe operating limits, and emergency instructions.
- Ensure that the plant is properly maintained.
- Keep adequate records of the most recent examination and any manufacturer's records supplied with new plant.

Here the safe operating units are basically the upper limits of pressure and temperature for which the plant was designed to be operated safely.

Emergency procedures

Emergencies include controllable events leading to safe shutdown of the plant and uncontrollable emergencies. The first can escalate into more serious and uncontrollable emergencies. Emergency planning is necessary to cover a wide range of eventualities. Within the UK procedures are required under the Management of Heath and Safety at Work Regulations 1992 and, in specific cases, the COMAH Regulations 1999.

Minor emergencies

- Personal injuries (refer to Table 13.8 for chemicals).
- Small fires (refer to Table 6.15).
- Overpressures, even if vented.
- Spillages, etc, e.g. onto land, into drains, into controlled water.
- Process 'excursions', e.g. temperature, pressure.
- Equipment failure.
- Disruption of essential services or computer networks.

Major emergencies

- Fire with or without potential for escalation.
- Explosion.
- Toxic releases.
- Flammables release (unignited).
- Large spillage.
- Pollution episode, on or off site.
- Person trapped in vessel/confined space.
- Multiple casualties.
- Emergency on adjacent site (mutual aid).
- Others (credible events, coincident emergencies).

However, most detailed planning tends to relate to major incidents, or those which might affect the public or neighbouring sites.

Any on-site emergency plan will be site- and business-specific. It should provide simple and logical procedures for effective direction and coordination of an emergency incident in a way which

- safeguards personnel including rescuers, fire-fighters, etc.;
- deals with casualties;

- minimizes damage to plant and equipment, buildings, and materials/stock;
- prevents escalation;
- triggers off-site responses, e.g. fire service, ambulance service, police, specialized staff;
- minimizes off-site effects of pollution; and
- restores the site to normal operation as rapidly as possible.

An inherent part of planning is to ensure that all personnel fully understand their own role, and appreciate the roles of others, in dealing with an emergency. In some situations it will be important to recover records, equipment and samples for subsequent investigation.

Warnings

A primary consideration is the method of warning that an emergency has occurred. Whether this involves bells, sirens, horns etc.,

- The alarm must be audible to everyone.
- Everyone must know what it means, and the correct action to take.

Visual alarms may be required in locations with a high level of background noise. Different alarms may be provided to warn of different emergencies, e.g. a minor fire, a toxic release or an incident requiring complete evacuation.

On a small site an emergency is likely to affect the complete site; hence a common warning system is required. On a very large site a two-tier warning system may be used, i.e. a local warning which differs from a warning to the whole complex.

Incident categorization

There are four typical categories of emergency.

A *minor incident* is one capable of being dealt with fairly effectively and quickly by personnel on the spot using the emergency equipment, e.g. fire extinguishers and monitors, on hand. The situation can be dealt with by the workforce, possibly including a Works Fire Brigade, fairly rapidly. It is reasonably localized and no other unit is affected.

A *Category 1 incident* is one requiring additional resources, but which can be dealt with by the site operator and internal fire brigade. Some thought needs, however, to be given as to whether, and when, the public emergency services should be notified. (The incident could escalate and result in the personnel on site requiring assistance quickly. If the total site forces are fully occupied dealing with a protracted incident, back-up resources may be needed to guard against a second incident.)

A *Category 2 incident* is one clearly recognizable from the outset as beyond the capability of the forces to hand. The emergency services should be informed immediately of the magnitude of the incident and the specific location. The resources sent in response, e.g. the number of fire appliances, ambulances, police, need to be predetermined by good pre-planning.

A *Category 3 incident* is an emergency of major proportions. It may have escalated from a smaller incident or arise instantaneously, e.g. due to a confined or unconfined vapour cloud explosion or a catastrophic failure of a storage vessel. The response from the emergency services will again need to be predetermined by pre-planning. Extensive evacuation of the general public may be required.

Calling in staff

When an emergency develops it may be necessary to call in key personnel, e.g.

- Staff able to continue operation of the plant or shut it down safely.
- Management who can take top-level decisions.
- Staff able to deal with the aftermath of casualties.
- Public relations staff to deal with the media.

Pre-planning may therefore include preparation of a 'call-in' list with the proviso that 90% of personnel will respond when called.

Responsibilities and pre-planning

Each person in the emergency team should understand clearly, and be well-practised in, the function that they have to perform.

One person, e.g. a Shift Manager, will have responsibility for responding immediately to an incident and for taking charge.

The priorities are:

1 Rescue of personnel, where appropriate; assessment of whether there are casualties or whether people may be missing and should be searched for.
2 Carrying out appropriate process action so that the situation does not escalate further.
3 Ensuring that the fire brigade, or other personnel trained in initial first aid/fire-fighting, are present.
4 Initiating some form of 'head-count'. In some situations provision is obviously necessary for wardens or properly equipped rescuers to search areas to ensure evacuation is complete.

Accounting for staff can be difficult; hence the advantage of site logging-in procedures. Instructions should be posted, stating

- Where personnel should go, by which route/alternative routes in an emergency.
- To whom they should report on arrival.
- What to do if they are unable to go to the nominated assembly point. (Different assembly points may be appropriate for different incidents, e.g. a fire in a specific area, a potential explosion or a spillage.)

Effective management of an emergency depends upon pre-planning, including:

- Provision of an adequately equipped emergency control centre of suitable design located outside potential danger zones.
- Procedures for the rapid assessment of the likely consequences in the event of, e.g., loss of containment of a hazardous material (identify the nature and quantity of material lost, check wind direction, determine from a map probable areas to be affected and the likely population density threatened).
- Provision of sets of process manuals, flowsheets, plot plans etc., ready for immediate use.
- Lists of mechanical equipment available, e.g. diggers, cranes, lifting equipment, and of emergency supplies, e.g. sand, wood, sandbags, drainseals, pipeblockers and absorbents for spillages on the ground and in drainage systems; floating booms for immiscible liquids (s.g. <1.0) that enter waters, and skimmers.

- Provision for monitoring and sampling, e.g. of oxygen levels, combustible gases and airborne toxic substances and the taking of samples from air, water and ground.
- Provision of action lists, list of emergency contacts and phone numbers, details of location of specialist services and equipment.
- Provision of simple refreshment facilities.
- Provision for dealing with emergency services and personnel responding from outside, e.g. designated locations for ambulances, fire appliance reception points, emergency services reporting points, vehicle parking.
- External liaison, e.g. to give essential personnel and the emergency services easy access to the site by the correct route, and to deal with the media.

Communications

Provisions for primary and emergency communications (internal, e.g. with staff, emergency personnel, and external, e.g. with the press, emergency services, hospital, employees' relatives, telephone links and radios) require careful assessment. If there is a flammable hazard on site, radios may need to be intrinsically safe or flameproof.

Training and exercises

Every part of an emergency plan has to be understood by all personnel involved, including the emergency services. Detailed instructions therefore have to be issued to all those who will act to control, or mitigate the effects of, a major incident. Everyone on site needs to understand the plan and the procedures in it; this includes:

- Familiarity with the different alarms, e.g. so as not to confuse a process plant alarm for a fire/ toxic gas release signal.
- Familiarity with the actions to take, escape routes, assembly areas.

Thorough training, including realistic practical exercises, with participation of outside services, is advisable. A full-scale rehearsal at least once a year is recommended. (Modifications can subsequently be introduced to cater for reported errors and omissions.)

Off-site requirements

Plans are also necessary to deal with the possible impact of accidents off-site. An indication of what these may entail is provided by the minimum information, listed in Table 13.5, which the occupier of a COMAH site must supply to the local authority to enable them to draw up an off-site emergency plan.

Spillage

All chemical spillages should be dealt with immediately since they may:

- Pose corrosive, or skin absorbent hazards.
- Pose flammable or toxic hazards (e.g. by inhalation or ingestion), or affect the eyes, mucous membranes or skin, especially if spillage involves volatile liquids in confined spaces, or if dust can become airborne.

Table 13.5 Minimum information to be supplied to local authority for an off-site emergency plan under the COMAH Regulations

- Details of site, location, nearby roads, site access.
- Site plan.
- Staffing levels.
- Off-site areas likely to be affected by major accidents and estimates of levels of harm which might result.
- Details of dangerous substances on site covered by the COMAH Regulations and of other hazardous materials on site.
- Details of technical advice the company can provide to assist an emergency response.
- Technical details of equipment and other resources normally available on-site and which may be available to assist the off-site emergency services.
- The functions of key posts with duties in the emergency response, their location and identification.
- An outline of the initial actions and procedures in the on-site emergency plan.

- Create slippery conditions (e.g. detergents, oils).
- Contaminate other materials/processes.
- Be symptomatic of more serious deficiencies in safety.
- Create risk of pollution.
- Be of nuisance value.

In general, systems of work should prevent spillages; nevertheless accidents will occur. As with accidents requiring first aid (see page 429), arrangements for dealing with spillages should be planned prior to the introduction of chemicals into the site. Bunds should be provided to contain spillage of hazardous substances. Procedures for cleaning up dangerous spillages will be dictated by circumstances (e.g. for mercury see page 157). Actions include:

- Assess the risk (e.g. identify the chemical and its hazardous properties, note the scale and prevailing circumstances); where necessary, raise the alarm, evacuate the area, obtain assistance etc.
- Don appropriate personal protection and, where it is safe to do so, contain the spillage (e.g. with sheets, pillows, booms, sand); if necessary, provide ventilation.
- Clean the spillage using appropriate clean-up material (sand, clay granules, cloth rags, synthetic sorbents, oil sorbents, general-purpose sorbents of high liquid-retention capability).
- Reassess risk, if necessary by environmental monitoring, swab tests.

Liquid spillages may be sucked up by pump and non-toxic solids can be vacuumed or brushed up (after wetting down where appropriate). Only small quantities of inert, water-soluble waste should be discarded to drains; acids and alkalis should first be neutralized.

Equipment to deal with spillages is summarized in Table 13.6. Treatment may require some combination of the measures given in Table 13.7.

Emergencies involving spillages during transport are mentioned in Chapter 15.

Table 13.6 Equipment for treatment of 'small' spillages, e.g. in a laboratory

Rubber or plastic gloves, face shields or goggles, rubber boots or over-shoes, face masks and an 'all-purpose' respirator
Household dustpan (rubber or polythene), brush and large bucket (preferably polythene), an ordinary steel shovel, stiff bristle brush and a soft brush, for sweeping up and containing broken glass, and industrial cotton mops, plastic foam mops or squeegees
Soap or detergent or special non-flammable dispersing agent to reduce fire hazards from flammable spillage and to reduce nuisance from lachrymators and strong-smelling fluids; disinfectants for biological smells
Sand in bucket or absorbents

Table 13.7 Procedure to treat spillages in a laboratory

Instruct others to keep a safe distance
Remove contaminated clothing and wash the hands before putting on protective clothing
Shut off all sources of ignition (for spillage of flammable fluid); do not operate electric switches in the vicinity
Ventilate the area well (for at least 10 min) to reduce irritant vapours and biological aerosols, and to evaporate remaining liquid (bear in mind that ventilation may transfer the danger to other parts of the premises)
Areas with sealed windows should have controlled ventilation (areas that cannot be adequately ventilated are unsuitable for use with hazardous materials)
Turn off the source of the leak and limit the spread of fluid by means of mop, dusters, rags, sand or diversion
Absorb spillage and shovel into buckets, and remove to a safe place, dilute greatly with running water and/or dilute with dispersing agent and water (as with ammonia solution, ethanol, methanol and formaldehyde solution)
Neutralize acid with soda ash, chalk or similar materials kept for the purpose, mop up cautiously with water (in case local areas of concentrated acid have been missed)
Use hypochlorite solutions in excess scattered over spillage of cyanide solutions and wearing respirator and gloves, mop up and collect in buckets. Allow to stand for 24 hours before diluting greatly and running to waste
Flood with a phenolic disinfectant any biologically contaminated area for 30 to 60 min and then clean up with water and allow to dry. After cleaning up wash hands with a suitable skin disinfectant (such as 0.5% Chlorhexidine in 70% methylated spirit)
Use forceps or a brush and pan with broken glass; never use the fingers, even with gloves. Disinfect broken glass arising from biological spills
Dispose of, by special arrangements, chemicals which cannot be admitted to the public sewerage system, e.g. flammable liquids and reagents with high toxicity

First aid

In Great Britain the Health and Safety (First Aid) Regulations 1981 require an employer to provide such equipment and facilities as are adequate and appropriate in the circumstances to enable first aid to be rendered to his employees if they are injured, or become ill, at work. An Approved Code of Practice gives more specific details on the number of first-aid personnel and their training, and the type of equipment.

Emergency first aid

A qualified first-aider, or nurse, should be called immediately to deal with any injury – however slight – incurred at work.

Any person on the spot may have to act immediately to provide first-aid treatment to prevent deterioration in the injured person's condition until assistance arrives. The aims are:

- To sustain life.
- To minimize danger.
- To relieve pain and distress.

Check the situation for danger to rescuers, then act as follows:

The patient:	is in danger	Remove from danger, or remove the danger from the patient
	is not breathing	If competent to do so, give artificial ventilation. Otherwise send for help without delay
	has no pulse	Start cardiac compressions
	is bleeding	Stop bleeding

is unconscious	Check the mouth for any obstruction. Open the airway by tilting the head back and lifting the chin using the tips of two fingers. If this does not keep the airway open, turn the casualty into the recovery position i.e. turn on side and ensure they cannot roll over; lift the chain
feels faint	Lie him or her down
has broken bones	Immobilize. Do not move the patient unless he/she is in a position which exposes them to immediate danger. Obtain expert help
has open wounds	After washing your hands, if possible cover with a dressing from the first-aid box. Seek appropriate help
has thermal or chemical burns	Immerse or flood copiously with cold water for ≥10 min
has minor injuries	Ignore these if there are more serious ones
is poisoned	Small amounts of water may be administered, more if the poison is corrosive. Administer a specific antidote if one exists. Do not induce vomiting.

Cuts

All minor cuts should be cleaned thoroughly and covered with a suitable dressing. After controlling bleeding, if there is a risk of a foreign body in the wound do not attempt to remove it, but cover loosely and take patient to a doctor or hospital, as should be done if there is any doubt about the severity of the wound.

Burns/scalds

Burns may arise from fire, hot objects/surfaces, radiant heat, very cold objects, electricity or friction. Scalds may arise from steam, hot water, hot vapour or hot or super-heated liquids.

The affected area should be cooled by holding in cold, clean running water. Swelling is liable to occur so jewellery or clothing likely to cause constriction must be removed. The area should then be covered with a sterile dressing, care being taken to apply the dressing without it sticking to the burned area. Blisters should not be pricked or damaged and cream or lotions should be avoided.

The patient may suffer from shock, in proportion to the extent of the injury. Give small drinks and keep warm: do not overheat. In all cases, speed of treatment is crucial to limit the effects of burns. Flowcharts which summarize the initial procedures for electrical, thermal and chemical burns respectively are shown in Figure 13.5.

Chemical splashes, poisoning by ingestion

Refer to Table 13.8.

All cases of ingestion should be referred to a doctor and/or hospital without delay.

Antidotes

Effective antidotes for use in an emergency are not common. Examples include:

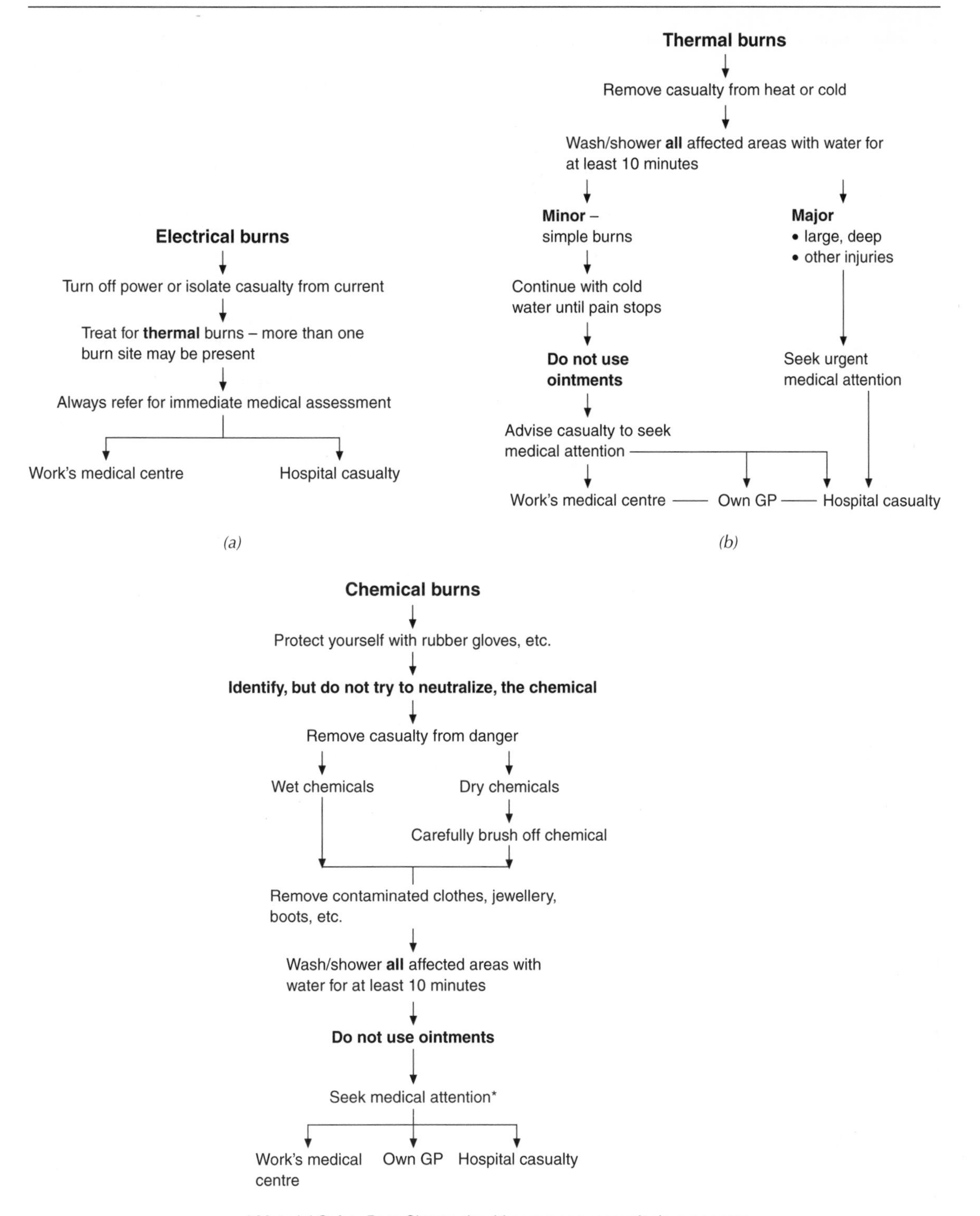

Figure 13.5 *Initial procedures for dealing with burn injuries*

Table 13.8 Standard first aid treatments for chemical exposure

Splashes on the skin	Flood the splashed surface thoroughly with large quantities of running water for ≥10 min or until satisfied that no chemical remains in contact with the skin Use soap to help remove splashes of solvents, solutions and chemicals known to be insoluble in water Remove all contaminated clothing, taking care not to contaminate yourself in the process If necessary, arrange for transport to hospital or refer for medical advice to nearest doctor Information to accompany the casualty: Chemical involved Details of treatment already given *(Special procedures apply to certain chemicals, e.g. anhydrous hydrogen fluoride, aqueous hydrofluoric acid, phenol and related compounds)*
Splashes in the eye	Flood the eye thoroughly with large quantities of clean, cool water or sterile liquid from a sealed container and continue for 10–15 min Ensure that the water bathes the eyeball by gently prising open the eyelids and keeping them apart until treatment is completed. Do not attempt to remove anything that is embedded All eye injuries from chemicals require medical advice. Apply an eye pad and arrange transport to hospital Information to accompany the casualty: Chemical involved Details of treatment already given
Inhalation of gas	Remove the casualty from the danger area after first ensuring your own safety Loosen clothing; administer oxygen if available If the casualty is unconscious, place in the recovery position and watch to see if breathing stops If breathing has stopped, apply artificial respiration by the mouth-to-mouth method; if no pulse is detectable, start cardiac compressions If necessary, arrange transport to hospital Information to accompany the casualty: Gas involved Details of treatment already given *(Special procedures apply to certain chemicals, e.g. hydrogen cyanide)*
Ingestion of poisonous chemical	If the chemical has been confined to the mouth, give large quantities of water as a mouthwash; ensure that the mouthwash is not swallowed If the chemical has been swallowed, small amounts of water may be administered, more if the chemical is corrosive; administer a specific antidote if one exists Do not induce vomiting Arrange transport to hospital Information to accompany the casualty: Chemical swallowed Details of treatment already given Estimate of quantity/concentration of chemical consumed

- *for cyanide poisoning.* Intravenous injection of:
  - 3% Sodium nitrite and 50% sodium thiosulphate.
  - Cobalt EDTA.
- *for HF on the skin.* Application of magnesium oxide paste with injection of calcium gluconate below the affected area.
- *for splashes of phenols.* Apply alcohol-soaked swabs.

Where there is a specific antidote suitable for emergency use it should be kept available and appropriate personnel trained in its use. Specific training should be given to first-aiders over and above their general training if they may need to administer oxygen or deal with incidents involving hydrogen cyanide, hydrofluoric acid or other special risks.

Records

Any injuries or cases of illness treated should be recorded. The minimum details are:

- Date, time and place of incident/treatment.
- Name and job of casualty.
- Details of injury/illness and of treatment.
- Subsequent movements of casualty (i.e. returned to work, sent home, sent to hospital, etc.).
- Name and signature of person providing treatment.

Personal protection

Because personal protection is limited to the user and the equipment must be worn for the duration of the exposure to the hazard, it should generally be considered as a last line of defence. Respiratory protection in particular should be restricted to hazardous situations of short duration (e.g. emergencies, maintenance, or temporary arrangements while engineering control measures are being introduced). Occasionally, personal protection may be the only practicable measure and a legal requirement. If it is to be effective, its selection, correct use and condition are of paramount importance. (Therefore within the UK the Personal Protective Equipment at Work Regulations 1992 require the provision of suitable p.p.e. except where and to the extent that a risk to health and safety is adequately controlled by other means. This has to be maintained, which covers: replacement or cleaning and keeping in an efficient state, in efficient working order and in good repair.)

Respiratory protection

Recommendations for the selection, use and maintenance of respiratory protective equipment are provided in the UK by BS 4275 1997 and BS EN 132. The two basic principles are:

- purification of the air breathed (respirator) *or*
- supply of oxygen from uncontaminated sources (breathing apparatus).

If the oxygen content of the contaminated air is deficient (refer to page 72), breathing apparatus is essential. The degree of protection required is determined by the level of contamination, the hygiene standard for the contaminant(s), the efficiency of any filter or adsorber available, and the efficiency with which the facepiece of the device seals to the user's face (this is reduced by beards, spectacles etc.). The level of comfort and user acceptability are further considerations.

The nominal protection factor (NPF) describes the level of protection afforded by a specific respirator:

$$\text{NPF} = \frac{\text{Concentration of contaminant in air}}{\text{Concentration of contaminant inside facepiece}}$$

The British Standard also uses the Assigned Protection Factor, i.e. the level of respiratory protection that can realistically be expected to be achieved in the workplace by 95% of adequately trained and supervised workers, using a properly functioning and correctly fitted respiratory protective device (see Table 13.9)

Table 13.9 Assigned Protection Factors for Respirators and Breathing Apparatus

Type	*APF (depending upon detailed equipment specification*
1 Filtering devices Half or quarter mask and filter Filtering half-mask without inhalation valves Valved filtering half masks Filtering half masks Full face mask and filter Power assisted filtering device with full, half or quarter masks Power filtering device with helmet or hood	**4–40**
2 Breathing apparatus Light duty construction air line Fresh air hose and compressed air line Self contained	**10–2000**

Gas masks (canister or cartridge mask)

Respirators for gases and vapours comprise a facepiece and a container filled with a specific adsorbent for the contaminant. Care must be taken to select the correct type. More than one canister can be attached.

The useful life of a canister should be estimated based on the probable concentration of contaminant, period of use, breathing rate and capacity of the canister.

Dust and fume masks

Dust and fume masks consist of one or two cartridges containing a suitable filter (e.g. paper or resin-impregnated wool) to remove particulate contaminant. The efficiency of the filters against particles of various sizes is quoted in manufacturers' literature and national standards. Such masks do not remove vapour from the air.

Facepiece fit is the limiting factor on the degree of protection afforded. Efficiency tends to increase with use, i.e. as the filter becomes loaded, but the resistance to breathing also increases. 'Paper cup' type masks are also available.

Powered dust masks

Masks are available with battery-powered filter packs which supply filtered air to a facepiece from a haversack filter unit. Another type comprises a protective helmet incorporating an electrically operated fan and filter unit complete with face vizor and provision for ear muffs.

Breathing apparatus

Compressed airline system: a facepiece or hood is connected to a filter box and hand-operated regulator valve which is provided with a safety device to prevent accidental complete closure. Full respiratory, eye and facial protection is provided by full-facepiece versions. The compressed air is supplied from a compressor through a manifold or from cylinders.

Self-contained breathing apparatus is available in three types:

- Open-circuit compressed air.

- Open-circuit oxygen-cylinder, liquid or solid-state generation.
- Regenerative oxygen.

All respiratory protective systems should be stored in clean, dry conditions but be readily accessible. They should be inspected and cleaned regularly, with particular attention to facepiece seals, non-return valves, harnesses etc. Issue on a personal basis is essential for regular use; otherwise the equipment should be returned to a central position. Records are required of location, date of issue, estimated duration of use of canisters etc.

Guidance on the choice of respiratory protection for selected environments is given in Figure 13.6. All persons liable to use such protection should be fully trained; this should cover details of hazards, limitations of apparatus, inspection, proper fitting of facepiece, testing, cleaning etc.

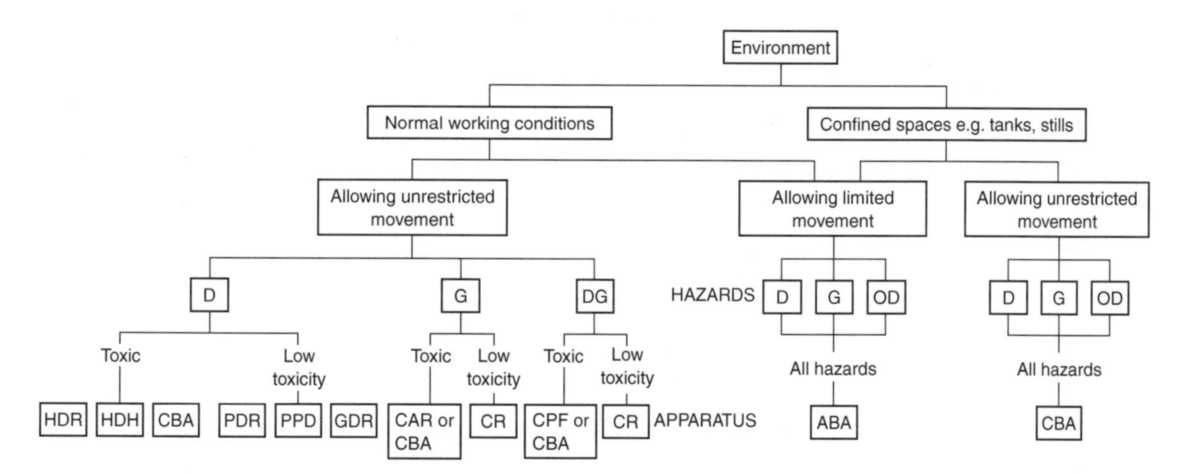

Figure 13.6 *Guide to the selection of respiratory protection*

Hazards	
D	*Dust*
G	*Gas*
DG	*Dust and gas*
OD	*Oxygen deficiency*

Equipment	
ABA	*Air line breathing apparatus*
CAR	*Canister respirator*
CBA	*Self-contained breathing apparatus*
CPF	*Canister respirators with particulate filters*
CR	*Cartridge respirator*
GDR	*General purpose dust respirators*
HDH	*High efficiency dust respirators*
HDR	*Positive pressure direct dust hoods and blouses (high efficiency)*
PDR	*Positive pressure powered dust respirators*
PPD	*Positive pressure powered dust hood and blouses*

Eye protection

The common-sense approach to the use of eye protection, includes:

- Eye protectors and/or fixed shields provided according to the nature of the process.
- Eye protectors on personal issue.
- Replacement of lost, defective or unsuitable eye protectors.
- Eye protectors suitable for individual needs.
- Eye protectors and shields that conform to an appropriate, approved specification and are marked to indicate intended use.

There are also requirements relating to the proper use and care of eye protectors/shields.

The normal range of personal eye protection is summarized in Table 13.10. In the UK the appropriate standard is BS EN 167, 168, 175 and BS 7028. Fixed shields can be of polycarbonate plastic to guard against splashing and projectiles, or of toughened glass or Perspex for protection against splashing only. Wrap-around of the hazard is required. If the need for access behind a shield cannot be eliminated personal protection is still necessary.

Table 13.10 Normal range of personal eye protection

Safety spectacles	A variety of styles and colours are available with different side-shields to protect against lateral hazards Attention must be given to both the lens (available toughened/tinted/to personal prescription for wearers requiring corrected vision) and the frame
Cup and wide-vision goggles	Tend to be more versatile and cheaper and offer more protection and, according to design, can protect against fine dust, fumes, liquid splashes and impact from flying particles Tend to be less comfortable than safety glasses and usually cannot be worn over ordinary spectacles Often mist up and as a rule are not fitted with prescription lenses
Face shields	Intended to protect from the forehead to the neck Some are attached to headgear and equipped with a chin-guard to prevent upward splashing of acids/alkalis The transparent screen is commonly made of polycarbonate Heavier special versions are available, e.g. for welding

Head protection

Head protection is required against falling objects but also serves against heat, chemical splashes, or entanglement of hair in machinery. The appropriate standard is BS EN 397 (1995).

Common-sense guidelines can be deduced from the requirements for 'building operations' and 'works of engineering construction' summarized in Table 13.11. In special situations, e.g. for fire-fighting or protection against toxic chemicals, hoods and suits cover the head and neck and many incorporate protection.

Table 13.11 Head protection requirements for construction activities (Construction (Head protection) Regulations 1989)

Provision, maintenance and replacement whenever necessary
Regular checking/replacement
Protection to be adequate, e.g. to BS EN 397 or equivalent standard (Type 1 helmets with comfort options strongly recommeded)
Protection to be compatible with the work/activity of the wearer
Protection to be worn at all times unless there is no risk from falling objects or possibility of hitting the head against something
Wearing of safety helmets to be ensured

Hand protection

Protection of the hands may rely upon gloves or barrier creams. The choice of gloves depends upon:

- Hazard to be protected against, e.g. corrosive or irritant chemicals, heat, abrasion.
- Degree of resistance required.
- Sensitivity required.
- Area to be protected, e.g. to wrist level or extending over the forearms.

For applications where surgical gloves provide adequate protection, if these are of natural rubber latex then powder-free gloves with low/undetectable protein and allergen levels are advisable. (Alternatively the use of nitrile gloves eliminates the risk of latex allergy.)

General advice on the suitability and degree of resistance of glove materials are summarized in Table 13.12. In all cases, however, in addition to the necessary mechanical properties (i.e. resistance to abrasion, blade cut resistance, tear resistance and puncture resistance rated according to European Standard EN 388), consideration should be given to the resistance to the specific chemicals involved.

Chemical permeation is rated in tests under EN 374-3. The breakthrough time of a chemical through a glove is quoted as a permeation index:

Breakthrough time (min) Greater than	10	30	60	120	240	480
Permeation index	1	2	3	4	5	6

If the time <10 minutes the index is 0. Clearly the greater the index the better the resistance to permeation.

Some glove suppliers also provide data on, or a noting of, degradation, i.e. the physical effects which a chemical will have upon a particular glove material.

Resistance to penetration by micro-organisms is covered by EN 374-2 and protection from low temperatures in EN 511. Careful handling and regular inspection are essential since chemicals and abrasion will eventually cause deterioration of gloves. Rinsing in clean water and drying naturally will prolong their life.

Barrier creams are applied before work to provide a protective film. A different type of cream, cleansing cream, is applied after work to aid dirt removal and to condition skin with humectant.

Limitations of barrier creams

- They may become a reservoir for harmful chemicals.
- Water-soluble preparations should not be used for aqueous situations; water-repellant creams are available for such applications.
- There is a limit to the quantity of chemical a barrier cream can repel.
- Effectiveness diminishes as a result of abrasion and massage.
- They are unlikely to prevent an allergic contact dermatitis.

Protective clothing

Protective clothing includes overalls, bibs, duffle coats, aprons, complete one-piece suits with hoods, spats, armlets etc. It is chosen for protection against mechanical hazards, abrasion, extremes of temperature etc. as well as chemicals. The nature of the hazard and user comfort dictate the choice. The properties of a range of protective clothing materials are listed in Table 13.13. Further guidance on selection, use and maintenance of chemical protective clothing is given in BS 7184:1989.

Table 13.12 Glove material resistance rating (courtesy James North & Sons Ltd)

Chemical	*Glove material*				
	Natural rubber	*Neoprene*	*Nitrile*	*Normal PVC*	*High grade PVC*
Organic acids					
Acetic acid	E	E	E	E	E
Citric acid	E	E	E	E	E
Formic acid	E	E	E	E	E
Lactic acid	E	E	E	E	E
Lauric acid	E	E	E	E	E
Maleic acid	E	E	E	E	E
Oleic acid	E	E	E	E	E
Oxalic acid	E	E	E	E	E
Palmitic acid	E	E	E	E	E
Phenol	E	E	G	E	E
Propionic acid	E	E	E	E	E
Stearic acid	E	E	E	E	E
Tannic acid	E	E	E	E	E
Inorganic acids					
Arsenic acid	G	G	G	E	E
Carbonic acid	G	G	G	E	E
Chromic acid (up to 50%)	G	F	F	E(1)	G
Fluorosilicic acid	G	G	G	E	G
Hydrochloric acid (up to 40%)	G	G	G	E	G
Hydrofluoric acid	G	G	G	E(1)	G
Hydrogen sulphide (acid)	F	F	G	E	E
Hydrogen peroxide	G	G	G	E	E
Nitric acid (up to 50%)	NR	NR	NR	G(1)	F(1)
Perchloric acid	F	G	F	E(1)	G
Phosphoric acid	G	G	G	E	G
Sulphuric acid (up to 50%)	G	G	F	E(1)	G
Sulphurous acid	G	G	G	E	E
Saturated salt solutions					
Ammonium acetate	E	E	E	E	E
Ammonium carbonate	E	E	E	E	E
Ammonium lactate	E	E	E	E	E
Ammonium nitrate	E	E	E	E	E
Ammonium nitrite	E	E	E	E	E
Ammonium phosphate	E	E	E	E	E
Calcium hypochlorite	NR	G	G	E	E
Ferric chloride	E	E	E	E	E
Magnesium chloride	E	E	E	E	E
Mercuric chloride	G	G	G	E	E
Potassium chromate	E	E	E	E	E
Potassium cyanide	E	E	E	E	E
Potassium dichromate	E	E	E	E	E
Potassium halides	E	E	E	E	E
Potassium permanganate	E	E	E	E	E
Sodium carbonate	E	E	E	E	E
Sodium chloride	E	E	E	E	E
Sodium hypochlorite	NR	F	F	E	E
Sodium nitrate	E	E	E	E	E
Solutions of copper salts	G	G	G	E	E
Stannous chloride	E	E	E	E	E
Zinc chloride	E	E	E	E	E
Alkalis					
Ammonium hydroxide	E	E	E	E	E
Calcium hydroxide	E	E	E	E	E
Potassium hydroxide	E	G	G	E	E

Table 13.12 Cont'd

Chemical	*Natural rubber*	*Glove material* *Neoprene*	*Nitrile*	*Normal PVC*	*High grade PVC*
Sodium hydroxide	E	G	G	E	E
Aliphatic hydrocarbons					
Hydraulic oil	F	G	F	G	E
Paraffins	F	G	E	G	E
Petroleum ether	F	G	E	F	G
Pine oil	G	G	E	G	E
Aromatic hydrocarbons[2]					
Benzene	NR	F	G	F	G
Naphtha	NR	F	F	F	G
Naphthalene	G	G	E	G	E
Toluene	NR	F	G	F	G
Turpentine	F	G	R	F	G
Xylene	NR	F	G	F	G[1]
Halogenated hydrocarbons[2]					
Benzyl chloride	F	F	G	F	G
Carbon tetrachloride	F	F	G	F	G
Chloroform	F	F	G	F	G
Ethylene dichloride	F	F	G	F	G
Methylene chloride	F	F	G	F	G
Perchloroethylene	F	F	G	F	G
Trichloroethylene	F	F	G	F	G
Esters					
Amyl acetate	F	G	G	F	G
Butyl acetate	F	G	G	F	G
Ethyl acetate	F	G	G	F	G
Ethyl butyrate	F	G	G	F	G
Methyl butyrate	F	G	G	F	G
Ethers					
Diethyl ether	F	G	E	F	G
Aldehydes					
Acetaldehyde	G	E	E	E	E
Benzaldehyde	F	F	E	G	E
Formaldehyde	G	E	E	E	E
Ketones					
Acetone	G	G	G	F	G
Diethyl ketone	G	G	G	F	G
Methyl ethyl ketone	G	G	G	F	G
Alcohols					
Amyl alcohol	E	E	E	E	E
Butyl alcohol	E	E	E	E	E
Ethyl alcohol	E	E	E	E	E
Ethylene glycol	G	G	E	E	E
Glycerol	G	G	E	E	E
Isopropyl alcohol	E	E	E	E	E
Methyl alcohol	E	E	E	E	E
Amines					
Aniline	F	G	E	E	E
Butylamine	G	G	E	E	E
Ethylamine	G	G	E	E	E
Ethylaniline	F	G	E	E	E
Methylamine	G	G	E	E	E
Methylaniline	F	G	E	E	E
Triethanolamine	G	E	E	E	E

Table 13.12 Cont'd

Chemical	*Glove material* *Natural rubber*	*Neoprene*	*Nitrile*	*Normal PVC*	*High grade PVC*
Miscellaneous					
Animal fats	F	G	G	G	E
Bleaches	NR	G	G	G	E
Carbon disulphide	NR	F	G	F	G
Degreasing solution	F	F	G	F	G
Diesel fuel	NR	F	G	F	E
Hydraulic fluids	F	G	G	G	E
Mineral oils	F	G	E	G	E
Ozone	F	E	G	E	E
Paint and varnish removers	F	G	G	F	G
Petrol	NR	G	G	F	G
Photographic solutions	G	E	E	G	E
Plasticizers	F	G	E	G	E
Printing inks	G	G	E	G	E
Refrigerant solutions	G	G	E	F	G
Resin oil	F	G	G	G	E
Vegetable oils	F	G	G	G	E
Weed killers	G	E	E	G	E
White spirit	F	G	G	F	G
Wood preservatives	NR	G	G	F	G

E Excellent
F Fair
G Good
NR Not recommended
(1) Resistance not absolute, but the best available.
(2) Aromatic and halogenated hydrocarbons will attack all types of natural and synthetic gloves. Should swelling occur, switch to another pair, allowing the swollen gloves to dry and return to normal.

Impervious clothing is essential when handling corrosive chemicals, liquids liable to cause dermatitis, or chemicals toxic by skin absorption.

All protective clothing should be maintained in a sound condition, cleaned/washed/replaced regularly as appropriate, and be stored apart from everyday clothing. With toxic chemicals a double locker system is advisable. With highly toxic substances, e.g. lead compounds or asbestos-based substances, separate storage of personal clothing and protective clothing is specifically required.

Protective footwear

Protective footwear includes shoes, boots with steel toecaps, full boots, e.g. Wellingtons. The choice of material determines durability, acid resistance, oil resistance, heat resistance, non-slip characteristics, impact resistance etc.

Washing facilities

Wherever chemicals are handled, adequate washing facilities are required conveniently situated with respect to the workplace. These comprise wash-basins or troughs with a constant supply of hot and cold or warm water; soap or liquid hand cleanser; clean towels or hot-air driers or disposable paper towels. Typical minimum standards are:

Normal work	One wash-basin per 10 workers
Handling toxic or irritant chemicals	One wash-basin per 5 workers

Table 13.13 Properties of protective clothing materials

Material	*Advantages*	*Disadvantages*	
Cotton	Lightweight, reasonably hard-wearing, no static generation Resists penetration of direct splashes of corrosive Unaffected by oils	Liable to shrinkage unless treated More flammable than wool Vulnerable to hot splashes	Suitable for under-gloves
Wool	Resists rapid penetration of direct splashes (more effective than cotton) Resists penetration of dust High absorption and porosity absorbs perspiration	Not resistant to hot splashes Takes up water and dirt Difficult to wash	
Artificial fibres (nylon, Terylene)	Hardwearing Terylene has good acid resistance	Not resistant to hot splashes High initial cost Can allow dust to pass through Static electricity can cause rapid soiling	Ceramic coating of fibres can render dust-proof
PVC	Impervious Non-flammable Chemically and biologically resistant Abrasion resistant	Not resistant to hot splashes Causes sweating unless well ventilated	Used alone or impregnated on fabric For gloves armlets, hats, bibs, spats, suits
Paper	Hygienic (disposable) Fairly resistant to chemicals if treated with polythene film	Liable to wetting if not treated with polythene film No strength Flammable Not resistant to hot splashes	Possibly used for disposable underwear or for clothing for visitors Used under headgear (disposable)
Polythene	Hygienic May be disposable Impervious	Low abrasion resistance Not resistant to hot splashes Low melting point could cause adhesion to skin	Suitable for disposable gloves or headgear

In some circumstances, where clothing has to be changed, bathing facilities, e.g. showers, are required. The provision of suitable and sufficient washing facilities and showers, if required by the nature of the work, at readily accessible places is required in the UK under Reg. 21 of the Workplace (Health, Safety and Welfare) Regulations 1992. Specific needs are also covered in other legislation, e.g. the Control of Lead at Work Regulations, and the Control of Asbestos at Work Regulations 1987 (as amended 1998).

Facilities for rest and food

Rest facilities should be provided in readily accessible places. To avoid contamination of food, or accidental ingestion of chemicals, these should include facilities to eat meals in a separate location.

Medical screening

The medical background of workers must be considered for work involving certain chemicals, e.g. radioactive substances, sensitizers. (Refer to Chapter 5.)

Monitoring standards

Management must institute procedures to assess levels of compliance with agreed standards for safety. Techniques include environmental and/or biological monitoring, health surveillance, safety audits, safety inspections, and procedures for accident reporting, investigation and analysis. Communication is essential, e.g. by provision of information (on specific chemicals, processes, etc.), safety meetings, notices, safety bulletins etc.

Training

Education, training and supervision are essential for the safe handling of chemicals. Training requirements vary according to position within the organization, and hence responsibility.

Topics should embrace a knowledge of the hazards and precautions, including the use and maintenance of protective devices including personal protection, under both normal and abnormal operating conditions including emergencies.

14

Marketing

Chemicals vary from commodity substances in bulk to household cleaning products in small packages. Mixtures of chemicals, such as formulated products, are usually termed 'preparations' which constitute 95% of commercially-available dangerous chemicals. Manufacturers, suppliers and importers of all chemicals have a legal obligation to ensure that their products are fit for use, properly packed, labelled and transported, and to provide the user with information on the hazards and precautions to ensure they can be used safely and without harm to the environment. (See Table 14.1 for UK legislation.) The hazards of dangerous substances, as opposed to dangerous preparations, must also appear in advertisements for those chemicals.

Important aspects of classification labelling and packaging chemicals are regulated in the UK by the Chemicals (Hazard Information and Packaging for Supply) Regulations 1994 ('CHIP'). These are applicable to any substance dangerous for supply excluding specific categories, e.g. cosmetic products, controlled drugs, foods, medicines and medical products, munitions, wastes, radioactive substances.

Classification

If a chemical is hazardous it is first classified into an appropriate category of danger to assist in the provision of the correct information and packaging.

This can be done by:

- use of the classification given to it by the supplier. (A check is necessary that the classification is accurate and also that processing has not invalidated it);
- use of an Approved Supply List from the UK Health and Safety Executive which covers most chemicals but not mixtures; or
- classification by reference to relevant information.

A list of danger categories is given in Table 14.2. Note that chemicals may possess several hazards, e.g. nitric acid is classed as both an oxidizer and a corrosive. If a chemical is not in one of these categories it is not generally considered to be dangerous. If the hazards of a new chemical have not been established it should be labelled 'Caution – substance not yet fully tested'. Mixtures can be classified either from results from tests on the preparation, or by calculation to predict the health effects of the product based on the properties of individual components and their concentration in the mixture. Preparations need to be classified for both physico-chemical and health effects but, to date, not for environmental effects.

Under CHIP there is a requirement for suppliers to exercise all due diligence in classifying a

Table 14.1 Legislation affecting the classification, packaging, labelling and carriage of dangerous substances in Great Britain (including their movement in harbours and harbour areas)

The Chemicals (Hazard Information and Packaging for Supply) Regulations 1994 as amended	Control the classification, provision of labels and safety data sheets, and safe packaging of chemicals by supplier.
The Carriage of Dangerous Goods by Road Regulations 1996	Covers the mode of dispatch of dangerous goods by road including bulk and tanker transport; suitability of containers and vehicles; examination and testing; information requirements; loading and unloading; procedures for emergencies and parking; exemptions.
The Carriage of Dangerous Goods (Classification, Packaging and Labelling) and Use of Transportable Pressure Receptacles Regulations 1996	Governs the classification of dangerous goods; consignment in packages and their markings/labelling; the design, manufacture, modification and repair of transportable pressure receptacles and their approval and certification, marking and filling; the role of approved persons and the need for records.
The Carriage of Dangerous Goods by Road (Driver Training) Regulations 1996	Addresses the instruction, training and certification of drivers of road vehicles used for the carriage of dangerous goods.
The Carriage of Explosives by Road Regulations 1996	Scope includes mode of transport in both passenger vehicles and bulk carriage; suitability of vehicle and container; approval; quantity limits; mixed loads; information and documentation; safety and security during carriage; equipment; precautions against fire and explosion and in the event of accidents and emergencies; age limits for those engaged in the carriage.
The Carriage of Dangerous Goods by Rail Regulations 1996	Governs the transport of dangerous goods by rail in small containers, large containers, tanks and wagons. Covers classification of dangerous goods; suitability, examination and testing of container; information including labelling. Also included is training and instruction; loading and unloading; marshalling; prohibition of overfilling, mixed loads and temperature-controlled substances; security and emergency procedures and special requirements for carriage of explosives.
The Packaging, Labelling and Carriage of Radioactive Materials by Rail Regulations 1996	Covers package design, approval, test procedures, notification of consignment, information requirements (for package and transport container).
Radioactive Material (Road Transport) (Great Britain) Regulations 1996	Includes radiation limits, prior approval, QA programmes, testing procedures, information/documentation requirements, and marking/labelling/placarding.
The Health and Safety at Work etc. Act 1974 (Application to Environmentally Hazardous Substances) Regulations 1996 as amended	Extends the scope of Health and Safety at Work etc. Act 1974 to include environmentally-hazardous substances including transportation of such dangerous goods and the control of volatile organic carbon emissions from storage and distribution of petrol.
The Ionizing Radiations Regulations 1999	Apply to all work activities with radioactive materials, including transport. The main provisions relevant to transport are those relating to driver training and the need, under some circumstances, to prepare contingency plans for emergencies and enforced stoppages.
The Dangerous Substances in Harbour Areas Regulations 1987 ('DSHA')	Control the carriage, loading, unloading and storage of all classes of dangerous substances, including explosives, in all harbours and harbour areas in Great Britain. Movement of dangerous substances by road, rail and sea is covered.

Table 14.1 Cont'd

The Classification and Labelling of Explosives Regulations 1983 ('CLER')	Require that explosives shall not be transported unless they have been classified by the classifying authority (i.e. the HSE, and the MOD for military explosives, in the UK) (except for journeys made specifically for classification and then only under certain conditions).

substance. Checks which may be required if reliance is placed on classification by a manufacturer are:

- reference to the Approved Supply List;
- use of experience about the reliability and experience of the supplier;
- use of common sense and experience;
- comparison with similar substances or preparations;
- comparison with any classification for carriage;
- enquiries with the supplier or other competent persons;
- checking with information in reference textbooks.

Packaging

A hazardous chemical must be supplied/conveyed in a package (drum, keg, cylinder, bottle, etc.) suitable for the purpose. In particular:

- The receptacle and any associated packaging must be designed, constructed, maintained and closed so as to prevent the escape of any of the contents of the receptacle when subjected to the stresses and strains of normal handling. A suitable safety device (e.g. pressure relief valve) may be fitted.
- The receptacle and any associated packagings (in so far as they are likely to come into contact with the dangerous substance) must be made of materials (a) which are not liable to be adversely affected by the substance, and (b) which are not liable to react with the substance to form any other dangerous substance.
- Where the receptacle is fitted with a replaceable closure, the closure must be designed so that the receptacle can be repeatedly reclosed without the contents escaping.
- The packaging of certain dangerous chemicals (e.g. those labelled 'toxic', 'very toxic' or 'corrosive' must be fitted with child-resistant closures, or tactile danger warnings (for the blind) if sold to the public.

Suppliers should perform a hazard analysis on each product. For example, aerosols can pose fire, explosion or inhalation risk. Basic precautions with an aerosol dispenser are:

- Assume that the aerosol contains a liquid of flammability equivalent to that of petrol and treat it as such.
- Do not use an aerosol spray when smoking, or near any other source of ignition, e.g. an electric fire, hot-plate, open-flame heater or boiler etc.
- Do not throw used aerosol dispensers onto bonfires.
- Do not let young children have access to aerosol dispensers, full or empty.

Table 14.2 Indication of dangers with chemical products

	Category of danger	*Indication of danger*	*Symbol*	*Symbol letter*
Physico-chemical	Explosive	Explosive		E
	Oxidizing	Oxidizing		O
	Extremely flammable	Extremely flammable		F+
	Highly flammable	Highly flammable		F
	Flammable	Flammable		–
Health	Very toxic	Very toxic		T+
	Toxic	Toxic		T
	Harmful	Harmful		Xn
	Corrosive	Corrosive		C
	Irritant	Irritant		Xi
	Sensitizing	Harmful		Xn
		Irritant		Xi
	Carcinogenic Categories 1 and 2	Toxic		T
	Category 3	Harmful		Xn
	Mutagenic Categories 1 and 2	Toxic		T
	Category 3	Harmful		Xn
	Toxic for reproduction Categories 1 and 2	Toxic		T
	Category 3	Harmful		Xn
Environmental	Dangerous for the environment	Dangerous for the environment		N

It is the responsibility of the supplier to draw these precautions to the attention of the user in the best way practicable.

Labelling

If a dangerous chemical is supplied in a package then the package must be labelled. Clearly this is not practical if the chemical is delivered in bulk or via a pipeline.

The elements of a label are:

- Full name and address and phone number of supplier.
- Chemical name or, if a preparation, the trade name.
- Indication of danger and associated symbol.
- Risk and safety phrases.

Requirements for labelling of containers for supply may differ from those for conveyance. Key features of a supply label are to identify the substance (the chemical name in most cases) and any hazards and safety precautions. Supply labels are black on yellow/orange. In Europe the classification, packaging and labelling of dangerous substances is covered by Directive 67/548/EEC as amended. This requires labels to identify appropriate risk and safety phrases (Tables 14.3 and 14.4) depending upon product properties. A substance is considered dangerous if in Part 1A of an approved list or if it exhibits hazardous properties as defined in Schedule 1 for supply, or Schedule 2 for conveyance as shown in Tables 14.5 and 14.6. Substances not tested should be labelled 'Caution – substance not yet fully tested'. Criteria for risk phrases are provided, e.g. as in Table 14.7 for toxic compounds.

Explosive chemicals tend to be governed by separate legislation, e.g., in the UK, The Classification and Labelling of Explosives Regulations 1983. These require the HSE to classify any explosive before it may be supplied. Under the scheme, explosives are labelled according to a classification based on hazard division (Table 14.8), and on compatibility (Table 14.9), which takes into account their sensitivity, explosivity and chemical nature. Labels are diamond shaped: the top half is reserved for the pictograph and division number, the bottom half shows the hazard code and the classification number. Figure 14.1 shows the label for Class 1, Division 1.1, 1.2 or 1.3 explosive.

Figure 14.1 *Example of label for explosive (Division 1.1, 1.2 or 1.3)*

Table 14.3 'Risk' phrases from the Approved List

1. Explosive when dry
2. Risk of explosion by shock, friction, fire or other sources of ignition
3. Extreme risk of explosion by shock, friction, fire or other sources of ignition
4. Forms very sensitive explosive metallic compounds
5. Heating may cause an explosion
6. Explosive with or without contact with air
7. May cause fire
8. Contact with combustible material may cause fire
9. Explosive when mixed with combustible material
10. Flammable
11. Highly flammable
12. Extremely flammable
13. Extremely flammable liquefied gas
14. Reacts violently with water
15. Contact with water liberates extremely flammable gases
16. Explosive when mixed with oxidizing substances
17. Spontaneously flammable in air
18. In use, may form flammable/explosive vapour–air mixture
19. May form explosive peroxides
20. Harmful by inhalation
21. Harmful in contact with skin
22. Harmful if swallowed
23. Toxic by inhalation
24. Toxic in contact with skin
25. Toxic if swallowed
26. Very toxic by inhalation
27. Very toxic in contact with skin
28. Very toxic if swallowed
29. Contact with water liberates toxic gas
30. Can become highly flammable in use
31. Contact with acids liberates toxic gas
32. Contact with acids liberates very toxic gas
33. Danger of cumulative effects
34. Causes burns
35. Causes severe burns
36. Irritating to eyes
37. Irritating to respiratory system
38. Irritating to skin
39. Danger of very serious irreversible effects
40. Possible risk of irreversible effects
41. Risk of serious damage to eyes
42. May cause sensitization by inhalation
43. May cause sensitization by skin contact
44. Risk of explosion if heated under confinement
45. May cause cancer
46. May cause heritable genetic damage
47. May cause birth defects
48. Danger of serious damage to health by prolonged exposure
49. May cause cancer by inhalation
50. Very toxic to aquatic organisms
51. Toxic to aquatic organisms
52. Harmful to aquatic organisms
53. May cause long-term adverse effects in the aquatic environment
54. Toxic to flora
55. Toxic to fauna
56. Toxic to soil organisms
57. Toxic to bees
58. May cause long-term adverse effects in the environment
59. Dangerous for the ozone layer

Table 14.3 Cont'd

60. May impair fertility
61. May cause harm to the unborn child
62. Possible risk of impaired fertility
63. Possible risk of harm to the unborn child
64. May cause harm to breastfed babies
65. Harmful: may cause lung damage if swallowed
66. Repeat exposure may cause skin dryness or cracking
67. Vapours may cause drowsiness and dizziness

Examples of combination of particular risks

14/15	Reacts violently with water, liberating extremely flammable gases
15/29	Contact with water liberates toxic, extremely flammable gas
20/21	Harmful by inhalation and in contact with skin
20/21/22	Harmful by inhalation, in contact with skin and if swallowed
20/22	Harmful by inhalation and if swallowed
21/22	Harmful in contact with skin and if swallowed
23/24	Toxic by inhalation and in contact with skin
23/24/25	Toxic by inhalation, in contact with skin, and if swallowed
23/25	Toxic by inhalation and if swallowed
24/25	Toxic in contact with skin and if swallowed
26/27	Very toxic by inhalation and in contact with skin
26/27/28	Very toxic by inhalation, in contact with skin and if swallowed
26/28	Very toxic by inhalation and if swallowed
27/28	Very toxic in contact with skin and if swallowed
36/37	Irritating to eyes and respiratory system
36/37/38	Irritating to eyes, respiratory system and skin
36/38	Irritating to eyes and skin
37/38	Irritating to respiratory system and skin
39/23	Toxic: danger of very serious irreversible effects through inhalation
39/23/24	Toxic: danger of very serious irreversible effects through inhalation and in contact with skin
39/23/24/25	Toxic: danger of very serious irreversible effects through inhalation, in contact with skin and if swallowed
39/23/25	Toxic: danger of very serious irreversible effects through inhalation, and if swallowed
39/24	Toxic: danger of very serious irreversible effects in contact with skin
39/24/25	Toxic: danger of very serious irreversible effects in contact with skin and if swallowed
39/25	Toxic: danger of very serious irreversible effects if swallowed
39/26	Very toxic: danger of very serious irreversible effects through inhalation
39/26/27	Very toxic: danger of very serious irreversible effects through inhalation and in contact with skin
39/26/27/28	Very toxic: danger of very serious irreversible effects through inhalation, in contact with skin and if swallowed
39/26/28	Very toxic: danger of very serious irreversible effects through inhalation and if swallowed
39/27	Very toxic: danger of very serious irreversible effects in contact with skin
39/27/28	Very toxic: danger of very serious irreversible effects in contact with skin and if swallowed
39/28	Very toxic: danger of very serious irreversible effects if swallowed
40/20	Harmful: possible risk of irreversible effects through inhalation
40/20/21	Harmful: possible risk of irreversible effects through inhalation and in contact with skin
40/20/21/22	Harmful: possible risk of irreversible effects through inhalation in contact with skin and if swallowed
40/20/22	Harmful: possible risk of irreversible effects through inhalation and if swallowed
40/22	Harmful: possible risk of irreversible effects if swallowed
40/21	Harmful: possible risk of irreversible effects in contact with skin
40/21/22	Harmful: possible risk of irreversible effects in contact with skin and if swallowed
42/43	May cause sensitization by inhalation and skin contact
48/20	Harmful: danger of serious damage to health by prolonged exposure through inhalation
48/20/21	Harmful: danger of serious damage to health by prolonged exposure through inhalation and in contact with skin
48/20/21/22	Harmful: danger of serious damage to health by prolonged exposure through inhalation in contact with skin and if swallowed
48/20/22	Harmful: danger of serious damage to health by prolonged exposure through inhalation and if swallowed
48/21	Harmful: danger of serious damage to health by prolonged exposure in contact with skin
48/21/22	Harmful: danger of serious damage to health by prolonged exposure in contact with skin and if swallowed
48/22	Harmful: danger of serious damage to health by prolonged exposure if swallowed
48/23	Toxic: danger of serious damage to health by prolonged exposure through inhalation

Table 14.3 Cont'd

48/23/24	Toxic: danger of serious damage to health by prolonged exposure through inhalation and in contact with skin
48/23/24/25	Toxic: danger of serious damage to health by prolonged exposure through inhalation, in contact with skin and if swallowed
48/23/25	Toxic: danger of serious damage to health by prolonged exposure through inhalation and if swallowed
48/24	Toxic: danger of serious damage to health by prolonged exposure in contact with skin
48/24/25	Toxic: danger of serious damage to health by prolonged exposure in contact with skin and if swallowed
48/25	Toxic: danger of serious damage to health by prolonged exposure if swallowed
50/53	Very toxic to aquatic organisms, may cause long-term adverse effects in the aquatic environment
51/53	Toxic to aquatic organisms, may cause long-term adverse effects in the aquatic environment
52/53	Harmful to aquatic organisms, may cause long-term adverse effects in the aquatic environment

Table 14.4 'Safety' phrases from the Approved List

1. Keep locked up
2. Keep out of reach of children
3. Keep in a cool place
4. Keep away from living quarters
5. Keep contents under . . . (appropropriate liquid to be specified by the manufacturer)
6. Keep under . . . (inert gas to be specified by the manufacturer)
7. Keep container tightly closed
8. Keep container dry
9. Keep container in a well-ventilated place
12. Do not keep the container sealed
13. Keep away from food, drink and animal feeding stuffs
14. Keep away from . . . (incompatible materials to be indicated by the manufacturer)
15. Keep away from heat
16. Keep away from sources of ignition – No Smoking
17. Keep away from combustible material
18. Handle and open container with care
20. When using do not eat or drink
21. When using do not smoke
22. Do not breathe dust
23. Do not breathe gas/fumes/vapour/spray (appropriate wording to be specified by manufacturer)
24. Avoid contact with skin
25. Avoid contact with eyes
26. In case of contact with eyes, rinse immediately with plenty of water and seek medical advice
27. Take off immediately all contaminated clothing
28. After contact with skin, wash immediately with plenty of . . . (to be specified by the manufacturer)
29. Do not empty into drains
30. Never add water to this product
33. Take precautionary measures against static discharges
35. This material and its container must be disposed of in a safe way
36. Wear suitable protective clothing
37. Wear suitable gloves
38. In case of insufficient ventilation, wear suitable respiratory equipment
39. Wear eye/face protection
40. To clean the floor and all objects contaminated by this material use . . . (to be specified by the manufacturer)
41. In case of fire and/or explosion do not breathe fumes
42. During fumigation/spraying wear suitable respiratory equipment (appropriate wording to be specified by the manufacturers)
43. In case of fire, use . . . (indicate in the space the precise type of fire-fighting equipment. If water increases the risk, add – Never use water)
45. In case of accident or if you feel unwell, seek medical advice immediately (show the label where possible)
46. If swallowed seek medical advice immediately and show this container or label
47. Keep at temperature not exceeding . . . °C (to be specified by the manufacturer)
48. Keep wetted with . . . (appropriate material to be specified by the manufacturer)
49. Keep only in the original container
50. Do not mix with . . . (to be specified by the manufacturer)
51. Use only in well-ventilated areas

Table 14.4 Cont'd

52. Not recommended for interior use on large surface areas
53. Avoid exposure – obtain special instructions before use
56. Dispose of this material and its container to hazardous or special waste collection point
57. Use appropriate containment to avoid environmental contamination
59. Refer to manufacturer/supplier for information on recovery/recycling
60. This material and/or its container must be disposed of as hazardous waste
61. Avoid release to the environment. Refer to special instructions/safety data sheet
62. If swallowed, do not induce vomiting: seek medical advice immediately and show this container or label
63. In case of accident by inhalation: remove casualty to fresh air and keep at rest
64. If swallowed, rinse mouth with water (only if the person is conscious)

Examples of combination of safety precautions

1/2	Keep locked up and out of the reach of children
3/9/14	Keep in a cool well-ventilated place away from [incompatible materials to be indicated by manufacturer]
3/9/14/49	Keep only in the original container in a cool well-ventilated place away from [incompatible materials to be indicated by manufacturer]
3/9/49	Keep only in the original container in a cool well-ventilated place
3/14	Keep in a cool place away from [incompatible materials to be indicated by the manufacturer]
7/8	Keep container tightly closed and dry
7/9	Keep container tightly closed and in a well-ventilated place
7/47	Keep container tightly closed and at a temperature not exceeding [ ]°C [to be specified by manufacturer]
20/21	When using do not eat, drink or smoke
24/25	Avoid contact with skin and eyes
29/56	Do not empty into drains, dispose of this material and its container to hazardous or special waste collection point
36/37	Wear suitable protective clothing and gloves
37/38/39	Wear suitable protective clothing, gloves and eye/face protection
36/39	Wear suitable protective clothing and eye/face protection
37/39	Wear suitable gloves and eye/face protection
47/49	Keep only in the original container at temperature not exceeding [ ]°C [to be specified by manufacturer]

Information

Suppliers must also provide the customer with more detailed information on the hazards and safety precautions of their products. If the chemicals are to be used in connection with work this is usually in a Material Safety Data Sheet. Relevant information from Table 1.10 should be included and at least cover that listed in Table 14.10.

The information must be provided under the headings shown in Table 14.11 and enable the user to determine the precautions required when using the chemical.

Where necessary the manufacturer must carry out, or arrange for, safety testing. Many countries operate mandatory premanufacturing and premarketing notification schemes of which safety testing is the cornerstone. Within the European Community under Directive 67/548/EEC and its sixth amendment 79/831/EEC, Competent Authorities must be notified before new substances are supplied in the marketplace. In the UK this Directive is enacted by the Notification of New Substances Regulations 1993 (NONS). Exempt are:

- Substances listed on the European Inventory of Existing Chemical Substances (EINECS).
- Medicinal products.

Table 14.5 Classification of and symbols for substances dangerous for supply (Schedule 1 Part I)

Characteristic properties of the substance	*Classification and indication of general nature of risk*	*Symbol*
A substance which may explode under the effect of flame or which is more sensitive to shocks or friction than dinitrobenzene.	Explosive	
A substance which gives rise to highly exothermic reaction when in contact with other substances, particularly flammable substances.	Oxidizing	
A liquid having a flash point <0°C and a boiling point ≤35°C.	Extremely flammable[1]	
A substance which (a) may become hot and finally catch fire in contact with air at ambient temperature without any application of energy; (b) is a solid and may readily catch fire after brief contact with a source of ignition and which continues to burn or to be consumed after removal of the source of ignition; (c) is gaseous and flammable in air at normal pressure; (d) in contact with water or damp air, evolves highly flammable gases in dangerous quantities; or (e) is a liquid having a flash point <21°C.	Highly flammable[1]	
A substance which is a liquid having a flash point ≥21°C and ≤55°C, except a liquid which when tested at 55°C in the manner described in Schedule 2 to the Highly Flammable Liquids and Liquefied Petroleum Gases Regulations 1972 does not support combustion.	Flammable[1]	No symbol required
A substance which if it is inhaled or ingested or it penetrates the skin, may involve extremely serious acute or chronic health risks and even death.	Very toxic[2]	
A substance which if it is inhaled or ingested or it penetrates the skin, may involve serious acute or chronic health risks and even death.	Toxic[2]	
A substance which if it is inhaled or ingested or it penetrates the skin, may involve limited health risks.	Harmful[2]	
A substance which may on contact with living tissues destroy them.	Corrosive	
A non-corrosive substance which, through immediate, prolonged or repeated contact with the skin or mucous membrane, can cause inflammation.	Irritant	

[1] Preparations packed in aerosol dispensers shall be classified as flammable in accordance with Part III of this schedule.
[2] Substances shall be classified as very toxic, toxic or harmful in accordance with the additional criteria set out in Part II of this schedule.

Table 14.6 Classification of and hazard warning signs for substances dangerous for conveyance (Schedule 2 Part I)

Characteristic properties of the substance	*Classification*	*Hazard warning sign*
A substance which (a) has a critical temperature <50°C or which at 50°C has a vapour pressure of more than 3 bar absolute and (b) is conveyed by road at a pressure of more than 500 millibars above atmospheric pressure or in liquefied form other than a toxic gas or a flammable gas.	Non-flammable compressed gas	COMPRESSED GAS
A substance which has a critical temperature <50°C or which at 50°C has a vapour pressure of more than 3 bar absolute and which is toxic.	Toxic gas	TOXIC GAS
A substance which has a critical temperature <50°C or which at 50°C has a vapour pressure of more than 3 bar absolute and is flammable.	Flammable gas[1]	FLAMMABLE GAS
A liquid with a flash point ≤55°C except a liquid which (a) has a flash point ≥21°C and ≤55°C and (b) when tested at 55°C in the manner described in Schedule 2 to the Highly Flammable Liquids and Liquefied Petroleum Gases Regulations 1972 does not support combustion.	Flammable liquid[1,2]	FLAMMABLE LIQUID
A solid which is readily combustible under conditions encountered in conveyance by road or which may cause or contribute to fire through friction.	Flammable solid	FLAMMABLE SOLID
A substance which is liable to spontaneous heating under conditions encountered in conveyance by road or to heating in contact with air being then liable to catch fire.	Spontaneously combustible substance	SPONTANEOUSLY COMBUSTIBLE

Table 14.6 Cont'd

Characteristic properties of the substance	*Classification*	*Hazard warning sign*
A substance which in contact with water is liable to become spontaneously combustible or to give off a flammable gas.	Substance which in contact with water emits flammable gas	DANGEROUS WHEN WET
A substance other than an organic peroxide which, although not itself necessarily combustible, may by yielding oxygen or by a similar process cause or contribute to the combustion of other material.	Oxidizing substance	OXIDIZING AGENT
A substance which is (a) an organic peroxide and (b) an unstable substance which may undergo exothermic self-accelerating decomposition.	Organic peroxide	ORGANIC PEROXIDE
A substance known to be so toxic to man as to afford a hazard to health during conveyance or which, in the absence of adequate data on human toxicity, is presumed to be toxic to man.	Toxic substance	TOXIC
A substance known to be toxic to man or, in the absence of adequate data on human toxicity, is presumed to be toxic to man but which is unlikely to afford a serious acute hazard to health during conveyance.	Harmful substance	HARMFUL – STOW AWAY FROM FOODSTUFFS
A substance which by chemical action will (a) cause severe damage when in contact with living tissue; (b) materially damage other freight or equipment if leakage occurs.	Corrosive substance	CORROSIVE

Table 14.6 Cont'd

Characteristic properties of the substance	*Classification*	*Hazard warning sign*
A substance which is listed in Part 1A of the approved list and which may create a risk to the health or safety of persons in the conditions encountered in conveyance by road, whether or not it has any of the characteristic properties set out above.	Other dangerous substance	! DANGEROUS SUBSTANCE
Packages containing two or more dangerous substances which have different characteristic properties.	Mixed hazards	! DANGEROUS SUBSTANCE

(1) An aerosol which is flammable in accordance with paragraph 2 or Part III of Schedule 1 shall have the classification of a flammable gas. Other aerosols need not be classified as flammable gas or flammable liquid.
(2) Viscous preparations which comply with the conditions in Part III of this schedule shall not be required to be classified as flammable.

Table 14.7 Criteria for the toxicity classification of substances

Category(1)	*LD_{50} absorbed orally in rat* (mg/kg)	*LD_{50} absorbed percutaneously in rat or rabbit* (mg/kg)	*LC_{50} absorbed by inhalation in rat* (mg/l, 4 hr)
Very toxic	≤25	≤50	≤0.5
Toxic	>25 to 200	>50 to 400	>0.5 to 2
Harmful	>200 to 2000	>400 to 2000	>2 to 20

(1) 'Very toxic', 'toxic' or 'harmful', applied to a dangerous substance not classified in accordance with Schedules 3, 4 and 5, means that the substance has a toxicity falling within the range tabulated for that category.
Note
If a substance produces other effects that make it inadvisable to use the LD_{50} or LC_{50} value as the principal basis of classification, the substance should be classified according to the magnitude of these effects.

Table 14.8 Hazard classification of explosives (UK)*

Division	*Division No.*
Substances and articles that have a mass explosion hazard	1.1
Substances and articles that have a projection hazard but not a mass explosion hazard	1.2
Substances and articles that have a fire hazard and either a minor blast hazard or a minor projection hazard but not a mass explosion hazard	1.3
Substances and articles that present no significant hazard	1.4
Very insensitive substances that have a mass explosion hazard	1.5
Extremely insensitive articles which do not have a mass explosion hazard	1.6

*See also UN recommendations on transportation of dangerous goods and the IARTA Regulations (page 460).

Table 14.9 Compatibility groups for explosives (UK)

Compatibility group	*Designating letter*
Primary explosive substance	A
Article containing a primary explosive substance and not containing two or more independent safety features	B
Propellant explosive substance or other deflagrating explosive substance or article containing such explosive substance	C
Secondary detonating explosive substance or black powder or article containing a secondary detonating explosive substance, in each case without means of initiation and without a propelling charge, or article containing a primary explosive substance and containing two or more independent safety features	D
Article containing a secondary detonatng explosive substance, without means of initiation and with a propelling charge (other than a charge containing a flammable or hypergolic liquid)	E
Article containing a secondary detonating explosive substance, with means of initiation and either with a propelling charge (other than a charge containing a flammable or hypergolic liquid) or without a propelling charge	F
A substance that is an explosive substance because it is designed to produce an effect by heat, light, sound, gas or smoke or a combination of these as a result of non-detonative self-sustaining exothermic chemical reactions or an article containing such a substance which is explosive because it is capable by chemical reaction in itself of producing gas at such a temperature and pressure and at such a speed as could cause damage to surroundings and an illuminating, incendiary, lachrymatory or smoke-producing substance (other than a water-activated article or one containing white phosphorus, phosphide or a flammable liquid or gel)	G
Article containing both an explosive substance and white phosphorus	H
Article containing both an explosive substance and a flammable liquid or gel	J
Article containing both an explosive substance and a toxic chemical agent	K
Explosive substance or explosive article presenting a special risk needing isolation of each type	L
Substance or article so packed or designed that any hazardous effect arising from accidental functioning is confined within the package unless the package has been degraded by fire, in which case all blast or projection effects are limited to the extent that they do not significantly hinder or prohibit fire-fighting or other emergency response efforts in the immediate vicinity of the package	S

- Plant protection products.
- Cosmetics.
- Foodstuffs including additives and flavourings.
- Animal feeding stuffs.
- Radioactive substances.
- Waste substances.
- Substances in transit under customs supervision.
- Substances classed as 'no longer polymers'. These are listed polymers which were considered to be polymers under the NONS 82 definition but not by the NONS 93 version. Unlisted substances *may* also be exempt if they could have been considered under NONS 82, or are not a polymer under NONS 93, or were placed on the market between 1 January 1971 and 1 November 1993.

Table 14.10 Minimum information for a Material Safety Data Sheet

- *Identification*
  Identification of product by trade name. Supplier's name and address plus a telephone number (for emergencies).
- *Chemical composition*
  Chemical composition; details of ingredients, formulae and proportions, including measures to be taken for over-exposure to eyes/skin or by inhalation/ingestion/absorption. Information for medical treatment.
- *Disposal – normal and emergency spills*
  General precautions, disposal methods and statutory controls.
  Clean-up, neutralization, disposal methods for spills.
- *Fire and explosion hazards and precautions*
  Flammable limits, flash point, auto-ignition temperature etc.
  Reactivity, stability.
  Fire-fighting procedures.
- *Health hazards*
  Acute (short-term) effects and chronic (long-term) effects.
  Special hazards.
  First detectable signs of over-exposure by all means of entry.
- *Environmental hazards*
- *Test data*
- *Exposure limits*
  Threshold Limit Values or Occupational Exposure Limits (preferably with reference to their interpretation, i.e. not as 'safe' levels).
  For mixtures – reference to prediction, or substance used for basis.
- *Legal requirements*
  Any legislation applicable to safe handling, storage and use.
- *Control measures*
  Containment, ventilation, means of limiting exposure generally.
- *Personal protection*
  Identify requirements and circumstances for use.
  Emergency requirements.
- *Storage and handling*
  Conditions of storage, segregation, materials of construction.
  Types of container and handling precautions.
  Where relevant, procedures for disposal of containers, e.g. aerosol dispensers, drums when 'empty'.
- *References*
  References consulted and further sources of information.

Table 14.11 Headings for information supplied on a Safety Data Sheet

Identification of the substance/preparation and company
Composition/information on ingredients
Hazards identification
First-aid measures
Fire-fighting measures
Accidental release measures
Handling and storage
Exposure controls/personal protection
Physical and chemical properties
Stability and reactivity
Toxicological information
Ecological information
Disposal consideration
Transport information
Regulatory information
Other information

- Substances intended solely for scientific research and development in quantities less than 100 kg per year.
- Substances placed on the market for process-orientated research and development with a limited number of customers registered with the notifier.

The aim of the notification systems is to identify possible risks posed to people and the environment from placing new substances on the market. A notification comprises:

- data on the chemical identity of the substance;
- an estimate of the quantity of the substance to be placed on the EC market;
- details of the functions and uses of the substance;

Table 14.12 Notification bands*

Type of notification	*Information requirements*	*Quantity placed on the market*
Deemed to have been notified	– 0 kg	
REDUCED	Acute toxicity Indications for use Chemical identity	10 kg
REDUCED	Additional acute toxicity Acute daphnia toxicity Biodegradability Physico-chemical properties	100 kg/annum or 500 kg cumulative
BASE SET	Mutagenicity Toxicity to reproduction Toxicity to algae Acute daphnia and fish toxicity Abiotic and readily biotic degradability Additional physico-chemical properties	1 t/annum or 5 t cumulative
'Lower' LEVEL 1	Possibly some of the information from 'Upper' LEVEL 1	10 t/annum or 50 t cumulative
'Upper' LEVEL 1	Chronic toxicity Toxicity in soil and plants Additional mutagenicity Long-term toxicity Bioaccumulation Inherent biodegradability Additional abiotic degradability	100 t/annum or 500 t cumulative
LEVEL 2	Carcinogencity Additional chronic toxicity Additional environmentally dangerous properties Toxicity to birds Long-term toxicity in water and soil Degradability simulation tests Additional abiotic degradability Mobility in water, soil and air	1000 t/annum or 5000 t cumulative

*A complete list of the information required for each type of notification is given in Schedules 2 and 3 of the Regulations.

- physico-chemical, toxicological and ecotoxicological properties of the substance;
- recommended precautions, disposal and emergency procedures;
- proposals for classification and labelling.

The notification dossier may also contain a risk assessment prepared by the notifier. The precise contents of a notification will depend on, e.g., the quantity of the substance to be placed on the market and, to an extent, on the properties and uses of the substance. The thresholds for notification and a summary of the information requirements are given in Table 14.12.

All testing to support notification must be performed by methods specified in Annex V to Directive 79/831/EEC and in accordance with the principles of good laboratory practice (GLP). GLP is concerned with the organizational processes and conditions under which laboratory studies are planned, performed, monitored, recorded and reported.

Laboratories wishing to claim GLP compliance are normally registered with Competent Authorities who issue statements of compliance following successful periodic inspections of the premises to monitor compliance with the relevant legislation. In the UK compliance with the Good Laboratory Practice Regulations 1999 is audited by the Medicines Control Agency.

15
Transport of chemicals

Transportation of a chemical between the supplier and/or agent and the user may introduce special risks in addition to those which can arise from the inherent chemical and physical properties. Comprehensive labelling and identification procedures, and information supply, covered in Chapter 14, aim to avoid any confusion over the nature of the chemical and recommended handling, storage and emergency procedures. However, different parties have to oversee, and take responsibility for, the chemical at different stages during its movement between sites, e.g. loading/storage, transport by various methods, storage in transit, unloading. Moreover, third parties can be prone to injury, and property or the environment can be subjected to damage, if loss of containment occurs. Security is a concern throughout.

The transportation of chemicals is therefore highly regulated. In addition to general transport legislation numerous statutory provisions govern the carriage of hazardous substances by air, road, rail and sea. These tend to address the need for:

- documentation;
- suitability of container, examination, testing and certification of tanks;
- classification, packaging and labelling of dangerous goods;
- carriage information;
- duties of consignors, operators, drivers, etc.;
- adequate information and warning labels to be displayed on containers, tanks, and wagons;
- training of crews/drivers and others engaged in the transport process;
- loading and unloading;
- security and other safety measures including emergency arrangements;
- special requirements.

Some categories of hazardous materials are covered by specific regulations as exemplified in the UK by the Carriage of Explosives by Road Regulations 1996; the Radioactive Material (Road Transport) (Great Britain) Regulations 1996; and the Packaging, Labelling and Carriage of Radioactive Material by Rail Regulations 1996. It is the duty of all those engaged in the transportation of hazardous chemicals to identify the relevant legislation and to comply with all appropriate detailed requirements. In addition to local legislation and guidance, the transportation of chemicals by air, rail, road and sea is the subject of international conventions, e.g. the Intergovernmental Marine Consultative Organization Dangerous Goods Code, European Agreement concerning the International Carriage of Dangerous Goods by Road, The International Regulations concerning the Carriage of Dangerous Goods by Rail, and the International Air Transport Association Restricted Articles Regulations. Table 15.12 lists dangerous goods forbidden for transport by air.

This chapter provides an insight into general principles associated with the transportation of hazardous substances; reference is made to selected regulations for illustrative purposes.

Irrespective of the mode of carriage, the safe transport of chemicals is the responsibility of:

- the producer;
- any hauliers involved; and
- the recipient, e.g. during unloading.

Effective communication between them is a vital consideration.

Road transport

Dangerous substances

Typical division of responsibilities in a road transport operation are shown in Figure 15.1. Producers transporting their own products need to establish a system to ensure that all necessary information and instructions are provided for safe handling under all foreseeable conditions. This will require:

- The enforcement of codes of practice, working procedures and general instructions incorporating all possible safety factors.
- Specific training and instruction of transport staff on actions to be taken in emergencies, e.g. fire, spillage on land, spillage entering drains.
- The institution, or development in consultation with the customer, of a safe unloading procedure (e.g. as exemplified in Table 15.18 for non-pressurized hazardous liquids).

These need to comply with all statutory legislation relevant to the specific chemical (e.g. if dangerous substances are carried into a UK port the operator must, under the Dangerous Substances in Harbour Areas Regulations 1987, give prior notice to the Harbour Master and in some cases the berth operator). The vehicles must be of the correct design and construction for the load carried and be maintained in sound condition. Various regulations, codes of practice and guidelines cover the labelling of containers and vehicles to identify the substances and their hazards in an emergency.

In the UK, the conveyance by road of dangerous chemicals in road tankers of any capacity and in tanks exceeding 3 m^3 is controlled by the Carriage of Dangerous Goods by Road Regulations 1996 as amended. Dangerous substances are defined as those named in an approved list or falling into the classifications in a schedule. Except for a slightly different definition for 'other dangerous substances' and 'multi-load', the latter classification is similar to that in Table 14.1.

Conveyance includes loading, transportation, unloading, cleaning and purging. A check-list of selected operator duties (excluding duties in relation to the unloading of petroleum spirit, covered by Schedule 4) arising where appropriate under these regulations is given in Table 15.1. The regulations also provide instruction on the form of hazard warning labels (Figure 15.2). In the event of an emergency, drivers should normally carry out the actions listed in Table 15.2.

The provision of written instructions in the form of a TREMCARD developed by CEFIC is exemplified by Figure 15.3.

Explosives

For the carriage of explosives by road the nature of the substance, the suitability of both the container and vehicle, and the quantity of explosive carried are paramount considerations. Private

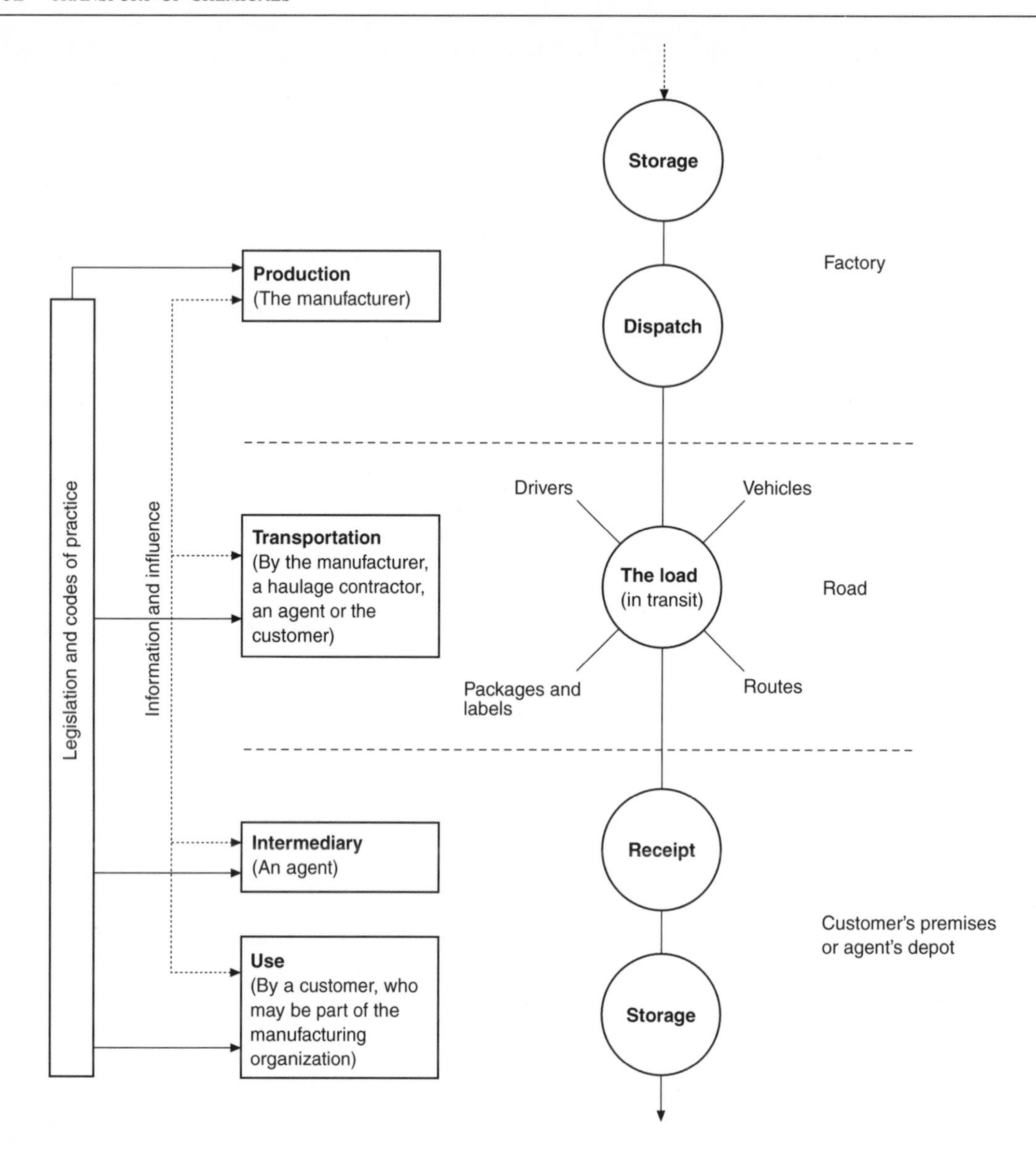

Figure 15.1 *Responsibilities for safe transport of chemicals by road*

light goods vehicles may be suitable only for small quantities of explosives (other than those in Compatibility Group S) and an ordinary goods vehicle is suitable for an intermediate quantity of most types of explosives. However, specially constructed vehicles are essential for more dangerous types of explosives or for the largest loads permitted under regulations. Cognizance must be taken of whether the vehicle is fuelled by diesel oil, petrol or LPG and whether or not it is an articulated vehicle. For the purposes of the UK Carriage of Explosives by Road Regulations 1996, road vehicles are classified as I, II or III (Table 15.3 – with type I defined as a vehicle other than type II or III). The characteristics of type II and III vehicles are also summarized in Table 15.3. (Whilst specialized, this illustrates the range of design features available for vehicles carrying thermally-unstable chemicals generally.) The selection of vehicle for a particular task will be influenced by

Table 15.1 Selected Operator duties under the Carriage of Dangerous Goods by Road Regulations 1996*

Operator of tanker or tank container	Ensure compliance with requirements specified in relevant approved documents and schedules to the regulations. Dangerous goods only transported if the letter Y appears in appropriate columns of Approved Carriage list and Schedules 5 and 6 are complied with, and tank pressures do not exceed the maximum working pressure as per certifications. Ensure specialist advice is always available on hazardous contents. Ensure the tank, container or vehicle is suitable for journey taking cognizance of the properties and volume of load. Ensure the vehicle has no more than one trailer or semi-trailer. Ensure packages of sensitive materials are suitably protected. Ensure tanks are certified by competent authority/approved person. Pass tank certificates to new owner when ownership changes. Obtain consignor's declaration where relevant. Obtain from consignor information listed in Section 13. Take reasonable steps to ensure goods are fit for carriage. Not permit driver to; • carry anyone other than crew member; • open package unless authorized; • bring sources of ignition into vehicle; • carry food unless adequately protected. Provide sub-contract operators and drivers relevant information from Section 14. Display information as per Schedule 10. Keep transport documentation relating to each journey for 3 months after completion of journey. Prevent carriage of materials which are liable to explosion. Ensure loading, stowage, unloading and cleaning of containers, tanks and containers creates no additional risk. Clean contaminated vehicles before re-loading. Establish emergency and parking arrangements as set out in Part VI.

* The full SI 1996 No. 2095 should be consulted for complete details.

the explosive Division Group (see Table 14.8), the Compatibility Group (see Table 14.9) and the quantity to be transported, as summarized by Table 15.4.

Radioactive substances

The transportation of radioactive material by road in the UK is prohibited:

Figure 15.2 *Hazard warning panels and labels*

Space 1 **Emergency action codes** *specified for that substance in Column 3 of Part 1 of the approved list*
Space 2 **Substance identification number and name** *as specified for that substance in Part 1 of the approved list*
Space 3 **Hazard warning sign for the classification** *of the substance as specified in Part 1 of the approved list*
Space 4 **Telephone number** *where expert advice can be obtained at all times when the substance is being conveyed by road*
Space 5 **Name of the manufacturer or owner** *of the substance*

Table 15.2 Emergency procedures for drivers

Arrange for the police and emergency services to be alerted
Arrange for assistance to be given to any person injured or in immediate danger
If considered safe to do so – having regard to the nature of the emergency, the substance and the emergency equipment available – follow a selection of the following procedures in an appropriate order:
- Stop the engine
- Turn off any battery isolating switch
- If there is no danger of ignition, operate the emergency flashing device
- Move the vehicle to a location where any leakage would cause less harm
- Wear appropriate protective clothing
- Keep onlookers away
- Place a red triangle warning device at the rear of the vehicle and near any spillage
- Prevent smoking and direct other vehicles away from any fire risk area

Upon the police/fire brigade taking charge:
- Show the written information, e.g. Tremcard, to them
- Tell them of action taken and anything helpful about the load, etc.

At the end of the emergency, inform the operator.
The written information given to the driver should include:
- The name of the substance
- Its inherent dangers and appropriate safety measures
- Action and treatment following contact/exposure
- Action in the event of fire and fire-fighting equipment to be used
- Action following spillage on the road
- How and when to use any special safety equipment

- in a public service vehicle;
- in a vehicle carrying explosives;
- generally in damaged packages;
- if inadequately labelled.

Radiation limits for individual packages for shipment by road are dictated by the type of package and the activity of the radionuclide. The Radioactive Material (Road Transport) (Great Britain) Regulations provide rules for calculating package limits. Radiation limits for road consignments in the UK under 'exclusive use' are given in Table 15.5. Instructions are provided on restrictions;

UN No. 1088 | CEFIC TEC (R)-687 | April 1974 | Class IIIa ADR

TRANSPORT EMERGENCY CARD
(Road)

Cargo
ACETAL
(Acetaldehyde Diethyl Acetal)
Colourless liquid with perceptible odour.
Immiscible with water, lighter than water.

Nature of hazard
Highly inflammable (flashpoint below 21°C).
Volatile.
The vapour is invisible, heavier than air and spreads along ground.
Can form explosive mixture with air particularly in empty uncleaned receptacles.
Heating will cause pressure rise, with risk of bursting and subsequent explosion.
The vapour has narcotic effect.
The substance has irritant effect on eyes.

Protective devices
Goggles giving complete protection to eyes.
Plastic or rubber gloves.
Eyewash bottle with clean water.

EMERGENCY ACTION – Notify police and fire brigade immediately

- Stop the engine
- No naked lights. No smoking
- Mark roads and warn other road users
- Keep public away from danger area
- Use explosion-proof electrical equipment
- Keep upwind

Spillage

- Shut off leaks if without risk
- Absorb in earth or sand and remove to safe place
- Subsequently flush road with water
- Prevent liquid entering sewers, basements and workpits, vapour may create explosive atmosphere
- Warn inhabitants – explosion hazard
- If substance has entered a water course or sewer or contaminated soil or vegetation, advise police

Fire

- Keep containers cool by spraying with water if exposed to fire
- Extinguish with water spray, dry chemical, alcohol foam or halons
- Do not use water jet

First aid

- Remove soaked clothing immediately and wash affected skin with plenty of water
- If substance has got into the eyes, immediately wash out with plenty of water for several minutes
- Seek medical treatment when anyone has symptoms apparently due to inhalation or contact with eyes

Additional information provided by manufacturer or sender

TELEPHONE
Prepared by CEFIC (CONSEIL EUROPEAN DES FEDERATIONS DE L'INDUSTRIE CHIMIQUE, EUROPEAN COUNCIL OF CHEMICAL MANUFACTURER'S FEDERATIONS)
Zurich, from the best knowledge available: no responsibility is accepted that the information is sufficient or correct in all cases. Obtainable from NORPRINT LIMITED, BOSTON, LINCOLNSHIRE
Applied only during road transport

English

Figure 15.3 *CEFIC TREMCARD*

Table 15.3 Characteristics of type II and type III vehicles for carriage of explosives

Feature	*Type II*	*Type III*
General construction requirements	• Compression ignition engine • Designed, constructed and equipped to protect explosive from external hazards and weather • Closed or securely sheeted • Trailers to be adequately braked or restrained upon detachment from towing vehicle	• Compression ignition engine • Vehicle design to be closed with loading surface continuous. Special attention to heat-resistant and fire-proofing properties of the body. Doors all lockable and placed and constructed as to overlap joints • Any trailers to be fitted with effective braking system acting on all wheels, actuated by the towing vehicle and automatically upon breakage of the coupling. Trailers equipped with only inertia braking devices limited to a maximum load of 50 kg net explosive mass
Electrical requirements	• Max. rated voltage of 24 V • Electrical installation of load-carrying compartment to be dust-protected (at least IP54 or equivalent) or in case of Compatibility Group J (see Table 14.9) flame-proof Ex d (at least IP65 or equivalent)	• Max. rated voltage of 24 V • Electrical installation of load-carrying compartment to be dust-protected (at least IP54 or equivalent) or in case of Compatibility Group J (see Table 14.9) flame-proof Ex d (at least IP65 or equivalent) *Wiring* • Conductors of sufficient size so as to prevent overheating, and adequately insulated. All circuits to be protected by fuses or automatic circuit breakers, except for specified battery connections • Cables securely fastened and positioned so that conductors are adequately protected against mechanical and thermal stresses *Battery master switch* • A switch to break electrical circuits placed as close as possible to the battery • Readily accessible, distinctively marked direct or indirect control devices installed, one in driver's cab and one on outside of vehicle. The cab version shall be within immediate reach, but protected against inadvertent operation • Switch should not constitute a fire risk in an explosive atmosphere and be capable of opening whilst engine is running, without causing any dangerous excess voltage • IP54 protection for cable connections on battery master switch • Battery terminals should be electrically insulated or covered by insulated battery box cover. Batteries not located under bonnet to be fitted in vented box

Table 15.3 Cont'd

Feature	*Type II*	*Type III*
		Tachograph • Electrical supply provided via safety barrier connected directly to battery and all the equipment certified to EN 50 020 *Electrical installations at rear of driver's cab* • Designed, constructed, installed and protected so as to prevent ignition or short-circuit under normal conditions of use, and risks minimized upon impact or deformation
Fire prevention	*Vehicle cab* • Driver's cab constructed of material not readily flammable *Fuel tank* • Tanks supplying the engine to be constructed so that fuel leakages drain to ground without contact of hot parts of vehicle or load *Engine* • Engine to be located forward of the load compartment (or under the load such that excess heat poses no hazard to the load) *Exhaust system* • Constructed and located such that excess heat poses no hazard to load *Combustion heaters* • Prohibited with certain specified exceptions for cabs	*Vehicle cab* • Driver's cab constructed of material not readily flammable *Fuel tank* • Tanks supplying the engine to be constructed so that fuel leakages drain to ground without contact of hot parts of vehicle or load *Engine* • Engine to be located forward of the load compartment (or under the load such that excess heat poses no hazard to the load) *Exhaust system* • Constructed and located such that excess heat poses no hazard to load *Vehicle endurance brake* • Systems emitting high temperatures placed behind the rear wall of the cab to be equipped with thermal shield appropriately fixed and located • The thermal shield should protect the braking system from outflow or leakage of the load *Combustion heaters* • Prohibited with certain specified exceptions for cabs

marking, labelling and placarding; contamination levels; prior approval; quality assurance programmes; test procedures; documentation; stowing and responsibilities. For example, during the transport of radioactive material the driver should:

- ensure none of the material is lost, escapes, or is unlawfully removed;
- not leave the vehicle unattended in a place to which the public has access;
- generally not park the vehicle for a continuous period of longer than one hour unless when parked there is clear space of at least 2 metres on both sides and at both ends of the vehicle;
- keep on the vehicle transport documentation until the package has been delivered.

The driver must notify the police and consignor in the event of an emergency such as material loss, escape or unlawful removal; if package has been opened or damaged, or if the vehicle has overturned, suffered serious damage, or is involved in a fire.

Table 15.4 Choice of vehicle to match the proposed load of explosive

Division (see Table 14.8)	*Compatibility group and form (see Table 14.9)*	*Special goods vehicle*	*Ordinary goods vehicle*	*Private light goods vehicle*
1.1	A	50 kg	Nil	Nil
	Substances of G	5 tonnes	50 kg	1 kg
	B	5 tonnes	500 kg	10 kg
	Substances of C or D	16 tonnes	500 kg	50 kg
	F	5 tonnes	50 kg	50 kg
	Articles of G	5 tonnes	5 tonnes	50 kg
	Articles of C, D, E or J	16 tonnes	5 tonnes	50 kg
	Articles of L	5 tonnes	5 tonnes	50 kg
1.2	Any	16 tonnes	5 tonnes	50 kg
1.3	Any	16 tonnes	5 tonnes	50 kg
1.4	B, C, D, E, F or G	16 tonnes	16 tonnes	50 kg
	S	Unlimited	Unlimited	Unlimited
1.5	D	16 tonnes	5 tonnes	50 kg
1.6	N	16 tonnes	5 tonnes	50 kg
*	*	500 kg	50 kg	10 kg

* Unclassified explosives carried solely in connection with an application for their classification.

Table 15.5 Radiation levels for a road consignment of radioactive material under 'exclusive use'

The radiation level for a consignment transported under exclusive use by a vehicle shall not exceed:

(a) 10 mSv/h at any point on the external surface of any package or overpack, and may only exceed 2 mSv/h provided that:
  (i) the vehicle by which the consignment is transported is equipped with an enclosure which, during routine transport, prevents the access of unauthorized persons to the interior of the enclosure;
  (ii) provisions are made to secure the package or overpackage so that its position within the vehicle remains fixed during routine transport; and
  (iii) there are no loading or unloading operations between the beginning and the end of the shipment;

(b) 2 mSv/h at any point on the outer surfaces of the vehicle, including the upper and lower surfaces, or in the case of an open vehicle, at any point on the vertical planes projected from the outer edges of the vehicle, on the upper surface of the load, and on the lower external surface of the vehicle; and

(c) 0.1 mSv/h at any point 2 m from the vertical planes represented by the outer lateral surfaces of the vehicle, or, if the load is transported in an open vehicle, at any point 2 m from the vertical planes projected from the outer edges of the vehicle.

Rail transport

Dangerous substances and explosives

The carriage of hazardous chemicals by rail in the UK is governed by, e.g., the Carriage of Dangerous Goods by Rail Regulations 1996. These regulations cover the transport in bulk in wagons and large containers, small containers and carriage in tanks. Mixed loading of certain goods is prohibited. Packages bearing a danger sign with a classification code specified in column 1 of Table 15.6 shall not be loaded in the same container or wagon with any package or small container bearing a danger sign with any classification codes specified in column 2. Small containers bearing danger signs and classifications in column 1 must not be loaded in the same container or wagon with packages or small containers bearing danger signs and classification codes in column

Table 15.6 Prohibition on mixed loads of dangerous goods (see Footnote and Table 14.8)

Column 1 *Classification code*	*Column 2* *Classification code*
1, 1.4, 1.5, 1.6 or 01	2, 3, 4.1, 4.2, 4.3, 5.1, 5.2, 6.1, 6.2, 7, 8 or 9
2	1, 1.4, 1.5, 1.6 or 01
3	1, 1.4, 1.5, 1.6 or 01
4.1	1, 1.4, 1.5, 1.6 or 01
4.2	1, 1.4, 1.5, 1.6 or 01
4.3	1, 1.4, 1.5, 1.6 or 01
5.1	1, 1.4, 1.5, 1.6 or 01
5.2	1, 1.4, 1.5, 1.6 or 01
6.1	1, 1.4, 1.5, 1.6 or 01
6.2	1, 1.4, 1.5, 1.6 or 01
8	1, 1.4, 1.5, 1.6 or 01
9	1, 1.4, 1.5, 1.6 or 01

*Consult the original regulations, regulations concerning the International Carriage of Dangerous Goods by Rail 1997, UN recommendations on transportation of dangerous goods, and IARTA Regulations (page 460).

Classification code	*Type*
1	Explosive substances and articles
2	Gases: compressed liquefied or dissolved under pressure
3	Flammable liquids
4.1	Flammable solids
4.2	Substances liable to spontaneous combustion
4.3	Substances when in contact with water emit flammable gases
5.1	Oxidizing substances
5.2	Organic peroxides
6.1	Toxic substances
6.2	Infectious substances
7	Radioactive materials
8	Corrosive substances
9	Miscellaneous dangerous substances

2. The carriage of temperature-controlled substances by rail is prohibited and special requirements are included for the carriage of explosives including:

- Prohibition of carriage in open containers.
- Restriction on the classification of explosive that can be transported and how, including the need for sheet steel spark-guards for carriage of certain explosives.
- Limitation on the quantity of material transported. Thus, Compatibility Group A materials are not permitted. Other than explosives in Division 1.4, the maximum quantity of explosive permitted in any container or wagon is 20 tonnes net explosive content. Materials in column 1 of Table 15.7 must not be transported in adjacent containers of wagons with materials in column 2 in amounts exceeding those quoted. (Where a train comprises more than one such group of adjacent containers or wagons the minimum separation distance between the nearest parts of each group is shown in column 3.) The regulations must be consulted for full details.
- Loading and stowage requirements, including the sealing of closed containers to prevent unauthorized access, and the need to complete journeys and unloading as quickly as practicable.
- Carriage on passenger trains is restricted.
- Security measures.

Table 15.7 Limits on quantities of explosives permitted for carriage by rail in containers and wagons, and separation distances

Column 1 *Type of explosive*	*Column 2* *Maximum net explosive content in tonnes in any group of adjacent containers or wagons*	*Column 3* *Minimum separation distance in metres between groups of containers or wagons*
1.1	40	80
Any combination of 1.1, 1.3 and 1.5	40 (Aggregate)	80
1.3 and 1.5 alone or mixed	120	40

Radioactive substances

UK regulations for the transport of radioactive materials set out the requirements for packaging, labelling and carriage of radioactive material including special forms of radioactive substances, and different types of package and their testing. Documentation requirements are set out together with advice on calculating maximum permissible radiation levels for packages of individual radionuclides and mixtures thereof. For excepted packages of limited activity the radiation level at any point on the external surface of the package must not exceed 5 μSv/h. Material which is not enclosed in, or forming a component part of, an instrument or manufactured article, may be transported in an excepted package if the activity of the material does not exceed the limit given in column 4 of Table 15.8.

Table 15.8 Activity limits for excepted packages (except mixtures of radionuclides)*

Column 1 *Physical state of contents*	*Column 2* *Instruments and articles* *Item limits*	*Column 3* *Instruments and articles* *Package limits*	*Column 4* *Materials* *Package*
Solids			
• Special form	$10^{-2} A_1$	A_1	$10^{-3} A_1$
• Other forms	$10^{-2} A_2$	A_2	$10^{-3} A_2$
Liquids	$10^{-3} A_2$	$10^{-1} A_2$	$10^{-4} A_2$
Gases			
• Tritium	$2 \times 10^{-2} A_2$	$2 \times 10^{-1} A_2$	$2 \times 10^{-2} A_2$
• Special form	$10^{-3} A_1$	$10^{-2} A_1$	$10^{-3} A_1$
• Other forms	$10^{-3} A_2$	$10^{-2} A_2$	$10^{-3} A_2$

*Values of A_1 and A_2 are given in the regulations for individual radionuclides (SI 1996 No. 1350).

Air transport

Dangerous substances

Air freighting of hazardous chemicals must comply with the IATA Dangerous Goods Regulations, based on the International Civil Aviation Organization. The regulations address packaging, marking and labelling, documentation, and handling arrangements. A section is devoted to radioactive materials. Detailed lists of individual articles and substances are included with their UN Classification, viz. explosives, gases, flammable liquids, flammable solids, oxidizing substances and organic peroxides, toxic and infectious substances, radioactive material, corrosives, and miscellaneous dangerous goods. Dangerous goods are assigned to the relevant packing group to reflect the degree of hazard they pose, viz.:

Packing Group I	Great danger
Packing Group II	Medium danger
Packing Group III	Minor danger

Explosive substances

Some classes of chemicals have been identified as too hazardous to be carried on aircraft under any conditions (Table 15.12). This includes the individual substances listed in Table 15.9. Other explosive materials have been forbidden under normal circumstances but may be carried with specific approval. Some are restricted to transport on cargo aircraft only but most can be transported on passenger aircraft provided certain requirements are met and the quantity is limited, e.g. the maximum net quantity of dangerous goods in inner packaging must be limited to that in Table 15.10. The regulations should be consulted for detailed advice. Both shippers and operators have responsibility to ensure the cargo is safe. Training is crucial in maintaining a safe regime; the minimum training requirements are summarized in Table 15.11.

Sea transport

Dangerous substances

In addition to a myriad of general safety maritime legislation (e.g. in the UK the Merchant Shipping (International Safety Management (ISM) Code) Regulations 1998; the Docks Regulations 1988; the Freight Containers (Safety Convention) Regulations 1984; the Merchant Shipping (Carriage of Cargoes) Regulations 1999) there is a raft of more specific regulations governing the carriage of hazardous chemicals at sea. Examples include the Merchant Shipping (Dangerous Goods and Marine Pollutants) Regulations 1997; the Explosives in Harbour Areas Regulations (Northern Ireland) 1995; and the Dangerous Substances in Harbour Areas Regulations 1987, plus international requirements and codes.

Those at risk from packaged hazardous goods inside a transport unit include packers/unpackers, land users, stevedores, customs officials, ships' crews, and passengers. Simple guidance for packers and drivers of transport units containing dangerous goods is given in Table 15.13.

Ships and their cargoes are subjected to stresses in all directions during handling, transfer, and the journey. Adequate stowage is crucial. In line with the transportation of chemicals on land systems to address include:

- Strength, integrity, labelling and marking of individual packages.
- Selection of appropriate container.
- Safe loading packages into the containing unit.
- Adequate securing of packages within the unit.

Table 15.9 Dangerous goods forbidden in aircraft under any circumstances (IATA)

Acetyl cyclohexanesulfonyl peroxide, more than 82%, wetted with less than 12% water
Acetylene (liquefied)
Acetylene silver nitrate
Acetone cyanohydrin, stabilized
Allyl chloroformate
Allyl isothiocyanate, inhibited
Aluminium dross, wet or hot
Ammonium azide
Ammonium bromate
Ammonium chlorate
Ammonium fulminate
Ammonium nitrite
Ammonium permanganate
Antimony sulphide and a chlorate, mixtures of
Arsenic sulphide and a chlorate, mixtures of
Arsenic trichloride
Ascaridole
Azaurolic acid (salt of), (dry)
Azidodithiocarbonic acid
Azidoethyl nitrate
Azido guanidine picrate (dry)
5-Azido-1-hydroxy tetrazole
Azido hydroxy tetrazole (mercury and silver salts)
3-Azido-1,2-propylene glycol dinitrate
Azotetrazole (dry)

Benzene diazonium chloride (dry)
Benzene diazonium nitrate (dry)
Benzene triozonide
Benzoxidiazoles (dry)
Benzoyl azide
Biphenyl triozonide
Boron tribromide
Bromine azide
Bromine solutions
4-Bromo-1,2-dinitrobenzene
Bromosilane
1,2,4-Butanetriol trinitrate
tert-Butoxycarbonyl azide
n-Butyl chloroformate
tert-Butyl hydroperoxide, >90% with water
n-Butyl isocyanate
terty-Butyl isocyanate
tert-Butyl peroxyacetate, >76% in solution
tert-Butyl peroxyisobutyrate, >77% in solution

Charcoal screenings, wet
Charcoal, wet
Chlorine azide
Chlorine dioxide
Chloroacetaldehyde
Chloroacetone (unstabilized)
Chloroacetonitrile
Chloroformates, n.o.s.
Chloroprene, uninhibited
Chlorosulphonic acid
Coal briquettes, hot
Coke, hot
Copper acetylide

Table 15.9 Cont'd

Copper amine azide
Copper tetramine nitrate
Crotonaldehyde, stabilized
Cyanogen bromide
Cyanuric triazide
Cyclotetramethylenetetranitramine (dry or unphlegmatized) (HMX)
Diacetone alcohol peroxides, >57% in solution with >9% hydrogen peroxide, <26% diacetone alcohol and <9% water; total active oxygen content >10% by weight
p-Diazidobenzene
1,2-Diazidoethane
1,1′-Diazoaminonaphthalene
Diazoaminotetrazole (dry)
Diazodinitrophenol (dry)
Diazodiphenylmethane
2-Diazo-1-naphthol-4-sulphochloride
2-Diazo-1-naphthol-5-sulphochloride
Diazonium nitrates (dry)
Diazonium perchlorates (dry)
1,3-Diazopropane
Dibenzyl peroxydicarbonate, >87% with water
Dibromoacetylene
2,2-Di-(tert-butylperoxy) butane, >55% in solution
Di-n-butyl peroxydicarbonate, >52% in solution
Di-(tert-butylperoxy) phthalate, >55% in solution
Dichloroacetylene
N,N′-Dichlorazodicarbonamidine (salts of), (dry)
Dichloroethyl sulphide
Dichlorovinylchloroarsine
2,2-Di-(4,4-di-tert-butylperoxycyclohexyl) propane, >42% with inert solid
Di-2,4-dichlorobenzoyl peroxide, >75% with water
Diethanol nitrosamine dinitrate (dry)
Diethyleneglycol dinitrate (dry)
Diethylgold bromide
Diethyl peroxydicarbonate, >27% in solution
1,8-Dihydroxy-2,4,5,7-tetranitroanthraquinone (chrysamminic acid)
Di-(1-hydroxytetrazole) (dry)
Diiodoacetylene
2,5-Dimethyl-2,5-dihydroperoxy hexane, >82% with water
Dimethylhexane dihydroperoxide, >82% with water
Di-(1-naphthoyl) peroxide
Dinitro-7,8-dimethylglycoluril (dry)
1,3-Dinitro-5,5-dimethyl hydantoin
1,3-Dinitro-4,5-dinitrosobenzene
1,1-Dinitroethane (dry)
1,2-Dinitroethane
Dinitroglycoluril
Dinitromethane
Dinitropropylene glycol
2,4-Dinitroresorcinol (heavy metal salts of) (dry)
4,6-Dinitroresorcinol (heavy metal salts of), (dry)
3,5-Dinitrosalicylic acid (lead salt), (dry)
Dinitrosobenzylamidine and salts of (dry)
2,2-Dinitrostilbene
1,4-Dinitro-1,1,4,4-tetramethylolbutanetetranitrate (dry)
2,4-Dinitro-1,3,5-trimethylbenzene
Di-(β-nitroxyethyl) ammonium nitrate
α, α'-Di-(nitroxyl) methylether
1,9-Dinitroxyl pentamethylene-2,4,6,8-tetramine (dry)
Dipropionyl peroxide, >28% in solution

Table 15.9 Cont'd

Ethanol amine dinitrate
Ethylene diamine diperchlorate
Ethylene glycol dinitrate
Ethyl hydroperoxide
Ethyl nitrate
Ethyl nitrite
Ethyl perchlorate

Fulminate of mercury (dry)
Fulminating gold
Fulminating mercury
Fulminating platinum
Fulminating silver
Fulminic acid

Galactsan trinitrate
Glycerol-1,3-dinitrate
Glycerol gluconate trinitrate
Glycerol lactate trinitrate
Guanyl nitrosaminoguanylidene hydrazine (dry)
Guanyl nitrosaminoguanyltetrazene

Hexamethylene triperoxide diamine (dry)
Hexamethylol benzene hexanitrate
Hexanitroazoxy benzene
2,2′,4,4′,6,6′-Hexanitro-3,3′-dihydroxyazobenzene (dry)
N,N′-(hexanitrodiphenyl) ethylene dinitramine (dry)
2,3′, 4,4′,6,6′-Hexanitrodiphenylether
Hexanitrodiphenyl urea
Hexanitroethane
Hexanitrooxanilide
HMX (dry or unphlegmatized)
Hydrazine azide
Hydrazine chlorate
Hydrazine dicarbonic acid diazide
Hydrazine perchlorate
Hydrazine selenate
Hydrogen cyanide, unstabilized
Hydroxyl amine iodide
Hyponitrous acid

Ignition element for lighter, containing pyrophoric liquid
Inositol hexanitrate (dry)
Inulin trinitrate (dry)
Iodine azide (dry)
Iodoxy compounds (dry)
Iridium nitratopentamine iridium nitrate
Isopropylcumyl hydroperoxide, >72% in solution
Isothiocyanic acid

Lead azide (dry)
Lead nitroresorcinate (dry)
Lead picrate (dry)
Lead styphnate (dry)
Lead trinitroresorcinate (dry)
Lighters (cigarettes) containing pyrophoric liquid
Lighters (cigarettes) with lighter fluids

Table 15.9 Cont'd

Magnesium dross, wet or hot
Mannitan tetranitrate
Mannitol hexanitrate (dry)
Mercurous azide
Mercury acetylide
Mercury iodide aquabasic ammonobasic (Iodide of Millon's base)
Mercury nitride
Methazoic acid
Methylamine dinitramine and dry salts thereof
Methylamine nitroform
Methylamine perchlorate (dry)
Methyldichloroarsine
Methylene glycol dinitrate
Methyl ethyl ketone peroxide, >50%
alpha-Methylglucoside tetranitrate
alpha-Methylglycerol trinitrate
Methyl nitramine (dry) metal salts of
Methyl nitrate
Methyl nitrite
Methyl picric acid (heavy metal salts of)
Methyl trimethylol methane trinitrate

Naphthalene diozonide
Naphthyl amineperchlorate
Nickel picrate
Nitrates of diazonium compounds
N-Nitroaniline
m-Nitrobenzene diazonium perchlorate
6-Nitro-4-diazotoluene-3-sulphonic acid (dry)
Nitroethyl nitrate
Nitroethylene polymer
Nitrogen trichloride
Nitrogen triiodide
Nitrogen triiodide monoamine
Nitroglycerin, liquid, not desensitized
Nitroguanidine nitrate
1-Nitro hydantoin
Nitro isobutane triol trinitrate
Nitromannite (dry)
Nitromethane
N-Nitro-*N*-methylglycolamide nitrate
2-Nitro-2-methylpropanol nitrate
m-Nitrophenyldinitro methane
Nitrosugars (dry)

1,7-Octadiene-3,5-diyne-1,8-dimethoxy-9-octadecynoic acid
Octogen (dry)

Pentaerythrite tetranitrate (dry)
Pentaerythritol tetratnitrate (dry)
Pentanitroaniline (dry)
Perchloric acid >72% strength
Peroxyacetic acid, >43% and with >6% hydrogen peroxide
PETN (dry)
m-Phenylene diaminediperchlorate (dry)
Phosphorus (white or red) and a chlorate, mixtures of
Potassium carbonyl
Pyridine perchlorate

Table 15.9 Cont'd

Quebrachitol pentanitrite

Security type attaché cases incorporating lithium batteries and/or pyrotechnic material
Selenium nitride
Silver acetylide (dry)
Silver azide (dry)
Silver chlorite (dry)
Silver fulminate (dry)
Silver oxalate (dry)
Silver picrate (dry)
Sodium picryl peroxide
Sodium tetranitride
Sucrose octanitrate (dry)

Tetraazido benzene quinone
Tetraethylammonium perchlorate (dry)
Tetramethylene diperoxide dicarbamide
Tetranitro diglycerin
2,3,4,6-Tetranitrophenol
2,3,4,6-Tetranitrophenyl methyl nitramine
2,3,4,6-Tetranitrophenylnitramine
Tetranitroresorcinol (dry)
2,3,5,6-Tetranitroso-1,4-dinitrobenzene
2,3,5,6-Tetranitroso nitrobenzene (dry)
Tetrazene (dry)
Tetrazine
Tetrazolyl azide (dry)
Thionyl chloride
Thiophosgene
Titanium tetrachloride
Trichloromethyl perchlorate
Triformoxime trinitrate
Trimethylacetyl chloride
Trimethylene glycol diperchlorate
Trimethylol nitromethane trinitrate
1,3,5-Trimethyl-2,4,6-trinitrobenzene
Trinitroacetic acid
Trinitroacetonitrile
Trinitroamine cobalt
2,4,6-Trinitro-1,3-diazobenzene
Trinitroethanol
Trinitroethylnitrate
Trinitromethane
2,4,6-Trinitrophenyl guanidine (dry)
2,4,6-Trinitrophenyl nitramine
2,4,6-Trinitrophenyl trimethylol methyl nitramine trinitrate (dry)
2,4,6-Trinitroso-3-methyl nitraminoanisole
Trinitrotetramine cobalt nitrate
2,4,6-Trinitro-1,3,5-triazido benzene (dry)
Tri-(β-nitroxyethyl) ammonium nitrate
Tris, bis-bifluoroamino diethoxy propane (TVOPA)

Vinyl nitrate polymer

p-Xylyl diazide

Table 15.10 Quantity limits of dangerous goods accepted in small amounts for transport by air

Packing group of the substance	*Packing group I*		*Packing group II*		*Packing group III*	
Class or division or primary or subsidiary risk[a]	*Packagings* *Inner*	*Outer*	*Packagings* *Inner*	*Outer*	*Packagings* *Inner*	*Outer*
1: Explosives			Forbidden			
2.1: Flammable gas			Forbidden			
2.2: Non-Flammable, non-toxic gas			See Note[b]			
2.3: Toxic gas			Forbidden			
3: Flammable liquid	30 ml	300 ml	30 ml	500 ml	30 ml	1l
4.1: Self reactive substances			Forbidden			
4.1: Other flammable solids	Forbidden		30 g	500 g	30 g	1 kg
4.2: Pyrophoric substances	Forbidden		Not Applicable		Not Applicable	
4.2: Spontaneously combustible substances	Not Applicable		30 g	500 g	30 g	1 kg
4.3: Water reactive substances	Forbidden		30 g or 30 ml	500 g or 500 ml	30 g or 30 ml	1 kg or 1 l
5.1: Oxidizers	Forbidden		30 g or 30 ml	500 g or 500 ml	30 g or 30 ml	1 kg or 1 l
5.2: Organic peroxides[c]	Not applicable		30 g or 30 ml	500 g or 500 ml	Not Applicable	
6.1: Toxic substances – inhalation	Forbidden		1 g or 1 ml	500 g or 500 ml	30 g or 30 ml	1 kg or 1 l
6.1: Toxic substances – oral	1 g or 1 ml	300 g or 300 ml	1 g or 1 ml	500 g or 500 ml	30 g or 30 ml	1 kg or 1 l
6.1: Toxic substances – dermal	1 g or 1 ml	300 g or 300 ml	1 g or 1 ml	500 g or 500 ml	30 g or 30 ml	1 kg or 1 l
6.2: Infectious substances			Forbidden			
7: Radioactive material[d]			Forbidden			
8: Corrosive materials[e]	Forbidden		30 g or 30 ml	500 g or 500 ml	30 g or 30 ml	1 kg or 1 l
9: Magnetized materials			Forbidden			
9: Other miscellaneous materials[f]	Not Applicable		30 g or 30 ml	500 g or 500 ml	30 g or 30 ml	1 kg or 1 l

[a] The more restrictive quantity required by either the Primary or Subsidiary Risk must be used.
[b] For inner packagings, the quantity contained in receptacle with a water capacity of 30 ml. For outer packagings, the sum of the water capacities of all the inner packagings contained must not exceed 1 litre.
[c] Applies only to Organic Peroxides when contained in a chemical kit or a first-aid kit.
[d] See 10.5.9.8.1, 10.5.9.8.2 and 10.5.9.7, radioactive material in excepted packages.
[e] UN 2803 and UN 2809 are not permitted in Excepted Quantities.
[f] For substances in Class 9 for which no packing group is indicated in the List of Dangerous Goods, Packing Group II quantities must be used.

Table 15.11 Minimum training requirements for staff engaged in the transport of hazardous chemicals by air

Relevant aspect with which staff should be familiar, as a minimum	*Type of staff* A	B	C	D	E	F	G	H
General philosophy	x	x	x	x	x	x	x	x
Limitations	x		x		x	x	x	x
General requirements	x		x			x		
Classification	x	x	x			x		
List of dangerous goods	x	x	x			x	x	
General packing requirements	x	x	x			x		
Packing instructions	x	x	x			x		
Labelling and marking	x	x	x	x	x	x	x	x
Shipper's declaration and other documentation	x		x			x		
Acceptance procedures			x					
Storage and loading procedures			x	x			x	
Pilot's notification			x	x			x	
Provisions for passengers and crew				x	x		x	x
Emergency procedures			x	x			x	x

Staff category

A Shippers, shipper's agents including operator's staff acting as shippers, operator's staff preparing dangerous goods
B Packers
C Cargo acceptance staff of operators and agencies acting on behalf of operators
D Staff of operators and agencies acting on behalf of operators engaged in the ground handling, storage and loading of cargo and baggage
E Passenger handling staff and security staff who deal with the screening of passengers and their baggage
F Staff of agencies other than operators involved in processing cargo
G Flight crew members
H Crew members (other than flight crew)

Depending upon responsibilities the training requirements will alter, e.g. if an operator carries only cargo those aspects pertaining to passengers may be omitted for his staff and flight crew.

Table 15.12 Classes of dangerous goods forbidden for transport by air

Explosives which ignite or decompose when subjected to a temperature of 75°C for 48 hours.
Explosives containing both chlorates and ammonium salts.
Explosives containing mixtures of chlorates with phosphorus.
Solid explosives which are classified as extremely sensitive to mechanical shock.
Liquid explosives which are classified as moderately sensitive to mechanical shock.
Any substance, as presented for transport, which is liable to produce a dangerous evolution of heat or gas under the conditions normally encountered in air transport.
Flammable solids and organic peroxides having, as tested, explosive properties and which are packed in such a way that the classification procedure would require the use of an explosives label as a subsidiary risk label.
Dangerous goods listed in Table 15.9

- Provision of relevant documentation/information, e.g. for packers, haulier, shipper, etc. For example, the ship's owner must be provided with:
  - the container packing certificate/vehicle declaration;
  - a signed dangerous goods declaration; and/or
  - a marine pollution declaration.

A checklist summarizing the range of responsibilities is given in Table 15.14.

Hazardous substances in ports are governed by the Dangerous Substances in Harbour Areas Regulations 1987. These provide for:

Table 15.13 Guidance relating to dangerous goods in cargo transport units (CTU) for transport by sea

	Packer	*Driver*
Do	Check individual packages for damage, leaks, etc. Check packages show: • proper shipping name; • UN number; • class label; • subsidiary risk label (where relevant); • marine pollutant mark (where relevant); • UN type approval mark Confirm nature of hazards from labels Visual check on condition of CTU before loading Check validity of safety approval plate of freight container Removal of irrelevant placards Follow loading plan when packing CTU Secure and brace packages inside CTU Securely close and seal CTU when loading is complete Ensure container packing certificate/vehicle declaration signed and forwarded to ship operator or master Ensure placards are present on sides and ends of CTU Ensure written information on potential hazards is passed from shipper to haulier	Check that goods and quantities are on dangerous goods declaration Obtain from packer the container packing certificate/vehicle declaration Obtain and carry transport documentation including TREMCARD, etc. Check that any freight container has valid safety approval plate Visually check the outside of the CTU for damage, etc. Ensure CTU doors are properly secured Learn emergency procedures Ensure availability of appropriate personal protective equipment as advised in transport documentation, and suitable fire extinguisher Check sides and ends of CTU are placarded appropriately Ensure irrelevant hazard data are removed Ensure controls against fire and explosion are taken during the journey When parking vehicle avoid creating risk to local people
Do not	Handle damaged, leaking or stained packages Load a freight container without valid safety approval plate Load packages of dangerous goods which are improperly labelled and marked Load packages into a dirty, wet or damaged CTU Place packages in a CTU with incorrect placards still visible Attempt to load packages without loading plan to ensure segregation of incompatibles Leave cargo in an insecure CTU Exceed the maximum gross mass of the CTU	Accept load or start journey without relevant hazard information from shipper Accept freight container without valid safety approval plate Allow CTU to travel with irrelevant hazard placards, marks or signs or if lacking in information Allow placards to become obscured or damaged Have irrelevant hazard warning information in the cab Start journey without checking availability of safety equipment in suitable condition for use Smoke or use other means of ignition Carry unauthorized passengers

- Entry of dangerous substances into harbours.
- Marking and navigation.
- Handling of dangerous substances.
- Liquid dangerous substances.
- Packaging and labelling.
- Emergency arrangements and untoward incidents.
- Storage of dangerous substances.
- Explosives.
- Ship/shore checklist.

Information on dangerous chemicals entering harbour areas from sea should be provided to the harbour master by the ship's master. Notification should also include the ship's name, call sign,

Table 15.14 Checklist of responsibilities for sea transportation of chemicals

Action	*Person responsible*
Dangerous chemical properly packaged and labelled	Manufacturer/supplier/shipper
Training and supervision of those packing	Packer
Health and safety of packers	Packer
Construction, maintenance and plating of freight container	Owner or lessee of container or others who use or permit use, e.g. packer, haulier
Visual inspection of container and packages	Packer
Container and vehicle loading – all packages securely stowed and braced	Packer
Provision of signed container/packing certificate – vehicle declaration	Packer
Placarding of cargo transport unit	Packer or shipper
Placarding of vehicle	Haulier
Provision of signed dangerous goods declaration	Shipper
Information of hazards to haulier	Shipper
Advanced notification to port	Haulier
Safety during transfer of dangerous goods by road	Haulier/driver

nationality, overall length, draught and beam, intended destination within the harbour, and estimated time of arrival. In addition, for marine tankers of 1600 tonnes gross, or carrying certain chemicals and gases or oil in bulk, the following information should be provided:

- nature and quantity of chemicals, gas or oil carried by the vessel;
- whether tanker is fitted with operational inert gas system;
- whether atmosphere of cargo tanks has been rendered non-flammable;
- information about tanker certificates;
- defects to hull, machinery or equipment which may affect safe manoeuvrability of ship, affect safety of other vessels, constitute a marine environment hazard;
- whether there is a hazard to property or people near to the harbour.

Controls for the safe transfer of chemicals between ship and shore will vary in detail. The escape of dust and vapour should be minimized when loading and unloading of bulk liquids or solids and persons should not be at risk. Precautions to avoid contact with water are needed when transferring water-incompatible materials (page 229). Precautions for handling explosives include:

- packages should be kept from berth until vessel/vehicle is ready to receive them;
- where necessary restrict the handling to a no-smoking secure area;
- warning signs should be displayed during the handling process;
- vehicle interiors should be clean;
- metal-free shoes should be used;
- radio transmitters should be banned within 50 metres during the transfer, except for low-power (25 watt) radios in mobile equipment such as cranes with aerials outside 2 metres of the explosives.

Table 15.15 summarizes the precautions to be followed by the berth operator for the transfer of dangerous bulk liquids.

Table 15.15 General precautions to be followed for the transfer of dangerous liquids in bulk*

- Pipelines/hoses should be properly designed, of adequate strength and of good construction, from sound and suitable material, properly maintained, appropriately protected against impact and used only for intended materials.
- Area in vicinity of pipeline berth should be adequately ventilated.
- Do not exceed safe working pressure; consider need for safety relief devices.
- Pipelines for flammable liquids/gases to be protected against arcing during connection and disconnection, e.g. hose strings and metal arms fitted with insulating flange or a single length of non-conducting hose to ensure discontinuity between vessel and shore. Seaward pipework should be electrically continuous to vessel, and landward piping electrically continuous to the jetty earthing system.
- Insulating flanges or single length non-conducting hose should not be short circuited with external metal.
- Insulating flanges to be inspected and tested at intervals not exceeding one month to ensure the insulation is clean and in good condition. Resistance should be measured between the metal pipe on shore side of flange and the end of the hose or metal arm when freely suspended. The measured value should be at least 1000 ohms.
- Cargo hoses with internal bonding between end flanges to be checked for electrical continuity before introduction into service and periodically thereafter.
- Berth operator should possess a supplier's certificate for any hose forming part of the pipeline confirming that tests have shown it is unlikely to burst in service.
- Hoses to be indelibly marked to indicate the substances for which it is intended, its safe working pressure, proof-test pressure, date tested, and maximum or minimum service temperature.
- Before first use hoses to be inspected visually internally and externally and tested at proof pressure. Repeat at least annually and include check on electrical resistance. The resistance of the complete assembly should be not more than 15 ohms, unless the hose is intended to be non-conducting when resistance should be at least 25 000 ohms. Inspect hose daily when in use. Hoses used at monobuoys should also be hydraulically tested.
- When rigged for use hoses should be under supervision.
- After use drain before disconnection.
- Close each end of hose until reconnected unless made safe by draining and purging.
- Check any cargo handling controls, emergency shutdown and alarms are working before commencing transfer.
- Prior to the transfer the berth operator and master should agree maximum loading or unloading rate. Berth operator to check periodically these are not exceeded.
- Communication should be established and maintained between people on the ship, the berth and at the storage installation during loading/unloading.
- Berth operator to take reasonably practicable steps to control flammable or toxic gas escapes (e.g. hose support, flange couplings liquid- or gas-tight, drip trays).
- Unloading or loading liquefied gas escapes should be limited to those vented via safety devices or authorized by the harbour master. Rate of transfer should ensure pipelines cool gradually; lines should be vented safely at the end of the transfer.
- Berth operator should ensure the unloading or loading of ship's stores does not endanger the transfer of dangerous chemicals or ballast water contaminated with dangerous substances, gas freeing, or tank cleaning.
- At the completion of the transfer of dangerous liquids the berth operator should render pipework, valves and associated equipment safe; valves and tanks should be closed and shore pipeline blanked off.

* The full SI 1987 No. 37 and Approved Code of Practice should be consulted for complete details.

Modes of transport for liquids, gases and solids

Liquids

Liquids may be transported in numerous ways depending upon the quantity and distance involved. The associated hazards are chemical-specific and also depend upon the physical condition, i.e.:

- liquid at atmospheric pressure and ambient temperature;
- liquefied gas under pressure and at ambient temperature; or
- liquefied gas at atmospheric pressure and at low temperature (i.e. fully-refrigerated transport).

Flammable liquid which spreads can result, on ignition, in a running liquid fire. If spilled onto water spreading will be more extensive, and vaporization will be more rapid, because of the increased rate of heat transfer. Unstable chemicals may pose an explosion risk. Toxic chemicals may be released as a liquid which spreads or as a vapour cloud. The risk of environmental damage is likely to be potentially serious in most cases.

The common means for transport are:

- In bottles, plastic drums, steel or resin-lined drums (e.g. of 210 litre capacity).
- Glass bottles which are used only for small quantities, e.g. 2.5 litres, but should be protected in specially designed carriers.
- Plastic drums, which must not be subjected to excessive loading and, if returnable, require checking for degradation, e.g. due to cracking, impact, distortion.
- By road or rail tanker.
- By ship, e.g. crude petroleum.
- By pipeline, e.g. LPG.

Accidents may be caused by impact, failure of container or pipeline, or during loading/unloading. The hazards arise from:

- Fire and/or explosion.
- Toxic release:
  - of a conventional type; or
  - of an ultratoxic; either instantaneous or prolonged.
- Environmental pollution.

These mirror those at fixed installations, but loss of containment due to the triggering event can occur anywhere en route. Thus accidents may occur in populated or environmentally sensitive locations, or where domino effects are less easily controlled. Common risks are also associated with all vehicular movements and mechanical/manual loading/unloading activities.

Problems may arise with switch-loading of road tankers, ships or pipelines and with the use of returnable containers. The important considerations for safety are:

- Compatibility of the chemical with the materials of construction.
- Adequacy of cleaning out, and removal of residues from, the previous chemical carried to avoid cross-contamination and potential reactive hazards on refilling.
- A sound information transfer system to avoid confusion of chemical identities and to ensure the specific risks of each load are identified and made known to the carrier/transporter.
- Adequacy of decontamination of 'empty' containers, tankers, etc. before return.

Gases

Gases are transported:

- Under pressure in cylinders or pressurized tanks subject to the Pressure Systems Safety Regulations 2000.
- By pipeline.

In general, loss of containment is more serious than with the majority of liquids (unless they are in a superheated state) since atmospheric dispersion will be immediate.

Release of a flammable gas or vapour may result in a jet or flash fire or any type of vapour cloud explosion (p. 178). In fire conditions there can be a BLEVE hazard from containers of liquefied gas. Release of a toxic gas or vapour always poses a potential risk of personal injury and possibly of environmental damage.

Whilst they will differ depending upon gas properties, the procedures and precautions appropriate for transport of cylinder gases are exemplified for LPG in Table 15.16.

Table 15.16 Procedures for safe transport of LPG cylinders by road

Transport

- Carry cylinders on open vehicles. Keep cylinders upright and adequately secured, e.g. with a rope.
- Keep a fire extinguisher, e.g. 1 kg dry power, in the cab to deal with any small fire, e.g. an engine fire.
- Do not leave cylinders on vehicles unsupervised.
- Ensure that the driver has received adequate training and instructions about the hazards of LPG, emergency procedures, driver duties, etc.
- Ensure that relevant information is readily available on the vehicle, e.g. on a clipboard in the cab. This written information, e.g. as a TREMCARD, should contain details of the nature of the load and the action to take in an emergency.

Duties of vehicle operator

- Check whether the Road Traffic (Carriage of Dangerous Packages, etc.) Regulations 1986 apply. Exceptions apply to cylinders <5 litres; cylinders which are part of equipment carried on the vehicles, e.g. burning gear, bitumen boilers; cylinders associated with vehicle operation, e.g. cooking, water heating.
- Ensure the vehicle is suitable, normally an open vehicle. Use of a closed vehicle should be restricted to a small number of cylinders with a load compartment having adequate ventilation.
- Ensure the driver has adequate information in writing, e.g. a TREMCARD.
- Ensure the driver is provided with adequate instruction and training and keeps necessary records.
- Ensure loading, stowage, unloading are performed safely. All cylinders should be packed, strapped, supported in frames, or loaded to avoid damage resulting from relative movement. Cylinders should be stowed with valves uppermost.
- Ensure all precautions are taken to prevent fire or explosion.
- Ensure suitable fire extinguishers are provided.
- Ensure the vehicle displays two orange plates if 500 kg of LPG is carried.
- Report any fire, uncontrolled release or escape of the LPG, to the appropriate authority.

Duties of the driver

- Ensure the relevant written information from the operator is always available during carriage. Destroy, remove or lock-away information about previous loads.
- Ensure loading, stowage and unloading are performed safely.
- Ensure all precautions against fire or explosion are taken during carriage.
- Display orange plates (when required) and keep them clean and free from obstruction.
- If >3 tonnes of LPG is carried, when the vehicle is not being driven, ensure parking is in a safe place or that it is supervised (by the driver or a competent person aged >18).
- On request provide appropriate information to persons authorized to inspect the vehicle and load.
- Report any fire, uncontrolled release or escape of LPG, to the operator.

The integrity of pipelines depends upon correct design, including materials selection, support and protection from mechanical damage. Depending upon the gas, routine inspection and maintenance may be supplemented by the provision of gas detection and alarm systems. Other considerations are exemplified by the safety-related controls on the transportation of domestic gas via pipeline systems summarized in Table 15.17.

Solids

A wide variety of containers of differing capacity and design are used for solids transport, i.e. fibreboard, metal, resin-lined metal, plastic drums; plastic, paper or hessian sacks; tote bins; bulk tanker; lorry-load. The material may then be stored in the containers as received, in hoppers or silos, or simply in piles depending upon its properties and value.

Table 15.17 Measures for the control of management of domestic gas through pipeline systems (the Gas Safety (Management) Regulations 1996)

Safety case – gas transporter

- Day-to-day management to ensure continuity of gas supply at the correct pressure and composition.
- Arrangement to deal with reports of gas leaks and suspected CO emissions.
- Arrangements for investigation of fire and explosion incidents.

Safety case – network emergency controller

- Arrangements to monitor the network – to identify any potential national gas supply emergency and to coordinate preventive action.
- Arrangements to direct transporters to reduce consumption if it is impossible to prevent a gas emergency developing.
- Procedures for the safe restoration of gas supply following an emergency.
- Arrangements for emergency services.

Bulk transport is often favoured since it reduces the requirement for manual handling and facilitates enclosed transfer into storage, thus reducing risks at the customer's factory. Stock inventories may, however, be increased.

Special considerations arise in the transport of:

- unstable chemicals (p. 235)
- chemicals prone to self-heating (p. 214). These may involve:
  - breakdown of stock into numerous smaller units;
  - blanketing of the material (e.g. to prevent ingress of air to oxidizable material or complete quenching, e.g. with water);
  - regular temperature measurements;
- reactive chemicals (p. 228). Segregation from incompatible chemicals is essential (p. 233)
- chemicals prone to generate combustible dust clouds (p. 220).

Loading and unloading

Tankers

Accidents during the loading/unloading of chemicals into/from road tankers, railway tankers, ships and barges may involve discharge of the wrong chemical, vehicle movement during transfer, failure of, or damage to, flexible transfer hoses, disconnection of transfer hoses whilst still under pressure, and overfilling. Loading/unloading should be in designated areas positioned at an appropriate distance from public roads, occupied buildings and – if relevant – possible ignition sources. A good level of ventilation should be ensured together with spillage control facilities. These should be backed up by standard operating procedures.

A basic safety audit covering design and system of work features to be covered for the loading/unloading of non-pressurized hazardous liquids to/from tankers of tank containers is given in Table 15.18.

Table 15.18 Basic safety audit checklist for loading/unloading non-pressurised hazardous liquids; tankers or tank containers

Site
- Are there adequate signs available at the site entrance to direct the traffic to the appropriate loading or off-loading point?
- Is there adequate access to the loading area?
- Where, and to whom, does the driver report?
- Is there a dedicated waiting area for the tanker?
- Are there 'no smoking' notices where appropriate?

Tanker bay
- Can the vehicles drive in and out without reversing?
- Check separation distances from equipment, buildings, other activities, etc.
- Does the ground slope? Check potential vehicle run-away.
- Check venting and ventilation facilities.
- Is vapour recovery provided?
- Are there notices for action in an emergency (e.g. in the event of spillage, do not attempt to drive away)?
- Is the lighting adequate? Is there back-up?
- Will spillages be contained?
- Would fire water be contained?
- Earthing provisions.

Tanker or tank container
- Is it a dedicated tanker or tank container?
- Who provides the pump and hoses and are these checked (integrity, compatibility)?
- How is the tanker filled – top or bottom?
- Are appropriate connections available (flanged, snap-on, self-sealing)?
- Check lids, pressure and vacuum relief.
- Is switch loading in operation? If so, have the hazards been recognized?
- Is tank dipping still necessary?
- Is the tanker labelled correctly?
- Check for cleanliness of tanker if loading.

Product
- Check hazards of substance (e.g. reactivity, flammability, toxicity, corrosivity, etc.) versus precautions.
- Check need for purging and inerting.
- Check COSHH regulations compliance.
- Check for provision of safety data sheet with paperwork.
- Are quality control provisions adequate?

Activities
- Correct, identified tank.
- Are up-to-date operating instructions available?
- What are the tanker driver's responsibilities?
- Check paperwork provided versus correct material and quantities. Delay if unconfirmed.
- Is a tanker run-away or drive-away protective system used?
- Check p.p.e. needs.
- Who has ultimate responsibility for rail cars, site management or the train driver?

Connection
- Earthing and bonding procedure.
- Approval to connect.
- Check provision of appropriate loading arms, connections, couplings.
- Is access available to tanker top for connections or operation of hatches?
- Are guard rails or safety harnesses provided for tanker top operations?

Loading/unloading
- Area cordoned off; restricted access.
- Quantities and levels (adequacy of).
- No simultaneous transfers.
- Is overfilling alarm and protection provided?
- Metering; calibration and reliability.
- Velocity of liquid in pipes versus static electricity hazards.
- Vapour venting and gas balancing.
- If vapour return is used, what pressure is required and is it below the tanker relief valve setting?
- Loading hot liquids; temperature checks, line heating and insulation.
- Procedure for blockages (and disposal of residues).

Safety and general management
- Timing of loading/unloading.
- The management policy for inclusion of safety, health and environment.
- A direct statement of the responsibility for the loading and unloading operation.
- Driver training and licence check.
- Operator training check.
- Need for barriers, chocks or brake interlocks.
- Drains and handling spillages.
- Other safety equipment.
- Control of ignition sources.
- Smoking policy and notices.
- Means of escape.
- Reporting of faults and incidents.

Disconnection
- Ensure lines clear and depressurized.
- Approvals.
- Spillages, containment and treatment.

Checking
- Product specification and sampling.
- Hoses, pumps, connections, lids, locked or capped valves, compartments.
- Contamination and cross-contamination.
- Quantities, weights or levels.
- Actions in the event of faults.
- Is the vehicle suitable for the material and in a roadworthy condition before loading?

Vehicle departure
- Tanker exterior cleaning?
- Security of openings, connections and valves.
- Procedure on overfilling.
- Approvals and paperwork. Complete documentation.

Table 15.18 Cont'd

- Tanker weighing.
- Exit route signs.
- Hazchem/Tremcards as necessary.
- Who checks the vehicle for integrity and cleanliness before despatch?

Emergencies
- Check fire-fighting capability and access for fire appliances. Foam make-up provisions.
- Check provision of fire alarms, extinguishers, emergency cut-offs.
- Provision of adsorbents.
- Water supplies.
- Emergency supplies.
- Emergency showers. Eyewash provisions. First aid.
- Communications.
- Liaison with emergency services.
- Procedures for dealing with the media.

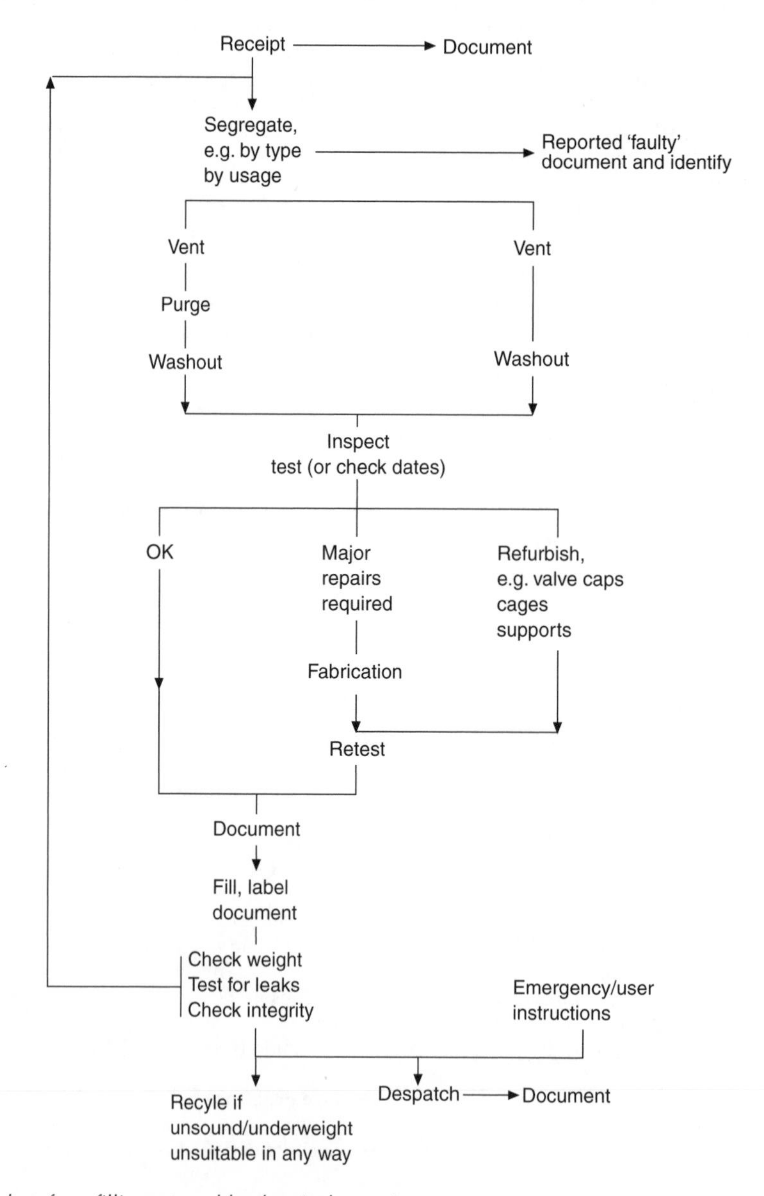

Figure 15.4 *Procedure for refilling returnable chemical containers*

Container filling/discharging

Considerations for the safe filling/discharging of containers will normally include:

- A segregated area adequately spaced from ignition sources, etc. with provision for spillage control and with appropriate protection from vehicle impact.
- Avoidance of overhead piping, e.g. process lines, water, steam.
- Adequate weather protection, both for operators and to avoid ingress of rainwater or other contaminants.
- Provision of information and instruction to personnel as to the hazards with the specific chemical, including possible reactions.
- Thorough inspection of containers for integrity, correct labelling, sound valves or closures.

Figure 15.4 shows a procedure for the refilling of returnable containers with a chemical of assured quality. (This will require some combination of documentation, sampling and analysis to ensure that the chemical is of the correct type, and in the intended condition.) This is applicable to drums and plastic containers, and bottles, but some modifications are necessary for pressure systems, i.e. gas cylinders, covered by the Pressure Systems Safety Regulations 2000 including:

- A check from marks on the cylinder indicating that it has undergone proper examination at appropriate intervals by a competent person.
- All appropriate safety checks including the provision of a correct valve, plug and protection cap, if appropriate.
- A check that it is not overfilled and is within its safe operating limits.

16

Chemicals and the environment: sources and impact

Protection of the environment from pollution caused by 'wastes' is a matter of increasing concern. It can be a technically-complex exercise because the variety of waste is extremely wide, the volumes may be large, and the levels 'permissible' for release may be very low.

Examples of industrial chemical waste are given in Table 16.1. Common industrial sources of pollution are given in Table 16.2. Since waste can result in pollution, the two terms are often used synonymously.

Table 16.1 Types of waste and forms of pollutant

Source	*Type of waste*	*Form of pollutant*[1]
Agriculture, horticulture		
	BOD waste (high and low)	L S A
	Disinfectant	A
	Pesticides, herbicides (see below)	PA
	Fertilizers, *see* Inorganic chemicals	A
Cement, bricks, lime		
	Chromium	L
	Dust	S
	Fluorides	G
	Sulphur dioxide	G
Coal distillation, coal tar, coke ovens		
	Ammonia	G A
	Aromatic hydrocarbons	G L
	Combustion products	G
	Cyanate	A
	Cyanides	G A
	Dust	S P
	Fluorides	G A
	Hydrocarbons (general)	G L A
	Hydrogen sulphide	G
	Phenols	L A
	Polycyclic hydrocarbons	G
	Sulphur dioxide	G
	Tar	G L S
	Thiocyanates	A
Construction, building, demolition		
	Combustion products	G P
	Dust	S P

Table 16.1 Cont'd

Source	*Type of waste*	*Form of pollutant*[1]
	Metals	S
	Rubble	S
	Timber	S
Electricity generation		
	Clinker	S
	Combustion products, e.g. smoke	G
	Cooling water	L
	Pulverized fuel ash	S P
	Sulphur dioxide, nitrogen dioxide	G
Fibres, textiles		
	Bleaches	A
	Cyanides	A
	Detergents	A
	Dyestuffs	A
	Grease	L S
	Oil	L A
	Resins	L A
	Silicones	A
	Speciality chemicals for fire-, rot- and waterproofing	L A
	Wax	PA
Food processing		
Animal	Abattoir waste	L S P A
	BOD waste (high and low)	A
	Disinfectants	A
	Grease	S A
	Oil	L A
Beverage	Alkali	A
	BOD waste (low)	A
	Carbon dioxide	G
	Cullet	S
	Detergents	A
Vegetable and fruit	Alkali	A
	Bleaches	A
	BOD waste (high and low)	P A
	Oil	L
	Solvents	L A
	Wax	S P A
Inorganic chemicals		
Chloralkali	Brine	A
	Calcium chloride	A
	Mercury	L
Desalination	Brines	A
Fertilizer	Ammonia	G A
	Nitrates	S
	Oxides of nitrogen	A
	Phosphates	S
Glass and ceramic	Arsenic	A
	Barium	S
	Manganese	S
	Selenium	S
Hydrofluoric acid	Calcium sulphate	P A
Nitric acid	Ammonia	G A
	Oxides of nitrogen	A
Phosphoric acid	Calcium sulphate	P A
	Hydrofluoric acid	A

Table 16.1 Cont'd

Source	*Type of waste*	*Form of pollutant*[(1)]
Pigments	Arsenic	S P
	Barium	S P
	Cadmium	S P
	Cobalt	S P
	Iron	S P A
	Lead	S P
	Manganese	S P
	Selenium	S P
	Titanium	S P
	Zinc	S P
Sulphuric acid	Sulphur dioxide	G P
Drycleaning	Chlorinated hydrocarbons	G L
	Hydrocarbon solvents	G L
General	Bleaches, e.g. hypochlorites	A
	Detergents	A
	Phosphates	A
	Sulphates	A
Metals		
Extraction and refining	Acid mine waters	A
	Combustion products	G P
	Carbon monoxide	G
	Chlorides	G S P
	Drosses	S
	Dust	S
	Fluorides	G S P
	Glass	S
	Spoil	S
	Sulphides	S P A
	Sulphur dioxide	G
Finishing/surface treatments		
Anodizing	Chromium	A
Degreasing	Chlorinated hydrocarbons	G L
	Detergents	A
	Grease; mineral-oil based	L S
	Solvents	G L
Electroplating	Alkali	A
	Boron	A
	Cadmium	A
	Chromium	P A
	Copper	A
	Cyanides	A
	Detergents	A
	Fluorides	A
	Iron	A
	Nickel	A P
	Organic complexing agents	A
	Phosphates	A
	Precious metals	A
	Silver	A
	Sulphates	A
	Tin	A
	Zinc	A
Foundries	Dust	S P
	Sands, some resin-bonded	S
Machine shops	Oils	L A
	Oil absorbents	S
	Semi-synthetic coolants	

Table 16.1 Cont'd

Source	*Type of waste*	*Form of pollutant*[1]
	Solvents	L G
	Swarf, various metals	S
	Synthetic coolants	L S A
Pickling	Acids	A
	Ferrous chloride	A
	Ferrous sulphate	A
	Hydrochloric acid	A
	Hydrofluoric acid	A
	Nitric acid	A
	Phosphoric acid	A
	Sulphuric acid	A
Pigments, *see* Inorganic chemicals		
Processing/engineering	Ammonia	G A
	Arsenic	G S A
	Cyanides	S A
	Emulsions	A
	Lubricating oils	L
	Phenols	L S A
	Soluble oils	A
	Thiocyanates	A
Products		
Batteries	Cadmium	S P
	Lead	S P
	Manganese	S
	Mercury	S
	Nickel	S
	Zinc	S
Catalysts	Cobalt	S
	Iron	S
	Manganese	S
	Mercury	S
	Nickel	S
	Organometallics	L S
	Platinum	S
	Silver	S
	Vanadium	S
Mining (excluding metals)		
	Spoil	S
	Dust	S P
Paint		
	Barium	S
	Cadmium	S
	Chromium	S
	Copper	S
	Lead	S
	Manganese	S
	Mercury	L S
	Selenium	S
	Solvents	G L
	Titanium	S
Paper, pulp		
	Bleaches	A
	Chlorine	G A
	Copper	A
	Fibres	S P A
	Lignin	A

Table 16.1 Cont'd

Source	*Type of waste*	*Form of pollutant*[(1)]
	Mercury	A
	Methanol	A
	Sulphides	A
	Sulphite liquor	A
	Titanium	S P A
	Wax	S P A
	Zinc	P A
Pesticides, herbicides		
	Arsenic	S P A
	Carbamates	S P A
	Chlorinated hydrocarbons	L S P A
	Copper	A
	Fluoride	G S A
	Lead	A
	Mercury	L A
	Organophosphorus compounds	P A
	Phenol	S A
	Polychlorinated biphenyls (PCB)	A
	Selenium	S
Petrochemicals		
Detergents	Boric acid	S A
	Phosphates	S A
	Sulphates	S A
Dyestuffs	Aniline	L A
	Chromium	L S P A
	Phenol	A
	Selenium	L S P A
General	Benzene, toluene, xylene	L
	Boric acid	A
	Chlorocarbons	G L A
	Fluorine	A
	Fluorocarbons	G L A
	Hydrocarbons	G L S P A
	Hydrochloric acid	A
	Hydrofluoric acid	A
	Phenols	L A
	Solvents	L
	Sulphuric acid	A
Miscellaneous	Polychlorinated biphenyls (PCB)	L
	Tetraethyl lead	G L P
Polymers, plastics, resins, rubber and fibres		
	Acids	A
	Alkali	A
	Cadmium	L S P A
	Cuprammonium compounds	A
	Detergents	A
	Dyestuffs	A
	Fibres	S P
	Formaldehyde	A
	Hydrocarbons	G L A
	Methanol	L A
	Phenols	L S A
	Phthalates	L A
	Polychlorinated byphenyls (PCB)	L A
	Solvents	L A

Table 16.1 Cont'd

Source	*Type of waste*	*Form of pollutant*[1]
	Sulphides	A
	Urea	S A
	Wood flour, various woods	S P
	Zinc	A
Pharmaceuticals		
	Drug intermediates and residues	L S A
	Solvents	L A
Refineries (Petroleum)		
	Alkali	A
	BOD waste	L P A
	Combustion products	G
	Emulsions	A
	Hydrocarbons	L A
	Mercaptans	G
	Mineral acids	A
	Phenols	L S P A
	Sulphides	G S P
	Sulphur	S P
	Tars	S P
Sewage treatment		
	Purified effluent	L
	Sewage sludge	S A
Tanneries		
	Arsenic	A
	Chromium	A
	Fibres	S P
	Hair	S P
	Lime	P A
Textiles, see Fibres		
Miscellaneous	Sulphides	A
Electrical, electronics	Copper	S
	Mercury	L
	Precious metals	S
	Selenium	S
Explosives, pyrotechnics	Barium	S
	Hydrocarbons	G L A
	Lead	S
	Manganese	S P
	Mercury	L S P
	Nitric acid	A
	Nitroglycerin	L A
	Phenol	S P
	Phosphorus	S
	Solvents	G L
	Strontium	S
	TNT	L A
Motor industry	Chromates	A
	Grease; mineral-oil based	L S
	Oil	L
	Paint	L S P
	Phosphates	S P A
	Solvents	L
Nuclear fuel and power	Radioactive substances	
	Radioisotopes	
Photography	Alkali	A

Table 16.1 Cont'd

Source	*Type of waste*	*Form of pollutant*[1]
	Cyanide	A
	Mercury	A
	Phenols	A
	Silver	P A
	Thiosulphate	A
Vehicle exhaust	Aromatic hydrocarbons	G
	Lead	S P
	Nitrogen oxides	G
Water treatment	Calcium salts	A P
	Filtered solids	L P A

[1] G Pollutant occurs as a gas
L Pollutant occurs as a liquid
S Pollutant occurs as a solid
P Pollutant occurs in particulate form
A Pollutant occurs in aqueous solution or suspension

Table 16.2 Common sources of pollutants

Chemical	*Form of pollutant*[1]	*Examples*	*Industrial source*
Acid Mineral	A	Hydrochloric acid	Pickling of metals
		Nitric acid	Chemical reagent
		Sulphuric acid	Byproducts, petrochemicals
Organic	A	Acetic acid	Petrochemicals
Aldehydes	A	Acetaldyhyde	Photochemical reaction in smog, incomplete combustion
			Petrochemicals
Alkali	P A	Sodium hydroxide	Electroplating
		Lime	Beverages production
			Photography
			Vegetable and fruit processing
Ammonia	G A	–	Coal distillation (coke ovens)
			Nitric acid production
			Urea and ammonium nitrate works
Aniline and related compounds	L A	–	Dyestuffs production
Aromatic hydrocarbons	G L	Benzene	Coal tar distillation
		Toluene	Vehicle exhausts
		Xylene	Petrochemicals
			Pesticides
			Herbicides
Arsenic	G S P A	Arsine	Pigment and dye
		Arsenous acid and salts	Pesticide and herbicide production
			Metallurgical processing of other metals
			Glass and ceramics industries
			Tanneries
Asbestos	S P	Chrysotile	Obsolete building products
		Amosite	Insulation removal operations
		Crocidolite	
Carbon dioxide	G A		Combustion
			Fermentation

Table 16.2 Cont'd

Chemical	*Form of pollutant*[1]	*Examples*	*Industrial source*
Carbon monoxide	G		Coke ovens Incomplete combustion generally Smelting Vehicle exhausts Metal extraction and refining
Chlorinated hydrocarbons			
Chemical	G L	Trichloroethylene; 1,1,1-trichloroethane (obsolescent)	Degreasing (engineering) Drycleaning Solvents
Pesticidal	L S	DDT BHC Aldrin Dieldrin	Pesticides Wood treatment
Chlorine and chlorides	G S P A		Chlorinated hydrocarbons Chloralkali Paper and pulp processing Petrochemicals Metal extraction and refining
Chromium and compounds	S P A	Chromic acid Sodium dichromate	Anodizing Cement Dyes Electroplating Paint Tanneries
Cobalt and compounds	S P A	Cobalt oxide	Catalysts Fibres Paint Paper and pulp processing Tungsten carbide binder
Copper and compounds	S P A	Copper sulphate Copper pyrophosphate Cuprammonium compounds	Electroplating Electrical and electronics Etching Pesticides
Cyanate	A		Coal distillation Oxidation of cyanide
Cyanide	S P A	Sodium cyanide Copper cyanide	Heat treatment of metal Coal distillation Electroplating Photographic Synthetic fibre
Disinfectants			Agriculture and horticulture Abattoirs Food processing
Fluorides	G S P A	Hydrogen fluoride Calcium fluoride	Cement Aluminium
Hydrocarbons, general			Coal distillation Petrochemicals Petroleum refineries
Iron and compounds	S P A	Iron oxide Ferrous chloride	Aluminium refining Steelworks Electronics Electroplating Pickling Pigments

Table 16.2 Cont'd

Chemical	*Form of pollutant*[1]	*Examples*	*Industrial source*
Lead and compounds	S P A	Lead oxide	Batteries
		Tetraethyl lead (TEL)	Explosives and pyrotechnics
			Paint
			Pesticides
			Petrochemicals
			Printing
			Refineries
			Vehicle exhausts
Manganese and compounds			Batteries
			Catalyst
			Glass
			Paint
			Pyrotechnics
Meat wastes	S A		Meat processing and preparation
			Abattoirs
			Dairies
			Tanneries
Mercaptans	G		Refineries
			Coke ovens
Mercury			Herbicides
Organic	S A	Methyl mercury	Bacterial activity on inorganic mercury
			Pesticides
Inorganic	L S A	Mercurous chloride	Batteries
			Catalysts
			Cement
			Chloralkali process
			Combustion of coal and oil
			Electrical and electronic
			Explosives
			Paints
			Paper and pulp
			Pesticides
			Pharmaceuticals
			Photographic
			Scientific instruments
Methanol	L A		Paper
			Resins
Nitrates	S P	Potassium nitrate	Metals heat treatment
			Water treatment
Nitrogen oxides	G	Nitrogen dioxide	Combustion processes and explosives
			Electricity generation
			Forage tower silos
			Nitric acid
Oil and soluble oil	L		Engineering
			Petrochemicals
			Refineries
Paraquat	L S		Herbicide
Pesticides (includes acaricides, avicides, bactericides, insecticides, molluskicides, nematocides, piscicides, rodenticides)		Chlorinated hydrocarbons (q.v.)	
		Carbamates (q.v)	
		Organophosphorus compounds (q.v.)	
Pharmaceuticals	L S P A	Aspirin	Pharmaceutical industry
		Penicillin	

Table 16.2 Cont'd

Chemical	*Form of pollutant*[1]	*Examples*	*Industrial source*
Phenol and related compounds	S P A	Phenol	Coal distillation
		Cresol	Dyestuffs
			Explosives
			Pesticides
			Petrochemicals
			Photographic
			Plastics manufacture
			Refineries
Phosphorus and compounds	S P A	Phosphoric acid	Boiler blowdown
			Corrosion protection
			Detergents
			Fertilizers
			Matches
			Metal finishing
Phthalates	L S P A	Dibutyl phthalate	Adhesives
Platinum and compounds	S P		Catalysts
Polychlorinated biphenyls (PCB)	L S A		Lubricants and hydraulic fluids
			Pesticides
			Plasticizer in paint and polymers
			Plasticizer (polymers)
			Transformer oils
Silicates	P		Cement
			Metal extraction and refining
Sulphur oxides	G	Sulphur dioxide	Coal distillation
		Sulphur trioxide	Combustion of coal and heavy fuel oil
			Detergents (sulphonation of alkyl benzenes)
			Electricity generation
Tar	L		Refineries
			Coal distillation
Thiocyanate	A		Coal distillation
Tin and compounds	G S P A		Tinplating
Titanium and compounds	S P A	Titanium dioxide	Astronautics
			Paint
			Paper
Vanadium and compounds	S P	Catalysts	Animal feed
Vegetable waste	L S P A		Breweries
			Natural rubber
			Starch
			Sugar refineries
			Vegetable and fruit processing and preparation
Wax	S		Fruit preserving
			Paper
			Refineries
			Textiles
Zinc and compounds	G S P A		Electroplating
			Galvanizing
			Paper and pulp
			Synthetic fibres

[1] G Pollutant occurs as a gas
L Pollutant occurs as a liquid
S Pollutant occurs as a solid
P Pollutant occurs in particulate form
A Pollutant occurs in aqueous solution or suspension

Conventionally, wastes are considered as being predominantly either solid, liquid or gaseous but as illustrated in Table 16.3, they may be multi-phase. Solid waste comprises liquid slurries, sludges, thixotropic solids and solids of varying particle sizes; it may be heterogeneous. Typical examples are given in Table 16.4.

Table 16.3 Some physical forms of waste

Phase	*Examples*	
Gas		
Gas–vapour	–	SO_2; NO_x; HCl; CO; hydrocarbons
Gas–liquid	Mist	Acid mist carryover; chromic acid; oil mists; tar fog
Gas–solid	Fume	Metal oxides, cement dust
Gas–liquid–solid	–	Paint spray
Liquid		
Liquid	Solution	Metal plating effluent; 'spent' acids; wash-waters
Liquid–gas	Foam	Detergent foam
Liquid–liquid	Emulsions	Oil-in-water (e.g. suds); water-in-oil
Liquid–solid	Slurry Suspension	Aqueous effluent from fume scrubbing
Solid		
Solid	–	Asbestos insulation; heat treatment salts, pulverized fuel ash; refuse
Solid–liquid	Sludge, wet solid	Filter cake Sewage sludge

Waste treatment prior to disposal may introduce phase changes which result in quite different pollution control considerations. For example, the gases generated by incineration of a solid waste can be scrubbed with liquid in order to meet an acceptable discharge criterion; hence, in addition to ash for disposal, a liquid effluent stream is produced and requires treatment. Other waste treatment processes may result in the liberation of flammable or toxic gaseous emissions as exemplified in Table 16.5.

Pollutants may enter the environment via air, water or land and prove:

- Damaging to the environment or public health and well-being.
- Damaging to buildings and materials of construction (Table 16.6).
- Wasteful of valuable resources.
- Illegal.
- Technically difficult and expensive to deal with (e.g. to clean up spillages and rectify damage).
- Harmful to company reputation.

The range of measures applicable to control are summarized in Chapter 17. Recycling and recovery are potentially attractive measures; hence solvents, mineral oils, metals, e.g. lead, copper, nickel, mercury, and glass are commonly recycled. However, recycling options generally depend upon favourable economics, particularly low collection costs.

The two different, but related, considerations in waste disposal are hazard control and loss prevention in the treatment and disposal operations, and the control of environmental hazards. With gas and liquid streams the control of on-site hazards arising from the chemical properties and processing operations generally follows the principles summarized in earlier chapters. The measures necessary with 'solid' wastes may, however, differ, particularly if they are heterogeneous in nature and disposed of on land.

Table 16.4 Major solid wastes: origin, quantities and destination

Type	*Origin*	*Method of disposal*	*Present uses*
Colliery spoil	Mining	Mainly tipping on land, some in sea Some used	As fill and in manufacture of bricks, cement and lightweight aggregate
China clay waste, overburden sand, micaceous residue	Quarrying	Mainly tipping and in lagoons Some used	As fine aggregate in concrete, in manufacture of bricks and blocks and as fill
Household refuse	Household	Mainly landfill Some incineration	Some resource and energy recovery
Pulverized fuel ash and furnace bottom ash	Waste from power stations burning pulverized coal	Some used Remainder in old workings or artificial lagoons	As fill and in manufacture of cement, concrete blocks, light-weight aggregate, bricks etc.
Blastfurnace slag	Iron smelting	Nearly all used	As roadstone, railway ballast, filter medium, aggregate for concrete, fertilizer and in manufacture of cement
Trade waste	Industry	Mainly landfill Some incineration	Unknown
Steel-making slag	Steel making	Some returned to blast-furnaces, remainder dumped or used as fill near steel works, or sold	As roadstone
Furnace clinker	Waste from chain grate power stations	All used	Concrete block making
Incinerator ash	Residue from direct incineration	Most dumped Minor usage	As fill and for covering refuse tips
Byproduct calcium sulphate	Manufacture of phosphoric acid and of hydrofluoric acid	Mainly in sea. Some dumped on land Some used	In manufacture of floor screeds
Waste glass	Waste glass within household refuse	No utilization in household refuse	If segregated, recycled
Slate waste	Mining and quarrying slate	Mainly tipping, some backfilling of old workings Minor usage	Inert filler, granules, expanded slate aggregate and filter medium, road building
Tin mine tailings	Tin mining	Minor utilization. Tailing lagoons and discharge into sea	Aggregate for concrete
Fluorspar mine tailings	Fluorspar mining	Minor utilization Tailing lagoons	Aggregate for roadmaking and concrete
Red mud	Production of alumina	Minor utilization, rest in lagoons	Pigment in paints and plastics
Copper slag	Smelting	Complete utilization	Grit blasting
Tin slag	Smelting	Major utilization Some in tips	Grit blasting and road building
Zinc-lead slags	Smelting of zinc and lead	Stockpiled and used locally	Bulk fill and some in pavement asphalt
Quarry wastes	Quarrying	Some utilization Remainder tipped	Roadmaking, brickmaking

Table 16.5 Examples of hazardous gases generation from waste pretreatment processes

Process	*Associated problems*	*Hazardous airborne contaminants liberated (in addition to the principal reagents)*
Neutralization of strong mineral acids from metal finishing trades (sulphide and hypochlorite contamination common)	Fierce reaction Possibility of mixing with water or organic materials	Chlorine Nitrogen dioxide Sulphur dioxide Hydrogen sulphide
Chlorination/oxidation of cyanide wastes from heat treatment plant	Mixing cyanide with acids liberates hydrogen cyanide	Hydrogen cyanide
Separation of oil and water mixtures from engineering and heat raising plant	Emulsion splitting may involve generation of heat, hydrogen and hydrogen sulphide	Hydrogen Hydrogen sulphide Phosphine
Detoxification of chromic acid and chromium salts from the plating industry	Use of sulphite	Sulphur dioxide
Detoxification of by-products from smelting	Water and weak acids liberate attack gas	Arsine Phosphine
Treatment of ammonia-bearing waste from chemical industry	Liberation of gaseous ammonia	Ammonium chloride Nitric acid
Removal of sulphides from leather industry waste	Generation of sulphide gas	Hydrogen sulphide

The common practice in waste management is either to recycle waste or to treat and dispose of it at the end of a process. However, many 'end-of-pipe' treatment processes simply transfer a waste, albeit sometimes in a different chemical form, from one environmental medium (i.e. air, land or water) to another often at high dilution.

It would obviously be desirable to eliminate the generation of waste. Practical alternatives are to minimize it by reduction at source or to recycle it. Table 16.7 illustrates the hierarchy of waste minimization practices. Source reduction includes increased process efficiency; economies in the use of energy are also relevant, e.g. it may result in a decrease in the consumption of fossil fuels:

- so reducing the air pollutants generated; or
- reducing boiler operation, with a consequent decrease in the discharge of cooling water and boiler-blowdown discharges.

'Accidental' pollution incidents arise from spillages, gas emissions, and liquid discharges sometimes in transportation. The main considerations are prevention, containment, and effective mitigatory and emergency action according to the principles outlined in earlier chapters. Effective clean-up and decontamination are then needed.

The more common requirement to control routine disposal and dispersion of solid, liquid or gaseous pollutants is based upon different criteria, e.g. their persistence in the environment (as with the effects attributed to ozone-depleting gases, or the problem of heavy metal contamination

Table 16.6 Effects of major pollutants on materials

Pollutant	*Primary materials attacked*	*Typical effect*
Carbon dioxide	Building stone e.g. limestone	Deterioration
Sulphur oxides	Metals	
	Ferrous metals	Corrosion
	Copper	Corrosion to copper sulphate (green)
	Aluminium	Corrosion to aluminium sulphate (white)
	Building materials (limestone, marble, slate, mortar)	Leaching, weakening
	Leather	Embrittlement, disintegration
	Paper	Embrittlement
	Textiles (natural/synthetic)	Reduced tensile strength, deterioration
Hydrogen sulphide	Metals	
	Silver	Tarnish
	Copper	Tarnish
	Paint	Leaded paint blackened by formation of lead sulphide
Ozone	Rubber and elastomers	Cracking, weakening
	Plastics	Degradation, weakening
	Textiles (natural/synthetic)	Weakening
	Dyes	Fading
Nitrogen oxides	Dyes	Fading
Hydrogen fluoride	Glass	Etching, becoming opaque
Solid particulates (soot, tars)	Building materials	Soiling
	Painted surfaces	Soiling
	Textiles	Soiling
Acid water (pH <6.5)	Cement and concrete	Slow disintegration
Ammonium salts		Slow to rapid disintegration
		If cement is porous, corrosion of steel reinforcement may occur
Fats, animal or vegetable oils		May cause slow disintegration
Calcium chloride		May cause steel corrosion if concrete is porous/cracked
Calcium sulphate (and other sulphates)		Disintegrates concrete of inadequate sulphate resistance
Solvents	Plastics, rubber	May cause changes due to penetration
Oxalic acid		Beneficial: reduces effects of CO_2, salt water, dilute acetic acid
Paraffin		Slow penetration, not harmful
Phenol (5% solution)		Slow disintegration
Caustic alkali		Not harmful <10–15% concentration

of land), synergistic effects (as with the enhanced toxicological effects of smoke in the presence of sulphur dioxide) or long-term cumulative effects. Thus the different considerations relating to occupational compared with environmental exposures to air pollutants are exemplified in Table 16.8.

Atmospheric emissions

Sources

Important atmospheric pollutants comprise smoke, dust, grit, fumes and gases. Types of emission are shown in Table 16.9.

Table 16.7 Waste minimization in the hierarchy of waste management practices

		Priority	
Elimination	Complete elimination of waste	Higher priority ↑	Waste minimization
Source reduction	Avoidance, reduction or elimination of waste, generally within the confines of the production unit, through changes in industrial processes or procedures		
Recycling	Use, reuse and recycling of wastes for the original or some other purpose, e.g. input material, materials recovery or energy production		
Treatment	Destruction, detoxification, neutralization, etc. of wastes into less harmful substances	Lower priority ↓	Treatment and safe disposal
Disposal	Discharge of wastes to air, water or land in properly controlled, or safe, ways such that compliance is achieved. Secure land disposal may involve volume reduction, encapsulation, leachate containment and monitoring techniques		

Table 16.8 Factors in exposure and effects of air pollutants

Exposure	*Occupational*	*Environmental*
Population	Adults (16–65) Fit for work; possibly monitored	All population including infants, aged, infirm hypersensitive subjects
Period	Basic working week, e.g. <40 hr/wk, 48 wk/yr plus overtime; therefore intermittent elimination and recovery times. Holidays excepted	Continual unless area vacated; therefore elimination and recovery depend on irregular periods of low/zero concentrations
Levels	Possibly measurable fractions of OESs (mg/m^3)	Normally very low: at limits of analytical/instrumental sensitivity (μg/m^3)
Dust, fume, gas, vapour, mist	Generally single pollutants Known origin Hazards probably known Recognized problem Personal protection and possibly health surveillance provided Probably freshly formed/released	Mixture of primary pollutants, from different sources and secondary pollutants. Possible additive or synergistic effects Origins may be difficult to prove Hazards not quantified Exposure largely unheeded
Concern	Specific effects, therefore increased costs, accident rate; reduced productivity	Decreased well-being Non-specific respiratory troubles Irritation of eyes, nose and throat Damage to property and vegetation Injury to animals Decrease in 'amenity' Long-term ecological effects

Combustion processes are the most important source of air pollutants. Normal products of complete combustion of fossil fuel, e.g. coal, oil or natural gas, are carbon dioxide, water vapour and nitrogen. However, traces of sulphur and incomplete combustion result in emissions of carbon monoxide, sulphur oxides, oxides of nitrogen, unburned hydrocarbons and particulates. These are 'primary pollutants'. Some may take part in reactions in the atmosphere producing 'secondary pollutants', e.g. photochemical smogs and acid mists. Escaping gas, or vapour, may

Table 16.9 Continuous/intermittent emissions to atmosphere

High level		
Routine	Vents	General ventilation (factory atmosphere) Local extraction (dust, fumes, odour)
	Flare stacks	Normal flaring
	Chimney	
Irregular	Plant maloperation	Process plant Dust and fume extraction plant
	Flare stack	Emergency/occasional flaring
	Plant failure	Process plant – emergency venting Extraction/collection plant (cyclones, precipitators, filters, scrubbers)
Low level		
Routine	Process equipment cleaning Materials handling (discharge, conveying, bagging) Waste handling/deposition	
Irregular	Plant maloperation (e.g. unauthorized 'venting') Plant failure – spillages, pipe joint failures Start-up/shutdown Dismantling/demolition Unauthorized waste incineration (rubbish burning) Fires	

also be associated with the storage or processing of volatile materials or 'accidents' which occur in handling and transportation.

Dust can be emitted wherever solids are mined (e.g. in quarries), processed (e.g. flour mills, woodworking factories, metal smelting and foundries), or handled/transported in particulate form. Construction and demolition operations also generate dust.

In Table 16.9 emissions are classified as 'persistent' or 'irregular', and as to the level (i.e. the height above ground) at which they are likely to be generated.

Effects

The effects of atmospheric pollution are chemical, and concentration, specific. However, additive, and sometimes synergistic, effects arising from mixtures are important. A summary of the potential effects of common air pollutants is given in Table 16.10.

As noted above (Table 16.8), in considering a limit on concentration at ground level to ensure that there is no significant risk to health of local residents it is essential to recognize the important difference from occupational exposure. Thus the Occupational Exposure Limits quoted in Chapter 5 cannot readily be extrapolated to assess or control non-occupational exposure. Damage to animals, vegetation and growing crops (which can occur at extremely low levels with some pollutants) and to buildings, etc. must also be considered.

Liquid effluents

Sources

The manner by which aqueous-based waste streams may arise in a factory from process uses of

Table 16.10 Major atmospheric pollutants and their effects

Contaminant	*Manmade source*	*Effects*
Sulphur dioxide	Combustion of coal, oil and other sulphur-containing fuels Petroleum refining, metal smelting, paper-making	Vegetation damage Sensory and respiratory irritation Corrosion Discoloration of buildings
Hydrogen sulphide	Various chemical processes Oil wells, refineries Sewage treatment	Odours, toxic Crop damage/reduced yields
Carbon monoxide	Vehicle exhaust and other combustion processes	Adverse health effects
Carbon dioxide	Combustion processes	'Greenhouse' effects
Nitrogen oxides	High-temperature reaction between atmospheric nitrogen and oxygen, e.g. during combustion Byproduct from manufacture of fertilizer	Adverse health effects Sensory irritants Reduced visibility Crop damage
Ammonia	Waste treatment	Odour Irritant
Fluorides	Aluminium smelting Manufacture of ceramics, fertilizer	Crop damage HF has adverse health effects on cattle fed on contaminated food
Lead	Combustion of leaded petrol Solder, lead-containing paint Lead smelting	Adverse health effects
Mercury	Manufacture of certain chemicals, paper, paint Pesticides Fungicides	Adverse health effects
Volatile hydrocarbons	Motor vehicles Solvent processes Chemical industry	Vegetation damage (especially unsaturated hydrocarbons) Some are irritants Adverse health effects
Particulates	Chemical processes Fuel combustion Construction Incineration processes Motor vehicles	Nuisance Adverse health effects Reduced visibility Deposition on buildings

water, cooling water, wash-rooms and canteens, and steam-raising is exemplified in Figure 16.1. Within the UK the majority are treated prior to discharge to sewers operated by a sewage undertaker. Following further treatment they are then discharged to natural waters or the open sea. Some effluents are subjected to treatment by the producer before similar final discharge.

Strict control is necessary over pollutant levels, acidity, temperature, turbidity, etc. because of their impact upon aquatic environments. Common pollutants in aqueous effluents are identified in Table 16.11.

Effects

The pH of rainwater is normally about 6 but can be reduced significantly by absorption of acidic exhaust gases from power stations, industrial combustion or other processes, and vehicles. Acids may also enter the waterways as a component of industrial effluent. In addition to the direct adverse effects on aquatic systems (Table 16.12) low pH can result in the leaching of toxic metals from land, etc.

The salts of the heavy metals beryllium, cadmium, chromium, copper, lead, mercury, nickel and zinc are all of high eco-toxicity. For example, the toxicity of some heavy metals to rainbow trout is demonstrated in Table 16.13; coarse fish are somewhat more resistant.

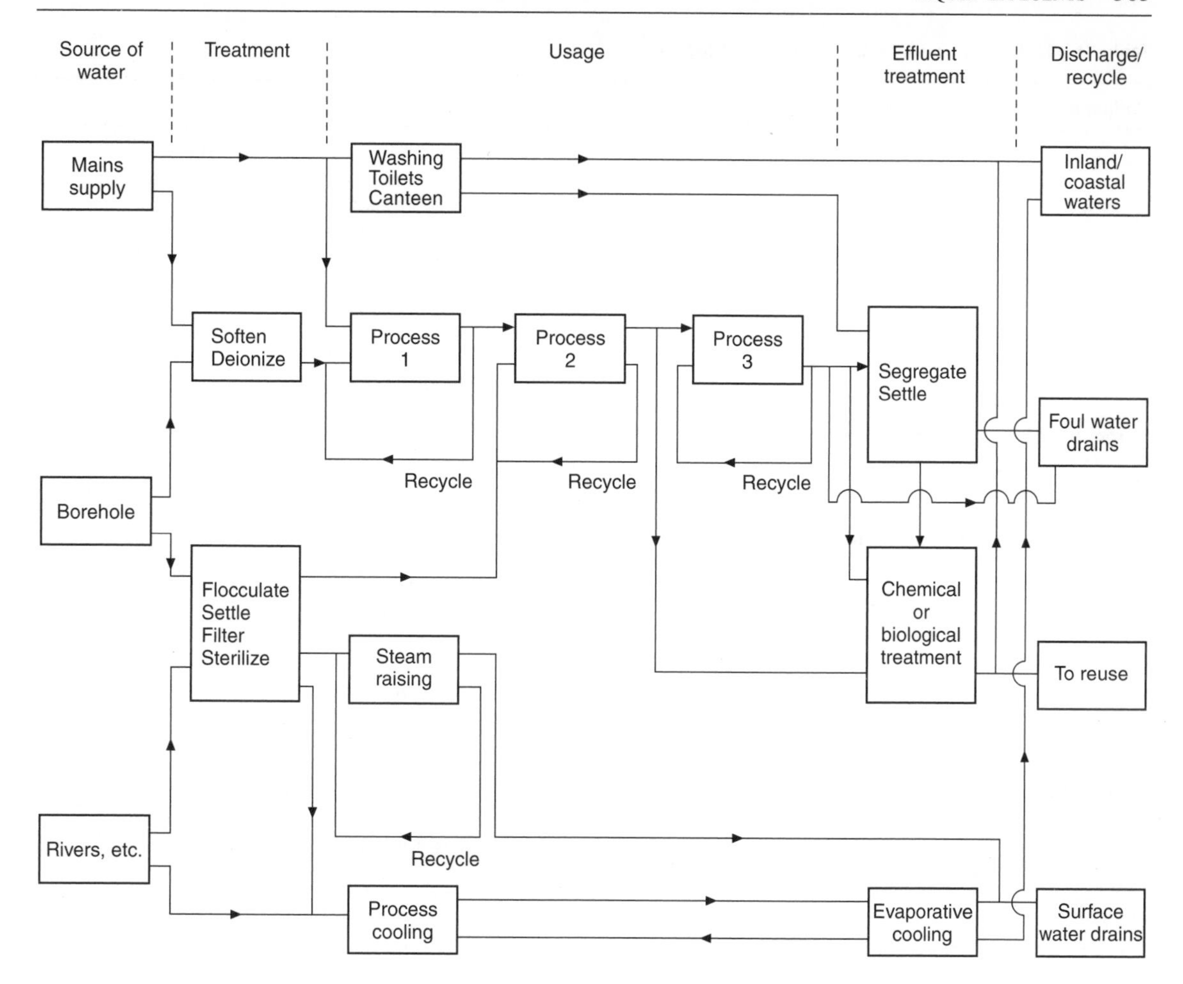

Figure 16.1 *Factory water supply: usage and disposal*

Fish are generally more susceptible to poisoning than microscopic plants or other animals; as such they are a good indicator species. A summary of the concentrations of selected substances at which toxic effects have been detected in bacteria, algae, crustacia and protozoa is given in Table 16.14.

Heavy metals may also be concentrated in passage up the food chain. Other pollutants, e.g. fungicides, pesticides, biocides, polychlorinated biphenyls or organic mercury compounds, are persistent and can therefore also bioaccumulate.

Cyanides can be fatal to fish at <1 ppm. Because of concern over the possible *in vivo* conversion of nitrate into carcinogenic nitrosamines, the nitrate content of drinking water must be strictly controlled. Nitrate and phosphate pollution can also cause eutrophication in still, or slow-moving, warm waters by stimulation of algae growth in the presence of strong sunshine. This chokes the water and results in depletion of light and oxygen content.

Trace contaminants such as sodium chloride, iron and phenols (especially if chlorinated) may also impart a characteristic taste if water is subsequently used for drinking purposes.

As discussed in Chapter 17, legislative controls including stringent consent conditions are applied in the UK to all discharges to specified sewers. (Discharge of effluent to surface water drains is prohibited.) Limits, or even total prohibitions, are placed upon certain chemicals to avoid

Table 16.11 Common pollutants of aqueous effluents

Pollutant	*Example*
Alkali	Sodium hydroxide (caustic soda)
	Potassium hydroxide (caustic potash)
	Calcium oxide (lime)
	Calcium hydroxide
	Sodium, potassium and calcium carbonates
	Ammonia (q.v.)
Ammonia	
Biocides	
Biodegradable waste	Sewage
	Food waste
	Organic chemicals
Boron, borates, fluoroborates	–
Bromine	–
Chloride	Sodium chloride
Chlorine	Hypochlorites
Chromic acid (hexavalent chromium)	Potassium dichromate
Cyanide	Copper cyanide
	Nickel cyanide
	Potassium cyanide
	Silver cyanide
	Sodium cyanide
	Zinc cyanide
Emulsified oil	Suds
Fluoride	–
Fluorine	–
Metal salts in alkaline solution	Cuprammonium complex
	Nickel and cobalt ammonia complex
	Cyanides (q.v.)
	Copper pyrophosphates
	Plumbites
	Zincates
Metal salts in acid solution	Most metals as acid salts, e.g.
	chloride
	nitrate
	sulphate
Mineral acid	Hydrobromic acid
	Hydrochloric acid
	Hydrofluoric acid
	Nitric acid
	Sulphuric acid
Miscible/soluble organic materials	Acetone
	Alcohol
	Acetic acid
Non-metallic inorganic dissolved compounds	Arsenic
	Selenium
Oil, grease, wax and immiscible organics	Lubricating oil
	Animal fat
	Chlorinated solvents
Organometallic compounds	–
pH	Acids
	Alkalis
Pharmaceuticals	–
Phenols and related compounds	Phenol
	Cresol
Phosphate	Detergents
Sewage	–
Sulphate	Calcium sulphate

Table 16.11 Cont'd

Pollutant	*Example*
Sulphite liquor	Sodium sulphite
Suspended particles	–
Total dissolved solids	Carbonate
	Chloride
	Nitrate
	Phosphate
	Sulphate

Table 16.12 Sensitivity of aquatic organisms to lowered pH

pH	*Effect*
6.0	Crustaceans, molluscs etc. disappear
	White moss increases
5.8	Salmon, char, trout and roach die
	Salamander eggs fail to hatch
	Sensitive insects, phytoplankton and zooplankton die
5.5	Whitefish and grayling die
5.0	Perch and pike die
4.5	Eels and brook trout die
4.0	Crickets, frogs and northern spring peepers die

Table 16.13 Toxic concentrations of heavy metals to rainbow trout

	Concentration (mg/l) which kills half the number of exposed fish in 48 hours (LC_{50})	
Metal	*Soft water*	*Hard water*
Copper	0.04	0.5
Cadmium	0.05	5.5
Zinc	0.5	3
Lead	1	–
Nickel	15	100

risks to sewermen, e.g. from the generation of toxic gases such as hydrogen sulphide, to avoid obstruction of flow, e.g. from suspended solids, to eliminate any risk of explosion or fire from volatile flammable liquids, and to ensure there is no subsequent interference with waste treatment processes.

The type of consent conditions imposed, where the effluent represents only a small proportion of the flow in a sewer, are exemplified in Table 16.15.

More stringent limits are imposed upon discharges to natural waters as exemplified in Table 16.16. In both cases the BOD, the biological oxygen demand, represents the demand for oxygen in order to decompose the organics in the effluent.

Solid wastes

Land pollution has arisen from the direct dumping of domestic and industrial solid waste, but this

Table 16.14 Concentrations, in ppm, of substances which show toxic effects in various organisms (in river water, pH 7.5)

	Bacteria–Escherichia coli	*Green algae–Scenedesmus quadricauda*	*Daphnia magna*	*Protozoa microregma*
Copper	0.08	0.15	0.1	0.05
Zinc	1.4 to 2.3	1 to 1.4	1.8	0.33
Chromium III		4 to 6	42	37
Cyanide	0.4 to 0.8	0.16	0.8	0.04
Cyanate	10	520	23	21
Sulphide	93	40	26	
Phenol		40	16	30
Toluene	200	120	60	
o-Xylene	500	40	16	10
m-Xylene		40	24	70
p-Xylene		40	10	50
Amyl alcohol		280	440	20
Formic acid		100	120	
Butyric acid		200	60	
Butyl acetate		320	44	20
Amyl acetate		180	120	40

Table 16.15 Type of consent conditions for discharge to sewers. (Other factors, e.g. the specific sewer, flowrate, total volume discharged, are also defined)

Parameter/substance	*Maximum allowed*
pH range	6–10 (permitted range)
Sulphate (as SO_3)	500–1000 mg/l
Free ammonia (as NH_3)	500 mg/l
Suspended solids	500–1000 mg/l
Tarry and fatty matter	500 mg/l
Sulphide (as S)	10 mg/1
Cyanide (as CN)	10 mg/l
Immiscible organic solvents	nil
Calcium carbide	nil
Temperature	45°C
Petroleum and petroleum spirit	nil
Total non-ferrous metals	30 mg/l
Soluble non-ferrous metals	10 mg/l
Separable oil and/or grease	300–400 mg/l

is not the sole cause. It may arise from excessive application of agrochemicals, and indirect contamination resulting from leaks or from leaching of hazardous components from liquid waste disposal sites, or from atmospheric fallout. Land may also become contaminated by chemicals processed, stored or dumped at the site, perhaps in the distant past. Such contamination may pose a health risk to workers on the site, those subsequently involved in building, construction or engineering works, or the public (e.g. arising from trespass), and to animals.

Soil surveys relating to construction work require samples from at least the depth of excavation; water seepage may cause cross-contamination of land. Some guidelines for the classification of contaminated soils are summarized in Table 16.17.

Table 16.16 Typical consent conditions for the discharge of industrial effluent to rivers and streams (UK)

	Maximum allowed
Fishing streams	
BOD (5 days at 20°)	20 mg/l
Suspended solids	30 mg/l
pH	5–9
Sulphide, as S	1 mg/l
Cyanide, as CN	0.1 mg/l
Arsenic, cadmium, chromium, copper, lead, nickel, zinc, individually or in total	1 mg/l
Free chlorine	0,5 mg/l
Oils and grease	10 mg/l
Temperature	30°C
Non-fishing streams	
BOD (5 days at 20°C)	40 mg/l
Suspended solids	40 mg/l
pH	5–9
Transparency of settled sample	≥100 mm
Sulphide, as S	1 mg/l
Cyanide, as CN	0.2 mg/l
Oils and grease	10 mg/l
Formaldehyde	1 mg/l
Phenols (as cresols)	1 mg/l
Free chlorine	1 mg/l
Tar	none
Toxic metals, individually or in total	1 mg/l
Soluble solids	7500 mg/l
Temperature	32.5°C
Insecticides or radioactive material	none

Sources

When classified in terms of handling and disposal methods, 'solid wastes' may comprise:

- liquids, e.g. in drums, slurries, sludges, thioxtropic solids, solids, as
- inert or 'hazardous' waste.

(The sources of solid wastes *per se* are summarized in Tables 16.1 and 16.4.) However, dealing with any of them will involve some combination of the activities shown in Figure 16.2, i.e. collection, segregation and identification, processing, recycling, transport and final disposal.

Effects

The effects from solid waste treatment and disposal depend upon the specific waste and the methods employed. The major disposal methods, depending upon the quantity and nature of the waste, are:

- disposal on land by landfill or land-raising;
- incineration, which requires air pollution control measures (Table 17.9) and a procedure for ash disposal; energy recovery may be practised;

Table 16.17 Guidelines for classification of contaminated soils: suggested range of values (ppm) on air dried soils

Parameter	*Typical values for uncontaminated soils*	*Slight contamination*	*Contaminated*	*Heavy contamination*	*Unusually heavy contamination*
pH (acid)	6–7	5–6	4–5	2–4	(<2)
pH (alkaline)	7–8	8–9	9–10	10–12	12
Antimony	0–30	30–50	50–100	100–500	500
Arsenic	0–30	30–50	50–100	100–500	500
Cadmium	0–1	1–3	3–10	10–50	50
Chromium	0–100	100–200	200–500	500–2500	2500
Copper (available)	0–100	100–200	200–500	500–2500	2500
Lead	0–500	500–1000	1000–2000	2000–1.0%	1.0%
Lead (available)	0–200	200–500	500–1000	1000–5000	5000
Mercury	0–1	1–3	3–10	10–50	50
Nickel (available)	0–20	20–50	50–200	200–1000	1000
Zinc (available)	0–250	250–500	500–1000	1000–5000	5000
Zinc (equivalent)	0–250	250–500	500–2000	2000–1.0%	1.0%
Boron (available)	0–2	2–5	5–50	50–250	250
Selenium	0–1	1–3	3–10	10–50	50
Barium	0–500	500–1000	1000–2000	2000–1.0%	1.0%
Beryllium	0–5	5–10	10–20	20–50	50
Manganese	0–500	500–1000	1000–2000	2000–1.0%	1.0%
Vanadium	0–100	100–200	200–500	500–2500	2500
Magnesium	0–500	500–1000	1000–2000	2000–1.0%	1.0%
Sulphate	0–2000	2000–5000	5000–1.0%	1.0–5.0%	5.0%
Sulphur (free)	0–100	100–500	500–1000	1000–5000	5000
Sulphide	0–10	10–20	20–100	100–500	500
Cyanide (free)	0–1	1–5	5–50	50–100	100
Cyanide	0–5	5–25	25–250	250–500	500
Ferricyanide	0–100	100–500	500–1000	1000–5000	5000
Thiocyanide	0–10	10–50	50–100	100–500	2500
Coal tar	0–500	500–1000	1000–2000	2000–1.0%	1.0%
Phenol	0–2	2–5	5–50	50–250	250
Toluene extract	0–5000	5000–1.0%	1.0–5.0%	5.0–25.0%	25.0%
Cyclohexane extract	0–2000	2000–5000	5000–2.0%	2.0–10%	10.0%

- recycling, a procedure now promoted – either by primary or secondary routes;
- pyrolysis – yielding useful by-products.

Any hazard which land deposition may create requires assessment with regard to the risk of injury or impairment of health to persons or animals, damage to vegetation, pollution of controlled waters including aquifers – either directly or because of water run-off, and of long-term accumulation, e.g. of heavy metals or persistent chemicals.

Landfill disposal of certain categories of solid waste may result in gas generation, mainly methane, and a highly polluted leachate. The methane may be drawn off, to avoid a flammable hazard on- or off-site. The leachate is pumped off for treatment.

A summary of the potential hazards associated with toxic waste deposition on land is given in Table 16.18.

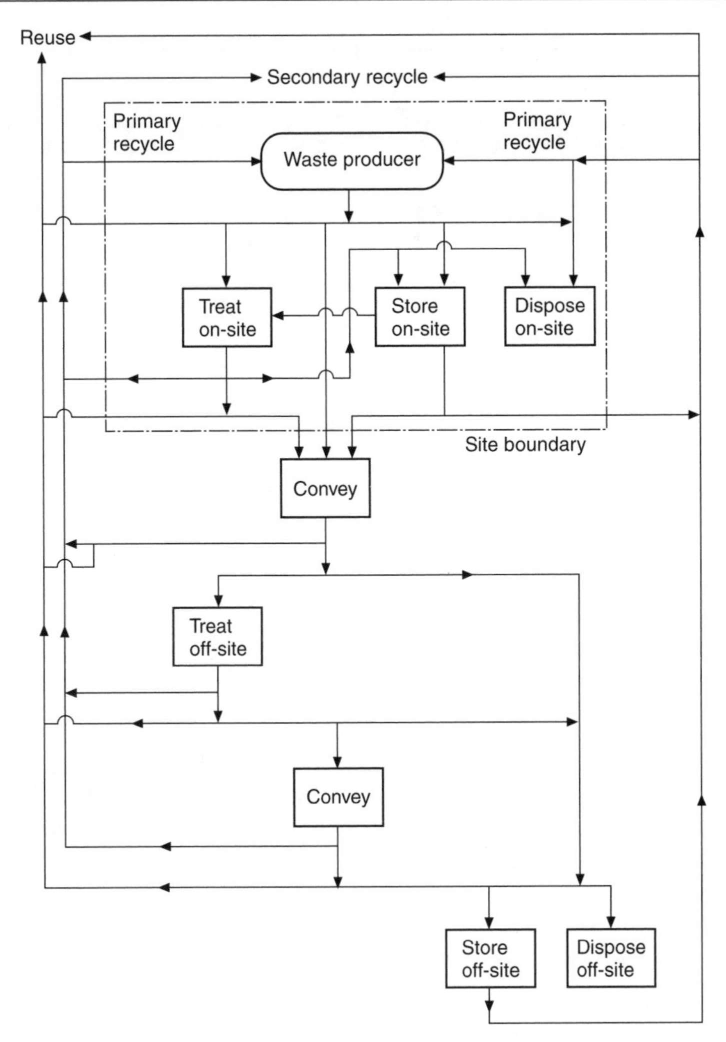

Figure 16.2 *Three basic techniques – treatment, storage, conveyance – to effect recycle or disposal of 'solid' waste either on- or off-site.*

Table 16.18 Potential hazards from toxic waste deposition

Air pollution	Dust, effluvia, smoke and fume, odour
	Toxic or flammable gas generation
Land pollution	Gross amenity damage
	Undermining of site stability
	Sterilization of surrounding land due to heavy metals
	pH changes etc.
	Permanent 'land contamination'
Water pollution	Deposited material or percolate escapes either by surface run-off or by underground movement, threatening streams, rivers, aquifers or the sea
	Direct 'poisoning' or eutrophication

17

Chemicals and the environment: monitoring and protection

Legislative control

Comprehensive pollution control legislation exists in most developed countries and there is increasing conformity within the EC.

The Environmental Protection Act (EPA) 1990

In the UK the Environmental Protection Act 1990 established an interlocking framework for pollution control. Integrated Pollution Control (IPC) introduced a new approach with waste minimization at its centre and a commitment to higher environmental standards. The latter is based upon selection of the Best Practical Environmental Option (BPEO), i.e.:

> the option which in the context of releases from a prescribed process, provides the most benefit or least damage to the environment as a whole, at acceptable cost, in the long term as well as the short term.

The reform of waste disposal aims to minimize waste generation and maximize recycling.

Under Part 1 of EPA 90, an IPC system controls emissions to air, land and water for the most polluting industrial and similar processes. All operators of prescribed processes, e.g. fuel and power, minerals, waste disposal and chemical, require prior authorization. They are required to use BATNEEC, the Best Available Technology Not Entailing Excessive Cost:

- for preventing the release of substances prescribed for any environmental medium, or where that is not practicable by such means;
- for reducing the release of such substances to a minimum and for rendering harmless any such substances which are released;
- for rendering harmless any other substances which might cause harm if released into any environmental medium.

Any authorization may contain specific conditions with details on how BATNEEC is to be achieved, e.g.:

> the mode of operation of the plant; the pollution abatement technology to be used; and, explicit limits on the amount and concentration of substances which may be discharged.

Some aspects of the process may not be covered explicitly by conditions, but there remains a residual duty on the operator to prevent or minimize the release of prescribed substances by the use of BATNEEC.

Local authority control of air pollution covers a second tier of less-polluting processes. Incinerators for waste chemicals, or waste plastic arising from their manufacture, and other waste incinerators dependent upon size are, however, subject to both the BATNEEC and BPEO requirements under the IPC regime.

Any IPC application will include, in addition to details of the operator:

- A description of the prescribed process.
- A list of prescribed substances, and other substances which are potentially environmentally harmful, used in or resulting from the process. The quantities that might be released with control technologies in place.
- Details of any proposed release of prescribed substances and an assessment of the environmental consequences of emissions. (For local authorities air pollution controlled processes this applies only to atmospheric emissions.)
- Information to show that BATNEEC is to be applied; a description of techniques for preventing, minimizing and rendering harmless the emissions. How these techniques will be monitored.
- Proposals for monitoring the release of such substances and the environmental consequences.

It must include an assessment to demonstrate that the process represents the BPEO.

Part II of EPA 90 deals with waste disposal, handling and management. The provisions impose duties on the producers of waste to ensure the safe disposal. There are strong licensing powers for local authorities, with continuing responsibilities for licensees to monitor and maintain sites after closure.

The Clean Air Act 1993 provides a comprehensive control mechanism for the protection of the environment from smoke, dust and fumes. The 1993 Act consolidates the previous provisions of the 1956 and 1968 Clean Air Acts.

Section 1 of the Act prohibits:

dark smoke emissions from a chimney of any building
dark smoke emissions from a chimney serving the furnace of a fixed boiler or industrial plant.

The Dark Smoke (Permitted Periods) Regulations, give details on circumstances and time limits when dark smoke is allowed.

Section 2 of this Act prohibits dark smoke from any industrial or trade premises (there is no requirement under Section 2 for the dark smoke to be emitted from a chimney).

The Water Industry Act 1991 and Water Resources Act 1991

The disposal of aqueous-based wastes, i.e. liquid effluents, is regulated in part by the Water Industry Act 1991 and the Water Resources Act 1991.

The discharge of any trade effluent into a public sewer requires a specific consent from the sewerage undertaker. If granted, conditions are attached to the consent relating to which sewer, flowrates and composition (see Table 16.15), metering of flowrate, and possibly automatic monitoring. For Special Category Effluents, e.g. mercury, pesticides, chlorinated organics, as listed in Schedule 1 of the Trade Effluents (Prescribed Processes and Substances) Regulations 1989 (Table 17.1), agreement must also be obtained from the Environment Agency.

Discharge to 'controlled waters' is controlled under the Water Resources Act 1991. These waters are defined and generally include territorial waters, coastal waters, inland waters and ground water. If the discharge is from a prescribed process, and of a prescribed substance under the Environmental Protection Act 1990, consent for discharge is required under an IPC authorization. All other discharges require consent from the Environment Agency under the Water Resources Act. Such consent is subject to conditions as to composition, flowrate, measures to minimize pollution, etc. A specimen consent form is given in Figure 17.1. Schedule 5 of the regulations details the prescribed substances (Red List). Where a discharge contains more than the stated amount of a prescribed substance (Table 17.2) it cannot be released to water without an authorization from the Environment Agency.

NATIONAL RIVERS AUTHORITY
WATER RESOURCES ACT 1991 – CONSENT TO DISCHARGE

Reference number ....

The National Rivers Authority, in pursuance of its powers under the above mentioned Act, HEREBY GIVES CONSENT to the discharge described hereunder subject to the terms and conditions set out below.

Name and Address of Applicant: ....

Date of Application: ....

Date of Consent: ....

Description of discharge: ....

Type: ....

From: ....

To: ....

National Grid Reference of
Discharge Point: ....

This consent shall not be taken as providing a statutory defence against a charge of pollution in respect of any poisonous, noxious or polluting constituents not specified herein.

....

Conditions
1. General
2. As to Outlet
3. As to Volume Discharged to watercourse
4. As to Discharge composition

NRA Region: ....

....

....

NRA Authorized Signature: ....

Figure 17.1 *Typical consent form*

Table 17.1 Prescribed substances

Aldrin, dieldrin, endrin and isodrin
Arsenic
Atrazine
Azinphos-ethyl
Azinphos-methyl
Boron
Cadmium and its compounds
Carbon tetrachloride
Chloroform
Chromium
Copper
Cyanide
Cyfluthrin
DDT
1,2-Dichloroethane
Dichlorvos
Dioxins
Endosulfan
Fenitrothion
Fenthion
Flucofuron
Hexachlorobenzene (HCB)
Hexachlorobutadiene (HCBD)
Hexachlorocyclohexane (Lindane and related compounds)
Iron
Lead
Malathion
Mercury and its compounds
Nickel
Parathion
Parathion-methyl
PCSDs
Pentachlorophenol (PCP)
Perchloroethylene
Permethrin
pH outside the range 5.5 to 9.0
Polychlorinated biphenyls
Simazine
Sulcofuron
Tetrachloroethylene
Tributyltin compounds
Trichlorobenzene
Trichloroethane
Trichloroethylene
Trifluralin
Triphenyltin compounds
Vanadium
Zinc

Whereas the Red List was produced by the UK government in 1989, and includes the most harmful and polluting substances, the EC in Directive 76/464/EEC introduced Black and Grey Lists, referred to in the Directive as Lists 1 and 2, respectively. National governments are required to eliminate pollution by the substances blacklisted and to reduce pollution by those in the Grey List. The 129 substances in the Black List were selected mainly on the basis of their toxicity, persistence and accumulation in living organisms and in sediment. They are generally considered to be more dangerous than those on the Grey List and are to be controlled by authorization in

Table 17.2 The Red List maximum discharge levels above which prior authorization is required

Prescribed substances	*Amount released above background in any 12 month period (grams)*	
Mercury and its compounds	200	(expressed as metal)
Cadmium and its compounds	1000	(expressed as metal)
All isomers of hexachlorocyclohexane	20	
All isomers of DDT	5	
Pentachlorophenol and its compounds	350	(expressed as PCP)
Hexachlorobenzene	5	
Hexachlorobutadiene	20	
Aldrin	2	
Dieldrin	2	
Endrin	1	
Polychlorinated biphenyls (PCBs)	1	
Dichlorvos	0.2	
1,2-Dichloroethane	2000	
All isomers of trichlorobenzene	75	
Atrazine*	350	
Simazine*	350	
Tributyltin compounds	4	(expressed as TBT)
Triphenyltin compounds	4	(expressed as TPT)
Trifluralin	20	
Fenitrothion	2	
Azinphos-methyl	2	
Malathion	2	
Endosulfan	0.5	

* Where both atrazine and simazine are released, the figure in aggregate is 350 grams.

accordance with European Community limit values or quality objectives to be set in daughter Directives on specific substances. They can be divided into the families shown in Table 17.3. The Grey List comprises the families of materials shown in Table 17.4. The contents of the Grey List may be added to the Black List in due course, and may thus be subject to the same tight controls.

Pesticides are also one of the main sources of inland water pollution and the EC has set limits for pesticide levels in surface waters. Most pesticides in surface waters come from diffuse losses soon after spraying. High levels of other specific pesticides often result from farmyard spillages. Any person using a pesticide must take all reasonable precautions to protect the health of human beings, creatures and plants, to safeguard the environment, and in particular to avoid pollution of water. They should be given instruction, training and guidance in the safe, efficient and humane use of pesticides and be competent in the duties they are called upon to perform. They are not permitted to combine or mix two or more pesticides nor use a pesticide in conjunction with an

Table 17.3 The Black List substances prohibited from discharge to ground water

Organohalogenated compounds and substances which may form such compounds in the aquatic environment
Organophosphorus compounds
Organotin compounds
Substances which possess carcinogenic properties
Mercury and its compounds
Cadmium and its compounds
Persistent mineral oils and hydrocarbons of petroleum origin
Persistent synthetic substances which may float, remain in suspension or sink and interfere with any use of the waters

Table 17.4 The Grey List substances for which authorization is required prior to discharge

Biocides and their derivatives not appearing in the black list
Substances which have a deleterious effect on the taste and/or smell of the products for human consumption derived from the aquatic environment
Toxic or persistent organic compounds of silicon
Inorganic compounds of phosphorus and elemental phosphorus
Non-persistent mineral oils and hydrocarbons of petroleum origin
Cyanides, fluorides
Substances which have an adverse effect on the oxygen balance, particularly ammonia, nitrates
The following metalloids and metals and their compounds:
Antimony
Arsenic
Barium
Beryllium
Boron
Chromium
Cobalt
Copper
Lead
Molybdenum
Nickel
Selenium
Silver
Tellurium
Thallium
Tin
Titanium
Uranium
Vanadium
Zinc

adjuvant except under authorized circumstances. A list of pesticides and pesticide residues as defined by EC Directive 90/642/EEC as amended by 94/30/EC is given in Table 17.5. Limits are to be set by subsequent Directives in due course.

Waste management

Under the EPA 90 a person shall not:

- deposit 'controlled waste', or knowingly cause or permit its deposit;
- treat, keep or dispose of controlled waste or permit it, etc.;

except under, and in accordance with, a waste management licence. Controlled waste is household, industrial, commercial or any such waste.

Extra duties are imposed on the producers of 'Special Waste', i.e. a waste that is on the European Hazardous Waste list reproduced in Table 17.6 and if it has one or more of the hazardous properties listed in Table 17.7 (reproduced from Part 2 of Schedule 2 of the Special Waste Regulations, 1996). Also Special Waste is any Controlled Waste which has one of the listed properties. Extra requirements are detailed for the safe transfer and management of such waste.

Section 34 of EPA 90 imposes a 'cradle to grave' philosophy for waste management. This places a duty of care upon anyone who has 'control' of controlled waste, i.e. importers, producers,

Table 17.5 Pesticides and pesticide residues

Pesticide	*Residue*
Aldrin and dieldrin	singly or combined, expressed as dieldrin (HEOD)
2-Aminobutane	2-aminobutane
Aminotriazole	aminotriazole
Azinphos-methyl	sum of azinphos-methyl and azinphos-ethyl
Bitertanol	bitertanol
Captafol	captafol
Captan	sum of captan and folpet
Carbaryl	carbaryl
Carbendazim	carbendazim (from use of benomyl, thiophanate-methyl and carbendazim)
Carbon disulphide	carbon disulphide
Carbon tetrachloride	carbon tetrachloride
Carbophenothion	sum of carbophenothion, its sulphoxide and its sulphone, expressed as carbophenothion
Chlordane	(1) for products of animal origin: sum of cis- and trans-isomers and oxychlordane expressed as chlordane; (2) for cereals, fruit and vegetables: sum of cis- and trans-isomers expressed as chlordane
Chlorfenvinphos	sum of E- and Z-isomers of chlorfenvinphos
Chlorpyrifos-methyl	chlorpyrifos-methyl
DDT	sum of pp'-DDT, op'-DDT, pp'-TDE and pp'-DDE expressed as DDT
Diazinon	diazinon
1,2-Dibromoethane	1,2-dibromoethane
Dichlofluanid	dichlofluanid
Dichlorvos	dichlorvos
Dicofol	Dicofol
Diflubenzuron	diflubenzuron
Dimethipin	dimethipin
Dimethoate	dimethoate
Dithiocarbamates	alkylenbisdithiocarbamates and alkylthiuramdisulphides and dialkyldithiocarbamates determined and expressed as carbon disulphide (CS_2)
Endosulfan	sum of alpha- and beta-isomers and of endosulfan sulphate, expressed as endosulfan
Endrin	endrin
Ethion	ethion
Etrimfos	sum of etrimfos, its oxygen analogue and 6-ethoxy-2-ethyl-4-hydroxypyrimidine
Fenitrothion	fenitrothion
Fluazifop	fluazifop and esters (including conjugates) of haloxyfop, expressed as free acid
Fluorochloridone	fluorochloridone
Haloxyfop	haloxyfop and esters (including conjugates) of haloxyfop, expressed as free acid
Hexachlorobenzene (HCB)	hexachlorobenzene
Hexachlorocyclohexane (HCH)	hexachlorocyclohexane alpha-isomer beta-isomer gamma-isomer
Heptachlor	sum of heptachlor and heptachlor epoxide, expressed as heptachlor
Hydrogen cyanide	cyanides expressed as hydrogen cyanide
Hydrogen phosphide	phosphides expressed as hydrogen phosphide
Imazalil	imazalil
Inorganic bromide	determined and expressed as total bromine from all sources
Ioxynil	ioxynil
Iprodione	sum of iprodione and all metabolites containing 3,5-dichloroaniline moiety, expressed as iprodione
Malathion	sum of malathion and malaoxon, expressed as malathion
Mercury compounds	determined as total mercury and expressed as mercury
Methacrifos	methacrifos
Methyl bromide	bromomethane
Mevinphos	sum of cis- and trans-mevinphos

Table 17.5 Cont'd

Pesticide	*Residue*
Omethoate	omethoate (from use of formothion, demethoate and omethoate)
Parathion	parathion
Parathion-methyl	parathion-methyl
Phosalone	phosalone
Phosphamidon	sum of phosphamidon (E- and Z-isomers) and N desethylphosphamidon (E- and Z-isomers) expressed as phosphamidon
Pirimiphos-methyl	pirimiphos-methyl
Pyrethrins	sum of pyrethrins I and II, cinerins I and II, jasmolins I and II
Quintozene	sum of quintozene, pentachloroaniline and methyl pentachlorophenylsulphide expressed as quintozene
Tecnazene	tecnazene
Thiabendazole	thiabendazole
Triazophos	triazophos
Trichlorfon	trichlorfon
2,4,5-T	2,4,5-T
Vinclozolin	sum of vinclozolin and all metabolites containing 3,5-dichloroaniline moeity, expressed as vinclozolin

carriers, keepers, treaters, disposers, recyclers and brokers. They are required to take all such measures as are reasonable in the circumstances:

- to prevent any person from contravening EPA 90, i.e. the prohibition placed upon unauthorized or harmful deposition, treatment or disposal of waste;
- to prevent the escape of waste; and
- on transfer, to secure that it is only to an authorized person, or for authorized transport, and with an adequate written description of the waste.

Duty of Care Regulations outline a transfer note system from which the Environment Agency can trace the destination of waste and previous holders. The information required on a transfer note is illustrated in Figure 17.2. In order to comply with the statutory duty of care in respect of waste the following procedures are advised in a Code of Practice:

- Identify and describe the waste:
  - Is it controlled waste?
  - What problems are associated with the waste and its disposal?
  - What should be included in the description?
- Keep the waste safely:
  - Are containers secure for both storage and transit?
  - What precautions have been taken to ensure security of waste?
- Transfer to the right person:
  - Will public authorities take the waste?
  - Is the carrier registered, or exempt from registration (under the Control of Pollution (Amendment) Act 1989 and the Controlled Waste (Registration of Carriers and Seizures of Vehicles) Regulations 1991)?
  - Is the waste manager properly licensed?
- Steps on receiving waste:
  - Who did the waste come from?
  - Does the waste tally with its description?
  - Are proper records being kept?

Table 17.6 European Hazardous Waste List

Code	*Description*
02	Waste from agricultural, horticultural, hunting, fishing and aquaculture primary production, food preparation and processing
03	Wastes from wood processing and the production of paper, cardboard, pulp, panels and furniture
04	Wastes from the leather and textile industries
05	Wastes from petroleum refining, natural gas purification and pyrolitic treatment of coal
06	Wastes from inorganic chemical processes
07	Wastes from organic chemical processes
08	Wastes from the manufacture, formulation, supply and use (MFSU) of coatings (paints, varnishes and vitreous enamels), adhesive, sealants and printing inks
09	Wastes from the photographic industry
10	Inorganic wastes from thermal processes
11	Inorganic waste with metals from metal treatment and the coating of metals; non-ferrous hydro-metallurgy
12	Wastes from shaping and surface treatment of metals and plastics
13	Oil wastes (except edible oils, 0500 and 1200)
14	Wastes from organic substances employed as solvents (except 0700 and 0800)
16	Wastes not otherwise specified in the catalogue
17	Construction and demolition waste (including road construction)
18	Wastes from human or animal health care and/or related research (excluding kitchen and restaurant wastes which do not arise from immediate health care)
19	Wastes from waste treatment facilities, off-site waste water treatment plants and the water industry
20	Municipal wastes and similar commercial, industrial and institutional wastes including separately collected fractions
02	Waste from agricultural, horticultural, hunting, fishing and aquaculture primary production, food preparation and processing
	0201 primary production waste
	020105 agrochemical wastes
03	Wastes from wood processing and the production of paper, cardboard, pulp, panels and furniture
	0302 wood preservation waste
	030201 non-halogenated organic wood preservatives
	030202 organochlorinated wood preservatives
	030203 organometallic wood preservatives
	030204 inorganic wood preservatives
04	Wastes from the leather and textile industries
	0401 wastes from the leather industry
	040103 degreasing wastes containing solvents without a liquid phase
	0402 wastes from textile industry
	040211 halogenated wastes from dressing and finishing
05	Wastes from petroleum refining, natural gas purification and pyrolitic treatment of coal
	0501 oily sludges and solid wastes
	050103 tank bottom sludges
	050104 acid alkyl sludges
	050105 oil spills
	050107 acid tars
	050108 other tars
	0504 spent filter clays
	050401 spent filter clays
	0506 waste from the pyrolytic treatment of coal
	050601 acid tars
	050603 other tars
	0507 waste from natural gas purification
	050701 sludges containing mercury
	0508 wastes from oil regeneration
	050801 spent filter clays
	050802 acid tars
	050803 other tars
	050804 aqueous liquid waste from oil regeneration
06	Wastes from inorganic chemical processes
	0601 waste acidic solutions

Table 17.6 Cont'd

Code		*Description*
	060101	sulphuric acid and sulphurous acid
	060102	hydrochloric acid
	060103	hydrofluoric acid
	060104	phosphoric and phosphorus acid
	060105	nitric acid and nitrous acid
	060199	waste not otherwise specified
	0602	alkaline solutions
	060201	calcium hydroxide
	060202	soda
	060203	ammonia
	060299	wastes not otherwise specified
	0603	waste salts and their solutions
	060311	salts and solutions containing cyanides
	0604	metal-containing wastes
	060402	metallic salts (except 0603)
	060403	wastes containing arsenic
	060404	wastes containing mercury
	060405	wastes containing heavy metals
	0607	wastes from halogen chemical processes
	060701	wastes containing asbestos from electrolysis
	060702	activated carbon from chlorine production
	0613	wastes from other inorganic chemical processes
	061301	inorganic pesticides, biocides and wood preserving agents
	061302	spent activated carbon (except 060702)
07	Wastes from organic chemical processes	
	0701	waste from the manufacture, formulation, supply and use (MFSU) of basic organic chemicals
	070101	aqueous washing liquids and mother liquors
	070103	organic halogenated solvents, washing liquids and mother liquors
	070104	other organic solvents, washing liquids and mother liquors
	070107	halogenated still bottoms and reaction residues
	070108	other still bottoms and reaction residues
	070109	halogenated filter cakes, spent absorbents
	070110	other filter cakes, spent absorbents
	0702	waste from the MFSU of plastics, synthetic rubber and man-made fibres
	070201	aqueous washing liquids and mother liquors
	070203	organic halogenated solvents, washing liquids and mother liquors
	070204	other organic solvents, washing liquids and mother liquors
	070207	halogenated still bottoms and reaction residues
	070208	other still bottoms and reaction residues
	070209	halogenated filter cakes, spent absorbents
	070210	other filter cakes, spent absorbents
	0703	waste from the MFSU for organic dyes and pigments (excluding 0611)
	070301	aqueous washing liquids and mother liquors
	070303	organic halogenated solvents, washing liquids and mother liquors
	070304	other organic solvents, washing liquids and mother liquors
	070307	halogenated still bottoms and reaction residues
	070308	other still bottoms and reaction residues
	070309	halogenated filter cakes, spent absorbents
	070310	other filter cakes, spent absorbents
	0704	waste from the MFSU for organic pesticides (except 020105)
	070401	aqueous washing liquids and mother liquors
	070403	organic halogenated solvents, washing liquids and mother liquors
	070404	other organic solvents, washing liquids and mother liquors
	070407	halogenated still bottoms and reaction residues
	070408	other still bottoms
	070409	halogenated filter cakes, spent absorbents
	070410	other filter cakes, spent absorbents

Table 17.6 Cont'd

Code	*Description*
0705	waste from the MFSU of pharmaceuticals
070501	aqueous washing liquids and mother liquors
070503	organic halogenated solvents, washing liquids and mother liquors
070504	other organic solvents, washing liquids and mother liquors
070507	halogenated still bottoms and reaction residues
070508	other still bottoms
070509	halogenated filter cakes, spent absorbents
070510	other filter cakes, spent absorbents
0706	waste from the MFSU of fats, grease, soaps, detergents, disinfectants and cosmetics
070601	aqueous washing liquids and mother liquors
070603	organic halogenated solvents, washing liquids and mother liquors
070604	other organic solvents, washing liquids and mother liquors
070607	halogenated still bottoms and reaction residues
070608	other still bottoms
070609	halogenated filter cakes, spent absorbents
070610	other filter cakes, spent absorbents
0707	waste from the MFSU of fine chemicals and chemical products
070701	aqueous washing liquids and mother liquors
070703	organic halogenated solvents, washing liquids and mother liquors
070704	other organic solvents, washing liquids and mother liquors
070707	halogenated still bottoms and reaction residues
070708	other still bottoms
070709	halogenated filter cakes, spent absorbents
070710	other filter cakes, spent absorbents
08	Wastes from the manufacture, formulation, supply and use (MFSU) of coatings (paints, varnishes and vitreous enamels), adhesive, sealants and printing inks
0801	wastes from MFSU of paint and varnish
080101	waste paints and varnish containing halogenated solvents
080102	waste paints and varnish free of halogenated solvents
080106	sludges from paint or varnish removal containing halogenated solvents
080107	sludges from paint or varnish removal free of halogenated solvents
0803	wastes from MFSU of printing inks
080301	waste ink containing halogenated solvents
080302	waste ink free of halogenated solvents
080305	ink sludges containing halogenated solvents
080306	ink sludges free of halogenated solvents
0804	wastes from MFSU of adhesive and sealants (including water-proofing products)
080401	waste adhesives and sealants containing halogenated solvents
080402	waste adhesives and sealants free of halogenated solvents
080405	adhesives and sealants sludges containing halogenated solvents
080406	adhesives and sealants sludges free of halogenated solvents
09	Wastes from the photographic industry
09101	wastes from the photographic industry
09101	water-based developer and activator solutions
09102	water-based offset plate developer solutions
09103	solvent-based developer solutions
09104	fixer solutions
09105	bleach solutions and bleach fixer solutions
09106	waste containing silver from on-site treatment of photographic waste
10	Inorganic wastes from thermal processes
1001	wastes from power station and other combustion plants (except 1900)
100104	oil fly ash
100109	sulphuric acid
1003	wastes from aluminium thermal metallurgy
100301	tars and other carbon-containing wastes from anode manufacture
100303	skimmings
100304	primary smelting slags/white drosses

Table 17.6 Cont'd

Code		*Description*
	100307	spent pot lining
	100308	salt slags from secondary smelting
	100309	black drosses from secondary smelting
	100310	waste from treatment of salt slags and black drosses treatment
	1004	wastes from lead thermal metallurgy
	100401	slags (1st and 2nd smelting)
	100402	dross and skimmings (1st and 2nd smelting)
	100403	calcium arsenate
	100404	flue gas dust
	100405	other particulates and dust
	100406	solid waste from gas treatment
	100407	sludges from gas treatment
	1005	wastes from zinc thermal metallurgy
	100501	slags (1st and 2nd smelting)
	100502	dross and skimmings (1st and 2nd smelting)
	100503	flue gas dust
	100505	solid waste from gas treatment
	100506	sludges from gas treatment
	1006	wastes from copper thermal metallurgy
	100603	flue gas dust
	100605	waste from electrolytic refining
	100606	solid waste from gas treatment
	100607	sludges from gas treatment
11	Inorganic waste with metals from metal treatment and the coating of metals; non-ferrous hydro-metallurgy	
	1101	liquid wastes and sludges from metal treatment and coating of metals (e.g. galvanic processes, zinc coating processes, pickling processes, etching, phosphatizing, alkaline degreasing)
	110101	cyanidic (alkaline) wastes containing heavy metals other than chromium
	110103	cyanidic (alkaline) wastes which do not contain heavy metals
	110104	cyanidic-free wastes containing chromium
	110105	acidic pickling solutions
	110106	acids not otherwise specified
	110107	alkalis not otherwise specified
	110108	phosphatizing sludges
	1102	wastes and sludges from non-ferrous hydrometallurgical processes
	110202	sludges from zinc hydrometallurgy (including jarosite, goethite)
	1103	sludges and solids from tempering processes
	110301	wastes containing cyanide
	110302	other wastes
12	Wastes from shaping and surface treatment of metals and plastics	
	1201	wastes from shaping (including forging, welding, pressing, drawing, turning, cutting and filing)
	120106	waste machining oils containing halogens (not emulsioned)
	120107	waste machining oils free of halogens (not emulsioned)
	120108	waste machining emulsions containing halogens
	120109	waste machining emulsions free of halogens
	120110	synthetic machining oils
	120111	machining sludges
	120112	spent waxes and fats
	1203	wastes from water and steam degreasing processes (except 1100)
	120301	aqueous washing liquids
	120302	steam degreasing wastes
13	Oil wastes (except edible oils, 0500 and 1200)	
	1301	waste hydraulic oils and brake fluids
	130101	hydraulic oils, containing PCBs or PCTs
	130102	other chlorinated hydraulic oils (not emulsions)
	130103	non-chlorinated hydraulic oils (not emulsions)
	130104	chlorinated emulsions
	130105	non-chlorinated emulsions

Table 17.6 Cont'd

Code		Description
	130106	hydraulic oils containing only mineral oil
	130107	other hydraulic oils
	130108	brake fluids
	1302	waste engine, gear and lubricating oils
	130201	chlorinated engine, gear and lubricating oils
	130202	non-chlorinated engine, gear and lubricating oils
	1303	waste insulating and heat transmission oils and other liquids
	130301	insulating or heat transmission oils and other liquids containing PCBs or PCTs
	130302	other chlorinated insulating and heat transmission oils and other liquids
	130303	non-chlorinated insulating and heat transmission oils and other liquids
	130304	synthetic insulating and heat transmission oils and other liquids
	130305	mineral insulating and heat transmission oils
	1304	bilge oils
	130401	bilge oils from inland navigation
	130402	bilge oils from jetty sewers
	130403	bilge oils from other navigation
	1305	oil/water separator contents
	130501	oil/water separator solids
	130502	oil/water separator sludges
	130503	interceptor sludges
	130504	desalter sludges or emulsions
	130505	other emulsions
	1306	oil waste not otherwise specified
	130601	oil waste not otherwise specified
14	Wastes from organic substances employed as solvents (except 0700 and 0800)	
	1401	wastes from metal degreasing and machinery maintenance
	140101	chlorofluorocarbons
	140102	other halogenated solvents and solvent mixes
	140103	other solvents and solvent mixes
	140104	aqueous solvent mixes containing halogens
	140105	aqueous solvent mixes free of halogens
	140106	sludges or solid wastes containing halogenated solvents
	140107	sludges or solid wastes free of halogenated solvents
	1402	wastes from textile cleaning and degreasing of natural products
	140201	halogenated solvents and solvent mixes
	140202	solvent mixes or organic liquids free of halogenated solvents
	140203	sludges or solid wastes containing halogenated solvents
	140204	sludges or solid wastes containing other solvents
	1403	wastes from the electronic industry
	140301	chlorofluorocarbons
	140302	other halogenated solvents
	140303	solvents and solvent mixes free of halogenated solvents
	140304	sludges or solid wastes containing halogenated solvents
	140305	sludges or solid wastes containing other solvents
	1404	wastes from coolants, foam/aerosol propellants
	140401	chlorofluorocarbons
	140402	other halogenated solvents and solvent mixes
	140403	other solvents and solvent mixes
	140404	sludges or solid wastes containing halogenated solvents
	140405	sludges or solid wastes containing other solvents
	1405	wastes from solvent and coolant recovery (still bottoms)
	140501	chlorofluorocarbons
	140502	halogenated solvents and solvent mixes
	140503	other solvents and solvent mixes
	140504	sludges containing halogenated solvents
	140505	sludges containing other solvents

Table 17.6 Cont'd

Code		*Description*
16		Wastes not otherwise specified in the catalogue
	1602	discarded equipment and shredder residues
	160201	transformers and capacitors containing PCBs or PCTs
	1604	waste explosives
	160401	waste ammunition
	160402	fireworks waste
	160403	other waste explosives
	1606	batteries and accumulators
	160601	lead batteries
	160602	Ni–Cd batteries
	160603	mercury dry cells
	160606	electrolyte from batteries and accumulators
	1607	waste from transport and storage tank cleaning (except 0500 and 1200)
	160701	waste from marine transport tank cleaning, containing chemicals
	160702	waste from marine transport tank cleaning, containing oil
	160703	waste from railway and road transport tank cleaning, containing oil
	160704	waste from railway and road transport tank cleaning, containing chemicals
	160705	waste from storage tank cleaning, containing chemicals
	160706	waste from storage tank cleaning, containing oil
17		Construction and demolition waste (including road construction)
	1706	insulation materials
	170601	insulation materials containing asbestos
18		Wastes from human or animal health care and/or related research (excluding kitchen and restaurant wastes which do not arise from immediate health care)
	1801	waste from natal care, diagnosis, treatment or prevention of disease in humans
	180103	other wastes whose collection and disposal is subject to special requirements in view of the prevention of infection
	1802	waste from research, diagnosis, treatment or prevention of disease involving animals
	180204	other wastes whose collection and disposal is subject to special requirements in view of the prevention of infection by discarded chemicals
19		Wastes from waste treatment facilities, off-site waste water treatment plants and the water industry
	1901	wastes from incineration or pyrolisis of municipal and similar commercial, industrial and institutional wastes
	190103	fly ash
	190104	boiler dust
	190105	filter cake from gas treatment
	190106	aqueous liquid waste from gas treatment and other aqueous liquid wastes
	190107	solid waste from gas treatment
	190110	spent activated carbon from flue gas treatment
	1902	wastes from specific physico-chemical treatments of industrial wastes (e.g. dechromatation, decyanidation, neutralization)
	190201	metal hydroxide sludges and other sludges from metal insolubilization treatment
	1904	vitrified wastes and wastes from vitrification
	190402	fly ash and other flue gas treatment wastes
	190403	fly ash and other flue gas treatment wastes non-vitrified solid phase
	1908	wastes from waste water treatment plants not otherwise specified
	190803	grease and oil mixture from oil/waste water separation
	190806	saturated or spent ion exchange resins
	190807	solutions and sludges from regeneration of ion exchangers
20		Municipal wastes and similar commercial, industrial and institutional wastes including separately collected fractions
	2001	separately collected fractions
	200112	paint, inks, adhesives and resins
	200113	solvents
	200117	photo chemicals
	200119	pesticides
	200121	fluorescent tubes and other mercury containing waste

Table 17.7 Hazard properties of Special Waste (see Table 17.6)

Reference	*Description*
H1	Explosive: substances and preparations which may explode under the effect of flame or which are more sensitive to shocks or friction than dinitrobenzene
H2	Oxidizing: substances and preparations which exhibit highly exothermic reactions when in contact with other substances, particularly flammable substances
H3-A	Highly flammable: liquid substances and preparations having a flash point below 21ºC (including extremely flammable liquids), or substances and preparations which may become hot and finally catch fire in contact with air at ambient temperature without any application of energy, or solid substances and preparations which may readily catch fire after brief contact with a source of ignition and which continue to burn or to be consumed after removal of the source of ignition, or gaseous substances and preparations which are flammable in air at normal pressure, or substances and preparations which, in contact with water or damp air, evolve highly flammable gases in dangerous quantities
H3-B	Flammable: liquid substances and preparations having a flash point equal to or greater than 21°C and less than or equal to 55°C
H4	Irritant: non-corrosive substances and preparations which, through immediate, prolonged or repeated contact with the skin or mucous membrane, can cause inflammation
H5	Harmful: substances and preparations which, if they are inhaled or ingested or if they penetrate the skin, may involve limited health risks
H6	Toxic: substances and preparations (including very toxic substances and preparations) which, if they are inhaled or ingested or if they penetrate the skin, may involve serious, acute or chronic health risks and even death
H7	Carcinogenic substances and preparations which, if they are inhaled or ingested or if they penetrate the skin, may induce cancer or increase its incidence
H8	Corrosive: substances and preparations which may destroy living tissue on contact
H9	Infectious: substances containing viable micro-organisms or their toxins which are known or reliably believed to cause disease in man or other living organisms
H10	Teratogenic: substances and preparations which, if they are inhaled or ingested or if they penetrate the skin, may induce non-hereditary congenital malformations or increase their incidence
H11	Mutagenic: substances and preparations which if they are inhaled or ingested or if they penetrate the skin, may induce hereditary genetic defects or increase their incidence
H12	Substances and preparations which release toxic or very toxic gases in contact with water, air or an acid
H13	Substances and preparations capable by any means, after disposal, of yielding another substance, e.g. a leachate, which possesses any of the characteristics listed above
H14	Ecotoxic: substances and preparations which present or may present immediate or delayed risks for one or more sectors of the environment

- Checking up on disposal.
- Report any suspicions to the Environment Agency. Change carrier/contractor.

Environmental Impact Assessment

An Environmental Impact Assessment is a technique by which information regarding the environmental effect of a project is collected, assessed and considered when reaching a decision on whether it should proceed. The assessment:

- provides a framework within which environmental considerations interact with design;
- can identify ways in which a project may be modified to minimize potential adverse environmental effects; and
- can assist in satisfying public interest that every possible impact has been considered.

Section A – Description of waste

1. Please describe the waste being transferred:
2. How is the waste contained?

   Loose ☐ Sacks ☐ Skip ☐ Drum ☐ Other ☐ → Please describe
3. What is the quantity of waste (number of sacks, weight, etc.):

Section B – Current holder of the waste

1. Full name (BLOCK CAPITALS):
2. Name and address of company:
3. Which of the following are you? (Please tick one or more boxes)

☐ producer of the waste ☐ holder of waste disposal or waste management licence Licence number: Issued by:	☐ importer of the waste ☐ exempt from requirement to have a waste disposal or waste management licence Give reason:	☐ waste collection authority ☐ registered waste carrier Registration number: Issued by:	☐ waste disposal authority (Scotland only) ☐ exempt from requirement to register Give reason:

Section C – Person collecting the waste

1. Full name (BLOCK CAPITALS):
2. Name and address of company:
3. Which of the following are you? (Please tick one or more boxes)

☐ waste collection authority ☐ holder of waste disposal or waste management licence Licence number: Issued by:	☐ waste disposal authority (Scotland only) ☐ exempt from requirement to have a waste disposal or waste management licence Give reason:	☐ exporter ☐ registered waste carrier Registration number: Issued by:	☐ exempt from requirement to register Give reason:

Section D

1. Address of place of transfer/collection point:
2. Date of transfer:
3. Time(s) of transfer (for multiple consignments, give 'between' dates):
4. Name and address of broker who arranged this waste transfer (if applicable):
5. Signed:
   Full name:
   Representing:

   Signed:
   Full name:
   Representing:

Figure 17.2 *Controlled Waste Transfer Note to comply with Duty of Care*

An assessment normally includes a description of the proposed development, its location, design and scale. Data are included to enable the main effects which the development is likely to have to be assessed. There is then a description of the likely significant effects, either directly or indirectly, upon the environment. This will cover its impact upon:

- water as per Tables 16.12–14 and 17.8;
- land as per Table 17.8;
- people;
- air;
- climate;
- landscape, including visual impact;
- material assets;
- cultural heritage.

Table 17.8 Considerations for impact upon water and land

Water	Flora and fauna, e.g. fisheries and shell beds Amenity, e.g. loss of recreation facilities Loss of potable water supplies (rivers, reservoirs, aquifers) Agricultural water supplies, for irrigation and animal feedstock Riverbanks, soils, sediments Direct interaction of the substance, followed by slow release into the water may give long-term problems The substance may simply collect in hollows and be gradually dispersed back into the main flow of water to give medium-term problems
Land	Danger to crops, flora and vegetation. Action may range from disposal of crops to merely monitoring the contamination. The substance may also be washed off the crops or be biodegraded, although such processes are slower on land than in water. Deposition, followed by run-off to water sources or via soil permeation Deposition followed by biodegradation, or persistence.

The consequential effects of transportation of materials to, and products and wastes from, the development are considered. For each significant potential impact identified, a description is required of the measures envisaged to mitigate the effects.

Control of atmospheric emissions

Solvents such as 1,1,1-trichloroethane (methyl chloroform), carbon tetrachloride and CFC 113 are well known as ozone-depleting chemicals. They are commonly used as cleaning agents in a number of industries including engineering, electronics and dry cleaning. Companies using these solvents need to consider the impact of phasing them out in accordance with the decisions of the Montreal Protocol (e.g. see EC Decision 88/540/EEC and Regulation 3093/94) and seek alternative cleaning processes as costs are likely to rise as the substances will become increasingly difficult to obtain.

Small solid particles, present in dust and grit emissions, have very low settling velocities (Table 4.4) The collection efficiencies of simple cyclones are therefore, as shown in Figure 17.3, relatively low. Fabric filters, electrostatic precipitators or wet scrubbers may be required to remove particles <5 μm in size with an acceptable efficiency. Therefore the cost of pollution control inevitably increases when dealing with particle size distributions skewed towards the lower end.

The complexity of equipment required to arrest mists and sprays also increases with a reduction

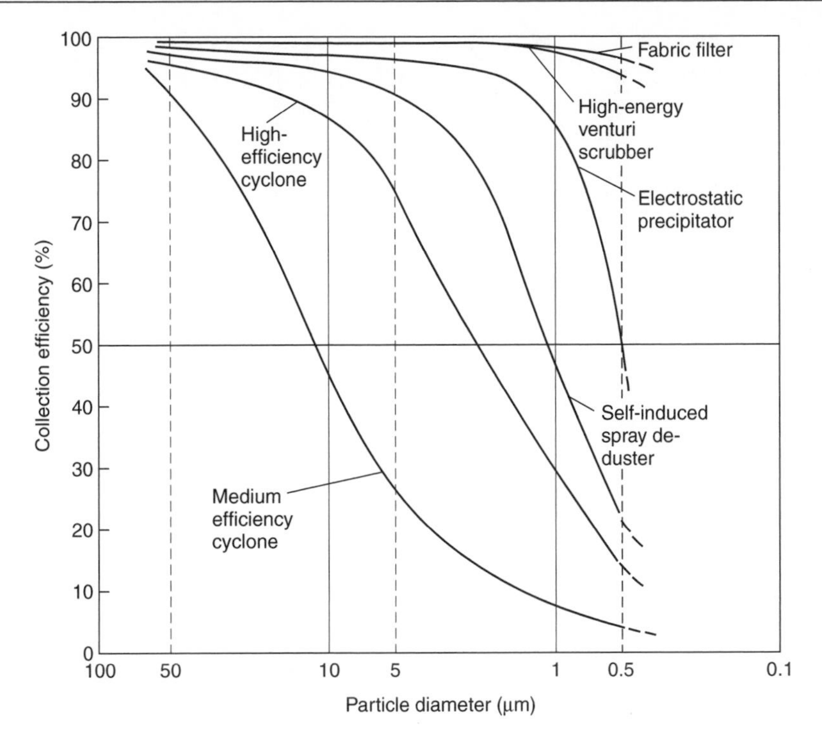

Figure 17.3 *Particle arrestment efficiencies for different equipment*

in droplet diameter. The range of droplet diameters associated with various processes, and the types of equipment used for their separation, are exemplified in Figure 17.4.

Processes for the control of atmospheric emissions are summarized in Table 17.9. Some of these involve the types of chemical engineering operations referred to in Chapter 7; the added complications are:

- the inlet pollutant concentrations may be quite low;
- the total stream flowrates may be large;
- the target outlet concentrations are probably subject to stringent regulation to very low levels.

As with pollution control equipment generally, a high degree of reliability is required.

Liquid effluent treatment operations

Any of the variety of chemical engineering operations or unit processes listed in Table 17.10 may be used to treat aqueous effluents. Oxidation includes both chemical and biological processes; the latter include trickling filters or an activated sludge bed.

Process operation and the storage and handling of effluents and chemicals involve potential chemical and biological hazards (Chapter 5). Safeguards of the type outlined in Chapters 12 and 13 are essential, particularly since the activities are often on a site's periphery and have low manning levels.

EC policy is that waste oils from use as lubricants should be recycled or used as fuel as far as possible. Some EC countries have a more regulated organizational framework for collection and

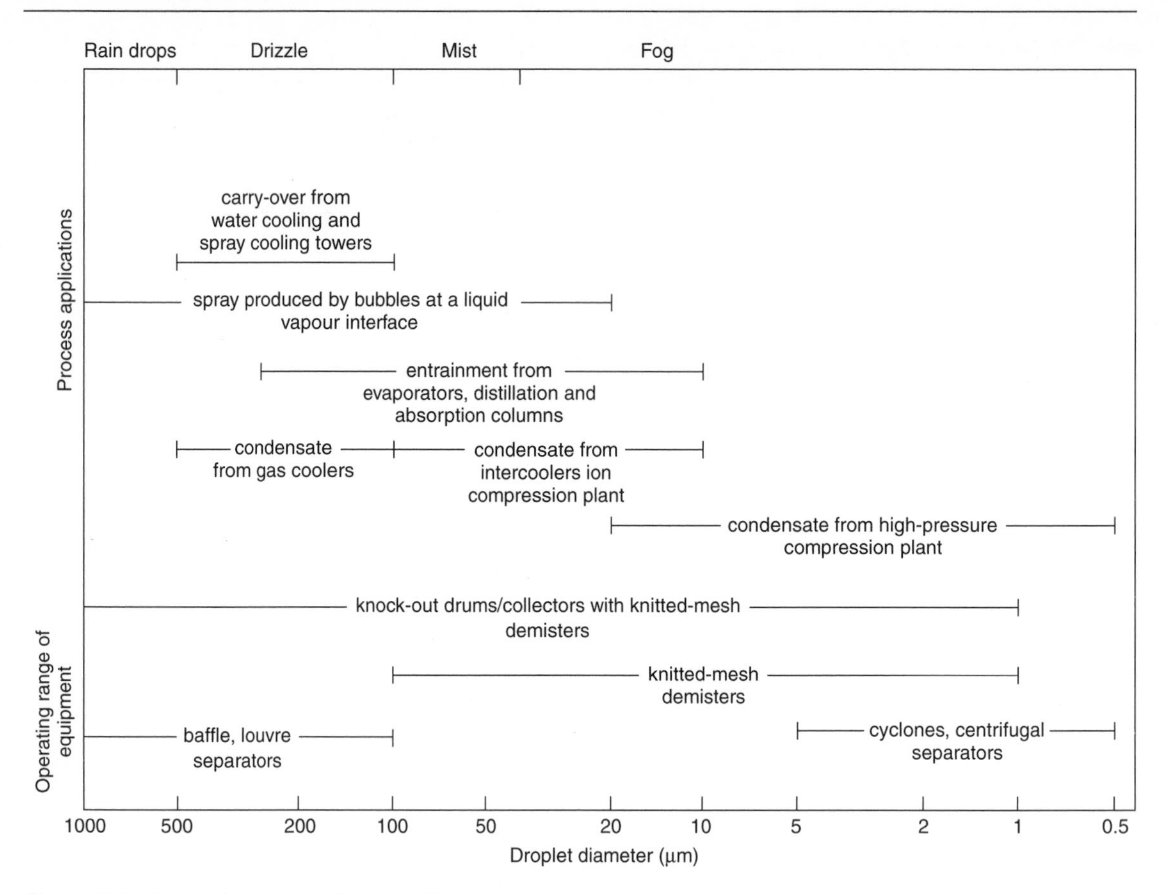

Figure 17.4 *Mists and sprays; applications of separators*

recycling of used oil. Used oil can be drained from machinery during routine servicing and should be stored in a central store using suitable leak-proof containers. Different types of oil should be stored separately. Used oils must not be disposed of into soakaways, watercourses, drains and sewers. The requirement of 'no visible oil' is often specified for releases to water, usually taken to mean no more than 20 ppm depending on droplet size. Used oils should only be disposed of to registered waste oil carriers. Alternatively waste oil can be burnt for heating provided that approval is given by the local authority Environmental Health Department for the installation of the burner.

The disposal of waste oils is largely governed in the UK by the EPA 90. Waste oils are a 'Special Waste' covered by Special Waste Regulations. Establishments handling more than 500 litres per annum have a duty to maintain records. There are also requirements for record keeping where waste oils contain toxic impurities and so become 'Special Waste'.

PCBs and PCTs are particularly troublesome liquids because of their toxicity and persistence in the environment. They are defined as polychlorinated biphenyls, polychlorinated terphenyls, monomethyl-dibromo-diphenyl methane, monomethyl-dichloro-diphenyl methane or monomethyl-tetrachlorodiphenyl methane. With low electrical conductivity and heat resistance they found wide use as dielectric fluids and were formerly used as hydraulic fluids. PCBs have not been made in the UK since 1977 and whilst most new uses for the substance are banned in most countries, around two-thirds of the 1.5 million tonnes manufactured in Europe and the US prior to 1985 still remain in equipment such as transformers. PCTs have been used in the past in a restricted range of specialist industrial applications.

Table 17.9 Treatment methods for gaseous waste

General methods	*Alternatives*
Centrifugal techniques	Centrifuge Cyclone
Coalescence	
Condensation	
Destruction	Chemical reaction Incineration
Direct recycle	
Dispersion	
Electrostatic precipitation	
Filtration	Needle bonded fabric Reverse jet Reverse pressure Shaker type
Gravity settlement	
Total enclosure	
Wet scrubbing	Absorption tower Fluidized bed scrubber Impingement scrubber Irrigated target scrubber Pressure spray scrubber Rotary scrubber Self-induced spray scrubber Spray tower Venturi scrubber

Waste gas treatment system

Dry methods	*Pollutant*	*Wet methods*
Absorption Destruction Direct recycling Dispersion Total enclosure	**Gas**	Wet scrubbing
Destruction	**Liquid**	Centrifugal techniques Coalescence Destruction Gravity settlement Total enclosure Wet scrubbing
Centrifugal techniques Destruction Direct recycling Electrostatic precipitation Filtration Gravity settlement Total enclosure	**Solid**	Electrostatic precipitation Wet scrubbing

Table 17.10 Treatment methods for liquid effluent

Treatment method	*Example*
Chemical	
Cementation	Copper recovery
Chlorination	Cyanide oxidation
Coagulation, *see* Flocculation	
Electrolytic processes	Metal recovery
Emulsion breaking	Soluble oil recovery
Flocculation	Sewage treatment
Hydrolysis	Cellulose waste
Incineration	Waste oils
Ion exchange	Metal recovery
Leaching	Metal-bearing sludges
Neutralization	Waste acid, waste alkali
Oxidation	Phenol removal
Ozonization	Cyanide oxidation
Precipitation	Metals
Reduction	Hexavalent chromium
Thermal decomposition	Recycling hydrochloric acid
Physical	
Adsorption	Removal of volatile organics
Cooling	Water reuse
Crystallization	Recovery of inorganic salts
Dewatering, *see* Filtration	
Dialysis	Desalination
Distillation	Solvent recovery
Drying	Pig manure
Electrodialysis	Desalination
Evaporation	Sulphuric acid recovery
Filtration	Sewage sludge
Flotation	Dairy wastes
Foam fractionation	Metal separation
Fractionation, *see* Distillation	
Freezing	Desalination
Heating	Demulsification
Phase separation	Oily wastes, hydrocarbons
Reverse osmosis	Desalination
Screening	Sewage
Sedimentation	Suspended solids removal
Solvent extraction	Metal recovery
Stripping	Ammonia removal, solvent removal
Ultrafiltration, *see* Dialysis, Reverse osmosis	
Biological	
Activated sludge	Sewage
Anaerobic digestion	Food wastes
Chemical production	Ethanol
Disinfection	Sewage plant effluent
High-rate filtration	Phenol removal
Oxidation, *see* Activated sludge, High-rate filtration, Trickling filter	
Reduction, *see* Anaerobic digestion	
Single-cell protein production	Organic waste
Trickling filter	Sewage

Under the requirements of the Environmental Protection (Disposal of Polychlorinated Biphenyls and other Dangerous Substances) (England and Wales) Regulations, SI 2000/1043, equipment containing PCBs is now required to be: identified, registered with the Environment Agency, labelled so that the PCB presence can be identified, and decontaminated without endangering health and safety to reduce the level of PCBs in equipment to less than 50 ppm (less than 0.005% in weight). It was, however, regarded as acceptable to achieve a level between 50 and 500 ppm and to dispose of equipment by 31 December 2000 unless the exemptions set out under the Environmental Protection (Disposal of Polychlorinated Biphenyls and other Dangerous Substances) Regulations, apply.

Control of solid waste

The overall stages in handling waste are summarized in Figure 16.2. The treatment methods applicable to it are listed in Table 17.11.

The range of precautions required during handling, transfer, and disposal, e.g. by landfill, depend upon the risk including the nature and degree of contamination and the work to be undertaken.

It will, however, encompass personal protective equipment: a high standard of personal hygiene; enclosure, possibly pressurization and regular cleaning of vehicle cabs; vehicle washing facilities; site security, and control of designated dirty areas. Air monitoring and medical surveillance may be required.

Monitoring and auditing

Auditing

The basic database for the control of environmental pollution is a pollution inventory, viz. a tabulation of emissions, discharges and wastes from the factory or activity covering all operating phases. This includes:

- atmospheric emissions; continuous, occasional and fugitive emissions during normal operation, start-up, shutdown, etc.;
- aqueous discharges; to sewers, direct to controlled waters, etc.;
- solid wastes; by type, origins, routes;
- noise; by reference to environmental noise surveys with identification of special features, tonal characteristics and intermittency;
- transport; normal plus any hazardous cargoes.

An environmental audit is a systematic, documented, periodic and objective evaluation of how well environmental organization, management and equipment are performing. The aim is to help safeguard the environment by:

- assessing compliance with company policies, including compliance with all regulatory requirements; and
- facilitating management control of environmental practices.

Table 17.11 Treatment methods for solid waste

Treatment method	*Examples/comments*
Chemical	
Calcination	Gypsum
Chlorination	Tin removal
Cooking	Inedible offal
Froth flotation	Coal recovery
	Glass
Hydrolysis	Household refuse
	Vegetable waste
Incineration	Household refuse, rubber tyres
Leaching	Gives an aqueous solution for treatment as liquid waste (Table 17.4)
Oxidation	Weathering
Pyrolysis	Household refuse
	Polystyrene
Sintering	Colliery spoil
	Millscale
Physical	
Adhesion	Household refuse
Agglomeration	Pulverized fuel ash
Ballistic separation	Household refuse
Baling	Cans, paper, cardboard
Centrifugation	Animal oil separation
	De-oiling swarf
Classification (air)	Household refuse
Classification (wet)	Plastic
Comminution	Mining wastes
	Motor vehicles
Compaction	Household refuse
Dewatering	Sewage sludge
Dissolution	Forms a liquid waste for treatment as in Table 17.4
Drying	Filter cake
Electrostatic separation	Household refuse
Foaming	Slag
Freezing	Meat products
Granulation	Slag
Impalement	Household refuse
Jigging	Household refuse
Magnetic treatment	Iron removal from slag
	Household refuse
Melting	Selective non-ferrous metal recovery
Pelletization	Iron and steel fines
Pulverization	Household refuse
	Swarf
Quenching	Incinerator residues
Screening	Clinker
Settlement	China clay wastes
Shape separation	Household refuse
Sliding separation	Household refuse, agricultural wastes
Biological	
Anaerobic digestion	Farm waste
Bacterial leaching	Low-grade copper ore
Composting	Household refuse
Degradation	Plastic
Fermentation	Requires solution or dilute slurry, *see* Table 17.4

Audits may differ. Some concentrate on particular facets, e.g. waste minimization, energy conservation. A management audit will probably cover organizational structures and relationships, communications, procedures acceptance and implementation, training and quality assurance. A technical audit will consider the performance of plant and equipment, monitoring and inspection programmes. Items for inclusion in auditing are listed in Table 17.12.

Table 17.12 Some items in a typical environmental pollution audit

- Checks on environmental policy, organization and responsibilities, legislation, training records, emergency plans, performance targets, means of investigating incidents and written work instructions.
- Listing gaseous emissions, concentrations, smoke characteristics; prevailing winds and exposed zones; toxicity or nuisance potential; effects of synergism or poor atmospheric dispersing conditions. Consent limits. Compliance record keeping.
- Querying the adequacy of chimneys and stacks, scrubbers and particulates collection equipment (e.g. filters, cyclones).
- Listing effluents, their analyses and discharge from processes. Querying flammability, corrosivity, toxicity, miscibility, reactivity. Checking the reliability of treatment and analysis. Consent limits. Compliance record keeping.
- Listing of 'solid' wastes, quantities, analyses and physical form (e.g. solid, slurry, suspension, sludge); 'toxicity' and flammability. Checking the reliability and legality of waste disposal options, including any transportation offsite. Record keeping.
- Bunding of storage areas, segregation of waste storage areas, security of landfill sites, etc.
- Soil contamination.
- Monitoring (preferably continuous). Record keeping (on- and off-site).
- Checking for land contamination, e.g. analysis of soil samples. Record keeping.
- Checks on waste removal. (Whilst a producer is under no obligation to audit his waste's final destination, such a visit is a prudent means of demonstrating the steps taken to prevent subsequent illegal treatment of the waste.)
- Checking noise levels (on- and off-site).

Specific attention is given to auditing for compliance with legislation, including the requirements of IPC, BATNEEC and BPEO.

Monitoring

Monitoring of emissions, discharges and disposals is a prerequisite of control. Depending upon the waste stream, sampling and analysis may be continuous or periodic. Measurement methods for physico-chemical properties and chemical analysis of waste streams are very diverse. The data obtained are used both for monitoring to ensure adequate control and compliance with legislation, and for process plant design and operation – including recycling and recovery. A selective list of methods for physico-chemical properties is given in Table 17.13 and for chemical analysis in Table 17.14.

All the parameters on which various consents (or permissions to dispose of, waste streams) are based must be reliably measured and recorded. This is easier to achieve with gaseous emissions (Chapter 10) and liquid effluents than with heterogeneous solid wastes. Systematic analysis of solid wastes will cover as a minimum the information in Table 17.15.

The gold standard for environment management systems is BS EN ISO 14001 as summarized by Table 17.16.

Table 17.13 Methods for measurement of physico-chemical properties
(R = routine test)

Gaseous wastes

Property	*For consideration*	*Methods*
Gases generally		
Calorific value	Incineration with/without heat recovery	Calculation from published data Calorimeters
Concentration of gaseous (R) pollutant	Disposal constraints Design of treatment/disposal/ recovery plant	Chemical analysis (q.v.) Chromatography
Density	Dispersion Design of disposal/treatment plant	Chancel flask, or effusiometer
Dewpoint	Design of treatment/disposal/ recovery plant Condensation	Tabulated Hygrometer ASTM method
Flammable limits	Fire and explosion hazard Control of combustion	Tabulated
Flammability	Fire and explosion hazard Control of combustion	*see* Calorific value and Flammable limits ASTM methods
Ignition temperature	Fire and explosion hazard Control of combustion	Adiabatic compression Concentric tube Tabulated
Moisture content	*see* Dewpoint Potential nuisance Disposal Necessity for treatment	Subjective testing ASTM methods Proprietary meters
Pressure	Design of treatment/disposal/ recovery plant	Gauges Transducer Manometer
Solubility of pollutant (R)	Design of separation/ treatment plant	Wide range ASTM methods Tabulated
Temperature (R)	Corrosion resistance Design of treatment/disposal plant Materials of construction	Thermometer Thermocouple Resistance treatment Pyrometer ASTM methods
Velocity (R)	Design of disposal/treatment plant	Pitot tube Anemometer Orifice plate, etc. Rotameter Gas meters ASTM method
Viscosity	Design of disposal/treatment plant	Calculated or tabulated
Liquids suspended in gases (*see also* Liquid wastes)		
Concentration of liquid pollutants (R)	Design of treatment/recovery plant Legislation	Sampling and analysis
Density of liquid	Design of treatment/recovery plant	ASTM methods Pyknometer Hydrometer Relative density balance Tabulated

Table 17.13 Cont'd

Property	*For consideration*	*Methods*
Electrical properties	Electrostatic precipitation	Direct testing
Flammability and related properties	Fire/explosion hazard Incineration	*see* Non-aqueous liquids
Droplet size distribution	Design of treatment/recovery plant Environmental impact	Numerous including: • Holography • Light scatter • ASTM methods
pH of liquid	Corrosion resistance Environmental impact	Indicators Test meter pH meter
Surface tension of liquid	Coalescence for removal Design of treatment/recovery plant	Tabulated ASTM methods
Viscosity of liquid	Design of treatment plant for removal	ASTM methods Viscometer Tabulated
Solids suspended in gases (*see also* Solid wastes)		
Concentration of solid in gas (R)	Design of treatment or recovery plant Legislation	Sampling Density meter
Density (bulk)	Design of treatment or recovery plant and handling equipment	Direct measurement ASTM methods
Density (particle)	Design of separation equipment	Displacement Pyknometer Tabulated ASTM methods
Electrical properties	Electrostatic precipitation	Direct testing
Flammability and related properties	Fire and explosion hazard Dust explosion hazard	ASTM methods *see* Solid wastes
Hygroscopicity	Collection and removal of solid	Direct measurement with exposure ASTM method
Particle geometry	Design of separation equipment Toxic hazard	Observation ASTM method
Particle size distribution (R)	Design of separation equipment Toxic hazard Environmental impact	Wide range including: • Microscopy • Holography • Light scatter • Sieving
Solubility (R)	Solubility for removal by dissolution	Wide range Tabulated
Wettability	Suitability for removal by scrubbing	Experimentation ASTM method
Liquid wastes		
Aqueous liquids		
Ash	Incineration	Ignition
BOD (Biological Oxygen Demand) (R)	Type of treatment required Legislation and controls	Dilution Manometry

Table 17.13 Cont'd

Property	*For consideration*	*Methods*
COD (Chemical Oxygen Demand)	Legislation and controls	Oxidation by dichromate
Colour (R)	Legislation and controls	Visual comparison ASTM methods
Density (bulk)	Extent of contamination Design of treatment or recovery plant	Hydrometer Pyknometer Relative density balance General
Density (liquid droplet)	Emulsion breaking Stability Design of treatment or recovery plant	Hydrometer Pyknometer Relative density balance Tabulation ASTM methods
Density (solid particle)	Design of treatment or recovery plant (gravity settling)	Displacement Pyknometer Tabulated ASTM methods
Flowrate (R)	Design of treatment or recovery plant	Wide range including: • Meters • Orifice plates • Pitot tubes • Venturi meters
Non-aqueous liquid content	Design of treatment or recovery plant	Breaking emulsion Solvent extraction ASTM methods
Odour	Legislation and controls Nuisance	Subjective, empirical
Particle size distribution	Design of treatment or recovery plant	*see* Suspensions in gases
Permanganate value	Legislation and controls	Oxidation with permanganate
pH (R)	Corrosion Legislation and controls	Test paper pH meter Titration
Surface tension	Foaming Design of treatment/recovery plant	ASTM methods Tabulated Capillary height Bubble pressure Drop
Suspended solids (R)	Design of treatment/recovery plant Legislation and controls	Filtration Centrifuge ASTM methods
Taste	Contamination of potable water	Subjective and/or empirical ASTM methods
Temperature (R)	Legislation and controls	Thermocouple Thermometer Resistance thermometers ASTM methods
Total dissolved solids (R)	Legislation and controls Design of treatment/recovery plant	Evaporation Electrical conduction ASTM methods
Turbidity	Legislation and controls	ASTM method Turbidity meter (*see also* Suspended solids)
Vapour pressure	Fire and explosion hazard Design of treatment/recovery plant Vapour-release (toxic, odorous)	Viscometer ASTM methods Tabulated

Table 17.13 Cont'd

Property	*For consideration*	*Methods*
Non-aqueous liquids		
Ash	Incineration with/without heat recovery	Ignition
BOD	Design of treatment plant Legislation and controls	Dilution Manometry Probe
Boiling point (R)	Design of separation equipment Fire and explosion hazard Handling and storage	Wide range including: • Conventional boiling point • Ebulliometer • Ramsay and Young • Tabulated (*see also* Vapour pressure and Distillation range)
Calorific value	*see* Heat of combustion	—
COD	Design of treatment plant Legislation and controls	Oxidation by acid dichromate
Density (bulk)	Design of treatment/recovery plant, e.g. gravity settling	Hydrometer Pyknometer Relative density balance ASTM methods
Density (particle)	Design of treatment/recovery plant	Displacement Pyknometer Tabulated ASTM methods
Distillation range	Design of recovery plant Incineration	Distillation ASTM methods
Flammable limits	Fire and explosion hazard Handling and storage	
Flash point (R)	Fire and explosion hazard Handling and storage	Abel Cleveland Pensky Martin Equilibrium apparatus (various)
Flowrate (R)	Design of treatment/recovery plant	*see* Gases
Heat of combustion	Incineration with/without heat recovery	Bomb calorimeter Tabulated
Interfacial tension	Design of treatment/recovery plant	*see* Surface tension ASTM methods
Miscibility with water	Design of treatment/recovery plant Evaporation rate Spillage control	Wide range including: • Hill • Hertz Tabulated
Odour	Legislation and controls	Subjective, empirical ASTM methods
Temperature (R)	Design of treatment/recovery plant Fire and explosion hazard Handling and storage	Resistance thermometry Thermocouple Thermometer ASTM methods
Vapour pressure (R)	Design of treatment/recovery plant Evaporation rate Fire and explosion hazards Handling and storage	Isoteriscope Micro Reid Tabulated *see also* Boiling point ASTM methods
Viscosity	Design of treatment/ disposal/recovery plant	Viscometer ASTM methods Tabulated

Table 17.13 Cont'd

Property	*For consideration*	*Methods*
Volatility	Fire and explosion hazard Handling and storage Vapour release	*see* Distillation range and Vapour pressure
Solid wastes		
Angle of repose	Handling and storage Disposal	Direct measurement
Ash	Incineration	ASTM methods Radiation methods *see* Loss on ignition
Calorific value	Incineration	Bomb calorimeter ASTM methods
Combustibility (R)	Fire hazard	
Compressibility	Disposal	Abitrary methods relating changes in bulk density ASTM methods
Density (actual)	Design of treatment/disposal/ recovery plant	Pyknometer Displacement Tabulated ASTM methods
Density (bulk)	Handling and storage Disposal	Direct measurement of weight and volume Bulk density/ash meter ASTM methods
Elasticity	Handling and storage Disposal	Tabulated Experimentation
Flowrate (R)	Design of treatment/recovery plant Handling	Meters Impact flow meter Radiation
Hardness	Handling Design of treatment/recovery plant	Tabulated ASTM methods Experimentation
Hygroscopicity	Disposal Handling and storage	*see* Water absorption
Loss on ignition	Incineration	Ignition ASTM methods
Mineralogy	Disposal Potential reuse	Examination
Oil content	Disposal Incineration Fire and explosion hazard	Extraction with solvent
Particle size distribution	Handling and storage Disposal Design of treatment/recovery plant Environmental impact (atmosphere dispersion, settling)	Wide range including: • Holography • Microscopy Sieving ASTM methods
pH	Corrosion Design of treatment/recovery plant	Test paper pH meter Titration
Physical composition	Design of treatment/recovery plant Handling Disposal	Examination and observation
Porosity	Disposal	Calculated from bulk density (q.v.)

Table 17.13 Cont'd

Property	*For consideration*	*Methods*
Solubility	Design of treatment/recovery plant Environmental impact (leaching)	Wide range including: • Campbell • Weyl Tabulated
Volatile matter	Fire and explosion hazard Incineration	Heating ASTM methods *see also* Distillation range
Water content	Handling and storage Disposal	Drying to constant weight BS and ASTM methods Range of methods including: • Microwave attenuation • Capacitance • Resistivity • Infra-red absorption • Neutron moderation
Water absorption	Disposal Handling and storage	Experimentation ASTM methods
Wettability	Design of treatment/recovery plant	Angle of contact related to surface tension (q.v.)

Sources of test methods: ASTM (American Society of Testing and Materials); IP (Institute of Petroleum); BS (British Standard).

Table 17.14 Methods of chemical analysis

Method	*Application*						
					Phase		
	Metals	*Other elements*	*Inorganic chemicals*	*Organic chemicals*	*Gas*	*Liquid*	*Solid*
Amperometry	+	+	(+)	+		+	
Atomic absorption spectrometry	+					+	+
Atomic emission spectrometry	+	(+)	(+)			+	
Atomic fluorescence spectrometry	+	(+)	(+)			+	
Conductometry	+	(+)	+			+	
Coulometry	+	(+)	(+)			+	
Electrogravimetry	(+)					+	
Electrophoresis				+		+	+
Flame photometry	+					+	(+)
Gas chromatography	(+)			+	+	+	(+)
Fluorimetry	(+)			+		+	+
Helium plasma	+	+			+	+	+
Infra-red spectrometry			(+)	+	+	+	+
Isotope dilution	+	+			+	+	+
Mass spectrometry				+	+	+	
Nuclear magnetic resonance spectrometry	(+)	+	(+)	+		+	
Polarimetry				+			+
Polarography	+		+	+		+	
Potentiometry	+		+			+	
Radioactivity	+	+			+	+	+
Raman spectroscopy			+	+	+	+	+
Spark (emission) spectrograph		+					+
Specific ion electrode	+	+	+		+	+	+
Thin layer chromatography				+		+	+
Ultraviolet spectrometry			(+)	+	+	+	
Visible light spectrometry			+		+	+	
X-ray diffraction spectrometry			+	+			+
X-ray fluorescence spectrometry	+	+					+

Table 17.15 Minimum analysis for solid waste sample

For inorganic wastes	Reaction with water Reaction with acids Reaction with alkalis Effect of heat pH, total solids Presence of sulphides, total cyanide, ammonium compounds Concentration of metals, particularly Pb, Zn, Cd, Hg, Sn, As, Cu, Cr, Ni	For mixtures of inorganic and organic wastes all these analyses should be completed
For organic wastes	Calorific value Flash point Miscibility with water (or other wastes) Viscosity at various temperatures Halogen, sulphur, nitrogen content Ash content Analysis of ash Organic content by BOD, COD, PV or total carbon methods	

Table 17.16 Elements of ISO 14001

1 Scope
2 Normative references
3 Definitions
4 Environmental management system requirements
- 4.1 General requirements
- 4.2 Environmental policy
- 4.3 Planning
  - 4.3.1 Environmental aspects
  - 4.3.2 Legal and other requirements
  - 4.3.3 Objectives and targets
  - 4.3.4 Environmental management programmes
- 4.4 Implementation and operation
  - 4.4.1 Structure and responsibility
  - 4.4.2 Training, awareness and competence
  - 4.4.3 Communication
  - 4.4.4 Environmental management system documentation
  - 4.4.5 Document control
  - 4.4.6 Operational control
  - 4.4.7 Emergency preparedness and response
- 4.5 Checking and corrective action
  - 4.5.1 Monitoring and measurement
  - 4.5.2 Non-conformance and corrective and preventive action
  - 4.5.3 Records
  - 4.5.4 Environmental management system audit
- 4.6 Management review

18

Conversion tables and measurement data

Frequently used atomic weights

Element	*Symbol*	*Atomic weight*	*Element*	*Symbol*	*Atomic weight*
Aluminium	Al	26.98	Magnesium	Mg	24.31
Antimony	Sb	121.75	Manganese	Mn	54.94
Argon	Ar	39.95	Mercury	Hg	200.59
Arsenic	As	74.92	Molybdenum	Mo	95.94
Barium	Ba	137.34	Nickel	Ni	58.71
Beryllium	Be	9.01	Nitrogen	N	14.01
Bismuth	Bi	208.98	Oxygen	O	16.00
Boron	B	10.81	Phosphorus	P	30.97
Bromine	Br	79.90	Platinum	Pt	195.09
Cadmium	Cd	112.40	Potassium	K	39.10
Calcium	Ca	40.08	Selenium	Se	78.96
Carbon	C	12.01	Silicon	Si	28.09
Chlorine	Cl	35.45	Silver	Ag	107.87
Chromium	Cr	52.00	Sodium	Na	22.99
Cobalt	Co	58.93	Strontium	Sr	87.62
Copper	Cu	63.55	Sulphur	S	32.06
Fluorine	F	19.00	Tantalum	Ta	189.95
Gold	Au	196.97	Tin	Sn	118.69
Helium	He	4.00	Titanium	Ti	47.90
Hydrogen	H	1.01	Tungsten	W	183.85
Iodine	I	126.90	Uranium	U	238.03
Iron	Fe	55.85	Vanadium	V	50.94
Lead	Pb	207.19	Zinc	Zn	65.37
Lithium	Li	6.94	Zirconium	Zr	91.22

SI Units

Quantity	*Unit*	*Symbol*	*Equivalent*
Base units			
Length	metre	m	–
Mass	kilogramme	kg	–
Time	second	s	–
Electric current	ampere	A	–
Temperature	kelvin	K	–
Luminous intensity	candela	cd	–
Supplementary units			
Plane angle	radian	rad	–
Solid angle	steradian	sr	–
Derived units			
Frequency	hertz	Hz	s^{-1}
Force	newton	N	kg m/s^2
Work, energy, quantity of heat	joule	J	Nm
Power	watt	W	J/s
Electric charge, quantity of electricity	coulomb	C	As
Electric potential, electromotive force	volt	V	W/A
Electric capacitance	farad	F	C/V
Electric resistance	ohm	Ω	V/A
Magnetic flux	weber	Wb	Vs
Magnetic flux density	tesla	T	Wb/m^2
Inductance	henry	H	Wb/A
Luminous flux	lumen	lm	cd sr
Illumination	lux	lx	lm/m^2

Other metric units

Quantity	*Unit*	*Symbol*	*Equivalent*
Length	angstrom	Å	10^{-10} m
	micrometre	μm	10^{-6} m
Area	are	a	10^2 m^2
Volume	litre	l	10^{-3} m^3
Mass	tonne	t	10^3 kg
Force	dyne	–	10^{-5} N
Pressure	pascal	Pa	1 N/m^2
	bar	bar	$10^5 N/m^2$
Energy	erg	–	10^{-7} J
Viscosity (dynamic)	centipoise	cP	10^{-3} Ns/m^2 (= 1 mPa s)
Viscosity (kinetic)	centistokes	cSt	10^{-6} m^2/s
Electrical conductivity	siemens	S	1 Ω^{-1}
Magnetic field strength	oersted	–	$10^3/4\pi A/m$
Magnetomotive force	gilbert	–	$10/4\pi A$
Magnetic flux density	gauss	–	10^{-4} T
Magnetic flux	maxwell (or line)	–	10^{-8} Wb
Luminance	nit	nt	1 cd/m^2
	Stilb	–	10^4 cd/m^2

Multiples and sub-multiples of units

Factor	*Preflix*	*Symbol*
10^{18}	exa	E
10^{15}	peta	P
10^{12}	tera	T
10^{9}	giga	G
10^{6}	mega	M
10^{3}	kilo	k
10^{2}	hecto	h
10	deca	da
10^{-1}	deci	d
10^{-2}	centi	c
10^{-3}	milli	m
10^{-6}	micro	μ
10^{-9}	nano	n
10^{-12}	pico	p
10^{-15}	femto	f
10^{-18}	atto	a

Temperature

°C	°C + 273.15 = K
	°C × (1.8) + 32 = °F
	°C × (1.8) + 491.58 = °R
°F	°F + 459.48 = °R
	(°F – 32) × 5/9 = °C
	(°F – 32) × 5/9 + 273.15 = K
K	K – 273.15 = °C
	(K – 273.15) × 1.8 + 32 = °F
°R	°R – 459.48 = °F
	°R – 491.58 × 5/9 = °C
	(°R – 491.58) × 5/9 + 273.15 = K

Length

	To obtain: m	cm	mm	μm	Å	in	ft
Multiply no. of							
m *by*	1	100	1000	10^{6}	10^{10}	39.37	3.28
cm *by*	0.01	1	10	10^{4}	10^{8}	0.394	0.0328
mm *by*	0.001	0.1	1	10^{3}	10^{7}	0.0394	0.00328
μ *by*	10^{-6}	10^{-4}	10^{-3}	1	10^{4}	3.94×10^{-5}	3.28×10^{-6}
Å *by*	10^{-10}	10^{-8}	10^{-7}	10^{-4}	1	3.94×10^{-9}	3.28×10^{-10}
in *by*	0.0254	2.540	25.40	2.54×10^{4}	2.54×10^{8}	1	0.0833
ft *by*	0.305	30.48	304.8	304 800	3.048×10^{9}	12	1

1 angstrom Å = 0.1 nm
= 0.000 0001 mm
= 0.000 000 003 937 in
1 m = 1.094 yd
1 yd = 0.914 m
1 km = 0.621 mile
1 mile = 5280 ft
= 1760 yd
= 1.609 km

Area

	To obtain: m^2	in^2	ft^2	cm^2	mm^2
Multiply no. of					
m^2 *by*	1	1550	10.76	10 000	10^6
in^2 *by*	6.452×10^{-4}	1	6.94×10^{-3}	6.452	645.2
ft^2 *by*	0.0929	144	1	929.0	92 903
cm^2 *by*	0.0001	0.155	0.001	1	100
mm^2 *by*	10^{-6}	0.001 55	0.000 01	0.01	1

1 cm^2 = 0.155 in^2
1 m^2 = 1.196 yd^2
1 km^2 = 247.104 acres
= 0.386 square mile
1 yd^2 = 0.836 m^2
1 square mile = 259.0 hectares

Mass

	To obtain: g	kg	gr	oz	lb
Multiply no. of					
g *by*	1	0.001	15.432	0.035 27	0.002 20
kg *by*	1000	1	15 432	35.27	2.205
gr *by*	0.0648	6.480×10^{-5}	1	2.286×10^{-3}	1.429×10^{-4}
oz *by*	28.35	0.028 35	437.5	1	0.0625
lb *by*	453.59	0.4536	7000	16	1

1 short ton = 2000 lb
= 0.892 857 long ton
= 907.1846 kg
= 0.907 1846 t

1 long ton = 2240 lb
= 1.12 short tons
= 1016.047 kg
= 1.016 047 t

1 tonne (t) = 1000 kg
= 2204.623 lb
= 1.102 31 short tons
= 0.984 2107 long ton

Volume

To obtain:	ft^3	Imp gal	litres	cm^3	m^3
Multiply no. of					
ft^3 *by*	1	6.229	28.32	28 320	0.0283
Imp gal *by*	0.1605	1	4.546	4546	4.55×10^{-3}
litres *by*	0.035 31	0.220	1	1000	1×10^{-3}
cm^3 *by*	3.531×10^{-5}	2.20×10^{-4}	0.001	1	10^{-6}
m^3 *by*	35.31	220	1000	10^6	1

1 cm^3 = 0.061 023 7 in^3 = 1 ml
= 0.035 1951 fl oz
= 0.033 814 US fl oz

1 m^3 = 264.172 US gal

1 litre = 0.264 172 US gal
= 35.195 fl oz
= 33.814 US fl oz

1 in^3 = 16.3871 cm^3 or ml
= 0.576 744 fl oz
= 0.554 113 US fl oz

1 ft^3 = 1728 in^3

1 fl oz = 1.7339 in^3
= 28.413 cm^3 or ml
= 0.960 76 US fl oz

1 Imp gal = 160 fl oz
= 277.42 in^3
= 1.2010 US gal
= 153.72 US fl oz
= 4546.1 cm^3 or ml
= 4.546 litres

1 US gal = 231 in^3
= 128 US fl oz
= 133.23 UK fl oz
= 0.832 67 Imp gal
= 3785.41 cm^3 or ml
= 3.7854 litres

Density

To obtain:	g/cm^3	lb/ft^3	lb/Imp gal	lb/US gal
Multiply no. of				
g/cm^3 *by*	1	62.43	10.02	8.345
lb/ft^3 *by*	0.016 02	1	0.1605	0.1337
lb/Imp gal *by*	0.100	6.229	1	1.2010
lb/US gal *by*	0.1198	7.481	0.832 67	1

1 mg/m^3 = 1000 mg/litre
= 35.314 mg/ft^3

1 mg/ft^3 = 28.32 mg/litre
= 0.028 32 mg/m^3

1 lb/Imp gal = 1.2010 lb/US gal
1 lb/US gal = 0.832 67 lb/Imp gal

Viscosity

Dynamic (20°C)		
centipoise	× 0.001	= N s/m^2 = Pa s
newton second per m^2	× 1000	= centipoise
poise	× 0.1	= N s/m^2 = Pa s
pound force second per ft^2	× 47 880	= centipoise
	× 47.880	= Pa s
Kinematic (20°C)		
centistoke	× 1	= mm^2/s
ft^2/hr	× 25.807	= centistokes
	× 0.092 90	= m^2/hr
m^2/hr	× 10.764	= ft^2/hr

Pressure

	To obtain: bar	MPa (N/mm^2)	atm	mm Hg	psi (lb/in^2)	lb/ft^2
Multiply no. of						
bar *by*	1	0.1	1.013 25	7.5006×10^2	14.5038	2088.54
MPa *by*	10	1	10.1325	7.5006×10^3	145.0377	20 885.4
atm *by*	0.986 9233	0.098 6923	1	760.0	14.696	2116
mm Hg *by*	133.322×10^{-5}	133.322×10^{-6}	0.0013	1	0.019 34	2.785
psi *by*	$6.894\ 76 \times 10^{-2}$	$6.894\ 76 \times 10^{-3}$	0.068	51.71	1	144
lb/ft^2 *by*	4.788×10^{-4}	0.4788×10^{-4}	4.72×10^{-4}	0.359	0.0069	1

Radiant energy units

	To obtain: erg	joule	W s	µW s	g cal
Multiply no. of					
erg *by*	1	10^{-7}	10^{-7}	0.1	2.39×10^{-8}
joule *by*	10^7	1	1	10^6	0.239
W s *by*	10^7	1	1	10^6	0.239
µW s *by*	10	10^{-6}	10^{-6}	1	2.39×10^{-7}
g cal *by*	4.19×10^7	4.19	4.19	4.19×10^6	1

Flow rates

Multiply no. of / *To obtain:*	litres/min	m³/hr	Imp gal/min	US gal/min	ft³/min	ft³/s
litres/min *by*	1	0.06	0.2200	0.2642	0.0353	5.89×10^{-4}
m^3/hr *by*	16.67	1	3.7	4.4	0.588	9.89×10^{-3}
Imp gal/min *by*	4.55	0.273	1	1.2010	0.1605	2.67×10^{-3}
US gal/min *by*	3.78	0.227	0.832 67	1	0.1338	2.23×10^{-3}
ft^3/min *by*	28.32	1.699	6.23	7.50	1	0.016 67
ft^3/s *by*	1699	102	373.7	448.8	60	1

Energy/unit area (dose units)

Multiply no. of / *To obtain:*	erg/cm²	joule/cm²	W s/cm²	µW s/cm²	g cal/cm²
erg/cm^2 *by*	1	10^{-7}	10^{-7}	0.1	2.39×10^{-8}
joule/cm^2 *by*	10^7	1	1	10^6	0.239
W s/cm^2 *by*	10^7	1	1	10^6	0.239
µW s/cm^2 *by*	10	10^{-6}	10^{-6}	1	2.39×10^{-7}
g cal/cm^2 *by*	4.19×10^7	4.19	4.19	4.19×10^6	1

Concentration

mg/litre in water (specific gravity = 1.0)	× 1.0	= ppm or g/m³
	× 0.001	= g/litre
	× 0.070157	= grains per Imp gal
	× 0.058418	= grains per US gal
	× 6.95	= lb per million Imp gal
	× 8.35	= lb per million US gal
mg/kg	× 0.002	= lb/ton (short)
	× 0.0018	= lb/ton (long)

Emission rates

	To obtain: **g/s**	**g/min**	**kg/hr**	**kg/day**	**lb/min**	**lb/hr**	**lb/day**
Multiply no. of							
g/s *by*	1.0	60.0	3.6	86.40	0.132 28	7.9367	190.48
g/min *by*	0.016 667	1.0	0.06	1.4400	2.2046×10^{-3}	0.132 28	3.1747
kg/hr *by*	0.277 78	16.667	1.0	24.000	0.036 744	2.2046	52.911
kg/day *by*	0.011 574	0.694 44	0.041 667	1.0	1.5310×10^{-3}	9.1860×10^{-2}	2.2046
lb/min by	7.5598	453.59	27.215	653.17	1.0	60.0	1440
lb/hr *by*	0.126 00	7.5598	0.453 59	10.886	1.6667×10^{-2}	1.0	24.0
lb/day *by*	5.2499×10^{-3}	0.314 99	1.8900×10^{-2}	0.453 59	6.9444×10^{-4}	4.1667×10^{-2}	1.0

Heat, energy or work

	To obtain: **joule**	**ft lb**	**kW hr**	**hp hr**	**kcal**	**cal**	**Btu**
Multiply no. of							
joules *by*	1	0.737	2.773×10^{-7}	3.725×10^{-7}	2.39×10^{-4}	0.2390	9.478×10^{-4}
ft lb by	1.356	1	3.766×10^{-7}	5.05×10^{-7}	3.24×10^{-4}	0.3241	1.285×10^{-3}
kW hr by	3.6×10^{6}	2.66×10^{6}	1	1.341	860.57	860 565	3412
hp hr by	2.68×10^{6}	1.98×10^{6}	0.7455	1	641.62	641 615	2545
kcal *by*	4184	3086	1.162×10^{-3}	1.558×10^{-3}	1	1000	3.9657
cal by	4.184	3.086	1.162×10^{-6}	1.558×10^{-6}	0.001	1	0.003 97
Btu *by*	1055	778.16	2.930×10^{-4}	3.93×10^{-4}	0.252	252	1

Velocity

	To obtain: **cm/s**	**m/s**	**km/hr**	**ft/s**	**ft/min**	**mph**
Multiply no. of						
cm/s *by*	1	0.01	0.036	0.0328	1.968	0.022 37
m/s *by*	100	1	3.6	3.281	196.85	2.237
km/hr *by*	27.78	0.2778	1	0.9113	54.68	0.6214
ft/s *by*	30.48	0.3048	18.29	1	60	0.6818
ft/min *by*	0.5080	0.005 08	0.0183	0.0166	1	0.011 36
mph *by*	44.70	0.4470	1.609	1.467	88	1

1 km/hr = 0.277 778 m/s
= 0.621 371 miles/hr
= 0.911 344 ft/s

1 ft/s = 0.681 818 miles/hr

19

Bibliography

It is impractical to catalogue all relevant publications but the following is a selection of useful sources of information relating to hazardous substances. The scope of legislation is restricted to industrial sources of hazardous chemicals; excluded are pollution by domestic fires, emissions from road vehicles and other means of transport, etc. Some proposals for legislation will have now become enacted whereas others may be dropped. Some statutory instruments will have been repealed by subsequent versions. When investigating incidents, complaints etc. it is important to be aware of the legislation/guidance available at the time of the events. For this reason the selected legislation listed demonstrates the requirements in vogue at, or after, the time of publication of the first edition, even though many will have since been updated or repealed.

Selected textbooks

Inorganic chemistry

Barnett, J., and Malati, M.A. (1997) *Fundamentals of Inorganic Chemistry*, Horwood Publishing, Chichester.
Buchel, K.H. *et al.* (2000) *Industrial Inorganic Chemistry*, Wiley-VCH, Weinheim.
Cotton, F.A. *et al.* (1999) Advanced Inorganic Chemistry, Wiley, Chichester.
Cotton, F.A. *et al.* (1999) *Basic Inorganic Chemistry*, Wiley, Chichester.
Lee, J.D. (1998) *Concise Inorganic Chemistry*, Blackwell Science, Oxford.
Striver, D.F., and Atkin, P.W. (1999) *Inorganic Chemistry*, Oxford University Press, Oxford.

Organic chemistry

Brown, W., and Foote, C. (1997) *Organic Chemistry*, Saunders College Publishing, London.
Finar, I.L. (1996) *Organic Chemistry*, Longman Higher Educational, Harlow.
Groundwater, P.W., and Taylor, G.A. (1997) *Organic Chemistry for Health and Life Sciences*, 4th edn Longman, Harlow.
Housecroft, C.E., and Constable, E.E. (1997) *Chemistry, an Integrated Approach*, Longman, Harlow.
McMurry, J., and Castellion, M.E. (1997) *Fundamentals of Organic Chemistry*, Brooks Cole.
Peach, J., and Hornby, M. (2000) *Foundations of Organic Chemistry: Worked Examples*, Oxford University Press, Oxford.
Tiwari, K.S., and Vishmoi, S.N. (1999) *A Textbook of Organic Chemistry*, Sangam Books.
Widom, J.M., and Edelstein, S.J. (1981) *Chemistry, An Introduction to General, Organic and Biological Chemistry*, Freeman, San Francisco.

Physical chemistry

Astarita, E. (1989) *Thermodynamics*, Kluwer Academic/Plenum, The Hague.
Atkin, P.W. (1998) *Physical Chemistry*, Oxford University Press, Oxford.
Isaacs, N.S. (1995) *Physical Organic Chemistry*, Longman Higher Educational, Harlow.

Kettle, S.F.A. (1998) *Physical Inorganic Chemistry*, Oxford University Press, Oxford.
Ladd, M.F.C. (1998) *Introduction to Physical Chemistry*, Cambridge University Press, Cambridge.
Lawrence, C. *et al.* (1996) *Foundations of Physical Chemistry*, Oxford University Press, Oxford.
Mortimer, R.G. (2000) *Physical Chemistry*, Academic Press, London.
Mozunder, A. (1999) *Fundamentals of Radiation Chemistry*, Academic Press, Kingston-upon Thames.
Wright, M.R. (1999) *Fundamentals of Chemical Kinetics*, Horwood Publishing, Hemel Hempstead.

Safety and environment

Alderson, M.R. (1986) *Occupational Cancer*, Butterworths, London.
Allen, D.T. and Rosselot, K.S. (1996) *Pollution Prevention for Chemical Processes*, Wiley, New York.
American Conference of Governmental Industrial Hygienists (2000) *Threshold Limit Values for Chemical Substances in the Work Environment*, ACGIH, Cincinnati.
Armour, M.A. (1996) *Hazardous Laboratory Chemicals Disposal Guide*, CRC Press, Lewis Publishers.
Ashford, N., and Miller, C. (1998) *Chemical Exposures, High Levels and High Stakes*, 2nd edn, Van Nostrand Reinhold, New York.
Bartknecht, W. (1981) *Explosions*, Springer-Verlag, New York.
Barton, J., and Rogers, R. (ed.) (1966) Chemical Reaction Hazards, *Institution of Chemical Engineers*, Rugby.
Bendito, D.P. *et al.* (1998) *Environmental Analytical Chemistry*, Elsevier Science Ltd, Amsterdam.
Boleij, J.S.M. *et al.* (1994) *Occupational Hygiene of Chemical and Biological Agents*, Elsevier Science Ltd, Amsterdam.
Bond, J. (1991) *Sources of Ignition – Flammability Characteristics of Chemicals and Products.* Butterworth-Heinemann, Oxford.
Bourdeau, P., and Green, G. (1989) *Methods for Assessing and Reducing Injury from Chemical Accidents*, Wiley, Chichester.
Bretherick, L. (1990) *Handbook of Reactive Chemical Hazards*, 4th edn, Butterworth-Heinemann, Oxford.
British Cryogenics Council (1991) *Cryogenics Safety Manual*, 3rd edn, Butterworth-Heinemann, Oxford.
British Occupational Hygiene Society (1985) *The Selection and Use of Personal Sampling Pumps*, Technical Guide No. 5, BOHS, Norwood, Science Reviews.
British Occupational Hygiene Society (1988) *Controlling Airborne Contaminants in the Workplace*, Technical Guide No. 7, BOHS, Norwood, Science Reviews.
British Occupational Hygiene Technology Committee (1975) *Guide to the Design and Installation of Laboratory Fume Cupboards*.
Britton, L.G. (2000) *Avoiding Static Ignition Hazards in Chemical Processes*, American Institute of Chemical Engineers.
Browning, E. (1965) *Toxicity and Metabolism of Industrial Solvents*, Elsevier, Amsterdam.
Burgess, W.A. (1995) *Recognition of Health Hazards in Industry*, 2nd edn, Wiley, New York.
Campbell, M. (ed.) (1996) *Sensor System for Environmental Monitoring*, Kluwer Academic Publishers, The Hague.
Carson, P.A., and Dent, N.J. (ed.) (1990) *Good Laboratory and Clinical Practices*, Heinemann Newnes, Oxford.
Carson, P.A., and Mumford, C.J. (1988) *The Safe Handling of Chemicals in Industry* (Vols 1 and 2), Longman Scientific and Technical, Harlow.
Carson, P.A., and Mumford, C.J. (1996) *The Safe Handling of Chemicals in Industry* (Vol. 3), Longman Scientific and Technical, Harlow.
Center for Chemical Process Safety, American Institute of Chemical Engineers
Guidelines for Chemical Transport Risk Analysis (1995)
Guidelines for Design Solutions to Process Equipment Failures Safety (1997)
Guidelines for Engineering Design for Process Safety (1993)
Guidelines for Evaluating the Consequences of Vapour Cloud Explosions, Flash Fires, and BLEVEs (1994)
Guidelines for Implementing Process Safety Management Systems (1994)
Guidelines for Improving Plant Reliability through Equipment Data Collection and Analysis (1998)
Guidelines for Investigating Chemical Process Incidents (1992)
Guidelines for Post Release Mitigation in the Chemical Process Industry (1997)
Guidelines for Process Safety in Batch Reaction Systems (2000)
Guidelines for Safe Automation of Chemical Processes (1993)
Guidelines for Safe Warehousing of Chemicals (1997)
Guidelines for Technical Management of Chemical Process Safety (1989)
Chemical Industries Association, London
Allergy to Chemicals at Work (1990)
An Approach to the Categorisation of Process Plant Hazard and Control Building Design (1990)
Be Prepared for an Emergency–Guidance for Emergency Planning (1991)

Be Prepared for an Emergency: Training and Exercises (1992)
Chemicals in a Sustainable World (1993)
Chemical Exposure Treatment Cards (1991)
Chemsafe – Assistance in Chemical Distribution Emergencies (1996)
Control of Substances Hazardous to Health Regulations (COSHH). Guidance on Exposure Limits for Mixtures (Regulation 7). Revised (1993)
Control of Substances Hazardous to Health Regulations (COSHH). Guidance on Setting In-House Occupational Exposure Limits (Regulation 7) (1990)
Control of Substances Hazardous to Health Regulations (COSHH). Guidance on Monitoring of Exposure (Regulation 10) (1993)
Control of Substances Hazardous to Health Regulations (COSHH). Guidance on Information, Instruction and Training (Regulation 12) (1990)
COSHH: COSHH in distribution. Some aspects of the application of the Control of Substances Hazardous to Health Regulations 1988 (COSHH) to Product Distribution (1992)
COSHH: Guidance on Record Keeping (1991)
COSHH: Guidance on Collection and Evaluation of Hazard Information (Regulation 2) (1991)
Determination of Formaldehyde in the Workplace Atmosphere (1993)
Flame Retardant Products and their Uses (1990)
Formaldehyde: Questions and Answers (1993)
A Guide to Hazard and Operability Studies (1993)
Guidance on Transfer Connections for the Safe Handling of Anhydrous Ammonia in the UK (1990)
Guidance on Carcinogens in the Workplace. Revised (1993)
Guidelines for Bulk Handling of Ethylene Oxide (1996)
Guidelines for Certification to ISO 9001 – Health, Safety and Environmental Management Systems (and BS7750 – Environmental Management Systems) in the Chemical Industry (1992)
Guidelines on Employers' Liability and Public Liability Risk Assessment in the Chemical Industry (1992)
Guidelines on Material Damage Risk Assessment in the Chemical Industry (3rd edn) (1992)
Guidelines on Product Liability Risk Assessment in the Chemical Industry (1992)
Guidance on Safety, Occupational Health and Environmental Protection Auditing (1991)
Guidance on Prevention or Control of Exposure (Regulation 7) (1998)
Guidelines for Safe Warehousing (1990)
Management of Wastes at Production Sites (1995)
Plastics Containers for the Carriage of Hazardous Materials. An Outline of Recommended Quality Control Standards (1991)
Protection of the Eyes (3rd edn) (1990)
Responsible Care Management Systems for Health, Safety and Environment (2nd edn) (1995)
Responsible Care Management System – Self Assessment (1998)
Responsible Care (1996)
Responsible Care Management Systems – Guidance (3rd edn) (1998)
Responsible Care Management Systems – Additional Guidance (1999)
Risk Reduction for Existing Substances (1995)
Risk-Benefit Analysis of Existing Substances (1995)
Road Transport of Dangerous Substances in Bulk. Guidance on Procedures for Organising and Monitoring (1990)
Safe Handling of Colourants (1990)
Safe Handling of Colourants 2: Hazard Classification and Selection of Occupational Hygiene Strategies (1993)
Safe Handling of Formaldehyde (2nd edn) (1995)
Steel Containers for the Carriage of Hazardous Materials. An Outline of Recommended Quality Control Standards (1991)
Terms and Phrases used in the Chemical Exposure Treatment Cards (1992)
Timber Pallets for the Carriage of Hazardous Substances. An Outline of Recommended Quality Control Standards (1992)
Use of Waste Disposal Contractors (1995)
What is Toxicology (1995)

Chemical and Process Industries Training Board (1977) *Permit to Work Systems*, CAPITB-16, London.

Cheremisinoff, P.N. (1992) *Waste Incineration Handbook*, Butterworth-Heinemann, Oxford.

Cheremisinoff, P.N., and Graffia, M. (1995) *Safety Management Practices for Hazardous Materials*, Marcel Dekker, New York.

Chou, J. (2000) *Hazardous Gas Monitors: A Practical Guide to Selection, Operation and Applications*. McGraw-Hill Book Company, New York.

Clayton, G.D., and Clayton, F.E. (ed.) (1982) *Patty's Industrial Hygiene and Toxicology,* 3rd edn, Wiley, New York.
Clarke, A.G. (ed.) (1997) *Indoor Air Pollution Monitoring*, Kluwer Academic Publishers, The Hague.
Connor, K.J., and McLintock, I.S. (1995) *Radiation Protection: Handbook for Laboratory Workers*, H&H Scientific Consultants.
Copin, A. (ed.) (1995) *Environmental Behaviour of Pesticides and Regulatory Aspects*, Royal Society of Chemistry, London.
Cox, S.J., and Tait, N.R.S. (1991) *Reliability, Safety and Risk Management: An Integrated Approach*, Butterworth-Heinemann, Oxford.
Cralley, L.V., and Cralley, L.L. (ed.) (1983/84) *Industrial Hygiene Aspects of Plant Operations*, 3 vols, Macmillan, New York.
Croner Publications Ltd (1986) *Dangerous Chemicals: Emergency First-aid Guide*, 2nd edn.
Cullis, C.F., and Firth, J.G. (ed.) (1988) *Detection and Measurement of Hazardous Gases*, Heinemann, London.
Daugherty, J., (1997) *Assessment of Chemical Exposures: Calculation Methods for Environmental Professionals*, Lewis Publishing, London.
Davison, G., and Hewitt, C.N. (ed.) (1997) *Air Pollution in the United Kingdom*, Royal Society of Chemistry, London.
Dow Chemical Company (1993) *Dow's Fire and Explosion Index Hazard Classification Guide American Institute of Chemical Engineers.*
Drake, J.A.G. (1994) *Integrated Pollution Control*, Royal Society of Chemistry, London.
Down, R. (2000) *Handbook of Environmental Instrumentation*, McGraw-Hill Publishing Co, New York.
Duffins, J.J. (ed.) (1997) *Carcinogenicity of Inorganic Substances*, Royal Society of Chemistry, London.
Environmental Protection Agency (1995) *Landfill Manuals: Landfill Monitoring.*
Everett, K., and Hughes, D. (1981) *A Guide to Laboratory Design*, 2nd edn, Butterworths, London.
Fawcett, H.H., and Woods, W.S. (ed.) (1982) *Safety and Accident Prevention in Chemical Operations*, 2nd edn, Wiley, Chichester.
Fire Protection Association (1977a) *Fire and Related Properties of Industrial Chemicals*, London.
Fire Protection Association (1977b) *Guide to Fire Precautions with Outside Contractors*, London.
Forsberg, K., and Keith, L.H. (1999) *Chemical Protective Clothing Performance Index*, Wiley, Chichester.
Foster, S.S.D. *et al.* (2000) *Groundwater Quality Monitoring*, Spon Press.
Furr, A.K. (ed.) (2000) *CRC Handbook of Laboratory Safety*, 5th edn, CRC.
Gangoli, S. (ed.) (1999) *The Dictionary of Substances and their Effects* (DOSE), Royal Society of Chemistry, London.
Gibbons, R. (1994) *Statistical Methods for Groundwater Monitoring*, Wiley.
Gilbert, R.O. (1987) *Statistical Methods for Environmental Pollution Monitoring,* Wiley, Chichester.
Grandjean, P. (1990) *Skin Penetration, Hazardous Chemicals at Work*, Taylor and Francis, London.
Greenberg, M.I. (ed.) (1997) *Occupational, Industrial and Environmental Toxicology*, Mosby, St Louis.
Griffiths, R. (ed.) (1995) *Sulphur Trioxide, Oleum and Sulphuric Acid Mist*, Institution of Chemical Engineers, Rugby.
Harrington, J.M. and Gardiner, K. (1995) *Occupational Hygiene*, 2nd edn, Blackwell Science, Oxford.
Harrison, R.M. (ed.) (1996) *Pollution Causes, Effects and Control*, Royal Society of Chemistry, London.
Harrison, R.M. (1999) *Understanding Our Environment: An Introduction to Environmental Chemistry and Pollution*, Royal Society of Chemistry, London.
Hathaway, G.J., Proctor, N.H., Hughes, J.P., and Fishman, M.L. (1991) *Chemical Hazards of the Workplace*, 3rd edn, Van Nostrand Reinhold, New York.
Harvey, B. (ed.) (1990) *Handbook of Occupational Hygiene*, Croner Publications, New Malden.
Hazell, R.W. (ed.) (1996) *COSHH in Laboratories*, Royal Society of Chemistry, London.
Henning, H. (ed.) (1993) *Solvent Safety Sheets: A Compendium for the Working Chemist*, Royal Society of Chemistry, London.
High Pressure Technology Association (1975) *High Pressure Safety Code*, Leeds.
Hughes, J.R. (1970) *Storage and Handling of Petroleum Liquids – Practice and Law*, 2nd edn, Griffin, London.
Institution of Chemical Engineers (1983) *The Preparation of Plant for Maintenance*, London.
Institution of Chemical Engineers (1987) *User Guide for the Safe Operation of Centrifuges*, London.
Institution of Chemical Engineers (1990) *Prevention of fires and Explosions in Dryers: A User Guide*, 2nd edn, London.
International Atomic Energy Agency, *Safe Handling of Radioisotopes*, Safety Series No. 1, Vienna.
Isman, W.E., and Carlson, G.P. (1980) *Hazardous Materials*, Glencoe, Encino, California.
Jahnke, J.A. (2000) *Continuous Emission Monitoring*, Wiley, Chichester.
Jones, C.W. (1999) *Applications of Hydrogen Peroxide and Derivatives*, Royal Society of Chemistry, London.
Kim, U., and Falkenbury, J. (1996) *Environmental and Safety Auditing; Program Strategies for Legal, International and Financial Issues*, Science Behaviour Books.
Kimber, I. (1997) *Toxicology of Chemical Respiratory Hypersensitivity*, Taylor and Francis, London.
Kletz, T.A. (1999) *HAZOP and HAZAN Identifying and Assessing Process Industry Hazards*, 4th edn, Institutional Chemical Engineers, Rugby.

Kletz, T.A. (1989) *What Went Wrong? Case Histories of Process Plant Disasters*, Gulf Publishing, Houston.
Kletz, T.A. (1998) *Process Plants: A Handbook of Inherently Safer Design*, Taylor and Francis Inc.
Kletz, T.A. (1984) *Cheaper, Safer Plants or Wealth and Safety at Work: Notes on Inherently Safer and Simpler Plants*, Institution of Chemical Engineers Loss Prevention Information Exchange Scheme – Hazard Workshop Module.
Kletz, T.A. (1993) *Lessons from Disasters: How Organisations Have No Memory and Accidents Recur*, Institutional Chemical Engineers, Rugby.
King, R.W., and Magid, J. (1979) *Industrial Hazard and Safety Handbook*, Newnes-Butterworth, London.
Kneip, T.J., and Crable, J.V. (ed.) (1988) *Methods for Biological Monitoring*, American Public Health Assoc., Washington DC.
Lees, F.P., and Ang, M.L. (1989) *Safety Cases Within the Control of Industrial Major (CIMAH) Regulations 1984*, Butterworth, London.
Lees, F.P. (1996) *Loss Prevention in the Process Industries*, 2nd edn, Butterworth-Heinemann, Oxford.
Lefevre, M.J. (ed.) (1989) *First Aid Manual for Chemical Accidents*, 2nd edn, Wiley, Chichester.
Leonelli, J., and Althouse, M.L. (eds) (1999) *Air Monitoring and Detection of Chemical and Biological Agents*, SPIE Society of Photo-optical Instrumentation Engineers.
Lewis, P. (1993) *Health Protection from Chemicals in the Workplace*, Ellis Horwood Ltd, Chichester.
Lewis, R.J. (2000) *Rapid Guide to Hazardous Chemicals in the Workplace*, Wiley, Chichester.
Lioy, P.J., and Lioy, M.J.Y. (1989) *Air Sampling Instruments*, ACGIH, Cincinnati.
Lipton, S., and Lynch, J. (1994) *Handbook of Health Hazard Control in the Chemical Process Industry*, Wiley, New York.
Lunn, G., and Sansone, E.B. (1994) *Destruction of Hazardous Chemicals in the Laboratory*, Wiley, Chichester.
Lunn, G.A. (1984) *Venting Gas and Dust Explosions – A Review*, Instn of Chem. Eng., London.
Luxon, S.G. (ed.) (1992) *Hazards in the Chemical Laboratory*, 5th edn, Royal Society of Chemistry, London.
Major, D.W., and Fitchko, J. (1992) *Hazardous Waste Treatment On-site and In Situ*, Butterworth-Heinemann, Oxford.
Manufacturing Chemists Association (1970) *Laboratory Waste Disposal Manual*, Washington, DC.
Manufacturing Chemists Association (1972) *Guide for Safety in the Chemical Laboratory*, Van Nostrand Reinhold, New York.
McBean, E.A., and Rovers, F. (1998) *Statistical Procedures for Analysis of Environmental Monitoring Data and Risk Assessment,* Prentice Hall, New York.
McConnell, R.L., and Abel, D.C. (1998) *Environmental Issues: Measuring, Analyzing and Evaluating*, Prentice Hall, New York.
Martin, K., and Bastock, T.W. (ed.) (1994) *Waste Minimisation: A Chemist's Approach*, Royal Society of Chemistry, London.
Martin, W., Lippitt, J.M., and Prothero, T.G. (1992) *Hazardous Waste Handbook for Health and Safety*, Butterworth-Heinemann, Oxford.
Mecklenburgh, J.C. (ed.) (1985) *Process Plant Layout*, 2nd edn, Longman, Harlow.
Meidl, J.H. (1970) *Explosive and Toxic Hazardous Materials,* Glencoe, Westerville, OH.
Meidl, J.H. (1978) *Flammable Hazardous Materials*, Glencoe, Encino, California.
Nagda, N.L. (1999) *Guidelines for Monitoring Indoor Air Quality*, Taylor and Francis Inc.
National Institute for Occupational Safety and Health (1984) *Manual of Analytical Methods*, US Department of Health and Human Services, Washington, DC.
Nowel, L.H. *et al.* (1999) *Pesticides in Stream Sediment and Aquatic Biota*, CRC Press.
Oil Industry Advisory Committee (1996) *Management of Occupational Health Risks in the Offshore Oil and Gas Industry*, HSE, Bootle.
Oil Insurance Association, *General Recommendations for Spacing in Refineries, Petrochemical Plants, Gasoline Plants, Terminals, Oil Pump Stations and Offshore Properties* (No. 361), USA.
Palmer, K.N. (1973) *Dust Explosions and Fires*, Chapman and Hall, London.
Paul, J., and Pradier, C. (1994) *Carbon Dioxide Chemistry: Environmental Issues*, Royal Society of Chemistry, London.
Pipitone, D.A. (ed.) (1991) *Safe Storage of Laboratory Chemicals*, 2nd edn, Wiley, New York.
Pohanish, R.P., and Greene, S.A. (1997) *Rapid Guide to Chemical Incompatibilities*, Wiley, New York.
Pohanish, R.P., and Greene, S.A. (1998) *Electronic and Computer Industry Guide to Chemical Safety and Environmental Compliance*, Wiley, New York.
Porteous, A. (ed.) (1985) *Hazardous Waste Management Handbook*, Butterworths, London.
Printing Industry Advisory Committee (1995) *Chemical Safety in the Printing Industry*, H.S.E. Bootle.
Quevauviller, P. (ed.) (1995) *Quality Assurance in Environmental Monitoring*, Wiley-VCH, Weinheim.
Quint, M. *et al.* (ed.) (1995) *Environmental Impact of Chemicals: Assessment and Control*, Royal Society of Chemistry, London.
Radojavic, M., and Bashkin, V.N. (1999) *Practical Environmental Monitoring*, Royal Society of Chemistry, London.
Raffle, P.A.B., Adams, P.H., Baxter, P.J., and Lee, W.R. (ed.) (1994) *Hunter's Diseases of Occupations*, 8th edn, Edward Arnold, London.

Rao, S. (1999) *Impact Assessment of Hazardous Aquatic Contaminants*, CRC Press.
Reinhardt, R.A. (ed.) *et al.* (1996) *Pollution Prevention and Waste Minimization in Laboratories*, Lewis Publishing.
Richardson, M.L. (ed.) (1992) *Risk Management of Chemicals*, Royal Society of Chemistry, London.
Richardson, M.L. (ed.) (1991) *Chemistry, Agriculture and the Environment*, The Royal Society of Chemistry, London.
Richardson, M.L. (ed). (1986) *Toxic Hazard Assessment of Chemicals*, The Royal Society of Chemistry, London.
Richardson, M.L. (1986) *Risk Assessment for Hazardous Installations,* Pergamon, Oxford.
Richardson, M.L. (ed.) (1988) *Risk Assessment of Chemicals in the Environment*, The Royal Society of Chemistry, London.
Ridley, J.R.R., and Channing, J. (1999) *Safety at Work*, 5th edn, Butterworth-Heinemann, Oxford.
Royal Society of Chemistry (1977) *Code of Practice for Chemical Laboratories*, London.
Royal Society of Chemistry (1989) *COSHH in Laboratories*, London.
Rubber Industry Advisory Committee – *A New Practical Guide to Complying with COSHH in the Rubber Industry*
Dust Control in Powder Handling and Weighing
Dust and Fume Control in Rubber Mixing and Milling
Control of Solvents in the Rubber Industry
Sadhra, S.S., and Rampal, K.G. (1999) *Occupational Health Assessment and Management*, Blackwell Science, Oxford.
Sanders, R.E. (1999) *Chemical Process Plant Safety*, American Institute of Chemical Engineers.
Saunders, G.T. (1993) *Laboratory Fume Hoods: A User's Manual*, Wiley.
Schafer, K. (ed.) (1999) *Spectroscopic Atmospheric Environmental Monitoring Techniques*, SPIE Society of Photo-optical Instrumentation Engineers.
St John Ambulance Association (1982) *First Aid*, 3rd edn, St Andrews Ambulance Association and The British Red Cross Society, London.
Sax, N.I., and Lewis, R.J. (1988) *Dangerous Properties of Industrial Materials*, 7th edn, Van Nostrand Reinhold. New York.
Schieler, L., and Pauze, D. (1976) *Hazardous Materials*, Van Nostrand Reinhold, Wokingham.
Scohfield, D. (1984) *User Guide to Explosion Prevention and Protection, Pt 2*, Instn of Chem. Eng., London.
Shapton, D.A., and Board, R.G. (1972) *Safety in Microbiology*, Academic Press, London.
Simpson, D., and Simpson, W.G. (ed.) (1991) *The COSHH Regulations: A Practical Guide*, The Royal Society of Chemistry, Cambridge.
Sittig, M. (1981) *Handbook of Toxic and Hazardous Chemicals*, Noyes Publications, Park Ridge, NJ.
Stricoff, R.S., and Walters, D.B. (1995) *Handbook of Laboratory Health and Safety*, 2nd edn, Wiley, Chichester.
Strong, C.B., and Irvin, T.R. (1995) *Emergency Response and Hazardous Chemical Management: Principles and Practices* (Advances in Environmental Management Series), St Lucie Press.
Tapp, J. *et al.* (ed.) (1996) *Toxic Impact of Wastes on the Aquatic Environment*, Royal Society of Chemistry, London.
Taylor, F. *et al.* (2000) *Analytical Methods for Environmental Monitoring*, Prentice Hall, New York.
T'So, P.O., and Di Paulo, J.A. (1974) *Chemical Carcinogenesis*, Marcel Dekker, New York.
Warren, P.J. (ed.) (1985) *Dangerous Chemicals, Emergency Spillage Guide*, Walters Samson (UK), New Malden, Surrey.
Wallace, I.G. (1994) *Developing Effective Safety Systems*, Institution of Chemical Engineers, Rugby.
Wentz. C.A. (1999) *Safety, Health and Environmental Protection*, McGraw-Hill Publishing Company (ISE edn).
Whiston, J. (ed.) (1991) *Safety in Chemical Production*, Blackwell Science, Oxford.
Young, J.A. (ed.) (1991) *Improving Safety in the Chemical Laboratory: A Practical Guide*, Wiley, Chichester.

UK legislation on dangerous substances

Selected acts of parliament

Environmental Protection Act 1990
Radioactive Material (Road Transport) Act 1991
Radioactive Substances Act 1993
The Environment Act 1995
The Health and Safety at Work etc. Act 1974

Selected statutory instruments

SI 1962/2645	Radioactive Substances (Exhibitions) Exemption Order
SI 1962/2646	Radioactive Substances (Storage and Transit) Exemption Order
SI 1962/2648	Radioactive Substances (Phosphatic Substances, Rare Earths, etc.) Exemption Order
SI 1962/2649	Radioactive Substances (Lead) Exemption Order
SI 1962/2710	Radioactive Substances (Uranium and Thorium) Exemption Order
SI 1962/2711	Radioactive Substances (Prepared Uranium and Thorium Compounds) Order
SI 1962/2712	Radioactive Substances (Geological Specimens) Exemption Order
SI 1963/1831	Radioactive Substances (Waste Closed Sources) Exemption Order
SI 1963/1832	Radioactive Substances (Schools, etc.) Exemption Order
SI 1963/1836	Radioactive Substances (Precipitated Phosphate) Exemption Order
SI 1963/1832	Radioactive Substances (Schools, etc.) Exemption Order
SI 1967/1797	Radioactive Substances (Electronic Valves) Exemption Order
SI 1980/953	Radioactive Substances (Smoke Detectors) Exemption Order
SI 1980/1248	Control of Lead at Work Regulations
SI 1982/1357	The Notification of Installations Handling Hazardous Substances Regulations
SI 1983/1649	Asbestos (Licensing) Regulations
SI 1983/1140	The Classification and Labelling of Explosives Regulations
SI 1984/1890	Freight Containers (Safety Convention) Regulations
SI 1985/1047	Radioactive Substances (Gaseous Tritium Light Devices) Exemption Order
SI 1985/1048	Radioactive Substances (Luminous Articles) Exemption Order
SI 1985/1049	Radioactive Substances (Testing Instruments) Exemption Order
SI 1985/2042	Asbestos Product (Safety) Regulations
SI 1986/1002	Radioactive Substances (Substances of Low Activity) Exemption Order
SI 1986/1510	Control of Pesticides Regulations
SI 1987/1979	Asbestos Products (Safety) (Amendment) Regulations
SI 1987/37	Health and Safety – The Dangerous Substances in Harbour Areas Regulations
SI 1987/2115	Control of Asbestos at Work Regulations
SI 1988/819	Collection and Disposal of Waste Regulations
SI 1988/1655	Health and Safety – The Docks Regulations
SI 1989/1156	Trade Effluents (Prescribed Processes and Substances) Regulations
SI 1990/556	Control of Asbestos in Air Regulations
SI 1990/2504	Radioactive Substances (Appeals) Regulations
SI 1990/2512	Radioactive Substances (Hospitals) Exemption Order
SI 1991/447	Radioactive Substances (Smoke Detectors) Exemption (Amendment) Order
SI 1991/836	Environmental Protection (Prescribed Processes and Substances) Regulations
SI 1992/31	Environmental Protection (Controls on Injurious Substances) Regulations
SI 1992/43	Road Traffic (Carriage of Dangerous Substances in Road Tankers and Tank Containers) Regulations
SI 1992/647	Radioactive Substances (Substances of Low Activity) Exemption (Amendment) Order
SI 1992/656	Planning (Hazardous Substances) Regulations
SI 1992/1583	Environmental Protection (Controls on Injurious Substances) (No. 2) Regulations
SI 1992/1685	Radioactive Substances (Records of Convictions) Regulations
SI 1992/2617	The Environmental Protection Act 1990 (Modification of Section 112) Regulations
SI 1992/2997	Public Information for Radiation Emergencies Regulations
SI 1992/3067	Asbestos (Prohibitions) Regulations
SI 1992/3068	Control of Asbestos at Work (Amendment) Regulations
SI 1992/3217	The Genetically Modified Organisms (Contained Use) Regulations
SI 1992/3280	The Genetically Modified Organisms (Deliberate Release) Regulations
SI 1993/1	Environmental Protection (Controls on Injurious Substances) Regulations
SI 1993/15	The Genetically Modified Organisms (Contained Use) Regulations
SI 1993/152	The Genetically Modified Organisms (Deliberate Release) Regulations
SI 1993/1643	Environmental Protection (Controls on Injurious Substances) (No. 2) Regulations
SI 1993/3031	Transfrontier Shipment of Radioactive Waste Regulations
SI 1994/199	Environmental Protection (Non-Refillable Refrigerant Containers) Regulations
SI 1994/232	Batteries and Accumulators (Containing Dangerous Substances) Regulations
SI 1994/2844	Dangerous Substances and Preparations (Safety) (Consolidation) Regulations
SI 1994/3247	Chemical (Hazard Information and Packaging for Supply) Regulations

SI 1995/304	The Genetically Modified Organisms (Deliberate Release) Regulations
SI 1995/2395	Radioactive Substances (Hospitals) Exemption (Amendment) Order
SI 1996/967	The Genetically Modified Organisms (Contained Use) (Amendment) Regulations
SI 1996/1092	Chemicals (Hazard Information and Packaging for Supply) (Amendment) Regulations
SI 1996/1106	The Genetically Modified Organisms (Risk Assessment) (Records and Exemptions) Regulations
SI 1996/1350	Radioactive Material (Road Transport) (Great Britain) Regulations
SI 1996/2075	Health and Safety at Work etc. Act 1974 (Application to Environmentally Hazardous Substances) Regulations
SI 1996/2089	The Carriage of Dangerous Goods by Rail Regulations
SI 1996/2090	Packaging, Labelling and Carriage of Radioactive Material by Rail Regulations
SI 1996/2092	Carriage of Dangerous Goods (Classification, Packaging and Labelling) and Use of Transportable Pressure Receptacles Regulations
SI 1996/2093	Health and Safety – The Carriage of Explosives by Road Regulations
SI 1996/2094	The Carriage of Dangerous Goods by Road (Driver Training) Regulations
SI 1996/2095	The Carriage of Dangerous Goods by Road Regulations
SI 1996/2635	Dangerous Substances and Preparations (Safety) (Consolidation) (Amendment) Regulations
SI 1997/1460	Chemicals (Hazard Information and Packaging for Supply) (Amendment) Regulations
SI 1997/1900	The Genetically Modified Organisms (Deliberate Release and Risk Assessment) (Amendment) Regulations
SI 1997/2367	Marine Pollution – Merchant Shipping – Safety – The Merchant Shipping (Dangerous Goods and Marine Pollutants) Regulations
SI 1998/545	Environmental Protection (Control of Hexachoroethane) Regulations
SI 1998/767	Environmental Protection (Prescribed Processes and Substances) (Amendment) (Hazardous Waste Incineration) Regulations
SI 1998/1548	The Genetically Modified Organisms (Contained Use) (Amendment) Regulations
SI 1998/2885	The Carriage of Dangerous Goods (Amendment) Regulations
SI 1998/3106	The Chemical (Hazard Information and Packaging for Supply) (Amendments) Regulations
SI 1998/3233	Asbestos (Licensing) (Amendment) Regulations
SI 1998/3235	Control of Asbestos at Work (Amendment) Regulations
SI 1999/40	Health and Safety at Work etc. Act 1974 (Application to Environmentally Hazardous Substances) (Amendment) Regulations
SI 1999/197	The Chemical (Hazard Information and Packaging for Supply) (Amendment) Regulations
SI 1999/257	The Transport of Dangerous Goods (Safety Advisers) Regulations
SI 1999/303	Carriage of Dangerous Goods (Amendment) Regulations
SI 1999/336	Merchant Shipping – Safety – The Merchant Shipping (Carriage of Cargos) Regulations
SI 1999/396	Environmental Protection Act 1990 (Extension of Section 140) Regulations
SI 1999/437	The Control of Substances Hazardous to Health Regulations 1999
SI 1999/743	The Control of Major Accident Hazards Regulations
SI 1999/981	Planning (Control of Major Accident Hazards) Regulations
SI 1999/2084	Dangerous Substances and Preparations (Safety) (Consolidation) (Amendment) Regulations
SI 1999/2373	Asbestos (Prohibitions) (Amendment) Regulations
SI 1999/2977	Asbestos (Prohibitions) (Amendment) (No. 2) Regulations
SI 1999/2978	Road Vehicles (Brake Linings Safety) Regulations
SI 1999/3244	Environmental Protection (Controls on Injurious Substances) Regulations
SI 1999/3106	Health and Safety – The Good Laboratory Practice Regulations
SI 1999/3165	Chemicals (Hazard Information and Packaging for Supply) (Amendment) (No. 2) Regulations
SI 1999/3193	Dangerous Substances and Preparations (Safety) (Consolidation) (Amendment) (No. 2) Regulations
SI 1999/3194	Chemicals (Hazard Information and Packaging for Supply) (Amendment) (No. 3) Regulations
SI 1999/3232	Health and Safety – The Ionising Radiations Regulations
SI 2000/128	Pressure Systems Safety Regulations
SI 2000/1043	Environmental Protection (Disposal of Polychlorinated Biphenyls and other Dangerous Substances) (England and Wales) Regulations

European legislation on dangerous substances

Selected regulations

2455/92	Regulation on the export and import of dangerous chemicals
3600/92	Regulation on the implementation of the first stage of the programme referred to in Directive 91/414/EEC on the placing of plant protection products on the market
793/93	Regulation on the evaluation of existing substances
933/94	Regulation on the laying down of active substances of plant protection products and designating rapporteur Member States for the implementation of Regulation 3600/92
1179/94	Regulation on priority substances under Regulation 793/93
1488/94	Regulation on the assessment of risks in accordance with Regulation 793/93
3135/94	Regulation amending 2455/92 on the export and import of dangerous chemicals
2230/95	Regulation amending Regulation 933/94 on the laying down of active substances of plant protection products and designating rapporteur Member States for the implementation of Regulation 3600/92
1492/96	Regulation amending Regulation 2455/92 on the import and export of dangerous chemicals
1199/97	Regulation amending Regulation 3600/92 on the implementation of the first stage of the programme referred to in Directive 91/414/EEC on the placing of plant protection products on the market
1237/97	Regulation amending Annex II to Regulation 2455/92 concerning the export and import of certain dangerous chemicals
2247/98	Regulation amending Regulation 2455/92 concerning the export and import of certain dangerous chemicals
1493/93	Regulation 1493/93 on shipments of radioactive substances between Member States
1972/99	Regulation amending Regulation (EEC) No. 3600/92 on the implementation of the first stage of the programme of work referred to in Directive 91/414/EEC concerning the placing of plant protection products on the market
645/2000	Regulation on rules for provisions of Directive 86/362/EEC and Directive 90/642/EEC on the maximum levels of pesticide residues

Selected directives

67/548/EEC	Directive on the classification, etc. of dangerous substances
73/404/EEC	Directive relating to detergents
75/439/EEC	Directive on the disposal of waste oils
76/116/EEC	Directive on the approximation of the laws relating to fertilizers
76/403/EEC	Directive on PCBs
76/464/EEC	Directive on dangerous substances discharged into water
76/769/EEC	Directive on the marketing and use of dangerous substances
76/895/EEC	Directive on pesticide residues in and on fruit and vegetables
77/93/EEC	Directive on protective measures against the introduction of organisms harmful to plants or plant products and against their spread within the community
78/176/EEC	Directive on waste from the titanium dioxide industry
78/319/EEC	Directive on toxic and dangerous waste
79/831/EEC	Directive sixth amendment to Directive 67/548/EEC
82/176/EEC	Directive on Mercury (Chlor-alkali electrolysis industry)
82/501/EEC	Directive on major accident hazards (Seveso Directive)
82/806/EEC	Directive amending Directive 76/769/EEC (benzene)
83/478/EEC	Directive amending Directive 76/769/EEC (asbestos)
83/513/EEC	Directive on cadmium
83/477/EEC	Directive on the protection of workers from exposure to asbestos at work
84/156/EEC	Directive on mercury
84/491/EEC	Directive on hexachlorocyclohexane
85/467/EEC	Directive amending (PCBs) Directive 76/769/EEC relating to restrictions on the marketing and use of certain dangerous substances and preparations.
86/362/EEC	Directive on pesticide residues in and on cereals
86/363/EEC	Directive on pesticide residues in foodstuffs of animal origin

87/217/EEC Directive on pollution by asbestos
88/183/EEC Directive amending Directive 76/116/EEC in respect of fluid fertilizers
89/178/EEC Directive amending Directive 88/379/EEC relating to the classification, packaging and labelling of dangerous preparations
89/284/EEC Directive amending Directive 76/116/EEC in respect of the calcium, magnesium, sodium and sulphur content of fertilizers
89/530/EEC Directive amending Directive 76/116/EEC in respect of the trace elements contained in fertilizers
89/677/EEC Directive amending Directive 76/769/EEC relating to restrictions on the marketing and use of certain dangerous substances
89/684/EEC Directive on training for drivers of vehicles carrying dangerous goods
90/219/EEC Directive on the contained use of genetically modified micro-organisms
90/220/EEC Directive on the deliberate release into the environment of genetically modified organisms
90/394/EEC Directive on the protection of workers from the exposure from carcinogens at work
90/492/EEC Directive amending Directive 88/379/EEC relating to the classification, packaging and labelling of dangerous preparations.
90/642/EEC Directive on pesticide on certain products of plant origin
91/132/EEC Directive amending Directive 74/63/EEC on animal nutrition
91/155/EEC Directive defining the detailed arrangements for the system of specific information relating to dangerous preparations with respect to the implementation of 88/379/EEC
91/157/EEC Directive on batteries and accumulators
91/173/EEC Directive amending Directive 76/769/EEC on marketing and use of dangerous substances and preparations
91/188/EEC Directive amending 79/117/EEC on plant protection products
91/338/EEC Directive amending Directive 76/769/EEC on the marketing and use of dangerous substances
91/339/EEC Directive amending Directive 76/769/EEC on the marketing and use of dangerous substances
91/382/EEC Directive amending Directive 83/477/EEC on the protection of workers from exposure to asbestos at work
91/414/EEC Directive on plant protection products (marketing)
91/689/EEC Directive on hazardous waste
92/32/EEC Directive amending Directive 67/548/EEC
92/3/EURATOM Directive on the supervision and control of shipments of radioactive waste between Member States and into and out of the Community
93/18/EEC Directive amending Directive 88/379/EEC relating to the classification, packaging and labelling of dangerous preparations
93/69/EEC Directive adapting to technical progress Directive 76/116/EEC on the approximation of the laws relating to fertilizers
93/71/EEC Directive amending Directive 91/414/EEC on plant protection products
93/75/EEC Directive on minimum requirements of tankers in Community ports
93/86/EEC Directive adapting to technical progress 91/157/EEC on batteries
94/15/EC Directive adapting to technical progress Directive 90/220/EC on the deliberate release into the environment of genetically modified organisms
94/27/EC Directive amending 76/769/EEC on the marketing of dangerous substances
94/29/EC Directive amending 86/362/EEC and 86/363/EEC on pesticide residues
94/30/EC Directive amending 90/642/EEC on pesticide on products of plant origin
94/51/EC Directive adapting to technical progress Directive 90/219/EC on the contained use of genetically modified micro-organisms
94/63/EC Directive on the control of volatile organic compounds (VOC) emissions resulting from the storage of petrol and its distribution from terminals to petrol stations
94/48/EC Directive amending for the 13th time Directive 76/769/EEC
94/55/EC Directive on the transport of dangerous goods by road
94/60/EC Directive amending for the 14th time 76/769/EEC
94/69/EC Directive adapting to technical progress 67/548/EEC dangerous substances
94/79/EC Directive amending Directive 91/414/EEC on plant protection products
95/35/EC Directive amending Directive 91/414/EEC concerning the placing of plant protection products on the market
95/36/EC Directive amending Directive 91/414/EEC concerning the placing of plant protection products on the market
95/38/EC Directive amending Directive 90/642/EEC on pesticides in products of plant origin
95/39/EC Directive amending 86/362/EEC and 86/363/EEC on pesticides in foodstuffs of animal origin

95/50/EC	Directive on uniform procedures for checks on the transport of dangerous goods by road
95/61/EC	Directive amending Annex II to Directive 90/642/EEC relating to the fixing of maximum levels for pesticide residues in and on certain products of plant origin
95/543/EC	Directive on the request from the UK to exempt certain transport operations from the application 89/684/EEC
96/12/EC	Directive amending Directive 91/414/EEC concerning the placing of plant protection products on the market
96/28/EC	Directive on adaptation of laws relating to fertilizers
96/29/EURATOM	Directive laying down basic safety standards for the protection of workers and the general public against dangers arising from ionizing radiation
96/32/EC	Directive amending the Pesticide Residue Directive (60/642/EEC)
96/33/EC	Directive amending Directives 86/362/EEC and 86/363/EEC on pesticide residues
96/35/EC	Directive on appointing safety advisers for the transport of dangerous goods
96/46/EC	Directive amending Directive 91/414/EEC concerning the placing of plant protection products on the market
96/49/EC	Directive on the transportation of dangerous goods by rail
96/54/EC	Directive adapting to technical progress on the classification of dangerous substances Directive 67/548/EEC
96/55/EC	Directive adapting to technical progress Annex I to Directive 76/769/EEC
96/56/EC	Directive amending Directive 67/548/EEC on the approximation of the laws relating to the classification, packaging and labelling of dangerous substances
96/59/EC	Directive on the disposal of polychlorinated biphenyls and polychlorinated terphenyls (PCB/PCT)
96/65/EC	Directive on dangerous substances adapting the technical progress
96/82/EC	Directive on the control of major-accident hazards involving dangerous substances
96/86/EC	Directive adapting to technical progress Directive 94/55/EC on the approximation of the laws with regard to the transport of dangerous goods by road
96/94/EC	Directive establishing a value for Directive 80/1107/EEC on the protection of workers from the risks from exposure to chemical, physical and biological agents
97/10/EC	Directive adapting to technical progress Annex I to Directive 76/769/EEC on the approximation of the laws relating to restrictions on the marketing and use of certain dangerous substances and preparations
97/16/EC	Directive amending Directive 76/769/EEC on restrictions on the marketing and use of dangerous substances and preparations
97/35/EC	Directive adapting (for the second time) to technical progress Directive 90/220/EC on the deliberate release into the environment of genetically modified organisms
97/42/EC	Directive amending Directive 90/394/EEC on the protection of workers from exposure to carcinogens at work
97/56/EC	Directive amending for the 16th time Directive 76/769/EEC on dangerous substances and preparations
97/57/EC	Directive establishing Annex VI for Directive 91/414/EEC regarding plant protection products
97/64/EC	Directive amending Directive 79/769/EEC on dangerous substances lamps oils
97/69/EC	Directive on the laws, regulations and administrative provisions relating to the classification, packaging and labelling of dangerous substances
97/71/EC	Directive amending 86/362/EEC, 86/363/EEC and 90/642/EEC on the fixing of maximum levels for pesticide residues, etc.
97/865/EC	Directive on the possible inclusion of various substance into Annex 1 of Directive 91/414 on plant protection products
98/3/EC	Directive adapting to technical progress 76/116/EEC on the approximation of the laws relating to fertilizers
98/8/EC	Directive concerning the placing of biocidal products on the market
98/55/EC	Directive amending 93/75/EEC on the minimum requirements for vessels bound for or leaving Community ports, carrying dangerous or polluting goods
98/73/EC	Directive adapting to technical progress for the 24th time 67/548/EEC on dangerous substances
98/74/EC	Directive amending 93/75/EEC on the minimum requirements for vessels bound for or leaving Community ports, carrying dangerous or polluting goods
98/91/EC	Directive on trailers for the transport of dangerous goods
98/98/EC	Directive adapting to technical progress for the 25th time 67/548/EEC on dangerous substances
98/101/EC	Directive adapting to technical progress 91/157/EEC on batteries and accumulators containing dangerous substances
98/676/EC	Directive on the possible inclusion of various substances into Annex 1 of Directive 91/414 on plant protection products

99/1/EC Directive including the active substance kresoxim-methyl in Annex I of 91/414/EEC concerning plant protection products

99/13/EC Directive on limitation of emissions of volatile organic compounds (VOCs) due to the use of organic solvents in certain activities and installations

99/43/EC Directive amending for the 17th time Directive 76/769/EEC on the marketing and use of certain dangerous substances and preparations

99/45/EC Directive concerning the approximation of the laws, regulations and administrative provisions of the Member States relating to the classification, packaging and labelling of dangerous preparations

99/57/EC Directive adapting Annex I to Directive 76/769/EEC on the marketing and use of certain dangerous substances and preparations (tin, PCP and cadmium)

99/65/EC Directive amending Directives 86/362/EEC and 90/642/EEC on the maximum levels for pesticide residues in and on cereals and certain products of plant origin including fruit and vegetables respectively

99/73/EC Directive including spiroxamine in Annex I to Council Directive 91/414/EEC for the placing of plant protection products on the market

99/80/EC Directive including an active substance (azimsulfuron) in Annex I to Council Directive 91/414/EEC concerning the placing of plant protection products on the market

99/47/EC Directive adapting for the second time to technical progress Council Directive 94/55/EC on the transport of dangerous goods by road

99/48/EC Directive adapting for the second time to technical progress Directive 96/49/EC on the transport of dangerous goods by rail

99/53/EC Directive amending Annex III to Directive 77/93/EEC on protective measures against the introduction into the Community of organisms harmful to plants or plant products and against their spread within the Community

99/77/EC Directive adapting for the sixth time Annex I to Directive 76/769/EEC on the marketing and use of certain dangerous substances and preparations (asbestos)

2000/10/EC Directive on the inclusion of fluroxypyr in Annex 1 of Directive 91/414/EEC on plant protection products

2000/21/EC Directive concerning the list of legislation referred to in Directive 67/548/EEC on the classification, packaging and labelling of dangerous substances

2000/18/EC Directive on minimum examination requirements for safety advisers for the transport of dangerous goods by road, rail or inland waterway

2000/32/EC Directive adapting for the 26th time Directive 67/548/EEC on the classification, packaging and labelling of dangerous substances

2000/33/EC Directive adapting for the 27th time Directive 67/548/EEC on the classification, packaging and labelling of dangerous substances

Selected Decisions

94/785/EC Decision on UK exemption from 89/684/EEC on training of drivers

98/269/EC Decision on the withdrawal of Dinoterb.

98/270/EC Decision on the withdrawal of plant protection products containing Fenvalerate.

98/433/EC Decision regarding a harmonized criteria for dispensations from 96/82/EC on major accident hazards

98/685/EC Decision on the Convention on the Transboundary Effects of Industrial Accidents

99/164/EC Decision concerning the non-inclusion of DNOC in Annex I of Directive 91/414/EEC on plant protection products

99/237/EC Decision on the possible inclusion of oxasulfuron in Annex I to Council Directive 91/414/EEC concerning the placing of plant-protection products on the market

99/392/EC Decision recognizing the dossiers for the possible inclusion of mesotrione, iodosulfuron-methyl-sodium, silthiopham and gliocladium catenulatum in Annex I to Directive 91/414/EEC concerning the placing of plant-protection products on the market

99/462/EC Decision 99/462/EC on the view of the possible inclusion of alanycarbe in Annex I to Directive 91/414/EEC concerning the placing of plant-protection products on the market

99/610/EC Decision on the view of the possible inclusion of carvone in Annex I to Council Directive 91/414/EEC concerning the placing of plant-protection products on the market

99/636/EC Decision authorizing the Member States to permit temporarily the marketing of forest reproductive material not satisfying the requirements of Directive 66/404/EEC

2000/166/EC	Decision extending the possible time period for provisional authorizations of the new active substance quinoxyfen
2000/180/EC	Decision extending the possible time period for provisional authorizations of the new active substance pseudomonas chlororaphis
2000/181/EC	Decision on the view of the possible inclusion of thiacloprid, forchlorfenuron, thiamethoxam in Directive 91/414/EEC on the placing of plant-protection products on the market
2000/210/EC	Decision on the view of the possible inclusion of spinosad in Annex I to Council Directive 91/414/EEC concerning the placing of plant-protection products on the market
2000/233/EC	Decision on the withdrawal of authorizations for plant-protection products containing pyrazophos and non-inclusion in Directive 91/414/EEC on the placing of plant-protection products on the market
2000/234/EC	Decision on the withdrawal of authorizations of plant protection products containing monolinuron and non-inclusion in Directive 91/414/EC
2000/251/EC	Decision on the possible inclusion of RPA407213 (fenamidone) in Directive 91/414/EEC on the placing of plant-protection products on the market
2000/320/EC	Decision authorizing Member States to permit temporarily the marketing of forest reproductive material not satisfying the requirements of Council Directive 66/404/EEC
2000/358/EC	Decision extending time period for provisional authorizations of the new active substances flupyrsulfuron methyl, carfentrazone ethyl, prosulfuron, flurtamone, isoxaflutole
2000/390/EC	Decision on the inclusion of EXP60707B (acetamiprid) in Annex I to Council Directive 91/414/EEC concerning the placing of plant-protection products on the market
2000/412/EC	Decision on the possible inclusion of IKF 916 (cyazofamid) in Annex I to Council Directive 91/414/EEC concerning the placing of plant protection products on the market

Selected Proposed Legislation

COM(91)373	Proposal on PCBs
COM(93)349	Proposal concerning ionizing radiation
COM(94)573	Proposal for a Directive on the approximation of the laws with regard to the transport of dangerous goods by rail
COM(94)579	Proposal for a Regulation concerning the creation of a supplementary protection certificate for plant protection products
COM(95)272	Proposal to amend Directives 76/895/EEC, 86/362/EEC, 86/363/EEC and 90/642/EEC on the fixing of maximum pesticide residues in and on certain specified foodstuffs
COM(95)240	Proposal on major-accident hazards involving dangerous substances
COM(95)636	Proposal to amend Directive 67/546/EEC on dangerous substance classification
COM(96)347	Proposal on the classification, packaging and labelling of dangerous preparations
COM(96)513	Proposal for a Directive amending Directive 79/769/EEC on dangerous substances and preparations
COM(96)594	Proposal for a Directive amending Directives 76/116/EEC, 80/876/EEC, 89/284/EEC and 89/530/EEC on the laws relating to fertilizers
COM(97)214	Proposal to amend the Basel Convention
COM(97)330	Proposal on the transboundary effects of industrial accidents.
COM(97)738	Proposal amending 76/769/EEC on the laws relating to restrictions on the marketing and use of certain dangerous substances and preparations
COM(97)604	Proposal to amend 94/67/EC on incineration of hazardous waste
COM(98)174	Proposal for a Directive on the examination requirements for safety advisers for the transport of dangerous goods
COM(99)157	Proposal for a Directive amending Directive 96/49/EC on the transport of dangerous goods by rail
COM(99)158	Proposal for a Directive amending Directive 94/55/EC on the transport of dangerous goods by road

Other documents

OJ: C361/3/94	Communication pursuant to 85/71/EEC and 67/548/EEC
96/199/EC	Recommendation on the 1996 pesticide residue monitoring programme
OJ: C22/10/98	On reference numbers for the notification of the export of certain dangerous chemicals

OJ: L294/1/98	An Annex to Council Directive 96/49/EC with regard to the transport of dangerous goods by rail
99/829/Euratom	Recommendation on the application of Article 37 of the Euratom Treaty

UK legislation on air pollution

Selected Acts of Parliament

Health and Safety Act 1974 (Part 1)
Highway Act 1980 Sections 161 and 161A
Environmental Protection Act 1990
Clean Air Act 1993
Environment Act 1995

Selected Statutory Instruments

SI 1958/498	Dark Smoke (Permitted Periods) Regulations
SI 1969/411	Clean Air (Height of Chimneys) (Exemption) Regulations
SI 1969/1262	Clean Air (Arrestment Plant) (Exemption) Regulations
SI 1969/1263	Clean Air (Emissions of Dark Smoke) (Exemption) Regulations
SI 1971/161	Clean Air (Measurement of Grit and Dust from Furnaces) Regulations
SI 1971/162	Clean Air (Emission of Grit and Dust from Furnaces) Regulations
SI 1977/17	Control of Atmosphere Pollution (Appeals) Regulations
SI 1977/18	Control of Atmospheric Pollution (Exempted Premises) Regulations
SI 1983/943	Health and Safety (Emissions into the Atmosphere) Regulations
SI 1989/317	Air Quality Standards Regulations
SI 1989/319	Health and Safety (Emissions into the Atmosphere) (Amendment) Regulations
SI 1990/556	Control of Asbestos in Air Regulations
SI 1991/472	Environmental Protection (Prescribed Processes and Substances) Regulations
SI 1991/507	Environmental Protection (Application, Appeals and Registers) Regulations
SI 1991/513	Environmental Protection (Authorization of Processes) (Determination Periods) Order
SI 1991/1282	Smoke Control Areas (Authorized Fuels) Regulations
SI 1992/72	Smoke Control Areas (Authorized Fuels) (Amendment) Regulations
SI 1992/2225	Notification of Cooling Towers and Evaporative Condensers Regulations
SI 1993/2499	Smoke Control Areas (Authorized Fuels) (Amendment) Regulations
SI 1994/199	Environmental Protection (Non-Refillable Refrigerant Containers) Regulations
SI 1994/440	Ozone Monitoring and Information Regulations
SI 1995/2644	Statutory Nuisance (Appeals) Regulations
SI 1995/3146	Air Quality Standards (Amendment) Regulations
SI 1996/506	Environmental Protection (Controls on Substances that Deplete the Ozone Layer) Regulations
SI 1996/1145	Smoke Control Areas (Authorized Fuels) (Amendment) Regulations
SI 1996/2678	The Environmental Protection (Prescribed Processes and Substances, etc.) (Amendment) (Petrol Vapour Recovery) Regulations
SI 1997/2658	Smoke Control Areas (Authorized Fuels) (Amendment) Regulations
SI 1997/3043	Air Quality Regulations
SI 1998/767	Environmental Protection (Prescribed Processes and Substances) (Amendment) (Hazardous Waste Incineration) Regulations
SI 1998/2154	Smoke Control Areas (Authorized Fuels) (Amendment) Regulations
SI 1998/3096	Smoke Control Areas (Authorized Fuels) (Amendment No. 2) Regulations
SI 1999/3107	Motor Fuel (Composition and Content) Regulations
SI 2000/928	The Air Quality (England) Regulations
SI 2000/1077	Smoke Control Areas (Authorized Fuels) (Amendment) (England) Regulations
SI 2000/1460	The Sulphur Content of Liquid Fuels (England and Wales) Regulations

European legislation on air pollution

Selected regulations

3093/94	Regulation replacing Regulation 594/91 on ozone depleting substances

Selected directives

77/312/EEC	Directive on the biological screening of the population for lead
80/779/EEC	Directive on air quality standards – sulphur dioxide
82/884/EEC	Directive on air quality standards – lead concentrations
84/360/EEC	Directive on air pollution from industrial plants
85/203/EEC	Directive on air quality standards – nitrogen dioxide
85/210/EEC	Directive on lead content of petrol
87/217/EEC	Directive on pollution by asbestos
88/609/EEC	Directive on large combustion plants
89/369/EEC	Directive on new municipal waste incineration plants
89/429/EEC	Directive on existing municipal waste incineration plants
92/55/EEC	Directive on motor vehicle exhaust emission tests
92/73/EEC	Directive on air pollution by ozone
93/12/EEC	Directive on sulphur content of liquid fuels
93/389/EEC	Council decision on a monitoring mechanism of community CO_2 and other greenhouse gas emissions
94/63/EC	Directive on the control of volatile organic compounds (VOCs) emissions resulting from the storage of petrol and its distribution from terminals to petrol stations
94/66/EC	Directive amending Directive 88/609/EEC on the limitation of emissions of certain pollutants into the air from large combustion plants
94/67/EC	Directive on the incineration of hazardous waste
96/61/EC	Directive concerning integrated pollution prevention and control
96/62/EC	Directive on ambient air quality assessment and management
98/70/EC	Directive amending 93/12/EC on the sulphur content of liquid fuels
99/13/EC	Directive on limitation of emissions of volatile organic compounds (VOCs) due to the use of organic solvents in certain activities and installations
99/30/EC	Directive relating to limit values for sulphur dioxide, nitrogen dioxide and oxides of nitrogen, particulate matter and lead in ambient air
99/32/EC	Directive relating to a reduction in the sulphur content of certain liquid fuels and amending Directive 93/12/EEC

Selected decisions

82/459/EEC	Decision on air pollution monitoring
88/540/EEC	Decision on Montreal Protocol and the Vienna Convention
93/389/EEC	Decision on a monitoring mechanism of Community greenhouse gas emissions
94/69/EEC	Decision on the UN Convention on Climate Change
94/563/EC	Decision on the quantities of CFCs allowed for essential uses
95/107/EC	Decision allocating production and import quotas for methyl bromide, import quotas for hydrobromofluorocarbons and consumption quotas for hydrochlorofluorocarbons for the period 1 January to 31 December 1995
95/324/EC	Decision on the quantities of controlled substances allowed for essential uses in the Community in 1996 under Regulation 3093/94 on substances that deplete the ozone layer
95/555/EC	Decision on companies authorized to use controlled substances for essential uses in 1996
97/414/EC	Decision on the essential uses of CFCs for 1997
97/461/EC	Decision on CFCs import quotas for 1997
98/67/EC	Decision on controlled substances allowed for essential uses in 1998 under Regulation 3093/94 on substances that deplete the ozone layer
98/27/EC	Decision on import quotas for chlorofluorocarbons (CFCs) for 1998

98/184/EC — Decision concerning a questionnaire for Member States' reports on the implementation of Directive 94/67/EC on the incineration of hazardous waste

98/686/EC — Decision on the Protocol to the 1979 Convention on Transboundary Air Pollution on Further Reduction of Sulphur Emissions

99/58/EC — Decision allocating the 1999 import quotas of substances ozone depleting substances

99/59/EC — Decision on controlled substances allowed for essential uses in 1999 under Regulation 3093/94 on substances that deplete the ozone layer

99/296/EC — Decision amending Decision 93/389/EEC for a monitoring mechanism of Community CO_2 and other greenhouse gas emissions

2000/22/EC — Decision on the allocation of quantities of controlled substances allowed for essential uses in the Community in 2000 under Council Regulation (EC) No. 3093/94 on substances that deplete the ozone layer

2000/379/EC — Decision on import quotas for halogenated chlorofluorocarbons, halons, carbon tetrachloride, 1,1,1-trichloroethane, hydrobromofluorocarbons and methyl bromide and allocating placing on the market quotas for hydrochlorofluorocarbons for the period 1 January to 31 December 2000

Selected proposed legislation

COM(91)268 — Proposal on the Geneva Convention on Long Range Air Pollution

COM(94)109 — Proposal on ambient air quality assessment and management

COM(97)604 — Proposal to amend 94/67/EC on incineration of hazardous waste

COM(98)398 — Proposal for a regulation on substances which deplete the ozone layer

COM(98)415 — Proposal to amend 88/609/EEC on the limitation of emissions of certain pollutants into the air from large combustion plants

COM(98)558 — Proposal on waste incineration

COM(98)591 — Proposal for a Directive on limit values for benzene and carbon monoxide in ambient air

COM(99)67 — Amended proposal for a Regulation on substances which deplete the ozone layer

UK water legislation

Selected acts of parliament

Public Health Act 1936

Water Act 1945

Rivers (Prevention of Pollution) Act 1951

Countryside Act 1968

Control of Pollution Act 1974

Salmon and Freshwater Fisheries Act 1975

Environmental Protection Act 1990

Land Drainage Act 1991

Water Industry Act 1991

Water Resources Act 1991

Environment Act 1995

Water Industry Act 1999

Selected statutory instruments

SI 1976/958 — Control of Pollution (Discharges to Sewers) Regulations

SI 1978/564 — Detergents (Composition) Regulations

SI 1983/1182 — Control of Pollution (Exemption of Certain Discharges from Control) Order

SI 1984/864 — The Control of Pollution (Consents for Discharges) (Notices) Regulations

SI 1986/1623 — Control of Pollution (Exemption of Certain Discharges from Control) (Variations) Order

SI 1987/1782 — The Control of Pollution (Exemption of Certain Discharges from Control) (Variation) Order

SI 1989/1147 — Water Supply (Water Quality) Regulations

SI 1989/1149	Controlled Water (Lakes and Ponds) Order
SI 1989/1152	Water and Sewerage (Conservation Access and Recreation) (Code of Practice) Order
SI 1989/1156	Trade Effluent (Prescribed Processes and Substances) Regulations
SI 1989/1378	The River (Prevention of Pollution) Act 1951 (Continuation of Bylaws) Order
SI 1989/1384	Water Supply (Water Quality) (Amendment) Regulations
SI 1989/1968	Water (Consequential Amendments) Regulations
SI 1989/2286	Surface Waters (Dangerous Substances) (Classification) Regulations
SI 1990/1629	Trade Effluents (Prescribed Processes and Substances) (Amendment) Regulations
SI 1991/324	Control of Pollution (Silage, Slurry and Agricultural Fuel Oil) Regulations
SI 1991/472	Environmental Protection (Prescribed Processes and Substances) Regulations
SI 1991/507	Environmental Protection (Applications, Appeals and Registers) Regulations
SI 1991/1597	Bathing Waters (Classification) Regulations
SI 1991/1837	Water Supply (Water Quality) (Amendment) Regulations
SI 1991/2790	Private Water Supplies Regulations
SI 1992/337	Surface Waters (Dangerous Substances) (Classification) Regulations
SI 1992/339	Trade Effluent (Prescribed Processes and Substances) Regulations
SI 1994/1057	Surface Waters (River Ecosystem) (Classification) Regulations
SI 1994/1729	Nitrate Sensitive Areas Regulations
SI 1994/2841	Urban Waste Water Treatment (England and Wales) Regulations
SI 1995/1708	The Nitrate Sensitive Areas (Amendment) Regulations
SI 1995/2095	The Nitrate Sensitive Areas (Amendment) (No. 2) Regulations
SI 1996/888	Protection of Water Against Agricultural Nitrate Pollution (England and Wales) Regulations
SI 1996/2044	The Control of Pollution (Silage, Slurry and Agricultural Fuel Oil) (Amendment) Regulations
SI 1996/2075	Health and Safety at Work etc. Act 1974 (Application to Environmentally Hazardous Substances) Regulations
SI 1996/2971	Control of Pollution (Applications, Appeals and Registers) Regulations
SI 1996/3001	Surface Waters (Abstraction for Drinking Water) (Classification) Regulations
SI 1996/3105	The Nitrate Sensitive Areas (Amendment) Regulations
SI 1997/547	The Control of Pollution (Silage, Slurry and Agricultural Fuel Oil) (Amendment) Regulations
SI 1997/990	The Nitrate Sensitive Areas (Amendment) Regulations
SI 1997/1331	Surface Waters (Fishlife) (Classification) Regulations
SI 1997/1332	Surface Waters (Shellfish) (Classification) Regulations
SI 1997/2560	The Surface Waters (Dangerous Substances) (Classification) Regulations
SI 1998/314	Food Protection (Emergency Prohibitions) (Oil and Chemical Pollution of Fish) Order 1997 (Revocation) Order
SI 1998/389	Surface Water (Dangerous Substances) (Classification) Regulations
SI 1998/79	Nitrate Sensitive Areas (Amendments) Regulations
SI 1998/892	The Mines (Notice of Abandonment) Regulations
SI 1998/1202	Action Programme for Nitrate Vulnerable Zones (England and Wales) Regulations
SI 1998/1649	Control of Pollution (Channel Tunnel Rail Link) Regulations
SI 1998/2138	Nitrate Sensitive Areas (Amendments) (No. 2) Regulations
SI 1998/2746	The Groundwater Regulations
SI 1998/3084	Water (Prevention of Pollution) (Code of Practice) Order
SI 1999/1006	Anti-Pollution Works Regulations
SI 1999/1148	Water Supply (Water Fittings) Regulations
SI 1999/1506	Water Supply (Water Fittings) (Amendment) Regulations
SI 1999/1524	The Water Supply (Water Quality) (Amendment) Regulations
SI 1999/3440	Water Industry Act 1999 (Commencement No. 2) Order
SI 1999/3441	Water Industry (Charges) (Vulnerable Groups) Regulations
SI 1999/3442	Water Industry (Prescribed Conditions) Regulations
SI 2000/477	Water and Sewerage (Conservation, Access and Recreation) (Code of Practice) Order
SI 2000/519	Water Industry (Charges) (Vulnerable Groups) (Amendment) Regulations
SI 2000/1297	The Drinking Water (Undertakings) (England and Wales) Regulations
SI 2000/1842	Water and Sewerage Undertakers (Inset Appointments) Regulations

European legislation on water pollution

Selected regulations

No Regulations in this section

Selected directives

73/404/EEC	Directive on detergents
73/405/EEC	Directive on methods of testing the biodegradability of anionic surfactants
75/439/EEC	Directive on waste oils
75/440/EEC	Directive on surface water quality
76/160/EEC	Directive on bathing water quality
76/464/EEC	Directive on the discharge of dangerous substances (Framework Directive)
77/795/EEC	Directive on information on water pollution
78/176/EEC	Directive on titanium dioxide waste
78/659/EEC	Directive on water standards for freshwater fish
79/115/EEC	Directive on pilotage of sea vessels
79/869/EEC	Directive concerning the sampling and analysis of drinking water
79/923/EEC	Directive on water standards for shellfish waters
80/68/EEC	Directive on groundwater
80/778/EEC	Directive on drinking water quality
82/176/EEC	Directive on mercury (chlor-alkali electrolysis industry)
82/242/EEC	Directive on the manufacture and marketing of detergents
82/883/EEC	Directive on monitoring of environments concerned by waste from the titanium dioxide industry
83/513/EEC	Directive on cadmium discharges
84/156/EEC	Directive on mercury from non-chlor-alkali electrolysis industry
84/491/EEC	Directive on hexachlorocyclohexane discharges
86/278/EEC	Directive on sewage sludge used in agriculture
86/280/EEC	Directive on limits for certain dangerous substances included in 76/464/EEC
87/217/EEC	Directive on pollution by asbestos
88/347/EEC	Directive amending Directive 86/280/EEC on certain dangerous substances in 76/464/EEC
90/415/EEC	Directive amending 86/280/EEC on certain dangerous substances included in 76/464/EEC
91/271/EEC	Directive on urban waste water treatment
91/676/EEC	Directive on nitrates from agricultural sources
92/112/EEC	Directive on the reduction and eventual elimination of pollution caused from the titanium dioxide industry
93/75/EEC	Directive on minimum requirements of tankers in Community ports
94/57/EC	Directive on ship inspection maritime administrations
94/58/EC	Directive on the minimum level of training for seafarers
95/21/EC	Directive concerning the enforcement of international standards for ship safety, pollution prevention and shipboard living and working conditions
96/61/EC	Directive concerning integrated pollution prevention and control
98/15/EC	Directive amending 91/271/EEC on urban waste water treatment
98/25/EC	Directive amending 95/21/EC on the enforcement of shipping using community ports, etc.
98/35/EC	Directive amending 94/58/EC on the minimum level of training for seafarers
98/42/EC	Directive amending 95/21/EC on the enforcement of shipping using Community ports
98/55/EC	Directive amending 93/75/EEC on the minimum requirements for vessels bound for or leaving Community ports, carrying dangerous or polluting goods
98/74/EC	Directive amending 93/75/EEC on the minimum requirements for vessels bound for or leaving Community ports, carrying dangerous or polluting goods
98/83/EC	Directive on the quality of water intended for human consumption
98/249/EC	Directive on the conclusion of the convention for the protection of the north east Atlantic
99/97/EC	Directive amending Directive 95/21/EC concerning the enforcement of international standards for ship safety, pollution prevention and shipboard living and working conditions

Selected decisions

95/308/EC	Decision on the protection and use of transboundary watercourses and international lakes

Selected proposed legislation

COM(93)680	Proposal on the ecological quality of water
COM(94)36	Proposal on bathing water
COM(97)49	Proposal for a Directive establishing a framework for Community action in the field of water policy
COM(97)416	Proposal for a Directive amending Directive 95/21/EC on ship safety and pollution prevention
COM(97)585	Amended proposal for a Directive on the quality of bathing water
COM(98)76	Amended proposal for a Directive establishing a framework for action on water policy
COM(98)769	Proposal setting up a Community framework for co-operation with regard to accidental marine pollution

Selected other legislation

79/114/EEC	Recommendation on minimum requirements for tankers
OJ: C162/1/78	Resolution on hydrocarbons at sea
98/480/EC	Recommendation on good environmental practice for household laundry detergents

UK legislation on land pollution

Selected acts

Environmental Protection Act 1990
Environment Act 1995

Selected statutory instruments

SI 1980/1709	Control of Pollution (Special Waste) Regulations 1980
SI 1988/1199	Town and Country Planning, England and Wales – The Town and Country Planning (Assessment of Environmental Effects)
SI 1990/1629	Water, England and Wales – The Trade Effluents (Prescribed Processes and Substances)(Amendment) Regulations 1990
SI 1990/1187	Water, England and Wales – Agriculture – The Nitrate Sensitive Areas (Designation) (Amendment) Order 1990
SI 1994/2286	Agriculture – Common Agricultural Policy – The Organic Products (Amendment) Regulations 1994
SI 1996/972	Environmental Protection – The Special Waste Regulations 1996
SI 1997/2778	Northern Ireland – The Waste and Contaminated Land (Northern Ireland) Order 1997
SI 1999/2977	Health and Safety – The Asbestos (Prohibitions)(Amendment)(No. 2) Regulations 1999
SI 2000/1973	Environmental Protection, England and Wales – The Pollution Prevention and Contrl (England and Wales) Regulations 2000
SI 2000/179	European Communities – Town and Country Planning – The Planning (Control of Major – Accident Hazards)(Scotland) Regulations 2000

Health and Safety Executive publications

Selected Codes of Practice

COP 2	Control of lead at work. Approved Code of Practice
COP 6	Plastic containers with nominal capacities up to 5 litres for petroleum spirit. Requirements for testing and marking or labelling
COP 14	Road tanker testing: Examination, testing and certification of the carrying tanks of road tankers and of tank containers used for the conveyance of dangerous substances by road
COP 20	Standards of training in safe gas installations. Approved Code of Practice
COP 23	Approved Code of Practice COP 23–Part 3: Exposure to radon. The Ionizing Radiations Regulations 1985 (Superseded by L 121: 2000)
COP 41	Control of substances hazardous to health in the production of pottery. The Control of Substances Hazardous to Health Regulations 1988. The Control of Lead at Work Regulations 1980 (superseded by COP 60)

HSE guidance notes

Agriculture

AIS 2	Zoonoses in Agriculture. Preventing the spread of disease to livestock handlers
AIS 3	Controlling Grain Dust on Farms
AIS 9	Preventing Access to Effluent Storage and Similar Areas on Farms
AIS 16	Guidance on Storing Pesticides for Farmers and other Professional Users
AIS 31	Safe Use of Rodenticides on Farms and Holdings
AS 5	Farmer's Lung
AS 25	Training in the Use of Pesticides
AS 26	Protective Clothing for Use with Pesticides
AS 27	Agricultural pesticides
AS 28	COSHH in Agriculture. Control of Substances Hazardous to Health Regulations 1988 (COSHH)
AS 29	Sheep Dipping
AS 30	COSHH in Forestry
AS 31	Veterinary Medicines – Safe Use by Farmers and Other Animal Handlers

Best practicable means

BPM 1	Best practicable means: general principles and practice
BPM 2	Mineral works (sand driers and coolers)
BPM 3	Petroleum works (PVC polymer plants)
BPM 4	Hydrofluoric acid works (HF manufacture)
BPM 5	Hydrochloric acid works
BPM 6	Mineral works (sintered aggregates)
BPM 7	Metal recovery works
BPM 8	Mineral works (plaster)
BPM 9	Aluminium (secondary) works
BPM 10	Amines works
BPM 11	Chemical incineration works
BPM 12	Iron works and steel works. Electric arc furnaces and associated processes
BPM 13	Mineral works (roadstone plants)
BPM 14	Cement production works
BPM 15	Mineral works (expansion of perlite)
BPM 16	Lead works
BPM 17	Sulphuric acid (Class II) works. Sulphuric acid manufacture
BPM 18	Chlorine works

BPM 20 Iron works and steel works. Basic oxygen steel plants
BPM 21 Iron works and steel works. Blast furnaces
BPM 22 Iron work and steel works. Machine scarfing
BPM 23 Iron works and steel works. Ore preparation

Chemical

CS 1 Industrial use of flammable gas detectors
CS 2 Storage of highly flammable liquids
CS 3 Storage and use of sodium perchlorate and other similar strong oxidants
CS 4 Keeping of LPG in cylinders and similar containers
CS 5 Storage and use of LPG at fixed installations
CS 6 Storage and use of LPG on construction sites
CS 7 Odorization of bulk oxygen supplies in shipyards
CS 8 Small scale storage and display of LPG at retail premises
CS 9 Bulk storage and use of liquid carbon dioxide: hazards and precautions
CS 10 Fumigation using methyl phosphine
CS 11 Storage and use of LPG at metered estates
CS 12 Fumigation using methyl bromide
CS 15 The cleaning and gas freeing of tanks containing flammable residues
CS 16 Chlorine vaporizers
CS 18 Storage and handling of ammonium nitrate
CS 19 Storage of approved pesticides: guidance for farmers and other professional users
CS 20 Sulphuric acid use in agriculture
CS 21 Storage and handling of organic peroxides
CS 22 Fumigation
CS 23 Disposal of waste explosives
CS 24 The interpretation and use of flashpoint information

Environmental hygiene

EH 1 Cadmium: health and safety precautions
EH 2 Chromium and its inorganic compounds: health and safety precautions
EH 4 Aniline: health and safety precautions
EH 6 Chromic acid concentrations in air (superseded by MDHS 52/3 (1998))
EH 7 Petroleum based adhesives in building operations
EH 8 Arsenic: toxic hazards and precautions
EH 9 Spraying of highly flammable liquids
EH 10 Asbestos: exposure limits and measurement of airborne dust concentrations
EH 11 Arsine: health and safety precautions
EH 12 Stibine: health and safety precautions
EH 13 Beryllium: health and safety precautions
EH 16 Isocyanates: toxic hazards and precautionary measures
EH 17 Mercury and its inorganic divalent compounds
EH 19 Antimony and its compounds: health hazards and safety precautions
EH 20 Phosphine: health and safety precautions
EH 21 Carbon dust: health and safety precautions
EH 22 Ventilation of the workplace
EH 23 Anthrax: health hazards
EH 24 Dust and accidents in malthouses
EH 25 Cotton dust sampling
EH 26 Occupational skin diseases: health and safety precautions
EH 27 Acrylonitrile: personal protective equipment
EH 28 Control of lead: air sampling techniques and strategies
EH 29 Control of lead: outside workers
EH 31 Control of exposure to polyvinyl chloride dust
EH 33 Atmospheric pollution in car parks

EH 34 Benzidine-based dyes: health and safety precautions
EH 35 Probable asbestos dust concentrations at construction processes. Revised 1989 (superseded by EH 71)
EH 36 Cement (superseded by EH 71)
EH 37 Work with asbestos insulating board (superseded by EH 71)
EH 38 Ozone: health hazards and precautionary measures
EH 40 Occupational exposure limits (annual)
EH 41 Respiratory protective equipment for use against asbestos
EH 42 Monitoring strategies for toxic substances
EH 43 Carbon monoxide
EH 44 Dust: general principles of protection
EH 45 Carbon disulphide: control of exposure in the viscose industry
EH 46 Man-made mineral fibres
EH 47 Provision, use and maintenance of hygiene facilities for work with asbestos insulation and coatings
EH 49 Nitrosamines in synthetic metal cutting and grinding fluids (withdrawn)
EH 50 Training operatives and supervisors for work with asbestos insulation and coatings
EH 51 Enclosures provided for work with asbestos insulation, coatings and insulating board
EH 52 Removal techniques and associated waste handling for asbestos insulation, coatings and insulating board
EH 53 Respiratory protective equipment for use against airborne radioactivity
EH 54 Assessment of exposure to fume from welding and allied processes
EH 55 The control of exposure to fume from welding, brazing and similar processes
EH 56 Biological monitoring for chemical exposures in the workplace
EH 57 The problems of asbestos removal at high temperatures
EH 58 Carcinogenicity of mineral oils
EH 59 Respirable crystalline silica
EH 60 Nickel and its inorganic compounds: health hazards and precautionary measures
EH 62 Metalworking fluids: health precautions
EH 63 Vinyl chloride: toxic hazards and precautions
EH 64 Summary criteria for occupational exposure limits 1996 (with updates/changes 1997, 1998, 1999)
EH 65/1 Trimethylbenzenes: criteria document for an OEL
EH 65/2 Pulverized fuel ash: criteria document for an OEL
EH 65/3 N,N-dimethylacetamide: criteria document for an OEL
EH 65/4 1,2-dichloroethane: criteria document for an OEL
EH 65/5 4,4'-methylene dianiline: criteria document for an OEL
EH 65/6 Epichlorohydrin: criteria document for an OEL
EH 65/7 Chlorodifluoromethane: criteria document for an OEL
EH 65/8 Cumene: criteria document for an OEL
EH 65/9 1,4-dichlorobenzene: criteria document for an OEL
EH 65/10 Carbon tetrachloride: criteria document for an OEL
EH 65/11 Chloroform: criteria document for an OEL
EH 65/12 Portland cement dust: criteria document for an OEL
EH 65/13 Kaolin: criteria document for an OEL
EH 65/14 Paracetamol: EH 65/13 criteria document for an OEL
EH 65/15 1,1,1,2-Tetrafluoroethane (HFC 134a): criteria document for an OEL
EH 65/16 Methyl methacrylate: criteria document for an OEL
EH 65/17 p-Aramid respirable fibres: criteria document for an OEL
EH 65/18 Propranolol: criteria document for an OEL
EH 65/19 Mercury and its inorganic divalent compounds: criteria document for an OEL
EH 65/20 Ortho-toluidine: criteria document for an OEL
EH 65/21 Propylene oxide: criteria document for an OEL
EH 65/22 Softwood dust: criteria document for an OEL
EH 65/23 Antimony and its compounds: criteria document for an OEL
EH 65/24 Platinum metal and soluble platinum salts: criteria document for an OEL
EH 65/25 Iodomethane: criteria document for an OEL
EH 65/26 Azodicarbonamide: criteria document for an OEL
EH 65/27 Dimethyl and diethyl sulphates: criteria document for an OEL
EH 65/28 Hydrazine: criteria document for an OEL
EH 65/29 Acid anhydrides: criteria document for an OEL

EH 65/30 Review of fibre toxicology
EH 65/31 Rosin-based solder flux fume: criteria document for an OEL
EH 65/32 Glutaraldehyde
EH 66 Grain dust
EH 67 Grain dust in maltings (maximum exposure limits)
EH 68 Cobalt: health and safety precautions
EH 69 How to handle PCBs without harming yourself or the environment
EH 70 The control of fire-water run-off from CIMAH sites to prevent environmental damage
EH 71 Working with asbestos cement and asbestos insulating board
EH 72/1 Phenylhydrazine: risk assessment document
EH 72/2 Dimethylaminoethanol: risk assessment document
EH 72/3 Bromoethane: risk assessment document
EH 72/4 3-Chloropropene: risk assessment document
EH 72/5 Alpha-chlorotoluene: risk assessment document
EH 72/6 2-Furaldehyde: risk assessment document
EH 72/7 1,2-Diaminoethane (ethylenediamine (EDA)): risk assessment document
EH 72/8 Aniline: risk assessment document
EH 72/9 Barium sulphate: risk assessment document
EH 72/10 N-Methyl-2-pyrrolidone: risk assessment document
EH 72/11 Flour dust: risk assessment document
EH 73 Arsenic and its compounds: health hazards and precautionary measures
EH 74/1 Dichloromethane: exposure assessment document
EH 74/2 Respirable crystalline silica: exposure assessment document
EH 74/3 Dermal exposure to non-agricultural pesticides: exposure assessment document
EH 75/1 Medium density fibreboard (MDF)

General

GS 3 Fire risk in the storage and industrial use of cellular plastics
GS 4 Safety in pressure testing (withdrawn, superseded by 1998 edition)
GS 5 Entry into confined spaces
GS 12 Effluent storage on farms
GS 16 Gaseous fire extinguishing systems: precautions for toxic and asphyxiating hazards (withdrawn)
GS 19 General fire precautions aboard ships being fitted out or under repair
GS 20 Fire precautions in pressurized workings
GS 26 Access to road tankers
GS 40 The loading and unloading of bulk flammable liquids and gases at harbours and inland waterways
GS 43 Lithium batteries
GS 46 *In-situ* timber treatment using timber preservatives

Health and safety

HSE 4 Employers' Liability (Compulsory Insurance) Act 1969. A short guide (superseded by HSE 36: 1998)
HSE 5 The Employment Medical Advisory Service and you
HSE 8 Fires and Explosions due to the misuse of oxygen
HSE 11 Reporting an Injury or a Dangerous Occurrence. (RIDDOR) The Reporting of Injuries, Disease and Dangerous Occurrences Regulations 1985
HSE 17 Reporting a Case of Disease. (RIDDOR) The Reporting of Injuries, Diseases and Dangerous Occurrences Regulations 1985
HSE 23 Health and safety legislation and trainees (a guide for employers)
HSE 25 The Health and Safety Executive and you
HSE 26 The Health and Safety Executive working with employers
HSE 31 RIDDOR explained – Reporting of Injuries, Diseases and Dangerous Occurences Regulations
HSE 33 RIDDOR 95 – Offshore
HSE 36 Employers' Liability (Compulsory Insurance) Act 1969 – A guide for employees and their representatives (supersedes HSE 4: 1990)

HS(G)1	Safe use and storage of flexible polyurethane foam in industry
HS(G)3	Highly flammable materials on construction sites
HS(G)4	Highly flammable liquids in the paint industry
HS(G)5	Hot work: welding and cutting on plant containing flammable materials
HS(G)7	Container terminals: safe working practices
HS(G)10	Cloakroom accommodation and washing facilities
HS(G)11	Flame arresters and explosion reliefs
HS(G)13	Electrical testing: safety in chemical testing
HS(G)14	Opening processes: cotton and allied fibres
HS(G)16	Evaporating and other ovens
HS(G)20	Guidelines for occupational health services
HS(G)21	Safety in the cotton and allied fibres industry: cardroom processes
HS(G)22	Electrical apparatus for use in potentially explosive atmospheres
HS(G)25	The Control of Industrial Major Accident Hazards Regulations 1984 (CIMAH): further guidance on emergency plans
HS(G)26	Transport of dangerous substances in tank containers
HS(G)27	Substances for use at work: the provision of information. Revised 1989 (superseded by L 62)
HS(G)28	Safety advice for bulk chlorine installations
HS(G)30	Storage of anhydrous ammonia under pressure in the United Kingdom
HS(G)34	Storage of LPG at fixed installations
HS(G)36	Disposal of explosives waste and the decontamination of explosives plant
HS(G)37	An introduction to local exhaust ventilation
HS(G)39	Compressed air safety
HSG40	Safe handling of chlorine from drums and cylinders
HS(G)41	Petrol filling stations: construction and operation
HS(G)49	The examination and testing of portable radiation instruments for external radiations
HS(G)50	The storage of flammable liquids in fixed tanks (up to 10 000 m^3 total capacity)
HS(G)51	The storage of flammable liquids in containers
HS(G)52	The storage of flammable liquids in fixed tanks (exceeding 10 000 m^3 total capacity)
HS(G)53	Respiratory protective equipment. A practical guide for users
HSG 54	Maintenance, examination and testing of local exhaust ventilation
HS(G)61	Health surveillance at work
HS(G)62	Health and safety in tyre and exhaust fitting premises
HS(G)63	Radiation protection offsite for emergency services in the event of a nuclear accident
HS(G)64	Assessment of fire hazards from solid materials and the precautions required for their safe storage and use
HSG65	Successful health and safety management
HS(G)66	Protection of workers and the general public during development of contaminated land
HS(G)70	The control of legionellosis including legionnaires' disease
HS(G)71	Storage of packaged dangerous substances
HSG 71	Chemical warehousing – storage of packaged dangerous substances
HS(G)72	Control of respirable silica dust in heavy clay and refractory processes
HS(G)73	Control of respirable crystalline silica in quarries
HS(G)74	Control of silica dust in foundries
HS(G)77	COSHH and peripatetic workers (withdrawn)
HSG 78	Dangerous goods in cargo transport units – packing and carriage for transport by sea
HS(G)86	Veterinary medicines: safe use by farmers and other animal handlers
HS(G)91	A framework for the restriction of occupational exposure to ionizing radiation
HS(G)92	Safe use and storage of cellular plastics
HS(G)93	Assessment of pressure vessels operating at low temperature
HS(G)94	Safety in the design and use of gamma and electron irradiation facilities
HS(G)97	A step by step guide to COSHH assessment
HS(G)103	Safe handling of combustible dusts: precautions against explosions
HS(G)108	CHIP for everyone
HS(G)110	Seven steps to successful substitution of hazardous substances
HS(G)113	Lift trucks in potentially flammable atmospheres
HS(G)114	Conditions for the authorization of explosives in Great Britain
HS(G)117	Making sense of NONS – a guide to the Notification of New Substances Regulations 1993
HS(G)122	New and expectant mothers at work: a guide for employees

HS(G)123	Working together on firework displays – a guide to safety for firework display organizers and operators
HS(G)124	Giving your own firework display – how to run and fire it safely
HS(G)125	A brief guide on COSHH for the offshore oil and gas industry
HS(G)126	CHIP 2 for everyone
HS(G)129	Health and safety in engineering workshops
HS(G)131	Energetic and spontaneously combustible substances: identification and safe handling
HS(G)132	How to deal with sick building syndrome – guidance for employers, building owners and building managers
HS(G)135	Storage and handling of industrial nitrocellulose
HS(G)137	Health risk management: a practical guide for managers in small and medium-sized enterprises
HS(G)139	The safe use of compressed gases in welding, flame cutting and allied processes
HS(G)140	The safe use and handling of flammable liquids
HS(G)143	Designing and operating safe chemical reaction processes
HS(G)146	Dispensing petrol – assessing and controlling the risk of fire and explosion at sites where petrol is stored and dispensed as a fuel
HS(G)158	Flame arresters – preventing the spread of fires and explosions in equipment that contains flammable gases and vapours
HS(G)160	The carriage of dangerous goods explained – Part 1: Guidance for consignors of dangerous goods by road and rail (classification, packaging, labelling and provision of information)
HS(G)161	The carriage of dangerous goods explained – Part 2: Guidance for road vehicle operators and others involved in the carriage of dangerous goods by road
HS(G)162	The carriage of dangerous goods explained – Part 4: Guidance for operators, drivers and others involved in the carriage of explosives by road
HS(G)163	The carriage of dangerous goods explained – Part 3: Guidance for rail operators and others involved in the carriage of dangerous goods by rail
HS(G)164	The carriage of dangerous goods explained – Part 5: Guidance for consignors, rail operators and others involved in the packaging, labelling and carriage of radioactive material by rail
HS(G)166	Formula for health and safety – guidance for small and medium-sized firms in the chemical industry
HS(G)167	Biological monitoring in the workplace – a guide to its practical application to chemical exposure
HS(G)168	Fire safety in construction work: guidance for clients, designers and those managing and carrying out construction work involving significant risks
HS(G)173	Monitoring strategies for toxic substances
HSG 174	Anthrax – safe working and the prevention of infection
HS(G)176	The storage of flammable liquids in tanks
HSG 178	The spraying of flammable liquids
HSG 181	Assessment principles for offshore safety cases
HS(G)183	Five steps to risk assessment – case studies
HS(G)	Health and safety in excavations. Be safe and shore
HSG 186	The bulk transfer of dangerous liquids and gases between ship and shore
HSG 187	Control of diesel engine exhaust emissions in the workplace
HSG 188	Health risks management – a guide to working with solvents
HSG 189/1	Controlled asbestos stripping techniques for work requiring a licence
HSG 189/2	Working with asbestos cement
HSG 190	Preparing safety reports – Control of Major Accident Hazards Regulations 1999
HSG 191	Emergency planning for major accidents – Control of Major Accident Hazards Regulations 1999
HSG 193	Datasheets: COSHH essentials – easy steps to control chemicals
HSG 202	General ventilation in the workplace. Guidance for employers
HSG204	Health and safety in arc welding
HS(G)248	Solder fume and you

Health and safety regulations

HS(R)1	Packaging and labelling of dangerous substances. Regulations and guidance notes
HS(R)6	A guide to the HSW Act
HS(R)12	A guide to the Health and Safety (Dangerous Pathogens) Regulations 1981
HS(R)13	A guide to the Dangerous Substances (Conveyance by Road in Road Tankers and Tank Containers) Regulations 1981

HS(R)14 A guide to the Notification of New Substances Regulations. Revised 1988 (obsolete – withdrawn)
HS(R)15 Administrative guidance on the European Community 'Explosive Atmospheres' Directives (76/117/EEC AND 79/196/EEC) and related Directives
HS(R)16 A guide to the Notification of Installations Handling Hazardous Substances Regulations 1982
HS(R)17 A guide to the Classification and Labelling of Explosives Regulations 1983
HS(R)21 A guide to the Control of Industrial Major Accident Hazards Regulations 1984
HS(R)22 A guide to the Classification, Packaging and Labelling of Dangerous Substances Regulations 1984
HS(R)23 A guide to the Reporting of Injuries, Diseases and Dangerous Occurrences Regulations 1985
HS(R)27 A guide to the Dangerous Substances in Harbour Areas Regulations 1987
HS(R)29 Notification and marking of sites. The Dangerous Substances (Notification and Marking of Sites) Regulations 1990
HS(R)30 A guide to the Pressure Systems and Transportable Gas Containers Regulations 1989

Industrial groups

IND(G)3L(Rev) First-aid needs in your workplace: your questions answered. Revised 1994
IND(G)4(P) General guidance for inclusion in first-aid boxes
IND(G)24L The Gas Regulations: for everybody's safety
IND(G)35L Hot work on tanks and drums
IND(G)39L(Rev) Permits-to-work and you. (An introduction for workers in the petroleum industry). Revised 1991
IND(G)54L Asbestos: does your company work with asbestos
IND(G)55P Bitumen boilers in construction fire hazards
IND(G)56P Flammable liquids on construction sites
IND(G)57L Review your occupational health needs. (An employer's guide)
IND(G)60L Down with dust. (A guide for employers)
IND(G)62L Protecting your health at work (good health is good business). Revised 4/1993
IND(G)64L Introducing assessment. (A simplified guide for employers). Control of Substances Hazardous to Health Regulations 1988 (COSHH)
IND(G)65L Introducing COSHH. A brief guide for all employers to the new requirements for controlling hazardous substances in the workplace introduced in the Control of Substances Hazardous to Health Regulations 1988 (COSHH)
IND(G)67L Hazard and risk explained. Control of Substances hazardous to Health Regulations 1988 (COSHH)
IND(G)68L Rev Do you use a steam/water pressure cleaner? You could be in for a shock!
IND(G)72L Health hazards to painters
IND(G)74L Need advice on occupational Health? A practical guide for employers
IND(G)76L Safe systems of work
IND(G)78L Transport of LPG cylinders by road
IND(G)79L Gas appliances – Get them checked – Keep them safe
IND(G)82L Your health and reactive dyes
IND(G)84L Leptospirosis – are you at risk?
IND(G)86(L) Contained use of genetically modified organisms
IND(G)91(rev2) Drug misuse at work – A guide for employers
IND(G)92L Dangerous substances on site: notification and warning signs
IND(G)93L Solvents and you
IND(G)95L Save your breath: respiratory sensitizers. (A guide for employers)
IND(G)96L Are you involved in the transport of dangerous substances by road (replaced by IND(G)234L)
IND(G)97L COSHH and Section 6 of the Health and Safety at Work Act
IND(G)98L Permit-to-work systems. (Chemical manufacturing)
IND(G)100L An important notice to all gas fitters
IND(G)101L AIDS: HIV infection and diving?
IND(G)104L Petrol filling stations: control and safety guidance for employees
IND(G)107(L) Asbestos and you – Asbestos dust kills – Are you at risk? (superseded by IND(G)289)
IND(G)111L Chemical manufacturing: personal protective equipment
IND(G)112L Chemical manufacturing: controlling contractors
IND(G)114L Are you complying with COSHH?
IND(G)115L New explosives controls – Do they affect you? (The Control of Explosives Regulations 1991)
IND(G)116L What your doctor needs to know
IND(G)123L Radon in the workplace

INDG 136	COSHH: a brief guide to the regulations
IND(G)139L	Electric storage batteries safe charging and use
IND(G)140L	Grain dust in non-agricultural workplaces
IND(G)141(rev 1)	Reporting incidents of exposure to pesticides and veterinary medicines – what to do if you think people, animals or the environment have been harmed by exposure to pesticides or veterinary medicines
IND(G)151L	The complete idiot's guide to CHIP, or How to amaze your friends with your knowledge of CHIP without really trying
IND(G)157L	Getting to grips with manual handling: pharmaceutical: a ahort guide for employers
IND(G)158L	Health surveillance in the pharmaceutical industry (COSHH)
IND(G)161	Emergency action for burns
IND(G)163(rev 1)	Five steps to risk assessment
IND(G)167L	Health risks from metalworking fluids – Aspects of good machine design
IND(G)168L	Management of metalworking fluids: a guide to good practice for minimizing risks to health
INDG 169	Metalworking fluids and you
IND(G)172L	Breathe freely: a workers' information card on respiratory sensitizers
IND(G)181(L)	The complete idiot's guide to CHIP 2
INDG 186	Chemical sites and safety reports: what you need to know (COMAH)
IND(G)182L	Why do I need a safety data sheet?
IND(G)184L	Signpost to the Health and Safety (Safety Signs and Signals) Regulations 1996
IND(G)186L	Read the label – How to find out if chemicals are dangerous
IND(G)187L	Asbestos dust – The hidden killer! – Are you at risk? Essential advice for building maintenance, repair and refurbishment workers (superseded by IND(G)289)
IND(G)189(L)	Safety zones around offshore oil and gas installations in waters around the UK – Notice to fishermen
IND(G)195L	It takes your breath away – Health advice to the plastics industry
IND(G)196L	Safety reports: how HSE assesses them: the Control of Industrial Major Accident Hazards Regulations 1984
IND(G)197L	Working with sewage – The health hazards – A guide for employers
IND(G)198L	Working with sewage – The health hazards – A guide for employers
IND(G)214	First aid at work – your questions answered
IND(G)215 (rev)	Basic advice on first aid at work
IND(G)218(L)	A guide to risk assessment requirements – Common provisions in health and safety law
IND(G)227(L)	Safe working with flammable substances
IND(G)230L	Storing and handling ammonium nitrate (replaces CS 18)
IND(G)233(L)	Preventing dermatitis at work – advice for employers and employees
IND(G)234	Are you involved in the carriage of dangerous goods by road or rail?
IND(G)235L	A guide to information, instruction and training – common provisions in health and safety law
IND(G)238	Gas appliances – get them checked – keep them safe
IND(G)239L	Play your part! – how you can help improve health and safety offshore
IND(G)245	Biological monitoring in the workplace – Information for employees on its application to chemical exposure
IND(G)246L	Prepared for . . . emergency
IND(G)248	Solder fume and you
IND(G)249L	Controlling health risks from rosin (colophony) based solder fluxes
IND(G)250L	How HSE assesses offshore safety cases
IND(G)251	WASP – Quality assurance for chemical analysis
IND(G)252(L)	Solvents and you – Working with solvents – Are you at risk? (revised 1997)
IND(G)253L	Controlling legionella in nursing and residential care homes
IND(G)254	Chemical reaction hazards and the risk of thermal runaway
IND(G)255	Asbestos dust kills – keep your mask on – guidance for employees on wearing respiratory protective equipment for work with asbestos (superseded by 1999 version)
IND(G)257	Pesticides – Use them safely
IND(G)258	Safe work in confined spaces (supersedes CS 15 'Confined spaces')
IND(G)259	An introduction to health and safety
IND(G)261	Pressure systems – Safety and you
IND(G)264	Selecting respiratory protective equipment for work with asbestos
IND(G)272	Health risks management – A guide to working with solvents
IND(G)273	Working safely with solvents – A guide to safe working practices
IND(G)286	Diesel engine exhaust emissions

IND(G)288 Selection of suitable respiratory protective equipment for work with asbestos
IND(G)289 Working with asbestos in buildings – Asbestos the hidden killer! Are you at risk?
INDG 297 Safety in gas welding, cutting and similar processes
INDG 300 Skin cancer caused by oil (replaces MS(B)5)
INDG 304 Understanding health surveillance at work – An introduction for employers
INDG 305 (rev1) Lead and you – A guide to working safely with lead
INDG 307 Hydrofluoric acid poisoning – recommendations on first aid procedures
INDG 308 The safe use of gas cylinders
INDG 311 Beryllium and you – working with beryllium – Are you at risk?

Legal series

L 5 General COSHH ACOP (Control of Substances Hazardous to Health) and Carcinogens ACOP (Control of Carcinogenic Substances) and Biological Agents ACOP (Control of Biological Agents) Approved Code of Practice
L 7 Approved Code of Practice – Part 4: Dose limitation – Restriction of exposure. Additional guidance on Regulation 6 of the Ionizing Radiations Regulations 1985 (superseded by L 121: 2000)
L 8 The Prevention or Control of Legionellosis (including Legionnaires' Disease) Approved Code of Practice
L 9 Safe use of pesticides for non-agricultural purposes – Control of Substances Hazardous to Health Regulations 1994 Approved Code of Practice
L 10 A guide to the Control of Explosives Regulations 1991
L 11 A guide to the Asbestos (Licensing) Regulations 1983
L 13 A guide to the Packaging of Explosives for Carriage Regulations 1991
L 21 Management of health and safety at work. Management of Health and Safety at Work Regulations 1999. Approved Code of Practice
L 22 Safe use of work equipment. Provision and Use of Work Equipment Regulations 1998. Approved Code of Practice and Guidance (revised)
L 24 Workplace health, safety and welfare. Workplace (Health, Safety and Welfare Regulations) 1992. Approved Code of Practice and Guidance
L 25 Personal protective equipment at work – Guidance on Regulations
L 27 The control of asbestos at work. Control of Asbestos at Work Regulations 1987 – Approved Code of Practice 3rd edition (revised)
L 28 Work with asbestos insulation, asbestos coating and asbestos insulating board – Control of Asbestos at Work Regulations 1987 – Approved Code of Practice 3rd edition (revised)
L 29 A guide to the Genetically Modified Organisms (Contained Use) Regulations 1992, as amended in 1996
L 30 A guide to Offshore Installations (Safety Case) Regulations 1992
L 37 Safety data sheets for substances and preparations dangerous for supply (superseded by L 62)
L 45 Explosives at coal and other safety-lamp mines
L 53 Approved methods for the classification and packaging of dangerous goods for carriage by road and rail – Carriage of Dangerous Goods by Road and Rail (Classification, Packaging and Labelling) Regulations 1994 (superseded by L 88 1996)
L 55 Preventing asthma at work: how to control respiratory sensitizers
L 56 Safety in the installation and use of gas systems and appliances – Gas Safety (Installations and Use) Regulations 1998 Approved Cope of Practice and guidance
L 57 Approved carriage list – Information approved for the classification, packaging and labelling of dangerous goods for carriage by road and rail – Carriage of Dangerous Goods by Road and Rail (Classification, Packaging and Labelling) Regulations 1994 (superseded by L 88: 1996)
L 58 Approved Code of Practice – The Protection of persons against ionizing radiation arising from any work activity – The Ionizing Radiations Regulations 1985 (superseded by L 121: 2000)
L 60 Approved Code of Practice – Control of Substances Hazardous to Health in the Production of Pottery (formerly COP 41) (reprinted with amendments 1998)
L 62 Safety datasheets for substances and preparations dangerous for supply
L 65 Approved Code of Practice and guidance – Prevention of Fire and Explosion, and Emergency Response on Offshore Instalations (Prevention of Fire and Explosion and Emergency Response) Regulations 1995
L 66 A guide to the Placing on the Market and Supervision of Transfers of Explosives Regulations 1993 (POMSTER)
L 67 Control of vinyl chloride at work
L 70 A guide to the Offshore Installations and Pipeline Works (Management and Administration) Regulations 1995

L73	RIDDOR explained – A guide to the Reporting of Injuries, Diseases and Dangerous Occurrences Regulations 1995
L 74	First aid at work. Approved Code of Practice and guidance
L 80	A guide to the Gas Safety (Management) Regulations 1996
L 82	A guide to the Pipelines Safety Regulations 1996
L 83	A guide to the Installation, Verification and Miscellaneous Aspects of Amendments by the Offshore Installations and Wells (Design and Construction, etc) Regulations 1996 to the Offshore Installations (Safety Case) Regulations 1992
L 84	A guide to the well aspects of amendments of the Offshore Installations and Wells (Design and Construction, etc.) Regulations 1996
L 85	A guide to the integrity, workplace environment and miscellaneous aspects of the Offshore Installations and Wells (Design and Construction, etc) Regulations 1996
L 86	Control of substances hazardous to health in fumigation operations
L 88	Approved requirements and test methods for the classification and packaging of dangerous goods for carriage
L 89	Approved vehicle requirements
L 90	Approved carriage list
L 91	Suitability of vehicles and containers and limits on quantities for the carriage of explosives by road
L 92	Approved requirements for the construction of vehicles intended for the carriage of explosives by road
L 93	Approved tank requirements: the provisions for bottom loading and vapour recovery systems of mobile containers carrying petrol
L 94	Approved requirements for the packaging, labelling and carriage of radioactive material by rails
L 95	A guide to the Health and Safety (Consultation with Employees) Regulations 1996
L 96	A guide to the Work in Compressed Air Regulations 1996
L 100	Approved guide to the classification and labelling of substances and preparations dangerous for supply
L 101	Safe working in confined spaces – Approved Code of Practice, Regulations and guidance – Confined Spaces Regulations 1997
L 110	A guide to the Offshore Installations (Safety Representatives and Safety Committees) Regulations 1989 – guidance on Regulations
L 111	A guide to the Control of Major Accident Hazards Regulations 1999 – COMAH – guidance on Regulations
L 118	Health and safety at quarries
L 121	Work with ionizing radiation. Ionizing Radiations Regulations 1999. Approved Code of Practice and guidance
L 122	Safety of pressure systems. Pressure Systems Safety Regulations 2000. Approved Code of Practice

Dangerous substances (general) – HSE

Major hazard aspects of the transport of dangerous substances 1991
Recommendations for training users of non-agricultural pesticides
Health surveillance under COSHH. *Guidance for employers*
Anaesthetic agents. *Controlling exposure under COSHH*
Risk assessments of notified new substances
Asthmagens. Critical assessments of the evidence for agents implicated in occupational asthma 1997.
The technical basis for COSHH essentials. *Easy steps to control chemicals*

Methods for determination of hazardous substances

MDHS 1	Acrylonitrile in air (charcoal adsorption tubes)
MDHS 2	Acrylonitrile in air (porous polymer adsorption tubes)
MDHS 3	Generation of test atmospheres of organic vapours by the syringe injection technique
MDHS 4	Generation of test atmospheres of organic vapours by the permeation tube method
MDHS 5	On-site validation of sampling methods
MDHS 6/3	Lead and inorganic compounds of lead in air (atomic absorption spectrometry)
MDHS 7	Lead and inorganic compounds of lead in air (X-ray fluorescence spectrometry)
MDHS 8	Lead and inorganic compounds of lead in air (colorimetric field method)
MDHS 9	Tetra alkyl lead compounds in air (personal monitoring)
MDHS 10/2	Cadmium and inorganic compounds in air (atomic absorption spectrometry)

MDHS 11 Cadmium and inorganic compounds of cadmium in air (X-ray fluorescence spectroscopy)
MDHS 12/2 Chromium and inorganic compounds of chromium in air (atomic absorption spectrometry)
MDHS 13 Chromium and inorganic compounds of chromium in air (X-ray fluorescence spectroscopy)
MDHS 14/3 General methods for sampling and gravimetric analysis of respirable and inhalable dust
MDHS 15 Carbon disulphide in air
MDHS 16 Mercury vapour in air (hopcalite adsorbent tubes)
MDHS 17 Benzene in air (charcoal adsorbent tubes)
MDHS 18 Tetra alkyl lead compounds in air (continuous on-site monitoring)
MDHS 19 Formaldehyde in air
MDHS 20 Styrene in air
MDHS 21 Glycol ether and glycol ether acetate vapours in air (charcoal adsorbent tubes)
MDHS 22 Benzene in air (porous polymer adsorbent tubes)
MDHS 23 Glycol ether and glycol ether acetate vapours in air (Tenax adsorbent tubes)
MDHS 24 Vinyl chloride in air
MDHS 25/3 Organic isocyanates in air
MDHS 26 Ethylene oxide in air
MDHS 27 Protocol for assessing the performance of a diffusive sampler
MDHS 28 Chlorinated hydrocarbon solvent vapours in air
MDHS 29/2 Beryllium and inorganic compounds of beryllium in air
MDHS 30/2 Cobalt and cobalt compounds in air – Laboratory method using flame atomic absorption spectrometry
MDHS 31 Styrene in air
MDHS 32 Dioctyl phthalates in air
MDHS 33/2 Sorbent tube standards
MDHS 34 Arsine in air
MDHS 35/2 Hydrogen fluoride and inorganic fluorides in air
MDHS 36 Toluene in air
MDHS 37 Quartz in respirable airborne dusts (direct method)
MDHS 38 Quartz in respirable airborne dusts (KBr disc technique)
MDHS 39/4 Asbestos fibres in air
MDHS 40 Toluene in air
MDHS 41/2 Arsenic and inorganic compounds of arsenic (except arsine) in air
MDHS 42/2 Nickel and inorganic compounds of nickel in air (except nickel carbonyl)
MDHS 43 Styrene in air (porous polymer diffusive samplers, thermal desorption and gas chromatography)
MDHS 44 Styrene in air (charcoal diffusive samplers, solvent desorption and gas chromatography)
MDHS 45 Ethylene dibromide in air
MDHS 46/2 Platinum metal and soluble platinum compounds in air
MDHS 47/2 Determination of rubber process dust and rubber fume (measured as cyclohexane-soluble material) in air
MDHS 48 Newspaper print rooms: measurement of total particulates and cyclohexane soluble material in air
MDHS 49 Aromatic isocyanates in air
MDHS 50 Benzene in air (porous polymer diffusion samplers, thermal desorption and gas chromatography)
MDHS 51 Quartz in respirable airborne dusts (X-ray diffraction)
MDHS 52/3 Hexavalent chromium in chromium plating mists (supersedes EH 6 and MDHS 52/2: 1990)
MDHS 53 1,3-Butadiene in air
MDHS 54 Protocol for assessing the performance of a pumped sampler for gases and vapours
MDHS 55 Acrylonitrile in air
MDHS 56/2 Hydrogen cyanide in air
MDHS 57 Acrylamide in air
MDHS 58 Mercury vapour in air (diffusive samplers)
MDHS 59 Man-made mineral fibre
MDHS 60 Mixed hydrocarbons (C3 to C10) in air
MDHS 61 Total hexavalent chromium compounds in air
MDHS 62 Aromatic carboxylic acid anhydrides in air
MDHS 63 1,3-Butadiene in air (molecular sieve diffusive samplers, thermal desorption and gas chromatography)
MDHS 64 Toluene in air (charcoal diffusive samplers, solvent desorption and gas chromatography)
MDHS 66 Mixed hydrocarbons (C5 to C10) in air
MDHS 67 Total (and speciated) chromium in chromium plating mists
MDHS 68 Coal tar pitch volatiles: measurement of particulates and cyclohexane soluble material in air
MDHS 69 Toluene in air

MDHS 70	General methods for sampling airborne gases and vapours
MDHS 71	Analytical quality in workplace air monitoring
MDHS 72	Volatile organic compounds in air
MDHS 73	Measurement of air change rates in factories and offices
MDHS 74	*n*-Hexane in air
MDHS 75	Aromatic amines in air and on surfaces
MDHS 76	Cristobalite in respirable airborne dust
MDHS 77	Asbestos in bulk materials – Sampling and identification by polarized light microscopy (PLM)
MDHS 78	Formaldehyde in air
MDHS 79	Peroxodisulphates in air
MDHS 80	Volatile organic compounds in air
MDHS 81	Dustiness of powders and materials
MDHS 82	The dust lamp – A simple tool for observing the presence of airborne particles
MDHS 83	Resin acids in rosin (colophony) solder flux fume
MDHS 84	Measurement of oil mist from mineral oil-based metalworking fluids
MDHS 85	Triglycidyl isocyanurate (and coating powders containing triglycidyl isocyanurate) in air
MDHS 86	Hydrazine in air
MDHS 87	Fibres in air
MDHS 88	Volatile organic compounds in air
MDHS 89	Dimethyl sulphate and diethyl sulphate in air
MDHS 91	Metals and metalloids in workplace air by X-ray fluorescence spectrometry
MDHS 92	Azodicarbonamide in air
MDHS 93	Glutaraldehyde in air
MDHS 94	Pesticides in air and/or on surfaces
MDHS 95	Measurement of personal exposure of metalworking machine operators to airborne water-mix metalworking fluid
MDHS 96	Volatile organic compounds in air

Medical

MS 3	Skin tests in dermatitis and occupational chest disease
MS 4	Organic dust surveys (obsolete – withdrawn)
MA 5	Health and Safety Executive's Medical Division: agency nurses and occupational health nursing
MA 6	Occupational health aspects of pregnancy
MS 9	Byssinosis (obsolete)
MS 12	Mercury: medical guidance notes
MS 13	Asbestos: medical guidance notes
MS 15	Welding
MS 17 (rev)	Medical aspects of work-related exposures to organophosphates
MS 21	Precautions for the safe handling of cytotoxic drugs
MS 22	The medical monitoring of workers exposed to platinum salts
MS 23	Health aspects of job placement and rehabilitation – Advice to employers
MS 24	Medical aspects of occupational skin disease
MS 25	Medical aspects of occupational asthma
MSA 1	Lead and you – A guide to working safely with lead
MS(A)7	Cadmium and you – working with cadmium – Are you at risk?
MS(A)8	Arsenic and you. Arsenic is poisonous – Are you at risk?
MS(A)9	What you should know about cyanide poisoning. (A guide for employers)
MS(A)11	Working with MbOCA
MS(A)12	Acid and you (withdrawn)
MSA 13	Benzene and you – Working with benzene – Are you at risk? (revision 1)
MS(A)14	Nickel and you – Working with nickel – Are you at risk?
MS(A)15	Silica dust and you
MS(A)16	Chromium and you
MS(A)17	Cobalt and you – Working with cobalt – Are you at risk?
MS(A)18	Beryllium and you – Working with beryllium – Are you at risk?
MS(A)19	PCBs and you – Do you know how to work safely with PCBs?

MSA 21 MbOCA and you – Do you use MbOCA?
MS(B)16 Save your breath: occupational lung disease. (A guide for employers)
MS(B)3 Anthrax (information card)
MS(B)4 Skin cancer caused by pitch and tar
MS(B)5 Skin cancer caused by oil (replaced by INDG 300)
MS(B)7 Poisoning by pesticides. First aid
MS(B)9 Save your skin: occupational contact dermatitis

Offshore

OT1442 A fibre optic sensor for flexible pipeline and riser integrity monitoring
OTI521 Improving inherent safety
OTI586 Representative range of blast and fire scenarios
OTI587 The prediction of single and two-phase release rates
OTI588 Legislation, codes of practice and certification requirements
OTI591 Gas/vapour build up on offshore structures
OTI592 Confined vented explosions
OTI593 Explosions in highly congested volumes
OTI594 The prediction of the pressure loading on structures resulting from an explosion
OTI595 Possible ways of mitigating explosions on offshore structures
OTI596 Oil and gas fires
OTI597 Behaviour of oil and gas fires in the presence of confinement and obstacles
OTI606 Passive fire protection
OTI607 Availability and properties of passive and active fire protection systems
OTI634 Jet-fire resistance test of passive fire protection materials

Plant and machinery

PM 75 Glass reinforced plastic vessels and tanks: advice to users
PM 77 Fitness of equipment used for medical exposure to ionising radiation
PM 81 Safe management of ammonia refrigeration systems – food and other workplaces
PM 82 The selection, installation and maintenance of electrical equipment for use in and around buildings containing explosives
PML 52 7 steps to successful substitution of hazardous substances

Plastics processing

PPIS 1 Fire and explosion risks from pentane in expandable polystyrene (EPS)
PPIS 2 Plastics recycling
PPS 4 Safety at injection moulding machines
PPS 5 Safety at blow moulding machines
PPS 6 Safety at thermoforming machines
PPS 9 Plastics processing sheet No. 9: safety at compression moulding machines
PPS 10 Safety at granulators

Research papers

RP 28 Grain dust: some of its effects on health
RP 30 Occupational health advice as part of primary health care nursing
RP 31 The fracture behaviour of polyethylene – A literature review
RP 32 First aid retention of knowledge survey

Cautionary notices

SHW 2125 Warning: Danger of explosion in oil fuel storage tanks fitted with immersed heaters

SHW 396 Effects of lemon and orange peeling on the skin
SHW 397 (rev) Effects of mineral oil on the skin
SHW 987 Celluloid fire dangers. Warning to workers. (Leaflet)

Specialist inspector reports

SIR 3 Control of exhaust fumes from vehicles working in enclosed spaces
SIR 5 Avoiding water hammer in steam systems
SIR 9 The fire and explosion hazards of hydraulic accumulators
SIR 10 Sick building syndrome
SIR 14 Dust control at woodworking processes
SIR 16 Low-volume high-velocity extraction systems
SIR 19 Fire and explosion hazards associated with the storage and handling of hydrogen peroxide
SIR 20 Spraying and the design of spray booths
SIR 21 Assessment of the toxicity of major hazard substances
SIR 22 Reliability of airflow measurements in assessing ventilation systems' performance
SIR 25 Cobalt – a review of properties, use and levels of exposure
SIR 26 A review of respirable crystalline silica – Exposure and control
SIR 27 Some occupational hygiene aspects of man-made mineral fibres and new technology fibres
SIR 31 Safe handling requirements during explosive, propellant and pyrotechnic manufacture
SIR 37 Simplified calculations of blast induced injuries and damage
SIR 41 Laboratory work with chemical carcinogens and oncogenes
SIR 42 Rosin (colophony): a review
SIR 48 Occupational hygiene aspects on the safe use and selection of refrigerant fluids
SIR 51 Remediation of contaminated land, occupational hygiene aspects on the safe selection and use of new soil clean-up techniques
SIR 56 A study of occupational dermatitis in further education training hairdressing establishments in Scotland

Textiles

TIS 1 Dyes and chemicals in textile finishing: an introduction
TIS 2 Non-dyestuff chemicals: safe handling in textile finishing
TIS 3 Dyestuffs: safe handling in textile finishing
TIS 4 Hazards from dyes and chemicals in textile finishing: a brief guide for employees
TIS 5 Reactive dyes: safe handling in textile finishing

Toxicity reviews

TR 1 Styrene
TR 2 Formaldehyde
TR 3 Carbon disulphide
TR 4 Benzene
TR 5 Pentachlorophenol
TR 6 Trichloroethylene
TR 7 Cadmium and its compounds
TR 8 Trimellitic anhydride (TMA) 4,4′-Methylenebis (2-chloroaniline) (MbOCA) *N*-Nitrosodiethanolamine (NDELA)
TR 9 1,1,1-Trichloroethane
TR 10 Glycol ethers
TR 11 1,3-Butadiene and related compounds
TR 12 Dichloromethane
TR 13 Vinylidene chloride
TR 14 Phthalate esters
TR 15 Carcinogenic hazard of wood dusts. Carcinogenicity of crystalline silica
TR 16 Inorganic arsenic compounds

TR 17	Tetrachloroethylene
TR 18	*n*-Hexane
TR 19	The toxicity of nickel and its inorganic compounds
TR 20	Toluene
TR 21	The toxicity of chromium and inorganic chromium compounds
TR 22	Bis (chloromethyl) ether
TR 23	Part 1: Carbon tetrachloride. Part 2: Chloroform
TR 24	Part 1: Ammonia. Part 2: 1-Chloro-2,3-epoxypropane (epichlorohydrin). Part 3: Carcinogenicity of cadmium and its compounds
TR 25	Cyclohexane, cumene, para-dichlorobenzene (p-DCB), chlorodifluoromethane (CFC22)
TR 26	Xylenes
TR 27	Triglycidyl isocyanurate, beryllium and beryllium compounds
TR 28	Cancer epidemiology in coal tar pitch volatile-associated industries
TR 29	Cobalt and cobalt compounds
TR 30	Phosphorus compounds (phosphoric acid, phosphorous pentoxide, phosphorous oxychloride, phosphorous pentachloride, phosphorous pentasulphide)
TR 31	Dimethylcarbamoyl chloride

Woodworking

WIS 1	Wood dust: hazards and precautions
WIS 6	COSHH and the woodworking industries
WIS 10	Glue spreading machines
WIS 11	Hardwood dust survey
WIS 12	Assessment and control of wood dust: use of the dust lamp
WIS 14	Selection of respiratory protective equipment suitable for use with wood dust
WIS 19	Health risks during furniture stripping using dichloromethane (DCM)
WIS 23	LEV – general principles of system design
WIS 24	LEV – dust capture at sawing machines
WIS 25	LEV – dust capture at fixed belt sanding machines
WIS 26	LEV – dust capture at fixed drum and disc sanding machines
WIS 29	Occupational hygiene and health surveillance at industrial timber pre-treatment plants
WIS 30	Toxic woods
WIS 32	Safe collection of woodwaste: prevention of fire and explosion
WIS 33	Health surveillance and wood dust
WIS 34	Health and safety priorities for the woodworking industry

Miscellaneous

Supplements to carriage of dangerous goods explained (HSG series)

SUPP 05	Supplement to Carriage of Dangerous Goods (HSG 160) Explained – Part 1 – Guidance for Consignors of Dangerous Goods by Road and Rail. Classification, Packaging, Labelling and Provision of Information
SUPP 06	Supplement to Carriage of Dangerous Goods Explained (HSG 161) – Part 2 – Guidance for Vehicle Operators and Others Involved in the Carriage of Dangerous Goods by Road
SUPP 07	Supplement to Carriage of Dangerous Goods Explained (HSG 163 which updates L 51) – Part 3 – Guidance for Rail Operators and Others Involved in the Carriage of Dangerous Goods by Rail
SUPP 08	Supplement to Carriage of Dangerous Goods Explained (HSG 164) – Part 5 – Guidance for Consignors, Rail Operators and Others Involved in the Packaging, Labelling and Carriage of Radioactive Material by Rail

Publications by the Environment Agency relating to chemicals

Prevention of pollution series

PPG1	General guide to the prevention of pollution of controlled waters
PPG2	Above ground oil storage tanks
PPG3	The use and design of oil separators
PPG4	Disposal of sewage where no mains drainage is available
PPG5	Works in, near or liable to affect water courses
PPG6	Working at demolition and construction sites
PPG7	Fuelling stations: construction and operation
PPG8	Safe storage and disposal of used oils
PPG9	Pesticides
PPG10	Highway depots
PPG11	Industrial sites
PPG12	Sheep dip
PPG13	The use of high pressure water and steam cleaners
PPG14	Boats and marinas
PPG15	Retail premises
PPG16	Schools and other educational establishments
PPG17	Dairies and other milk handling operations
PPG18	Managing fire water and major spillages
PPG19	Garages and vehicle service centres
PPG20	Dewatering underground ducts and chambers
PPG22	Dealing with spillages on highways

Masonry bunds for oil storage tanks
Concrete bunds for oil storage tanks
Chemical pollution and how to avoid it
River pollution and how to avoid it
Pollution from your home and how to avoid it
Solvent pollution and how to avoid it
Silt pollution and how to avoid it
Farm pollution and how to avoid it
Making the right connection
Follow the oil care code
Oil care at home
Oil care at work
Oil care on your boat
Oil tank sticker
Domestic oil tank sticker
Control of Pollution (Slurry, Silage and Agricultural Fuel Oil) Regulations 1991

Waste management

Duty of care – waste and your duty of care
The registration of waste carriers – who needs to register, how to apply and what it costs
New packaging regulations – how do they affect you?
Producer responsibility obligations 1997 – guidance on evidence of compliance and accreditation of re-processors
A new waste management licensing system – what it means – how it affects you

Special waste regulations

How they affect you
Classification of special waste

Use of the consignment note
Obtaining and sending consignment notes

Selected British Standards

BS 341: PART 1	AMD 1 Transportable gas container valves. Part 1: Industrial valves for working pressures up to and including 300 bar (AMD 7641) dated 15 May 1993: Partially replaced BS 341. Part 1: 1962 and BS 341: Part 2: 1963
BS 476: PART 12	Fire tests on building materials and structures. Part 12: Method of test for ignitability of products by direct flame impingement. Replaced BS 476: Part 5: 1979
BS 476: PART 20	AMD 1 Fire tests on building materials and structures. Part 20: Method for determination of the fire resistance of elements of construction (general principles) (AMD 6487) dated 30 April 1990. Replaced BS 476: Part 8: 1972
BS 476: PART 21	Fire tests on building materials and structures. Part 21: Methods for determination of the fire resistance of loadbearing elements of construction. Replaced BS 476: Part 8: 1972
BS 476: PART 22	Fire tests on building materials and structures. Part 22: Methods for determination of the fire resistance of non-loadbearing elements of construction. Replaced BS 476: Part 8: 1972
BS 476: PART 23	AMD 1 Fire tests on building materials and structures. Part 23: Methods for determination of the contribution of components to the fire resistance of a structure (AMD 9458) dated January 1998. Replaced BS 476: Part 8: 1972
BS 499: PART 2	(Withdrawn) 1980 AMD 2 Welding terms and symbols. Part 2: Specification for symbols for welding (AMD 7439) dated 15 November 1992. Withdrawn, superseded by BS EN 24063: 1992 and BS EN 22553: 1995
BS 1542	(Withdrawn) 1982 Equipment for eye, face and neck protection against non-ionizing radiation arising during welding and similar operations. Withdrawn, superseded by BS EN 175:1997
BS 1547	(Withdrawn) 1959 Flameproof industrial clothing (materials and design) . . . Withdrawn, superseded by BS EN 469: 1995 and BS EN 531: 1995
BS 1870: PART 1	AMD 1 Safety footwear. Part 1: Safety footwear other than all-rubber and all-plastics moulded types (AMD 6273) dated 28 February 1990. Superseded by BS EN 345: 1993 and BS EN 346: 1993
BS 1870: PART 2	AMD 4 Safety footwear. Part 2: Lined rubber safety boots (AMD 4741) dated 31 January 1985. Superseded by BS EN 345: 1993
BS 1870: PART 3	AMD 1 Safety footwear. Part 3: Polyvinyl chloride moulded safety footwear (AMD 4742) dated 31 January 1985. Superseded by BS EN 345: 1993
BS 2091	(Withdrawn) 1969 AMD 2 Respirators for protection against harmful dust, gases and schedule agricultural chemicals. Withdrawn, superseded by BS EN 136,140,141,143,371, and 372
BS 2092	(Withdrawn) 1987 AMD 1 Eye-protectors for industrial and non-industrial uses (AMD 7061) dated 15 October 1993. Withdrawn, partially superseded by BS EN 167 and 168: 1995
BS 2653	(Withdrawn) 1955 AMD 3 Protective clothing for welders. Withdrawn, superseded by BS EN 470-1: 1995
BS 2782	(Withdrawn) 1993 Methods of testing plastics. Method 140B: Determination of the burning behaviour of flexible vertical specimens in contact with a small-flame ignition source. Withdrawn, superseded by BS EN ISO 9773: 1999
BS 2782	(Withdrawn) 1993 Methods of testing plastics. Method 140C: Determination of the combustibility of specimens using a 125 mm flame source. Withdrawn, superseded by BS EN 60695-11-20: 1999
BS 2782	Methods of testing plastics. Method 140D: Flammability of a test piece 550 mm × 35 mm of thin polyvinyl chloride sheeting (laboratory method). Supersedes BS 2782: Part 1: Method 140D: 1980
BS 3042	(Withdrawn) 1992 Test probes to verify protection by enclosures. Withdrawn, superseded by BS EN 61032: 1998
BS 3156	Methods for the analysis of fuel gases
BS 3202: PART 1	Laboratory furniture and fittings. Supersedes BS 3202: 1959
BS 3841: PART 1	Determination of smoke emission from manufactured solid fuels for domestic use. Part 1: General method for determination of smoke emission rate. Superseded BS 3841: 1972
BS 3841: PART 2	Determination of smoke emission from manufactured solid fuels for domestic use. Part 2: Methods for measuring the smoke emission rate. Superseded BS 3841: 1972

BS 3951	(Withdrawn) 1989 AMD 2 Freight containers
BS 4078: PART 1	AMD 1 Powder actuated fixing systems. Part 1: Code of practice for safe use (AMD 9899) dated April 1998. Superseded BS 4078: 1966
BS 4422: PART 6	Glossary of terms associated with fire. Part 6: Evacuation and means of escape. Superseded BS 4422: Part 3: 1972
BS 4434	AMD 1 Specification for safety and environmental aspects in the design, construction and installation of refrigerating appliances and systems (AMD 9383) dated 15 March 1997. Supersedes BS 4434: 1989
BS 4555	(Withdrawn) 1970 High efficiency dust respirators. Withdrawn, superseded by BS EN 136: 1990, and BS EN 143: 1991
BS 4558	(Withdrawn) 1970 Positive pressure, powered dust respirators. Withdrawn, superseded by BS EN 136: 1990, BS EN 143: 1991 and BS EN 147
BS 4680	Clothes lockers. Superseded BS 4680: 1971
BS 4771	(Withdrawn) 1971 Positive pressure, powered dust hoods and blouses. Withdrawn, superseded by BS EN 143: 1991 and BS EN 146: 1992
BS 4972	AMD 4 Women's protective footwear (AMD 6374) dated 31 July 1991. Superseded by BS EN 346: 1993
BS 5045: PART 3	(Withdrawn) 1984 AMD 4 Transportable gas containers. Part 3: Seamless aluminium alloy gas containers above 0.5 litre water capacity and up to 300 bar charged pressure at 15°C (AMD 6320) dated 31 January 1991. Withdrawn, superseded by BS EN 1975: 2000 and BS 5045-8: 2000
BS 5111: PART 1	(Withdrawn) 1974 AMD 1 Laboratory methods of test for determination of smoke generation characteristics of cellular plastics and cellular rubber materials. Part 1: Method for testing a 25 mm cube test specimen of low density material (up to 130 kg/metres cubed) to continuous flaming conditions (AMD 7688) dated 15 July 1993. Withdrawn, superseded by BS ISO 5659-2: 1994
BS 5045: PART 3	(Withdrawn) 1984 AMD 4 Transportable gas container. Superseded by BS EN 1975: 2000 and BS 5045-8: 2000
BS 5111	(Withdrawn) 1974 AMD 1 Laboratory methods of test for determination of smoke generation characteristics of cellular plastics and cellular rubber materials. Superseded by BS ISO 5659-2: 1994
BS 5240: PART 1	(Withdrawn) 1987 Industrial safety helmets. Part 1: Construction and performance. Withdrawn, superseded by BS EN 397: 1995
BS 5266-7	Lighting applications. Emergency lighting. Also BS EN 1838: 1999 partially superseded BS 5266-1: 1988
BS 5274	(Withdrawn) 1985 Fire hose reels (water) for fixed installations. Withdrawn, superseded by BS EN 671-1: 1995
BS 5345: PART 1	(Withdrawn) 1989 AMD 3 Code of practice for selection, installation and maintenance of electrical apparatus for use in potentially explosive atmospheres (other than mining applications or explosives processing and manufacture). Part 1: General recommendations (AMD 7871) dated 15 September 1993. Withdrawn, superseded by BS EN 60079-14: 1997
BS 5345: PART 6	(Withdrawn) 1978 AMD 1 Code of practice for selection, installation and maintenance of electrical apparatus for use in potentially explosive atmospheres (other than mining applications or explosives processing and manufacture). Part 6: Recommendations for type of protection. Increased safety (AMD 5557) dated 30 November 1989. Withdrawn, superseded by BS EN 60079
BS 5423	(Withdrawn) 1987 AMD 3 Portable fire extinguishers (AMD 8585) dated 15 December 1995. Withdrawn, superseded by BS 7863: 1996, BS 7867: 1997 and BS EN 3-1 to 6
BS 5445: PART 1	(Withdrawn) 1977 Components of automatic fire detection systems. Part 1: Introduction. Withdrawn, superseded by BS EN 54-1: 1996
BS 5445: PART 5	AMD 1 Components of automatic fire detection systems. Part 5: Heat sensitive detectors; point detectors containing a static element (AMD 5762) dated 31 August 1988. Superseded BS 3116: Part 1: 1970
BS 5499: PART 1	AMD 1 Fire safety signs, notices and graphic symbols. Part 1: Fire safety signs (AMD 7444) dated 15 January 1993. Superseded BS 2560: 1968
BS 5499: PART 2	Fire safety signs, notices and graphic symbols. Part 2: Self-luminous fire safety signs. Superseded BS 4218: 1978
BS 5499: PART 3	Fire safety signs, notices and graphic symbols. Part 3: Internally illuminated fire safety signs. Superseded BS 2560: 1978

BS 5588: PART 1 AMD 1 Fire precautions in the design, construction and use of buildings. Part 1: Code of practice for residential buildings (AMD 7840) dated 15 September 1993. Superseded BS 5502: Sec 1.3: 1986 and BS 5588: Section 1.1: 1984 and CP 3: Chapter IV: Part 1

BS 5588: PART 11 AMD 2 Fire precautions in the design, construction and use of buildings. Part 11: Code of practice for shops, offices, industrial storage and other similar buildings (AMD 10213) dated January 1999. Supersedes BS 5588: Part 2: 1985 and BS 5588: Part 3: 1983

BS 5588: PART 2 (Withdrawn) 1985 AMD 3 Fire precautions in the design, construction and use of buildings. Part 2: Code of practice for shops (AMD 6478) dated 31 August 1990. Withdrawn, superseded by BS 5588: Part 11: 1997

BS 5588: PART 3 (Withdrawn) 1983 AMD 3 Fire precautions in the design, construction and use of buildings. Part 3: Code of practice for office buildings (AMD 6160) dated 31 October 1989. Withdrawn, superseded by BS 5588: Part 11: 1997

BS 5588: PART 8 Fire precautions in the design, construction and use of buildings. Part 8: Code of practice for means of escape for disabled people. Supersedes BS 5588-8: 1988, which is withdrawn

BS 5726: PART 4 AMD 1 Microbiological safety cabinets. Part 4: Recommendations for selection, use and maintenance (AMD 7784) dated 15 June 1993. Supersedes BS 5726: 1979

BS 6467 AMD 1 Electrical apparatus with protection by enclosure for use in the presence of combustible dusts. Superseded by BS EN 50281-1-1 and 1-2 but remains current

BS 6535: PART 3 (Withdrawn) 1989 Fire extinguishing media.. Part 3: Powder. Withdrawn, superseded by BS EN 615: 1995

BS 6535 (Withdrawn) 1990 Fire extinguishing media. Part 2: Halons. Section 2.1: Halon 1211 and halon 1301. Withdrawn, now BS EN 27201-1: 1994

BS 6535: SEC 2.2 (Withdrawn) 1989 Fire extinguishing media. Part 2: Halons. Section 2.2: Code of practice for safe handling and transfer procedures. Withdrawn, superseded by BS EN 27201-2: 1994

BS 6575 (Withdrawn) 1985 AMD 1 Fire blankets (AMD 8938) dated 15 December 1995. Withdrawn, superseded by BS 7944: 1999

BS 6941 Electrical apparatus for explosive atmospheres with type of protection N. Superseded by BS EN 50021: 1999 but remains current

BS 6967 (Withdrawn) 1988 Glossary of terms for personal eye-protection. Withdrawn, superseded by BS EN 165: 1996

BS 7355 (Withdrawn) 1990 Full face masks for respiratory protective devices. Withdrawn, superseded by BS EN 136: 1998

BS 7356 (Withdrawn) 1990 AMD 1 Half masks and quarter masks for respiratory protective devices (AMD 7113) dated 15 November 1992. Withdrawn, superseded by BS EN 140: 1999

BS 7535 Guide to the use of electrical apparatus complying with BS 5501 or BS 6491 in the presence of combustible dusts

BS 7750 (Withdrawn) 1994 Environmental management systems. Withdrawn, superseded by BS EN ISO 14001: 1996

BS 7863 AMD 2 Recommendations for colour coding to indicate the extinguishing media contained in portable fire extinguishers (AMD 9740) dated October 1997. With BS EN 3: Parts 1 to 6, supersedes BS 5423: 1987

BS 7944 Type 1 heavy duty fire blankets and type 2 heavy duty heat protective blankets. Superseded BS 7944: 1985

BS EN 2 Classification of fires. Superseded BS 4547: 1972

BS EN 3-1 Portable fire extinguishers. Description, duration of operation, Class A and B fire test. With BS EN 3-2 to 6 and BS 7863: 1996, Superseded BS 5423: 1987 (still current)

BS EN 3-2 Portable fire extinguishers. Tightness, dielectric test, tamping test, special provisions. With BS EN 3-1 and 3 to 6 and BS 7863: 1996, superseded BS 5423: 1987 (current)

BS EN 3-3 Portable fire extinguishers. Construction, resistance to pressure, mechanical tests. With BS EN 3 Parts 1 to 6, BS 7863, superseded BS 5423: 1987 (still current)

BS EN 3-4 Portable fire extinguishers. Charges, minimum required fire. With BS EN 3-1, 3 and 5 to 6 and BS 7963: 1996, superseded BS 5423: 1987 (current)

BS EN 3-5 AMD 1 Portable fire extinguishers. Specification and supplementary tests (AMD 9618) dated October 1997. With BS EN 3-1 to 4 and 6 and BS 7863: 1996, superseded BS 5423: 1987 (current)

BS EN 3-6 AMD 1 Portable fire extinguishers. Provisions for the attestation of conformity of portable fire extinguishers in accordance with EN 3 Parts 1 to 5 (AMD 10494) dated September 1999. With BS EN 3-1 to 5 and BS 7863: 1996, superseded BS 5423: 1987

BS EN 54-1 Fire detection and fire alarm systems. Introduction, supersedes BS 5445: Part 1: 1977

BS EN 54-2	Fire detection and fire alarm systems. Control and indicating equipment. With BS EN 54-4: 1997, superseded BS 5839: Part 4: 1998 which remains current
BS EN 54-4	Fire detection and fire alarm systems. Power supply equipment. With BS EN 54-2: 1997, supersedes BS 5839: Part 4: 1988 which remains current
BS EN 132	Respiratory protective devices – Definitions of terms and pictograms. Superseded BS 6927: 1988
BS EN 133	Respiratory protective devices – Classification superseded BS 6928: 1988
BS EN 135	Respiratory protective devices – List of equivalent terms. Supersedes BS 6930: 1988
BS EN 136	Respiratory protective devices – Full face masks – Requirements, testing, marking. Supersedes BS EN 136-10: 1992 and BS 7355: 1990
BS EN 137	AMD 1 Respiratory protective devices: self-contained open-circuit compressed air breathing apparatus (AMD 8167) dated 15 May 1994. Supersedes BS 7004: 1988 and BS 4667: Part 2: 1974
BS EN 139	AMD 2 Respiratory protective devices – Compressed air line breathing apparatus for use with a full face mask, half mask or a mouthpiece assembly – Requirements, testing, marking (AMD 10774) dated January 2000. With BS EN 138, BS EN 269 and BS EN 270, superseded BS 4667-3: 1974
BS EN 145	Respiratory protective devices – Self-contained closed-circuit breathing apparatus, compressed oxygen or compressed oxygen–nitrogen type. Requirements, testing, marking. Supersedes BS 7170: 1990 and BS EN 145-2: 1993
BS EN 149	Filtering half masks to protect against particles. Superseded BS 6016: 1980
BS EN 161	AMD 2 Automatic shut-off valves for gas burners and gas appliances (AMD 9796) dated January 1998. With BS 7461, superseded BS 5963: 1981
BS EN 165	Personal eye-protection – Vocabulary. Superseded BS 6967: 1988
BS EN 166	Personal eye-protection – Specifications. With BS EN 167: 1995 and BS EN 168: 1995. Superseded BS 2092: 1987
BS EN 167	Personal eye-protection – Optical test methods. Partially superseded BS 2092: 1987
BS EN 168	Personal eye-protection – Non-optical test methods. Partially superseded BS 2092: 1987
BS EN 175	AMD 1 Personal protection – Equipment for eye and face protection during welding and allied processes (AMD 9902) dated January 1998. Superseded BS 1542: 1982
BS EN 203-1	AMD 3 Gas heated catering equipment. Safety requirements (AMD 10573) dated December 1999. Superseded BS 5314
BS EN 345	(Withdrawn) 1993 Safety footwear for professional use. Withdrawn, now BS EN 345-1: 1993
BS EN 345-1	Safety footwear for professional use. Specification. Previously known as BS EN 345: 1993
BS EN 346	(Withdrawn) 1993 Protective footwear for professional use. Withdrawn, now known as BS EN 346-1: 1993
BS EN 346-1	Protective footwear for professional use. Specification. Previously BS EN 346: 1993, read with BS EN 344-1: 1993
BS EN 347	(Withdrawn) 1993 Occupational footwear for professional use. Withdrawn, now known as BS EN 347-1: 1993
BS EN 367	AMD 1 Protective clothing – Protection against heat and fire – Method of determining heat transmission on exposure to flame (AMD 7667) dated 15 May 1993. Superseded BS 3791: 1970
BS EN 372	SX gas filters and combined filters against specific named compounds used in respiratory protective equipment. Superseded BS 2091: 1969
BS EN 388	Protective gloves against mechanical risks. Superseded BS 1651: 1986
BS EN 397	Specification for industrial safety helmets. Superseded BS 5240: Part 1: 1987
BS EN 400	Respiratory protective devices for self rescue – Self-contained closed circuit breathing apparatus – Compressed oxygen escape apparatus – Requirements, testing and marking. Superseded BS 4667: Part 5: 1990
BS EN 407	Protective gloves against thermal risks (heat and/or fire). Superseded BS 1651: 1986
BS EN 420	AMD 1 General requirements for gloves (AMD 8515) dated 15 February 1995. Superseded BS 1651: 1986
BS EN 443	Helmets for firefighters. Supersedes BS 3864: 1989
BS EN 470-1	AMD 2 Protective clothing for use in welding and allied processes. General requirements (AMD 10318) dated February 1999. Superseded BS 2653: 1955
BS EN 1596	Specification for dedicated liquefied petroleum gas appliances – Mobile and portable non-domestic forced convection direct fired air heaters. Superseded BS 4096: 1967

BS EN 1751 Ventilation for buildings – Air terminal devices – Aerodynamic testing of dampers and valves. Superseded BS 6821: 1988

BS EN 1838 Lighting applications – Emergency lighting. Also BS 5266-7: 1999

BS EN 1869 Fire blankets. Partially. Superseded BS 6565: 1985

BS EN 25923 Fire protection – Fire extinguishing media – Carbon dioxide. Previously BS 6535: Part 1: 1990

BS EN 27201-1 Fire protection – Fire extinguishing media – Halogenated hydrocarbons. Halon 1211 and Halon 1301. Previously BS 6535: Section 2.1: 1990

BS EN 27201-2 Fire protection – Fire extinguishing media – Halogenated hydrocarbons. Code of practice for safe handling and transfer procedures. Supersedes BS 6535: Section 2.2: 1989

BS EN 12673 Water quality – Gas chromatographic determination of some selected chlorophenols in water. Also BS 6068-2.65: 1999

BS EN 12941 Respiratory protective devices – Powered filtering devices incorporating a helmet or a hood – Requirements, testing, marking. Superseded BS EN 146: 1992

BS EN 50014 AMD 2 Electrical apparatus for potentially explosive atmospheres – General requirements (AMD 10552) dated August 1999. Partially superseded BS EN 50014: 1993 which remains current

BS EN 50014 AMD 1 Electrical apparatus potentially explosive atmospheres – General requirements (AMD 8132) dated 15 March 1994. Partially superseded by BS EN 50014: 1998 (remains current)

BS EN 50015 Electrical apparatus for potentially explosive atmospheres – Oil immersion 'o'. Supersedes BS EN 50015: 1994 which remains current

BS EN 50017 Electrical apparatus for potentially explosive atmospheres – Powder filling 'q'. Superseded BS EN 50017: 1994 which remains current

BS EN 50020 AMD 1 Electrical apparatus for potentially explosive atmospheres – Intrinsic safety 'i' (AMD 10040) dated June 1998. Read with BS EN 50014: 1993

BS EN 50281-1-1 AMD 1 Electrical apparatus for use in the presence of combustible dust. Part 1: Electrical apparatus protected by enclosures – Construction and testing (AMD 10764) dated October 1999. Partially superseded BS 6467-1 and 2 which remain current

BS EN 50281-1-2 AMD 1 Electrical apparatus for use in the presence of combustible dust. Part 1: Electrical apparatus protected by enclosures – Selection, installation and maintenance (AMD 10763) dated October 1999. Inc.Cor.1(10856)Part.s/s BS 6467-1: 1985 and BS 6467-2: 1988, remain current

BS EN 531 AMD 1 Protective clothing for workers exposed to heat (AMD 10302) dated January 1999. Partially superseded BS 1547: 1959

BS EN 533 Protective clothing – Protection against heat and flame – Limited flame spread materials and material assemblies. Superseded BS 6249: Part 1: 1982

BS EN 615 Fire protection – Fire extinguishing media – Specifications for powders (other than class D powders). Superseded BS 6535: Part 3: 1989

BS EN 671-1 Fixed firefighting systems – Hose systems. Hose reels with semi-rigid hose. Supersedes BS 5274: 1985

BS EN 779 AMD 1 Particulate air filters for general ventilation – Requirements, testing, marking (AMD 8367) dated 15 November 1994. Superseded BS 6540: Part 1: 1985

BS EN ISO 6326-3 Natural gas – Determination of sulfur compounds. Determination of hydrogen sulfide, mercaptan sulfur, and carbonyl sulfide sulfur by potentiometry. Also BS 3156: Subsection 11.4.3: 1994

BS EN ISO 6326-5 Natural gas – Determination of sulfur compounds. Lingener combustion method. Also BS 3156: Subsection 11.4.5: 1994

BS EN ISO 9000-1 Quality management and quality assurance standards. Guidelines for selection and use. Previously BS 5750: Section 0.11: 1987

BS EN ISO 9001 Quality systems – Model for quality assurance in design, development, production, installation and servicing. Previously BS 5750: Part 1: 1987

BS EN ISO 9002 Model for quality assurance in production, installation and servicing. Previously BS 5750: Part 2: 1987

BS EN ISO 9004-1 Quality management and quality assurance standards. Guidelines. Previously BS 5750: Section 0.2: 1987

BS EN ISO 10101-1 Natural gas – Determination of water by the Karl Fischer method. Also BS 3156 (10053)

BS EN ISO 14001 Environmental management systems – Specification with guidance for use. Superseded BS 7750: 1994 which remains current

BS EN 60079-10 AMD 1 Electrical apparatus for explosive gas atmospheres. Classification of hazardous areas (AMD 9340) dated 15 November 1996. Superseded BS 5345: Part 2: 1983

BS EN 60079-14	AMD 1 Electrical apparatus for explosive gas atmospheres. Electrical installations in hazardous areas (other than mines). Corrigendum No. 1 (Corr. 10013) dated April 1998, superseded BS 5345: Part 1: 1989, Parts 2 and 7: 1979, Part 4: 1977, Part 5: 1983, Part 6: 1978 and Part 8: 1980
BS EN 60309-2	Plugs, socket-outlets and couplers for industrial purposes. Dimensional interchangeability requirements for pin and contact-tube accessories. Superseded BS EN 60309-2: 1998, which remains current
BS EN 60432-1	AMD 1 Incandescent lamps – Safety specifications. Tungsten filament lamps for domestic and similar general lighting purposes (AMD 9775) dated January 1998. Superseded by BS EN 60432-1: 2000, but remains current
BS EN 60529	AMD 1 Degrees of protection provided by enclosures (IP code) (AMD 7643) dated 15 July 1993. Superseded BS 5490: 1977
BS EN 60598-2-22	Luminaires. Part 2: Particular requirements. Particular requirements – Luminaires for emergency lighting. Incorporating Corrigendum No. 1 (AMD 10562) dated December 1999. Inc.Corl, Superseded BS 4533: Section 102.22: 1990, which remains current
BS EN 60695-11-20	Fire hazard testing. Part 11: Test flames – 500 W flame test methods. Superseded BS 2782 – 140C: 1993
BS EN 60695-2-2	AMD 1 Fire hazard testing. Part 2: Test methods. Needle-flame test (AMD 9176) dated 15 November 1996. Previously known as BS 6458: Section 2.2: 1993
BS EN 60695-4	Fire hazard testing. Terminology concerning fire tests. Superseded BS 6458: Part 1: 1990
BS EN 60707	Flammability of solid non-metallic materials when exposed to flame sources – List of test methods. Superseded BS 6334: 1983
BS EN 60721-3-1	Classification of environmental conditions. Part 3: Classification of groups of environmental parameters and their severities. Storage. Superseded BS EN 60721-3-1: 1993
BS EN 60721-3-2	Classification of environmental conditions. Part 3: Classification of groups of environmental parameters and their severities. Transportation. Superseded BS EN 60721-3-2: 1993
BS EN 60721-3-3	AMD 1 Classification of environmental conditions. Part 3: Classification of groups of environmental parameters and their severities. Stationary use at weather protected locations (AMD 9514) dated 15 June 1997. Superseded BS EN 60721-3-3: 1993. Previously BS 7527: Section 3.3: 1991
BS EN 60721-3-4	AMD 1 Classification of environmental conditions. Part 3: Classification of groups of environmental parameters and their severities. Stationary use at non-weather protected locations (AMD 9513) dated 15 June 1997. Previously BS 7527: Section 3.4: 1991
CP 15/1	Code of Practice 15/1: The safe re-rating of existing BS 5045: Part 1: 1982 Containers to amendment AMD 5145: 1986. Part 1: Permanent gas containers (excluding hydrogen trailer service)
CP 15/2	Code of Practice 15/2: The safe re-rating of existing BS 5045: Part 1: Cylinders to amendment AMD 5145: 1986. Part 2: Containers for hydrogen trailer service
DD 180	Guide for the assessment of toxic hazards in fire in buildings and transport. Superseded by BS 7899: Parts 1, 2 and 3 but remains current
DD 189	(Withdrawn) 1990 Safety of condensing boilers (2nd and 3rd family gases). Withdrawn, superseded by BS EN 677: 1998
DD 211	(Withdrawn) 1992 Method for detection and enumeration of Legionella organisms in water and related materials. Withdrawn, superseded by BS 6068-4.12: 1998

Appendix

Selected UK legislation relevant to environmental protection and occupational health and safety in relation to chemicals

The general principles outlined in this work for the identification and control of hazardous chemicals are independent of legislation. However, regulatory obligations forcing compliance with national and/or local requirements should be consulted. Space prevents encyclopaedic coverage but the following brief selection of relevant UK legislation serves to illustrate the scope. It is supplemented by the lists of legislation in the Bibliography.

General health and safety

The minimum standards for health and safety are those outlined in numerous pieces of statutory legislation. These include the:

Factories Act 1961
Health and Safety at Work etc. Act 1974
Management of Health and Safety at Work Regulations 1992
Management of Health and Safety at Work (Amendment) Regulations 1994
Manual Handling Operations Regulations 1992

All of these contain requirements of relevance to work with hazardous chemicals.

Selected legislation relevant to the control of hazardous chemicals

The following legislation is of direct relevance.

Asbestos (Prohibitions) Regulations 1992
Cover the prohibition of the importation, supply or new use of amphibole asbestos or products containing it. Also prohibit the supply and use of a range of products containing chrysotile asbestos.

Carriage of Dangerous Goods by Rail Regulations 1996
The scope includes carriage of dangerous substances by wagons, large containers, small containers and tanks. Cover the documentation requirements, the mode of transport, the information needs, loading/unloading, security, safety

and emergency requirements with special provisions for explosives. The carriage of temperature-controlled substances is prohibited.

Carriage of Dangerous Goods by Road Regulations 1996

Cover the carriage of dangerous goods by road in bulk, in tanks and in containers. The requirements for documentation, information, loading/unloading, and emergencies and parking are included.

Carriage of Dangerous Goods by Road (Driver Training) Regulations 1996

Cover the instruction, training and certification of drivers of motor vehicles engaged in the carriage of dangerous goods.

Carriage of Dangerous Goods (Classifications, Packaging and Labelling) and Use of Transportable Pressure Receptacles Regulations 1996

Cover the classification, packaging and labelling of dangerous goods. The requirements for design, manufacture, modification, repair, approval, certification and marking of transportable pressure containers are included. The role of 'approved persons' to comply with the regulations is explained.

Carriage of Explosives by Road Regulations 1996

Prohibit the carriage of certain explosives. Concerned with the carriage of explosives in vehicles used to carry passengers and carriage in bulk. The suitability of both vehicle and container, vehicle approval, types of vehicles, and quantity limits for loads are regulated. Cover the requirements for information, and safety and security during transit. This includes route of carriage, parking, load integrity, equipment, precautions against fire and explosion, and the requirements in the event of accidents and emergencies.

Chemicals (Hazard Information and Packaging for Supply) Regulations 1994

Regulate the classification, provision of safety data sheets, labelling and packaging of substances and preparations dangerous for supply. There are specific exceptions, e.g. medicines, pesticides, wastes, radioactive substances or preparations.

Confined Spaces Regulations 1997

Requirements for work in confined spaces, including avoidance of entry if reasonably practicable, provision of a safe system of work and establishment of adequate emergency arrangements.

Control of Asbestos at Work Regulations 1987 (amended 1992 and 1999)

Requirements for the control of asbestos exposures at work. Exposure is to be prevented or, if this is not reasonably practicable, reduced to the lowest level reasonably practicable by measures other than the use of respiratory protective equipment.

Control of Explosives Regulations 1991

Concerned mainly with the security of explosives and restricted substances. Applicable to the acquisition, keeping, handling and control of explosives, e.g. blasting explosives, detonators, fuses, ammunitions, propellants, pyrotechnics and fireworks.

Control of Lead at Work Regulations 1980

Requirements for protecting the health of workers exposed to lead including lead alloys, lead compounds and lead containing substances. A risk assessment is necessary.

Provision of adequate information, instruction and training; control against exposures by measures other than the use of personal protective equipment; provision of adequate washing facilities; prohibition of eating, drinking and smoking in contaminated areas; and health surveillance are covered.

Control of Major Accident Hazard Regulations 1999

Regulate the design and operation of defined major hazard installations. All necessary measures are required to prevent and limit the consequences of major accidents. Operators of upper-tier sites must produce a detailed Safety Report; those of lower-tier sites must prepare a Major Accident Prevention Policy.

Control of Pesticides Regulations 1986

Prohibit the advertisement, supply, storage and use of pesticides unless they have been approved. 'Pesticides' includes herbicides, fungicides, wood preservatives, plant growth hormones, soil sterilants, bird or animal repellants, masonry biocides and anti-foul boat paints. Conditions are applied to supply, storage and use. Competence training and certification is required for commercial use in agriculture generally.

Control of Substances Hazardous to Health Regulations 1999
Cover the control of substances classified as very toxic, toxic, harmful, corrosive, sensitizing or irritant under the Chemicals (Hazard Information and Packaging for Supply) Regulations 1994 (as amended) and to substances which have MELs or OESs. Also other substances that have chronic or delayed effects and biological agents. Special provisions are included for carcinogens.

Dangerous Substances in Harbour Areas Regulations 1987
Requirements for prior notice of arrival of dangerous substances into ports.

Dangerous Substances (Notification and Marking of Sites) Regulations 1990
Requirements for notification and appropriate marking of any site containing large quantities of dangerous substances. Principally aimed at the safety of fire officers attending incidents.

Fire Precaution Act 1971
Provides for the control of fire safety in all designated occupied premises, by ensuring that adequate general fire precautions are taken and appropriate means of escape and related precautions are present. Specifies the requirements for a fire certificate for various premises.

Fire Precautions (Workplace) Regulations 1997, as amended
Apply to all workplaces, unless specifically excepted, and require a fire risk assessment; where necessary, appropriate fire-fighting equipment with detectors and alarms; measures for fire-fighting; emergency routes and exits; maintenance of equipment provided.

Gas Safety (Installation and Use) Regulations 1994
Requirements for any employer of, or self-employed, persons to be members of the Council for Registered Gas Installers (CORGI). Individual gas fitters must possess certificates of competence.

Health and Safety (First-Aid) Regulations 1981
Requirements for every employer to provide equipment and facilities which are adequate and appropriate in the circumstances for administering first aid to employees. Adequate numbers of suitable persons are needed to administer first aid.

Health and Safety (Safety Signs and Signals) Regulations 1996
Cover means of communicating health and safety information in all workplaces. Include illuminated signs, alarms, verbal communication, fire safety signs, marking of pipework, etc.

Highly Flammable Liquids and Liquefied Petroleum Gases Regulations 1972
Cover the storage, handling and use of highly flammable liquids, viz. liquids with a flashpoint $<32°C$ and which support combustion when tested in the prescribed ways. Also cover the manner of storage and the marking of storage accommodation for LPG, viz. commercial propane, commercial butane and any mixture of the two.

Ionizing Radiations Regulations 1999
Cover all work with ionizing radiations, including exposure to naturally occurring radon gas. Special arrangements are required to protect outside workers, e.g. contractors. Specified practices need authorization and the regulatory authority must receive prior notification of specified work. Govern risk assessment, exposure, personal dosimetry, engineering controls, personal protective equipment, contingency plans, management controls, designated areas, control of radioactive substances.

Notification of New Substances Regulations 1993
Requirements for the notification of new substances, irrespective of their potential for harm, before they can be placed on the market. Aim to ensure that a chemical placed on the market can be validly marketed in any EU state without duplicating authorization.

Packaging, Labelling and Carriage of Radioactive Material by Rail Regulations 1996
Govern the need for documentation, design for special forms of radioactive material, approval, packaging, loading/unloading, and shipment including the requirement for prior inspections. Also cover security measures and procedures in the event of emergencies plus obligations on provision of information, emission evaluation and certification.

Personal Protective Equipment at Work Regulations 1992
Cover all aspects of the provision, maintenance and use of personal protective equipment at work and in other circumstances.

Petroleum Consolidation Act 1928
Regulates the use, transport or storage of petroleum spirit and – as extended by other legislation – mixtures, e.g. adhesives, thinners or lacquers containing petroleum, and non-petroleum based solvents with a flash point <73°F.

Provision and Use of Work Equipment Regulations 1998
Govern the provision and use of all work equipment (i.e. any machinery, appliance, apparatus, tool or installation for use at work) including mobile and lifting equipment. Cover general requirements relating to the selection, inspection and maintenance of equipment, protection against specific risks (e.g. dangerous parts of machinery, high or very low temperature, equipment failure scenarios, discharge of any article or substance from equipment) and the provision of instruction, information and training.

Radioactive Substance Act 1993
Governs the keeping and use of radioactive substances and the storage and disposal of radioactive waste.

Reporting of Injuries, Diseases and Dangerous Occurrences Regulations (RIDDOR) 1995
Requirements for the reporting of certain categories of injury and disease sustained at work, and specific dangerous occurrences or gas incidents, to the enforcing authority.

Social Security (Industrial Injuries) (Prescribed Diseases) Regulations 1985
List the diseases prescribed for the payment of disablement benefit, if related to specific occupations. Conditions due to chemical agents, e.g. poisoning by any of a range of chemicals and certain carcinomas, and miscellaneous conditions, e.g. pneumoconiosis, asthma, diffuse mesothelioma, non-infective dermatitis are included.

Transport of Dangerous Goods (Safety Advisers) Regulations 1999
Requirements for provision of adequate information, time and other resources to safety advisers to fulfil their functions. These include monitoring compliance with rules relating to transportation of dangerous goods; advising on transportation; implementing emergency procedures; investigating serious accidents, incidents or infringements and implementing measures to avoid a recurrence, etc.

Workplace (Health, Safety and Welfare Regulations) 1992
Aim to protect the health and safety of everyone in the workplace and ensure that adequate welfare facilities are provided. Covers e.g. general ventilation, temperature in indoor workplaces, lighting, cleanliness, space requirements, condition of floors and traffic routes, measures against falls/falling objects, washing facilities.

Environmental protection legislation

Clean Air Act 1993
Sets out ancillary controls related to air pollution control including regulation of smoke, grit, dust and fume emissions from non-prescribed industrial processes; provision of a lower level of control over some smaller combustion plants not covered by IPC or LAAPC; prohibition of the emission of ‘dark smoke’ from any chimney or industrial premises.

Environment Act 1995
Created the Environment Agency (EA) and the SEA for Scotland. Contains detailed provisions for dealing with a range of environmental issues including air quality; contaminated land; reinforcing the ‘polluter pays’ principle; water quality, with the EA empowered to require action to prevent water pollution and to require polluters to clean up after any pollution episode.

Environmental Protection (Duty of Care) Regulations 1991

Establish a system of transfer notes and record keeping of waste transfers to assist 'waste holders' comply with their duty of care under Section 34 of the EPA 1990.

They must take all reasonable steps to ensure waste is collected, treated, transported and disposed of by licensed operators.

Environmental Protection Act 1990

Covers eight aspects of the control of pollution and protection of the environment; Integrated Pollution Control (IPC) and Air Pollution Control identifying industrial processes scheduled for control by the Environment Agency or local authorities (LAS); Waste on Land, imposing a duty of care on anyone who imports, carries, keeps, treats or disposes of waste on land; Statutory Nuisances and Clean Air; Litter; amendments to the Radioactive Substances Act 1960; Genetically Modified Organisms, dealing with the prevention, or minimization of any damage to the environment from the escape or release of GMOs; Nature Conservation; miscellaneous provisions.

Pollution Prevention and Control Act 1999

Implements a new system of pollution control, Integrated Pollution Prevention and Control (IPPC), which will eventually replace IPC and LAPC under the Environmental Protection Act 1990.

Waste Management Licensing Regulations 1994 (as amended)

Implement in detail the regulatory system for a waste management licence under EPA 1990.

Water Industry Act 1991

Sets out the powers and duties of Statutory Undertakers (SUs) with regard to water supply, the provision of sewerage services and consent requirements for the receipt of trade effluents.

Water Resources Act 1991

Sets out the duties and powers of the NRA, part of the Environment Agency, in relation to water resources management, abstraction and impounding, control of pollution of water resources, flood defence.

Index